Lecture Notes in Computer Science 4815

Commenced Publication in 1973
Founding and Former Series Editors:
Gerhard Goos, Juris Hartmanis, and Jan van Leeuwen

Ashish Ghosh   Rajat K. De
Sankar K. Pal (Eds.)

Pattern Recognition and Machine Intelligence

Second International Conference, PReMI 2007
Kolkata, India, December 18-22, 2007
Proceedings

 Springer

Volume Editors

Ashish Ghosh
Rajat K. De
Sankar K. Pal
Machine Intelligence Unit
Indian Statistical Institute
203 B. T. Road, Kolkata 700108, India
E-mail: ash, rajat, sankar@isical.ac.in

Library of Congress Control Number: 2007940552

CR Subject Classification (1998): I.4, F.1, I.2, I.5, J.3, C.2.1, C.1.3

LNCS Sublibrary: SL 6 – Image Processing, Computer Vision, Pattern Recognition, and Graphics

ISSN   0302-9743
ISBN-10   3-540-77045-3 Springer Berlin Heidelberg New York
ISBN-13   978-3-540-77045-9 Springer Berlin Heidelberg New York

Springer is a part of Springer Science+Business Media

springer.com

© Springer-Verlag Berlin Heidelberg 2007
Printed in Germany

Typesetting: Camera-ready by author, data conversion by Scientific Publishing Services, Chennai, India
Printed on acid-free paper   SPIN: 12197972   06/3180   5 4 3 2 1 0

Message from the General Chair

Pattern recognition and machine intelligence form a major area of research and developmental activities that encompass the processing of multimedia information obtained from the interaction among science, technology and society. An important motivation behind the spurt of activity in this field is the desire to design and make intelligent machines capable of performing certain tasks that we human beings do. Potential applications exist in forefront research areas like computational biology, data mining, Web intelligence, brain mapping, global positioning systems, medical imaging, forensic sciences, and man – machine communication, besides other classical research problems.

There have been several conferences around the world organized separately in these two areas of pattern recognition and artificial intelligence, but hardly any meeting combining these two, although both communities share similar objectives. Therefore, holding an international conference covering these domains is very appropriate and timely, considering the recent trends in information technology, which is a key vehicle for the economic development of any country. Based on this promise, the first such conference, integrating these two topics (abbreviated as PReMI 2005), was held at the Indian Statistical Institute (ISI), Calcutta in December 2005. The event was very successful and well attended.

In December 2007, the second meeting (PReMI 2007) was organized under the aforesaid theme, which was planned to be held every alternate year. Its objective was to present the state-of-the-art scientific results, encourage academic and industrial interaction, and promote collaborative research, in pattern recognition, machine intelligence and related fields, involving scientists, engineers, professionals, researchers and students from India and abroad. The conference was an ideal platform for the participants to share their views and experiences. Particular emphasis in PReMI 2007 was placed on computational biology, data mining and knowledge discovery, soft computing applications, case-based reasoning, biometry, as well as various upcoming pattern recognition/image processing problems. There were tutorials, keynote talks and invited talks, delivered by speakers of international repute from both academia and industry.

PReMI 2007 had a special significance, as it coincided with the year of the Platinum Jubilee Celebration of ISI. The ISI was founded on December 17, 1931 by Prof. Prasanta Chandra Mahalanobis, a great visionary and a great believer of interdisciplinary research integrating statistics with natural and social sciences. In ISI, the research activity on computer science, in general, and pattern recognition and learning, in particular, started in the 1960s as one such outcome of interdisciplinary research. The institute, having a long tradition of conducting basic research in statistics/mathematics and related areas, has been able to develop profoundly the activities in pattern recognition and machine learning in its different facets and dimensions with various real-life applications under the institute's motto "Unity in Diversity." As evidence and justification of the inter-disciplinary research comprising statistics and computer science, one may note that statistics provides one of the best paradigms for learning, and it has become an integral part of the theories/paradigms of machine learning, e.g., artificial intelligence, neural networks, brain mapping, data mining, and search machines on the Internet. Zadeh, the founder of fuzzy set theory, has observed that there are three essential ingredients for dramatic success in computer applications, namely, a fuzzy model of data, Bayesian inference and genetic algorithms for optimization. Similarly, statistical science will be a part, in many ways, of the validation of the tentative model of the human brain, its functions and properties, including consciousness.

As a mark of the significant achievements in these activities in ISI, special mention may be made of the DOE-sponsored KBCS Nodal Center of ISI in the 1980s and the Center for Soft Computing Research of ISI recently established in 2004 by the DST, Government of India. The soft computing center is the first national initiative in the country in this domain, and has many important objectives like providing a six-month value addition certificate course for post-graduates, enriching national institutes, e.g., NITs through funding for research in soft computing, establishing linkage to premier institutes/industries, organizing specialized courses, apart from conducting fundamental research.

The conference proceedings of PReMI 2007, containing rigorously reviewed papers, is published by Springer, in the prestigious series of *Lecture Notes in Computer Science* (LNCS). Different professional sponsors and funding agencies (both national and international) came forward to support this event for its success. These include, DST-Sponsored Center for Soft Computing Research at ISI, Kolkata; International Association of Pattern Recognition (IAPR); Web Intelligence Consortium (WIC); International Center for Pure and Applied Mathematics (CIMPA), France; Yahoo! India Research & Development; and Philips Research Asia - Bangalore.

October 2007   Sankar K. Pal

Preface

It is our pleasure to welcome you all to the Proceedings of the Second International Conference on Pattern Recognition and Machine Intelligence (PReMI 2007), held in the Indian Statistical Institute Kolkata, India during December 18–22, 2007. PReMI is gradually gaining its popularity as a premier conference to present state-of-the-art research findings. The conference is also successful in encouraging academic and industrial interaction, and promoting collaborative research and developmental activities in pattern recognition, machine intelligence and related fields, involving scientists, engineers, professionals, researchers and students from India and abroad. The conference is scheduled to be held every two years to make it an ideal platform for people to share their views and experiences in the said areas.

The conference had one plenary talk, two keynote speeches and eight invited lecturers, all by very eminent and distinguished researchers from around the world. The conference had a very good response in terms of paper submission. It received 241 submissions from about 20 countries spanning six continents. Each paper was critically reviewed by experts in the field, after which 76 papers were accepted for inclusion in this proceedings. Accepted papers are divided into eight groups, although there could be some overlapping. Articles written by the plenary, keynote and invited speakers are also included in the proceedings.

We wish to express our appreciation to the Program Committee and additional reviewers, who worked hard to ensure the quality of the contributions of this volume. We take this opportunity to express our gratitude to A. Skowron for becoming the Plenary speaker of the conference. We are also thankful to N. Nasrabadi and N. Zhong for accepting our invitation to be the keynote speakers in this conference. We thank P. Bhattacharya, L. Bruzzone, S. Chaudhuri, V.D. Gesu, S.B. Cho, G. Wang, B. Lovell and M. Nachtegael for accepting our invitation to deliver invited lectures in this conference. We would also like to express our gratitude to Alfred Hofmann, Executive Editor in the Computer Science Editorial at Springer, for his co-operation in the publication of the PReMI 2007 proceedings. Finally we would like to thank all the contributors for their enthusiastic response.

Thanks are due to Sanjoy Kumar Das and Indranil Dutta for taking the major load of the secretarial jobs.

PReMI 2007 was academically very productive and believe that you will find the proceedings to be a valuable source of reference for your ongoing and future research.

December 2007

Ashish Ghosh
Rajat K. De
Sankar K. Pal

Organization

PReMI 2007 was organized by the Machine Intelligence Unit, Indian Statistical Institute (ISI) in Kolkata during December 18–22, 2007.

PReMI 2007 Conference Committee

General Chair   S.K. Pal (ISI, Kolkata, India)
Program Chairs   A. Ghosh (ISI, Kolkata, India)
                       R.K. De (ISI, Kolkata, India)
Organizing Chair   D.P. Mandal (ISI, Kolkata, India)
Tutorial Chair   S. Mitra (ISI, Kolkata, India)
Finance Chair   C.A. Murthy (ISI, Kolkata, India)
Industrial Liaison   M.K. Kundu (ISI, Kolkata, India)
                       B. Umashankar(ISI, Kolkata, India)
International Liaison   Simon C.K. Shiu (Hong Kong Polytechnic
                       University, Hong Kong)

Advisory Committee

A.K. Jain, USA
A. Skowron, Poland
B.L. Deekshatulu, India
D. Dutta Majumder
H. Bunke, Switzerland
H.-J. Zimmermann, Germany
H. Tanaka, Japan
J. Keller, USA

L.A. Zadeh, USA
L. Kanal, USA
M. Jambu, France
R. Chellappa, USA
S.I. Amari, Japan
S. Grossberg, USA
W. Pedrycz, Canada

Program Committee

A. Abraham, Korea
A. Dengel, Germany
A.K. Majumder, India
A. Lahiri, India
B. Lovell, Australia
B. Yegnanarayana, India
D.W. Aha, USA
D. Zhang, Hong Kong
H.A. Abbass, Australia
H.R. Tizhoosh, Canada

I. Bloch, France
J.K. Udupa, USA
J.T.L. Wang, USA
L. Bruzzone, Italy
L. Hall, USA
L. Wang, Singapore
M.J. Zaki, USA
M.K. Pakhira, India
M. Girolami, UK
M.N. Murty, India

M. Nachtegael, Belgium
N.M. Nasrabadi, USA
O. Nasraoui, USA
R. Kruse, Germany
R. Kothari, India
R. Setiono, Singapore
Santanu Chaudhuri, India
Subhasis Chaudhuri, India
S.B. Cho, Korea
S. Kundu, India
S. Mukhopadhyay, India

S.K. Mitra, India
S. Ray, Australia
S. Tsutsui, Japan
T. Nakashima, Japan
T. Heskes, Netherlands
T. Pham, Australia
T.K. Ho, USA
U. Seiffert, Germany
Y. Hayashi, Japan

Reviewers

A. Abraham
A. Adhikary
A. Basu
A. Pal
A. Ghosh
A.K. Mazumdar
A. Dengel
A. Skowron
A. Haldar
A. Bagchi
A. Konar
A. Roychoudhuri
A. Mohani
A. Dengel
B. Lovell
B. Umashankar
B. L. Narayanan
B. Chanda
C. Fyfe
C.I.J
C. Bhattacharyya
C.A. Murthy
C. Kamath
C.V. Jawahar
D.W. Aha
D. Bimbo
D.K. Pratihar
D.P. Mukherjee
D.K. Bhattacharyya
D.S. Guru

D.P. Mandal
D. Sen
D. Kaller
D. Dasgupta
F. Herrera
F. Gomide
F.C. Ferrera
G. Chakraborty
H.A. Abbass
H. Iba
H. Ishibuchi
H. Banka
I. Bloch
I. Hayashi
I. J. Cox
J.K. Sing
J.T.L. Wang
J. Wang
J.K. Udupa
Jayadeva
K. Ghosh
L. Wang
L. Hall
L. Bruzzone
L. Dey
L. Behera
M. Giorlami
M. Nachtegael
M.K. Kundu
M.N. Murthy

M. Mitra
M. K. Pakhira
M.A. Zaveri
M.T. Kundu
M. Acharyya
M. Nasipuri
M. Harvey
M. Zaki
N. M. Nasrabadi
N. Tagdur
N.R. Pal
O. Nasraoui
P. Mitra
P. Maji
P.K. Nanda
P. Bhattacharyya
P. Siarry
P. Chan
P. Nagabhusan
P. Pal
P. Sierra
R. Kruse
R. Setiono
R. Kothari
R.K. Mudi
R. Das
S. Chowdhuri
S.K. Mitra
S. Mukhopadhyay
S. Chaudhury

S. Tsutsui
S.B. Cho
S.K. Saha
S.K. Naik
S. Dehuri
S.K. Meher
S. Shevade
S. Basu
S. Chandran
S.S. Ray
S. Saha
S. Padhy
S. Bhandari
S.N. Biswas
S.P. Maity
S. Biswas
S. Bose
S. Ray
S. Kundu

S. Karmakar
S. Rakshit
S. Sural
S.C.K. Shiu
S.K. Parui
S. Purkayastra
S. Banerjee
S. Palit
S.C. Nandy
S. Devi
S. Shevade
S. Mitra
S. Saha
S.K. Saha
S.C. Kremer
S. DuttaRoy
S. Banerjee
S. Deb
S. Dehuri

T. Nakashima
T.K. Ho
T. Heskes
T. Pham
T.S. Dillon
T. Samanta
T. Murata
T. Nagabhushan
U. Chakraborty
U. Garain
U. Pal
U. Bhattacharyya
U. Seifert
V.D. Gesu
W. Pedrycz
W.H. Hsu
Y. Hayashi
Y.Y. Tang

Sponsoring Organizations from Academia

- Center for Soft Computing Research: A National Facility, ISI, Kolkata
- International Association for Pattern Recognition (IAPR)
- International Center for Pure and Applied Mathematics (CIMPA), France
- Web Intelligence Consortium (WIC)

Sponsoring Organizations from Industry

- Yahoo! India Research & Development
- Philips Research Asia - Bangalore

Table of Contents

Pattern Recognition

Image Analysis

Soft Computing and Applications

Data Mining and Knowledge Discovery

Bioinformatics

Signal and Speech Processing

Document Analysis and Text Mining

Biometrics

Video Analysis

Ensemble Approaches of Support Vector Machines for Multiclass Classification

Jun-Ki Min, Jin-Hyuk Hong, and Sung-Bae Cho

Department of Computer Science, Yonsei University
Biometrics Engineering Research Center
134 Shinchon-dong, Sudaemoon-ku, Seoul 120-749, Korea
{loomlike, hjinh}@sclab.yonsei.ac.kr, sbcho@cs.yonsei.ac.kr

Abstract. Support vector machine (SVM) which was originally designed for binary classification has achieved superior performance in various classification problems. In order to extend it to multiclass classification, one popular approach is to consider the problem as a collection of binary classification problems. Majority voting or winner-takes-all is then applied to combine those outputs, but it often causes problems to consider tie-breaks and tune the weights of individual classifiers. This paper presents two novel ensemble approaches: probabilistic ordering of one-vs-rest (OVR) SVMs with naïve Bayes classifier and multiple decision templates of OVR SVMs. Experiments with multiclass datasets have shown the usefulness of the ensemble methods.

Keywords: Support vector machines; Ensemble; Naïve Bayes; Multiple decision templates; Cancer classification, Fingerprint classification.

1 Introduction

Support Vector Machine (SVM) is a relatively new learning method which shows excellent performance in many pattern recognition applications [1]. It maps an input sample into a high dimensional feature space and tries to find an optimal hyperplane that minimizes the recognition error for the training data by using the non-linear transformation function [2]. Since the SVM was originally designed for binary classification, it is required to devise a multiclass SVM method [3].

Basically, there are two different trends for extending SVMs to multiclass problems. The first considers the multiclass problem directly as a generalization of the binary classification scheme [4]. This approach often leads to a complex optimization problem. The second decomposes a multiclass problem into multiple binary classification problems that can be solved by an SVM [5]. One-vs-rest (OVR) is a representative decomposition strategy, while winner-takes-all or error correcting codes (ECCs) is reconstruction schemes to combine the multiple outputs [6]. It has been pointed out that there is no guarantee on the decomposition-based approach to reach the optimal separation of samples [7]. There are several reasons for this, such as unbalanced sample sizes. However, these could be complemented by the proper selection of a model or a decision scheme.

A. Ghosh, R.K. De, and S.K. Pal (Eds.): PReMI 2007, LNCS 4815, pp. 1–10, 2007.

In this paper, we present two novel ensemble approaches for applying OVR SVMs to multiclass classification. The first orders OVR SVMs probabilistically by using naïve Bayes (NB) classifier with respect to the subsumption architecture [8]. The latter uses multiple decision templates (MuDTs) of OVR SVMs [9]. It organizes the outputs of the SVMs with a decision profile as a matrix, and estimates the localized template from the profiles of the training set by using the K-means clustering algorithm. The profile of a test sample is then matched to the templates by a similarity measure. The approaches have been validated on the GCM cancer data [10] and the NIST4 fingerprints data [11]. Experimental results have shown the effectiveness of the presented methods.

2 Background

2.1 Multiclass Classification Using Binary SVMs

Since SVM is a basically binary classifier, a decomposition strategy for multiclass classification is required. As a representative scheme, the OVR strategy trains M (the number of classes) SVMs, where each SVM classifies samples into the corresponding class against all the others. The decision function of the jth SVM replaces the class label of the ith sample, c_i, with t_i as:

$$t_i = +1 \quad \text{if} \quad c_i = j, \quad \text{otherwise} \quad t_i = -1. \tag{1}$$

After constructing multiple SVMs, a fusion strategy is required to combine their outputs. Popular methods for combination such as winner-takes-all and ECCs [6] are widely employed. Winner-takes-all method classifies a sample into the class that receives the highest value among the L classifiers for the M-class problem. ECCs method generates a coding matrix $E \in \{-1, 0, 1\}^{M \times L}$ where $E_{i,j}$ represents an entry in the ith row and jth column of E. $E_{i,j} = -1$ (or 1) indicates that the points in class i are regarded as negative (or positive) examples when training the jth classifier. If $E_{i,j} = 0$, class i is not used when the jth classifier is trained. A test point is classified into the class whose row in the coding matrix has the minimum distance to the vector of the outputs of the classifiers.

2.2 Cancer Classification Based on Gene Expression Profiles

Recently developed DNA microarray technologies produce large volume of gene expression profiles and provide richer information on diseases. It simultaneously monitors the expression patterns of thousands of genes under a particular experimental environment. Especially the classification of cancers from gene expression profiles which commonly consists of feature selection and pattern classification has been actively investigated in bioinformatics [12]. Gene expression profiles provide useful information to classify different forms of cancers, but the data also include useless information for the classification. Therefore, it is important to find a small subset of genes sufficiently informative to distinguish cancers for diagnostic purposes [13].

2.3 Fingerprint Classification

Fingerprint classification is a technique that classifies fingerprints into a predefined category [14]. It is useful for an automated fingerprint identification system (AFIS) as a preliminary step of the matching process to reduce searching time. Henry system is the most widely used system for the fingerprint classification. It categorizes a fingerprint into one of five classes: whorl (W), right loop (R), left loop (L), arch (A), and tented arch (T) according to its global pattern of ridges. There are many different ways to extract and represent ridge information [15]. Especially, FingerCode proposed by Jain in 1999 [11] is a representative fingerprint feature extraction algorithm among them. This method tessellated a given fingerprint image into 48 sectors and transformed the image using the Gabor filters of four directions ($0°$, $45°$, $90°$, and $135°$). Standard deviation was then computed on 48 sectors for each of the four transformed images in order to generate the 192-dimensional feature vector.

3   Novel Ensemble Methods for SVMs

3.1   PO-SVMs: Probabilistic Ordering of SVMs with Naïve Bayes Classifier

The method first estimates the probability of each class by using NB, and then organizes OVR SVMs as the subsumption architecture according the probability as shown in Fig. 1(a). Samples are sequentially evaluated with OVR SVMs. When a SVM is satisfied, the sample is classified into the corresponding class, while it is assigned with the class of the highest probability when no SVMs are satisfied. Fig. 1(b) shows the pseudo code of the proposed method.

The NB classifier estimates the posterior probability of each class given the observed attribute values for an instance, and the class with the highest posterior probability is finally selected [16]. In order to calculate the posterior probability, the classifier must be defined by the marginal probability distribution of variables and by a set of conditional probability distributions of each attribute A_i given each class c_j. They are estimated from the training set. When A_i is the ith state of the attribute A, which has a parent node B, and $count(A_i)$ is the frequency of cases in which the attribute A appears with the ith states, a priority probability $P(A_i)$ and a conditional probability $P(A_i|B_j)$ are estimated as follows:

$$P(A_i) = count(A_i)/n_T \ , \ P(A_i \mid B_j) = count(A_i, B_j)/count(B_j) \ . \tag{2}$$

Bayes' theorem yields the posterior probability of each class given n features as evidence over a class C and $F_1 \sim F_n$ by

$$P(C \mid F_1,...,F_n) = P(C)P(F_1,...,F_n \mid C)/P(F_1,...,F_n) \ . \tag{3}$$

In practice, we are only interested in the numerator of the fraction, since the denominator does not affect C. Feature F_i is conditionally independent from the other feature F_j for $j \neq i$, so the probability of a class is calculated by equation (4):

$$P(C)P(F_1,...,F_n \mid C) = P(C)P(F_1 \mid C)P(F_2 \mid C)...P(F_n \mid C) = P(C)\prod_{i=1}^{n}P(F_i \mid C). \quad (4)$$

Since there will be a heavy computation when using high dimensional features for NB, we adopt different features from SVMs. In the following subsections, feature selection and extraction methods used for NB classifier are described.

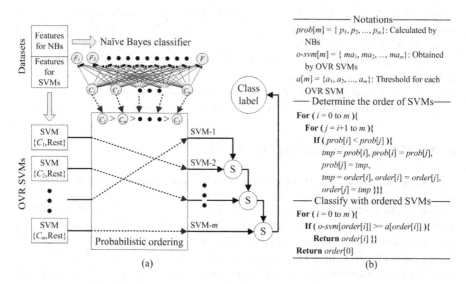

Fig. 1. (a) Overview of the PO-SVMs and (b) its pseudo code

Feature Selection for Cancer Classification. A subset of informative genes is selected by using the feature selection process based on Pearson correlation. We define two ideal markers that represent a standard of good features, and utilize the features by scoring the respective similarity with each ideal marker. Two ideal markers are negatively correlated to represent two different aspects of classification boundaries. The first marker l_1 is high in class A and low in class B, while the second marker l_2 is low in class A and high in class B. Since this feature selection method is originally designed for binary classification, we select features based on the OVR scheme. Ten genes are selected for each class: the first five for the l_1 and the rest for the l_2. When there are M classes, total $M \times 10$ genes are used to construct NB classifier.

The similarity between an ideal marker l and a gene g can be regarded as a distance, while the distance represents how far they are located from one another. A gene is regarded as an informative gene if the distance is small. Pearson correlation is used to measure the similarity as follows:

$$PC = \sum_{i=1}^{n}(l_i g_i) - \frac{\sum_{i=1}^{n}l_i \sum_{i=1}^{n}g_i}{n} \left/ \sqrt{\left(\sum_{i=1}^{n}l_i^2 - \left(\sum_{i=1}^{n}l_i\right)^2 \middle/ n\right) \times \left(\sum_{i=1}^{n}g_i^2 - \left(\sum_{i=1}^{n}g_i\right)^2 \middle/ n\right)}\right. \quad (5)$$

Feature Extraction for Fingerprint Classification. Two representative features of fingerprints called the singular points (core and delta points) [14] and the pseudo ridge [17] are considered in order to construct a NB classifier. Fig. 2(a) shows an example. Locations and distances between the core C and other points are parameterized as shown in Fig. 2(b). If there are two core points, the nearest core to the center is denoted as C (if there is no core, C represents the center of the image). In this paper, as shown in Table 1, the number of cores and deltas, the location and distance between them, and the location and distance between cores and the end points of pseudo ridges are used for the NB classifier.

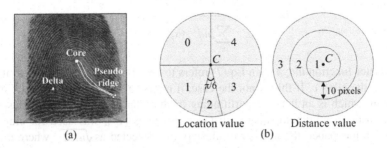

Fig. 2. (a) Fingerprint features which are used in this paper, (b) parameterized position and distance

Table 1. Features for the naïve Bayes classifier

Feature	Definition	State
Num_C, Num_D	Number of core points and delta points	0, 1, 2
Loc_{D1}, Loc_{D2}	Location of delta points ($D1$ and $D2$)	0, 1, 2, 3, 4, *absent*
Loc_{R1}, Loc_{R2}	Location of the end point of the pseudo ridge ($R1$ and $R2$)	0, 1, 2, 3, 4, *turn*
Dst_{D1}, Dst_{D2}	Distance between C and delta points	1, 2, 3, *absent*
Dst_{R1}, Dst_{R2}	Distance between C and the end point of the pseudo ridge	1, 2, 3, *turn*

3.2 MuDTs-SVMs: Multiple Decision Templates of SVMs

The original decision template (DT) is a classifier fusion method proposed by Kuncheva [18] which generates the template of each class (one per class) by averaging decision profiles for the training samples. The profile is a matrix that consists of the degree of support given by classifiers. The templates are estimated with the averaged profiles of the training data. In the test stage, the similarity between the decision profile of a test sample and each template is calculated. The class label refers to the class with the most similar templates.

The multiple decision templates (MuDTs) allow for clustering of each class and generate localized templates which are able to model intra-class variability and inter-class similarity of data. An overview of proposed method is shown in Fig. 3. For the M-class problem, a decision profile with the outputs of OVR SVMs is organized as represented in Eq (6), where $d_m(x_i)$ is the degree of support given by the mth classifier for the sample x_i.

$$DP(x_i) = \begin{bmatrix} d_1(x_i) \\ \vdots \\ d_M(x_i) \end{bmatrix}, \tag{6}$$

Since SVM is a binary classifier, we represented the profile to be one column matrix with positive or negative values. In order to generate multiple decision templates, profiles of training samples are clustered for each class. And then the localized template of the kth cluster in the class c, $DT_{c,k}$, is estimated by

$$DT_{c,k} = \frac{\sum_{i=1}^{n} ind_{c,k}(x_i) d_c(x_i)}{\sum_{i=1}^{n} ind_{c,k}(x_i)}. \tag{7}$$

An indicator function $ind_{c,k}(x_i)$ in Eq (7) refers to one if a sample x_i belongs to the kth cluster in the class c. If this is not the case, it refers to zero. In this paper, K-means algorithm, which is an iterative partitioning method that finds K compact clusters in the data using a minimum squared distance measures [19], was used for clustering method. In this paper, the number of clusters K is selected as $\sqrt{n/M}$, where n is the number of training samples and M is the number of classes.

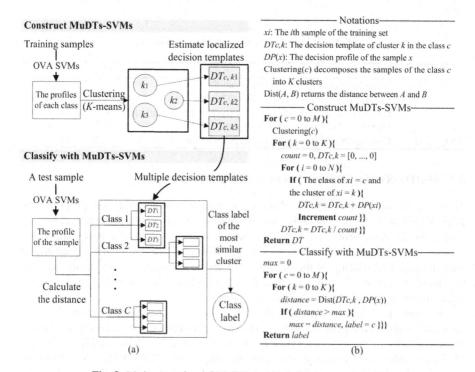

Fig. 3. (a) An overview of MuDTs method, (b) the pseudo code

In order to classify a new input sample, similarity between its profile and localized template of each cluster are calculated as:

$$dst_{c,k}(x) = \left| DT_{c,k} - U_c \cdot DP(x) \right|,$$ (8)

where

$$U_c = \begin{bmatrix} u_1 & \cdots & u_M \end{bmatrix}, \ u_m = 1 \ \text{if} \ m = c, \ \text{otherwise} \ u_m = 0.$$ (9)

The sample is then classified into the class that contains the most similar cluster.

4   Experimental Results

We have verified the proposed methods on the GCM cancer data set [10] and the NIST4 fingerprint data set [11], summarized in Table 2.

Table 2. Summary of the datasets used

Data set	# Features	# Classes	# Samples	SVM kernel (parameter)
GCM	16,063 for SVMs 140 for NB	14	198	Linear
NIST4	192 for SVMs 10 for NB	5	3,937	Gaussian (σ^2=0.0625)

4.1   Experiments on GCM Cancer Data

GCM cancer dataset contains 144 training samples and 54 test samples with 16,063 gene expression levels. In this paper, all genes are used for SVMs and 140 genes are selected for NB based on Pearson correlation. There are 14 different tumor categories including breast adenocarcinoma (Br), prostate (Pr), lung adenocarcinoma (Lu), colorectal adenocarcinoma (Co), lymphoma (Ly), bladder (Bl), melanoma (Me), uterine adenocarcinoma (Ut), leukemia (Le), renal cell carcinoma (Re), pancreatic adenocarcinoma (Pa), ovarian adenocarcinoma (Ov), pleural mesothelioma (Pl), and central nervous system (Ce). Since the dataset provides only a few samples but lots of features, it is a challenging task for many machine learning researchers to construct a competitive classifier. Table 3 shows the performances of the previous works on this dataset.

Comparison results of the proposed method with several traditional approaches are presented in Table 4. OVA SVMs with the winner-takes-all strategy produced 77.8% classification accuracy, and NB yielded an accuracy of 74.8%, individually. PO-SVMs showed higher performance than the others with a classification accuracy of 81.5%, while MuDTs produced an accuracy of 77.8% (which is the same with that of winner-takes-all SVMs and ECCs). This may be because the clustering was not effective enough since there were only a few samples in the GCM dataset. The confusion matrices of proposed methods are presented in Table 5. Rows and columns denote the true class and the assigned class, respectively.

Table 3. Related works on the GCM cancer dataset

Author	Method	Accuracy (%)
Ramaswamy et al. (2001) [10]	OVR SVMs	77.8
Deutsch (2003) [20]	GA/SVMs	77.9
Li et al. (2004) [21]	SVMs random	63.3

Table 4. Classification accuracy for the GCM cancer dataset

Mehod	Product (SVMs+NB)	ECCs (SVMs)	DTs (SVMs)	PO-SVMs (SVMs+NB)	MuDTs-SVMs (SVMs)
Accuracy (%)	66.7	77.8	72.2	81.5	77.8

Table 5. Confusion matrices for the GCM cancer dataset (left: PO-SVMs, right: MuDTs-SVMs)

Left (PO-SVMs):

	Br	Pr	Lu	Co	Ly	Bl	Me	Ut	Le	Re	Pa	Ov	Pl	Ce
Br	3	0	0	0	0	0	0	0	0	0	1	0	0	0
Pr	0	5	0	0	0	0	0	0	0	0	1	0	0	0
Lu	0	0	4	0	0	0	0	0	0	0	0	0	0	0
Co	0	0	0	4	0	0	0	0	0	0	0	0	0	0
Ly	0	0	1	0	5	0	0	0	0	0	0	0	0	0
Bl	0	0	1	0	0	2	0	0	0	0	0	0	0	0
Me	0	0	0	0	0	0	1	0	0	0	1	0	0	0
Ut	0	0	0	0	0	0	0	2	0	0	0	0	0	0
Le	0	1	0	0	0	0	0	0	5	0	0	0	0	0
Re	0	0	0	0	0	0	0	1	0	2	0	0	0	0
Pa	0	0	0	1	0	1	0	0	0	0	1	0	0	0
Ov	0	0	0	1	0	0	1	0	0	0	0	2	0	0
Pl	0	0	0	0	0	0	0	0	0	0	0	0	3	0
Ce	0	0	0	0	0	0	0	0	0	0	0	0	0	4

Right (MuDTs-SVMs):

	Br	Pr	Lu	Co	Ly	Bl	Me	Ut	Le	Re	Pa	Ov	Pl	Ce
Br	2	0	0	0	0	0	0	0	0	0	1	1	0	0
Pr	0	4	0	0	0	0	1	0	0	0	1	0	0	0
Lu	0	0	2	0	0	1	1	0	0	0	0	0	0	0
Co	0	0	0	4	0	0	0	0	0	0	0	0	0	0
Ly	0	0	0	0	6	0	0	0	0	0	0	0	0	0
Bl	0	0	0	0	0	2	1	0	0	0	0	0	0	0
Me	0	0	0	0	0	0	1	0	0	0	1	0	0	0
Ut	0	0	0	0	0	0	0	2	0	0	0	0	0	0
Le	0	1	0	0	0	0	0	0	5	0	0	0	0	0
Re	0	0	0	0	0	0	0	0	0	3	0	0	0	0
Pa	1	0	0	0	0	0	0	0	0	0	2	0	0	0
Ov	0	0	0	0	0	0	1	0	0	0	1	2	0	0
Pl	0	0	0	0	0	0	0	0	0	0	0	0	3	0
Ce	0	0	0	0	0	0	0	0	0	0	0	0	0	4

4.2 Experiments on NIST4 Fingerprint Data

This set consists of 4,000 scanned images (at 512×512 resolution) obtained from two impressions (F and S) of 2,000 fingerprints. Fingerprints were equally distributed into 5 classes, whorl (W), right loop (R), left loop (L), arch (A), and tented arch (T). Due to the ambiguity in fingerprint images, 350 fingerprints (17.5%) were cross-referenced with two classes. The first label was only considered in training phase while both labels were used in the test. In the experiment, the fingerprints of the first impression were used as the training set (F0001~F2000), and the other fingerprints constructed the test set. The FingerCode features was used for SVMs after normalization from −1 to 1 where some rejected images were included in the training set (1.4%) and the test set (1.8%) [11]. Singularities and pseudo ridge information was used for NB. Table 6 represents the performances of published fingerprint classification methods.

Several schemes for generating and combining multiple SVMs were compared with the proposed method (Table 7). OVA SVMs with the winner-takes-all method produced an accuracy of 90.1%, and NB yielded an accuracy of 85.4%, individually. Although the NB obtained lower accuracy than the SVMs, PO-SVMs (which combined NB and SVMs dynamically) showed an accuracy of 90.8% which is higher than the others.

Table 6. Related works on the NIST4 fingerprint dataset with FingerCode feature

Author	Method	Accuracy (%)
Jain et al. (1999) [11]	KNN+NN: Two stage classifiers	90.0
Yao et al. (2002) [22]	ECCs SVMs with RNN feature	90.0

Table 7. Classification accuracy for the NIST4 fingerprint dataset

Mehod	Product (SVMs+NB)	ECCs (SVMs)	DTs (SVMs)	PO-SVMs (SVMs+NB)	MuDTs-SVMs (SVMs)
Accuracy (%)	90.2	90.1	89.8	90.8	90.4

Table 8. Confusion matrices for the NIST4 fingerprint dataset (left: PO-SVMs, right: MuDTs-SVMs)

	W	R	L	A	T
W	373	10	10	0	0
R	4	374	0	6	15
L	5	0	377	8	9
A	0	6	4	365	40
T	1	8	15	39	295

	W	R	L	A	T
W	380	6	7	0	1
R	9	369	1	5	17
L	8	0	366	14	10
A	1	4	1	356	50
T	1	10	6	38	304

MuDTs-SVMs also achieved higher accuracy of 90.4% than the conventional methods. The confusion matrices of proposed methods are shown in Table 8.

5   Concluding Remarks

Multiclass classification is a challenging task in pattern recognition, where various approaches have been investigated especially using SVMs. Since SVM is a binary classifier, it is necessary to formulate decomposition and combination methods. In this paper, two novel ensemble approaches for OVR SVMs were proposed. The one uses probabilistic ordering of OVR SVMs with naïve Bayes classifier (PO-SVMs) and the other uses multiple decision templates of OVR SVMs (MuDTs-SVMs). Two benchmark data sets of GCM cancer data and NIST4 fingerprint data were used to verify these methods. As the future work, we will demonstrate the proposed methods with other popular benchmark datasets of multiclass, and study about the parameters for better performance.

Acknowledgments. This work was supported by the Korea Science and Engineering Foundation (KOSEF) through the Biometrics Engineering Research Center (BERC) at Yonsei University.

References

1. Cortes, C., Vapnik, V.: Support-Vector Networks. Machine Learning 20(3), 273–297 (1995)
2. Cristianini, N., Shawe-Taylor, J.: An Introduction to Support Vector Machines. Cambridge University Press, Cambridge (2000)

3. Bredensteiner, E., Bennett, K.: Multicategory Classification by Support Vector Machines. Computational Optimization and Applications. 12(1), 53–79 (1999)
4. Arenas-Garcia, J., Perez-Cruz, F.: Multi-Class Support Vector Machines: A New Approach. In: 2003 IEEE Int. Conf. Acoustics, Speech, and Signal Processing, pp. II-781–784 (2003)
5. Hsu, C., Lin, C.: A Comparison of Methods for Multiclass Support Vector Machines. IEEE Trans. Neural Networks. 13(2), 415–425 (2002)
6. Rifkin, R.M., Klautau, A.: In Defense of One-Vs-All Classification. J. Machine Learning Research. 5, 101–141 (2004)
7. Lee, Y., Lin, Y., Wahba, G.: Multicategory Support Vector Machines. Tech. Rep. 1043, Dept. Statistics, Univ. of Wisconsin (2001)
8. Hong, J.-H., Cho, S.-B.: Multi-Class Cancer Classification with OVR-Support Vector Machines Selected by Naive Bayes Classifier. In: King, I., Wang, J., Chan, L., Wang, D. (eds.) ICONIP 2006. LNCS, vol. 4234, pp. 155–164. Springer, Heidelberg (2006)
9. Min, J.-K., Hong, J.-H., Cho, S.-B.: Effective Fingerprint Classification by Localized Models of Support Vector Machines. In: Zhang, D., Jain, A.K. (eds.) Advances in Biometrics. LNCS, vol. 3832, pp. 287–293. Springer, Heidelberg (2005)
10. Ramaswamy, S., et al.: Multiclass Cancer Diagnosis using Tumor Gene Expression Signatures. Proc. National Academy of Science. 98(26), 15149–15154 (2001)
11. Jain, A.K., Prabhakar, S., Hong, L.: A Multichannel Approach to Fingerprint Classification. IEEE Trans. Pattern Analysis and Machine Intelligence. 21(4), 348–359 (1999)
12. Cho, S.-B., Ryu, J.: Classifying Gene Expression Data of Cancer using Classifier Ensemble with Mutually Exclusive Features. Proc. of the IEEE. 90(11), 1744–1753 (2002)
13. Hong, J.-H., Cho, S.-B.: The Classification of Cancer based on DNA Microarray Data That Uses Diverse Ensemble Genetic Programming. Artificial Intelligence in Medicine. 36(1), 43–58 (2006)
14. Maltoni, D., Maio, D., Jain, A.K., Prabhakar, S.: Handbook of Fingerprint Recognition. Springer, Heidelberg (2003)
15. Yager, N., Amin, A.: Fingerprint Classification: A Review. Pattern Analysis and Application. 7(1), 77–93 (2004)
16. Liu, J., Li, B., Dillon, T.: An Improved Naïve Bayesian Classifier Technique Coupled with a Novel Input Solution Method. IEEE Trans. Systems, Man, and Cybernetics-Part C: Applications and Reviews 31(2), 249–256 (2001)
17. Zhang, Q., Yan, H.: Fingerprint Classification based on Extraction and Analysis of Singularities and Pseudo Ridges. Pattern Recognition. 37(11), 2233–2243 (2004)
18. Kuncheva, L.I., Bezdek, J.C., Duin, R.P.W.: Decision Templates for Multiple Classifier Fusion: An Experimental Comparison. Pattern Recognition. 34(2), 299–314 (2001)
19. Jain, A.K., Dubes, R.C.: Algorithms for Clustering Data. Prentice-Hall, Englewood Cliffs (1988)
20. Deutsch, J.: Evolutionary Algorithms for Finding Optimal Gene Sets in Microarray Prediction. Bioinformatics. 19(1), 45–52 (2003)
21. Li, T., Zhang, C., Ogihara, M.: A Comparative Study of Feature Selection and Multiclass Classification Methods for Tissue Classification based on Gene Expression. Bioinformatics. 20(15), 2429–2437 (2004)
22. Yao, Y., Marcialis, G.L., Pontil, M., Frasconi, P., Roli, F.: Combining Flat and Structured Representations for Fingerprint Classification with Recursive Neural Networks and Support Vector Machines. Pattern Recognition. 36(2), 397–406 (2003)

Robust Approach for Estimating Probabilities in Naive-Bayes Classifier

B. Chandra[1], Manish Gupta[2], and M.P. Gupta[1]

[1]Indian Institute of Technology, Delhi
Hauz Khas, New Delhi, India 110 016
bchandra104@yahoo.co.in
[2]Institute for Systems Studies and Analyses
Metcalfe House, Delhi, India 110 054

Abstract. Naive-Bayes classifier is a popular technique of classification in machine learning. Improving the accuracy of naive-Bayes classifier will be significant as it has great importance in classification using numerical attributes. For numeric attributes, the conditional probabilities are either modeled by some continuous probability distribution over the range of that attribute's values or by conversion of numeric attribute to discrete one using discretization. The limitation of the classifier using discretization is that it does not classify those instances for which conditional probabilities of any of the attribute value for every class is zero. The proposed method resolves this limitation of estimating probabilities in the naive-Bayes classifier and improve the classification accuracy for noisy data. The proposed method is efficient and robust in estimating probabilities in the naive-Bayes classifier. The proposed method has been tested over a number of databases of UCI machine learning repository and the comparative results of existing naive-Bayes classifier and proposed method has also been illustrated.

1 Introduction

Classification has wide application in pattern recognition. In classification, a vector of attribute values describes each instance. Classifier is used to predict the class of the test instance using training data, a set of instances with known classes. Decision trees [13], k- nearest neighbor [1], naive- Bayes classifier [6,7,8] etc. are the commonly used methods of classification. Naive-Bayes classifier (NBC) is a simple probabilistic classifier with strong assumption of independence. Although attributes independence assumption is generally a poor assumption and often violated for real data sets, Langley et. al. [10] found that NBC outperformed an algorithm for decision-tree induction. Domingos et.al. [4] has also found that this limitation has less impact than might be expected. It often provides better classification accuracy on real time data sets than any other classifier does. It also requires small amount of training data. It is also useful for high dimensional data as probability of each attribute is estimated independently. There is no need to scale down the dimension of the data as required in some popular classification techniques.

A. Ghosh, R.K. De, and S.K. Pal (Eds.): PReMI 2007, LNCS 4815, pp. 11–16, 2007.
© Springer-Verlag Berlin Heidelberg 2007

NBC has a limitation in predicting the class of instances for which conditional probabilities of each class are zero i.e. conditional probability of any of the attribute value for every class is zero. To rectify this problem, the Laplace-estimate [3] is used to estimate the probability of the class and M-estimate [3] is used to estimate conditional probability of any of the attribute value. The results obtained from these estimates have more error of classification for noisy data i.e more number of classes or more number of attributes. A novel approach based on the maximum occurrence of the number for which conditional probabilities of any of the attributes are zero for a given instance, is proposed in the paper. The proposed method has been tested over a number of databases of UCI machine learning repository [12]. In proposed approach, the classification accuracy is much better even for noisy data as compared to that of basic approach of estimating probabilities in NBC and of Laplace-estimates and M-estimates.

The overview of NBC along with the estimation of probabilities in NBC is given in section 2 of the paper. The proposed approach has been described in section 3 with limitation of NBC using discretization and overcoming the limitation by proposed approach. Section 4 presents results and discussion of all approaches over several databases from UCI machine learning repository. Comparative evaluation of proposed approach with the existing basic and estimate approach are also present in this section. Concluding remarks is given in the last section of the paper.

2   Naive-Bayes Classifier (NBC)

NBC [6,7,8] is a simple probabilistic inductive algorithm with strong attribute independence assumption. NBC learns from training data and then predicting the class of the test instance with the highest posterior probability. Let C be the random variable denoting the class of an instance and let $\mathbf{X} < X_1, X_2, \ldots, X_m >$ be a vector of random variables denoting the observed attribute values. Let c represent a particular class label and let $\mathbf{x} < x_1, x_2, \ldots, x_m >$ represent a particular observed attribute value vector. To predict the class of a test instance $\mathbf{x}$, Bayes' Theorem is used to calculate the probability

$$p(C = c \,|\mathbf{X} = \mathbf{x}) = \frac{p(C = c)p(\mathbf{X} = \mathbf{x} \,|C = c)}{p(\mathbf{X} = \mathbf{x})} \tag{1}$$

Then, predict the class of test instance with highest probability. Here $\mathbf{X} = \mathbf{x}$ represents the event that $X_1 = x_1 \wedge X_2 = x_2 \wedge \ldots X_m = x_m$. $p(\mathbf{X} = \mathbf{x})$ can be ignored as it is invariant across the classes, then equation (1) becomes

$$p(C = c \,|\mathbf{X} = \mathbf{x}) \propto p(C = c)p(\mathbf{X} = \mathbf{x}|C = c) \tag{2}$$

$p(C = c)$ and $p(\mathbf{X} = \mathbf{x}|C = c)$ are estimated from the training data. As attributes $X_1, X_2, \ldots, X_m$ are conditionally independent of each other for given class then equation (2) becomes

$$p(C = c \,|\mathbf{X} = \mathbf{x}) \propto p(C = c)\prod_{i=1}^{m} p(X_i = x_i|C = c) \tag{3}$$

which is simple to compute for test instances and to estimate from training data. Classifiers using equation (3) are called naive-Bayes classifier.

2.1   Estimation of Probabilities in Naive-Bayes Classifier

NBC can handle both categorical and numeric attributes. For each discrete attribute, $p(X_i = x_i|C = c)$ in equation (3) is modeled by a single real number between 0 and 1, and the probabilities can be estimated with reasonable well accuracy from the frequency of instances with $C=c$ and the frequency of instances with $X_i = x_i \wedge C = c$ in the training data. We call this approach of estimating probabilities as basic approach. Laplace-estimate [3] and M-estimate [3] are also used to compute the probabilities in equation (3). In Laplace-estimate $p(C = c) : (n_c + k)/(N + n * k)$ where n_c is the number of instances satisfying $C=c$, N is the number of training instances, n is the number of classes and $k=1$. In M-estimate $p(X_i = x_i|C = c) : (n_{ci}+m*p)/(n_c+m)$ where n_{ci} is the number of instances satisfying $X_i = x_i \wedge C = c$, n_c is the number of instances satisfying $C = c$, p is the prior probability $p(X_i = x_i)$ (estimated by the Laplace-estimate), and $m = 2$.

In contrast to discrete attribute, for each numeric attribute the probability $p(X_i = x_i|C = c)$ is either modeled by some continuous probability distribution [8] over the range of that attribute's values or by conversion of numeric attribute to discrete one using discretization [11,14,15]. Equal width discretization (EWD) [2,5,9] is a popular approach to transform numeric attributes into discrete one in NBC. A discrete attribute X_i^c is formed for each numeric attribute X_i and each value of X_i^c corresponds to an interval $(a_i, b_i]$ of X_i. If $x_i \in (a_i, b_i]$, then $p(X_i = x_i|C = c)$ in equation (3) is estimated by

$$p(X_i = x_i|C = c) \approx p(a_i < x_i \le b_i|C = c) \tag{4}$$

It is estimated same as for discrete attribute mentioned above. Using equation (4), equation (3) becomes as

$$p(C = c \,|\mathbf{X} = \mathbf{x}) \propto p(C = c) \prod_{i=1}^{m} p(a_i < x_i \le b_i|C = c) \tag{5}$$

Thus for a numeric attribute of a test instance $\mathbf{x} < x_1, x_2, , x_m >$, the probability is computed by equation (5).

3   Proposed Approach

The limitation of NBC using basic approach for estimating probabilities is that if the training instance with $X_i = x_i \wedge C = c$ does not present in the training data, then $p(X_i = x_i|C = c)$ is zero which ultimately leads to $p(C = c \,|\mathbf{X} = \mathbf{x})$ as zero from equation (3). It means that for any instance, $p(C = c \,|\mathbf{X} = \mathbf{x})$ will be zero for all classes if any one of $p(X_i = x_i|C = c)$ is zero for each class. Thus, NBC cannot predict the class of such test instances. Laplace and M-estimate methods are also not well enough for noisy as well as large databases and classify the noisy data with high error of classification. To reduce the error of classification for noisy data and to overcome the problems of both the estimation of probabilities in NBC, a simplistic novel approach based on the maximum occurrence of the number for which conditional probabilities of any of the attributes are zero for a given instance has been proposed in the paper. The proposed approach is given as follows.

For given test instance $\mathbf{x} < x_1, x_2, \ldots, x_m >$, if $p(C = c \,|\mathbf{X} = \mathbf{x})$ for each class is zero, then for each class, count the occurrences of attribute values (say, n_i) for which $p(a_i < x_i \leq b_i | C = c) = 0$. Here, n_i signifies the number of attributes for which training instance with $X_i = x_i \wedge C = c$ does not present in the training data. The greater is the number n_i of a class c, the lesser is the probability that test instance $\mathbf{x}$ belong to that class c. n_i also depends upon the probability of that class in the training data. Therefore, instead of taking n_i as the significant number in deciding the class of such test instance, we compute for every attribute

$$N_i = n_i / p(C = c) \tag{6}$$

Now, N_i captures the dependency of $p(C = c)$ in the training data. Instead of taking $p(C = c \,|\mathbf{X} = \mathbf{x}) = 0$ for each class, we take $p(C = c \,|\mathbf{X} = \mathbf{x})$ is equal to $p(C = c)$ for a particular class for which N_i is minimum. It means, test instance $\mathbf{x}$ with $p(C = c \,|\mathbf{X} = \mathbf{x}) = 0$ for each class would be classified to the class for which N_i is minimum.Thus,N_i is computed using equation (6) for each test instances of such type.

This new approach of estimating probabilities will solve the problem in basic approach of NBC and also performs better than Laplace and M-estimate methods for noisy and large datasets. The proposed method is simple, efficient and robust for noisy data. We observe reasonably well reduction in error of classification on several datasets using the proposed approach. The comparative results of proposed approach with the existing basic approach and estimate approach are present in the subsequent section of the paper. The result shows that proposed approach outperforms the existing approaches of estimating probabilities using discretization for NBC.

4   Results and Discussion

To determine the robustness of our approach to real world data, we selected 15 databases from the UCI machine learning repository [12]. Table 1 gives mean accuracies of the ten-fold cross validation using different approaches of estimating probabilities in naive-Bayes classifier. For each datasets, Size is the number of instances, Feature is the number of attributes, Class is the number of classes, Basic, Estimate and Proposed represent probability estimation using basic approach,Laplace and M-estimate approach and proposed approach respectively and last two columns show the significance level of paired t test that proposed approach is more accurate than basic and estimate. For each datasets, we used ten-fold cross validation to evaluate the accuracy of the three different approaches i.e. basic, estimate and proposed.

Table 1 shows that the classification accuracy of proposed approach was much better than basic approach in 8 out of the 15 databases. The proposed approach was also significantly better than estimate approach in 4 of the 15 databases. The result also indicates that proposed approach improved the results significantly in case of basic approach of estimating the probabilities in NBC whereas in case of estimate approach, it outperformed for noisy databases such as sonar, pendigits, letter recognition and segmentation. It is important to note that all the databases where proposed approach outperformed estimate approach were large in size and had more number of classes than any other datasets.

Table 1. Mean Accuracies of the ten cross validation using different approaches of estimating probabilities in Naive-Bayes Classifier

Dataset	Size	Feature	Class	Basic	Estimate	Proposed	Proposed Better with Basic?	Proposed Better with Estimate?
Letter	20000	16	26	70.78	70.75	70.82	Yes(97.8%)	Yes(97.4%)
Pendigit	10992	16	10	87.65	87.43	87.65	Equal	Yes(99%)
Segmention	2310	19	7	90.61	89.96	91.08	Yes(99.9%)	Yes(95%)
Vowel	990	10	11	69.49	70.51	70.00	Yes(97.4%)	Equal
Vehicle	846	18	4	62.71	62.12	62.47	Equal	Equal
P-I-Diabetes	768	8	2	75.32	75.58	75.06	Equal	Equal
Wdbc	569	30	2	91.4	94.21	93.33	Yes(99.1%)	Equal
Ionosphere	351	34	2	85.43	90.57	88.00	Yes(99.8%)	Equal
Liver	345	6	2	64.71	64.71	65.00	Equal	Equal
New-Thyoroid	215	5	3	90.00	92.86	90.00	Equal	Equal
Glass	214	10	7	50.95	55.24	55.71	Yes(97.4%)	Equal
Sonar	208	60	2	70.00	65.24	74.29	Yes(97.9%)	Yes(100%)
Wpbc	198	30	2	73.16	70.00	73.16	Equal	Equal
Wine	178	13	3	81.11	96.67	91.67	Yes(100%)	No(99.8%)
Iris	150	4	3	94.67	95.33	95.33	Equal	Equal

Our experiments show that proposed approach is better than basic and estimate approaches and aims at reducing NBC's error of classification. It is observed from Table 1 that the proposed approach is especially useful in classifying noisy datasets.

5   Concluding Remarks

In this paper, a robust approach for estimating probabilities in naive-Bayes classifier based on the maximum occurrence of the number for which conditional probabilities of any of the attributes are zero for a given instance has been proposed in order to overcome the limitation of existing approaches of estimating probabilities in NBC. The effectiveness of the proposed approach over the existing approaches has been illustrated using different databases of UCI machine learning repository. The proposed approach performs remarkably well in terms of classification accuracy for large and noisy datasets as compared to other estimates. The approach can play an important role for wider variety of pattern recognition and machine learning problems by estimating the probabilities for naive-Bayes classifier.

References

1. Aha, D., Kibler, D.: Instance-based learning algorithms. Machine Learning 6, 37–66 (1991)
2. Catlett, J.: On changing continuous attributes into ordered discrete attributes. In: Proceedings of the European Working Session on Learning, pp. 164–178 (1991)
3. Cestnik, B.: Estimating probabilities: A crucial task in machine learning. In: Proceedings of the 9th European Conference on Artificial Intelligence, pp. 147–149 (1990)
4. Domingos, P., Pazzani, M.: On the optimality of the simple Bayesian classifier under zero-one loss. Machine Learning. 29, 103–130 (1997)
5. Dougherty, J., Kohavi, R., Sahami, M.: Supervised and unsupervised discretization of continuous features. In: Proceedings of the Twelfth International Conference on Machine Learning, pp. 194–202 (1995)
6. Duda, R.O., Hart, P.E.: Pattern classification and scene analysis. John Wiley and Sons, New York (1973)
7. Friedman, N., Geiger, D., Goldszmidt, M.: Bayesian network classifiers. Machine Learning 29, 131–163 (1997)
8. John, G.H., Langley, P.: Estimating continuous distributions in Bayesian classifiers. In: Proceedings of the 11th Conference on Uncertainty in Artificial Intelligence, pp. 338–345 (1995)
9. Kerber, R.: Chimerge: Discretization for numeric attributes. In: National Conference on Artificial Intelligence AAAI Press, pp. 123–128 (1992)
10. Langley, P., Iba, W., Thompson, K.: An analysis of Bayesian classifiers. In: Proceedings of the Tenth National Conference on Artificial Intelligence, pp. 223–228 (1992)
11. Lu, J., Yang, Y., Webb, G.I.: Incremental Discretization for Nave-Bayes Classifier. In: Proceedings of the Second International Conference on Advanced Data Mining and Applications, pp. 223–238 (2006)
12. Newman, D.J., Hettich, S., Blake, C.L., Merz, C.J.: UCI Repository of machine learning databases. University of California, Irvine, CA, Department of Information and Computer Science. (1998),
http://www.ics.uci.edu/mlearn/MLRepository.html
13. Quinlan, R.: C4.5: Programs for Machine Learning. Morgan Kaufmann Publishers, San Mateo, CA (1993)
14. Yang, Y., Webb, G.: A Comparative Study of Discretization Methods for Naive-Bayes Classifiers. In: Proceedings of the Pacific Rim Knowledge Acquisition Workshop, Tokyo, Japan, pp. 159–173 (2002)
15. Yang, Y., Webb, G.: On why discretization works for naive-Bayes classifiers. In: Proceedings of the 16th Australian Joint Conference on Artificial Intelligence (AI) (2003)

Weighted k-Nearest Leader Classifier for Large Data Sets

V. Suresh Babu and P. Viswanath

Department of Computer Science and Engineering,
Indian Institute of Technology–Guwahati, Guwahati-781039, India
{viswanath,vsbabu}@iitg.ernet.in

Abstract. Leaders clustering method is a fast one and can be used to derive prototypes called leaders from a large training set which can be used in designing a
classifier. Recently nearest leader based classifier is shown to be a faster version
of the nearest neighbor classifier, but its performance can be a degraded one since
the density information present in the training set is lost while deriving the prototypes. In this paper we present a generalized weighted k-nearest leader based
classifier which is a faster one and also an on-par classifier with the k-nearest
neighbor classifier. The method is to find the relative importance of each prototype which is called its weight and to use them in the classification. The design
phase is extended to eliminate some of the noisy prototypes to enhance the performance of the classifier. The method is empirically verified using some standard
data sets and a comparison is drawn with some of the earlier related methods.

Keywords: weighted leaders method, k-NNC, noise elimination, prototypes.

1 Introduction

Nearest neighbor classifier (NNC) and its variants like k-nearest neighbor classifier
(k-NNC) are popular because of their simplicity and good performance [1]. It is shown
that asymptotically (with infinite number of training patterns) k-NNC is equivalent to
the Bayes classifier and the error of NNC is less than twice the Bayes error[2,1]. It
has certain limitations and shortcomings as listed below. (1) It requires to store the
entire training set. So the space complexity is $O(n)$ where n is the training set size.
(2) It has to search the entire training set in order to classify a given pattern. So the
classification time complexity is also $O(n)$. (3) Due to the curse of dimensionality effect
its performance can be degraded with a limited training set for a high dimensional data.
(4) Presence of noisy patterns in the training set can degrade the performance of the
classifier.

Some of the various remedies for the above mentioned problems are as follows. (1)
Reduce the training set size by some editing techniques where we eliminate some of
the training patterns which are redundant in some sense [3]. For example, Condensed
NNC [4] is of this type. (2) Use only a few selected prototypes from the training set [5].
Prototypes can be selected by partitioning the training set by using some clustering
techniques and then taking a few representatives for each block of the partition as the
prototypes. Clustering methods like *Leaders* [6], *k-means* [7], *etc.,* can be used to derive

A. Ghosh, R.K. De, and S.K. Pal (Eds.): PReMI 2007, LNCS 4815, pp. 17–24, 2007.

the prototypes. (3) Reduce the effect of noisy patterns. Preprocessing the training set and removing the noisy patterns is a known remedy.

With data mining applications where typically the training set sizes are large, the space and time requirement problems are severe than the curse of dimensionality problem. Using only a few selected prototypes can reduce the computational burden of the classifier, but this can result in a poor performance of the classifier. *Leaders-Subleaders* method [5] applies the leaders clustering method to derive the prototypes. The classifier is to find the nearest prototype and assign its class label to the test pattern. While the *Leaders-Subleaders* method reduces the classification time when compared with NNC or k-NNC which uses the entire training set, it also reduces the classification accuracy.

This paper attempts at presenting a generalization over the *Leaders-Subleaders* method where along with the prototypes we derive its importance also which is used in the classification. Further, we extend the method to k nearest prototypes based classifier instead of 1-nearest prototype based one as done in the *Leaders-Subleaders* method. To improve the performance, noisy prototypes, which is appropriately defined as done in some of the density based clustering methods, are eliminated.

The rest of the paper is organized as follows. Section 2, first describes the leaders clustering method briefly and then its extension where the *weight* of each leader is also derived. It also describes regarding the noise elimination preprocessing. Section 3 describes the proposed method, *i.e.,* the k nearest leaders classifier. Section 4 gives the empirical results and finally Section 5 gives some of the conclusions.

2   Leaders and Weighted Leaders

2.1   Notation and Definitions

1. *Classes:* There are c classes *viz.*, $\omega_1, \ldots, \omega_c$.
2. *Training set:* The given training set is $\mathcal{D}$. The training set for class ω_i is $\mathcal{D}_i$, for $i = 1$ to c. The number of training patterns in class ω_i is n_i, for $i = 1$ to c. The total number of training patterns is n.
3. *Apriori probabilities:* Apriori probability for class ω_i is $P(\omega_i)$, for $i = 1$ to c. If this is not explicitly given, then $P(\omega_i)$ is taken to be n_i/n, for $i = 1$ to c.

2.2   Leaders Method

Leaders method [6] finds a partition of the given data set like most of the clustering methods. Its primary advantage is its running time which is linear in the size of the input data set. For a given threshold distance τ, leaders method works as follows. It maintains a set of leaders L, which is initially empty and is incrementally built. For each pattern x in the data set, if there is a leader $l \in L$ such that distance between x and l is less than τ, then x is assigned to the cluster represented by l. In this case, we call x as a *follower* of the leader l. Note that even if there are many such leaders, only one (the first encountered one) is chosen. If there is no such leader then x itself becomes a new leader and is added to L. The algorithm outputs the set of leaders L. Each leader can be seen as a representative for the cluster of patterns which are grouped with it. The leaders method is modified as enumerated below.

1. For each class of training patterns, the leaders method is applied separately, but the same distance threshold parameter, *i.e.,* τ is used. The set of leaders derived for class ω_i is denoted by L_i, for $i = 1$ to c.
2. Each leader l has an associated weight denoted by *weight(l)* such that $0 \leq weight(l) \leq 1$.
3. For a training pattern x which belongs to the class ω_i, if there are p leaders that are already derived such that their distance from x is less than τ, then weight of all these p leaders is updated. Let l be a leader among these p leaders, then its new weight is found by $weight(l) = weight(l) + 1/(p \cdot n_i)$. If $p = 0$, then x itself becomes a new leader whose weight is $1/n_i$.

The modified leaders method called *Weighted-Leaders* is given in Algorithm 1.

Algorithm 1. Weighted-Leaders($\mathcal{D}_i, \tau$)

$L_i = \emptyset$;
for each $x \in \mathcal{D}_i$ **do**
    Find the set $P = \{l \mid l \in L_i, ||l - x|| < \tau\}$;
    if $P \neq \emptyset$ **then**
        for each l such that $l \in P$ **do**
            $weight(l) = weight(l) + 1/(\mid P \mid \cdot n_i)$.
        end for
    else
        $L_i = L_i \cup \{x\}$;
        $weight(x) = 1/n_i$;
    end if
end for
Output L_i which is a set of tuples such that each tuple is in the form $< l, weight(l) >$ where l is a leader and $weight(l)$ is its weight.

The leaders and their respective weights derived depends on the order in which the data set is scanned. For example, for a pattern x, there might be a leader l such that $||l - x|| < \tau$ which is created in a later stage (after considering x) and hence the weight of l is not updated. By doing one more data set scan these kind of mistakes can be avoided. But since the method is devised to work with large data sets, and as the training set size increases the effect these mistakes diminishes, they are ignored.

We assume that the patterns are from a Euclidean space and Euclidean distance is used.

2.3 Eliminating Noisy Prototypes

Since noise (noisy training patterns) can degrade the performance of nearest neighbor based classifiers, we propose to eliminate noisy prototypes in this section.

A leader l which belongs to the class ω_i is defined as a noisy prototype if (1) the class conditional density at l is less than a threshold density and (2) there are no neighbors for l which are dense (*i.e.,* the class conditional density at each of these neighbors is less than the threshold density). This definition is similar to that used in density based clustering methods like DBSCAN [8] and its variants [9].

This process is implemented as follows. For a given ϵ distance, for each leader l (say this belongs to the class ω_i), we find all leaders from class ω_i which are at a distance less than ϵ. Let the set of these leaders be S. Let the cumulative weight of these leaders be W. Then we say that l is a non-dense prototype if $W < \delta$, for a predefined threshold weight δ. If all leaders in the set S are non-dense, then we say l is a noisy prototype and is removed from the respective set of leaders.

Since any two leaders that belongs to a class are separated by distance of at least τ (the threshold used in deriving the leaders), ϵ is normally taken to be larger than τ. Section 4 describes about how these parameters are chosen.

The process of eliminating noise can take time atmost $O(|L|^2)$ where L is the set of all leaders. Since $|L| << n$, where n is the data set size, the noise elimination process will not take much time.

3   Weighted k-Nearest Leaders Classifier

This section describes the proposed classifier. Let L_i be the set of leaders obtained after eliminating noisy leaders for class ω_i, for $i = 1$ to c. Let L be the set of all leaders. That is, $L = L_1 \cup \ldots \cup L_c$. For a given query pattern q, the k nearest leaders from L is obtained. For each class of leaders among these k leaders, their respective cumulative weight is found. Let this for class ω_i be W_i, for $i = 1$ to c.

Let the k nearest leaders are from the region R at q. Approximately class conditional density at q for class ω_i is:

$$\hat{p}(q \mid \omega_i) = \frac{m_i}{n_i \cdot V}$$

where m_i is the number of patterns that are present in the region R that belongs to the class ω_i and n_i is the total number of training patterns for the class ω_i, and V is the volume of the region R. Asymptotically as $n_i \to \infty, m_i \to \infty, m_i/n_i \to 0$ and $V \to 0$, it can be shown that $\hat{p}(q \mid \omega_i) \to p(q \mid \omega_i)$ [1].

It is easy to see that

$$W_i \approx \frac{m_i}{n_i}$$

and hence is proportionate to the $\hat{p}(q \mid \omega_i)$. So, $W_i \cdot P(\omega_i)$ is proportionate to the posterior probability $\hat{P}(\omega_i \mid q)$ where $P(\omega_i)$ is the apriori probability for class ω_i.

The classifier chooses the class according to $argmax_{\omega_i}\{W_1 P(\omega_1), \ldots, W_c P(\omega_c)\}$. If $P(\omega_i)$ is not explicitly given then it is taken to be n_i/n where n_i is the number of training patterns from class ω_i and n is the total number of training patterns.

From the above argument, it is clear that the proposed method is approximately doing the Bayes classification as done by the k-nearest neighbor classifier.

The proposed k nearest leader classifier is given in Algorithm 2.

4   Experimental Results

Experimental studies are done with one synthetic data set and two standard data sets, viz., *Covtype.binary* available at the URL: *htpp://www.csie.ntu.edu.tw/ cjlin/libsvmtools*

Algorithm 2. k-Nearest-Leader(L, q)

{L is the set of all leaders derived from all classes. q is the query pattern to be classified}

Find k nearest leaders of q from L.

Among the k nearest leaders find the cumulative weight of leaders that belongs to each class. Let this be W_i for class ω_i, for $i = 1$ to c.

Class label assigned for $q = argmax_{\omega_i}\{W_1 P(\omega_1), \ldots, W_c P(\omega_c)\}$.

/datasets/binary.html and *Pendigits* available at *http://www.ics.uci.edu/mlearn/MLRepository.html.*

A two dimensional *synthetic* data for a two class problem is generated as follows. First class having 60000 patterns were *i.i.d.* drawn from a normal distribution having mean as $(0, 0)^T$ and covariance matrix as $I_{2\times2}$(*i.e.*,identity matrix). Second class also is of 60000 patterns which is also *i.i.d.* drawn from a normal distribution with mean $(2.56, 0)^T$ and covariance matrix $I_{2\times2}$. The Bayes error rate for this synthetic data set is 10%. The data set is divided randomly into two parts consisting of 80000 and 40000 patterns which are used as training and testing sets respectively.

Covtype.binary is a large data set consisting of 581012 patterns of 54 dimensions which belongs to two classes. The data set is divided randomly into two parts consisting of 400000 and 181012 patterns which are used as training and test sets respectively.

Table 1. Synthetic Data Set

Classifier	Threshold	#Leaders	#Noisy leader	Design time	k	classification time	CA(%)
1-NNC	Nil				1	4692	85.17
k-NNC	Nil				74	7884	89.61
1-NLC	0.06	7947		4.75	1	36.97	73.93
	0.05	10327		6.81	1	53.74	75.52
	0.04	13949		11.98	1	73.9	77.35
	0.03	20164		22.53	1	105.2	79.80
	0.02	31743		46.41	1	151.15	82.43
k-NLC	0.06	7947		4.75	25	152.57	87.42
	0.05	10327		6.81	25	184.71	87.65
	0.04	13949		11.98	25	250.26	88.48
	0.03	20164		22.53	25	357.73	88.92
	0.02	31743		46.4	25	561.84	89.24
wk-NLC	0.06	7947		4.95	25	182.23	89.55
	0.05	10327		7.35	25	240.16	89.56
	0.04	13949		12.50	25	322.75	89.57
	0.03	20164		23.07	25	444.96	89.60
	0.02	31743		47.33	25	714.25	89.50
wk-NLC with noise elimination	0.06	5147	2800	10.61	25	112.97	89.56
	0.05	7067	3260	19.61	25	159.35	89.59
	0.04	10123	3826	37.99	25	233.45	89.62
	0.03	15789	4375	74.95	25	364.46	89.63
	0.02	26784	4959	163.98	25	607.32	89.56

Table 2. Covtype Data Set

Classifier	Threshold	#Leaders	#Noisy leader	Design time	k	classification time	CA(%)
1-NNC	Nil				1	1373066	94.89
1-NLC	0.3	4129		90.41	1	538.15	68.81
	0.25	6775		168.65	1	880.93	72.76
	0.2	12385		380.57	1	1577.86	77.36
	0.15	25746		1050.15	1	4137.72	82.78
	0.1	63433		3395.42	1	11927.2	88.72
k-NLC	0.3	4129		90.41	23	765.46	70.28
	0.25	6775		168.65	25	1311.98	73.28
	0.2	12385		380.57	10	2177.86	77.87
	0.15	25746		1050.15	7	4623.61	83.58
	0.1	63433		3395.42	3	12527.4	89.62
wk-NLC	0.3	4129		92.79	11	674.90	71.95
	0.25	6775		171.56	7	1011.11	74.75
	0.2	12385		385.67	5	1744.37	78.49
	0.15	25746		1059.80	3	4453.59	84.06
	0.1	63433		3415.32	2	12154.6	90.46
wk-NLC with noise elimination	0.3	3753	376	103.3	11	627.9	72.02
	0.25	6293	482	235.45	7	965.6	74.95
	0.2	11772	613	565.63	5	1684.28	79.14
	0.15	24844	902	1210.9	3	4086.28	84.98
	0.1	61877	1556	3812.85	2	11657.66	91.65

Pendigits is a medium sized data set consisting of 7494 training patterns and 3498 test patterns. The dimensionality is 16 and the number of classes is 10.

The classifiers chosen for the comparative study are: (1) the nearest neighbor classifier(NNC), (2) the k-nearest neighbor classifier(k-NNC), (3) the nearest leader classifier(NLC), (4) the k-nearest leader classifier(k-NLC) and (5) the weighted k-nearest leader classifier(wk-NLC) which is the proposed one in this paper. NLC and k-NLC are similar to NNC and k-NNC, except that, instead of nearest neighbor(s), nearest leader(s) are taken in to consideration. For wk-NLC experiments are done with noise elimination preprocessing and without it.

The experiments are conducted for various leader's threshold *i.e.,* τ values. For *Synthetic* data set the τ values chosen are $\{0.02, 0.03, 0.04, 0.05, 0.06\}$, for *Covtype.binary* data set the τ values chosen are $\{0.1, 0.15, 0.2, 0.25, 0.3\}$, and for *Pendigits* data set these are $\{20, 30, 40, 50, 60\}$. The ϵ value for noise elimination are 0.12, 0.6 and 120 for *Synthetic* data set, *Covtype.binary* data set and *Pendigits* data set respectively. The parameter δ used as weight threshold to eliminate noise is chosen as 5% of the average weight of the leaders in the respective data sets. Similarly, the k value for k-NNC, k-NLC and wk-NLC are chosen from a three fold cross validation from $\{1, 2, \ldots, 40\}$.

The results are summarized in Tables 1, 2,and 3.

From the results it can be seen that the leader based classifies are considerably faster than NNC and k-NNC. The classification time and classification accuracy(CA) of the

Table 3. Pendigits Data Set

Classifier	Threshold	#Leaders	#Noisy leader	Design time	k	classification time	CA(%)
1-NNC	Nil				1	302.6	97.74
k-NNC	Nil				3	320.8	97.83
1-NLC	60	385		0.10	1	0.34	91.76
	50	631		0.11	1	0.53	94.19
	40	1165		0.13	1	0.94	95.85
	30	2306		0.19	1	1.78	97.17
	20	4821		0.40	1	4.70	97.48
k-NLC	60	385		0.10	4	0.40	92.28
	50	631		0.11	4	0.63	94.72
	40	1165		0.13	5	1.19	95.11
	30	2306		0.19	3	2.16	96.04
	20	4821		0.40	1	4.70	97.48
wk-NLC	60	385		0.10	2	0.37	93.39
	50	631		0.11	2	0.58	95.48
	40	1165		0.13	3	1.13	95.19
	30	2306		0.19	2	2.20	96.68
	20	4821		0.40	2	5.48	97.57
wk-NLC with noise elimination	60	363	22	0.10	2	0.37	93.37
	50	606	25	0.13	2	0.57	95.43
	40	1133	32	0.21	3	1.09	95.23
	30	2260	46	0.47	2	2.11	96.71
	20	4753	68	1.48	2	5.43	97.57

leader based classifiers depends on the threshold τ. As the value τ reduces, the classification time increases, and also the CA increases. With $\tau = 0$ each distinct training pattern becomes a leader and hence NLC is same as NNC and k-NLC, wk-NLC both are same as k-NNC. With $\tau = 0.03$ for *synthetic* data set, with $\tau = 0.1$ for *Covtype.binary* data set, and with $\tau = 20$ for *Pendigits* data set, the CA of wk-NLC is almost similar to the CA of k-NNC but with much reduced classification time. The classification time of wk-NLC when compared with that of k-NNC and NNC are less than 6%,1% and 2% for Synthetic, Covtype.binary and Pendigits data sets respectively. With noise elimination wk-NLC shows some improvement with respect to classification accuracy over wk-NLC without noise elimination. Also with noise elimination wk-NLC can be slightly faster than wk-NLC without noise elimination.

5   Conclusions

In this paper an improvement over using leaders as prototypes is given. For each of the leaders an associated weight is found which gives its relative importance. These weights are used in the k nearest leader based classifier called weighted k nearest leader classifier. A preprocessing step to eliminate noisy prototypes is also presented. The proposed method is experimentally compared with nearest neighbor classifier(NNC), k nearest neighbor classifier(k-NNC), nearest leader classifier and k nearest leader classifier. With suitable parameters, the proposed method can achieve classification accuracy similar to k-NNC and is superior than the other methods, but is a much fast classifier than k-NNC or NNC. Hence the proposed method is a suitable one to be used with large data sets as in data mining applications.

References

1. Duda, R.O., Hart, E., Stork, P.: Pattern Classification, 2nd edn. John Wiley & Sons, Chichester (2000)
2. Cover, T., Hart, P.: Nearest neighbor pattern classification. IEEE Transactions on Information Theory 13, 21–27 (1967)
3. Dasarathy, B.V.: Nearest neighbor (NN) norms: NN pattern classification techniques. IEEE Computer Society Press, Los Alamitos, California (1991)
4. Hart, P.: The condensed nearest-neighbor rule. IEEE Transactions on Information Theory IT-4, 515–516 (1968)
5. Vijaya, P., Murty, M.N., Subramanian, D.K.: Leaders-subleaders: An efficient hierarchical clustering algorithm for large data sets. Pattern Recognition Letters 25, 505–513 (2004)
6. Spath, H.: Cluster Analysis Algorithms for Data Reduction and Classification. Ellis Horwood, Chichester, UK (1980)
7. Jain, A., Dubes, R., Chen, C.: Bootstrap technique for error estimation. IEEE Transactions on Pattern Analysis and Machine Intelligence 9, 628–633 (1987)
8. Ester, M., Kriegel, H.P., Xu, X.: A density-based algorithm for discovering clusters in large spatial databases with noise. In: Proceedings of 2nd ACM SIGKDD, Portand, Oregon, pp. 226–231 (1996)
9. Viswanath, P., Pinkesh, R.: l-dbscan: A fast hybrid density based clustering method. In: Proceedings of the 18th Intl. Conf. on Pattern Recognition (ICPR 2006), Hong Kong, vol. 1, pp. 912–915. IEEE Computer Society, Los Alamitos (2006)

Hybrid Approaches for Clustering

Laxmi Kankanala and M. Narasimha Murty

Indian Institute of Science, Bangalore, India
{laxmi,mnm}@csa.iisc.ernet.in

Abstract. Applications in various domains often lead to very large and frequently high-dimensional data. Successful algorithms must avoid the curse of dimensionality but at the same time should be computationally efficient. Finding useful patterns in large datasets has attracted considerable interest recently. The primary goal of the paper is to implement an efficient Hybrid Tree based clustering method based on CF-Tree and KD-Tree, and combine the clustering methods with KNN-Classification. The implementation of the algorithm involves many issues like *good accuracy*, *less space* and *less time*. We will evaluate the time and space efficiency, data input order sensitivity, and clustering quality through several experiments.

1 Introduction

Classification process attempts to generate the description of the classes, and those descriptions helps to classify the unknown data points. Data clustering identifies the sparse and the crowded places, and hence discovers the overall distribution patterns of the dataset. Besides, the derived clusters can be visualized more efficiently and effectively than the original dataset.[1]

A pattern classifier has two phases. They are: *design phase* where abstractions are created; and *classification phase*, where the classification of test patterns is done using these abstractions. Corresponding to these two phases, we have *design time* and *classification time*. In classification based on neighbourhood classifiers, there is no design phase; so, zero design time. However, the classification phase could be computationally expensive.

With an increasing number of new database applications dealing with very large high dimensional data sets, data mining on such data sets has emerged as an important research area. There are a number of different clustering algorithms that are applicable to very large data sets, and a few that address large high dimensional data.

In this paper we propose two algorithms. One is a tree based clustering algorithm which combines CF-Tree[6] and KD-Tree[2] to get moderate space and time complexities. And the other one is combination of clustering with classification using CF-Tree, KD-Tree and KNNC.

1.1 Review of Literature

In this section, we briefly discuss the two tree based clustering algorithms, CF-Tree and KD-Tree. And also a classification algorithm, KNNC.

A. Ghosh, R.K. De, and S.K. Pal (Eds.): PReMI 2007, LNCS 4815, pp. 25–32, 2007.

CF-Tree. ClusterFeature-Tree is based on the principle of agglomerative clustering, that is, at any given stage there are smaller subclusters and the decision at the current stage is to merge subclusters based on some criteria. It maintains a set of *cluster features*(CF) for each subcluster in the form of tree.

Each CF is given by a vector (n, ls, ss); where n is the number of data objects in CF, ls is the linear sum of the data objects, and ss is the square sum of the data objects in CF. CF-Tree is a height-balanced tree with two parameters: branching factor B and threshold T. Each nonleaf node contains at most B entries of the form $[CF_i, child_i]$, where $i \in 1, 2, \cdots, B$, $child_i$ is a pointer to its ith child node, and CF_i is the CF of the subcluster represented by this child. A nonleaf node represents a cluster made up of all the subclusters represented by its entries. And leaf node contains at most L entries, each of the form $[CF_i]$, where $i = 1, 2, \cdots, L$. In addition, each leaf node has two pointers, "prev" and "next" which are used to chain all leaf nodes together for efficient scans. A leaf node also represents a cluster made up of all the subclusters represented by its entries. But all entries in a leaf node must satisfy a threshold requirement, with respect to a threshold value T the diameter has to be less than T.

CF-Tree is a data dependent structure and the parameters T and B need to be tuned. This generally causes increased design times. It has been observed that it takes considerably more space than many other approaches.

KD-Tree. *KD-Tree* is a binary tree structure for storing a finite set of points from a k-dimensional space, generally to handle spatial data, in a simple way. Associated with each internal node N there is a coordinate x and a value v based on which the points are divided into either of the children. Like binary search trees, all points to the left of N will have the x-coordinate less than v and the points to the right will have the greater than (or equal to) that. To construct a KD-Tree for a given set of points, we start with a coordinate and continue until each leaf node satisfies a given set of constraints like all dimensions are used, all the points in the node are similar,etc.

This approach takes lot of time as for every decision we need to scan the whole data.

KNNC. K-Nearest Neighbor (KNN)[5] classification is a very simple, yet powerful classification method. The key idea behind KNN classification is that similar observations belong to similar classes. Thus, one simply has to look for the class designators of a certain number of the nearest neighbors and weigh their class numbers to assign a class number to the unknown. The weighing scheme of the class numbers is often a majority rule, but other schemes are conceivable. k should be kept small, since a large k tends to create misclassifications unless the individual classes are well-separated. It can be shown that the performance of a KNN classifier can be same as at least half of the best possible classifier for a given problem, under certain conditions.

One of the major drawbacks of KNN classifier is that the classifier needs all available data. This may lead to considerable overhead, if the training data set is large.

So in this paper, we combine the above approaches to overcome the drawbacks.

1.2   Outline of the Paper

The paper is organized as follows. Section 2 gives the details about the new clustering algorithm based on CF-Tree and KD-Tree. Section 3 gives description about KNN classification with CF-Tree and KD-Tree clustering. Section 4 has experimental results of our implementations of the algorithms. And we conclude in Section 5.

2   CFKD-Tree Algorithm

CF-Tree algorithm takes more space and less time compared to KD-Tree. So to achieve advantages of both the approaches we propose to combine CF-Tree and KD-Tree algorithms into a new algorithm, CFKD-Tree Algorithm. CFKD-Tree Algorithm has two phases, first phase builds level restricted CF-Tree and the second phase builds a KD-Tree for each leaf node of the CF-Tree. Level restriction can be done by tuning the branching factor(B) and threshold(T). As the leaves of the CF-Tree form smaller clusters, KD-Tree can perform better in terms of running time and at the same time we can achieve space savings. In the KD-Tree phase, we have used variance as the decision parameter for dimension selection. And we take mean as the basis for splitting. See Figure 1.

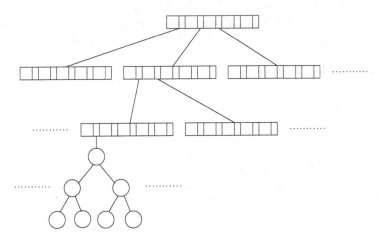

Fig. 1. CFKD-Tree

\ * CFKD-Tree construction for dataset $\mathbb{D}$ * \
CFKD($\mathbb{D}$, B, T)
{

1. Construct Tree using CF-Tree($\mathbb{D}$, B, T)
2. for each cluster C of Tree
   − construct KD-Tree(C)

}

CFKD-Tree algorithm improves the time and accuracy of KD-Tree algorithm, and takes lesser space than CF-Tree.

CFKD-Tree algorithm can be done in other way also. First we construct a constrained KD-Tree and after getting the clusters, we construct CF-Tree for each cluster. It will take lesser time than former but sometimes gives poor accuracy.

KD-Tree can be constrained by restricting the number of points in each cluster, by fixing the radius of the cluster and taking only few dimensions for splitting.

3   CF-KNNC and KD-KNNC Algorithms

We also propose a combination of a tree based clustering and KNNC with a view to combine advantages of KNNC, CF-Tree and KD-Tree algorithms. KNNC essentially provides good accuracy, however takes longer to run.

3.1   CF-KNNC Algorithm

This algorithm has two phases, design and classification. Where the design phase constructs level restricted CF-Tree, and in classification phase it runs KNNC on the selected leaves of the CF-Tree. To improve the execution time we added a heuristic where KNNC is avoided if all the patterns in a cluster have the same class label. We will find KNNs for only those patterns which belong to border clusters i.e., the clusters which contains data points of multiple classes.
\ * Design phase of CF-KNNC * \
CF-KNNC($\mathbb{D}$, B, T)
{

1. Construct CF-Tree($\mathbb{D}$, B, T)

}

\ * Classification phase of CF-KNNC * \
\ * Classifying pattern p using the *tree* constructed in design phase of CF-KNNC, and k is the number of nearest neighbors to be find * \
Classify(p, *tree*, k)
{

1. Find appropriate cluster C in *tree*
2. if all points of C are in same class
   − return the label of that class

3. else

    – apply KNNC on C

    – return the label

}

3.2 KD-KNNC Algorithm

This algorithm also has two phases, design and classification. In the first phase we construct a constrained KD-Tree by restricting the number of points in the cluster, by fixing the radius and taking only few dimensions for decision making. In the second phase, we find KNNs for the border clusters as in CF-KNNC.
\ * Design phase of KD-KNNC * \
KD-KNNC($\mathbb{D}$)
{

  1. Construct constrained KD-Tree($\mathbb{D}$)

}

\ * Classification phase of KD-KNNC * \
\ * Classifying pattern p using the *tree* constructed in design phase of KD-KNNC, and k is the number of nearest neighbors to be find * \
Classify(p, *tree*, k)
{

  1. Find appropriate cluster C in *tree* for p

  2. if C is a border cluster

    – apply KNNC on C

    – return the label

  3. else

    – return the label of that C

}

For the above example, Figure 2, we apply KNNC only on clusters 3 and 4.

4 Experimental Results

4.1 Setup

All the experiments are performed on a Intel(R) Pentium(R) 4 CPU 3.20GHz PC with 1GB main memory.

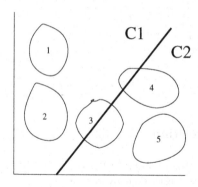

Fig. 2. KNNC with Clustering

Table 1 gives the details of the data sets which have been used for our experiments. KDD is 10% of Network Intrusion Detection Data used in KDDCUP'99. OCR is a digital handwritten data. Sonar is a sonar signals data. Covtype is forest cover type data.

Table 1. Description of Datasets

Dataset	Training Patterns	Testing Patterns	Attributes	Classes
OCR [4]	6670	3333	192	10
SONAR [3]	104	104	60	2
Covtype [3]	15120	565892	54	7
KDD [3]	494021	311029	36	5

4.2 Results

Tables 2, 3, 4 gives experimental results of CFKD, CF-KNN, KD-KNN respectively.

Table 2. Comparison of CF-Tree, KD-Tree and CFKD-Tree Algorithms

Dataset	CF			KD			CF-KD		
	Space (KB)	Time (sec)	Acc. (B, T)	Space (KB)	Time (sec)	Acc.	Space (KB)	Time (sec)	Acc. (B, T)
OCR	40965.5	4.72	85.5386 (18, 2.9)	5060.68	3.13	45.6046	45938.5	14.52	82.4782 (20, 2.5)
SONAR	236.17	0.013	79.8077 (5, 0.69)	28.82	0.012	62.5000	220.43	0.096	78.8462 (35, 0.65)
Covtype	30702.1	403.49	63.2009 (140, 0.3)	3839.0	15.39	63.9056	28850.3	504.74	66.7035 (160, 0.2)
KDD	368301	130.019	97.1986 (11, 1.3)	80787.6	976.22	96.9944	316437	149.37	97.1837 (12,1.12)

Table 3. Comparison of CF-Tree, KNNC and CF-KNNC Algorithms

Dataset	CF			KNNC		CF-KNNC		
	Space (KB)	Time (sec)	Acc. (B, T)	Time (sec)	Acc.	Space (KB)	Time (sec)	Acc. (B, T)
OCR	40965.5	4.72	85.5386 (18, 2.9)	41.37	92.4992	40965.5	4.72	85.6886 (18, 2.9)
SONAR	224.16	0.016	82.6923 (11, 0.65)	0.006	80.7600	128.02	0.035	84.6154 (22, 1.10)
Covtype	30702.1	403.49	63.2009 (140, 0.3)	5298.49	64.1391	27960.6	371.99	63.683 (190, 0.35)
KDD	368301	130.01	97.1986 (11, 1.3)	17271.2	97.1656	362338	156.72	97.4431 (12, 0.7)

Table 4. Comparison of KD-Tree, KNNC and KD-KNNC Algorithms

Dataset	KD			KNNC		KD-KNNC		
	Space (KB)	Time (sec)	Acc.	Time (sec)	Acc.	Space (KB)	Time (sec)	Acc.
OCR	5060.68	3.13	45.6046	41.37	92.4992	5035.4	23.00	85.7486
SONAR	28.82	0.012	62.5000	0.006	80.7600	24.84	0.0026	83.6538
Covtype	3839.0	15.39	63.9056	5298.49	64.1391	3252.99	421.781	64.5084
KDD	80787.6	976.22	96.9944	17271.2	97.1656	75279.2	1026.11	96.9944

5 Conclusions

In this paper, we discussed CF-Tree, KD-Tree and CFKD-Tree clustering algorithms; KNNC, CF-KNNC, KD-KNNC classification algorithms. It is seen that CFKD-Tree takes lesser space than CF-Tree and gives better accuracy than KD-Tree especially for large datasets. KNNC with Tree based clustering reduces the classification time of KNNC. It takes lesser space and provides better accuracy. CF-KNNC gives better accuracy than CF-Tree and much lesser time than KNNC. KD-KNNC gives better accuracy than KD-Tree and much lesser time than KNNC. If the training data is very large, CF-Tree or KD-Tree takes lot of space. That can be avoided at the cost of accuracy. So to handle that type of datasets we can use CF-KNNC or KD-KNNC, which take moderate time and space and with good accuracy. Our results suggest that Hybrid methods perform moderately compared to individual algorithms.

References

1. Jain, A.K., Murty, M.N., Flynn, P.J.: Data clustering: a review. ACM Comput. Surv. 31(3), 264–323 (1999)
2. Moore, A.W.: An intoductory tutorial on kd-trees (October 8, 1997)
3. Newman, D.J., Hettich, S., Blake, C.L., Merz, C.J.: UCI repository of machine learning databases (1998)
4. Viswanath, P., Narasimha, M., Bhatnagar, S.: Partition based pattern synthesis technique with efficient algorithms for nearest neighbor classification. Pattern Recognition Letters 27(14), 1714–1724 (2006)
5. Zhang, B., Srihari, S.N.: Fast K-nearest neighbor classification using cluster-based trees. IEEE Trans. Pattern Analysis and Machine Intelligence 26(4), 525–528 (2004)
6. Zhang, T., Ramakrishnan, R., Livny, M.: BIRCH: An efficient data clustering method for very large databases. In: Jagadish, H.V., Mumick, I.S. (eds.) Proceedings of the 1996 ACM SIGMOD International Conference on Management of Data, Montreal, Quebec, Canada, pp. 103–114 (June 4–6, 1996)

Recognizing Patterns of Dynamic Behaviors Based on Multiple Relations in Soccer Robotics Domain

Huberto Ayanegui and Fernando Ramos

ITESM Campus Cuernavaca
Reforma 182-A, Col Lomas De Cuernavaca,
62589, Temixco Morelos, Mexico
{huberto.ayanegui,fernando.ramos}@itesm.mx

Abstract. This work is focused on the recognition of team patterns represented by different formations played by a soccer team during a match. In the soccer domain, the recognition of formation patterns is difficult due to the dynamic and real time conditions of the environment as well as the multiple interactions among team mates. In this work, some of these multiple interactions are modeled as relations represented by a topological graph which is able to manage the dynamic changes of structures. Thus, the topological graph serves to recognize apparent changes of formations from real changes of them. The proposed model has been tested with different teams in different matches of the Robocup Simulation League. The results have shown that the model can recognize the different main formations used by a team during a match even the multiple changes of the players due to the dynamic nature of a match.

Keywords: Pattern recognition, robotic soccer, formations, dynamic behavior.

1 Introduction

Formations are the way a soccer team lines up its defense, midfield, and attack line during a match. When talking about formations, defenders are listed first and then midfielders and forwards. For example, a code 4:4:2 represents a formation composed by four defenders, four midfielders, and two forwards. As in the real soccer game the goalkeeper is not considered as part of the formation. Certainly, the dynamic conditions of the soccer game difficult the task of building adequate representations able to facilitate the recognition of formation patterns.

Usually, teams playing in strategic and organized ways search for respecting predefined patterns or formations [2][3]. The purpose of this work is the recognition of patterns represented by formations played by a team during a match. If the recognition task is performed on an opponent team, a soccer team can obtain advantages over it [8]. Visser and colleagues [3] recognize formations of opponent teams using neural networks. This work feeds the observed player positions into a neural network and tries to classify them into a predefined set of formations.

A. Ghosh, R.K. De, and S.K. Pal (Eds.): PReMI 2007, LNCS 4815, pp. 33–40, 2007.

If a classification can be done, the appropriate counter formation is looked up and communicated to the players. Due to the fact that Visser does not represent the multiple relations between players, so the neural network has a low level of accuracy for some cases of soccer teams.

Riley and colleagues in [4] use a model to identify 'home areas' of players to recognize formations. A home area specifies the region of the field in which the agent should generally be. Thus, they propose that identifying home areas, the agents can infer a role in the team (defender, midfielder or forward players). A drawback of this approach is that due to dynamic conditions of the world, the player movements can generate such a wide range extending considerably the home areas, which difficult the task of determining the role of a player, therefore a correct formation.

Kuhlmann and colleagues [11] learn team formations similar to Riley and Veloso [4], using home positions. They model the formation as a home position (X,Y) and introduce a ball attraction vector (BX,BY) for each player. The X and Y values are calculated as the average x and y coordinates of the observed player during the course of the game. Values for BX and BY were handpicked for each position and were found through brief experimentation. A weakness of this work is that the home positions have to be adjusted, for some cases, manually.

It is presented in this work an efficient model to recognize patterns of formations based on a representation that takes into account multiple relations among defender, midfielder and forward players. The test domain for this research is simulated robotic soccer, specifically, the Soccer Server System [6], used in the Robot World Cup Initiative [7], an international AI and robotics research initiative. The system is a rich multiagent environment including fully distributed team agents in two different teams composed of eleven agents.

2   The Multiple Relation Model

The focus of this work is on teams that play following patterns of high level of abstraction based on a distribution of zones named Defensive (D), Middle (M) and Attack (A), as in classic soccer game. These patterns will be represented as follows: D:M:A. Due to the dynamic conditions of the soccer game, the players are in constant movement and temporally breaking the alignment of players belonging to a zone. To handle the constant changes without an expressive representation of the relations between players can result in an inefficient way of recognizing formations submitted to a dynamic environment. In the next sections will be explained how the zones and the players belonging to them are recognized.

2.1   Recognition of Team Zones

As in human soccer domains the players in robotic soccer should tend to be organized [1],[9]. That is, each player has a strategic position that defines its movement range in the soccer field. The role of a player is quite related with a predefined area within which an individual player can play basically in the field. Any behaviours of a player depend on its current role. According to the position of the player, roles in robotic soccer can be divided into four types: goalkeeper, defenders, midfielders and forwards. Different roles are associated with different positions and different behaviours that players assume. However, due to the dynamic changing conditions of a match, a defender could become a forward temporarily as his team is trying to attack. As well, a forward could become a defender temporarily when his team is being attacked. So the roles of a player are dynamically changing and a player can have dynamic behaviours in a match. Consequently, the recognition of formation patterns is difficult due to the dynamic and real time conditions of the environment. It is needed to determine, first of all, the belonging of players to a specific zone. This first approach tells us the clustering of players to a zone. In this work, the clustering algorithm, K-means [10], is applied to meet this first stage. K-means classifies a given data set through a certain number of clusters (assume k clusters) fixed a priori. In this work, k=3 such that three zones will be defined: defensive, middle and attack zones (D:M:A). From the log file (game film), the data from one team is extracted and K-means is applied in each simulation cycle of the game. The positions of each player, with respect to the x axis, are taken as the input of the clustering algorithm and the output of clustering is the classification, according to their x position, of all players of the team in the three clusters. Due to the continuously moving of players, it is not feasible to conclude what players are in each zone of a team from a single cycle only, a period of time should be considered.

Clustering algorithm is useful to determine the three zones of a team but it is not able to represent the multiple relations between players of each zone. Given that patterns of formations are based on relations that determine structures then an additional model is crucial for the recognition of patterns of formations. Such model has to be sufficiently robust in order to manage the constant positional changes of players. The next section describes how a set of triangular sub-graphs connected together, that uses as input the result of K-means algorithm, can be used as an adequate representational model able to facilitate the recognition of formation patterns.

2.2   Representation of Multiple Relations

A formation is represented by a set of relations between players. Thus, the relations represent the structure that supports a formation. So, a change of relations between players entails a change of formation. It is needed at least the change of one relation to transform one structure into another one. Constant changes of relations could occur because the multiple relations in a formation and the dynamic nature of a match. Figure 1(a) illustrates the relations of each one of the players with the rest of their teammates. A total of 90 relations are obtained by given by the formula: $n(n-1)$, where n represents the number of players. This formula considers two relations by each pair of players. Thus, one relation is represented by the link from player A to player B and the second one from player B to player A. For practical reasons it is considered just one of these

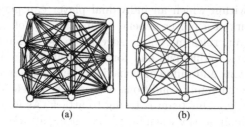

Fig. 1. All posible relations between players of a soccer team. (a) 90 relations and (b) 45 relations.

relations. Thus, the total of relations is $\frac{n(n-1)}{2} = 45$. Figure 1(b) illustrates these 45 relations.

On the one hand, the control of such number of relations becomes very difficult to be managed because any change of relations would produce a change of structure. In addition, it could happen that several changes of relations occur at the same time then the problem of detecting what relations are provoking changes of structures becomes much more difficult to be managed. On the other hand, the 45 relations are not relevant in a real match, because a relevant relation is the one in which a player uses to exchange passes and positions in a strategic way. Thus, a player stays related with his closer neighbor belonging to his zone and the closer neighbor belonging to the neighbor zone. In this work, the goal is to build a simple but robust structure based on relevant relations able to manage the dynamic nature of the game in a topological way, based on triangular sub-graphs that are built as indicated below:

- *Step 1.* Let A $= \{a_i < b_i < c_i < ... < n_i\}$ be finite number of nodes (players) belonging to a zone Z_i. Where player a_i is located at the top of the zone and player n_i at the bottom of it. The rest of the nodes are located at intermediate positions. Thus, a_i is linked with b_i, b_i is linked with c_i, and so on (see Fig. 2(a)).
- *Step 2.* Once the nodes of the zones have been linked, nodes of neighbor zones based on minimal distances are linked until a planar graph is built (see Fig. 2(b)).

Figure 2(c) shows the planar graph represented by triangular sub-graphs as result of applying the previous two steps.

The total number of relations of a graph, which has been built based on the method described above, is given by $N_m + 15$; where N_m is the number of nodes of the middle zone (Due to the lack of space the deduction of this formula is not described in this work). For instance, for a formation 4:4:2, the number of relations will be 19, because $N_m = 4$. The advantages of this method are expressed below:

- The number of relations has been reduced from 45 to 19 for the formation 4:4:2. Then, 26 relations have been eliminated.

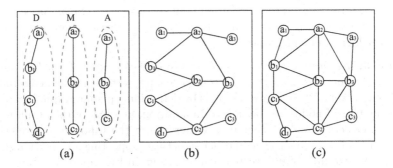

Fig. 2. (*a*) *Step 1.* Neighbor nodes of the same zone are linked. (*b*) *Step 2.* Neighbor nodes of neighbor zones are linked. (*c*) Planar graph obtained from step 1 and step 2.

- In a formation the minimal number of relations of a node is 2 meanwhile the maximal number is 9. An example of this kind of formation is 9:1:1. For this formation the total number of relations is 16.
- Triangular subgraphs are able to assume a topological behavior. That is, even if a structure is deformed because positional changes of nodes, the topological property of the triangular graphs helps to preserve the structure.

3 Pattern Recognition Process

Fig. 3 shows the process to recognize patterns of formations and changes of structures that support the formations. The first module serves to determine the zones by using a clustering algorithm; the second module builds the multiple relations which are expressed by a topological graph and finally in the third module the changes of structures are detected if topological properties of a defined structure have been broken.

Module 1. *Recognition of team zones.* The algorithm of clustering is performed during the first cycles of the match and it is stopped until the number of players in each group does not change. In this way, the three zones of a team, defensive, middle and attack zones are recognized.

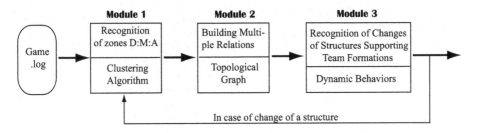

Fig. 3. Process to recognize pattern formations

Module 2. *Building multiple relations and a topological graph.* Based on the three zones recognized by the clustering algorithm, relevant multiple relations and a topological planar graph are built.

Module 3. *Recognition of Changes of Structures that support Team Formations.* Changes of structures are detected if topological properties of a defined structure have been broken. A topological graph is by definition a planar graph. In a planar graph any pair of nodes can be linked. In addition, any link of the graph should not be intersected by any other link. Otherwise, if the topological property of the graph has been broken then another structure supporting a formation should be built. An example of intersected links is shown in Figure 4(b). Intersections occur when players change their roles in order to build a new formation or due to reactive behavior in response to the opponent. If intersections of links occur during a short period of time, thus they have not considered as a change of formation. But, if they occur during a long period of time, clustering algorithm should redefine the zones and a new topological graph should be built. Fig. 4(b).

4   Experimental Results

In this section, important experimental results are shown. The results to be shown are derived from a match of the FC Portugal soccer team of the RoboCup Simulation Championship. This team has won several world RoboCup championships. However, in order to valid the approach presented in this work a vast number of matches has been analyzed.

Fig. 4 shows the results of the clustering algorithm derived from the first 1500 cycles of a match. As you can see, the clustering algorithm recognized the three zones of a team. 10 different classifications (formations) have been detected. However, formation 4.3:3 has been recognized more times than others. In particular, it has been recognized most frequently during the first 1000 cycles. The formation 4:2:2 was recognized in the next 500 cycles. On the contrary,

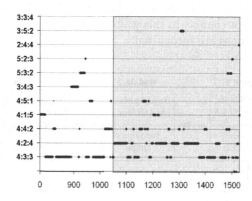

Fig. 4. Clustering results of the first 1500 simulation cycles

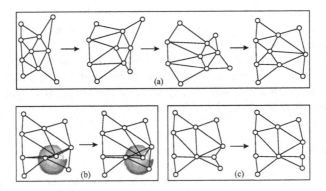

Fig. 5. (*a*) Four frames showing a plannar graph during the first 1000 simulation cycles. (*b*) Intersection of relations (shadow region). (*c*) Building of a new plannar graph for a new formation.

formations such as 3:5:2, 2:4:4, 5:2:3, and 5:3:2 have been detected during a very short period of time.

Fig. 5(a) shows the 9 triangles of the planar graph of some simulation cycles during the first 1000 simulation cycles. It is observed from this sequence of graphs that the structured supporting the formation 4:3:3 is maintained. However, the clustering algorithm recognized different formations even that the relations between agents have not changed. The team started with the formation 4:3:3 and changed to 4:2:4 at the cycle 1050, just after scoring the second goal. A possible cause could be that the score was in favour of this team, so the team has possibly decided to experiment a more offensive behaviour. Fig. 5(b) shows intersections of relations that indicates a change of formation. Once intersections of links have been detected the relations are redefined and a new formation is determined as describe by module 2 discussed before. Fig. 5(c) shows the new planar graph built because the intersections shown in Fig. 5(b).

5   Conclusion

In this work, we have been concerned with the recognition of pattern dynamic behaviors within complex, competitive and real time domains, such as soccer robotic games. The patterns to be recognized are formations that can change because strategic reasons or due to reactive behavior in response to the opponent. These facts bring about multiple interactions between agents of a team, and make difficult the task of recognizing formation patterns. In this work, we presented an efficient model to recognize formations based on a rich representation that takes into account multiple relations among players, including the neighbourhood relation between players of the same zone and the neighbourhood relation between players belonging to neighbour zones. In particular, a topological model, based on triangular sub-graphs, has been built, which is able to manage the deformations of structures due to the dynamic changes and the multiple relations. A vast number of matches have been analyzed to test the model. The experiments have shown that the model is able to recognize formations and apparent and real changes of them.

As future work variants around a formation will be modelled. A variant is defined as a temporal change around a given formation during a short period of time. We consider that variants are, in most of the times, the strategic key of the teams that decide the final result of a match.

References

1. Kaminka, G., A., Fidanboylu, M., Allen, C., Veloso, M.: Learning the Sequential Coordinated Behavior of Teams from Observations. In: Proceedings of the RoboCup-2002 Symposium, Fukuoka, Japan (June 2002)
2. Kuhlmann, G., Stone, P., Lallinger, J.: The Champion UT Austin Villa 2003 Simulator Online Coach Team. In: Polani, D., Browning, B., Bonarini, A., Yoshida, K. (eds.) RoboCup 2003. LNCS (LNAI), vol. 3020, Springer, Heidelberg (2004)
3. Ubbo, V., Christian, D., Sebastian, H., Esko, S.: Weland Hans-Georg: Recognizing Formations in Opponent Teams. In: Stone, P., Balch, T., Kraetzschmar, G.K. (eds.) RoboCup 2000. LNCS (LNAI), vol. 2019, Springer, Heidelberg (2001)
4. Patrick, R., Manuela, V., Gal, K.: An empirical study of coaching. In: Distributed Autonomous Robotic Systems 6, Springer, Heidelberg (2002)
5. Crelle, A.L.: Sammlung mathematischer Aufstze und Bemerkungen, vol. 1. Maurer Berlin, p. 176 (1821)
6. Itsuki, N., lan, F.: Investigating the Complex with Virtual soccer. In: Heudin, J.-C. (ed.) VW 1998. LNCS (LNAI), vol. 1434, pp. 241–253. Springer, Heidelberg (1998)
7. Kitano, H., Tambe, M., Stone, P., Veloso, M., Coradeschi, S., Osawa, E., Matsubara, Noda, I., Asada, M.: The RoboCup Synthetic Agent Challenge 1997. In: Proceedings of IJCAI 1997, Nagoya, Japan (August 1997)
8. David, C., Shaul, M.: Incorporating Opponent Models into Adversary Search. In: Thirteenth National Conference on Artificial Intelligence, Portland Oregon, AAAI Press (1996)
9. Bo, Y., Qinghua, W.: Agent brigade in dynamic formation of robotic soccer. In: Proceedings of the 3rd World Congress on Intelligent Control and Automation, vol. 1, pp. 174–178 (2000)
10. MacQueen, J.B.: Some Methods for classification and Analysis of Multivariate Observations. In: Proceedings of 5th Berkeley Symposium on Mathematical Statistics and Probability, Berkeley, vol. 1, pp. 281–29. University of California Press (1967)
11. Kuhlmann, G., Stone, P., Lallinger, J.: The UT Austin Villa 2003 Champion Simulator Coach: A Machine Learning Approach. In: Nardi, D., Riedmiller, M., Sammut, C., Santos-Victor, J. (eds.) RoboCup 2004. LNCS (LNAI), vol. 3276, pp. 636–644. Springer, Heidelberg (2005)

An Adaptive Algorithm for Failure Recovery During Dynamic Service Composition*

Xingzhi Feng, Huaimin Wang, Quanyuan Wu, and Bin Zhou

School of Computer, National University of Defense Technology,
410073 Changsha, China
billytree@gmail.com

Abstract. During the execution of Web Service Composition, if one component service fails or becomes overloaded not to be accessed, a failure recovery mechanism is needed to ensure that the running process is not interrupted and the failed service can be replaced quickly and efficiently. Recent researches on this problem have some disadvantages. They don't consider the influence of the number of service candidates or the connection state of the overlay network, so the algorithms are easily disabled. In this paper, we present an adaptive algorithm to find replacement path locally by virtue of the old path during dynamic service composition. We go backward along the execution path to find the branch node, and then construct the sub-graph by the predefined rules. Finally we choose the best path with the highest total utility to replace the failed one. The test's result shows the algorithm performs very well in the vigorousness and availability to dynamic adaptation.

1 Introduction and Related Work

Dynamic Web Service Composition (DWSC) organizes the component services across the different autonomic domains to finish complex application requirements by business planning. But for the incontrollable features of service providers and network connection of the component services in the Internet, the component services along execution path easily can fail during the runtime course while we have laid out the business process well in the design stage. So the running process is interrupted and couldn't be finished along the original execution path. A failure recovery mechanism is needed to ensure that a replacement path is found quickly and the business process can continue. Also the dynamic replacement problem is often considered in software upgrading and maintenance, because dynamic upgrading with minimal disruption to the consumers is very important feature for high-availability, mission-critical, and hard real-time systems.

---

* This work was supported by the National High-Tech Research and Development Plan of China under Grant Nos. 2003AA115210, 2003AA115410, 2005AA112030; the National Grand Fundamental Research 973 of China under Grant No.2005CB321800; the National Natural Science Foundation of China under Grant Nos. 60603063, 90412011.

A. Ghosh, R.K. De, and S.K. Pal (Eds.): PReMI 2007, LNCS 4815, pp. 41–48, 2007.

The QoS values of the single component services used for computation may be estimations in turn, declared by each service provider or obtained by computing statistics during previous executions of the service. At run-time, the actual measured QoS values may deviate from the estimate ones or, simply, some of the services may not be available (fail or become overloaded not to be accessed). Thus the composite service may have to be re-planned, so to still meet the constraints and maximize the QoS. In this time, quite a few component services perhaps have executed. If we just find an entirely new execution path over again. Although the new path is globally optimal but needs longer time to search, higher cost and delay unfit for the real-time application requirements. In this paper we consider the replacement solution in local path; assure the new path has the largest overlap with the old one with multiple QoS constraints. Thus the replacement process can be quickly finished with lower delay and cost. We select the optimal service candidates with higher utility to construct new replacement path. It makes sure that new path has the largest total utility with multi-dimensional QoS constraints assurance.

Many researchers have worked on the adaptive and dynamic service composition problem. The eFlow system is enacted by a centralized process engine, supports specification, enactment and management of composite services [1]. Service processes can adapt to environment changes and dynamically modify the process definition with minimal or no user intervention. However, eFlow does not have the rules for automatically switching service when service failure occurs. The reconfiguration (process definition modification) requires the user intervention and no quality issues are considered. In [2], authors develop Web Service Offering Infrastructure (WSOI) and proposed the idea of service offering similar to service levels defined in [3][4]. They address dynamic adaptation of service composition using manipulation of service offerings or classes via five mechanisms: switching, deactivation, reactivation, deletion, and creation. These methods are simple and fast but the ability is very limited. If we consider the end-to-end quality constraint issues of business processes, it is usually not enough to just switch among the service offerings within one Web service. The project in [5] majors in dynamic software upgrading with minimal disruption to consumers. In [6] they present the CARIS architecture and introduce application to dynamic composition of on-demand security associations. They address dynamic service composition supports business agility, flexibility, and availability. Same as eFlow, no algorithm has been designed to select replacing services. However, this is one of the most fundamental and important problems for the dynamic adaptation in Web service composition.

In [7-9], they study the end-to-end quality issue of business processes, model the system as a Directed Acyclic Graph (DAG) and convert the service composition problem to the problem finding the highest utility path in the graph theory under the multiple QoS constraints. The business process constructed can produce the highest user-defined utility as well as satisfy user's functional and quality requirements. In [10] [9], they present a replacement path algorithm to find the shortest path from source to target in graph $G-\{e_i\}$ (G without e_i), where e_i is an edge in the original shortest path of graph G. The algorithm proposed in [10] can produce n replacement paths with the same time complexity as the single-source shortest path algorithm. Our algorithm is similar to the algorithm in [9]. But our algorithm has no special requirement for the network topology, so it has more flexibility and availability than the algorithm in [9].

2 QoS Evaluation Model and QSC Problem

During the process of service composition, it should consider not only the functionality and behavior of service but also the nonfunctional properties, especially about the QoS metrics, which normally include response time, service time, availability, reliability, service cost and loss probability etc.

Some researchers have studied the replacement algorithm in local path. SpiderNet [8] removes all the other service candidates in the same Service Class that the old service performs well in the execution path. It is not suitable if the output degree of the immediate predecessor of the failed service equals to 1. QCWS [9] provides a backup path for each service node in the original execution path. The immediate predecessor of a failed service may quickly switch to a predefined backup path, so the running process is not interrupted. But this predefined backup path is unfit for the dynamic business process. They assume that if the output of service s_i is used as input of service s_j, add directed edges from all service level nodes in s_i to all service level nodes in s_j. It is too rigorous for each service node in the graph must be included in at least two paths. It's reasonable that not all service level nodes in s_i must be connected with all service level nodes in s_j.

In this paper we propose an algorithm that can solve the above shortages of other algorithms. If the component service fails while the composite service has executed half way, we will find the branch node backward along the execution path. Then we use current branch node as new source node, prune out the irrelevant service nodes by the algorithm and remove all edges whose start or end nodes reside outside of the sub-graph. According to this sub-graph, we construct the composition sub-tree. There is a list in each branch node recording the paths begin from it. We select the optimal path with highest utility as the replacement path with multi-dimensional QoS constraints assurance. We use the backward chaining algorithm to construct the sub-graph. And we make use of the utility function as the evaluation criteria.

The utility function is the evaluation criteria of QoS during service selection. The computation of the utility function should consider each QoS metric of the candidates in order to reflect the QoS of the service candidates deeply and thoroughly, such as response time, service time, availability, reliability, service cost and loss probability, etc. QoS metrics such as availability and reliability are benefit property, which need to be normalized by maximization. And other metrics such as response time and service time are cost property, which need to be normalized by minimization. For there exist some differentiations such as metrology unit and metric type between each QoS metric. In order to rank the Web Services fairly we must do some mathematics operations to normalize the metrics in the range [0,1]. Subsequently we can get a uniform measurement of service qualities independent of units.

Definition 1 (Service Candidate Utility Function). *Suppose there are α QoS metric values to be maximized and β QoS metric values to be minimized. The utility function for candidate k in a Service Class is defined as following:*

$$\mathcal{F}(k) = \sum_{i=1}^{\alpha} w_i \times \left(\frac{q_{ai}(k) - q_{ai\,min}}{d_i}\right) + \sum_{j=1}^{\beta} w_j \times \left(\frac{q_{bj\,max} - q_{bj}(k)}{d_j}\right) \tag{1}$$

$d_i = q_{ai\,max} - q_{ai\,min}$, if $q_{ai\,max} - q_{ai\,min} \neq 0$, else $(q_{ai}(k) - q_{ai\,min})/d_i = 1$,

$d_j = q_{bj\,max} - q_{bj\,min}$, if $q_{bj\,max} - q_{bj\,min} \neq 0$, else $(q_{bj\,max} - q_{bj}(k))/d_j = 1$.

Where w is the weight for each QoS metric set by user and application requirements $(0 < w_i, w_j < 1; \sum_{i=1}^{\alpha} w_i + \sum_{j=1}^{\beta} w_j = 1, \alpha + \beta = m)$. d_i, d_j *are the difference between maximum value and minimize value in the column. The concrete QoS metrics are defined in the matrix QoS= $[q_{i,j}]_{n \times m}$. $q_{ai\,max}$ ($q_{bj\,max}$) is the maximum value among all values on column i (j) in submatrix $[q_{a\,k,i}]_{n \times \alpha}$ ($[q_{b\,i,j}]_{n \times \beta}$) and $q_{ai\,min}$ ($q_{bj\,min}$)is the minimum value among all values on column i (j) in submatrix $[q_{a\,k,i}]_{n \times \alpha}$ ($[q_{b\,i,j}]_{n \times \beta}$).*

Each row in QoS matrix represents a Web service candidate sl_i ($1 \leq i \leq n$, and n represents the total number of candidates) while each column represents one of the QoS metrics q_v ($1 \leq v \leq m$, and m represents the total number of QoS metrics). As shown in matrix, QoS metrics of a Web Service candidate can be modeled as *m-dimensional* vectors; each Web service candidate can be represented as a point in m-dimensions.

We also consider the network QoS attributes (QoS metrics of the edge) such as bandwidth, network delay and loss probability. Because the QoS attributes of the network are only estimated, we can use indeterminacy reasoning with the degree of confidence and plausibility. Thus each service node has different utility value with different service link.

Definition 2 (Edge Utility Function). *Let LP denotes loss probability, D denotes network delay, BR denotes bandwidth ratio. The utility function of the edge $e_i = (u,v)$ in service graph G is defined as follows:*

$$\mathcal{F}(e_i) = w_1 \times \frac{\ln 1/1 - LP_{e_i}}{\ln 1/1 - LP_e^{min}} + w_2 \times \frac{D_{e_i}}{D^{max}} + w_3 \times \frac{BR_{e_i}}{BR^{max}} \tag{2}$$

where $\sum_{i=1}^{3} w_i = 1, 0 \leq w_i \leq 1, i = 1,2,3$ *the loss probability metric LP^{min} denotes the minimum value among all values. And* $D_{e_i} \leq D^{max}$, $BR_{e_i} \leq BR^{max}$.

Based on the above two definitions, the QoS-aware Service Composition (QSC) problem is defined as follows:

Definition 3 (QoS-aware Service Composition (QSC) problem). *Given overlay network G = (V,E), where V denotes the set of |V| nodes v_i and E denotes the set of |E| overlay links e_i. Let $Fn(s_i)$ denotes the business function provided by the service candidate s_i, s_i/v_j denote the service candidate s_i provided by the node v_j, and e_i/p_j denote the service edge e_i instantiated on the overlay path p_j. Each node $v \in V$ is associated with m non-negative QoS values; each edge $e \in E$ is associated with n* *non-negative QoS values. Given a user request $<\xi, Q^{req}>$, the QSC problem is to find the best service graph $\lambda \square \{\lambda_s, \lambda_e\}$ such that*

$$max\,(\mu_s \times \sum_{s_i / v_j \in \lambda_s} \mathcal{F}(s_i) + \mu_e \times \sum_{e_i / p_j \in \lambda_e} \mathcal{F}(e_i)\,) \qquad (3)$$

Subject to $\forall F_k \in \xi, \exists s_i \in \lambda_s, Fn(s_i) = F_k, q_i^\lambda \leq q_i^{req}, 1 \leq i \leq m$. Where $\mathcal{F}(s_i)$ and $\mathcal{F}(e_i)$ are the utility values of service candidate and edge, which can be computed by Def. 1 and Def. 2 respectively. μ_s and μ_e are their weights respectively.

A composite service may have several execution paths. The overall QoS value of the execution path can be calculated by the sum of the QoS values of selected component services. Therefore the optimal execution path is the maximum one.

Since the Service Flow Model is not related to any concrete Web Service, we must assign appropriate services to the corresponding activities to construct execution paths. And this is also the procedure to generate weighted multistage graph. We use the compatible rules to build up the service composition graph that we presented in another reviewing paper. The QoS value of service candidate and edge are used to set the weight of edges in the DAG.

3   QoS-Based Algorithm for Finding Replacement Service

During runtime an established service path can become broken or violate QoS constraints, particularly for the long-lived application session. When the service instance (or link) experiences outage or significant quality degradations, the dynamic service failure recovery mechanism is used to recover all the affected sessions. We define the Dynamic Failure Recovery (DFR) Problem as follows:

Definition 4 (Dynamic Failure Recovery (DFR) Problem). *Let λ^{old} and λ^{new} denote the broken service graph and the new service graph, respectively. Let $|\lambda_s^{old} \cap \lambda_s^{new}|$ denote the number of common service candidates between the old service graph and the new service graph. Given the overlay network $G=(V,E)$ and the user request $<\xi, Q^{req}>$, the DFR problem is to find a new service graph λ^{new} such that maximize $|\lambda_s^{old} \cap \lambda_s^{new}|$, subject to λ^{new} satisfies $\forall F_k \in \xi, \exists s_i \in \lambda_s, Fn(s_i) = F_k$, $q_i^\lambda \leq q_i^{req}, 1 \leq i \leq m$.*

The following algorithm is designed according to the above DFR problem. Our algorithm decomposes the candidate graph as composition sub-graph, and then transforms it into composition sub-tree. The replacement path begins from the branch node, backward along the original execution path of the failed service. We consider the special case: If the output degree of the predecessor s_p of the failed service s_f is equal to 1, i.e. $d_{out}(s_p) = 1$, then the algorithm would continue to find the branch node s_b backward sequentially. While this occasion would make the algorithms break down in [8][9]. The WSCPR algorithm is showed in Algorithm 1 in great details.

Algorithm 1. WSCPR(Web Service Composition Partial Re-composition) Algorithm

Step 1. From the failed node s_f backward to find the branch node s_b whose output degree $d_{out}(s_b) > 1$;
  Else no feasible node found, stop.
Step 2. For each Service Class S_i from branch node to last Service Class
    If $s_f \in S_f$
    // remain all other service nodes in the Service Class of the failed service s_f
      continue;
    Else if the predecessor s_p of s_f, $d_{out}(s_p) = 1$ **or**
        the successor node s_s of s_f, $d_{in}(s_s) = 1$
      continue;
    Else
      Remove all the other service nodes in the class S_i but not in the original execution path;
    Remove the failed node;
Step 3. Construct the sub-graph;
Step 4. Transform the sub-graph into composition sub-tree;
    Add all paths in the sub-tree into $p\_list(s_b)$, the list of the branch node s_b;
    Select the best optimal path according to computation of the utility function as the new execution path.
    $\mathcal{F}_{max} = 0$;
    for each $p \in p\_list(s_b)$
      $\mathcal{F}(p) = \sum \mathcal{F}(s)$, s is all the services along the path p;
      $\mathcal{F}_{max} = max\{\mathcal{F}_{max}, \mathcal{F}(p)\}$;
Return the path p with the highest utility in $p\_list(s_b)$ as optimal replacement path.

Using WSCPR algorithm, we suppose the number of QoS requirements is m, the number of Service Classes is N and each class has l candidates. Then the worst case time complexity for WSCPR is $O(N^2 lm)$, it performs very well in practice.

4   Validation of the Algorithm

We have performed the simulations and experiments to see the performance of our WSCPR algorithm. We also compare the running time between our WSCPR algorithm and CSPB algorithm in [9].

We use the degree-based Internet topology generator *nem* (network manipulator) [11] to generate a power-law random graph topology with 4000 nodes to represent the Internet topology. Then we randomly select 50~500 (depends on different test cases) nodes as the service candidate nodes in a business process. Some service nodes belong to the same Web service, representing different service levels. Several Web services group together to form a Service Class and several Service Classes in sequential order form an execution path. Each service node maintains the following information: *Service Class, service* and *execution path* it belongs to. The graph is

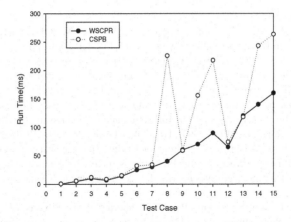

Fig. 1. Performance comparison between WSCPR and CSPB algorithm

constructed by randomly generating edges among service nodes. The generated graph is a DAG. The generated graph contains one or more execution paths.

The performance comparison between WSCPR and CSPB algorithm is showed in Fig. 1. We can see that the running time of WSCPR algorithm is reduced at a rather extent in contrast with CSPB algorithm. CSPB algorithm just uses another backup path when some node fails. And CSPB algorithm doesn't consider the errors in the dynamic environment, such as: the Internet connection fails during the SOAP message round trip; the WS times out because of a network connection failure; the Web Service returns a failure message because of data inconsistency. So the CSPB algorithm is easily disabled as showed in our test. Our WSCPR algorithm excludes quite a number of nodes independent of the final solution from the set of the service candidates. Thus the search space can be reduced at great certain extent.

References

1. Casati, F., Ilnicki, S., Jin, L., et al.: Adaptive and Dynamic Service Composition in eFlow. In: Wangler, B., Bergman, L.D. (eds.) CAiSE 2000. LNCS, vol. 1789, Springer, Heidelberg (2000)
2. Tosic, V., Ma, W., Pagurek, B.: On the Dynamic Manipulation of Classes of Service for XML Web Services. In: Proceedings of the 10th Hewlett-Packard Open View University Association (HP-OVUA) Workshop, Geneva, Switzerland (July 2003)
3. H.G., C., Yu, T., Lin, K.-J.: QCWS: An implementation of QoS-capable multimedia web services. In: Proceedings of IEEE 5th International Symposium on Multimedia Software Engineering, Taiwan, pp. 38–45 (December 2003)
4. Tosic, V., Mennie, D., Pagurek, B.: Software Configuration Management Related to Management of Distributed Systems and Services and Advanced Service Creation. In: Westfechtel, B., van der Hoek, A. (eds.) SCM 2001 and SCM 2003. LNCS, vol. 2649, pp. 54–69. Springer, Heidelberg (2003)
5. Feng, N., Ao, G., White, T., et al.: Dynamic Evolution of Network Management Software by Software Hot-Swapping. In: Proceedings of the 7th IFIP/IEEE International Symposium on Integrated Network Management (IM 2001), Seattle, USA, pp. 63–76 (May 2001)

6. Tosic, V., Mennie, D., Pagurek, B.: On dynamic Service Composition and Its Applicability to E-business Software Systems. In: Proceedings of the WOOBS 2001 (Workshop on Object-Oriented Business Solutions) workshop (at ECOOP 2001), pp. 95–108 (2001)
7. Yu, T., Lin, K.J.: Service Selection Algorithms for Web Services with End-to-end QoS Constraints. In: Proceedings of the IEEE International Conference on E-Commerce Technology (CEC 2004), San Diego, California, pp. 129–136 (2004)
8. Gu, X., Nahrstedt, K., Yu, B.: SpiderNet: An Integrated Peer-to-Peer Service Composition Framework. In: Proceedings of 13th IEEE International Symposium on High performance Distributed Computing(HPDC 2004), Honolulu, Hawaii, pp. 110–119 (June 2004)
9. Yu, T., Lin, K.-J.: Adaptive Algorithms for Finding Replacement Services in Autonomic Distributed Business Processes. In: Proceedings of the 7th International Symposium on Autonomous Decentralized Systems (ISADS 2005), Chengdu, China, pp. 427–434 (April 2005)
10. Hershberger, J., Suri, S.: Vickrey Prices and Shortest Paths: What is an Edge Worth? In: FOCS 2001. Proceedings of the 42nd IEEE symposium on Foundations of Computer Science, pp. 252–259. IEEE Computer Society, Los Alamitos (2001)
11. Magoni, D., Pansiot, J.-J.: Internet Topology Modeler Based on Map Sampling. In: ISCC 2002. Proceedings of the Seventh International Symposium on Computers and Communications, Taormina, 1-4 July 2002, pp. 1021–1027. IEEE Computer Society, Los Alamitos (2002)

Neuromorphic Adaptable Ocular Dominance Maps

Priti Gupta, Mukti Bansal, and C.M. Markan

VLSI Design Technology Lab
Department of Physics & Computer Science
Dayalbagh Educational Institute
AGRA – 282005
markan_cm@hotmail.com

Abstract. *Time staggered winner-take-all* (ts-WTA) is a novel analog CMOS neuron cell [8], that computes 'sum of weighted inputs" implemented as floating gate pFET 'synapses'. The cell behavior exhibits competitive learning (WTA) so as to refine its weights in response to stimulation by input patterns staggered over time such that at the end of learning, the cell's response favors one input pattern over others to exhibit feature selectivity. In this paper we study the applicability of this cell to form feature specific clusters and show how an array of these cells when connected through an RC-network, interacts diffusively so as to form clusters similar to those observed in cortical ocular dominance maps. Adaptive feature maps is a mechanism by which nature optimize its resources so as to have greater acuity for more abundant features. Neuromorphic feature maps can help design generic machines that can emulate this adaptive behavior.

Keywords: Floating Gate pFET, competitive learning, WTA, Feature maps, ocular dominance.

1 Introduction

Interconnectivity patterns between hierarchically organized cortical layers are known to extract different sensory features from a sensory image and map them over the cortical sheet. Higher cortical layers successively extract more complex features from less complex ones represented by lower layers. In fact it has been shown that different sensory cortices are also an outcome of a mechanism by which a generic cortical lobe adapts to the nature of stimulus it receives so as to extract sensory features embedded in it. [1]. Thus feature extraction and hence formations of feature maps are fundamental underlying principles of parallel and distributed organization of information in the cortex. Any effort towards an artificial or neuromorphic realization of cortical structure will have to comprehend these basic principles before any attempt is made to derive full benefit of cognitive algorithms that are active in the brain. Neuromorphic realization of cortical map finds useful application in the area of robotic vision where potential improvement are possible by employing mechanisms observed in a living brain [2]. Dedicated adaptive hardware can help design generic machines that conserve resources by acquiring greater sensory acuity to more abundant features at the cost of others

A. Ghosh, R.K. De, and S.K. Pal (Eds.): PReMI 2007, LNCS 4815, pp. 49–56, 2007.

depending on the environment they are nurtured. Another emerging area of interest is that of neural prostheses wherein implants such as retinal or cochlear artificially stimulate sensory nerves to overcome blindness or deafness. Such implants are effective only when cortical infrastructure (feature maps) to interpret inputs from these implants is intact [3]. Animals born with defunct sensory transducers find their representative cortical area encroached upon by competing active senses. In such animals or in those who have a damaged or diseased cortex, sensory implants are ineffective as cortical apparatus (maps) to interpret inputs from these implants is not in place. The only option in such cases is to revive the cortical feature maps either biologically or through prosthetic neuromorphic realization [3].

In this paper we build a framework for adaptive neurmorphic cortical feature maps. In section 2 we present a basic building block for competitive learning (ts-WTA). In section 3 we apply *ts*-WTA to extract visual features e.g. ocular dominance and investigate its ability to cluster on the basis of their feature preference i.e. form feature maps.

2 'Time Staggered WTA' Circuit

In building adaptive neuromorphic structures, the biggest bottleneck has always been implementing adaptable connection strengths (weights) in hardware. While capacitor storage is volatile, the non-volatile digital storage is bugged with severe overheads of analog/digital domain crossing. However, improved quantum mechanical charge transfer processes across the thin oxide now offer a floating gate pFET whose threshold voltage and hence conductivity adapts much like Hebbian learning in a 'synapse' [4]. In fact, recently a competitive learning rule similar to Kohonen's unsupervised learning was implemented using floating gate pFETs [5].

2.1 Floating Gate 'Synapses'

The trapped charge on the floating gate of a pFET is altered using two antagonistic quantum mechanical transfer processes [4]:

Tunneling: Charge can be added to the floating gate by removing electrons from it by means of Fowler-Nordheim tunneling across oxide capacitor. The tunnel current is expressed in terms of terminal voltages across oxide capacitor i.e. tunnel voltage V_T, and floating gate voltage V_{fg} as [6]

$$I_{tunnel} = F_T(V_T, V_{fg}) = I_{to} \exp\left(-\frac{V_f}{V_T - V_{fg}}\right) \tag{1}$$

Injection: Charge is removed from the floating gate by adding electrons to it by current hot-electron injection (IHEI) from channel to the floating gate across the thin gate oxide. The injection current is expressed as a semi-empirical relationship in terms of pFET terminal voltages i.e. source (V_s), drain (V_d), floating gate voltage (V_{fg}) and source current I_s as [6]

$$I_{injection} = F_I(V_d, Is, V_{fg}) = \eta.Is.\exp\left(-\frac{\beta}{(V_{fg} - V_d + \delta)^2} + \lambda.V_{sd}\right) \tag{2}$$

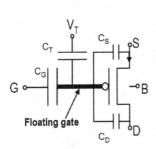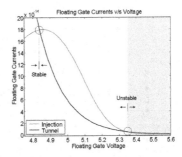

Fig. 1. (*Left*) A floating gate pFET, (*right*) Variation of Injection and Tunnel current as a function of Vfg for fixed tunnel voltage VT and drain voltage Vd

With floating gate capacitatively linked to input gate, drain, tunnel and source terminals of pFET (see figure 1(left)), the overall effect of continuous floating gate currents is expressed using Kirchoff's Current Law at the floating gate node:

$$C_F \frac{\partial V_{fg}}{\partial t} = C_G \frac{\partial V_G}{\partial t} + C_D \frac{\partial V_D}{\partial t} + C_T \frac{\partial V_T}{\partial t} + C_S \frac{\partial V_S}{\partial t} + I_{tunnel} - I_{injection}$$

2.2 Adaptation in Floating Gate 'Synapses'

Keeping V_G constant and variation in V_S small and by increasing the capacitances at the drain terminal, the variation of charge on floating gate can be reduced to

$$C_F \frac{\partial V_{fg}}{\partial t} = I_{tunnel} - I_{injection} = F_T(V_T, V_{fg}) - F_I(V_d, I_s, V_{fg})$$

With pFET in saturation its source current is largely expressed as a function of V_{fg}, so that above equation is rewritten with magnitude of I_s absorbed in F_I

$$C_F \frac{\partial V_{fg}}{\partial t} = I_{tunnel} - I_{injection} = F_T(V_T, V_{fg}) - F_I(V_d, V_{fg}).x \tag{3}$$

where binary x is 1 when I_s current flows and 0 otherwise.

Variation of I_{tunnel} and $I_{injection}$ is plotted as a function V_{fg} in figure 1(right). Two salient points (i) stable and (ii) unstable point characterize the adaptation dynamics of floating gate pFET. V_{fg} above unstable point increases uncontrollably as $I_{tunnel} > I_{injection}$. Below unstable point, V_{fg} always gravitates towards the stable point. Any variation in V_T and V_d merely relocates both stable and unstable point, though latter is affected less substantially. Thus we have a 'synapse' with an adaptable conductivity or weight which grows on stimulation (*injection*) and saturates to an upper limit (*stable point*). In absence of stimulation the self decay term (*tunneling*) prunes its weights till it finally goes dead from where it cannot recover (*unstable point*).

2.3 'Time Staggered Winner Take All'

We now build a competitive learning cell by connecting two such weighted inputs or 'synapses' as shown in figure 2(left) such that the voltage at common source terminal

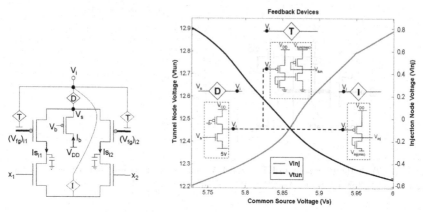

Fig. 2. (Left) Actual circuit of learning cell, $(V_{fg})_{i1}$, $(V_{fg})_{i2}$, shows the floating gate based weighted connection. I, T and D are dependent voltage sources for injection, tunnel and activation node. x_1, x_2 are inputs and node voltage V_S is common source voltage of the cell. (Right) Shows Feedback devices T(green), I(blue). Both the devices act in conjunction with buffer device 'D' and their equivalent circuits are shown as inset. The graph shows variation of common tunnel node voltage V_{tun} and common Injection node V_{inj} w.r.t. V_S.

(V_s) is expressed as a weighted sum of the two branch currents, $V_s = V_{DD} - I_b.R_b$ where R_b is the saturation resistance of pFET with a fixed bias voltage V_b, and source current I_b. With $x_j=1(0) \Rightarrow j^{th}$ branch nFET is ON(OFF), i.e. $V_{gate} = V_{DD}(0)$ we write

$$I_{b_i} = \sum_j Is_{ij}.x_j = Is_{i1}.x_1 + Is_{i2}.x_2 \tag{4}$$

$$V_s = \left[V_{DD} - R_b. \sum_j Is_{ij}.x_j \right] \tag{5}$$

Assuming floating gate pFET is in saturation, source current largely depends on its gate-to-source voltage. Also with one input applied at a time, source voltage (V_s) itself depends on floating gate voltage of stimulated branch and so it won't be improper to express source current as a function of floating gate voltage i.e. $Is_{ij} = f(V_{fg,ij}).x_j$. Rewriting (5), after substitution and suppressing constants, we have

$$V_s = \left[\sum_j f(V_{fg})_{ij}.x_j \right]$$

2.4 Feedback

Let us express tunnel feedback $V_T = T(V_i)$ and Injection feedback $V_d = I(V_i)$ as monotonically varying functions of activation node voltage V_i which itself is linked to common source voltage V_s through a buffer device D as $V_i = D(V_s)$ see figure 2. *In our design of feature selective cell we are guided by the basic principle "to accomplish WTA a self-excitation (injection) must accompany global nonlinear inhibition (tunneling)"* [7].

Rewriting adaptation eqn (3) after substitution and suppressing the constant terms

$$\frac{d(V_{fg})_{ij}}{dt} = F_T\left[T\left(\sum_j f(V_{fg})_{ij}x_j\right)(V_{fg})_{ij}\right] - F_I\left[I\left(\sum_j f(V_{fg})_{ij}x_j\right)(V_{fg})_{ij}\right]x_j \tag{6}$$

With two inputs stimulated one at a time, there are two input patterns $\mathbf{x_1}$: $x_1=1$ & $x_2=0$ or $\mathbf{x_2}$: $x_1=0$ & $x_2=1$. If in every epoch all patterns occur once in random order (*random-inside-epoch*) then at the end of an epoch the adaptation rate of V_{fg} will be

$$\frac{d(V_{fg})_{i1}}{dt} = \frac{d(V_{fg})_{i1}}{dt}\bigg|_{x_1} + \frac{d(V_{fg})_{i1}}{dt}\bigg|_{x_2} \quad ; \text{and} \quad \frac{d(V_{fg})_{i2}}{dt} = \frac{d(V_{fg})_{i2}}{dt}\bigg|_{x_1} + \frac{d(V_{fg})_{i2}}{dt}\bigg|_{x_2} \quad ;$$

In order to achieve WTA it would be necessary that adaptation rates of the two floating gates, in the above equation, are of opposite sign. This is achieved by ensuring that overall gain of our feedback devices is greater than unity [8]. Initially if $(V_{fg})_{i1} > (V_{fg})_{i2}$ and the two arms are stimulated with equal probability i.e. as random-inside-epoch[8] then this circuit will iteratively amplify the difference $\Delta(V_{fg})$ so that $(V_{fg})_{i1}$ rises to eventually switch off its pFET while $(V_{fg})_{i2}$ falls to reach the stable point so that its pFET stays on. Adaptability in floating gates with time is shown in the figure 3.

Though our circuit, which we call as *time-staggered WTA* (ts-WTA), is topologically similar to Lazzaro's WTA(*l*-WTA) circuit [9][10], yet it has some subtle and fundamental differences. *l*-WTA achieves positive feedback through common source terminal keeping gate voltages constant. In *ts*-WTA +ve feedback is achieved through special devices (I, T) that act on floating gates to achieve WTA. In *l*-WTA positive feedback through common source terminal require all inputs be stimulated simultaneously and the competition is instantaneous, whereas *ts*-WTA requires stimulation of one input at a time and competition is staggered over several epochs. For clustering in self-organizing feature maps (SOFM), cells are required to communicate their feature selectivity through their output node. This cannot happen if all inputs are stimulated at the same time as required by *l*-WTA. Therefore for map formation it is necessary that inputs be applied one at a time over all the participating cells. This clearly distinguishes the importance of '*ts*-WTA' w.r.t. *l*-WTA as a feature selective cell.

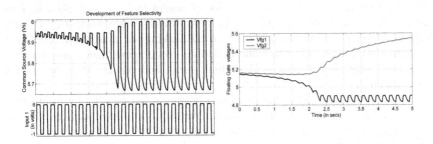

Fig. 3. (Left-top) Development of feature selectivity observed as response of the cell (Vs) to alternate input patterns. (Left-bottom) Input x1, x2 is invert of it. (Right) ts-WTA evolution of Vfgs. Here we have used alternate stimulation instead of random-inside-epoch.

3 Visual Feature Extraction

Having described the basic competition learning cell as ts-WTA we now illustrate its applicability to one of the least complex of the visual features i.e. ocular dominance.

Cells of primary visual cortex of mammals have been observed to be dominated by inputs from one eye or the other, termed as ocular dominance. This domination is not genetically predetermined at birth rather develops in response to stimulus received from the two eyes. In animals whose one eye has been sutured or closed immediately after birth, cortical cells normally belonging to the closed eye are taken over or dominated by inputs from the active eye. This behavior is explained by the fact that at birth cortical cell receive nearly equal inputs from both eyes, slowly competition (similar to *ts*-WTA) between inputs from the two eyes leads to ocular domination. Thus ts-WTA cell can be directly used as ocular dominance feature selective cell provided its inputs represent inputs from the two eyes. At the end of learning, ts-WTA cell responds only to the input that wins the competition i.e. it is either left eye dominated or right eye dominated.

Map formation through diffusive interaction: Feature selectivity maps require that cells must cluster into groups with similar feature preferences such that there is both variety and periodicity in spatial arrangement of these cells across the cortical sheet. The basic mechanism that leads to clustering is based on the fact that neighboring cells have overlapping receptive fields and hence receive similar inputs. If somehow these neighboring cells can also be made to have similar responses, then because of Hebbian learning, individual cell's receptive field will also develop similarly. This similarity will extend over a portion of cortex where a majority of cells have similar initial feature biases. With randomly chosen initial feature biases, clusters of feature selective cells will emerge distributed over the cortical sheet similar to those observed in cortical feature selectivity maps.

Thus an important property a learning cell must posses to achieve clustering is the ability to develop its feature selectivity under influence of its neighbors. More explicitly, it should exhibit three types of adaptive behavior i.e. (i) if cell's bias is same as its neighbors, it must strengthen it, (ii) if cell's bias is opposite to its neighbors, it must reverse it, and (iii) if cell has equal number of neighbors with biases in favor and against, it must become unbiased i.e. respond equally to both inputs. Though theoretical models employ these principles in different ways, yet not all are easy to realize in hardware. A class of model that is relatively easier to implement are those based on Turing's reaction-diffusion framework. Biologically such an interaction occurs in the form of chemicals leaking out of an active cell which enter neighboring cells to lower their threshold thus encouraging them to also become active so that neighboring cells, with similar inputs and outputs, develop similar feature selectivity.

We model the diffusive interaction between feature selective cells by means of a RC network. Cells are connected to RC-grid through a diffusion device 'D', see figure 4. Device 'D' serves (i) to isolate neighboring cells from directly altering activation of a cell, (ii) to drive the tunnel device that operates at higher voltage than voltages available at activation node. We test this hypothesis for two cases (i) a row of 10 learning cells that are diffusively coupled on a 1-D RC network, see

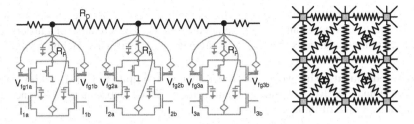

Fig. 4. Shows diffusive interaction between learning cells implemented in actual circuit by means of RC network. (left) 1-D, (right) 2-D, where every grey square represents ts-WTA cell.

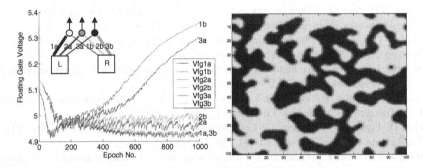

Fig. 5. (Left) Shows development of V_{FG} for three cells of figure 4(left). (1) left eye dominated cell (white), (2) binocular cell (gray), (3) right eye dominated cell (black). Lower VFG implies stronger connection strength. A dominated cell develops to have a large difference between two VFGs, but a binocular cell has nearly equal VFGs. (Right) Simulation results for 2D array of 100x100 ts-WTA cells of ocular dominance. White (Black) represent cells dominated by left (right) eye, grey represents equal domination or binocular cell.

figure 4(left), and (ii) a 2-D RC grid of 100 x 100 cells, see figure 4(right). Here R_R and R_D refer to resistance values that balance the effect of Reaction term (activation of a cell to given input) and Diffusion term (effect of activation of neighboring cell) respectively. To avoid any boundary effects we take periodic boundary conditions. Results for 1-D are shown in Figure 5(left) and for 2-D in figure 5(right). Initially all cells are given faint random biases as either left or right dominated cells. During development all the cells are stimulated with identical patterns $\{(0,1), (1,0)\}$ in an order chosen randomly in every epoch. Under this *random inside epoch* stimulation cells perform 'time staggered WTA' i.e. strengthen their biases. Since these cells develop under influence of diffusive interaction they begin to clusters into left or right dominated cells, also cells which lie at the boundary of opposite biases turn binocular i.e. they respond equally to both left and right input. This is reflected in the Figure 5(right) as grey cells. The graph (figure 5(left)) shows three cells chosen such that cells at the two ends represent clusters of opposite biases and the cell in the centre favors one of them. After learning, cells at the two corners develop according to their own biases while the cell in the centre becomes unbiased with equal weights.

The pattern of ocular dominance produced with this array of RC-coupled ts-WTA cells is shown in figure 5 showing well distributed clusters of left (black) and right eye (white) dominated cells. Observe the smooth change from one cluster to other occurs due to development of grey (binocular) cells. The creation of clusters and existence of binocular cells are the major strengths of this diffusive-Hebbian model.

Acknowledgements. This work was funded by research grant to CMM, (III.6 (74)/99-ST(PRU)) under SERC Robotics & Manufacturing PAC, Department of Science and Technology, Govt. of India.

References

1. Horng, S.H., Sur, M.: Visual activity and cortical rewiring: activity-dependent plasticity of cortical networks. Progress in Brain Research 157, 3–11 (2006)
2. Choi, T.Y.W., et al.: Neuromorphic Implementation of Orientation Hypercolumns. IEEE Tran. on Circuit & Systems-I 52(6), 1049–1060 (2005)
3. Merabet, L.B., et al.: What blindness can tell us about seeing again: merging neuroplasticity and neuroprostheses. Nature Reviews Neurosci. 6, 71–77 (2005)
4. Diorio, C., Hasler, P., Minch, B.A., Mead, C.A.: A Single-Transistor Silicon Synapse. IEEE Transactions on Electron Devices 43(11), 1972–1980 (1996)
5. Hsu, D., Figueroa, M., Diorio, C.: Competitive learning with floating-gate circuits. IEEE Transactions on Neural Networks 13(3), 732–744 (2002)
6. Rahimi, K., Diorio, C., Hernandez, C., Brockhausen, M.D.: A simulation model for floating gate MOS synapse transistors. In: ISCAS (2002)
7. Grossberg, S.: Adaptive pattern classification and universal recoding: I. Parallel development and coding of neural feature detectors. Biol. Cybern. 23, 121–134 (1988)
8. Bansal, M., Markan, C.M.: Floating gate Time staggered WTA for feature selectivity. In: Proc. of Workshop on Self-organizing Maps, WSOM 2003, Kitakyushu, Japan (2003)
9. Lazzaro, J., Ryckbusch, S., Mahowald, M.A., Mead, C.A.: Winner-take-all Networks of O(N) complexity, NIPS 1 Morgan Kaufman Publishers, San Mateo, CA, pp. 703–711 (1989)
10. Kruger, W., Hasler, P., Minch, B., Koch, C.: An adaptive WTA using floating gate technology. In: Advances in NIPS 9, pp. 713–719. MIT Press, Cambridge (1997)

Fault Diagnosis Using Dynamic Time Warping

Rajshekhar[2], Ankur Gupta[2], A.N. Samanta[2], B.D. Kulkarni[1,*],
and V.K. Jayaraman[1,*]

[1] Chemical Engineering Division, National Chemical Laboratory, Pune-411008, India
{bd.kulkarni,vk.jayaraman}@ncl.res.in
[2] Department of Chemical Engineering, Indian Institute of Technology,
Kharagpur-721302, India

Abstract. Owing to the superiority of Dynamic Time Warping as a
similarity measure of time series, it can become an effective tool for
fault diagnosis in chemical process plants. However, direct application
of Dynamic Time Warping can be computationally inefficient, given the
complexity involved. In this work we have tackled this problem by employing a warping window constraint and a Lower Bounding measure. A
novel methodology for online fault diagnosis with Dynamic Time Warping has been suggested and its performance has been investigated using
two simulated case studies.

1 Introduction

The process deviation from the normal operating range leads to deterioration in
product quality and can be source of potential hazard. The control of such deviations comes under abnormal event management (AEM) in chemical process
industry. The first step in AEM consists of timely detection and diagnosis of
fault, so that it can lead to situation assessment and planning of supervisory decisions to bring the process back to a normal and safe operating state. However
due to the size and complexity involved in the modern process plants, traditional method of complete reliance on human operators has become insufficient
and unreliable. The advent of computer based control strategies and its success in process control domain has lead to several automated fault diagnosis
methodologies.

Currently available fault diagnosis techniques can be classified into three broad
categories: quantitative model based, qualitative model based and process history based approaches. In this work, a novel process history based approach
for fault detection has been proposed. It employs the concept of Dynamic time
warping (DTW) for the similarity measurement. Direct application of DTW
leads to poor computational efficiency of the methodology. This problem has
been rectified in this work by using window warping constraint in DTW with
the application of lower bounding technique. We demonstrate the efficiency of
our proposed methodology by performing online fault diagnosis on two simulated
case studies.

---

* Corresponding author.

A. Ghosh, R.K. De, and S.K. Pal (Eds.): PReMI 2007, LNCS 4815, pp. 57–66, 2007.
© Springer-Verlag Berlin Heidelberg 2007

2   Dynamic Time Warping

Consider two multivariate time series Q and C, of length n and m respectively,

$$Q = q_1, q_2, q_3, \ldots\ldots, q_i, \ldots\ldots, q_n \tag{1}$$

$$C = c_1, c_2, c_3, \ldots\ldots, c_j, \ldots\ldots, c_m \tag{2}$$

such that, $q_i, c_j \in \mathbb{R}^p$. Since the DTW measure is symmetric with respect to the order of the two time series, without any loss of generality we can assume that $n \geq m$ in our work. To align these two sequences using DTW, we construct a $n - by - m$ matrix where the (i^{th}, j^{th}) element of the matrix corresponds to the squared distance,

$$d(i,j) = \sum_{r=1}^{r=p} (q_{i,r} - c_{j,r})^2 \tag{3}$$

In order to find the best match between these two sequences, one finds a path through the matrix that minimizes the total cumulative distance between them. Such a path will be called a warping path. A warping path W is a contiguous set of matrix elements that characterizes a mapping between Q and C. The kth element of W is defined as $W(k) = (i, j)_k$. The time-normalized distance [1] between the two time series is defined over the path as,

$$DTW(Q, C) = \min_{W} \left[\sqrt{\frac{\sum\limits_{k=1}^{k=K} d(W(k)) \cdot \phi(k)}{\sum\limits_{k=1}^{k=K} \phi(k)}} \right] \tag{4}$$

Where, $\phi(k)$ is the non-negative weighting co-efficient and K is the length of the warping path W, which satisfies the condition,

$$\max(m, n) \leq K \leq m + n - 1 \tag{5}$$

The normalization is done to compensate for K, the number of steps in the warping path W, which can be different for different cases. The symmetric normalization, owing to its established superiority [1] has been used for purpose, given as,

$$\phi(k) = (i(k) - i(k-1)) + (j(k) - j(k-1)) \tag{6}$$

The warping path however, is constrained to the following conditions [2]:

Boundary conditions: The path must start at $W(1) = (1, 1)$ and end at $W(K) = (n, m)$.

Continuity and monotonic condition: From a point (i, j) in the matrix, the next point must be either $(i, j + 1), (i + 1, j + 1)$ or $(i + 1, j)$.

Warping Window condition: In case of equal length time series, the path as we know intuitively should not wander too far from the diagonal. For the case

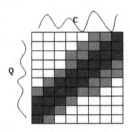

Fig. 1. A sample warping window band. Note the two regions in the band, in black and grey.

of unequal length time-series, a unique diagonal does not exist, thus we define a region analogous to the diagonal, which is bounded by the two lines whose equations are given by (7).

$$i = j \ \& \ i = j + n - m \text{ where } 1 \le j \le m \text{ for } n \ge m \tag{7}$$

Figure 1 shows an example of such a region for the case of $n \ge m$. This region, analogous to the diagonal has been shown in darker (black) shades in the figure. It should be noted that this region reduces to a diagonal, in case the two time series are of equal length. In order to restrict the warping path from straying away from the diagonal (or defined-region for unequal length time-series), we limit the distance by which it can wander away from the diagonal/defined-region. This limits the warping path to a warping-window or a band, which allows the warping path a distance (say) R_i to wander directly above and to the right of the diagonal/defined-region. This definition of band makes it symmetric with respect of the diagonal/defined-region. This is a generalization of the warping window originally proposed by [2].

We now mathematically define the band as,

$$j - R_i \le i \le j + R_i + (n - m) \text{ for } n \ge m \tag{8}$$

Where, $R_i = d$ such that $0 \le d \le max(m)$, and $1 \le i \le n$. max(m) is the length of the longest time-series in the data available, and R_i is the permissible range of warping above and to the right of the region defined in (7). In the case where the time series are univariate and of equal length i.e. $n = m$, the band reduces to the R-K band as defined in [2].

In general, the above definition of band allows the formulation of any arbitrary shaped band. However, we are employing the bands for online fault diagnosis purpose and hence the temporal variation in bandwidth should be avoided. Hence, for current work, R_i has been considered independent of i. In order to solve the optimization problem given by (4), we employ the dynamic programming technique as done in [1,2]. For this cumulative distance matrix is calculated [1,3],

$$D(i,j) = \min \begin{cases} D(i-1,j) + d(i,j) \\ D(i-1,j-1) + 2 \cdot d(i,j) \\ D(i,j-1) + d(i,j) \end{cases} \tag{9}$$

This takes care of the continuity and monotonic conditions along with the path normalization. This recursive formula for dynamic programming enables us to find the warping path W, along which the values of $d(i,j)$ can be summed up, and the final value of DTW distance can be found out as given by (4). However, when the warping window constraint is employed, (9) admits only those points which lie inside the warping window or the band. Thus, only those points are eligible for consideration in (9) which satisfy (8). We can see that application of warping window constraint drastically reduces the number of paths that are needed to be considered, thus speeding up the DTW calculation. For a detailed introduction to DTW, lower bounding measures, warping windows and their application to classification we refer the readers to [1,2,3].

3   Lower Bounding Measures

Computation of DTW distances is computationally very expensive, consequently a classification algorithm based on DTW distance as a similarity measure is bound to be computationally inefficient. This problem has been circumvented by using a fast lower bounding measure, which saves unnecessary calculations by pruning off the time-series which cannot be nearest to the given time-series [3]. The lower bounding measure is obviously computationally cheaper than the actual DTW calculation, but gives a fairly tight approximation of the DTW distance. The original method has already been successfully applied to the case of time series of equal length [2]. In this work, we present a modification of the lower bounding measure which can be applied to time series of unequal lengths, which we will call as LB_UMV (stands for Lower-Bound Unequal Multivariate).

3.1   Lower-Bounding Measure (LB_UMV)

Let us consider the two time-series Q and C defined in (1) and (2). Using the global constraints on the warping path given by (8), we construct an envelope using the warping window across Q bounded by two time-series U and L given as,

$$u_{j,r} = \max(q_{j-R,r} : q_{j+R+(n-m),r}) \tag{10}$$

$$l_{j,r} = \min(q_{j-R,r} : q_{j+R+(n-m),r}) \tag{11}$$

Using the above definitions of U and L, we define LB_UMV as,

$$LB\_UMV(Q,C) = \sqrt{\frac{1}{(m+n-1)} \sum_{j=1}^{j=m} \sum_{r=1}^{r=p} \begin{cases} (c_{j,r} - u_{j,r})^2 & if\ c_{j,r} > u_{j,r} \\ (l_{j,r} - c_{j,r})^2 & if\ c_{j,r} < l_{j,r} \\ 0 & otherwise \end{cases}} \tag{12}$$

It is important to note that the lower bounding measure defined in (12) for the two time series will always be lower than the corresponding DTW distance. A mathematical proof of this has been provided in [3] for the case of equal length time series. The same can be proved for the current definition of band by following a similar approach.

4   Learning Multiple Bands for Classification Using Heuristics

A lower value DTW distance implies that the given time series are similar to each other, while a large value implies dissimilarity between the two. In a classification task, one usually compares a test sample with the samples from the reference dataset, and then assigns a label to the test sample of the class to which it resembles the most. In this work, both the test and reference samples are time series and we utilize the concept of DTW distance to define the similarity between them. Each class in the dataset will have a warping window or a band of its own, which will be utilized when the test time-series is compared to a reference time-series of that class. Our aim is to automatically find these bands using the reference dataset. Similar to approach employed by [2], we pose this problem as a search problem and make use of the generic heuristic techniques to obtain bands. The search can be performed as a either forward or backward hill-climbing search algorithm, with the aim of maximizing a heuristic function. An intuitive selection for the heuristic function is the accuracy of classification, which is defined as the percentage of correct classifications obtained by using a given set of bands. A backward search begins with maximum bandwidth , above and to the right of the region defined in (7); the corresponding accuracy is evaluated. The bandwidth of these bands is reduced and the accuracy is re-evaluated. The change is accepted if an improvement in accuracy is registered. The reduction in bandwidth is continued until a reduction in accuracy is observed. This is done until we reach zero bandwidth.

5   Online Fault Diagnosis

We follow the methodology described in Section 4 to obtain a set of trained bands (one for each class) by using the reference dataset, which is then used to compute DTW distances wherever necessary. A process plant continuously provides us with a multivariate time-series in the form of measured variables at some time intervals. We utilize the DTW concept for the purpose of online fault diagnosis by using a window wise pattern matching approach. We select a window from the current sequence of measured variables, which is used as a test sample to compare with windows (of nearly equal length) from the reference dataset. This allows us to predict the nature of the current sequence of variables to be faulty or not as well as the type of fault. The selection of the current and reference windows has been illustrated in Figure 2.

The approach consists of three major steps:

1. *Selection of Current Window:* A window is selected from the current batch of measured variables obtained from process plant, which is then matched against similar windows from the reference batches. There are three parameters involved in this step: Initial point (IP), Window Width (WW) and Window Shift (WS). We start selecting windows from an Initial Point (IP).

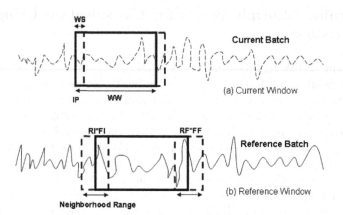

Fig. 2. (a) Selection of Current Window, the current window shifts by WS. (b) Selection of Reference Window, the reference window can be of various sizes due to variation in the neighborhood on both sides.

Initial Point decides the reliability of initial transient part in identifying the fault type. Window Width is the size of the current window and Window Shift is the number of time points by which the window shifts. Every current window acts as a test sample which is then compared with windows from the reference dataset in order to determine the fault.

2. *Selection of Reference Window:* The main parameter involved in this step is the window location in a particular reference time series. We will identify this window with initial point RI and final point RF. However, in order to counter the effect of different or uneven sampling time, these parameters will be updated by two other parameters, FI and FF respectively. A more effective independence from the sampling variation can be obtained by allowing the RI and RF to deviate in the neighborhood. This can be achieved by introducing two parameters, number of neighboring points (N) by which the RI and RF can deviate and the time (n) by which these points are apart. In a particular batch of the reference dataset, we consider all the possible windows as a match against the current window and nearest window is selected as the reference window. This will also update the FI and FF parameters.

3. *Fault Diagnosis:* After the current window and corresponding reference windows are selected, we rank the reference windows in increasing order of their distances from the current windows, such that the nearest window is ranked first. The current batch is expected to be of the same class to which its nearest neighbor belongs to. However, instead we follow a more robust probability based approach to identify the class of the current batch as described subsequently. If the assigned class belongs to a fault type, the process is likely to be faulty. But if it belongs to normal class, the process is considered to be operating normally and we modify the current window by shifting it by WS, and repeat the classification procedure. However, it is possible that

a current window is falsely assigned a faulty class. In order to suppress such false alarms, another parameter viz. Fault Count (FC) has been introduced. This parameter requires the current batch to be assigned a faulty class for a given number of times (given by FC) before it is declared faulty. It should be noted that the reference dataset is first used to generate trained bands and then later to compare with current windows, which are obtained from the test dataset.

Similarity Measure: To make the similarity search more robust, a probability based method has been used. The probability value will indicate the chances of current time series to be a faulty or not. The similarity measurement has been performed by finding 50 best matched reference windows and based on this set a probability function was defined as,

$$P_i = \frac{\sum\limits_{\substack{j=1 \\ s.t.\, j \in class\, i}}^{50} \frac{1}{r_j \times d_j}}{\sum\limits_{j=1}^{50} \frac{1}{r_j \times d_j}} \tag{13}$$

Where, P_i is the probability of the current batch to belong to class i (i.e. particular fault or normal type), r_j and d_j are rank and distance measure with current time series of jth reference time series respectively.

6   Simulated Case Studies

6.1   Case Study 1: Batch Acetone-Butanol Fermentation

This case study employs the mathematical model of batch acetone-butanol fermentation, originally proposed by Votruba et al. [4]. This model has been further explored by Singhal [5]. He introduced different operating parameters consisting of one normal operation and four abnormal operations.

Generation of historical database: Each abnormal operating condition was characterized by an abnormal value of a cell physiology parameter. The magnitude of the abnormal parameter varied randomly from batch to batch. The duration of each batch was 30 h and sampling interval was randomly varied from 9 to 12 minutes. The operating conditions and parameter ranges can be found in [5]. Each of the five operating conditions was simulated 60 times to provide a historical database of 300 batches. Gaussian measurement noise was added to the measured variables so that the signal-to-noise ratio for a normal batch run was equal to ten. These batch profiles were divided into two subsets; a set of 250 profiles (50 profiles per class) was taken as the reference dataset and another set of 50 profiles (10 profiles per class) was used for the online diagnosis. The reference dataset was first used to generate trained bands for each class, and then later to compare with the current window in online fault diagnosis. The test dataset is used to generate the current windows, and the proposed methodology was investigated. In order to exploit the temporal nature of the batch profiles, we considered each profile in the reference dataset to contain a transient, steady state and an intermediate part. Thus, each profile was split into three equal time-series, and a new dataset was created which contained 750 time-series.

Diagnosis Result: The various parameters defined above were tuned with an objective of achieving maximum diagnosis efficiency. The chosen values of these parameters are shown in Table 1.

When the fault count parameter was set at 1, the classification accuracy is 100% for all the classes except the normal class. This is because many of the test batch profiles were declared faulty due to false alarms. These were controlled when fault count was increased to 2. Furthermore, we determine the reaction time of the proposed methodology in detecting a fault. The detection times for the four faults lie between 985 minutes and 1308 minutes. We compare our results with the LLE-SVDD approach, has been found to be superior to the PCA approach for the given case study, as given by Kumar et al. [8]. We can see that both the methods are able to perform perfect diagnosis efficiency; however LLE-SVDD approach detects a fault at an earlier stage than the proposed method. The LLE-SVDD approach however performs only fault detection, while our proposed methodology performed fault diagnosis. Thus, our proposed approach is advantageous in cases where it is imperative to diagnose a fault.

Table 1. Fault Diagnosis Results for a particular set of parameters

Parameters	Parameter Values	Fault Type	Diagnosis Efficiency(%)	
			Fault Count=1	Fault Count=2
Initial Point	51	Normal	80	100
Window Width	50	Fault 1	100	100
Window Shift	10	Fault 2	100	100
N	2	Fault 3	100	100
n	2	Fault 4	100	100

6.2   CSTR Batch Profiles

In the second case study, we have performed fault diagnosis of a jacketed CSTR in which an irreversible, exothermic, first order reaction ($A \rightarrow B$) is taking place using the proposed methodology. The system is equipped with three control loops, controlling the outlet temperature, the reactor holdup and the outlet concentration. A detailed description of mathematical model has been provided by Luyben [6]. This model has been further explored by Venkatasubramanian et al. [7] in their work on application of neural networks for fault diagnosis. They have introduced different ranges of operating parameters resulting in one normal operation and six abnormal operations. The variables that cause malfunctions are inlet flowrate, inlet temperature and inlet concentration of the reactant. Deviations beyond ±5% of the normal values of these variables were considered as faults, while values within ±5% were considered normal.

Generation of the historical database: The normal operations were simulated by varying any of the input variables in the neighborhood of the perfectly normal operation. This variation was kept in the range of 2.0%. The magnitudes of the input variable were varied randomly in each simulation. The duration of each simulation was 4 hours and sampling interval was randomly varied from 2 minutes. Each of the seven operating conditions was simulated 50 times to provide a historical database of 350 batches. Gaussian measurement noise was added to the measured variables so that the signal-to-noise ratio for the CSTR profile was approximately equal to ten. These batch profiles were divided into two subsets; a set of 175 profiles (25 profiles per class) was taken as the reference dataset and another set of 175 profiles (25 profiles per class) was used for the online diagnosis.

Diagnosis Result: Various parameters defined above were tuned with an objective of achieving maximum diagnosis efficiency, as shown in Table 2.

Table 2. Fault Diagnosis Results for a particular set of parameters

Parameters	Parameter Values	Fault Type	Diagnosis Efficiency(%) Fault Count=2
Initial Point	11	Normal	68
Window Width	50	Fault 1	100
Window Shift	10	Fault 2	88
N	2	Fault 3	80
n	2	Fault 4	96
		Fault 5	92
		Fault 6	80

We can see that the overall classification accuracy is 86.2857% for the above case. Very high diagnosis efficiency is obtained for Fault1, Fault4 and Fault5 operations, while the diagnosis efficiency for the case of Normal is quite less. We compare the results of the proposed algorithm the SVM-based classification technique, where one-against-all strategy is emplyed to classify the test profiles. An overall classification accuracy of 80% is achieved by using this approach, which is significantly less than the classification accuracy of 86.2857% obtained by the DTW approach. Thus, our method performs better in cases where fault diagnosis is required.

7   Result

In this work we have proposed a novel methodology for the purpose of online fault diagnosis employing the concept of Dynamic Time Warping as a superior distance measure between two time series. Warping window constraints and lower bounding measures have been utilized for the case of unequal length time-series, which substantially reduce the computational expense required. The proposed methodology is capable of handling time-series of unequal lengths with different sampling times. This allows the application of our methodology to time-series with missing time points. A moving window approach has been demonstrated for the purpose of fault diagnosis in two simulated case studies. In both the cases, a simulated reference dataset is used to train bands for different classes using a heuristic hill-climbing approach, which were then employed for similarity measurement. The diagnosis efficiency is heavily dependent on the parameter values, however for appropriate set of parameters, the methodology is found to be satisfactorily efficient in detecting the faults in the process and can predict the type of the fault also.

Acknowledgements

We gratefully acknowledge financial assistance provided by Department of Science and Technology, New Delhi, India.

References

1. Sakoe, H., Chiba, S.: Dynamic-Programming Algorithm Optimization for Spoken Word Recognition. IEEE Trans Acoust Speech Signal Process 26, 43–49 (1978)
2. Ratanamahatana, C.A., Keogh, E.: Making Time-series Classification More Accurate Using Learned Constraints. In: Jonker, W., Petković, M. (eds.) SDM 2004. LNCS, vol. 3178, pp. 11–22. Springer, Heidelberg (2004)
3. Rath, T.M., Manmatha, R.: Lower-Bounding of Dynamic Time Warping Distances for Multivariate Time Series. Technical Report MM-40, Center for Intelligent Information Retrieval, University of Massachusetts Amherst (2002)
4. Vortruba, J., Volesky, B., Yerushalmi, L.: Mathematical model of a batch acetone-butanol fermentation. Biotechnol. Bioeng. 28, 247–255 (1986)
5. Singhal, A.: Pattern-matching in multivariate time-series data. Ph.D. dissertation, Univ. of California, Santa Barbara (2002)
6. Luyben William, L.: Process modeling, simulation and control for Chemical Engineers. McGraw Hill, New York (1973)
7. Venkatasubramanian, V., Vaidyanathan, R., Yamamoto, Y.: Process fault detection and diagnosis using Neural Networks-I. Steady-state processes. Computers Chem. Engg. 14(7), 699–712 (1990)
8. Kumar, R., Jade, A.M., Jayaraman, V.K., Kulkarni, B.D.: A Hybrid Methodology For On-Line Process Monitoring. IJCRE 2(A14) (2004)

Kernel-Based Spectral Matched Signal Detectors for Hyperspectral Target Detection

Nasser M. Nasrabadi

U.S. Army Research Laboratory, 2800 Powder Mill Road, Adelphi, MD 20783, USA
nnasraba@arl.army.mil

Abstract. In this paper, we compare several detection algorithms that are based on spectral matched (subspace) filters. Nonlinear (*kernel*) versions of these spectral matched (subspace) detectors are also discussed and their performance is compared with the linear versions. Several well-known matched detectors, such as matched subspace detector, orthogonal subspace detector, spectral matched filter and adaptive subspace detector (adaptive cosine estimator) are extended to their corresponding kernel versions by using the idea of kernel-based learning theory. In kernel-based detection algorithms the data is implicitly mapped into a high dimensional kernel feature space by a nonlinear mapping which is associated with a kernel function. The detection algorithm is then derived in the feature space which is *kernelized* in terms of the kernel functions in order to avoid explicit computation in the high dimensional feature space. Experimental results based on real hyperspectral imagery show that the kernel versions of these detectors outperform the conventional linear detectors.

1 Introduction

Detecting signals of interest, particularly with wide signal variability, in noisy environments has long been a challenging issue in various fields of signal processing. Among a number of previously developed detectors, the well-known matched subspace detector (MSD) [1], orthogonal subspace detector (OSD) [2], spectral matched filter (SMF) [3], and adaptive subspace detectors (ASD) also known as adaptive cosine estimator (ACE) [4], have been widely used to detect a desired signal (target).

Matched signal detectors, such as spectral matched filter and matched subspace detectors (whether adaptive or non-adaptive), only exploit second order correlations, thus completely ignoring nonlinear (higher order) spectral inter-band correlations that could be crucial to discriminate between target and background. In this paper, our aim is to introduce nonlinear versions of MSD, OSD, SMF and ASD detectors which effectively exploits the higher order spectral inter-band correlations in a high (possibly infinite) dimensional feature space associated with a certain nonlinear mapping via kernel-based learning methods [5]. A nonlinear mapping of the input data into a high dimensional feature space is often expected to increase the data separability and reduce the complexity of the corresponding data structure.

This paper is organized as follows. Section 2 provides the background to the kernel-based learning methods and kernel trick. Section 3 introduces a linear matched subspace and its kernel version. The orthogonal subspace detector is defined in Section 4 as well

A. Ghosh, R.K. De, and S.K. Pal (Eds.): PReMI 2007, LNCS 4815, pp. 67–76, 2007.

as its kernel version. In Section 5 we describe the conventional spectral matched filter ad its kernel version in the feature space and reformulate the expression in terms of the kernel function using the kernel trick. Finally, in Section 6 the adaptive subspace detector and its kernel version are introduced. Performance comparison between the conventional and the kernel versions of these algorithms is provided in Section 7 and conclusions are given in Section 8.

2 Kernel-Based Learning and Kernel Trick

Suppose that the input hyperspectral data is represented by the data space ($\mathcal{X} \subseteq \mathcal{R}^l$) and $\mathcal{F}$ is a feature space associated with $\mathcal{X}$ by a nonlinear mapping function Φ

$$\Phi : \mathcal{X} \to \mathcal{F}, \mathbf{x} \mapsto \Phi(\mathbf{x}), \tag{1}$$

where $\mathbf{x}$ is an input vector in $\mathcal{X}$ which is mapped into a potentially much higher – (could be infinite) – dimensional feature space. Due to the high dimensionality of the feature space $\mathcal{F}$, it is computationally not feasible to implement any algorithm directly in feature space. However, kernel-based learning algorithms use an effective kernel trick given by Eq. (2) to implement dot products in feature space by employing kernel functions [5]. The idea in kernel-based techniques is to obtain a nonlinear version of an algorithm defined in the input space by implicitly redefining it in the feature space and then converting it in terms of dot products. The kernel trick is then used to implicitly compute the dot products in $\mathcal{F}$ without mapping the input vectors into $\mathcal{F}$; therefore, in the kernel methods, the mapping Φ does not need to be identified.

The kernel representation for the dot products in $\mathcal{F}$ is expressed as

$$k(\mathbf{x}_i, \mathbf{x}_j) = \Phi(\mathbf{x}_i) \cdot \Phi(\mathbf{x}_j), \tag{2}$$

where k is a kernel function in terms of the original data. There are a large number of Mercer kernels that have the kernel trick property, see [5] for detailed information about the properties of different kernels and kernel-based learning. Our choice of kernel in this paper is the Gaussian RBF kernel and the associated nonlinear function Φ with this kernel generates a feature space of infinite dimensionality.

3 Linear MSD and Kernel MSD

3.1 Linear MSD

In this model the target pixel vectors are expressed as a linear combination of target spectral signature and background spectral signature, which are represented by subspace target spectra and subspace background spectra, respectively. The hyperspectral target detection problem in a p-dimensional input space is expressed as two competing hypotheses $\mathbf{H_0}$ and $\mathbf{H_1}$

$$\mathbf{H}_0 : \mathbf{y} = \mathbf{B}\zeta + \mathbf{n}, \qquad\qquad\qquad \text{Target absent} \tag{3}$$

$$\mathbf{H}_1 : \mathbf{y} = \mathbf{T}\theta + \mathbf{B}\zeta + \mathbf{n} = \begin{bmatrix} \mathbf{T} & \mathbf{B} \end{bmatrix} \begin{bmatrix} \theta \\ \zeta \end{bmatrix} + \mathbf{n}, \qquad \text{Target present}$$

where $\mathbf{T}$ and $\mathbf{B}$ represent orthogonal matrices whose p-dimensional column vectors span the target and background subspaces, respectively; θ and ζ are unknown vectors whose entries are coefficients that account for the abundances of the corresponding column vectors of $\mathbf{T}$ and $\mathbf{B}$, respectively; $\mathbf{n}$ represents Gaussian random noise ($\mathbf{n} \in \mathcal{R}^p$) distributed as $\mathcal{N}(0, \sigma^2 \mathbf{I})$; and $\begin{bmatrix} \mathbf{T} & \mathbf{B} \end{bmatrix}$ is a concatenated matrix of $\mathbf{T}$ and $\mathbf{B}$. The numbers of the column vectors of $\mathbf{T}$ and $\mathbf{B}$, N_t and N_b, respectively, are usually smaller than p ($N_t, N_b < p$).

The generalized likelihood ratio test (GLRT) for the model (3) was derived in [1], given as

$$\mathbf{L}_2(\mathbf{y}) = \frac{\mathbf{y}^T (\mathbf{I} - \mathbf{P_B}) \mathbf{y}}{\mathbf{y}^T (\mathbf{I} - \mathbf{P_{TB}}) \mathbf{y}} \underset{H_0}{\overset{H_1}{\gtrless}} \eta \tag{4}$$

where $\mathbf{P_B} = \mathbf{B}(\mathbf{B}^T\mathbf{B})^{-1}\mathbf{B}^T = \mathbf{BB}^T$ is a projection matrix associated with the N_b-dimensional background subspace $< \mathbf{B} >$; $\mathbf{P_{TB}}$ is a projection matrix associated with the ($N_{bt} = N_b + N_t$)-dimensional target-and-background subspace $< \mathbf{TB} >$

$$\mathbf{P_{TB}} = \begin{bmatrix} \mathbf{T} & \mathbf{B} \end{bmatrix} \begin{bmatrix} [\mathbf{T}\ \mathbf{B}]^T \begin{bmatrix} \mathbf{T}\ \mathbf{B} \end{bmatrix} \end{bmatrix}^{-1} \begin{bmatrix} \mathbf{T}\ \mathbf{B} \end{bmatrix}^T. \tag{5}$$

3.2 Linear MSD in the Feature Space and Its Kernel Version

The hyperspectral detection problem based on the target and background subspaces can be described in the feature space $\mathcal{F}$ as

$$\mathbf{H}_{0_\Phi} : \Phi(\mathbf{y}) = \mathbf{B}_\Phi \zeta_\Phi + \mathbf{n}_\Phi, \qquad\qquad \text{Target absent} \tag{6}$$

$$\mathbf{H}_{1_\Phi} : \Phi(\mathbf{y}) = \mathbf{T}_\Phi \theta_\Phi + \mathbf{B}_\Phi \zeta_\Phi + \mathbf{n}_\Phi = \begin{bmatrix} \mathbf{T}_\Phi & \mathbf{B}_\Phi \end{bmatrix} \begin{bmatrix} \theta_\Phi \\ \zeta_\Phi \end{bmatrix} + \mathbf{n}_\Phi, \quad \text{Target present}$$

where $\mathbf{T}_\Phi$ and $\mathbf{B}_\Phi$ represent full-rank matrices whose column vectors span target and background subspaces $< \mathbf{B}_\Phi >$ and $< \mathbf{T}_\Phi >$ in $\mathcal{F}$, respectively; θ_Φ and ζ_Φ are unknown vectors whose entries are coefficients that account for the abundances of the corresponding column vectors of $\mathbf{T}_\Phi$ and $\mathbf{B}_\Phi$, respectively; $\mathbf{n}_\Phi$ represents Gaussian random noise; and $\begin{bmatrix} \mathbf{T}_\Phi & \mathbf{B}_\Phi \end{bmatrix}$ is a concatenated matrix of $\mathbf{T}_\Phi$ and $\mathbf{B}_\Phi$. Using a similar reasoning as described in the previous subsection, the GLRT of the hyperspectral detection problem depicted by the model in (6) is given by

$$\mathbf{L}_2(\Phi(\mathbf{y})) = \frac{\Phi(\mathbf{y})^T (\mathbf{P_{I_\Phi}} - \mathbf{P_{B_\Phi}}) \Phi(\mathbf{y})}{\Phi(\mathbf{y})^T (\mathbf{P_{I_\Phi}} - \mathbf{P_{T_\Phi B_\Phi}}) \Phi(\mathbf{y})}, \tag{7}$$

where $\mathbf{P_{I_\Phi}}$ represents an identity projection operator in $\mathcal{F}$; $\mathbf{P_{B_\Phi}} = \mathbf{B}_\Phi (\mathbf{B}_\Phi^T \mathbf{B}_\Phi)^{-1} \mathbf{B}_\Phi^T = \mathbf{B}_\Phi \mathbf{B}_\Phi^T$ is a background projection matrix; and $\mathbf{P_{T_\Phi B_\Phi}}$ is a joint target-and-background projection matrix in $\mathcal{F}$

$$\mathbf{P_{T_\Phi B_\Phi}} = \begin{bmatrix} \mathbf{T}_\Phi & \mathbf{B}_\Phi \end{bmatrix} \begin{bmatrix} [\mathbf{T}_\Phi\ \mathbf{B}_\Phi]^T \begin{bmatrix} \mathbf{T}_\Phi\ \mathbf{B}_\Phi \end{bmatrix} \end{bmatrix}^{-1} \begin{bmatrix} \mathbf{T}_\Phi\ \mathbf{B}_\Phi \end{bmatrix}^T$$

$$= \begin{bmatrix} \mathbf{T}_\Phi & \mathbf{B}_\Phi \end{bmatrix} \begin{bmatrix} \mathbf{T}_\Phi^T \mathbf{T}_\Phi & \mathbf{T}_\Phi^T \mathbf{B}_\Phi \\ \mathbf{B}_\Phi^T \mathbf{T}_\Phi & \mathbf{B}_\Phi^T \mathbf{B}_\Phi \end{bmatrix}^{-1} \begin{bmatrix} \mathbf{T}_\Phi^T \\ \mathbf{B}_\Phi^T \end{bmatrix}. \tag{8}$$

The kernelized GLRT (7) is drived in [6] and is given by

$$L_{2K} = \frac{K(\mathbf{Z_{TB}},\mathbf{y})^T \Delta\Delta^T K(\mathbf{Z_{TB}},\mathbf{y}) - K(\mathbf{Z_B},\mathbf{y})^T \mathcal{B}\mathcal{B}^T K(\mathbf{Z_B},\mathbf{y})}{K(\mathbf{Z_{TB}},\mathbf{y})^T \Delta\Delta^T K(\mathbf{Z_{TB}},\mathbf{y}) - [K(\mathbf{Z_T},\mathbf{y})^T \mathcal{T} \; K(\mathbf{Z_B},\mathbf{y})^T \mathcal{B}] \Lambda_1^{-1} \begin{bmatrix} \mathcal{T}^T K(\mathbf{Z_T},\mathbf{y}) \\ \mathcal{B}^T K(\mathbf{Z_B},\mathbf{y}) \end{bmatrix}},$$

(9)

where $\Lambda_1 = \begin{bmatrix} \mathcal{T}^T K(\mathbf{Z_T},\mathbf{Z_T})\mathcal{T} & \mathcal{T}^T K(\mathbf{Z_T},\mathbf{Z_B})\mathcal{B} \\ \mathcal{B}^T K(\mathbf{Z_B},\mathbf{Z_T})\mathcal{T} & \mathcal{B}^T K(\mathbf{Z_B},\mathbf{Z_B})\mathcal{B} \end{bmatrix}$.

In the above derivation (9) we assumed that the mapped input data was centered in the feature space by removing the sample mean. However, the original data is usually not centered and the estimated mean in the feature space can not be explicitly computed, therefore, the kernel matrices have to be properly centered. The resulting centered $\hat{\mathbf{K}}$ is shown in [5] to be given by

$$\hat{\mathbf{K}} = (\mathbf{K} - \mathbf{1}_N \mathbf{K} - \mathbf{K}\mathbf{1}_N + \mathbf{1}_N \mathbf{K}\mathbf{1}_N),$$

(10)

where the $N \times N$ matrix $(\mathbf{1}_N)_{ij} = 1/N$. The empirical kernel maps $K(\mathbf{Z_T},\mathbf{y})$, $K(\mathbf{Z_B},\mathbf{y})$, and $K(\mathbf{Z_{TB}},\mathbf{y})$ have also to be centered by removing their corresponding empirical kernel map mean. (e.g. $\hat{K}(\mathbf{Z_T},\mathbf{y}) = K(\mathbf{Z_T},\mathbf{y}) - \frac{1}{N}\sum_{i=1}^{N} k(\mathbf{y}_i,\mathbf{y})$, $\mathbf{y}_i \in \mathbf{Z_T}$.)

4 OSP and Kernel OSP Algorithms

4.1 Linear Spectral Mixture Model

The OSP algorithm [2] is based on maximizing the SNR (signal-to-noise ratio) in the subspace orthogonal to the background subspace and only depends on the noise second-order statistics. It also does not provide any estimate of the abundance measure for the desired end member in the mixed pixel. A linear mixture model for pixel $\mathbf{y}$ consisting of p spectral bands is described by

$$\mathbf{y} = \mathbf{M}\alpha + \mathbf{n},$$

(11)

where the $(p \times l)$ matrix $\mathbf{M}$ represent l endmembers spectra, α is a $(l \times 1)$ column vector whose elements are the coefficients that account for the proportions (abundances) of each endmember spectrum contributing to the mixed pixel, and $\mathbf{n}$ is an $(p \times 1)$ vector representing an additive zero-mean Gaussian noise with covariance matrix $\sigma^2\mathbf{I}$ and $\mathbf{I}$ is the $(p \times p)$ identity matrix.

Assuming now we want to identify one particular signature (e.g. a military target) with a given spectral signature $\mathbf{d}$ and a corresponding abundance measure α_l, we can represent $\mathbf{M}$ and α in partition form as $\mathbf{M} = (\mathbf{U} : \mathbf{d})$ and $\alpha = \begin{bmatrix} \gamma \\ \alpha_l \end{bmatrix}$ then model (11) can be rewritten as

$$\mathbf{r} = \mathbf{d}\alpha_l + \mathbf{B}\gamma + \mathbf{n},$$

(12)

where the columns of $\mathbf{B}$ represent the undesired spectral signatures (background signatures or eigenvectors) and the column vector γ is the abundance measures for the undesired spectral signatures. The OSP operator that maximizes the signal to noise ratio is given by

$$q_{OSP}^T = d^T(I - BB^\#) \tag{13}$$

which consists of a background signature rejecter followed by a matched filter. The output of the OSP classifier is now given by

$$D_{OSP} = q_{OSP}^T r = d^T(I - BB^\#)y. \tag{14}$$

4.2 OSP in Feature Space and Its Kernel Version

The mixture model in the high dimensional feature space $\mathcal{F}$ is given by

$$\Phi(r) = M_\Phi \alpha + n_\Phi, \tag{15}$$

where M_Φ is a matrix whose columns are the endmembers spectra in the feature space and n_Φ is an additive zero-mean Gaussian noise with covariance matrix $\sigma^2 I_\Phi$ and I_Φ is the identity matrix in the feature space. The model (15) can also be rewritten as

$$\Phi(r) = \Phi(d)\alpha_p + B_\Phi \gamma + n_\Phi, \tag{16}$$

where $\Phi(d)$ represent the spectral signature of the desired target in the feature space and the columns of B_Φ represent the undesired background signatures in the feature space.

The output of the OSP classifier in the feature space is given by

$$D_{OSP_\Phi} = q_{OSP_\Phi}^T r = \Phi(d)^T(I_\Phi - B_\Phi B_\Phi^\#)\Phi(r). \tag{17}$$

This output (17) is very similar to the numerator of (7). It is easily shown in [7] that the kernelized version of (17) is given by

$$D_{KOSP} = K(Z_{Bd}, d)^T \Upsilon \Upsilon^T K(Z_{Bd}, y) - K(Z_B, d)^T \mathcal{B} \mathcal{B}^T K(Z_B, y) \tag{18}$$

where $Z_B = [x_1 \ x_2 \ \dots \ x_N]$ correspond to N input background spectral signatures and $\mathcal{B} = (\beta^1, \beta^2, \dots, \beta^{N_b})^T$ are the N_b significant eigenvectors of the centered kernel matrix (Gram matrix) $K(Z_B, Z_B)$ normalized by the square root of their corresponding eigenvalues [5]. $K(Z_B, r)$ and $K(Z_B, d)$, are column vectors whose entries are $k(x_i, y)$ and $k(x_i, d)$ for $x_i \in Z_B$, respectively. $Z_{Bd} = Z_B \cup d$ and Υ is a matrix whose columns are the N_{bd} eigenvectors ($v_1, v_2, \dots, v_{N_{bd}}$) of the centered kernel matrix $K(Z_{Bd}, Z_{Bd}) = (K)_{ij} = k(x_i, x_j)$, $x_i, x_j \in Z_{Bd}$ with nonzero eigenvalues, normalized by the square root of their associated eigenvalues. Also $K(Z_{Bd}, y)$ is the concatenated vector $[K(Z_B, r)^T \ K(d, y)^T]^T$ and $K(Z_{Bd}, d)$ is the concatenated vector $[K(Z_B, d)^T \ K(d, d)^T]^T$.

5   Linear SMF and Kernel Spectral Matched Filter

5.1   Linear Spectral Matched Filter

In this section, we introduce the concept of linear SMF. The constrained least squares approach is used to derive the linear SMF. Let the input spectral signal $\mathbf{x}$ be $\mathbf{x} = [x(1), x(2), \ldots, x(p)]^T$ consisting of p spectral bands. We can model each spectral observation as a linear combination of the target spectral signature and noise

$$\mathbf{x} = a\mathbf{s} + \mathbf{n}, \tag{19}$$

where a is an attenuation constant (target abundance measure). When $a = 0$ no target is present and when $a > 0$ target is present, vector $\mathbf{s} = [s(1), s(2), \ldots, s(p)]^T$ contains the spectral signature of the target and vector $\mathbf{n}$ contains the added background clutter noise.

Let us define $\mathbf{X}$ to be a $p \times N$ matrix of the N mean-removed background reference pixels (centered) obtained from the input image. Let each centered observation spectral pixel to be represented as a column in the sample matrix $\mathbf{X}$

$$\mathbf{X} = [\mathbf{x}_1 \ \mathbf{x}_2 \ \ldots \ \mathbf{x}_N]. \tag{20}$$

We can design a linear matched filter such that the desired target signal $\mathbf{s}$ is passed through while the average filter output energy is minimized. The solution to this minimization problem is given by

$$y_{\mathbf{r}} = \mathbf{w}^T \mathbf{r} = \frac{\mathbf{s}^T \hat{\mathbf{C}}^{-1} \mathbf{r}}{\mathbf{s}^T \hat{\mathbf{C}}^{-1} \mathbf{s}} \tag{21}$$

where $\hat{\mathbf{C}}$ is the estimated covariance matrix.

5.2   SMF in Feature Space and Its Kernel Version

Consider the linear model of the input data in a kernel feature space which is equivalent to a non-linear model in the input space

$$\Phi(\mathbf{x}) = a_\Phi \Phi(\mathbf{s}) + \mathbf{n}_\Phi, \tag{22}$$

where Φ is the non-linear mapping that maps the input data into a kernel feature space, a_Φ is an attenuation constant (abundance measure), the high dimensional vector $\Phi(\mathbf{s})$ contains the spectral signature of the target in the feature space, and vector $\mathbf{n}_\Phi$ contains the added noise in the feature space.

Using the constrained least squares approach it can easily be shown that the output of the desired matched filter for the input $\Phi(\mathbf{r})$ is given by

$$y_{\Phi(\mathbf{r})} = \frac{\Phi(\mathbf{s})^T \hat{\mathbf{C}}_\Phi^{-1} \Phi(\mathbf{r})}{\Phi(\mathbf{s})^T \hat{\mathbf{C}}_\Phi^{-1} \Phi(\mathbf{s})} \tag{23}$$

where $\hat{\mathbf{C}}_\Phi$ is the estimated covariance of pixels in the feature space.

We now show how to kernelize the matched filter expression (23) where the resulting non-linear matched filter is called the kernel matched filter. The pseudoinverse (inverse) of the estimated background covariance matrix can be written in terms of its eigenvector decomposition as

$$\hat{\mathbf{C}}_\Phi^\# = \mathbf{X}_\Phi \mathcal{B} \Lambda^{-1} \mathcal{B}^T \mathbf{X}_\Phi^T \tag{24}$$

where $\mathbf{X}_\Phi = [\Phi(\mathbf{x}_1)\ \Phi(\mathbf{x}_2)\ \ldots\ \Phi(\mathbf{x}_N)]$ is a matrix whose columns are the mapped background reference data in the feature space and $\mathcal{B} = [\beta^1\ \beta^2\ \ldots\ \beta^{N_b}]$ are the nonzero eigenvectors of the centered kernel matrix (Gram matrix) $\mathbf{K}(\mathbf{X}, \mathbf{X})$ normalized by the square root of their corresponding eigenvalues.

Inserting Equation (24) into (23) it can be rewritten as

$$y_{\Phi(\mathbf{r})} = \frac{\Phi(\mathbf{s})^T \mathbf{X}_\Phi \mathcal{B} \Lambda^{-1} \mathcal{B}^T \mathbf{X}_\Phi^T \Phi(\mathbf{r})}{\Phi(\mathbf{s})^T \mathbf{X}_\Phi \mathcal{B} \Lambda^{-1} \mathcal{B}^T \mathbf{X}_\Phi^T \Phi(\mathbf{s})}. \tag{25}$$

Also using the properties of the Kernel PCA [5], we have the relationship $\mathbf{K}^{-1} = \mathcal{B}\Lambda^{-1}\mathcal{B}$. We denote $\mathbf{K} = \mathbf{K}(\mathbf{X}, \mathbf{X}) = (\mathbf{K})_{ij}$ an $N \times N$ Gram kernel matrix whose entries are the dot products $< \Phi(\mathbf{x}_i), \Phi(\mathbf{x}_j) >$. Finally, the kernelized version of SMF as shown in [8] is now given by

$$y_{\mathbf{K_r}} = \frac{\mathbf{K}(\mathbf{X}, \mathbf{s})^T \mathbf{K}^{-1} \mathbf{K}(\mathbf{X}, \mathbf{r})}{\mathbf{K}(\mathbf{X}, \mathbf{s})^T \mathbf{K}^{-1} \mathbf{K}(\mathbf{X}, \mathbf{s})} = \frac{\mathbf{K_s}^T \mathbf{K}^{-1} \mathbf{K_r}}{\mathbf{K_s}^T \mathbf{K}^{-1} \mathbf{K_s}} \tag{26}$$

where the empirical kernel maps $\mathbf{K_s} = \mathbf{K}(\mathbf{X}, \mathbf{s})$ and $\mathbf{K_r} = \mathbf{K}(\mathbf{X}, \mathbf{r})$.

6 Adaptive Subspace Detector and Kernel Adaptive Subspace Detector

6.1 Linear ASD

In this section, the GLRT under the two competing hypotheses ($\mathbf{H}_0$ and $\mathbf{H}_1$) for a certain mixture model is described. The subpixel detection model for a measurement $\mathbf{x}$ (a pixel vector) is expressed as

$$\mathbf{H}_0 : \mathbf{x} = \mathbf{n}, \qquad\qquad \text{Target absent} \tag{27}$$
$$\mathbf{H}_1 : \mathbf{x} = \mathbf{U}\theta + \sigma\mathbf{n}, \qquad\qquad \text{Target present}$$

where $\mathbf{U}$ represents an orthogonal matrix whose column vectors are the eigenvectors that span the target subspace $< \mathbf{U} >$; θ is an unknown vector whose entries are coefficients that account for the abundances of the corresponding column vectors of $\mathbf{U}$; $\mathbf{n}$ represents Gaussian random noise distributed as $\mathcal{N}(0, \mathbf{C})$.

In the model, $\mathbf{x}$ is assumed to be a background noise under $\mathbf{H}_0$ and a linear combination of a target subspace signal and a scaled background noise, distributed as $\mathcal{N}(\mathbf{U}\theta, \sigma^2\mathbf{C})$, under $\mathbf{H}_1$. The background noise under the two hypotheses is represented by the same covariance but different variances because of the existence of subpixel targets under $\mathbf{H}_1$. The GLRT for the subpixel problem as described in [4] (so called ASD) is given by

$$D_{ASD}(\mathbf{x}) = \frac{\mathbf{x}^T \hat{\mathbf{C}}^{-1} \mathbf{U}(\mathbf{U}^T \hat{\mathbf{C}}^{-1} \mathbf{U})^{-1} \mathbf{U}^T \hat{\mathbf{C}}^{-1} \mathbf{x}}{\mathbf{x}^T \hat{\mathbf{C}}^{-1} \mathbf{x}} \underset{H_0}{\overset{H_1}{\gtrless}} \eta_{ASD}, \qquad (28)$$

where $\hat{\mathbf{C}}$ is the MLE (maximum likelihood estimate) of the covariance $\mathbf{C}$ and η_{ASD} represents a threshold. Expression (28) has a constant false alarm rate (CFAR) property and is also referred to as the adaptive cosine estimator because (28) measures the angle between $\tilde{\mathbf{x}}$ and $< \tilde{\mathbf{U}} >$ where $\tilde{\mathbf{x}} = \hat{\mathbf{C}}^{-1/2}\mathbf{x}$ and $\tilde{\mathbf{U}} = \hat{\mathbf{C}}^{-1/2}\mathbf{U}$.

6.2   ASD in the Feature Space and Its Kernel Version

We define a new subpixel model by assuming that the input data has been implicitly mapped by a nonlinear function Φ into a high dimensional feature space $\mathcal{F}$. The model in $\mathcal{F}$ is then given by

$$\mathbf{H}_{0_\Phi} : \Phi(\mathbf{x}) = \mathbf{n}_\Phi, \qquad \text{Target absent} \qquad (29)$$
$$\mathbf{H}_{1_\Phi} : \Phi(\mathbf{x}) = \mathbf{U}_\Phi \theta_\Phi + \sigma_\Phi \mathbf{n}_\Phi, \qquad \text{Target present}$$

where $\mathbf{U}_\Phi$ represents a full-rank matrix whose M_1 column vectors are the eigenvectors that span target subspace $< \mathbf{U}_\Phi >$ in $\mathcal{F}$; θ_Φ is unknown vectors whose entries are coefficients that account for the abundances of the corresponding column vectors of $\mathbf{U}_\Phi$; $\mathbf{n}_\Phi$ represents Gaussian random noise distributed by $\mathcal{N}(0, \mathbf{C}_\Phi)$; and σ_Φ is the noise variance under $\mathbf{H}_{1_\Phi}$. The GLRT for the model (29) in $\mathcal{F}$ is now given by

$$D(\Phi(\mathbf{x})) = \frac{\Phi(\mathbf{x})^T \hat{\mathbf{C}}_\Phi^{-1} \mathbf{U}_\Phi (\mathbf{U}_\Phi^T \hat{\mathbf{C}}_\Phi^{-1} \mathbf{U}_\Phi)^{-1} \mathbf{U}_\Phi^T \hat{\mathbf{C}}_\Phi^{-1} \Phi(\mathbf{x})}{\Phi(\mathbf{x})^T \hat{\mathbf{C}}_\Phi^{-1} \Phi(\mathbf{x})}, \qquad (30)$$

where $\hat{\mathbf{C}}_\Phi$ is the MLE of $\mathbf{C}_\Phi$.

The kernelized expression of (30) was derived in [9] and is given by

$$D_{KASD}(\mathbf{x}) = \frac{\mathbf{K}_\mathbf{x}[\mathcal{T}^T \mathbf{K}(\mathbf{X}, \mathbf{Y})^T \mathbf{K}(\mathbf{X}, \mathbf{X})^{-1} \mathbf{K}(\mathbf{X}, \mathbf{Y})\mathcal{T}]^{-1} \mathbf{K}_\mathbf{x}^T}{\mathbf{k}(\mathbf{x}, \mathbf{X})^T \mathbf{K}(\mathbf{X}, \mathbf{X})^{-1} \mathbf{k}(\mathbf{x}, \mathbf{X})} \qquad (31)$$

where $\mathbf{K}_\mathbf{x} = \mathbf{k}(\mathbf{x}, \mathbf{X})^T \mathbf{K}(\mathbf{X}, \mathbf{X})^{-1} \mathbf{K}(\mathbf{X}, \mathbf{Y})\mathcal{T}$, background spectral signatures is denoted by $\mathbf{X} = [\mathbf{x}_1 \ \mathbf{x}_2 \ \dots \ \mathbf{x}_N]$, target spectral signatures are denoted by $\mathbf{Y} = [\mathbf{y}_1 \ \mathbf{y}_2 \ \dots \ \mathbf{y}_M]$ and $\mathcal{T} = [\alpha^1 \ \alpha^2 \ \dots \ \alpha^{M_1}]$, $M_1 < M$, is a matrix consisting of the M_1 eigenvectors of the kernel matrix $\mathbf{K}(\mathbf{Y}, \mathbf{Y})$.

7   Experimental Results

In this section, the kernel-based matched signal detectors, such as the kernel MSD (KMSD), kernel ASD (KASD), kernel OSP (KOSP) and kernel SMF (KSMF) as well

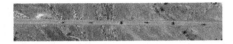

Fig. 1. A sample band image from the DR-II data

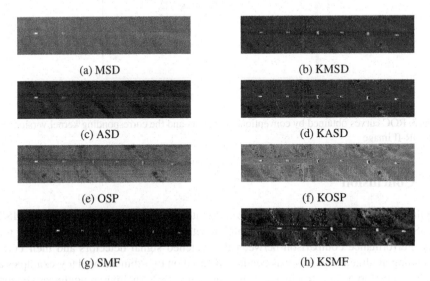

(a) MSD

(b) KMSD

(c) ASD

(d) KASD

(e) OSP

(f) KOSP

(g) SMF

(h) KSMF

Fig. 2. Detection results for the DR-II image using the conventional detectors and the corresponding kernel versions

as the corresponding conventional detectors are implemented. The Gaussian RBF kernel, $k(\mathbf{x}, \mathbf{y}) = \exp(\frac{-\|\mathbf{x}-\mathbf{y}\|^2}{c})$, was used to implement the kernel-based detectors. c represents the width of the Gaussian distribution and the value of c was chosen such that the overall data variations can be fully exploited by the Gaussian RBF function. In this paper, the values of c were determined experimentally. HYDICE (HYperspectral Digital Imagery Collection Experiment) image from the Desert Radiance II data collection (DR-II) was used to compare detection performance between the kernel-based and conventional methods. The HYDICE imaging sensor generates 210 bands across the whole spectral range $(0.4 - 2.5 \ \mu m)$ which includes the visible and short-wave infrared (SWIR) bands. But we only use 150 bands by discarding water absorption and low signal to noise ratio (SNR) bands; the spectral bands used are the 23rd–101st, 109th–136th, and 152nd–194th for the HYDICE images. The DR-II image includes 6 military targets along the road, as shown in the sample band images in Fig. 1. The detection performance of the DR-II image was provided in both the qualitative and quantitative – the receiver operating characteristics (ROC) curves – forms. The spectral signatures of the desired target and undesired background signatures were directly collected from the given hyperspectral data to implement both the kernel-based and conventional detectors. Figs. 2-3 show the detection results including the ROC curves generated by applying the kernel-based and conventional detectors to the DR-II image.

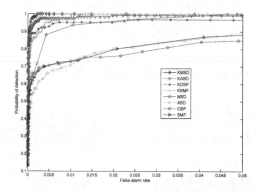

Fig. 3. ROC curves obtained by conventional detectors and the corresponding kernel versions for the DR-II image

8 Conclusion

In this paper, nonlinear versions of several matched signal detectors, such as KMSD, KOSP, KSMF and KASD have been implemented using the kernel-based learning theory. Performance comparison between the matched signal detectors and their corresponding nonlinear versions was conducted based on two-dimensional toy-examples as well as a real hyperspectral image. It is shown that the kernel-based nonlinear versions of these detectors outperform the linear versions.

References

1. Scharf, L.L., Friedlander, B.: Matched subspace detectors. IEEE Trans. Signal Process 42(8), 2146–2157 (1994)
2. Harsanyi, J.C., Chang, C.I.: Hyperspectral image classification and dimensionality reduction: An orthogonal subspace projection approach. IEEE Trans. Geosci. Remote Sensing 32(4), 779–785 (1994)
3. Robey, F.C., Fuhrmann, D.R., Kelly, E.J.: A cfar adaptive matched filter detector. IEEE Trans. on Aerospace and Elect. Syst. 28(1), 208–216 (1992)
4. Kraut, S., Scharf, L.L., McWhorter, T.: Adaptive subspace detectors. IEEE Trans. Signal Process. 49(1), 1–16 (2001)
5. Schölkopf, B., Smola, A.J.: Learning with Kernels. MIT Press, Cambridge (2002)
6. Kwon, H., Nasrabadi, N.M.: Kernel matched subspace detectors for hyperspectral target detection. IEEE Trans. Pattern Anal. Machine Intell. 28(2), 178–194 (2006)
7. Kwon, H., Nasrabadi, N.M.: Kernel orthogonal subspace projection for hyperspectral signal classification. IEEE Trans. Geosci. Remote Sensing 43(12), 2952–2962 (2005)
8. Kwon, H., Nasrabadi, N.M.: Kernel spectral matched filter for hyperspectral imagery. Int. J. of Computer Vision 71(2), 127–141 (2007)
9. Kwon, H., Nasrabadi, N.M.: Kernel adaptive subspace detector for hyperspectral target detection. 3(2), 178–194 (2006)

A Multiscale Change Detection Technique Robust to Registration Noise

Lorenzo Bruzzone, Francesca Bovolo, and Silvia Marchesi

Department of Information and Communication Technologies
Via Sommarive, 14
I-38050 Trento, Italy
`lorenzo.bruzzone@ing.unitn.it`

Abstract. This paper addresses the problem of unsupervised change detection in multitemporal very high geometrical resolution remote sensing images. In particular, it presents a study on the effects and the properties of the registration noise on the change-detection process in the framework of the polar change vector analysis (CVA) technique. According to this study, a multiscale technique for reducing the impact of residual misregistration in unsupervised change detection is presented. This technique is based on a differential analysis of the direction distributions of spectral change vectors at different resolution levels. The differential analysis allows one to discriminate sectors associated with residual registration noise from sectors associated with true changes. The information extracted is used at full resolution for computing a change-detection map where geometrical details are preserved and the impact of residual registration noise is strongly reduced.

Keywords: Change detection, change vector analysis, registration noise, multi-temporal images, very high geometrical resolution images, multiscale techniques, remote sensing.

1 Introduction

The availability of images acquired by satellite sensors on the same geographical area at different times makes it possible to identify and label possible changes occurred on the ground. In this context, the new generation of passive sensors (e.g., Ikonos, Quickbird, SPOT-5), which can acquire images with very high geometrical resolution, potentially can identify changes occurred on the ground with a high geometrical precision. However, the change-detection problem on VHR images is more complex than in the case of medium resolution images, and requires the development of proper advanced change-detection techniques.

In the literature, several unsupervised change-detection methods have been proposed for the analysis of medium resolution multispectral images [1]. Among them, a widely used unsupervised technique is the change vector analysis (CVA). Although CVA was used in several different application domains, only recently a rigorous and theoretically well defined framework for polar CVA has been introduced in the literature [2]. This framework is based on the analysis of the distribution of

A. Ghosh, R.K. De, and S.K. Pal (Eds.): PReMI 2007, LNCS 4815, pp. 77–86, 2007.

spectral change vectors (SCVs) in the polar domain and makes it possible to better analyze the properties of the change-detection problem. A basic assumption in the CVA technique is that multitemporal images acquired at different times are similar to each other, but for the presence of changes occurred on the ground. Unfortunately, this assumption is seldom completely satisfied due to differences in atmospheric and sunlight conditions, as well as in the acquisition geometry. In order to fulfill the similarity assumption, radiometric and geometric corrections, co-registration, and noise reduction are required. However, it is not possible to reach the ideal conditions with the pre-processing. This is particularly critical in the co-registration procedure, which usually results in a residual misalignment between images (registration noise) which affects the accuracy of the change-detection process leading to a sharp increase of the number of false alarms [2-5]. The effects of the registration noise (RN) are very critical in presence of multitemporal images acquired by VHR sensors. This is due to the complexity of the co-registration of VHR images, which mainly depends on the fact that last generation sensors acquire data with different view angles, thus resulting in multitemporal images with different geometrical distortions.

This paper analyzes the properties of the residual RN in multitemporal VHR images in the context of the polar framework for CVA. Then, according to this theoretical analysis, an adaptive multiscale technique is presented that aims at reducing the impact of the registration noise in the final change-detection map.

2 Notation and Background

Let us consider two multispectral images $\mathbf{X_1}$ and $\mathbf{X_2}$ acquired over the same geographical area at different times t_1 and t_2. Let us assume that these images do not show significant radiometric differences and that they are perfectly coregistered. Let $\Omega = \{\omega_n, \Omega_c\}$ be the set of classes of unchanged and changed pixels to be identified. In greater detail, ω_n represents the class of unchanged pixels, and $\Omega_c = \{\omega_{c_1}, ..., \omega_{c_K}\}$ the set of the K possible classes (kinds) of changes occurred in the considered area. In order to analyze the effects of misregistration in change detection on VHR images and to define a proper technique for reducing the errors induced from registration noise on the change-detection map, we take advantage from the theoretical polar framework for unsupervised change detection based on the CVA proposed in [2]. As in [2], for simplicity, in the following the study is developed by considering a 2–dimensional feature space (however, the analysis can be generalized to the case of more spectral channels).

The CVA technique computes a multispectral difference image ($\mathbf{X}_D$) by subtracting the feature vectors associated to each corresponding spatial position in the two considered images. Each SCV is usually implicitly represented in *polar* coordinates, with its magnitude ρ and direction ϑ defined as:

$$\vartheta = \tan^{-1}\left(\frac{X_{1,D}}{X_{2,D}}\right) \quad \text{and} \quad \rho = \sqrt{(X_{1,D})^2 + (X_{2,D})^2} \tag{1}$$

where $X_{b,D}$ is the random variable representing the *b-th* component (spectral channel) of the multispectral difference image $\mathbf{X}_D$.

Let us define the magnitude-direction (*MD*) domain that includes all change vectors for the considered scene as a circle in the *Polar* representation:

$$MD = \{\rho \in [0, \rho_{max}] \text{ and } \vartheta \in [0, 2\pi)\} \tag{2}$$

where $\rho_{max} = \max \left\{ \sqrt{(X_{1,D})^2 + (X_{2,D})^2} \right\}$. The complete modeling of a change-detection problem requires to identify regions associated with: i) unchanged and ii) changed pixels. Thus the *MD* domain can be split into two parts: i) *circle* C_n of no-changed pixels; and ii) *annulus* A_c of changed pixels. This can be done according to the optimal Bayesian threshold value T that separates pixels belonging to ω_n from pixel belonging to Ω_c.

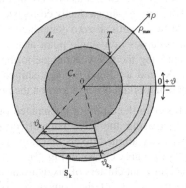

Fig. 1. Representation of the regions of interest for the CVA technique in the Polar coordinate system

Definition 1: the *Circle of unchanged pixels* C_n is defined as $C_n = \{\rho, \vartheta : 0 \leq \rho < T \text{ and } 0 \leq \vartheta < 2\pi\}$. C_n can be represented in the *polar* domain as a circle with radius T.

Definition 2: the *Annulus of changed pixels* A_c is defined as $A_c = \{\rho, \vartheta : T \leq \rho \leq \rho_{max} \text{ and } 0 \leq \vartheta < 2\pi\}$. A_c can be represented in the *Polar* domain as a ring with inner radius T and outer radius ρ_{max}.

The previous definitions have been based on the values of the magnitude, independently on the direction variable. A further important definition is related to *sectors* in the *Polar* domain, which are mainly related to the direction of the SCVs and therefore to the kind of changes occurred on the ground.

Definition 3: the *Annular sector* S_k of change $\omega_k \in \Omega_c$ is defined as $S_k = \{\rho, \vartheta : \rho \geq T \text{ and } \vartheta_{k_1} \leq \vartheta < \vartheta_{k_2}, 0 \leq \vartheta_{k_1} < \vartheta_{k_2} < 2\pi\}$. S_k can be represented in the *Polar* domain as a sector of change within A_c and bounded from two angular thresholds ϑ_{k_1} and ϑ_{k_2}.

The three regions are depicted in Fig. 1. For space constraint, we refer the reader to [2] for a complete set of definitions and for a discussion on them.

3 Properties of Registration Noise

In this section we analyze the effects and the properties of the RN in the polar CVA domain according to the use of a VHR reference image. To this purpose, let us consider a Quickbird image acquired in July 2006 on the Trentino area (Italy) (see Fig. 3.b and section V for further details on this image). From this image, two simulated multitemporal datasets are generated made up of the original image (X_1) and of one shifted version of it (X_2) (shifts of 2 and 6 pixels were considered). These two simulated data sets allow understanding the behavior of the RN components in a controlled experimental setup, as there are no true changes in the two images and there are no differences in their radiometric properties. Let us observe that the application of the CVA technique to X_1 and itself when images are perfectly co-registered leads to a multispectral difference image made up of SCVs with all zero components. Thus the representation in polar coordinates of SCVs collapses in a single point at the origin. An analysis of the histograms obtained after applying the CVA technique to the red and the near infrared spectral channels of image X_1 and its 2 and 6 pixels shifted versions (Fig. 2.a and b) allows one to derive two main important observations on the properties of the RN:

1. The RN results in a cluster in the circle of unchanged pixels. The variance of the cluster increases by increasing the RN. This effect is mainly related to the comparisons of non-perfectly aligned pixels that however correspond to the same object in the two images (see grey circles in Fig. 2).

2. The RN also results in significant clusters of pixels in the annulus of changed pixels. This is mainly due to the effects of the comparison of pixels which belong to different objects (pixels associated with detail and border regions) and is particularly significant in VHR images due to both the presence of regular structures and the presence of very high frequency regions. These clusters give raise to annular sectors having properties very similar to annular sectors of changed pixels (see black circles in Fig. 2), and are associated with specific misregistration components. We define these annular sectors as *annular sectors of registration noise* (S_{RN_i}). They are very critical because they cannot be distinguished from sectors of true changes resulting in a significant false alarm rate in the change-detection process.

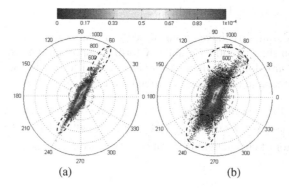

Fig. 2. Histograms in the Polar coordinate system obtained after applying CVA to the simulated data sets that show (a) 2 pixels of residual registration noise, and (b) 6 pixels of residual registration noise

4 Proposed Multiscale Technique for Registration Noise Reduction

In order to identify the annular sectors of registration noise and to reduce their impact on the change-detection process, we propose a multiscale technique based on the analysis of the behaviors of SCVs in the polar domain at different resolution levels. One of the main assumptions at the basis of the proposed technique is that we expect that true significant changes are associated with objects with a non-negligible size and that misregistration appears in the difference image with relatively thin structures. Therefore, if we reduce the resolution of images, we implicitly decrease the impact of the registration noise with respect to that on the original scene. The lower is the geometrical resolution, the lower is the probability of identifying in the polar representation annular sectors of registration noise. This means that at low resolution levels in the *annulus of changed pixels* mainly sectors due to the presence of true changes on the ground can be detected. However, in order to obtain a change-detection map characterized by a good geometrical fidelity, we should work at full resolution. On the basis of this analysis, we propose a multiscale strategy for reducing the impact of the registration noise on the change-detection map, which is based on the following two steps: i) multiscale decomposition of the multitemporal images; and ii) differential analysis of the statistical behaviours of the direction distribution of SCVs in the *annulus of changed pixels* at different scales.

The first step aims at computing a multiscale representation of the information present in the considered images. Irrespectively on the multiscale decomposition approach considered, the multiresolution decomposition step produces two sets of images $X_{MS_i} = \{ \mathbf{X}_i^0, ..., \mathbf{X}_i^n, ..., \mathbf{X}_i^{N-1} \}$, where the subscript i (i=1,2) denotes the acquisition date, and the superscript n (n=0,1,..., N-1) indicates the resolution level. Images in X_{MS_i} show different tradeoffs between registration noise and geometrical detail content.

In the second step, the set of multiscale difference images $X_D = \{ \mathbf{X}_D^0, ..., \mathbf{X}_D^n, ..., \mathbf{X}_D^{N-1} \}$ is computed by applying the CVA technique to each corresponding pair ($\mathbf{X}_1^n, \mathbf{X}_2^n$) of images in X_{MS_1} and X_{MS_2}. The distributions of the direction of SCVs at different resolution levels are then analyzed. In greater detail, the information extracted from the spectral difference image $\mathbf{X}_D^{N-1}$ at the lowest considered resolution level is exploited for reducing the effects of registration noise in the change-detection process performed at full resolution. Sectors associated with true changes on the ground are detected at lower scales and this information is exploited at full resolution for recovering change information, by properly processing sectors characterized by the presence of registration noise. It is expected that reducing the resolution the pixels associated with registration noise tend to disappear given their properties that usually result in small and thin structures. On the opposite, true changes (that usually have a non-negligible size) will be reduced in terms of size but not completely smoothed out. It follows that, as the resolution decreases, the marginal conditional density of the direction of pixels in A_c (i.e., pixels candidate to be true changed pixels) $p(\vartheta | \rho \geq T)$ [1] decreases in *annular sectors of registration noise* (S_{RN_i}) whereas it remains nearly constant in *annular sectors of true changes* (S_k).

---

[1] T is the threshold value that separates the circle of unchanged pixels from the annulus of changed pixels.

According to the previous analysis, in order to discriminate *annular sectors of registration noise* (S_{RN_i}) from annular *sectors of true changes* (S_k) we define the marginal conditional density of registration noise (*RN*) in the direction domain as:

$$\hat{p}^{RN}(\vartheta|\rho \geq T) = C[p^0(\vartheta|\rho \geq T) - p^{N-1}(\vartheta|\rho \geq T)] \tag{3}$$

where $p^0(\vartheta|\rho \geq T)$ and $p^{N-1}(\vartheta|\rho \geq T)$ are the marginal conditional densities of the direction of pixels in A_c at the full resolution and at the lowest considered resolution level N-1, respectively, and C is a constant defined such that $\int_{-\infty}^{+\infty} \hat{p}^{RN}(\vartheta|\rho \geq T) = 1$. The proposed procedure associates the sectors where $\hat{p}^{RN}(\vartheta|\rho \geq T)$ exceeds a predefined threshold value T_{RN} with sectors of registration noise S_{RN_i}, i.e.,

$$S_{RN_i} = \{\rho, \vartheta : \rho \geq T, \ \vartheta_{t_1} \leq \vartheta \leq \vartheta_{t_2} \text{ and } \hat{p}^{RN}(\vartheta|\rho \geq T) \geq T_{RN} \text{ with } 0 \leq \vartheta_{t_1} < \vartheta_{t_2} < 2\pi\}. \tag{4}$$

In this way we can apply the CVA technique to magnitude and direction variables in the polar domain and automatically reject patterns in the annulus of changed pixels associated with S_{RN_i}, thus reducing the false alarms due to the registration noise.

5   Experimental Results

In order to assess the reliability of the proposed analysis and the effectiveness of the proposed technique for reducing the impact of registration noise in the change-detection process, a multitemporal data set made of two images acquired by the Quikbird satellite on the Trentino area (Italy) in October 2005 and July 2006 was considered. In the pre-processing phase the two images were: i) pan-sharpened by applying the Gram–Schmidt procedure implemented in the ENVI software package [6] to the panchromatic channel and the four bands of the multispectral image; ii) radiometrically corrected; iii) co-registered. The registration process was carried out by using a polynomial function of order 2 according to 14 ground control points (GCPs), and applying a nearest neighbor interpolation. The final data set was made up of two pan-sharpened multitemporal and multispectral images of 984×984 pixels with a spatial resolution of 0.7 m that have a residual miregistration of about 1 pixel on GCPs. Fig. 3.a and b show the pan-sharpened near infrared channels of images X_1 and X_2, respectively. Between the two acquisition dates two kinds of changes occurred: i) new houses were built on rural area (ω_{c_1}, white circles in Fig. 3.b); and ii) some roofs in the industrial and urban area were rebuilt (ω_{c_2}, black and light gray circles in Fig. 3.b). In order to allow a quantitative evaluation of the effectiveness of the proposed method, a reference map (which includes 21248 changed pixels and 947008 unchanged pixels) was defined according to the available prior knowledge on the considered area (see Fig. 3.c).

Fig. 4.a shows the histograms in polar coordinates obtained after applying the CVA technique to the red and the near infrared spectral channels, which better emphasize the changes occurred in the considered test area. From the analysis of these histograms and

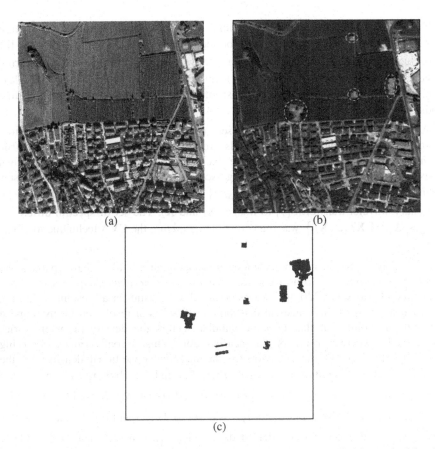

Fig. 3. Near infrared spectral channels of images of the Trentino area (Italy) acquired by the Quickbird VHR multispectral sensor in: (a) October 2005; and (b) July 2006 (changes occurred between the two acquisition dates appear in dashed white, black and gray circles). (c) Change-detection reference map.

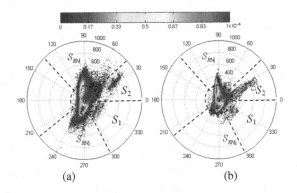

Fig. 4. Histograms in the Polar coordinate system of (a) $\mathbf{X}_D^0$ and (b) $\mathbf{X}_D^4$

of the available reference map, the annular sectors S_1 and S_2 associated with true changes ω_{c_1} and ω_{c_2}, respectively, can be identified (see Fig. 4.a). As expected, the two clusters associated with ω_{c_1} and ω_{c_2} show a high magnitude and a preferred direction. All the other annular sectors: i) can be considered statistically meaningless due to the low occurrence of patterns; or ii) are associated with the residual registration noise (S_{RN_i}).

As we are dealing with unsupervised change detection, all annular sectors including a significant amount of pixels are candidate to be associated with true changes on the ground, and can not be excluded *a priori*. In order to distinguish registration noise sectors from true change sectors, we computed two multiscale sets of images $X_{MS_1} = \{ \mathbf{X}_1^0, ..., \mathbf{X}_1^4 \}$ and $X_{MS_2} = \{ \mathbf{X}_2^0, ..., \mathbf{X}_2^4 \}$ applying the *Daubechies-4* stationary wavelet transform [7] to images $\mathbf{X}_1$ and $\mathbf{X}_2$. The set of spectral difference images $X_D = \{ \mathbf{X}_D^0, ..., \mathbf{X}_D^4 \}$ was then computed applying the CVA technique to X_{MS_1} and X_{MS_2}.

According to the proposed technique a decision threshold T that separates the *annulus of unchanged* pixels (A_c) from the *circle of no-changed* pixels (C_n) was computed (T was set to 220, which was assumed to be valid for all resolution levels). A visual analysis of the statistical distribution of SCVs at resolution levels 0 and 4 (see Fig. 4) points out that in some annular sectors the density of pixels varies significantly, whereas in others it is nearly stable. This is confirmed by observing Fig. 5.a, which shows the behaviors of the marginal conditional densities of the direction in A_c at resolution level 0 [$p^0(\vartheta | \rho \geq T)$] and 4 [$p^4(\vartheta | \rho \geq T)$]. In order to identify annular sectors S_{RN_i}, the behavior of $\hat{p}^{RN}(\vartheta | \rho \geq T)$ was derived (see Fig. 5.b). Accordingly, the decision threshold T_{RN} was set to 3×10^{-3} and the 2 annular sectors having registration noise were identified, i.e., $S_{RN_1} = \{ \rho, \vartheta : \rho \geq 220 \text{ and } 37° \leq \vartheta < 143° \}$, and $S_{RN_2} = \{ \rho, \vartheta : \rho \geq 220 \text{ and } 232° \leq \vartheta < 293° \}$ (see Fig. 4).

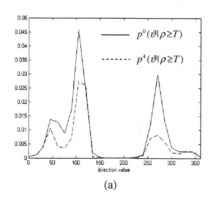
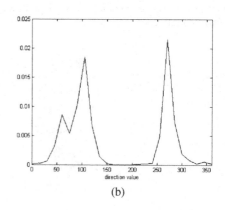

(a)   (b)

Fig. 5. Behaviors: (a) of the marginal conditional densities of the direction in A_c at full resolution $p^0(\vartheta | \rho \geq T)$ (level 0) and at the lowest resolution $p^4(\vartheta | \rho \geq T)$ (level 4) for the considered data set; and (b) of $\hat{p}^{DRN}(\vartheta | \rho \geq T)$ for the considered data set

Table 1. Change-detection results obtained with the proposed multiscale technique and the standard technique that does not consider the presence of residual registration noise

Technique	False alarms	Missed alarms	Overall error
Proposed	9138	3812	12950
Standard	55924	6354	62278

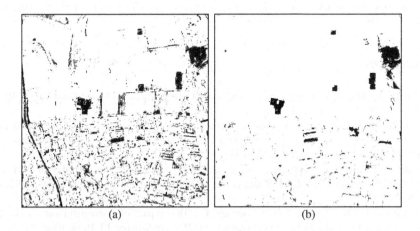

(a)   (b)

Fig. 6. Change-detection maps obtained with: (a) the standard approach that does not consider the presence of residual registration noise; (b) the proposed multiscale technique

The identified sectors were then used for computing the final change-detection map. SCVs that fall into C_n, S_{RN_2} and S_{RN_1} were classified as belonging to ω_n, whereas the other statistically significant clusters in A_c (i.e., associated with sectors S_1 and S_2) were classified as belonging to ω_{c_1} and ω_{c_2}, respectively. The obtained change-detection map was compared with the one yielded ignoring the presence of registration noise, which was computed assigning SCVs that fall into C_n to ω_n, and all statistically significant clusters in A_c to the class of changes. Numerical results reported in Tab. 1 point out the effectiveness of the proposed technique, which sharply reduces both the overall error and the false alarm rate. These results are confirmed by the change-detection maps reported in Fig. 6. It is worth noticing that the obtained change-detection map is computed at full resolution level, and thus exhibits a high geometrical detail content.

6   Conclusion

This paper has addressed the problem of unsupervised change detection in VHR multitemporal remote sensing images in the context of a polar framework for CVA. In particular, a study on the properties of registration noise on VHR images has been presented, and an adaptive multiscale technique for reducing the effects of residual registration noise in unsupervised change detection has been proposed. The proposed technique exploits both the magnitude and the direction information of SCVs in the polar domain at different resolution levels. *Annular sectors of registration noise* are identified in an automatic way according to a differential analysis of the marginal distributions of the direction in the *annulus of changed pixels* at different scales. The pixels included in these sectors are then rejected in the change-detection process at full resolution level. Experimental results, obtained on a pair of VHR multitemporal images acquired by the Quikbird sensor on a mixed urban and rural areas, confirmed the effectiveness of the proposed technique in reducing the effects of the registration noise.

References

1. Singh, A.: Digital change detection techniques using remotely-sensed data. Int. J. Rem. Sens. 10(6), 989–1003 (1989)
2. Bovolo, F., Bruzzone, L.: A Theoretical Framework for Unsupervised Change Detection Based on Change Vector Analysis in Polar Domain. IEEE Transactions on Geoscience and Remote Rensing 45(1), 218–236 (2007)
3. Dai, X., Khorram, S.: The effects of image misregistration on the accuracy of remotely sensed change detection. IEEE Transactions on Geoscience and Remote Sensing 36, 1566–1577 (1998)
4. Townshend, J.R.G., Justice, C.O., Gurney, C.: The impact of misregistration on change detection. IEEE Transactions on Geoscience and Remote Sensing 30, 1054–1060 (1992)
5. Bruzzone, L., Cossu, R.: An adaptive approach for reducing registration noise effects in unsupervised change detection. IEEE Transactions on Geoscience and Remote Sensing 41(11), 2455–2465 (2003)
6. ENVI User Manual. Boulder, CO: RSI (2003), http://www.RSInc.com/envi
7. Bovolo, F., Bruzzone, L.: A detail-preserving scale-driven approach to change detection in multitemporal SAR images. IEEE Transactions on Geoscience and Remote Sensing 43(12), 2963–2972 (2005)

Image Quality Assessment Based on Perceptual Structural Similarity

D. Venkata Rao and L. Pratap Reddy

Bapatla Engineering College, Bapatla, India
JNTU College of Engineering, Hyderabad, India
dv2002in@yahoo.co.in,pratplr@rediffmail.com

Abstract. We present a full reference objective image quality assessment technique which is based on the properties of the human visual system (HVS). It consists of two major components: 1) structural similarity measurement (SSIM) between the reference and distorted images, mimicking the overall functionality of HVS in a top down frame work. 2) A visual attention model which indicates perceptually important regions in the reference image based on the characteristics of intermediate and higher visual processes through the use of Importance Maps. Structural similarity in a region is weighted, depending on the perceptual importance of the region to arrive at Perceptual Structural Similarity Metric (PSSIM) indicative of the image quality.

Keywords: Objective image quality, HVS, structural distortion, perceptually important regions.

1 Introduction

The role of images in present day communication has been steadily increasing. In this context the quality of an image plays a very important role. Different stages and multiple design choices at each stage exist in any image processing system. They have direct bearing on the quality of the resulting image. Unless we have a quantitative measure for the quality of an image, it becomes difficult to design an ideal image processing system. Though subjective quality assessment is an alternative, it is not feasible to be incorporated into real world systems. Hence, objective quality metrics play an important role in image quality assessment.

In the last two decades a lot of objective metrics have been proposed [1-7] to assess image quality. The most widely adopted statistics feature is the Mean Squared Error (MSE). However, MSE and its variants do not correlate well with subjective quality measures because human perception of image distortions and artifacts is unaccounted for. MSE also not good because the residual image is not uncorrelated additive noise,it also contains components of the original image. A detailed discussion on MSE is given by Girod [8].

A major emphasis in recent research has been given to a deeper analysis of the Human Visual System (HVS) features [1]. There are lot of HVS characteristics [9] that may influence the human visual perception on image quality. Although

A. Ghosh, R.K. De, and S.K. Pal (Eds.): PReMI 2007, LNCS 4815, pp. 87–94, 2007.
© Springer-Verlag Berlin Heidelberg 2007

HVS is too complex to fully understand with present psychophysical means, the incorporation of even a simplified model into objective measures reportedly leads to a better correlation with the response of the human observers [1]. However, most of these methods are error sensitivity based approaches,explicitly or implicitly, and make a number of assumptions [10] which need to be validated. These methods suffer from the problems like natural image complexity problem, Minkowski error pooling problem, and cognitive interaction problem [10].

Structural similarity based methods [11, 12] of image quality assessment claim to account for the fact that the natural image signal samples exhibit strong dependencies amongst themselves, which is ignored by most of these methods. Structural similarity based methods replace the Minkowski error metric with different measurements that are adapted to the structures of the reference image signal, instead of attempting to develop an ideal transform that can fully decouple signal dependencies.

However, Vision models[13, 14, 15, 16] which treat visible distortions equally, regardless of their location in the image, may not be powerful enough to accurately predict picture quality in such cases. This is because we are known to be more sensitive to distortions in areas of the image to which we are paying attention than to errors in peripheral areas.

In this paper we present an image quality metric which integrates the notions of structural similarity measure mimicking the overall functionality of HVS and perceptually important regions based on the characteristics of intermediate and higher visual processes. We observed that the proposed index correlates effectively with subjective scores and found to posses superior performance when compared with other metrics discussed in this paper.

This paper is organized as follows. Section 2 explains the structural similarity method. Section 3 explains the perceptual importance map. Section 4 describes the computation of proposed quality index. Experimental results follow in Section 5. Finally, in Section 6, the conclusions of the paper are presented.

2   Structural SIMIlarity(SSIM)

Based on the assumption that the HVS is highly adapted to extract structural information from the viewing field, a new philosophy for image quality measurement$SSIM$ was proposed by Wang et al [12]. Let $x = \{x_i | i = 1, 2, 3, ...N\}$and $y = \{y_i | i = 1, 2, 3, ...N\}$ be two discrete non-negative signals been aligned with each other and let $\bar{x}, \sigma_x^2$,and σ_{xy} be the mean of x,variance of x,and the covariance of x and y respectively. $\bar{x}, \sigma_x^2$ are the estimates of luminance and contrast of x and σ_{xy} measures the tendency of x and y to vary together, which is an indication of structural similarity.$SSIM$ index is given in equation(1) where C_1, C_2 and C_3 are small constants introduced to avoid instability when the denominator is close to zero.

$$SSIM(x, y) = \frac{(2\mu_x \mu_y + C_1)(2\sigma_{xy} + C_2)}{(\mu_x^2 + \mu_y^2 + C_1)(\sigma_x^2 + \sigma_y^2 + C_2)} \tag{1}$$

3   Perceptual Importance Map

Visual attention is a process that locates features in an image that have high information content so that limited computational resources can be directed toward them. Cave [17] writes that attention "only allows a small part of the incoming sensory information to reach short-term memory and visual awareness, allowing us to break down the problem of scene understanding into rapid series of computationally less demanding, localized visual analysis problems".

Psycho visual studies reveal that human eye is very sensitive to the edge and contour information of the image. Texture is one of the important characteristics used in identifying objects or regions of interest in an image. Texture contains important information about the structural arrangement of surfaces. The HVS converts luminance into contrast at an early stage of processing. Regions which have a high contrast with their surrounds attract our attention and are likely to be of greater visual importance [18, 19]. The following explains the computation of perceptual importance map based on these three parameters.

1. Edges per unit area E is determined by detecting edges in an image using Canny extension of the Sobel operator[20] and then congregating the edges detected within an 8x8 block [21].The value of E is normalized to the range [0 1]. A block without edges will have a value of 0.

2. The texture content$N$, in a local region is quantified by computing the entropy [22]. From basic information theory, entropy is defined as in equation (2), where z_i is a random variable indicating intensity, $p(z_i)$ is the histogram of the intensity levels in a region,L is the number of possible intensity levels (0 to 255 for gray scale images).

$$N = -\sum_{i=0}^{L-1} p(z_i)\lg_2 p(z_i) \tag{2}$$

The value of N is normalized to fit in the range [0 1]. Image blocks with rich texture will have a value of 1.

3. Michaelson contrast C [23]is most useful in identifying high contrast regions, generally considered to be an important choice feature for human vision. Michaelson contrast is calculated as in equation (3) where l_m is the mean luminance within an 8 x 8 block and L_M is the overall mean luminance of the image. C is scaled to the range [0 1], 1 indicating highest contrast block and 0 indicating lowest contrast block.

$$C = \|(l_m - L_M)/(l_m + L_M)\| \tag{3}$$

For each block of the original image, the importance value is calculated as the normalized value of sum of the squares of the respective parameters, which forms the perceptual importance map P of the image.

4   Perceptual Structural Similarity

At first, the original and distorted images are divided into 8 x 8 non-overlapping blocks. The $SSIM$ for each block is computed using equation (1), to form a matrix S,as shown below where each element s_{ij} represents the structural similarity between corresponding blocks of the original and distorted images with coordinates $(i,j), 1 \leq i \leq m = \lfloor H/8 \rfloor, 1 \leq j \leq n = \lfloor W/8 \rfloor$, where H and W represent the height and width of the image respectively.

Secondly, the perceptual importance map P specified in section 4 is obtained for the original image as shown below. p_{ij} represents perceptual importance of each block with coordinates (i,j) as defined earlier. We define Perceptual

$$S = \begin{pmatrix} s_{11} & s_{12} & \cdots & s_{1n} \\ s_{21} & s_{22} & \cdots & s_{2n} \\ \cdot & \cdot & \cdots & \cdot \\ \cdot & \cdot & \cdots & \cdot \\ s_{m1} & s_{m2} & \cdots & s_{mn} \end{pmatrix} \qquad P = \begin{pmatrix} p_{11} & p_{12} & \cdots & p_{1n} \\ p_{21} & p_{22} & \cdots & p_{2n} \\ \cdot & \cdot & \cdots & \cdot \\ \cdot & \cdot & \cdots & \cdot \\ p_{m1} & p_{m2} & \cdots & p_{mn} \end{pmatrix}$$

Structural Similarity index $PSSIM$ as the weighted average of the structural similarity indices s_{ij} with coordinates (i,j), where each $s_{i,j}$ is weighted with the corresponding perceptual importance $p_{i,j}$. Eqaution(4) gives the expression for $PSSIM$.

$$PSSIM = \frac{\sum_{i=1}^{m} \sum_{j=1}^{n} [S][P]}{\sum_{i=1}^{m} \sum_{j=1}^{n} [P]} \qquad (4)$$

5   Experimental Results

The proposed quality index was tested using LIVE image database [24]. The database consists of twenty-nine high resolution 24-bits/pixel RGB color images (typically 768 x 512), distorted using five distortion types: JPEG2000, JPEG, White noise in the RGB components, Gaussian blur in the RGB components, and bit errors in JPEG2000 bit stream using a fast-fading Rayleigh channel model. Each image was distorted with each type, and for each type the perceptual quality covered the entire quality range. Difference Mean Opinion Score (DMOS) value for each distorted image was computed based on the perception of quality of the images by observers.

We tested the proposed method on all the images and distortions available in the LIVE database, after converting the color images to gray level images. In order to provide quantitative measures on the performance of the objective quality assessment models, different evaluation metrics were adopted in the Video Quality Experts Group (VQEG) Phase-I test [27]. We performed non-linear mapping between the objective and subjective scores, using 4-parameter logistic function of the form shown in Equation (5).

$$y = a/(1.0 + e^{-(x-b)/c}) + d \qquad (5)$$

After the non-linear mapping, the Correlation Coefficient (CC), the Mean Absolute Error (MAE), and the Root Mean Squared Error (RMS) between the subjective and objective scores are calculated as measures of prediction accuracy. The prediction consistency is quantified using the outlier ratio (OR), which is defined as the percentage of the number of predictions outside the range of ± 2 times the standard deviation. Finally, the prediction monotonicity is measured using the Spearman rank-order-correlation coefficient (ROCC).

To evaluate the performance of the proposed metric, we considered two image quality assessment models, PSNR and $SSIM$. Table 1 shows the evaluation results for the models being compared with that of the $PSSIM$ for different types of distortions.For each of the objective evaluation criteria, $PSSIM$ outperforms the other models being compared across different distortion types. Fig. 1 shows the scatter plots of DMOS versus $PSSIM$ for different kinds of distortions.

Table 1. Performance comparison of image quality assessment models on LIVE image database [18]. CC: non-linear regression correlation coefficient; ROCC: Spearman rank-order correlation coefficient; MAE: mean absolute error; RMS: root mean square error; OR: outlier ratio. (a) JPEG2000 (b) JPEG (c) White noise (d) Gaussian blur (e) Fast fading.

Model	CC	ROCC	MAE	RMS	OR
PSNR	0.859	0.851	6.454	8.269	5.917
SSIM	0.899	0.894	5.687	7.077	2.366
PSSIM	0.941	0.935	4.426	5.442	2.958
			(a)		
Model	CC	ROCC	MAE	RMS	OR
PSNR	0.842	0.828	6.636	8.622	6.285
SSIM	0.891	0.863	5.386	7.236	5.714
PSSIM	0.930	0.893	4.262	5.871	6.285
			(b)		
Model	CC	ROCC	MAE	RMS	OR
PSNR	0.922	0.938	4.524	6.165	5.555
SSIM	0.94	0.914	4.475	5.459	2.777
PSSIM	0.964	0.952	3.514	4.247	3.472
			(c)		
Model	CC	ROCC	MAE	RMS	OR
PSNR	0.744	0.725	8.395	10.50	3.448
SSIM	0.947	0.940	3.992	5.027	3.448
PSSIM	0.969	0.964	3.240	3.871	2.758
			(d)		
Model	CC	ROCC	MAE	RMS	OR
PSNR	0.857	0.859	6.383	8.476	6.896
SSIM	0.956	0.945	3.806	4.799	5.517
PSSIM	0.967	0.959	3.328	4.189	4.827
			(e)		

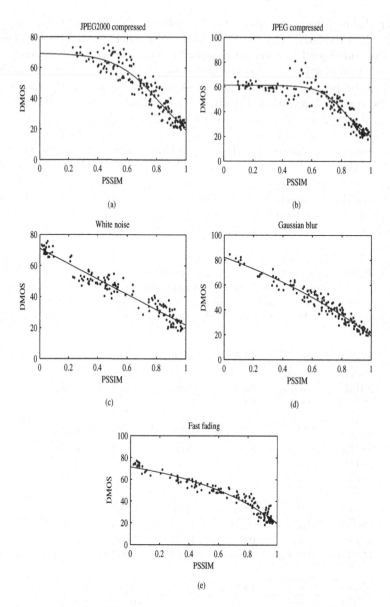

Fig. 1. Scatter plots for DMOS versus model prediction for (a) JPEG2000 (b) JPEG (c) White noise (d) Gaussian blur (e) Fast fading distorted images

6   Conclusions

In this paper we present an image quality assessment technique which is based on the properties of the human visual system (HVS). It combines the notions of structural similarity with visual attention model. The results prove the fact that human eye is sensitive to important image features like edges, texture, and contrast. The results also justify the visual attention model built on these three parameters. Statistical indices of performance as set by VQEG for the proposed quality index indicate that the index matches well with the Human Visual System obviating the need for subjective tests and proves to be a better choice than other indices mentioned in the paper. The index is found to have good sensitivity across all the distortion types mentioned.

References

1. Eskicioglu, A.M., Fisher, P.S.: Image quality measures and their performance. IEEE Transactions on Communications 43(12), 2959–2965 (1995)
2. Karunasekera, S.A., Kingsbury, N.G.: A distortion measure for blocking artifacts in images based on human visual sensitivity. IEEE Transactions on Image Processing 4(6), 713–724 (1995)
3. Mill, N.B.: A visual model weighted cosine transform for image compression and quality assessment. IEEE Transactions on Communications 33(6), 551–557 (1985)
4. Saghri, J.A.: Image quality measure based on a human visual system model. Optical Engineering 28(7), 813–818 (1989)
5. Daly, S.: The visible differences predictor: an algorithm for the assessment of image fidelity. In: Watson, A.B. (ed.) Digital Images and Human Vision, Ch. 14, pp. 179–206. MIT press, Cambridge (1993)
6. Lubin, J.: A visual discrimination model for imaging system design and evaluation. In: Peli, E. (ed.) Vision Models for Target Detection and Recognition, Ch.10, pp. 245–283. World Scientific Publishing (1995)
7. Watson, A.B.: DCT quantization matrices visually optimize for individual images. In: Human Vision, Visual Processing and Digital Display IV, Proc. SPIE, vol. 1913, pp. 202–216 (1993)
8. Girod, B.: What's wrong with mean-squared error. In: Watson, A.B. (ed.) Digital Images and Human Vision, pp. 207–220. MIT Press, Cambridge (1993)
9. Wandell, B.A.: Foundations of Vision, Sinauer Associates, Inc. (1995)
10. Wang, Z., Sheikh, H.R., Bovik, A.C.: Objective video quality assessment. In: Furht, B., Marques, O. (eds.) The Handbook of Video Databases: Design and Applications, pp. 1041–1078. CRC press (September 2003)
11. Wang, Z., Lu, L., Bovik, A.C.: Video quality assessment based on structural distortion measurement, Signal Processing: Image Communication. special issue on objective video quality metrics 19 (January 2004)
12. Wang, Z., Bovik, A.C., Sheikh, H.R., Simocelli, E.P.: Image quality assessment: From error measurement to structural similarity. IEEE Trans. Image Processing 13(4), 600–612 (2004)
13. Geri, G.A., Zeevi, Y.Y.: Visual assessment of variable-resolution imagery. Journal of the Optical Society of America 12(10), 2367–2375 (1995)
14. Kortum, P., Geisler, W.: Implementation of a foveated image coding system for image bandwidth reduction. In: SPIE - Human Vision and Electronic Imaging, vol. 2657, pp. 350–360 (February 1996)
15. Stelmach, L.B., Tam, W.J., Hearty, P.J.: Static and dynamic spatial resolution in image coding: An investigation of eye movements. In: Proceedings of the SPIE, San Jose, vol. 1453, pp. 147–152 (1991)
16. Yarbus, A.L.: Eye Movements and Vision Press, New York (1967)

17. Cave, R.: The feature Gate model of visual selection. Psychological research 62, 182–194 (1999)
18. Findlay, J.: The visual stimulus for saccadic eye movement in human observers. Perception 9, 7–21 (1980)
19. Senders, J.: Distribution of attention in static and dynamic scenes. In: Proceedings SPIE, San Jose, vol. 3016, pp. 186–194 (February 1997)
20. Canny, J.: A Computational Approach to Edge Detection. IEEE Trans. Pattern Analysis and Machine Intelligence 8(6), 679–698 (1986)
21. Richards, W., Kaufman, L.: Centre-of-Gravity Tendencies for Fixations and Flow Patterns. Perception and Psychology 5, 81–84 (1969)
22. Gonzalez, Woods.: Digital Image Processing. Prentice Hall, Englewood Cliffs (2002)
23. Mannan, S.K., Ruddock, K.H., Wooding, D.S.: The Relationship between the Locations of Spatial Features and Those of Fixations Made during Visual Examination of Briefly Presented Images. Spatial Vision 10(3), 165–188 (1996)
24. Sheikh, H.R., Bovik, A.C., Cormack, L., Wang, Z.: LIVE Image Quality Assessment Database (2004), http://live.ece.utexas.edu/research/quality
25. Corriveau, P., et al.: Video quality experts group: Current results and future directions. In: presented at the SPIE Visual Communication and Image Processing, vol. 4067 (June 2000)
26. Van Dijk, A.M., Martens, J.B., Watson, A.B.: Quality assessment of coded images using numerical category scaling. In: Proc. SPIE, vol. 2451, pp. 90–101 (March 1995)
27. VQEG: Final report from the video quality experts group on the validation of objective models of video quality assessment (March 2004), http://www.vqeg.org/

Topology Adaptive Active Membrane

Sitansu Kumar Das and Dipti Prasad Mukherjee

Electronics and Communication Sciences Unit
Indian Statistical Institute, Kolkata, India

Abstract. Segmentation of multiple objects especially when the objects are touching each other in an image is a challenging problem. In this paper we propose a parametric active membrane, which can change its topology to detect multiple objects present in the image. The membrane evolves in image space and also along the image intensity surface and if requires, splits into multiple membranes. The methodology is tested for a number of real images that demonstrates the efficacy of the proposed scheme.

Keywords: Active membrane, topology adaptive parametric model.

1 Introduction

Multiple object detection in a scene using a membrane like surface is always a challenging problem. The topology adaptive parametric active contours [1], [2], [3] and region-based surface fitting approach [4] were used to detect multiple objects in a scene. However, for the conjoint or touching objects sharing common boundary, topology adaptive snakes fail to identify the objects separately. We have designed a dynamic evolving membrane to identify the conjoint objects efficiently.

In our membrane model we assume that the membrane moves from the top of the image surface to approximate the image intensity such that the sum of internal energy and external energy of the approximated membrane is minimum. The membrane internal energy is responsible for maintaining the smoothness of the approximated surface while the image intensity influenced external energy drives the membrane towards the image surface. The force governing membrane evolution is applied along the image plane. Once the deformation of membrane exceeds the allowable deformation limit that membrane can sustain, tearing of membrane occurs in the high-stressed zone and then the membrane pieces continue propagation once again. The membrane fitting process continues until the deformation process of membrane comes to equilibrium when internal and external energy of the membrane balance each other. The membrane at the equilibrium state defines multiple objects present in the image.

The membrane is modeled using finite element method [5], [6] and the partial differential stress equation is solved at the finite element vertices [7], [8], [3]. In [1] topology adaptive active contour is used where the image surface is tessellated using triangular elements and the splitting and joining of parametric active

A. Ghosh, R.K. De, and S.K. Pal (Eds.): PReMI 2007, LNCS 4815, pp. 95–102, 2007.
© Springer-Verlag Berlin Heidelberg 2007

contour(s) depends on the sign of vertices at the triangular element. This is similar to the signed distance function based level set approach [9] and computationally expensive. Moreover, the approach in [1] only relies on edge force and therefore, cannot segment the conjoint objects. Conjoint objects are distinguishable from one another because each of them is defined by a specific band of intensities compared to their adjoining neighborhood objects. Our dynamic membrane approach segments the image within different bands of intensities and therefore can segment conjoint objects efficiently. This is one of the specific contributions of our proposed work. We have presented the result and compared our approach with [1] in Section 3 followed by conclusion. But before that in the next section we present our model of active membrane and its evolution.

2   Methodology

For the proposed 3D membrane we consider an image $I(x, y) : \Re \times \Re \to \Re$ as a 3D surface where an image pixel value is the height of the surface along z-axis at an image location (x, y). Different types of the undulations are viewed on the image surface due to intensity difference created by the presence of objects in the image.

Simulation of membrane for image segmentation consists of two major parts. The first part deals with driving the membrane toward the image surface. In the second part the elongation of the membrane due to the deformation constrained by the image characteristics is calculated and tearing of membrane is carried out depending on the stress map of the membrane. While first part requires solution of PDE for the membrane, in the second part calculated stress is compared with preset value. In the next section we formulate the construction and evolution of membrane model.

2.1   Design of Membrane

We define the membrane as $v(r, s, t)$, where r and s are 2D co-ordinates on the membrane surface and t is the time parameter for membrane evolution. The membrane proceeds towards the image surface by minimizing the energy functional,

$$E(v) = \int \int \left[\alpha(I) \left(\left\| \frac{\partial v}{\partial r} \right\|^2 + \left\| \frac{\partial v}{\partial s} \right\|^2 \right) + \beta(I)P(I) \right] dr ds \qquad (1)$$

In (1) the first two terms in the right hand side are the internal energy terms dependent on the amount of deformation of the membrane. The last term is an external energy term dependent on the image characteristics. The weights for the internal and external energies are $\alpha(I)$ and $\beta(I)$ respectively which depends on the image property. The external energy $P(I)$ depends on the image gradient and also on the difference of local heights of membrane and image intensity surface. Minimization of $E(v)$ is done by

$$\frac{\partial v}{\partial t} + \frac{\partial E(v)}{\partial v} = 0 \qquad (2)$$

where the evolution of the membrane v has to be carried out in 3D both in image space (x, y) and along intensity axis z. Next we present the numerical implementation scheme of (2).

2.2 Numerical Implementation

Solving (2) for the membrane needs the assumption of the existence of an imaginary grid over the membrane. The grid tessellates the membrane into a mesh of small triangular elements. The grid points over the membrane are represented in a matrix form referred as grid matrix. If the membrane is tessellated into a mesh of $2mn$ triangular elements, then there exist $m+1$ rows and $n+1$ columns in the grid matrix. One typical element of the membrane is denoted by e and we assume that we know the properties of the surface in the three vertices of e [7], [8]. We use linear basis function using weighted combinations of the element vertices since only first order PDE is used to evolve the membrane. Following [6] the basis function is given by,

$$p_e = ar + bs + c, a = \frac{1}{h}(v_{2,1} - v_{1,1}), b = \frac{1}{h}(v_{1,2} - v_{1,1}), c = v_{1,1} \qquad (3)$$

where $v_{1,1}$, $v_{1,2}$ and $v_{2,1}$ are the vertices of the element as shown in the Fig. 1. Discrete version of (1) using (3) is given by,

$$E(V) = \frac{h^2}{2} \sum_{i=1}^{m+1} \sum_{j=1}^{n+1} \alpha(I(x_{i,j}, y_{i,j}))[(v_{i+1,j} - v_{i,j})^2 + (v_{i,j+1} - v_{i,j})^2]$$
$$+ \beta(I(x_{i,j}, y_{i,j}))P(I(x_{i,j}, y_{i,j})) \qquad (4)$$

where $(x_{i,j}, y_{i,j})$ is the image location at (i, j)th grid point. The discrete vertices of the elements are $V = (v_{1,1}, v_{2,1}, ..., v_{i,j}, ..., v_{m+1,n+1})$ and $v_{i,j} = (x_{i,j}, y_{i,j}, z_{i,j})$

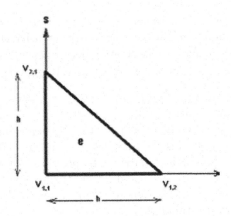

Fig. 1. A triangular finite element

denotes the co-ordinate of the (i,j)th grid point. Differentiating $E(V)$ with respect to V implies differentiating (4) with respect to $v_{i,j}$ for $i = 1, ..., m+1$ and $j = 1, ..., n+1$ and we get $(m+1)(n+1)$ number of equations. However these equations are nonlinear due to the dependability of α and β on $I(x_{i,j}, y_{i,j})$. Therefore, to facilitate linear solution, we assume α and β are constants in each iteration and they are obtained from the membrane at previous iteration. So differentiating (4) with respect to $v_{i,j}$ gives,

$$\alpha(I(x_{i,j}^{t-1}, y_{i,j}^{t-1}))(4v_{i,j}^t - v_{i-1,j}^t - v_{i+1,j}^t - v_{i,j-1}^t - v_{i,j+1}^t)$$
$$+\beta(I(x_{i,j}^{t-1}, y_{i,j}^{t-1}))\left(\frac{\partial P(I(x_{i,j}^{t-1}, y_{i,j}^{t-1}))}{\partial x}, \frac{\partial P(I(x_{i,j}^{t-1}, y_{i,j}^{t-1}))}{\partial y}, \frac{\partial P(I(x_{i,j}^{t-1}, y_{i,j}^{t-1}))}{\partial z}\right) = 0 \quad (5)$$

Three discrete equations can be obtained from (5),

$$\alpha(I(x_{i,j}^{t-1}, y_{i,j}^{t-1}))(4x_{i,j}^t - x_{i-1,j}^t - x_{i+1,j}^t - x_{i,j-1}^t - x_{i,j+1}^t) = \beta(I(x_{i,j}^{t-1}, y_{i,j}^{t-1}))\frac{\Delta P}{\Delta x}$$

$$\alpha(I(x_{i,j}^{t-1}, y_{i,j}^{t-1}))(4y_{i,j}^t - y_{i-1,j}^t - y_{i+1,j}^t - y_{i,j-1}^t - y_{i,j+1}^t) = \beta(I(x_{i,j}^{t-1}, y_{i,j}^{t-1}))\frac{\Delta P}{\Delta y} \quad (6)$$

$$\alpha(I(x_{i,j}^{t-1}, y_{i,j}^{t-1}))(4z_{i,j}^t - z_{i-1,j}^t - z_{i+1,j}^t - z_{i,j-1}^t - z_{i,j+1}^t) = \beta(I(x_{i,j}^{t-1}, y_{i,j}^{t-1}))\frac{\Delta P}{\Delta z}$$

The external force component in the image plane is Gaussian convolved image gradients $|\nabla G(I)|$. Given the weight of the external force $N(k)$ at the kth point out of total q number of discrete points in the inter-vertex distance, $N(k) = \left(1 - \frac{k}{q}\right)$, $k = 0, ..., q$, the external force component is given by,

$$\left(\frac{\Delta P}{\Delta x}\right)_{i,j} = \sum_{\substack{(\xi, \eta) = \\ v_{i+1,j}, \ v_{i-1,j} \\ v_{i,j+1}, \ v_{i,j-1}}} \sum_{k=0}^{q} N(k)\left|\nabla_x G\left(I\left(x_{i,j} + \frac{x_{i,j}-\xi}{q}k, y_{i,j} + \frac{y_{i,j}-\eta}{q}k\right)\right)\right|$$

$$\left(\frac{\Delta P}{\Delta y}\right)_{i,j} = \sum_{\substack{(\xi, \eta) = \\ v_{i+1,j}, \ v_{i-1,j} \\ v_{i,j+1}, \ v_{i,j-1}}} \sum_{k=0}^{q} N(k)\left|\nabla_y G\left(I\left(x_{i,j} + \frac{x_{i,j}-\xi}{q}k, y_{i,j} + \frac{y_{i,j}-\eta}{q}k\right)\right)\right|$$

$$(7)$$

The derivative of the external force in the z-direction essentially depends on the difference of membrane height and the image intensity. The force to drive the membrane towards image surface minimizes $\rho = (I(x_{i,j}, y_{i,j}) - z_{i,j})$. Therefore, we model $\left(\frac{\Delta P}{\Delta Z}\right)_{i,j} = -\rho(1 + exp(|\rho|))$. The weights $\alpha(I)$ (and similarly $\beta(I)$) are taken as linear function $\alpha(I(x,y)) = \alpha_{low} + \frac{I(x,y)}{255} \times (\alpha_{high} - \alpha_{low})$ where $(\alpha_{low}, \alpha_{high})$ are set experimentally.

Equation (6) can be rewritten as $\alpha \cdot *AX = \beta \cdot *Fx$, and $\alpha \cdot *AY = \beta \cdot *Fy$, $\alpha \cdot *AZ = \beta \cdot *Fz$ where A is the $(m+1)(n+1) \times (m+1)(n+1)$ pentadiagonal matrix referred as the stiffness matrix. The position vectors of the element vertices are X, Y and Z and Fx, Fy and Fz are the force vectors at the membrane node points. The operation '$\cdot*$' denotes element wise multiplication. We assume that we have a priori estimation X^{t-1}, Y^{t-1} and Z^{t-1} at iteration $(t-1)$ for the current iteration is t. Then, the final discrete form of (2) for the evolving membrane is given by,

$$\frac{X^t - X^{t-1}}{\Delta t} + \alpha \cdot *AX^t = \beta \cdot *Fx \Rightarrow (I + \Delta t \times \alpha \cdot *A)X^{t-1} = (X^{t-1} + \Delta t \times \beta \cdot *Fx)$$

$$\frac{Y^t - Y^{t-1}}{\Delta t} + \alpha \cdot *AY^t = \beta \cdot *Fy \Rightarrow (I + \Delta t \times \alpha \cdot *A)Y^{t-1} = (Y^{t-1} + \Delta t \times \beta \cdot *Fy) \, (8)$$

$$\frac{Z^t - Z^{t-1}}{\Delta t} + \alpha \cdot *AZ^t = \beta \cdot *Fz \Rightarrow (I + \Delta t \times \alpha \cdot *A)Z^{t-1} = (Z^{t-1} + \Delta t \times \beta \cdot *Fz)$$

where I and Δt denotes identity matrix and time step respectively. Next we describe the tear model based on which the membrane splits into pieces.

2.3   Tear Model

Tearing process involves deletion of connectivity between two neighbouring vertices. A vertex may have maximum four and minimum zero connected neighbours, and in the later case the vertex is deleted. As a result the stiffness matrix A is changed due to the vertex deletion. If (i, j)th vertex is connected to its (k, l)th neighbour and the connectivity is lost due to the deletion process then $R_{i,j}$ and $R_{k,l}$ rows of A is modified where $R_{i,j} = (i - 1)(n + 1) + j$ and $R_{k,l} = (k - 1)(n + 1) + l$. In A the diagonal elements of $R_{i,j}$ and $R_{k,l}$ rows is reduced by one and $(R_{i,j}, R_{k,l})$ and $(R_{k,l}, R_{i,j})$ elements are changed to zero. If (i, j)th vertex of the grid matrix is deleted then the row and column corresponding to $R_{i,j}$ is deleted and A is reduced in size by one both in row and column. Deletion of a connection deletes the two elements sharing the connection. Also, deletion of two elements in the two sides of a connection also deletes the connection. The criterion to delete a connection is the excessive elongation of the element governed by (8). When the elongation of a connection exceeds a preset distance it is disconnected. We keep this distance variable and make it dependent on the variation of local intensity. Taking preset distance as function of image intensity allows us to take variable distance for tearing the membrane in the different parts of image. We take the distance function as

$$d(x, y) = d_{low} + \frac{I(x, y)}{255} \times (d_{high} - d_{low}) \qquad (9)$$

where, d_{high} and d_{low} are application dependent preset constants.

So for each iteration we compare the inter vertex Euclidean distance between a pair of neighbouring nodes with its preset distance $d(x, y)$ and also for each connected link we check the existence of its two neighbourhood elements. If inter vertex distance exceeds preset limit or there exists no neighbourhood element in either side of the connection then connectivity between the vertices is deleted. After this we delete all the vertices having zero connected neighbour. Simultaneously, as mentioned earlier, we modify A deleting corresponding row column and start the next iteration. In the next section we discuss the results of our proposed method.

3   Results

We have implemented the proposed methodology to detect objects in a wide variety of real images, especially on some low-contrast images where multiple objects are present. But, first a set of examples to show that it works equally well for simple image segmentation problem.

For the examples of Figs. 2 and 3, the active membrane is evolved following (8). For poorly contrasted white blood cells, segmentation of cells having different shapes can be achieved from a single initialization of the membrane. In all the

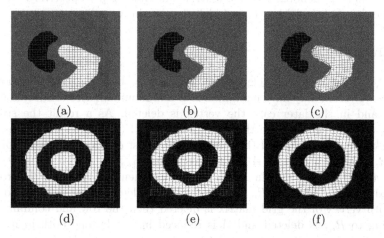

Fig. 2. (a), (d): Initial membrane on synthetic objects. (b), (e): Intermediate stages of membrane evolution after 20 iterations. (c), (f): Final segmentation using topology adaptive membrane after 150 iterations.

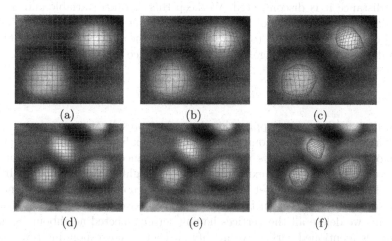

Fig. 3. (a), (d): Multiple white blood cells with initial active membrane. (b), (e): After 15 iterations of the membrane. (c), (f): After segmentation using 180 iterations.

above examples, we take $d_{high} = 5g_d$ and $d_{low} = 1.1g_d$ where g_d is the grid distance and $\alpha_{high} = 0.05$, $\alpha_{low} = 0.03$, $\beta_{high} = 0.001$ and $\beta_{low} = 0.01$. The entire approach was implemented in Matlab 6.5 in Pentium 4, 2.1 GHz PC. In the next section we compare our method with topology adaptive snake [1].

3.1   Comparison

Proposed method has two advantages over topology adaptive contour [1]. The first advantage is that the proposed approach uses automatic initialization, where as for [1] separate initialization is required for a group of objects. The second advantage is that the proposed approach can detect touching objects sharing a common boundary as it looks for both image gradients and image intensity surface as opposed to only image gradient information in [1].

The touching objects of Fig. 4(a) share common boundary with an initial membrane placed on the image. Fig. 4(b) shows membrane detects the two objects correctly and separately. The results of Fig. 4(c) and (d) show that the approach in [1] fails for this kind of situation. For Fig. 4(c) the initial contour is outside the object while the same for Fig. 4(d) is inside the object. The active contour of [1] was contracting in case of Fig. 4(c) while it was expanding in case of Fig. 4(d). In either case, the topology adaptive snake fails to capture conjoint object correctly and separately, which can be achieved using the proposed method.

To compare the performance of our method numerically we define a performance measure coefficient as $PMC = 1/\left(1 + \sum_{i=1}^{n} d_i\right)$, where n is the number of contour points and d_i is the distance between the ith contour point to its closest object boundary point. Therefore, the closer the value of PMC to 1 the better is the segmentation. Table 1 shows that as expected, the proposed approach performs much better than the topology adaptive contour [1].

| (a) | (b) | (c) | (d) |

Fig. 4. (a): Initial membrane on the conjoint objects. (b) After correct segmentation of the touching objects after 200 iterations. (c)-(d) Topology adaptive active contour evolution [1] for different initialization after 250 iterations.

Table 1. PMC of the proposed approach and topology adaptive contour [1]

	Synthetic image	Blood cell image	Conjoint object image
Proposed approach	0.0109	0.0044	0.0011
Topology adaptive contour [1]	0.0041	0.00031	0.000017

4   Conclusions

We have developed a topology adaptive membrane to segment multiple conjoint objects. The proposed approach is initialization independent where an initial membrane without any a priori knowledge of the objects evolves and adapts to the shape of objects present in the image. Comparison of the proposal with similar approach shows promise. Our goal is to use this technique in a tracking framework constraining the membrane evolution with shape and motion information in a video.

References

1. McInerney, T., Terzopoulous, D.: T-snake: Topology adaptive snakes. Medical Image Analysis 4, 73–91 (2000)
2. Kass, M., Witkin, A., Terzopoulous, D.: Snakes: Active contour models. Int. J. Computer Vision 1(4), 321–331 (1988)
3. Cohen, L., Cohen, I.: Finite-element methods for active contour models and balloons for 2-d and 3-d images. IEEE Trans. Pattern Analysis Machine Intelligence 15(11), 1131–1147 (1993)
4. Besl, P., Jain, R.: Segmentation through variable order surface fitting. IEEE Trans. Pattern Analysis Machine Intelligence 10(2), 167–192 (1988)
5. McInerney, T., Terzopoulous, D.: Non-rigid motion tracking. In: IEEE proceedings, vol. 4, pp. 73–91 (2000)
6. Terzopoulous, D.: The computation of visible-surface representations. IEEE Trans. Pattern Analysis Machine Intelligence 10(4), 417–438 (1988)
7. Desai, C.: Elementary Finite Element Method. Prentice-Hall, Englewood Cliffs (1979)
8. Hinton, E., Owen, D.: An Introduction To Finite Element Computations. Pineridge Press Limited (Swansea) (1981)
9. Mukherjee, D., Ray, N., Acton, S.: Level set analysis for cell detection and tracking. IEEE Trans. on Image Processing 13(4), 562–572 (2004)

Bit Plane Encoding and Encryption[*]

Anil Yekkala and C.E. Veni Madhavan

[1] Philips Electronics India Ltd., Bangalore
anil.yekkala@philips.com
[2] Indian Institute of Science, Bangalore
cevm@csa.iisc.ernet.in

Abstract. The rapid growth in multimedia based Internet systems and applications like video telephony, video on demand and also tele-medicine has created a great need for multimedia security. One of the important requirements for multimedia security is encryption. Owing to the size of multimedia data and real time requirements, lightweight encryption schemes are important. Lightweight encryption schemes are based on taking the structure of multimedia data into consideration and partially encrypting the content. Even though several lightweight encryption schemes are existing for lossy compression schemes, their do not exist any lightweight encryption scheme for lossless compression. In this paper we present a lossless compression scheme for image, and we show how this scheme can be also used for supporting scalable lightweight encryption.

1 Introduction

The rapid growth in multimedia based Internet systems and applications like video telephony, video on demand, network based DVD recorders and IP television have created a substantial need for multimedia security. One of the important requirements for multimedia security is transmission of the digital multimedia content in a secure manner using encryption for protecting it from eavesdropping. The simplest way of encrypting multimedia content is to consider the two-dimensional/three-dimensional image/video stream as an one-dimensional stream and to encrypt the entire content using standard block ciphers like AES, DES, IDEA or RC4 or using a stream cipher. The method of encrypting the entire multimedia content is considered as a naive encryption approach.

Even though the naive encryption approach provides the desired security requirements, it imposes a large overhead on the multimedia codex. This is due to the size of the multimedia content, and also due to real time requirements of transmission and rendering. Hence, lightweight encryption schemes are gaining popularity for multimedia encryption. Lightweight encryption schemes are based on the principle "encrypt minimal and induce maximum noise". Lightweight encryption schemes are designed to take the structure of the multimedia content into consideration.

---

[*] This work was done by the author as part of his Msc(Engg) thesis at Indian Institute of Science, Bangalore.

A. Ghosh, R.K. De, and S.K. Pal (Eds.): PReMI 2007, LNCS 4815, pp. 103–110, 2007.
© Springer-Verlag Berlin Heidelberg 2007

Several lightweight encryption schemes exist for lossy compression [5],[1], [2],[3] and [7], as well as few schemes exist for uncompressed data [6],[4]. But we do not see any schemes for encrypting multimedia data within lossless compression. Even though lossy compression schemes are generally acceptable for most applications. But for some applications like medical imaging any form of loss is not acceptable. In these cases the images and video are generally stored in lossless or uncompressed format. The only approach currently available is to encrypt the content after lossless compression. In current standard lossless compression schemes, the compression factor is generally less then 2. Thus, encrypting the entire compressed content after lossless compression will still prove to be an overhead. In this section we propose a new lossless encoding scheme for images, which permits scalable lightweight encryption scheme.

2 Proposed Scheme

The proposed lossless encoding and encryption scheme for image uses two important properties of image data, one for designing the encoding scheme and one for designing the encryption scheme.

Firstly, compression within an image can be achieved by dividing it into bit planes, and using the spatial redundancy within each bit plane. This is due to the fact that the MSB bit planes will have longer runs of 1's or 0's. This fact can be observed from Table 1. From the table it can be observed except for baboon image, every other image have more than 60% of runs of length greater then equal to 8 in their $MSB0$ plane. Similarly for $MSB1$, more then 45% of runs are of length greater then equal to 8. Similarly, the percentage of run lengths greater then equal to 4 have been computed for the remaining bit planes. It is seen from the table that long runs are generally present for bit planes $MSB0$ to $MSB4$, and as expected the percentage of long runs decrease from bit plane $MSB0$ to bit plane $MSB7$. Hence, an efficient way of encoding the bit planes will be to encode the run's using a variable length encoding scheme.

Secondly, most of the information within an image is present in its most significant bit planes, and hence encrypting the most significant bit-planes is sufficient for purpose of confidentiality. The number of most significant bit planes to be

Table 1. Percentage of long runs in various bit planes

% of long runs in each bit plane	Images						
	Baboon	Bandon	Brandyrose	Lena	Opera	Peppers	Pills
% runs of length ≥ 8 in $MSB0$	34.6	80.7	86.2	66.7	62.2	75.7	81.8
% runs of length ≥ 8 in $MSB1$	18.6	70.4	60.0	45.3	46.3	53.8	57.1
% runs of length ≥ 4 in $MSB2$	27.3	78.4	66.2	54.8	58.2	52.6	64.9
% runs of length ≥ 4 in $MSB3$	17.3	64.2	45.6	34.6	35.9	32.5	48.6
% runs of length ≥ 4 in $MSB4$	13.2	45.0	24.1	21.0	20.2	17.2	32.9
% runs of length ≥ 4 in $MSB5$	12.4	29.7	13.7	13.6	13.1	12.6	19.5
% runs of length ≥ 4 in $MSB6$	12.4	25.5	12.5	12.6	12.6	12.6	13.5
% runs of length ≥ 4 in $MSB7$	12.5	24.1	12.4	12.5	12.5	12.4	12.6

Table 2. Amount of energy in the MSB bit planes

% of energy	Images						
	Baboon	Bandon	Brandyrose	Lena	Opera	Peppers	Pills
% energy in MSB 0	46.28	42.91	43.23	47.20	38.08	44.97	41.45
% energy in MSB 0-1	60.15	62.12	68.54	65.70	58.62	62.58	68.87
% energy in MSB 0-2	79.55	78.53	81.97	79.87	79.12	79.56	82.19
% energy in MSB 0-3	89.91	88.77	91.08	89.88	89.70	90.01	90.80
% energy in MSB 0-4	95.19	94.67	95.74	95.19	95.00	95.23	95.58
% energy in MSB 0-5	97.92	97.65	98.15	97.91	97.83	97.94	98.08
% energy in MSB 0-6	99.30	99.19	99.38	99.30	99.27	99.31	99.36
% energy in MSB 0-7	100.00	100.00	100.00	100.00	100.00	100.00	100.00

encrypted will depend upon the amount of security required. The amount of the information present in the most significant bit planes can be observed from the Table 2. The table shows the percentage of energy present in the most significant bit planes. From the Table 2 it can be clearly observed that approximately 40%–45% of the energy is present in the $MSB0$, whereas approximately 80% of the energy is present in the first 3 most significant bit planes, namely $MSB1$ and $MSB2$.

Based on the observation that most of the energy within an image is present in the MSB bits planes, a lightweight encryption scheme was proposed by Schmidth et al. in their work [6]. Schmidth et al. proposed to divided the entire image into 8 bit planes and encrypt the most significant bit planes. Unfortunately the scheme is limited to images in uncompressed domain and cannot be extended to images compressed in lossless domain. This is due to the fact that the statistics of the image gets disturbed after encryption and the image loses most of its spatial redundancy. Hence the amount of compression that can be gained on using a lossless compression on the encrypted image reduces drastically.

Hence, from the observations on importance of MSB planes for purpose of security and long runs in MSB planes for purpose of compression, an approach can be to divide the image into bit planes and compress each bit plane before applying encryption. Hence a lightweight encryption scheme can be incorporated with a lossless encoding scheme in following manner

1. Divide the image into bit planes. For 8-bit grey image the number of bit planes will be equal to 8, whereas for color images the number of bit planes will be 24, 8 bit planes for each of the three color code.
2. Encode each bit plane by encoding the run lengths.
3. Encrypt the MSB planes after encoding. The number of MSB planes to be encrypted will depend upon the level of security required.

2.1   Encoding Algorithm

Each of the bit planes are scanned row by row starting from the top-left pixel, and the run lengths are computed. For $MSB0$ and $MSB1$ bit planes, the maximum run length is restricted to 7. The run lengths are encoded using a Huffman coding. The Huffman codewords can be either predefined computed from several training images, or alternatively, it can be computed for the respective image, by computing the relative frequencies of run lengths of size 1-7 and run length of size greater then equal to 8, and subsequently computing the Huffman codeword. The relative frequencies and the corresponding Huffman codewords of run lengths generated from a Lena image for $MSB0$ and $MSB1$ bit planes are shown in Table 3.

For $MSB2$ to $MSB7$ bit planes, the maximum run length is restricted to 3. Similar to $MSB0$ and $MSB1$ bit planes the Huffman codewords for $MSB2$ to

Table 3. Frequencies and Huffman codewords for MSB0 and MSB1

Length of runs	Frequencies		Huffman codeword	
	MSB0	MSB1	MSB0	MSB1
Runs of length 1	0.105	0.179	011	011
Runs of length 2	0.064	0.114	000	001
Runs of length 3	0.046	0.078	0100	0101
Runs of length 4	0.036	0.058	0011	0001
Runs of length 5	0.034	0.048	0010	0000
Runs of length 6	0.025	0.038	01011	01001
Runs of length 7	0.024	0.032	01010	01000
Runs of length ≥ 8	0.667	0.453	1	1

Table 4. Frequencies for MSB2 to MSB7

Length of runs	Frequencies					
	MSB2	MSB3	MSB4	MSB5	MSB6	MSB7
Runs of length 1	0.227	0.347	0.441	0.492	0.497	0.497
Runs of length 2	0.137	0.192	0.226	0.248	0.251	0.252
Runs of length 3	0.088	0.115	0.122	0.124	0.126	0.125
Runs of length ≥ 4	0.548	0.346	0.210	0.136	0.126	0.125

Table 5. Compression achieved using bit plane encoding

Image	Original Image Size (bytes)	Compressed Image Size (bytes)	Compression Ratio
Baboon	262144	210643	1.244494
Bandon	245830	133377	1.843121
Brandyrose	385392	238625	1.615053
Lena	262144	176800	1.4827
Opera	407270	277194	1.469260
Peppers	262144	174890	1.498908
Pills	415200	249233	1.665911

$MSB7$ bit planes can either be pre-defined or can be defined from the relative frequencies of run lengths of the respective image. The relative frequencies of various run lengths computed from Lena image for bit planes $MSB2$ to $MSB7$ are shown in Table 4.

The amount of compression achieved for some standard images using run length encoding on bit planes is shown in Table 5.

2.2   Encryption Algorithm

The MSB planes can be encrypted using a standard stream cipher or preferably a standard block cipher in OFB mode. The number of MSB planes to be encrypted can be decided based on the security level required for the underlying content and the system. The scheme in terms of security and computational cost allow eight levels of scalability. At the first level of security, only $MSB0$ plane is encrypted, at the next level i.e. the second level of security $MSB0$ and $MSB1$ planes are encrypted. Finally at the eight level of security all the bit planes will be encrypted, which will be equivalent to a naive encryption approach.

2.3   Extension to Color Images

For color images the scheme can be extended by following the procedure similar to gray images on all the color components, namely R (red) component, G (green) component and B (blue) component.

3   Results

The amount of noise introduced by the encryption scheme at various levels of security measured in terms of PSNR and MAD are shown in Table 6 and Figure 1. From Table 6 it can be observed that the amount of additional noise introduced by encrypting $MSB2$ to $MSB7$ planes is negligible. Hence, it can be concluded that encrypting only $MSB0$ and $MSB1$ will provide sufficient security (it is to be noted that encrypting only $MSB0$ was found not to be secured by Schmidth et al. in their work [6]). But, for data with high security requirements it is recommended to encrypt atleast the first four MSB bit planes. The need for encrypting the first four MSB planes can be observed from Figure 2. The figure shows 8 sets of images of Lena. The first image is when all eight bit planes are present; the second image is formed by shifting each eight bit pixel of Lena image left by 1 bit (i.e. image is formed from $MSB1$ to $MSB7$). Finally the eight image is formed by left shifting each pixel by 7 bits (i.e. image is formed from only $MSB7$). From the figure it can be observed that the original image is visible to certain extent even when only $MSB4$ to $MSB7$ are available.

The advantage of compressing the bit planes before encrypting also results in saving of computational time. The computational time saved at various levels of security for Lena image of size 2097152 bits ($= 512 \times 512 \times 8$) by encoding the bit planes before encrypting them is shown in Table 7. First column shows

Table 6. MAD and PSNR for bit plane encryption scheme

Image		Level1	Level2	Level3	Level4	Level5	Level6	Level7	Level8
Lena	MAD	64.04	72.25	72.30	72.91	73.03	73.06	73.07	73.07
	PSNR	8.99	8.96	9.23	9.22	9.22	9.22	9.22	9.22
Baboon	MAD	63.99	68.00	70.25	70.78	70.93	70.96	70.97	70.97
	PSNR	8.99	9.58	9.54	9.53	9.53	9.53	9.53	9.53
Peppers	MAD	64.10	71.58	74.99	75.52	75.62	75.66	75.67	75.68
	PSNR	64.10	71.58	74.99	75.52	75.62	75.66	75.67	75.68
Opera	MAD	64.12	71.65	72.69	73.14	73.24	73.27	73.27	73.27
	PSNR	8.98	9.04	9.18	9.18	9.19	9.19	9.19	9.19
Bandon	MAD	64.04	79.25	86.12	88.98	90.31	90.87	91.13	91.25
	PSNR	8.99	8.10	7.60	7.38	7.27	7.22	7.20	7.19
Brandyrose	MAD	64.04	73.57	75.02	75.36	75.47	75.50	75.51	75.51
	PSNR	8.99	8.79	8.86	8.88	8.88	8.88	8.88	8.88
Pills	MAD	64.13	76.07	77.83	78.42	78.55	78.58	78.59	78.59
	PSNR	8.98	8.48	8.50	8.49	8.49	8.49	8.49	8.49

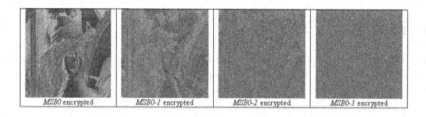

Fig. 1. Lena image encrypted using Bit plane encryption at various levels

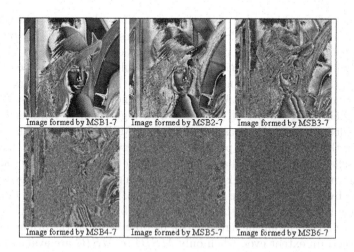

Fig. 2. Lena image formed from its LSBs

Table 7. Bit plane encryption: gain in computation time due to encoding

Security Level	% of total bits encrypted without compression	% of total bits encrypted with bit plane compression
Level 1	12.5%	3.4%
Level 2	25.0%	8.90%
Level 3	37.5%	15.23%
Level 4	50.0%	24.30%
Level 5	62.5%	34.72%
Level 6	75.0%	45.56%
Level 7	90.0%	56.50%
Level 8	100.0%	67.44%

the various security levels i.e from Level-1 to Level-8. Second column shows the percentage of bits that needs to be encrypted at a particular security level compared to encrypting the full uncompressed image when compression is not applied on bit planes. Finally, the third column shows percentage of bits that need to be encrypted at a particular security level using bit plane compression compared to encrypting the full uncompressed image. From the Table 7 it can be observed that using the proposed method only 24.30% of original uncompressed image needs to be encrypted for encrypting the first four MSB planes. Also from previous discussions it can be concluded that encrypting first four MSB planes provides very high level of security, since very minimal information is present in the first four LSB's.

4   Conclusions

In this paper we presented an lossless encoding scheme for images by dividing the image into bit planes and encoding each bit plane individually. The encoding scheme is designed in such a way that it can support partial encryption. The encryption scheme in addition to being secure also supports scalability at eight different levels. A very high level of security can be achieved by encrypting only first four MSB planes, which is equivalent to encrypting roughly only 24% of bits compared to encrypting the entire image. Moreover since the encryption is achieved by encrypting only the MSB planes, and the MSB planes can be compressed to higher extent compared to the LSB planes, the amount of data to be encrypted is very minimal. The compression achieved may be improved further by using an alternative variable length encoding scheme for encoding the runs of each bit plane. The scheme can be also easily extended for supporting lossless video coding and scalable encryption, by considering each frame of the video as an image. In addition the scheme can be also extended to support progressive encoding, since it can support encoding/encryption and decoding/decryption of the image and video in bit planes starting from MSB's.

References

1. Agi, I., Gong, L.: An Empirical Study of Secure MPEG Video Transmission. In: Proceedings of the Symposium on Network and Distributed Systems Security, IEEE, Los Alamitos (1996)
2. Aly, S.: Multimedia Security: Survey and Analysis, Multimedia and Networking Research Lab. CTI, DePaul University, Chicago, http://www.mnlab.cs.depaul.edu
3. Bhargava, B., Shi, C., Wang, Y.: MPEG Video Encryption Algorithms (August 2002), http://raidlab.cs.purdue.edu/papers/mm.ps
4. Choo, E., Lee, J., Lee, H., Nam, G.: SRMT: A Lightweight Encryption Scheme for Secure Real-time Multimedia Transmission. In: International Conference on Multimedia and Ubiquitous Engineering (MUE 2007) (2007)
5. Furht, B., Socek, D., Eskicioglu, A.M.: Chapter 3: Fundamentals of Multimedia Encryption Techniques. In: Multimedia Security Handbook, Published by CRC press LLC(December 2004)
6. Podesser, M., Schmidt, H.P., Uhl, A.: Selective Bitplane Encryption for Secure Transmission of Image Data in Mobile Environments. In: 5th Nordic Signal Processing Symposium on board, Hurtigruten, Norway (October 4-7, 2002)
7. Yekkala, A.K., Udupa, N., Bussa, N., Veni Madhavan, C.E.: Lightweight Encryption for Images. In: IEEE international Conference on Consumer Electronics (2007)

Segmenting Multiple Textured Objects Using Geodesic Active Contour and DWT

Surya Prakash and Sukhendu Das

VP Lab, Dept. of CSE, IIT Madras, Chennai-600 036, India
surya@cse.iitm.ernet.in, sdas@iitm.ac.in

Abstract. We address the issue of segmenting multiple textured objects in presence of a background texture. The proposed technique is based on Geodesic Active Contour (GAC) and can segment multiple textured objects from the textured background. For an input texture image, a texture feature space is created using scalogram obtained from discrete wavelet transform (DWT). Then, a 2-D Riemannian manifold of local features is extracted via the Beltrami framework. The metric of this surface provides a good indicator of texture changes, and therefore, is used in GAC algorithm for texture segmentation. Our main contribution in this work lie in the development of new DWT and scalogram based texture features which have a strong discriminating power to define a good texture edge metric which is used in GAC technique. We validate our technique using a set of synthetic and natural texture images.

Keywords: Snake, segmentation, texture, DWT, scalogram.

1 Introduction

Active contours are extensively used in the field of computer vision and image processing. In this paper, we present a texture object segmentation technique which is based on Geodesic Active Contour (GAC) [1] and discrete wavelet transform (DWT) based texture features, and can segment multiple textured objects from the textured background. Our algorithm is based on the generalization of the GAC model from 1-D intensity based feature space to multi-dimensional feature space [2]. In our approach, image is represented in a n-dimensional texture feature space which is derived from the image using scalograms [3] of the DWT. We derive edge indication function (stopping function) used in GAC from the texture feature space of the image, by viewing texture feature space as Riemannian manifold. Sochen et al. [4] showed that the images or image feature spaces can be described as Riemannian manifolds embedded in a higher-dimensional space, via the Beltrami framework. Their approach is based on the polyakov action functional which weights the mapping between the image manifold (and its metric) and the image features manifold (and its corresponding metric). In our approach, a 2-D Riemannian manifold of local features is extracted from the texture features via the Beltrami framework [4]. The metric of this surface provides a good indicator of texture changes, and therefore, is used

A. Ghosh, R.K. De, and S.K. Pal (Eds.): PReMI 2007, LNCS 4815, pp. 111–118, 2007.

in GAC for texture segmentation. The determinant of the metric of this manifold is interpreted as a measure of the presence of the gradient on the manifold.

Similar approaches where the GAC scheme is applied to some feature space of the images, were studied in [5,6,7]. The aim of our study is to generalize the intensity based GAC model and apply it to DWT and scalogram based wavelet feature space of the images. Our main contribution in this work lie in the development of new texture features which give a strong texture discriminating power and in turn, use of these features to define a good texture edge metric to be used in GAC algorithm.

2   Background

2.1   Geodesic Active Contour

Here, we briefly review of the GAC model presented in [1]. Let $C(q) : [0, 1] \rightarrow R^2$ be a parameterized curve, and let $I : [0, m] \times [0, n] \rightarrow R^+$ be the image where we want to detect the objects boundaries. Let $g(r) : [0, \infty] \rightarrow R^+$ be an inverse edge detector, so that $g \rightarrow 0$ when $r \rightarrow \infty$. g represents the edges in the image. Minimizing the energy functional proposed in the classical snakes [8] is generalized to finding a geodesic curve in the Riemannian space with a metric derived from the image by minimizing following functional:

$$L_R = \int g(|\nabla I(C(q))|)|C'(q)|dq \qquad (1)$$

where, L_R is a new length definition (called geodesic length) in the Riemannian space. It can be considered as a weighted length of a curve, where the Euclidian length is weighted by a factor $g(|\nabla I(C(q))|)$, which contains information regarding the edges in the image. To find this geodesic curve, steepest gradient descent is used which gives following curve evolution equation to get the local minima of L_R.

$$\frac{dC}{dt} = g(|\nabla I|)k\mathbf{N} - (\nabla g.\mathbf{N})\mathbf{N} \qquad (2)$$

where, k denotes Euclidian curvature and $\mathbf{N}$ is a unit inward normal to the curve. Let us define a function $u : [0, m] \times [0, n] \rightarrow R$ such that curve C is parameterized as a level set of u, i.e. $C = \{(x, y)|u(x, y) = 0\}$. Now, we can use the Osher-Sethian level sets approach and replace above evolution equation for the curve C with an evolution equation for the embedded function u as follows:

$$\frac{du}{dt} = |\nabla u|div\left(g(\nabla I)\frac{\nabla u}{|\nabla u|}\right) \qquad (3)$$

where, div is divergence operator. Stopping function $g(\nabla I)$ is generally given by $g(\nabla I(x, y)) = \frac{1}{1+|\nabla I(x,y)|^p}$, where p is an integer and usually equal to 1 or 2. The goal of $g(\nabla I)$ is to stop the evolving curve when it reaches to the object boundary. For an ideal edge, ∇I is very large so $g = 0$ at the edge and the curve stops $(u_t(x, y) = 0)$. The boundary is then given by $u(x, y) = 0$.

2.2   Discrete Wavelet Transform (DWT) and Scalogram

DWT analyses a signal based on its content in different frequency ranges. Therefore, it is very useful in analyzing repetitive patterns such as texture. DWT decomposes a signal into different bands (approximation and detail) with different resolution in frequency and spatial extent. Let $\xi(x)$ be the image signal and $\psi_{u,s}(x)$ be a wavelet function at a particular scale, then signal filtered at point u is obtained by taking the inner product of the two $< \xi(x), \psi_{u,s}(x) >$. This inner product is called *wavelet coefficient* of $\xi(x)$ at position u and scale s [9]. *Scalogram* [3] of a signal $\xi(x)$ is the variance of this wavelet coefficient:

$$w(u, s) = \mathbf{E}\{| < \xi(x), \psi_{u,s}(x) > |^2\} \qquad (4)$$

The $w(u, s)$ has been approximated by convolving the square modulus of the filtered outputs with a Gaussian envelop of a suitable width [3]. The $w(u, s)$ gives the energy accumulated in a band with frequency bandwidth and center frequency inversely proportional to scale. We use scalogram based discrete wavelet features to model the texture characteristics of the image in our work.

3   Texture Feature Extraction

In this section, we explain how DWT is used to extract texture features of the input image. It discusses the computational framework based on multi-channel processing. We use DWT-based dyadic decomposition of the signal to obtain texture properties. A simulated texture image shown in Fig. 1(a) is used to illustrate the computational framework with the results of intermediate processing. Modeling of texture features at a point in an image involves two steps: scalogram estimation and texture feature estimation. To obtain texture features at a particular point (pixel) in an image, a $n \times n$ window is considered around the point of interest (see Fig. 1(b)). Intensities of the pixels in this window are arranged in the form of a vector of length n^2 whose elements are taken column wise from the $n \times n$ cropped intensity matrix. Let this intensity vector (signal) be ξ. It represents the textural pattern around the pixel and is subsequently used in the estimation of scalogram.

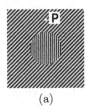
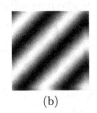
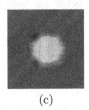

(a)   (b)   (c)

Fig. 1. (a) Synthetic texture image, (b) Magnified view of 21×21 window cropped around point P, shown in Fig. 1(a); (c) Mean texture feature image of Fig. 1(a)

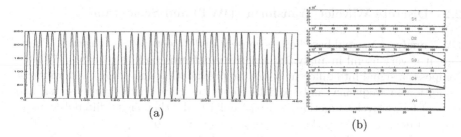

Fig. 2. (a) 1-D texture profile of Fig. 1(b); (b) Scalogram of the texture profile

3.1 Scalogram Estimation

1-D input signal ξ, obtained after arranging the pixels of $n \times n$ window as explained above, is used for the scalogram estimation. Signal ξ is decomposed using wavelet filter. We use orthogonal Daubechies 2-channel (with dyadic decomposition) wavelet filter. Daubechies filter with level-L dyadic decomposition, yields wavelet coefficients $\{A_L, D_L, D_{L-1}, .., D_1\}$ where, A_L represents approximation coefficient and D_i's are detail coefficients. The steps of processing to obtain scalogram from the wavelet coefficients are described in [10]. Fig. 2(b) presents an example of scalogram obtained for the signal shown in Fig. 2(a) using level-4 DWT decomposition.

3.2 Texture Feature Estimation

Once the scalogram of the texture profile is obtained, it is used for texture feature estimation. Texture features are estimated from the "energy measure" of the wavelet coefficients of the scalogram subbands. This texture feature is similar to the "texture energy measure" proposed by Laws [11].

Let E be the texture energy image for the input texture image I. E defines a functional mapping from 2-D pixel coordinate space to multi-dimensional energy space Γ, i.e. $E : [0, m] \times [0, n] \to \Gamma$. Let for the kth pixel in I, D_k be the set of all subbands of scalogram S and $E_k \in \Gamma$ be the texture energy vector associated with it. Texture energy space, Γ, can be created by taking the l_1 norm of each subband of the scalogram S. Then, Γ represents $L + 1$ dimensional energy space for level-L decomposition of the texture signal. Formally, i^{th} element of the energy vector E_k belonging to Γ, is given as follows:

$$E_{(k,i)} = \frac{1}{N} \left\{ \sum_j S_{(i,j)} \right\} \qquad (5)$$

where, i represents a scalogram subband of set D_k, $S_{(i,j)}$ is the jth element of the ith subband of scalogram S and N is the cardinality of the ith subband. Texture energy image computed using Eqn. 5 is a multi-dimensional image and provides good discriminative information to estimate the texture boundaries.

These texture energy measures constitute a texture feature image. Fig. 1(c) shows an image obtained by taking the mean of all bands of a texture energy image computed using Eqn. 5 for the texture image shown in Fig. 1(a).

One common problem in texture segmentation is the problem of precise detection of the boundary efficiently. A pixel near the texture boundary has neighboring pixels belonging to different textures. In addition, a textured image may contain a non-homogeneous, non-regular texture regions. This would cause the obtained energy measure to deviate from "expected" values. Hence, it is necessary that the obtained feature image be further processed to remove noise and outliers. To do so, we apply smoothing operation to the texture energy image in every band separately. In our smoothing method, the energy measure of the kth pixel in a particular band is replaced by the average of a block of energy measures centered at pixel k in that band. In addition, in order to reduce the block effects and to reject outliers, the p percentage of the largest and the smallest energy values with in the window block are excluded from the calculation. Thus, the smooth texture feature value of pixel k in i^{th} band of the feature image is obtained as:

$$F_{(k,i)} = \frac{1}{w^2(1 - 2 \times p\%)} \left\{ \sum_{j=1}^{(w^2)(1-2\times p\%)} E_{(k,j)} \right\} \qquad (6)$$

where, $E_{(k,j)}$s are the energy measures within the $w \times w$ window centered at pixel k of ith band of the texture energy image. The window size $w \times w$ and the value of p are chosen experimentally to be 10×10 and 10 respectively in our experiments. Texture feature image F, computed by smoothing the texture energy image E as explain above, is used in the computation of texture edges using inverse edge indicator function which is described in the next section.

4   Geodesic Active Contours for Texture Feature Space

We use GAC technique in the scalogram based wavelet texture feature space by using the generalized inverse edge detector function g proposed in [5]. GAC, in presence of texture feature based inverse edge detector g, is attracted towards texture boundary.

Let $X : \Sigma \to M$ be an embedding of Σ in M, where M is a Riemannian manifold with known metrics, and Σ is another Riemannian manifold with unknown metric. As proposed in [7], metric on Σ can be constructed using the knowledge of the metric on M using the pullback mechanism [4]. If Σ is a 2-D image manifold embedded in n-dimensional manifold of texture feature space $\vec{F} = (F^1(x,y), ..., F^n(x,y))$, metric $h(x,y)$ of 2-D image manifold can be obtained from the embedding texture feature space as follows [7]:

$$h(x,y) = \begin{pmatrix} 1 + \Sigma_i(F_x^i)^2 & \Sigma_i F_x^i F_y^i \\ \Sigma_i F_x^i F_y^i & 1 + \Sigma_i(F_y^i)^2 \end{pmatrix} \qquad (7)$$

Then, stopping function g used in GAC model for texture boundary detection can be given as the inverse of the determinant of metric h as follows [5]:

$$g(\nabla I(x,y)) = \frac{1}{1 + |\nabla I(x,y)|^2} = \frac{1}{det(h(x,y))} \tag{8}$$

where, $det(h)$ is the determinant of h. Eqn. 3, with g obtained in Eqn. 8, is used for the segmentation of the textured object from the texture background.

4.1   Segmenting Multiple Textured Objects

As GAC model is topology independent and does not require any special strategy to handle multiple objects, proposed method can segment multiple textured objects simultaneously. Evolving contours naturally split and merge allowing the simultaneous detection of several textured objects so number of objects to be segmented in the scene are not required to be known prior in the image. Section 5 presents some multiple textured objects segmentation results for the synthetic and natural images.

5   Experimental Results

We have used our proposed method on both synthetic and natural texture images to show its efficiency. For an input image, texture feature space is created using the DWT and scalogram, and Eqn. 5 is used for texture energy estimation. We use the orthogonal Daubechies 2-channel (with dyadic decomposition) wavelet filter for signal decomposition. The metric of the image manifold is computed considering the image manifold embedded in the higher dimensional texture feature space. This metric is used to obtain the texture edge detector function of GAC. Initialization of the geodesic snake is done using a signed distance function.

To start the segmentation, an initial contour is put around the object(s) to be segmented. Contour moves towards the object boundary to minimize objective function L_R (Eqn. 1) in the presence of new g (Eqn. 8). Segmentation results obtained using proposed technique are shown in Fig. 3 and Fig. 4 for synthetic and natural images respectively. Input texture images are shown with the initial contour (Fig. 3(a) and Fig. 4(a)). Edge maps of the input synthetic and natural

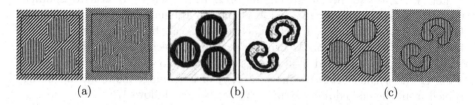

(a)   (b)   (c)

Fig. 3. Results on synthetic images: (a) Input images with initial contour, (b) Texture edge maps, (c) Segmentation results

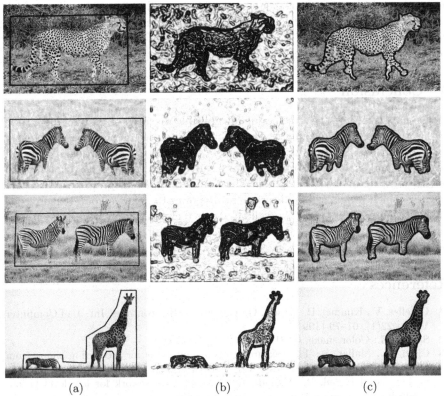

(a)   (b)   (c)

Fig. 4. Results on natural images: (a) Input images with initial contour, (b) Texture edge maps, (c) Segmentation results

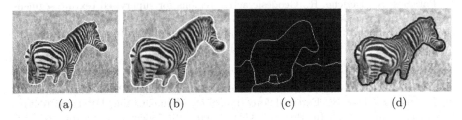

(a)   (b)   (c)   (d)

Fig. 5. Comparative study of results: (a) reproduced from [12], (b) reproduced from [13], (c) obtained using the method presented in [14], (d) proposed technique

texture images, computed using Eqn. 8, are shown in Fig. 3(b) and Fig. 4(b) respectively. Final segmentation results are shown in Fig. 3(c) and Fig. 4(c). Results obtained are quite encouraging and promising.

Fig. 5 shows comparative results for zebra example. We can carefully observe that our result is superior than the results of other techniques. The results reported in the literature show errors in any one of the following places of the object (mouth, back, area near legs etc.). The overall computation cost (which includes the cost of texture feature extraction and segmentation) of the proposed method lies in the range of 70 to 90 seconds on a P-IV, 3 GHz machine with 2 GB RAM for images of size 100×100 pixels.

6 Conclusion

In this paper, we present a technique for multiple textured objects segmentation in the presence of background texture. The proposed technique is based on GAC and can segment multiple textured objects from the textured background. We use DWT and scalogram to model the texture features. Our main contribution in this work lie in the development of new DWT and scalogram based texture features which have a strong discriminating power to define a good texture edge metric to be used in GAC. We validated our technique using various synthetic and natural texture images. Results obtained are quite encouraging and accurate for both types of images.

References

1. Caselles, V., Kimmel, R., Saprio, G.: Geodesic active contours. Int. J. of Computer Vision 22(1), 61–79 (1997)
2. Sapiro, G.: Color snake. CVIU 68(2), 247–253 (1997)
3. Clerc, M., Mallat, S.: The texture gradient equations for recovering shape from texture. IEEE Trans. on PAMI 24(4), 536–549 (2002)
4. Sochen, N., Kimmel, R., Malladi, R.: A general framework for low level vision. IEEE Trans. on Image Proc. 7(3), 310–318 (1998)
5. Sagiv, C., Sochen, N.A., Zeevi, Y.Y.: Gabor-space geodesic active contours. In: Sommer, G., Zeevi, Y.Y. (eds.) AFPAC 2000. LNCS, vol. 1888, pp. 309–318. Springer, Heidelberg (2000)
6. Paragios, N., Deriche, R.: Geodesic active regions for supervised texture segmentation. In: Proc. of ICCV 1999, pp. 926–932 (1999)
7. Sagiv, C., Sochen, N., Zeevi, Y.: Integrated active contours for texture segmentation. IEEE Trans. on Image Proc. 15(6), 1633–1646 (2006)
8. Kass, M., Witkin, A., Terzopoulos, D.: Snakes: Active contour models. Int. J. of Computer Vision 1(4), 321–331 (1988)
9. Mallat, S.: A Wavelet Tour of Signal Processing. Academic Press, London (1999)
10. Prakash, S., Das, S.: External force modeling of snake using DWT for texture object segmentation. In: Proc. of ICAPR 2007, ISI Calcutta, India, pp. 215–219 (2007)
11. Laws, K.: Textured image segmentation. PhD thesis, Dept. of Elec. Engg., Univ. of Southern California (1980)
12. Rousson, M., Brox, T., Deriche, R.: Active unsupervised texture segmentation on a diffusion based feature space. In: Proc. of CVPR 2003, pp. II-699–704 (2003)
13. Awate, S.P., Tasdizen, T., Whitaker, R.T.: Unsupervised texture segmentation with nonparametric neighborhood statistics. In: Leonardis, A., Bischof, H., Pinz, A. (eds.) ECCV 2006. LNCS, vol. 3952, pp. 494–507. Springer, Heidelberg (2006)
14. Gupta, L., Das, S.: Texture edge detection using multi-resolution features and self-organizing map. In: Proc. of ICPR 2006 (2006)

Projection Onto Convex Sets with Watermarking for Error Concealment

Chinmay Kumar Nayak, M. Jayalakshmi, S.N. Merchant, and U.B. Desai

SPANN Lab, Electrical Engineering Department
Indian Institute of Technology, Bombay, Powai, Mumbai-76, India
chinmay.kn@gmail.com, {jlakshmi,merchant,ubdesai}@ee.iitb.ac.in

Abstract. Transmission of images and video over unreliable channels produces visual artifacts due to either loss or corruption of bit streams. Error concealment schemes aim at reducing such degradation and improving the visual quality. Error concealment schemes usually conceal errors of a specific type. This points towards the possibility of combining more than one scheme, for efficient error concealment and better visual effects. In this paper, we propose a scheme for error concealment using a combination of watermarking and Projection Onto Convex Sets (POCS). Watermarking an image using the information derived from the host data itself conceals errors initially and those which are prominent even after this stage are concealed through POCS. The proposed method can be used for concealing errors in images and intra coded video by preserving average values through watermarking and edge information through POCS. Simulation results of POCS, watermarking and watermarking with POCS are given in the paper to show the advantage of combining both the methods.

1 Introduction

Data transmitted over channels are vulnerable to transmission errors. Imperfect transmission of block coded images and video results in loss of blocks. Error correction, error control and error concealment techniques have been developed for reducing these visual artifacts [1]. Error concealment techniques reduce image distortions through post processing at the decoder side. Hence error concealment techniques do not have access to the original information and are usually based on estimation and interpolation procedures that do not require additional information from the encoder.

Digital watermarking is basically a means of inserting some content into the original data which can be later recovered to prove authentication or copyright. Of late it has also been proposed for error concealment of images and video [2]. In this case some important information derived from the image itself is chosen as the watermark and is retrieved at the decoder as in the case of a blind watermarking scheme and used for concealing errors. The scheme proposed in [3] conceals block errors of various sizes by replacing the lost blocks with their average value. This is performed by embedding some important information

A. Ghosh, R.K. De, and S.K. Pal (Eds.): PReMI 2007, LNCS 4815, pp. 119–127, 2007.

extracted from the original image itself to introduce sufficient redundancy in the transmitted image. When there are too many losses, blockiness may become visible and some blocks may remain unconcealed. Thus it would be a good idea to combine this method with another method to remove this blockiness.

Projection onto Convex Sets (POCS) [4] utilizes correlated edge information from local neighborhood in images and intra frames of video to restore missing blocks. It is an iterative algorithm satisfying spatial domain and spectral domain constraints. Thus POCS method preserves edge, but is computationally complex and non real time as it requires much iteration to extrapolate missing pixels from the available correctly received surrounding blocks.

This paper proposes a combined error concealment algorithm using both watermarking and POCS. Watermarking ensures that each block is replaced by its average value. But this has disadvantage of high blockiness effect as the number of error blocks increases. This can be improved through POCS by recovering coefficients extrapolated from the neighboring correctly received blocks. POCS is an efficient algorithm to maintain edge continuity. Moreover, it gives superior concealment for blocks in smooth regions by exploiting the local texture information. POCS being a computationally expensive method, is applied after replacing error blocks with their DC values using watermarking method. This reduces the number of iterations for POCS. Most of the erroneous blocks are concealed using watermarking based method. So a low complexity error measure is worked out to select blocks on which POCS will be applied. This ensures better performance with reduced complexity. We have chosen Peak Signal to Noise Ratio (PSNR) and Structural Similarity (SSIM) [5] as two performance indices for evaluating our method. To show the superiority of the algorithm we have performed 8×8 and 16×16 block errors on three different images. The performance with POCS, watermarking and combined watermarking and POCS are tabulated to show the advantage of combining both the methods.

The paper is organized as follows. Section 2 gives a detailed description of the algorithm and the performance indices. Section 3 includes the simulation results. Section 4 gives conclusion.

2   Proposed Method

Error concealment of the corrupted image using watermarking is explained in the first part of this section. Second part gives the implementation of POCS in the subsequent stage of concealment on selected blocks. At the end of this section the performance indices chosen are explained.

2.1   Error Concealment Through Watermarking

We have used a wavelet based algorithm for this purpose and the gray level values of the approximate band is hidden in two selected sub-bands of the image. Before hiding these values, they are encrypted using a pseudo random sequence generated by using a key. So only a person who has the key can decrypt the

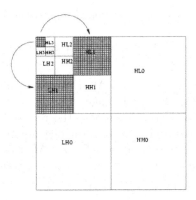

Fig. 1. 4-level Wavelet transform

hidden information for concealment purpose. Four level wavelet decomposition
of the original image is taken and the approximate band is selected as the information to be hidden for error concealment as in Fig 1. Consider a general
case of 4-level DWT which gives an approximate band of size $N \times N$. Then
bands LH1 and HL1 would be of size $4N \times 4N$. If we use 8-bit representation of
each coefficient in approximate band, we can have four copies of the bit stream
embedded into the horizontal and vertical detail bands which give redundancy.

To hide the approximate band, the transformed coefficients are first scaled to
0-255 gray scale. This scale factor should be known at the receiver side to retrieve
the approximate band exactly. Then 8×8 blocks from approximate band are
selected row wise and zig-zag scanned to get a one dimensional array. Converting
these gray levels to 8-bit binary representation, we generate a one dimensional
array of binary numbers. This generated bit stream is Ex-ORed with a pseudo
random sequence generated using a key to get the watermark. The bands LH1
and HL1 are also divided into 8×8 blocks row wise and then zig-zag scanned.
Each coefficient pixel from LH1 and HL1 would carry a single watermark bit.
A block error in LH1 would implicate the same error in HL1 also in the same
location. This may lead to loss of the same watermark bit in the redundant
streams. The bit streams embedded in LH1 and HL1 are spatially displaced
from one another so as to minimize the effect of same bits being affected by
an error. We have given a shift of 50 bits while embedding in LH1. A reversed
stream is embedded in HL1.

For embedding the watermark there are four different cases to be considered.
The watermark bit can be either 0 or 1 and the coefficient on which embedding
is to be performed can be positive or negative. We have followed the following
strategy for embedding under these four conditions.

- If watermark bit is 0 and the coefficient I is positive
  $I' = 8(floor(I/8)) + 1$
- If watermark bit is 0 and the coefficient I is negative
  $I' = 8(ceil(I/8)) - 1$

- If watermark bit is 1 and the coefficient I is positive
  $I' = 8(floor(I/8)) + 5$
- If watermark bit is 1 and the coefficient I is negative
  $I' = 8(ceil(I/8)) - 5$

Here I' is the watermarked coefficient. After the embedding, inverse transform is performed to get the watermarked image.

For retrieving the watermark, decompose the received image using wavelet transform. Then select HL1 and LH1 from the decomposed image. Group 8×8 blocks and then zig-zag scan each of these blocks. After that for each coefficient Y in the scanned pattern, we decide if a 0 or a 1 is embedded.

- If $(mod(Y, 8) \geq -2)$ or $(mod(Y, 8) \leq 2)$,
  the recovered bit is 0
- Else, the recovered bit is 1.

The range specified in the above case eliminates round off errors. We get 4 copies of the embedded stream from HL1 and LH1. The shift given to the bit stream is considered while reconstructing the four copies. Then out of the four bits representing a single bit in the watermark, we choose the one which occurs maximum number of times. The extracted stream is Ex-ORed with the pseudo random sequence generated using the key to get the bit stream corresponding to the approximate band. Convert groups of 8 consecutive bits into decimal values and place them in 8×8 blocks as in zig-zag scanning. The approximate band is recovered by multiplying this entire band by the scale factor chosen while embedding.

Once the approximate band is retrieved, the blocks in error are to be reconstructed using these values. The 4-level wavelet decomposition of the error image is performed. The error coefficients of the approximate band of the received image are replaced by the corresponding coefficients from the retrieved approximate band. After wards, we zoom in the approximate band to the original size. Zooming in at every stage is performed by simple replication of row and column at every level and averaging the four neighbors of each of the pixel. Once the zoomed in image is obtained, its 4-level wavelet decomposition is taken. Comparing the transformed coefficients of the zoomed in image and error image at every resolution,the error blocks are replaced from the transformed coefficients of the zoomed in image. Then inverse transform is performed to get the error corrected image.

2.2 Implementation of POCS

Watermarking based error concealment gives good performance when the number of blocks in error is less. As more number of blocks are lost during transmission, it becomes difficult to retrieve the correct copy of the embedded data. This affects the quality of the error concealed image. Thus, with a large number of error blocks, it is better to combine another method for further improvement in the image quality. In the proposed method, POCS is implemented after the previous stage of error concealment. So the first step to be performed is to identify those blocks which are still in error. The selection of error blocks is separately performed if it is an 8 × 8 or 16 × 16 block error.

POCS for 8 × 8 Blocks. Sum of Absolute Difference (SAD) of the boundary pixels with the surrounding good ones is taken as a measure for selection of 8 × 8 block errors. POCS is applied to only those blocks which are above some threshold. The threshold value has been decided by varying the number of erroneous blocks and considering the perceptual quality improvement.

The damaged block with correctly received 8 surrounding blocks is used to form a larger search area. This search area is classified as a monotone block or edge block. This is done using a Sobel edge detector. The local gradient components g_x and g_y for each pixel $x(i,j)$ in the correctly received pixels in search area is computed as follows.

$$g_x = x_{i+1,j-1} - x_{i-1,j-1} + 2x_{i+1,j} - 2x_{i-1,j} + x_{i+1,j+1} - x_{i-1,j+1}$$

$$g_y = x_{i-1,j+1} - x_{i-1,j-1} + 2x_{i,j+1} - 2x_{i,j-1} + x_{i+1,j+1} - x_{i-1,j-1} \qquad (1)$$

The magnitude of gradient, G and angular direction, θ are computed as follows.

$$G = \sqrt{g_x^2 + g_y^2}$$

$$\theta = tan^{-1}(\frac{g_y}{g_x}) \qquad (2)$$

For each pixel in the surrounding block, the actual edge direction is estimated with Sobel edge detector. The slope at each pixel is then determined as $slope = tan(\theta + 90)$. After wards, a line is drawn from the pixel with the calculated slope. If this line intersects the boundaries of the error block, θ is quantized to one of the eight directions in $0 - 180^0$ and the corresponding accumulator is incremented by the gradient value. If the line drawn is not touching the boundary pixels of the error block, classification process is shifted to the next pixel. This process is repeated exhaustively to all pixels in the neighborhood and the accumulator with highest value is compared against a threshold. This completes the classification of a particular error block as a monotone block or an edge block.

Once the classification process is performed, two projection operations are applied to restore the missing block. The first projection operation is adaptive to local image characteristics and it imposes smoothness and edge continuity constraints. In monotone areas of the image, the spectrum is nearly isotropic and has very low bandwidth. So for restoring missing blocks, classified as monotone areas, a low pass band-limited spectrum is imposed. Specifically, in the Fourier transform domain, all high frequency coefficients outside a particular radius are set to zero. In edge areas of the image, the spectrum has a bandpass characteristic in which energy is localized in transform coefficients that lie in a direction orthogonal to the edge. The remaining coefficients are usually very small. Thus for missing blocks classified as edge areas, we can impose the constraint that any

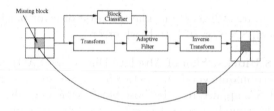

Fig. 2. Adaptive POCS iterative restoration process

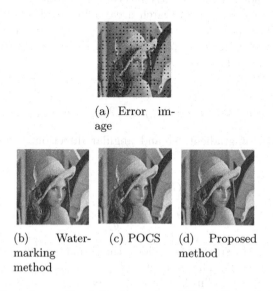

(a) Error image

(b) Watermarking method   (c) POCS   (d) Proposed method

Fig. 3. Error Image (8 × 8 block error) and Error Concealed Images

feasible restoration must have a bandpass spectrum oriented orthogonal to the classified edge. All the coefficients located outside this spectrum are set to zero.

The second projection operation imposes a range constraint and restricts the output values after the first projection to a range of 0–255. For neighboring blocks that are correctly received, the values are maintained unaltered. These two projection operations are iteratively applied until the block does not change anymore under further projections.

Selection of 16 × 16 Blocks for Applying POCS. With larger block sizes of errors, POCS does not give a very good result. Emphasis should be given to texture information while selecting the blocks to apply POCS. We choose three different criteria for deciding this.

- SAD of boundary pixels with surrounding good ones with a threshold value that removes the blocking effect
- SAD of boundary pixels with the surrounding good ones with a lower threshold, and finding only one strong edge

Table 1. PSNR and SSIM after concealment of 8 × 8 block errors

Image	Error blocks	before conceal	PSNR POCS	PSNR WM	PSNR WM+POCS	SSIM original	SSIM POCS	SSIM WM	SSIM WM+POCS
	50	24.59	42.42	40.94	43.21		0.994	0.9941	0.9957
	100	21.70	38.55	37.32	40.68		0.9855	0.9848	0.9903
Lena	150	19.73	37.14	36.04	38.33	0.9998	0.9829	0.9821	0.9869
	200	18.83	36.50	35.28	37.56		0.9794	0.9782	0.9839
	250	17.77	35.30	34.13	36.47		0.9734	0.9719	0.9796
	50	21.84	36.37	37.78	38.17		0.990	0.9920	0.9924
	100	19.15	33.55	34.85	35.35		0.9817	0.9846	0.9855
Aerial	150	16.84	31.21	32.80	33.06	0.9999	0.9691	0.9743	0.9755
	200	16.13	30.50	31.93	32.29		0.9635	0.9693	0.9710
	250	14.88	29.29	30.69	31.04		0.9518	0.9590	0.9615
	50	24.90	39.91	40.28	40.66		0.9925	0.9926	0.9933
	100	22.36	37.11	37.76	37.86		0.9864	0.9869	0.9879
Couple	150	20.69	35.53	36.20	36.45	0.9998	0.9807	0.9815	0.9830
	200	18.93	33.54	34.40	34.53		0.9699	0.9719	0.9740
	250	18.34	33.28	33.84	34.17		0.9664	0.9680	0.9709

- SAD of boundary pixels with surrounding good ones with higher threshold and four strong edges

This classifies the image texture and POCS has been applied to them. In case of multiple strong edges, bandpass filtering in three directions are applied in spectral domain.

2.3 Performance Measures

We have used two performance measures namely, Peak Signal to Noise Ratio and Structural Similarity (SSIM) in this work. PSNR is one of the simplest and most widely used full reference quality metrics. However, it is not very well matched to perceived visual quality. SSIM is a measure that compares local patterns of pixel intensities that have been normalized for luminance and contrast. It is based on extraction of structural information by Human Visual System [5]. This measure gives better consistency with perceived quality.

3   Simulation Results

In this section, the results of simulations are included and discussed. We have chosen three test images of size 512 × 512. We have considered block errors of size 8 × 8 and 16 × 16. Number of lost blocks was varied from 50 to 250 and 10 to 35 in case of 8 × 8 and 16 × 16 block errors respectively. The results of error concealment with POCS, watermarking and watermarking with POCS are tabulated to have a comparative study of all three methods.

(a) Error image

(b) Watermarking method   (c) POCS   (d) Proposed method

Fig. 4. Error Image (16 × 16 block error) and Error Concealed Images

Table 2. PSNR and SSIM after concealment of 16 × 16 block errors

Image	Error blocks	before conceal	PSNR POCS	PSNR WM	PSNR WM+POCS	SSIM original	SSIM POCS	SSIM WM	SSIM WM+POCS
	10	27.60	42.25	44.33	45.09	0.9999	0.994	0.9954	0.9955
	20	22.34	36.57	38.72	39.02		0.9878	0.9899	0.9904
Lena	30	21.08	34.56	36.97	37.13		0.9833	0.9857	0.9872
	35	20.04	33.79	36.62	36.29		0.9792	0.9839	0.9845
	10	22.89	36.53	39.71	39.74	0.9999	0.9923	0.9938	0.9939
	20	19.31	32.53	35.85	36.05		0.9825	0.9857	0.9859
Aerial	30	18.13	31.60	34.89	34.98		0.9770	0.9818	0.9819
	35	17.49	31.96	35.94	35.53		0.9734	0.9789	0.9791
	10	25.91	38.33	41.32	41.68	0.9998	0.9932	0.9944	0.9945
	20	23.37	37.69	39.56	39.76		0.9872	0.9891	0.9893
Couple	30	21.32	34.85	38.31	38.61		0.9796	0.9831	0.9832
	35	20.75	34.58	37.46	37.50		0.9772	0.9812	0.9813

4   Conclusion

In this paper, we propose to use a combination of watermarking and POCS for error concealment of images. Watermarking method conceals most of the errors in the first stage of concealment. The error blocks which degrade the quality of the image even after this preliminary stage will be concealed using POCS in the successive stage. This minimizes computational complexity due to POCS. Both improvement in performance and reduced complexity depend on the threshold values chosen which in turn depend on the content.

References

1. Wang, Y., Zhu, Q.: Error control and concealment for video communications. IEEE Trans. 86, 974–995 (1998)
2. Adsumilli, B.C.B., Farias, M.C.Q., Mitra, S.K., Carli, M.: A robust error concealment technique using data hiding for image and video transmission over lossy channels. IEEE Transaction on Circuit and Systems for Video Technology 15, 1394–1406 (2005)
3. M, J., Merchant, S.N., Desai, U.B., G, A., M, A., P, S., J, S.: Error concealment using digital watermarking. In: Proc. Asia Pacific Conference on Circuits and Systems (2006)
4. Sun, H., Kwok, W.: Concealment of damaged block transform coded images using projections onto convex sets. IEEE Transaction on Image Processing 4, 470–477 (1995)
5. Wang, Z., Bovik, A.C., Sheikh, R., Simoncelli, E.P.: Image quality assessment: From error visibility to structural similarity. IEEE Transaction on Image Processing 13, 600–612 (2004)

Spatial Topology of Equitemporal Points on Signatures for Retrieval

D.S. Guru, H.N. Prakash*, and T.N. Vikram

Dept of Studies in Computer Science,University of Mysore,
Mysore - 570 006, India
dsg@compsci.uni-mysore.ac.in, prakash_hn@yahoo.com

Abstract. In this paper, we address the problem of quick retrieval of online signatures. The proposed methodology retrieves signatures in the database for a given query signature according to the decreasing order of their spatial similarity with the query. Similarity is computed based on orientations of corresponding edges drawn in between sampled points of the signatures. We retrieve the best hypotheses in a simple yet efficient way to speed up the subsequent recognition stage. The runtime of the signature recognition process is reduced, because the scanning of the entire database is narrowed down to contrasting the query with a few top retrieved hypotheses. The experimentation conducted on a large MCYT_signature database [1] has shown promising results.

Keywords: Signature retrieval, Spatial similarity, Online Signature.

1 Introduction

Handwritten signature is one of the earliest biometrics used for general authentication. Its simplicity, ease to capture and the flexibility that it provides for human verification, makes it the most widely used biometric. Offline signature (conventional signature) is supplemented by other features like azimuth, elevation and pressure in case of online signature. The online signature is more robust as it stores additional features, other than just signature image. Any biometric identification problem [2] has two distinct phases: i) recognition and ii) verification. In verification, the query signature is contrasted with a limited set of signatures of the class whose identity is claimed. At the recognition phase, presence of an identity in the database is ascertained [3]. It involves matching stage that extends to entire dataset/database, which is more time consuming. Gupta and McCabe [4] have presented a review on online signature verification. Found et al, [5] investigated spatial properties of handwritten images through matrix analysis. Martinez et al., [6] compare support vector machines and multilayer perceptron for signature recognition.

Essentially any signature recognition system can be optimized when the query signature is matched with best hypotheses than the entire database. Hence signature retrieval mechanism that retrieves the best hypotheses from the database

---

* Corresponding author.

A. Ghosh, R.K. De, and S.K. Pal (Eds.): PReMI 2007, LNCS 4815, pp. 128–135, 2007.

attains importance. Efficient retrieval of handwritten signatures is still a challenging work if the signature database is large. Unlike fingerprint, palm print and iris, signatures have significant amount of intra class variations, making the research even more compelling.

In so far the only work on signature retrieval is by Han and Sethi [7]. They work on handwritten signatures and use a set of geometrical and topological features to map a signature onto 2D-strings [8]. However, 2D-strings are not invariant to similarity transformations and any retrieval systems based on them are hindered by many bottlenecks citenine. There are several approaches for perceiving spatial relationships such as nine-directional lower triangular matrix (9DLT) [12], triangular spatial relationship (TSR) [13] and Similarity measure (SIM_R) [10]. In order to overcome the said problem, we propose an online signature retrieval model in this paper based on SIM_R. The proposed methodology retrieves signatures quickly from the database for a given query in the decreasing order of their spatial similarity with the query. Consequently the proposed system can be used as a preprocessing stage which reduces the runtime of the recognition process as scanning of the entire database is narrowed down to contrasting the query with a top few retrieved hypotheses during recognition. Experimentation has been conducted on a MCYT_signature database [1] which consists of 8250 signatures and it has shown promising results.

The remaining part of the paper is organized as follows. The proposed methodology is explained in Section 2. The details of the experimental results are given in Section 3, and finally in Section 4 some conclusions are drawn.

2   Proposed Model

Our approach involves sampling of online signature at equitemporal interval for x, y coordinates, to get n sample points. The first sampled point is labeled as '1' and the second as '2' and so on and so forth until n, the last sampled point. A directed graph of n nodes is envisaged where directions originate from the node with smaller label to the one with larger label as shown in 1 for $n = 5$. A vector V consisting of the slopes of all the directed edges forms the symbolic representation of a signature and is given by:

$$V = \theta_{12}, \theta_{13}, \cdots, \theta_{1n}, \theta_{23}, \theta_{24}, \cdots, \theta_{n-1n} \tag{1}$$

Where θ_{ij} is the slope of the edge directed from node i to node j, $1 \leq i \leq n-1$, $2 \leq j \leq n$, and $i < j$.

Let S_1 and S_2 be two signatures and V_1 and V_2 be the corresponding vectors representing the slopes of the edges in S_1 and S_2. Now the similarity between S_1 and S_2 is analogous to the similarity between the vectors V_1 and V_2. Let

$$V_1 = \{{}^{s_1}\theta_{12}, {}^{s_1}\theta_{13}, \cdots, {}^{s_1}\theta_{1n}, {}^{s_1}\theta_{23}, {}^{s_1}\theta_{24}, \cdots, {}^{s_1}\theta_{ij}, \cdots, {}^{s_1}\theta_{n-1n}\} \tag{2}$$

$$V_2 = \{{}^{s_2}\theta_{12}, {}^{s_2}\theta_{13}, \cdots, {}^{s_2}\theta_{1n}, {}^{s_2}\theta_{23}, {}^{s_2}\theta_{24}, \cdots, {}^{s_2}\theta_{ij}, \cdots, {}^{s_2}\theta_{n-1n}\} \tag{3}$$

Let $\Delta V = |V_1 - V_2|$, i.e.

$$\Delta V = \{\Delta\theta_{12}, \Delta\theta_{13}, \cdots, \Delta\theta_{1n}, \Delta\theta_{23}, \Delta\theta_{24}, \cdots, \Delta\theta_{ij}, \cdots, \Delta\theta_{n-1n}\} \tag{4}$$

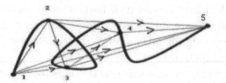

Fig. 1. Online signature with nodes and edges

Here, ΔV represents the vector of the absolute differences in the slopes of corresponding edges in signatures S_1 and S_2. The total number of edges is $(n)(n-1)/2$. Assuming a maximum possible similarity of 100, each edge contributes a value of $100.00/(n)(n-1)/2$ towards the similarity. If the difference in the corresponding edge orientations of the two signatures is zero then the computed similarity value is maximum. When the differences in corresponding edge orientations tend to be away from zero, then the similarity between the two signatures reduces. In this case contribution factor [10] towards similarity from each corresponding edges directed from node i to node j in S_1 and S_2 is

$$\frac{100}{n(n-1)/2} \left[\frac{1 + cos(\Delta\theta_{ij})}{2} \right] \tag{5}$$

where $\Delta\theta_{ij} = |^{s_1}\theta_{ij} - ^{s_2}\theta_{ij}|$, $1 \le i \le n-1$, $2 \le j \le n$, and $i < j$.
Consequently the similarity [10] between S_1 and S_2 due to all edges is

$$SIM(S_1, S_2) = \frac{100}{n(n-1)/2} \sum_{ij} \left[\frac{1 + cos(\Delta\theta_{ij})}{2} \right] \tag{6}$$

where $1 \le i \le n-1$, $2 \le j \le n$, and $i < j$.

Rotation invariance is achieved by aligning the first edge of the query signature with that of database signature before contrasting. The computation complexity of the proposed methodology is $O(n^2)$. During retrieval, the query signature is sampled and slopes of the edges are extracted to form a query vector. The query vector is contrasted with the training vectors in the database. Signatures are retrieved according to the similarity ranks and top K retrievals are selected for further matching for accurate recognition.

In the proposed methodology, we have considered the orientations of edges between the two corresponding sampled points of query and database signatures. For the sake of comparison with the proposed methodology, we have also considered the orientations of edges among three sampled points, forming triangles as shown in Fig.2 for six sampled points. The computation of triangular spatial relationship [9] among all possible triangles is $O(n^3)$ time complexity. Therefore we considered the orientation of edges of only successive triangles of query and database signatures for matching as it is of $O(n)$. We refer this method as successive triangle matching. Triangles formed among sampled points are $\Delta123$,

$\Delta 234$, $\Delta 345$ and $\Delta 456$. The corresponding triangles are contrasted sequentially between query and database signatures for similarity computation. In general let $1, 2, 3, \cdots, n$ be equitemporal sampled points of online signature. From n points, we can form $n - 2$ triangles by considering three successive points at a time. During matching process, the correspondence is drawn between sides of the respective triangles of query and database signatures.

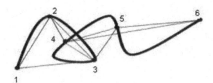

Fig. 2. Successive Triangle Matching

3   Experimental Results

The dataset: The MCYT_signature corpus [1] consists of 50 signatures; 25 are genuine and remaining 25 are forgeries for each of the 330 individuals. Totally it forms a signature database of 8250 (i.e. 330 × 25) genuine and 8250 (i.e. 330×25) forged online signatures. The online signature consists of x-y co-ordinate positions, pressure(P), azimuth(AZ) and elevation(EL) signals. An x-y plot of an online signature is shown in Fig. 3, along with pressure, azimuth and elevation information plots. For our experimentation only spatial relationships of x, y sampled points are considered. The comparison of retrieval performances of the

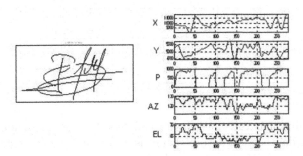

Fig. 3. Sample online signature from MCYT_signature corpus

proposed method and successive triangle method is made through a series of extensive experimentation in this section. Retrieval experiments are conducted for different number of sample points n: 10, 20 and 30. For each sampling, 10 signatures were considered as queries keeping the remaining 15 in the database,

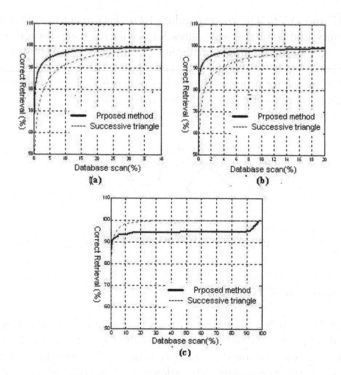

Fig. 4. Retrieval performance with 15 database signatures and 10 queries per class for different sample points: (a) for 10 sample points (b)for 20 sample points and (c)for 30 sample points

Table 1. Query and database signatures combination

Combination	Number of database signatures	Number of Query signatures
(a)	15	10
(b)	10	15
(c)	5	20

out of 25 genuine signatures per class. In all 3300 ($i.e.330 \times 10$) queries and 4950 ($i.e.330 \times 15$) database signatures comprised the test set for experimentation. The retrieval results are as shown in Fig. 4.

The output of the retrieval system is the top K hypotheses. In our experimentations we have set $K = 10$. We define the correct retrieval (CR) for the performance evaluation of retrieval system as

$$CR = (K_c/K_d) \times 100 \qquad (7)$$

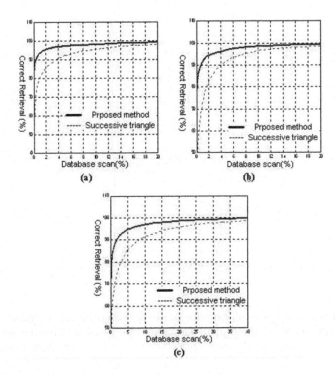

Fig. 5. Retrieval performance (correct retrieval v/s database scan) with 20 sample points for different number of queries and database signatures: (a) 15 database signatures and 10 queries per class. (b) 10 database signatures and 15 queries per class. (c) 05 database signatures and 20 queries per class.

where K_c is the number of correctly retrieved signatures, K_d is the number of signatures in the database. The retrieval performance is best for 20 sample points per signature (Fig. 4b).

We have conducted another set of experiments for different numbers of data base signatures for 20 sample points per signatures (Fig. 5). Experiments were carried out for the different numbers (see Table 1) of database signatures (out of 25 signatures) for each class and remaining signatures as queries to the system. The system shows the good retrieval performance for 15 database signatures per class (fig.5 (a)). That shows the best performance is obtained for higher number of database signatures. Correct retrieval is 98% for 5% database scan and correct retrieval 99% for 10% of database scan.

To evaluate the retrieval performance of the proposed methods for the combination of queries and database signatures as in Table 1, we compute the Precision and Recall ratios. For all these experiments, 20 sample points per signature is used. Proposed methodology based on edge orientation shows better precision compared to successive triangle matching method. The results are shown in Fig.6.

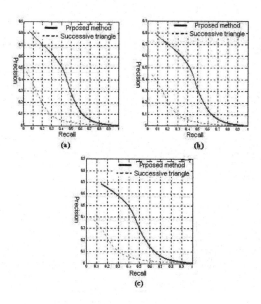

Fig. 6. Precision ratio v/s recall ratio for 20 sample points and for different number of queries and database signatures: (a) 15 database signatures and 10 queries per class. (b) 10 database signatures and 15 queries per class. (c) 05 database signatures 20 queries and per class.

4   Conclusions

Experiments were conducted for quick retrieval of online signatures and results are presented. The retrieval performance of the proposed method based on edge correspondence is compared with the retrieval method based on successive triangle matching. The proposed method is simple, efficient and outperforms the retrieval system based on successive triangle match with respect to all parameters (Precision, Recall and Correct Retrieval).

The MCYT_ signature dataset used here consists of signatures whose $x - y$ sample length varies from 400 points to around 6000 points. However, our retrieval system is fast as it employs just 20 sample points per signature with promising results in retrieving top K hypotheses. The minimum percentage of database scan required to retrieve relevant signatures for all queries is supposed to be fixed experimentally. This is essentially a K-nearest neighbor problem and K best hypotheses should be retrieved. An attempt has been made in the work of Ghosh [11] in this regard where the parameter K is fixed without experimentation. Hence, the decision of arriving at the optimal percentage of database scan where all the authentic queries find a match can be fixed up analytically.

Acknowledgement

Authors thank Dr. Julian Firrez Auguilar, Biometric Research Lab-AVTS, Madrid, Spain for providing MCYT_signature dataset.

References

1. http://atvs.ii.uam.es/mcyt100s.html
2. Ismail, M.A., Gad, S.: Off-line Arabic signature verification. Pattern Recognition 33, 1727–1740 (2002)
3. Lee, S., Pan, J.C.: Off-line tracing and representation of signatures. IEEE Transaction Systems man and Cybernetics 22, 755–771 (1992)
4. Gupta, G., McCabe, A.: Review of dynamic handwritten signature verification. Technical Report, James Cook University Townsville, Australia (1997)
5. Found, B., Rogers, D., Schmittat, R.: Matrix Analysis: A technique to investigate the spatial properties of handwritten images. Journal of Forensic Document Examination 11, 54–74 (1998)
6. Martinez, E.F., Sanchez, A., Velez, J.: Support vector machines versus Multilayer perceptrons for efficient off-line signature recognition. Artificial Intelligence 19, 693–704 (2006)
7. Ke, H., Sethi, I.K.: Handwritten signature retrieval and identification. Pattern Recognition Letters 17, 83–90 (1995)
8. Chang, S.K., Li, Y.: Representation of multi resolution symbolic and binary pictures using 2D-H strings. In: Proceedings of the IEEE Workshop on Languages for Automata, Maryland, pp. 190–195 (1998)
9. Guru, D.S., Punitha, P., Nagabhushan, P.: Archival and retrieval of symbolic images: An invariant scheme based on triangular spatial relationship. Pattern Recognition Letters 24(14), 2397–2408 (2003)
10. Gudivada, V.N., Raghavan, V.V.: Design and evaluation of algorithms for image retrieval by spatial similarity. ACM Transactions on Information Systems 13(2), 115–144 (1995)
11. Ghosh, A.K.: An optimum choice of k in nearest neighbor classification. Computational Statistics and Data Analysis 50(11), 3113–3123 (2006)
12. Chang, C.C.: Spatial match retrieval of symbolic pictures. Information Science and Engineering 7(3), 405–422 (1991)
13. Guru, D.S., Nagabhushan, P.: Triangular spatial relationship: A new approach for spatial knowledge representation. Pattern Recognition Letters 22(9), 999–1006 (2001)

Modified 9DLT Matrix for Similarity Retrieval of Line-Drawing Images

Naveen Onkarappa* and D.S. Guru

Department of Studies in Computer Science, University of Mysore,
Mysore - 570 006, India
naveen_msc@yahoo.com, dsg@compsci.uni-mysore.ac.in

Abstract. An attempt towards perception of spatial relationships existing among the generic components of line-drawing images is made for their
similarity retrieval. The proposed work is based on modified 9DLT matrix
representation. The conventional concept of 9DLT matrix has been tuned
to accommodate multiple occurrences of identical components in images.
A novel similarity measure to estimate the degree of similarity between
two 9DLT matrices representing images is introduced and exploited for
retrieval of line-drawing images. A database of 118 line-drawing images
has been created to corroborate the effectiveness of the proposed model
for similarity retrieval through an extensive experimentation.

1 Introduction

As the availability of information in the form of images is increasing, it has
become important to process, store and search the images based on their content. Here in this paper, we concentrate on the need of storing and searching
of line-drawing images (scanned copies of paper based drawings). Generally,
line-drawing images (which include flowcharts, flow diagrams and engineering
drawings such as electrical circuit diagrams, logic circuit diagrams and architec-
tural plans) are made up of primitive components such as straight lines, curve
segments, geometrical shapes and symbols. The possible dispersion of primitive
components leads to many different line-drawings. The crucial feature that we
can make out in discriminating line-drawing images is thus the spatial locations
of the primitive components with respect to the other primitive components. In
literature, few approaches for recognizing [18], matching [6,16], and interpreting
[15] of line-drawing images have been proposed. Though Fanti et al., (2000) consider the spatial knowledge globally, it cannot discriminate line-drawing images
considering the specific spatial relationships between specific components.

Thus an attempt to exploit the spatial knowledge, an important feature in discriminating images, for the purpose of similarity retrieval of line-drawing images is
made in this paper. There have been several theoretical works towards development
of effective spatial data structures, viz., nine-directional codes [1,2,9], triangular
spatial relationship[8], spatial orientation graph [7] and 2D-strings and its variants

---

* Corresponding author.

A. Ghosh, R.K. De, and S.K. Pal (Eds.): PReMI 2007, LNCS 4815, pp. 136–143, 2007.

[1,3,4,13,17]. Chang and Wu (1995) have proposed a technique for exact match retrieval based on 9DLT matrix [2] using principal component analysis. A survey of indexing and retrieval of iconic images can be found in [14]. The similarity retrieval [11,12] using 2D-strings and its variants involve subsequence matching which is non-deterministic polynomial complexity problem. Most of the previous representation and retrieval methods based on variants of nine-directional codes and 2D-string representations are limited by the number/type of components (primitive objects) in the database, in addition to being ineffectual in handling multiple similar spatial relationships between identical components in images.

In this paper, a similarity retrieval of line-drawing images from a database using modified 9DLT matrix representation is proposed. The conventional 9DLT matrix which supports a single existence of a component and exact match retrieval is tuned up to accommodate multiple components/spatial relationships and also to support similarity retrieval. The proposed similarity measure is demonstrated on a database of 118 synthetic line-drawing images containing primitive components such as lines, circles, triangles, rectangles and trapeziums. The overall contributions of the paper are: 1. Modification of 9DLT matrix to accommodate multiple occurrences of identical components/spatial relationships, 2. A novel similarity measure to estimate the similarity between two such 9DLT matrices, and 3. Exploitation of spatial topology for similarity retrieval of line-drawing images.

This paper is organized as follows. Section 2 gives an overview of 9DLT matrix representation of symbolic images. In section 3, proposed transformation of physical line-drawing images into symbolic representation and the corresponding retrieval schemes are presented. Section 4 shows the retrieval results of the proposed scheme on a database of 118 images. Finally section 5 follows with conclusion.

2   9DLT Matrix: An Overview

Consider an abstract image consisting of four components with labels L_1, L_2, L_3 and L_4 as shown in Fig.1(a). We may use nine directional codes shown in Fig.1(b) to represent the pair-wise spatial relationships between x, a referenced component and y, a contrasted component. The directional code say $d = 0$, represents that y is to the east of x, $d = 1$ represents that y is to the north-east of x, and so on. Thus, the 9DLT matrix T for the image of Fig.1(a) is as shown in Fig.1(c). Since each relationship is represented by a single triple (x, y, d), the 9DLT matrix is a lower triangular matrix.

The 9DLT matrix can now be formally defined as follows (Chang, 1991). Let $V = \{v_1, v_2, v_3, v_4, \cdots, v_m\}$ be a set of m distinct components/objects. Let Z consist of ordered components $z_1, z_2, z_3, , z_s$ such that, $\forall i = 1, 2, \cdots, s, z_i \in V$. Let C be the set of nine-directional codes as defined in Fig.1(b). Each directional code is used to specify the spatial relationship between two components. So, a 9DLT matrix T is an $s \times s$ matrix over C in which t_{ij}, the ith row and j^{th} column element of T is the directional code of Z_j to Z_i if $j < i$, and undefined otherwise. The matrix T is a 9DLT matrix according to the ordered set Z.

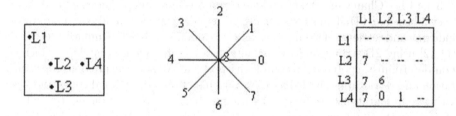

Fig. 1. (a):An abstract image, (b):The nine directional codes and (c):The 9DLT matrix of Fig.1(a)

3   Proposed Scheme

In this section, the method of transforming physical line-drawing images into their symbolic representation is presented. The symbolic representation is done using nine-directional spatial relationships. Using this representation, a scheme for retrieval of similar images is also proposed.

3.1   Symbolic Representation of Line-Drawing Images

The technique of measuring similarity between two images requires the images to be represented in some suitable form. Therefore it is necessary to transform the line-drawing images into symbolic form. In the transformation procedure, we first detect the straight lines using hough transform [5]. Using these detected straight lines, we find the possibility of constructing triangles, trapeziums and rectangles, and extract those possible constructs. After constructing the possible triangles, trapeziums and rectangles, the remaining non-contributing straight lines are generalized as line components. We also extract the circles present in the images using hough transform [10]. Using these extracted components (circles, triangles, trapeziums, rectangles and remaining straight lines) in an image, 9DLT matrices are constructed by considering components of particular type one after the other in a predefined sequence. That is, all line components first, followed by circles, rectangles, triangles and trapeziums. This constraint simplifies the similarity measure presented in section 3.2. While determining directional codes we are considering the area of components rather than just points (centroids) of components because such a representation is too sensitive in spatial reasoning. This is to decide the directional code in case of one component spreads over many directions with respect to another component. In such situations, the directional code is determined as the direction, where the maximum amount of referring component lies with respect to the other component. Further the triplets are generated for each 9DLT matrix corresponding to an image in the database. To generate the triplets, the components of particular type are given unique label. Each triplet of the form (L_1, L_2, d) indicates that the component with label L_1 is in d-direction to the component with label L_2. Thus the physical line-drawing images are converted to symbolic representation as sets of triplets. Fig.2(c) shows the 9DLT matrix for the line-drawing image shown in Fig.2(a) using the directional codes in Fig 1. Fig. 2(b) shows the different components identified in Fig.2(a), where L_i's represent line components, C_i's represent circle components and R_i's represent rectangle components. By giving the unique labels to the components of particular type (here say label 10 for lines, label 20 for circles, label 30 for rectangles), the set of triplets generated for the image in Fig.2(a) is $S = \{(10, 10, 0), (10, 10, 2), (10, 10, 3), (10, 10, 1), (10, 10, 2), (10, 10, 7), (20, 10, 1),$ $(20, 10, 3), (20, 10, 0), (20, 10, 3), (30, 10, 2), (30, 10, 2), (30, 10, 7), (30, 10, 4), (30, 20, 6)\}.$ This set of triplets generated for a line drawing image is then stored in the database as a representative of the image.

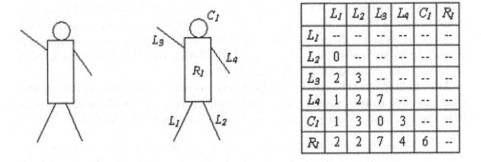

	L_1	L_2	L_3	L_4	C_1	R_1
L_1	--	--	--	--	--	--
L_2	0	--	--	--	--	--
L_3	2	3	--	--	--	--
L_4	1	2	7	--	--	--
C_1	1	3	0	3	--	--
R_1	2	2	7	4	6	--

Fig. 2. (a). A line drawing image, 2(b). A labeled image of 2(a) and 2(c).Symbolic representation 2(a).

Thus, the following is the algorithm designed for creating a symbolic database of line-drawing images.

Algorithm 1: Symbolic database creation.
Input: Line Drawing Images.
Output: Symbolic representation of all line-drawing images.
Method:
Repeat the following steps for each input line-drawing image.

1. Detect straight lines and circles using hough transform.
2. Extract possible constructs (rectangles, trapeziums and triangles) using the detected straight lines as explained in section 3.1.
3. Using the extracted components (lines, circles, rectangles, triangles and trapeziums), construct the 9DLT matrix by considering components of particular type one after the other (i.e., all line components first, followed by circles, rectangles, triangles and trapeziums).
4. Construct the set of triplets for the 9DLT matrix by giving a unique label for a particular type of component and store it as a representative.

Algorithm Ends

3.2   Similarity Retrieval

The similarity between two line-drawing images (i.e., between two symbolic representations) is obtained in terms of the number of triplets in common between the representative sets of triplets. In finding the number of triplets in common, the uniqueness of two triplets is defined as follows.

Let $P_1 = (L_{i1}, L_{i2}, d_i)$ and $P_2 = (L_{j1}, L_{j2}, d_j)$ be two triplets. The two triplets P_1 and P_2 are same if $((L_{i1} == L_{j1})$ and $(L_{i2} == L_{j2})$ and further $(d_i == d_j))$ in the case $(L_{i1} \neq L_{i2})$. In the case $(L_{i1} == L_{i2})$, i.e., while comparing spatial relationships between particular type components, in addition to the conditions $(L_{i1} == L_{j1})$ and $(L_{i2} == L_{j2})$, it needs to be checked that $(d_i == d_j)$ or d_i is oppositional (complementary) directional code of d_j. This is necessary because the components would have interchanged during the process of grouping and putting the components of particular type together at the time of 9DLT matrix construction. This complimentary directional check is not required in the case $(L_{i1} \neq L_{i2})$, because there is no chance of interchange in components of different types due to the constraint of grouping and listing out them in a predefined sequence while constructing 9DLT matrix.

After finding out the number of triplets in common between two sets of triplets corresponding to two images (query and a reference image), we propose to compute the similarity between a query and reference image as follows.

$$Sim = \frac{2 \times number\ of\ triplets\ in\ common\ between\ query\ and\ reference\ image}{number\ of\ triplets\ in\ query\ +\ number\ of\ triplets\ in\ reference} \quad (1)$$

The above proposed similarity measure ensures that the value of Sim lies in the range $[0, 1]$. The following algorithm is thus devised to retrieve similar images for a given line-drawing query image.

Algorithm 2: Retrieval Scheme.
Input: Q, a query line-drawing image.
Output: List of line-drawing images.
Method:

1. Obtain the symbolic representation for a given line-drawing query image using algorithm 1.
2. Compute the similarity of the query image with each of the image in the database using equation 1.
3. Retrieve the images in the database in descending order of their similarities.

Algorithm Ends

Since the similarity between two line-drawing images is found using the number of triplets in common, the time complexity is of $O(mn)$, where m, is the number of triplets in query and n is the number of triplets in referenced image. Thus the proposed methodology is computationally efficient when compared to exponential/non-polynomial computational times of few methods proposed in literature [3,4].

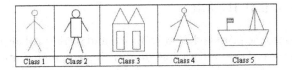

Fig. 3. Representative images of the five different classes

Query image	
Retrieved images for the query with similarity = 1	
Retrieved images for the query with similarity = 0.9333	
Retrieved images for the query with similarity = 0.8	
Retrieved images for the query with similarity = 0.5714	

Fig. 4. Retrieval results for a query image of class 1

Query image and itself the retrieved image with similarity = 1.0	
Retrieved images for the query with similarity = 0.5	
Retrieved images for the query with similarity = 0.2857	

Fig. 5. Retrieval results for a query image of class 2

4   Experimental Results

In order to validate the proposed representation and retrieval algorithms, we have created a database of 118 synthetic line-drawing images, representatives of which are shown in Fig.3. Out of the results obtained for many query images during experimentation, the retrieval results for only three query images are shown in Fig.4, Fig.5 and Fig.6. The top ten retrieved images with their similarity value for the two query images are given in Fig.4 and Fig.5. Fig.6 shows only four images retrieved for the query as our database contains only four images of that class. The experimental results on this database appear to be effectual and encouraging.

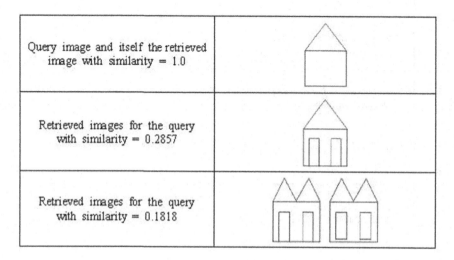

Fig. 6. Retrieval results for a query image of class 3

5   Conclusion

A similarity measure for retrieval of line-drawing images based on the modified 9DLT matrix representation is proposed. The proposed similarity retrieval scheme is invariant to translation, scale and can deal with multiple instances of components/relationships in images. The proposed method is validated on a database of 118 synthetic line-drawing images. The method can be extended to accommodate other primitives including curve segments. The focus is on extending the approach to automatic understanding and also towards indexing for efficient retrieval.

References

1. Chang, C.C., Wu, T.C.: An exact match retrieval scheme based upon principal component analysis. Pattern Recognition Letters 16(5), 465–470 (1995)
2. Chang, C.C.: Spatial Match Retrieval of Symbolic Pictures. J. Information Science and Engineering 7 17(3), 405–422 (1991)

3. Chang, S.K., Shi, Q.Y., Yan, C.W.: Iconic Indexing by 2D Strings. IEEE Tran. on Pattern Analysis and Machine Intelligence 9(3), 413–428 (1987)
4. Chang, S.K., Yan, C.W., Dimitroff, D.C., Arndt, T.: An Intelligent Image Database System. IEEE Tran. Software Engineering 14(3) (1988)
5. Duda, R.O., Hart, P.E.: Use of the Hough Transformation to Detect Lines and Curves in Pictures. Comm. ACM 15, 11–15 (1972)
6. Franti, P., Mednonogov, A., Kalviainen, H.: Hough transform for rotation invariant matching of line-drawing images. In: Proceedings of the International Conf. on Pattern Recognition (ICPR), pp. 389–392 (2000)
7. Gudivada, V.N., Raghavan, V.V.: Design and Evaluation of Algorithms for Image Retrieval by Spatial Similarity. ACM Transactions on Information Systems 13(2), 115–144 (1995)
8. Guru, D.S., Nagabhushan, P.: Triangular spatial relationship: a new approach for spatial knowledge representation. Pattern Recognition Letters 22(9), 999–1006 (2001)
9. Po-Whei, H., Chu-Hui, L.: Image Database Design Based on 9D-SPA Representation for Spatial Relations. IEEE Transactions on Knowledge and Data Engineering 16(12), 1486–1496 (2004)
10. Kerbyson, D.J., Atherton, T.J.: Circle detection using Hough transform filters. In: Proceedings of Fifth International Conf. on Image Processing and its Applications, pp. 370–374 (1995)
11. Lee, S.Y., Hsu, F.J.: Spatial reasoning and similarity retrieval images using 2D-C string knowledge representation. Pattern Recognition 25(3), 305–318 (1992)
12. Lee, S.Y., Shan, M.K., Yang, W.P.: Similarity retrieval of iconic image database. Pattern Recognition 22(6), 675–682 (1989)
13. Lee, S.Y., Hsu, F.J.: 2D C-String: A New Spatial Knowledge Representation for Image Database Systems. Pattern Recognition 23(10), 1077–1087 (1990)
14. Punitha, P.: IARS: Image Archival and Retrieval Systems, Ph. D. Thesis, Dept. of Studies in Computer Science, University of Mysore, India (2006)
15. Kasturi, R., Bow, S.T., El-Masri, W., Shah, J., Gattiker, J.R., Mokate, U.B.: A System for Interpretation of Line Drawings. IEEE Tran. on Pattern Analysis and Machine Intelligence 12(10), 978–992 (1990)
16. Tabbone, S., Wendling, L., Tombre, K.: Matching of graphical symbols in line-drawing images using angular signature information. International Journal on Document Analysis and Recognition 6(2), 115–125 (2003)
17. Ying-Hong, W.: Image indexing and similarity retrieval based on spatial relationship model. Information Sciences 154, 39–58 (2003)
18. Yu, Y., Samal, A., Seth, S.C.: A System for Recognizing a Large Class of Engineering Drawings. IEEE Tran. on Pattern Analysis and Machine Intelligence 19(8), 868–890 (1997)

Image Retrieval Using Fuzzy Relevance Feedback and Validation with MPEG-7 Content Descriptors

M. Banerjee* and M.K. Kundu

Machine Intelligence Unit, Center for Soft Computing Research
Indian Statistical Institute
203, B. T. Road, Kolkata 700 108, India
{minakshi_r, malay}@isical.ac.in

Abstract. Content-Based Image retrieval has emerged as one of the most active research directions in the past few years. In CBIR, selection of desired images from a collection is made by measuring similarities between the extracted features. It is hard to determine the suitable weighting factors of various features for optimal retrieval when multiple features are used. In this paper, we propose a relevance feedback frame work, which evaluates the features, from fuzzy entropy based feature evaluation index (FEI) for optimal retrieval by considering both the relevant as well as irrelevant set of the retrieved images marked by the users. The results obtained using our algorithm have been compared with the agreed upon standards for visual content descriptors of MPEG-7 core experiments.

Keywords: Content-Based image retrieval, fuzzy feature evaluation index, invariant moments, MPEG-7 feature descriptors.

1 Introduction

Digital images are widely used in many application fields such as biomedicine, education, commerce crime prevention, World Wide Web searching etc. This necessitates finding of relevant images from a database, by measuring similarities between the visual contents (color, texture, shape etc.) of the query images and those stored in the database and commonly known as Content-based image retrieval (CBIR) [1],[2], [3], [4].

Due to the vast activities in CBIR research over the last few years, Moving Pictures Expert Group (MPEG) has started the standardization activity for " Multimedia Content description Interface" in MPEG-7 [5], to provide standardization of feature descriptions for audiovisual data.

In a conventional CBIR, an image is usually represented by a set of features, where the feature vector is a point in the multidimensional feature space. However, images with high feature similarities may differ in terms of semantics. The

* M. Banerjee is grateful to the Department of Science and Technology, New Delhi, India, for supporting the work under grant (SR/WOS-A/ET-111/2003).

A. Ghosh, R.K. De, and S.K. Pal (Eds.): PReMI 2007, LNCS 4815, pp. 144–152, 2007.

discrepancy between the low level features like color, texture shape etc. and the high-level semantic concepts (such as sunset, flowers, outdoor scene etc.) is known as "semantic gap". To bridge this gap, users feedback may be used in an interactive manner which is popularly known as "relevance feedback" [6], [7], [8], [9]. The user rates the relevance of the retrieved images, from which the system dynamically learns the user's judgment to gradually generate better results. Owing to these facts, derivation and selection of optimal set of features, which can effectively model human perception subjectivity via relevance feedback, still remain a challenging issue.

Majority of the relevance feedback methods employ two approaches [10] namely, query vector moving technique and feature re-weighting technique to improve retrieval results. In the first approach, the query is reformulated by moving the vector towards positive / relevant examples and away from the negative examples, assuming that all positive examples will cluster in the feature space. Feature re-weighting method is used to enhance the importance of those components of a feature vector, that help in retrieving relevant images, while reducing the importance of the features that does not help. However in such cases, the selection of positive and negative examples, from a small number of samples having large number of features, still remains as a problem.

Relevance feedback techniques in CBIR, have mostly utilized information of the relevant images but have not made use of the information from irrelevant images. Zin et al., [11] have proposed a feature re-weighting technique by using both the relevant and the irrelevant information, to obtain more effective results. Recently, relevance feedback is considered as a learning and classification process, using classifiers like Bayesian classifiers [12], neural network [13]. However trained classifiers become less effective when the training samples are insufficient in number. To overcome such problems, active learning methods have been used in [14].

A fuzzy entropy based feature evaluation mechanism is provided for relevance feedback, combining information from both relevant and irrelevant images. The effectiveness of the proposed method is compared, with the results of Schema(XM) which uses the feature descriptors of MPEG-7 core experiments. The remaining sections are organized as follows : The section 2 describes the mathematical formulations for relevance feedback frame work. The experimental results and conclusion are described in section 3

2   Estimation of Relative Importance of Different Features from Relevance Feedback

Image retrieval using relevance feedback can be considered as a pattern classification problem. An effective relevance feedback system should be able to accumulate knowledge from small set of feedback images to form an optimal query. Each type of visual feature tends to capture only one aspect of the image property. To evaluate the importance of individual components for a particular query I_{qr}, the information from relevant and irrelevant images may be combined as follows :

Let an image database S_d be composed of d distinct images, $I=\{I_1, I_2, ..., I_d\}$ where $I \in S_d$. The image I is represented by a set of features $F = \{f_q\}_{q=1}^{N}$, where f_q is the qth feature component in the N dimensional feature space. The commonly used similarity function between the query image I_{qr} and other images I, is represented as,

$$D_{is}(I, I_{qr}) = \sum_{q=1}^{N} w_q \|f_q(I) - f_q(I_{qr})\| \tag{1}$$

where $\|f_q(I) - f_q(I_{qr})\|$ is the Euclidean distance between the qth component and w_q is the weight assigned to the qth feature component. The weights should be adjusted such that, the features have small variations over the relevant images and large variation over the irrelevant images. Let k similar images $I_s=\{I_1, I_2, ..., I_k\}$ where, $I_k \in I_s$, are returned to the user. The information from relevant(intraclass) images I_r and irrelevant (interclass) images I_{ir} are combined to compute fuzzy feature evaluation index (FEI) proposed by Pal et al., [15], [16] in pattern classification problems.

Feature evaluation index: The fuzzy measure (FEI) is defined from interclass and intraclass ambiguities and explained as follows. Let $C_1, C_2, ... C_j ... C_m$ be the m pattern classes in an N dimensional $(f_1, f_2, f_q, ...f_N)$ feature space where class C_j contains, n_j number of samples. It is shown that fuzzy entropy (H_{qj}) [17] gives a measure of 'intraset ambiguity' along the qth co-ordinate axis in C_j. The entropy of a fuzzy set, having n_j points in C_j is computed as,

$$H(A) = (\frac{1}{n_j \ln 2}) \sum_{i} S_n(\mu(f_{iqj})); i = 1, 2...n_j \tag{2}$$

where the Shannon's function $,(S_n\mu(f_{iqj}))=-\mu(f_{iqj})\ln\mu(f_{iqj})-\{1-\mu(f_{iqj})\}$ ln $\{1-\mu(f_{iqj})\}$

For computing H of C_j along qth component, a standard S-type function shown in Fig. 1 is considered. At b (cross over point), S(b;a,b,c)= 0.5. Similarly at c (shoulder point) $S(c; a, b, c)=1.0$ and at a (feet point) S(a;a,b,c)= 0.0.

The parameters are set as follows. $b = (f_{qj})av$, $c = b + max\{|(f_{qj})av - (f_{qj})max|, |(f_{qj})av - (f_{qj})min|\}$, $a = 2b - c$ where $(f_{qj})av$, $(f_{qj})max$, $(f_{qj})min$ denote the mean, maximum and minimum values respectively computed along the qth co-ordinate axis over all the n_j samples in c_j.

The values of H are 1.0 at $b = (f_{qj})av$ and would tend to zero when moved away from b towards either c or a of the S function, where $\mu(b) =\mu(f_{qj})av =0.5$, eqn. (2). Selecting $b= (f_{qj})av$ indicates that, the cross over point is near to the query feature component. Higher value of H, indicates more samples having $\mu(f)$ equal to 0.5. i.e., cluster around the mean value, resulting in less internal scatter within the class. After combining the classes C_j and C_k the mean, maximum and minimum values $(f_{qkj})av$, $(f_{qjk})max$, $(f_{qjk})min$ respectively of qth dimension over the samples $(n_j + n_k)$ are computed similarly to evaluate "interset ambiguity" H_{qjk}, where n_k are the samples in class C_k.

The criteria of a good feature is that, it should be nearly invariant within class, while having differences between patterns of different classes [15]. The value of H would therefore decrease, after combining C_j and C_k as the goodness of the qth feature in discriminating pattern classes C_j and C_k increases. Considering the two types of ambiguities, the proposed Feature evaluation index (FEI) for the qth feature is,

$$(FEI_q) = \frac{H_{qjk}}{H_{qj} + H_{qk}} \tag{3}$$

Lower the value of FEI_q, higher is the quality of importance of the qth feature in recognizing and discriminating different classes. The user marks the relevant and irrelevant set from 20 returned images. To evaluate the importance of the qth feature, the qth component of the retrieved images is considered. i.e., $I^{(q)}$ $= \{I_1^{(q)}, I_2^{(q)}, I_3^{(q)},, ...I_k^{(q)}\}$. Value of H_{qj} is computed from $I_r^{(q)} = \{I_{r1}^{(q)}, I_{r2}^{(q)}, I_{r3}^{(q)},,$ $....I_{rk}^{(q)}\}$. Similarly H_{qk} is computed from the set of images, $I_{ir}^{(q)} = \{I_{ir1}^{(q)}, I_{ir2}^{(q)}, I_{ir3}^{(q)},,$ $....I_{irk}^{(q)}\}$. H_{qkj} is computed combining both the sets.

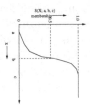

Fig. 1. S-type membership function

Effect of sample size on evaluating importance: To address the issue like, how to manage with limited number of returned images, we study the effect of image sample size on the FEI values. We combine two distinct categories of images. We mark category (1) as relevant and those from category (2) as irrelevant. We increase the sample size of the relevant and irrelevant images (double, triple, half, etc). We also test with other combinations. The FEI values computed from different combinations are shown in Table.1. As seen from the table, the order of the values obtained for $(FEI)_q$ is same for a particular query category.

Adjustment of weights: The marking of relevant and irrelevant images is subjective. One may emphasize more on similarity between one feature (eg., color) than some other features (eg., shape), when comparing two images. Accordingly the weights need to be adjusted. For the query feature vector F. The individual components of relevant images are expected to vary within a smaller range say (ϵ) and may be represented as .

$$I_r = \{I_j \in I_s : \frac{\delta f_q}{|F|} \le \epsilon\} \tag{4}$$

In the first pass, all features are considered to be equally important. Hence $w_1=w_2,...=w_q=1$. The feature spaces of the relevant images are therefore altered in a similar fashion after updating the components with w_q. As a result, the ranks of the relevant images are not affected much. For irrelevant images, one feature component may be very close to the query, whereas other feature component may be far away from the query feature. But the magnitude of the similarity vector may be close to the relevant ones. These images may be characterized as,

$$I_{ir} = \{I_j \in I_s : \frac{\delta f_{q1}}{|F|} \gg \epsilon \; and \frac{\delta f_{q2}}{|F|} \ll \epsilon\} \tag{5}$$

After the features of the query and the stored images (S_d) have been updated with the FEI values, the weighted components are expected to dominate over feature space such that, the rank of the irrelevant ones are pulled down. We have tested the results, from updating the weights with $w_q= FEI_q^2$, $\frac{1}{FEI_q}^2$ and obtained better results from $w_q= FEI_q^2$, in majority of the cases. Intuitively this depends on how the combination of important features dominate over the others for a particular query. Importance of a features is decided from the decreasing order of (FEI) values. In order to further enhance the effect of the important features, we introduce another multiplying factor t_{qf} to make the component more dominating The weights of the individual features for successive iterations are expressed as follows ; $w_{qf} = t_{qf} \times (FEI_{qf})^2$ where the value of t_{qf} are chosen as $\{1.0, 0.1\}$. to get better results.

Features used for characterizing an image: The features are computed from a set of moments invariant to rotation, translation and scaling. The moments m_{pq} of order p and q of a function $f(x,y)$ for discrete images, are usually approximated as,

$$m_{pq} = \sum_x \sum_y x^p y^q f(x,y) \tag{6}$$

The centralized moments are expressed as, $\mu_{pq} = \sum_x \sum_y (x - \bar{x})(y - \bar{y})f(x,y)$

where $\bar{x}=\frac{m_{10}}{m_{00}}$, $\bar{y}=\frac{m_{01}}{m_{00}}$

The normalized central moments are computed as, $\eta_{pq}= \frac{\mu_{pq}}{\mu_{00}^\gamma}$ where $\gamma = \frac{p+q}{2}+1$ for $p + q = 2, 3,....$ A set of seven moments invariant to translation, rotation and scale can be computed from η_{pq}, from which $\theta=\eta_{02}+\eta_{02}$ is considered. Let $I(x,y)_R, I(x,y)_G, I(x,y)_B$ represent the R,G,B component planes of the image matrix $I(x,y)$ and $I_s(x,y)_R$ $I_s(x,y)_G$ $I_s(x,y)_B$ represnt the component planes of the representative locations [18] shown in Fig.2. Six values of invariant moments (θ) are computed from the described component planes, to represent the feature vector.

3   Experimentation

The performance of image retrieval system is tested upon SIMPLIcity images which consists of 1000 images from 10 different categories. The average precision value from each category (randomly chosen queries), after retrieving (10,20,40) images are shown in Table 2. Altough the performance mostly depends uopn the choice of features, it is observed that, the results almost converged after two iterations. A retrieval result is shown in Fig.3.

Complexity. The time complexity T(n) for matching and sorting our results is represented as, T(n) = O(ND)+ O(NlogN). The complexity involved in computing Euclidean distance is O(ND) where D is the number of components within the feature vector. Sorting of (N) images with quick sort of O(NlogN). As (D=6). Hence $D \ll logN$. Therefore $T(n) \simeq O(NlogN)$.

Table 1. Feature evaluation index

Images(Intra, Inter)	$(FEI_1)^2$	$(FEI_2)^2$	$(FEI_3)^2$	$(FEI_4)^2$	$(FEI_5)^2$	$(FEI_6)^2$
(10,10)	0.207	0.115	0.199	0.351	0.351	0.351
(20,20)	0.207	0.115	0.199	0.351	0.351	0.351
(30,30)	0.220	0.121	0.195	0.350	0.350	0.350
(5,5)	0.309	0.151	0.247	0.349	0.349	0.349
(5,10)	0.289	0.164	0.265	0.306	0.306	0.306
(10,5)	0.433	0.213	0.401	0.534	0.534	0.534

(a)   (b)   (c)

Fig. 2. (a) Original image (b) Edge signature on which the high curvature region marked as (*) (c) corner signature

Comparison with MPEG-7: MPEG-7 [19] is an ISO/IEC standard, which provides a collection of specific, agreed upon standard (audio, visual) descriptors. MPEG-7 experimentation Model (XM) software [5] is the frame work of all the reference codes, and make use of MPEG-7 visual descriptors. These can serve as a test bed for evaluation and comparison of features in CBIR context. The features of XM commonly used as standard visual content descriptors for still images are listed in Table. 3.

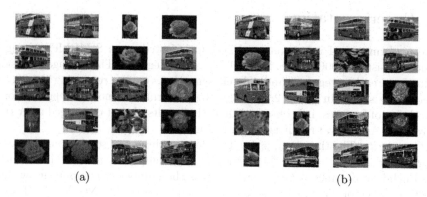

(a)   (b)

Fig. 3. Retrieved results using set(A) (a) first set of candidates (b) After feature evaluation with (FEI), (iteration 1), $t_i*FEI_{i=1}^N$ ={1 × 0.22, 1 × 0.32, 1 × 0.39, 0.1 × 0.36, 0.1 × 0.32, 0.1 × 0.37}. The top left image is the query image.

Table 2. Average precision % from our algorithm

Category	unweighted features			weighted features (iteration 1)		
	10 images	20 img.	40 img.	10 img.	20 img.	40 img.
Africa	68.20	60.25	55.03	70.50	61.00	58.00
Beach	70.00	55.23	50.60	71.48	56.23	52.58
Building	70.26	60.50	54.46	72.40	63.67	56.00
Bus	80.00	70.59	60.67	81.45	72.77	62.67
Dinosaur	100.0	95.0	90.7	100.0	95.0	92.0
Elephant	83.5	75.5	65.8	85.0	77.0	66.0
Flower	90.0	80.5	70.6	92.0	83.0	71.2
Horses	100.0	90.0	80.5	100.0	95.0	83.0
Mountains	70.5	65.8	60.9	72.0	68.0	62.0
Food	60.8	55.8	53.40	62.0	57.0	55.20

These features have been rigorously tested in the standardization process. We have seen the query results for all the classes available in Schema (Beaches, buildings, horses, cars, flowers etc.)which utilize the MPEG-7 visual descriptors. The performance is compared for the case of evaluating overall similarity between images from precision rate defined as, $P_r = (1/n_1) \sum r/n$ where, the system retrieves r images that belongs to the same class $C1$ from n retrieved images. n_1 is the number images queried from category $C1$. The images from other categories are taken as outliers. The results are shown in Figs. 4 and 5. Our results can be fairly compared with MPEG-7 visual descriptors, as seen from Figs. 4 and 5. Although the initial precision is found better for SchemaXM as shown in Fig. 4(a) and Fig.5(b) but our algorithm fairly catches the results.

Table 3. Standard Visual content descriptors of MPEG-7

Color Descriptors	Texture Descriptors	Shape Descriptors based
Dominant Colors	Edge Histogram	Region Based Shape
Scalable Color	Homogeneous Texture	Contour Based Shape
Color Layout		
Color Structure		

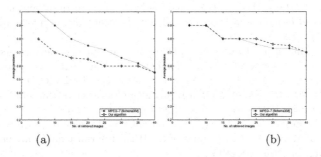

(a)   (b)

Fig. 4. Comparison with MPEG-7 visual descriptors (a) Category beaches (b) Flowers

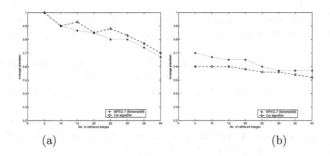

(a)   (b)

Fig. 5. Comparison with MPEG-7 visual descriptors (a) Category horse (b) Vehicles

Conclusion: In the current work, a fuzzy entropy based relevance feedback, framework for image retrieval is proposed. We intend to test the effectiveness of the system, using simple features , invariant moments in the present case. We plan to improve our scheme by associating text and other features, for better compatibility with MPEG-7 and better semantic modeling.

References

1. Smeulders, A.W.M., Worring, M., Santini, S., Gupta, A., Jain, R.: Content-based image retrieval at the end of the early years. IEEE Transactions on Pattern Analysis and Machine Intelligence 22(12), 1349–1380 (2000)
2. Kunttu, I., Lepisto, L., Rauhamaa, J., Visa, A.: Multiscale fourier descriptors for defect image retrieval. Pattern Recognition Letters 27(2), 123–132 (2006)

3. Gevers, T., Smeulders, A.W.M.: Combining color and shape invariant features for image retrieval. Image and Vision computing 17(7), 475–488 (1999)
4. Chen, Y., Wang, J.Z., Krovetz, R.: Clue: Cluster-based retrieval of images by unsupervised learning. IEEE Transactions on Image Processing 14(8), 1187–1201 (2005)
5. MPEG-7 Multimedia Description Schemes (XM). ISO/IEC JTC1/SC29/WG11 N3914 (2001)
6. Han, J., Ngan, K.N., Li, M., Zhang, H.J.: A memory learning framework for effective image retrieval. IEEE Transactions on Image Processing 14(4), 521–524 (2005)
7. Chang, F.C., Hang, H.M.: A relevance feedback image retrieval scheme using multi-instance and pseudo image concepts. IEICE Transactions on Information and Systems D(5), 1720–1731 (2006)
8. Yin, P.Y., Bhanu, B., Chang, K.C., Dong, A.: Integrating relevance feedback techniques for image retrieval using reinforcement learning. IEEE Transactions on Pattern Analysis and Machine Intelligence 27(10), 1536–1551 (2005)
9. Lim, J.H., Jin, J.S.: Combining intra-image and inter-class semantics for consumer image retrieval. Pattern Recognition 38(6), 847–864 (2005)
10. Rui, Y., Huang, T.S., Mehrotra, S.: Content-based image retrieval with relevance feedback in MARS. In: Proceedings of the IEEE International Conference on Image Processing, pp. 815–818 (1997)
11. Jin, Z., King, I., Li, X.Q.: Content-based retrieval by relevance feedback. In: Laurini, R. (ed.) VISUAL 2000. LNCS, vol. 1929, pp. 521–529. Springer, Heidelberg (2000)
12. ves, E.D., Domingo, J., Ayala, G., Zuccarello, P.: A novel bayesian framework for relevance feedback in image content-based retrieval systems. Pattern Recognition 39(9), 1622–1632 (2006)
13. Qian, F., Zhang, B., Lin, F.: Constructive learning algorithm-based rbf network for relevance feedback in image retrieval. In: Bakker, E.M., Lew, M.S., Huang, T.S., Sebe, N., Zhou, X.S. (eds.) CIVR 2003. LNCS, vol. 2728, pp. 352–361. Springer, Heidelberg (2003)
14. He, X., King, O., Ma, W., Li, M., Zhang, H.J.: Learning a semantic space from user's relevance feedback for image retrieval. IEEE transactions on Circuits and Systems for Video technology 2003 13(1) (2003)
15. Pal, S.K., Chakraborty, B.: Intraclass and interclass ambiguities (fuzziness) in feature evaluation. Pattern Recognition Letters 2, 275–279 (1984)
16. Pal, S.K., Chakraborty, B.: Fuzzy set theoretic measures for automatic feature evaluation. IEEE Transactions on Systems, Man and Cybernatics 16(5), 754–760 (1986)
17. Pal, S.K., Majumder, D.D.: Fuzzy mathematical Approach to Pattern Recognition. Willey Eastern Limited, New York (1985)
18. Banerjee, M., Kundu, M.K.: Content Based Image Retrieval with Multiresolution Salient points. In: ICVGIP 2004. Fourth Indian Conference Computer Vision, Graphics and Image Processing, India, pp. 399–404 (2004)
19. Manjunath, B.S., Salembier, P., S, T.: Introduction to MPEG-7: Multimedia Content description Interface. John Wiley and Sons, Inc, USA (2002)

Recognition of Isolated Handwritten Kannada Numerals Based on Image Fusion Method

G.G. Rajput and Mallikarjun Hangarge

Dept. Of Computer Science, Gulbarga University,Gulbarga
ggrajput@yahoo.co.in, mhangarge@yahoo.co.in

Abstract. This paper describes a system for isolated Kannada handwritten numerals recognition using image fusion method. Several digital images corresponding to each handwritten numeral are fused to generate patterns, which are stored in 8x8 matrices, irrespective of the size of images. The numerals to be recognized are matched using nearest neighbor classifier with each pattern and the best match pattern is considered as the recognized numeral.The experimental results show accuracy of 96.2% for 500 images, representing the portion of trained data, with the system being trained for 1000 images. The recognition result of 91% was obtained for 250 test numerals other than the trained images. Further to test the performance of the proposed scheme 4-fold cross validation has been carried out yielding an accuracy of 89%.

1 Introduction

Automatic recognition of handwritten digits has been the subject of intensive research during the last few decades. Digit identification is very vital in applications such as interpretation of ID numbers, Vehicle registration numbers, Pin Codes, etc.

Most of the published work has identified a number of approaches for handwritten character and numerals recognition having their own merits and demerits. In most published work, training samples are used for automated digit prototyping which are available as images stimulating direct application of several well investigated image processing techniques [1, 2]. A systematic survey of numerous automated systems can be found in [3, 4]. An overview of feature extraction methods for offline recognition of isolated characters has been proposed in [5]. A texture based approach using modified invariant moments for pin code script identification is reported in [6]. Unconstrained handwritten character recognition based on fuzzy logic is proposed in [7]. Multiple classifiers based on third order dependency for handwritten numeral recognition is presented in [8]. Unconstrained digit recognition using radon function which represents image as a collection of projections along various directions is reported in [9]. Font and size independent OCR system for printed Kannada documents using support vector machines has been proposed in [10]. Substantial research related to Indian scripts can be found in [11- 15]. From literature survey, it is clear that

A. Ghosh, R.K. De, and S.K. Pal (Eds.): PReMI 2007, LNCS 4815, pp. 153–160, 2007.

much work has been concentrated for recognition of characters rather than numerals. Moreover, in Indian context, it is evident that still handwritten numerals recognition research is a fascinating area of research to deign a robust optical character recognition (OCR), in particular for handwritten Kannada numeral recognition. Hence, we are motivated to design a simple and robust algorithm for handwritten Kannada numerals recognition system. In this paper, we propose a simple and effective method for feature extraction based on image fusion technique [16]. The 64 dimensional features are used to classify the handwritten Kannada numerals using basic nearest neighbour classifier and aimed to obtain encouraging results. The rest of the paper is organized as follows.

In Section 2, the brief overview of data collection and pre-processing is presented. Section 3 deals with the feature extraction and recognition system. The algorithm is presented in Section 4. The experimental results obtained are presented in Section 5. Conclusion is given in Section 6.

2 Data Collection and Preprocessing

The Kannada language is one of the four major south Indian languages. The Kannada alphabet consists of 16 vowels and 36 consonants. Vowels and consonants are combined to form composite letters. Writing style in the script is from left to right. It also includes 10 different symbols representing the ten numerals of the decimal number system shown in Fig. 1.

Fig. 1. Kannada numerals 0 to 9

The database of totally unconstrained Kannada handwritten numerals has been created for validating the recognition system, as the standard database is not available at the moment. Hence, samples of 100 writers were chosen at random from schools, colleges and professionals for collecting the handwritten numerals. The writers were not imposed by any constraint like type of pen and style of writing etc., and the purpose of data collection is not disclosed. Writers were provided with plain A4 normal papers and were asked to write Kannada numerals 0 to 9.

The collected documents are scanned using HP-Scan jet 2400 scanner at 300 dpi which usually yields a low noise and good quality document image. The digitized images are stored as binary images in bit map format. The noise has been removed using morphological dilate and erode operations. Further, it is assumed that the skew correction has been performed before pre-processing. The preprocessed image is inverted resulting in the background as black (binary 0) and numeral as white(binary 1). A total of 1000 binary images of handwritten Kannada numerals are obtained. A sample of Kannada handwritten numerals from this data set is shown in the Fig. 2.

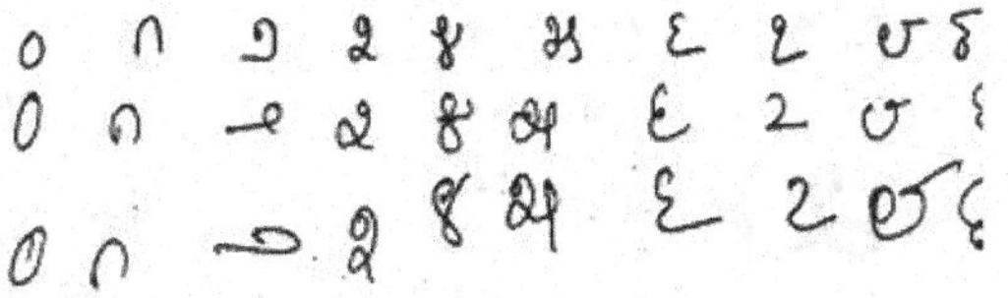

Fig. 2. Sample handwritten Kannada numerals 0 to 9

3   Features Extraction and Recognition

3.1   Feature Extraction

Region labeling is performed on the preprocessed image and a minimum rectangle bounding box is inserted over the numeral and the numeral is cropped. The normalization of the numerals is essential because of the varied writing styles of the writers in the shapes and sizes. Therefore, to bring uniformity among the input numerals, the cropped numeral is normalized to fit into a size of 32x32 pixels without disturbing the aspect ratio. (See Fig. 3(b)). Next, image thinning is performed on this numeral. The purpose of thinning is to reduce the image components to their essential information so that further analysis and recognition are facilitated. It is very important to extract features in such a way that the recognition of different numerals becomes easier on the basis of individual features of each numeral. Therefore, the binary image is divided into 64 zones of equal size, each zone being of size 4 x 4 pixels, so that the portions of numeral will lie in some of these zones. Obviously, there could be zones that are empty. A pattern matrix Mpt_j of size 8 x 8 is created, where j ranges from 0 to 9 representing the numerals. Next, for each zone of the image, if the number of on pixels is greater than 5% of total pixels in that zone, 1 is stored in Mpt_j, for that image. Thus, Mptj represents the reduced image (see Fig. 3 (c)) having the same features as that of its original image. The reduced images, Mpt_j's, of each numeral are fused to generate the pattern matrix ,P_{Mj}, of size 8x8, j ranging from 0 to 9, using the following recursive equation [16].

$$P_{Mj}^{New} = \frac{1}{num+2}(num * P_{Mj}^{Old} + Mpt_j), \ 0 \le j \le 9 \tag{1}$$

Here, P_{Mj}^{New} is the fused pattern matrix obtained after fusing training images contained in Mpt_j and P_{Mj}^{Old} (already stored in the pattern matrix table). P_{Mj}^{New} is copied back to the table along with num increased by 1. The content of each cell of fused pattern matrix represents the probability of occurrence of a white pixel that is mapped with the test image to a typical numeral.

3.2   Recognition

We use nearest neighbour classifier for recognition purpose. The test numeral feature vector is classified to a class, to which its nearest neighbour belongs to.

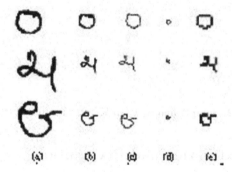

Fig. 3. (a) Original Image, (b) Normalized image, (c) Thinned image,(d) Reduced image, (e) Enlarged view of reduced image

Feature vectors stored priori are used to decide the nearest neighbour of the given feature vector. The recognition process is described below.

The extracted features form the knowledge base and are used to identify the test numerals. The test numeral which is to be recognized is processed in a similar way up to the image reduction step to generate the matrix Mp_{test}. Euclidean distance between Mp_{test} and , j ranging from 0 to 9, is computed using the formula given below.

$$D_j(x) = ||Mp_{test} - P_{Mj}|| \quad 0 \le j \le 9 \tag{2}$$

where Mp_{test} represents the pattern matrix of input test image and P_{Mj} is the fused pattern matrix of jth numeral in the database. The nearest neighbour technique assigns the test numeral to a class that has the minimum distance. The corresponding numeral is declared as recognized numeral.

4   Algorithm

Feature extraction algorithm is given below:

Input: Digitized binary Kannada numeral image set
Output: Pattern matrix of size 8 x 8 as feature vector
Method: Feature Extraction
Step 1: For each sample of Kannada numeral do

  i remove noise, if any
  ii invert the image to get background black and numeral white
  iii fit a bounding box for the numeral and crop the numeral
  iv resize the cropped numeral to 32 x 32 pixels without loosing aspect ration
  v apply thinning process to the normalized image
  vi divide the normalized image into 64 zones of equal size
  vii create a pattern matrix Mp_{tj} of size 8 x 8

viii For each zone if the number of on pixels is greater than 5'% of total pixels store 1 in Mp_{tj}

ix Fuse the reduced image Mp_{tj} with P_{Mj} using equation (1) and store back the result. P_{Mj} forms the feature matrix for that numeral

Step2. Repeat step 1 for all numerals.

Recognition algorithm is given below.

Input: Digitized binary Kannada numeral test image
Output: recognized numeral display
Method: Recognition

Step 1: For the test Kannada numeral perform feature extraction to obtain pattern matrix Mp_{test} of size 8 x 8.

Step 2: Compute the Euclidian distance between the test pattern Mp_{test} and each of the trained patterns P_{Mj} using equation (2).

Step 3: Choose the minimum distance as the best match and declare the corresponding numeral as the recognized numeral.

The proposed image recognition system was implemented with Matlab scripts as illustrated in Fig 4.

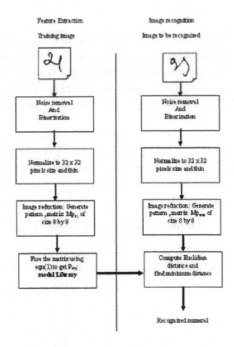

Fig. 4. Recognition system

5   Results and Discussion

The experiments were carried out on the data set discussed in section 2. To see how well the system represents the data it has been trained on, we extracted 500 numerals at random, 50 samples for each numeral, and checked the results. Further, the test data contained a separate set of 250 numerals other than the training set. The Table 1 shows the recognition results for both the cases.

Table 1. Experimental results

Digits	For a portion of trained data		For test data	
	No. of trained numerals	% of recognition	No. of test numerals	% of recognition
೦	50	100	25	96
೧	50	86	25	100
೨	50	100	25	100
೩	50	96	25	72
೪	50	96	25	92
೫	50	92	25	92
೬	50	98	25	76
೭	50	98	25	92
೮	50	98	25	100
೯	50	98	25	92
Average recognition		96.2%		91.2 %

I n the test set, a recognition rate of 91.2% was achieved. Understandably, the training set produced much higher recognition rate of 96.2% than the test data. Further, to validate the performance of the system trained for 1000 images, we carried out K-Fold Cross-validation on 1000 images. With K=4, for each of K experiments we used K-1 folds for training and the remaining one for testing. The recognition rate turned out to be 89%.

It is a difficult task to compare results for handwritten Kannada numeral recognition with other researchers in the literature due the differences in experimental methodology, experimental settings and the size of the database used. Table 2 presents the comparison of the proposed method with other methods in

Table 2. Comparative results

Method	Feature extraction method	Data set size	Classifier	Success rate
V. N. Manjunath Aradhya et. Al.[9]	Radon transform	1000	Nearest Neighbour	91.2
N. Sharma et. Al. [11]	chain code	2300	quadratic classifier	97.87
Dinesh Acharya U. et Al.[15]	Structural features	500	k- means classifier	90.5
Proposed	Image Fusion	1000	Nearest Neighbour	91.2

the literature dealing with Kannada numerals. The proposed method yields a comparable recognition rate of 91.2% at lesser computation cost even with the basic classifier. However, the higher recognition rate can certainly be achieved by using better classifiers.

6 Conclusion

A system capable of recognizing isolated handwritten Kannada numerals is presented. Image fusion method is used to generate patterns as feature matrices of 64 dimensions. The nearest neighbor classification is used to identify the test numeral. The novelty of this paper is that the proposed scheme is simple and effective. The results are discussed for portion of the trained data as well as for unseen instances of data collected from the writers other than that for trained data. The recognition rates are encouraging and supported by 4-Fold cross validation. Our future aim is to improve the recognition rate using fuzzy based classifier and to perform experiments using larger data set. Further, the proposed scheme may be tested for recognition of handwritten numerals and characters of other languages also.

Acknowledgement. The authors thank the referees for their comments and suggestions. The authors wish to acknowledge Dr. P. S. Hiremath, Dr. B. V. Dhandra, Department of Computer Science, Gulbarga University, Gulbarga and Dr. P. Nagabhushan, Dr. Hemantkumar and Dr. Guru, Department of Computer Science, University of Mysore, Mysore for their support.

References

1. Duda, R., Hart, P.: Pattern Classification and Scene Analysis. John Wiley, New York (1973)
2. Gader, P., Khabou, M.: Automated Feature Generation for Handwritten Digit Recognition. IEEE Transactions On Pattern Analysis And Machine Intelligence 18(12), 1256–1261 (1996)
3. Casy, R., Lecolinet, E.: A Survey Of Methods And Strategies In Character Segmentation. IEEE Transactions On Pattern Analysis And Machine Intelligence 18(7), 690–706 (1996)
4. Plamodin, R., Srihari, S.N.: Online and Offline Handwriting Recognition: A Comprehensive survey. IEEE Transaction on Pattern Analysis and Machine Intelligence 22(1), 63–84 (2000)
5. Trier, O., Jain, A.K., Taxt, T.: Feature Extraction Methods For Character Recognition - A Survey Pattern Recognition. 29(4), 641–662 (1996)
6. Nagabhushan, P., Angadi, S.A., Anami, B.S.: An Intelligent Pin Code Script Identification Methodology Based On Texture Analysis using Modified Invariant Moments. In: Proc. of ICCR, pp. 615–623. Allied Pub (2005)
7. Hanmandlu, M., Mohan, R.M., Chakraborty, S., Goyal, S., Choudhary, D.R.: Unconstrained Handwritten Character Recognition Based on Fuzzy Logic. Pattern Recognition 36, 603–623 (2003)

8. Kang, H.J.: Combining Multiple Classifiers Based on Third Order Dependency For Handwritten Numeral Recognition. Pattern Recognition Letters 24, 3027–3036 (2003)
9. Aradhya, V.N.M., Kumar, G.H., Noushath, S.: Robust Unconstrained Handwritten Digit Recognition Using Radon Transform. In: Proc. of IEEE-ICSCN 2007, IEEE Computer Society Press, Los Alamitos (2007)
10. Ashvin, T.V., Shastry, P.S.: A Font And Size Independent Ocr System For Printed Kannada Documents using Support Vector Machines. Sadhana 27(1), 23–24 (2002)
11. Sharma, N., Pal, U., Kimura, F.: Recognition of Handwritten Kannada Numerals. In: Proc. of 9th International Conference on Information Technology (ICIT 2006), pp. 133–136 (2006)
12. Hariharana, S.: Recognition of Handwritten Numerals Through Genetic Based Machine Learning. Jour. of the CSI 30(2), 16–22 (2000)
13. Bhattacharya, U., et al.: Recognition of Hand Printed Bangla Numerals Using Neural Network Models. In: Pal, N.R., Sugeno, M. (eds.) AFSS 2002. LNCS (LNAI), vol. 2275, pp. 228–235. Springer, Heidelberg (2002)
14. Raju, G.: Recognition Of Unconstrained Handwritten Malayalam Characters Using Zero Crossing of Wavelet Coefficients. In: Proc. of ADCOM, pp. 217–221 (2006)
15. Dinesh Acharya, U., Subbareddy, N.V., Krishnamoorthy,: Isolated Kannada Numeral Recognition Using Structural Features and K-Means Cluster. In: Proc. of IISN-2007, pp. 125–129 (2007)
16. Chakraborty, R., Sil, J.: Handwritten Character Recognition Systems Using Image Fusion And Fuzzy Logic. In: PReMI 2005. LNCS, vol. 3766, pp. 344–349. Springer, Heidelberg (2005)

Semi-supervised Learning with Multilayer Perceptron for Detecting Changes of Remote Sensing Images

Swarnajyoti Patra,[1] Susmita Ghosh[1,*], and Ashish Ghosh[2]

[1] Department of Computer Science and Engineering
Jadavpur University, Kolkata 700032, India
susmitaghoshju@gmail.com
[2] Machine Intelligence Unit and Center for Soft Computing Research
Indian Statistical Institute
203 B. T. Road, Kolkata 700108, India

Abstract. A context-sensitive change-detection technique based on semi-superv-ised learning with multilayer perceptron is proposed. In order to take contextual information into account, input patterns are generated considering each pixel of the difference image along with its neighbors. A heuristic technique is suggested to identify a few initial labeled patterns without using ground truth information. The network is initially trained using these labeled data. The unlabeled patterns are iteratively processed by the already trained perceptron to obtain a soft class label. Experimental results, carried out on two multispectral and multitemporal remote sensing images, confirm the effectiveness of the proposed approach.

1 Introduction

In remote sensing applications, change detection is the process of identifying differences in the state of an object or phenomenon by analyzing a pair of images acquired on the same geographical area at different times [1]. Such a problem plays an important role in many different domains like studies on land-use/land-cover dynamics [2], burned area assessment [3], analysis of deforestation processes [4], identification of vegetation changes [5] *etc.* Since all these applications usually require an analysis of large areas, development of automatic change-detection techniques is of high relevance in order to reduce the effort required by manual image analysis.

In the literature [2,3,4,5,6,7,8,9,10], several supervised and unsupervised techniques for detecting changes in remote-sensing images have been proposed. The supervised methods need "ground truth" information whereas the unsupervised approaches perform change detection without using any additional information, besides the raw images considered. Besides these two methods of learning, another situation may arise where only a few training patterns are available. The semi-supervised learning [11] comes into play in such a situation. In this article

---

* Corresponding author.

A. Ghosh, R.K. De, and S.K. Pal (Eds.): PReMI 2007, LNCS 4815, pp. 161–168, 2007.

we propose a context-sensitive semi-supervised change-detection technique based on multilayer perceptron (MLP) that automatically discriminates the changed and unchanged pixels of the difference image. In order to take care of the contextual information, the input patterns are generated considering each pixel of the difference image along with its neighbors. Initially the network is trained using a small set of labeled data. We suggest a technique to initially identify some labeled patterns automatically. The unlabeled patterns are iteratively processed by the MLP to obtain a soft class label for each of them.

2 Proposed Change Detection Technique

In this section we propose a context-sensitive semi-supervised technique that automatically discriminates the changed and unchanged pixels of the difference image. Few labeled patterns are identified by applying a suggested heuristic technique without using ground truth information. In the following subsections we will describe the steps involved in the proposed change detection technique.

2.1 Generation of Input Patterns

To generate the input patterns, we first produce the difference image by considering the multitemporal images in which the difference between the two considered acquisitions are highlighted. The most popular Change Vector Analysis (CVA) technique is used here to generate the difference image. Let us consider two co-registered and radiometrically corrected γ-spectral band images X_1 and X_2, of size $p \times q$, acquired over the same geographical area at two times T_1 and T_2, and let $D = \{l_{mn}, 1 \leq m \leq p, 1 \leq n \leq q\}$ be the difference image obtained by applying the CVA technique to X_1 and X_2. Then

$$l_{mn} = (int)\sqrt{\sum_{\alpha=1}^{\gamma}\left(l_{mn}^{\alpha(X_1)} - l_{mn}^{\alpha(X_2)}\right)^2}.$$

Here $l_{mn}^{\alpha(X_1)}$ and $l_{mn}^{\alpha(X_2)}$ are the gray values of the pixels at the spatial position (m, n) in α^{th} band of images X_1 and X_2, respectively.

After producing the difference image, the input patterns are generated corresponding to each pixel in the difference image D, considering its spatial neighborhood of order d. In the present case 2nd order neighborhood ($d = 2$) is considered, and the input vectors contain nine components considering the gray value of the pixel and the gray values of its eight neighboring pixels. So the pattern set $U = \{u(1), u(2), ..., u(N)\}$ contains N ($N = p \times q$) pattern vectors in nine-dimension feature space. The patterns generated by the above technique help us to produce better change detection map as they take some spatial contextual information from the difference image.

2.2   Description of the MLP

The multilayer perceptron [12] has one input, one output and one or more hidden layers. Neurons/nodes in one layer of the network are connected to all the neurons in the next layer. Let S be the number of layers in the network and $y_j^r(n)$ denote the output signal of the jth neuron in the rth layer for an input pattern $\boldsymbol{u}(n)$, where $n = 1, 2, ..., N$ and w_{ij}^r be the connection strength between the ith neuron in the $(r-1)^{th}$ layer and j^{th} neuron in the rth layer. For an input pattern vector $\boldsymbol{u}(n)$, the output value of neuron j in the input layer is defined as $y_j^0(n) = u_j(n)$, which is sent to the first hidden layer as an input signal. A neuron j in the rth ($r \geq 1$) layer takes input signal $v_j^r(n) = \sum_i y_i^{r-1}(n).w_{ij}^r + w_{0j}^r$, where w_{0j}^r is the connection strength between a fixed unit (bias) to neuron j, and produces an output $y_j^r(n) = f(v_j^r(n))$. The activation function $f(.)$ mapped the output sigmoidally between 0 to 1. The network is trained using backpropagation algorithm [12] that iteratively adjusts coupling strengths (weights) in the network to minimize the sum-square error $\sum_{n=1}^N \sum_{j=1}^C (y_j^{S-1}(n) - t_j(n))$, where $y_j^{S-1}(n)$ and $t_j(n)$ are the predicted and desired value of the output layer neuron j for input pattern $\boldsymbol{u}(n)$, respectively. As the generated patterns have nine features and belong to either changed class or unchanged class, the architecture of the MLP used here has nine neurons in the input layer and two neurons in the output layer.

2.3   Labeled (Training) Pattern Generation

MLP needs labeled patterns for learning. In this section we suggest a technique to automatically identify some patterns which either belong to changed or unchanged class without using ground truth information.

As component values of the generated pattern vectors contain gray values of pixels in the difference image, the patterns whose component values are very low belong to unchanged class and the patterns whose component values are very high belong to changed class. To identify these patterns automatically, K-means (K=2) clustering algorithm [13] is used. Let lc and uc be the two centroid obtained by K-means algorithm in nine-dimensional feature space. We also considered two other points lb $(0, ..., 0)$, the possible minimum component values of the patterns near to lc and ub $(255, ..., 255)$, the possible maximum component values of the patterns which is near to uc in the same feature space (see Fig. 1). A pattern is assigned to the unchanged class if it is inside the hypersphere whose center is at lb and radius is the distance between lb and lc; it is assigned to the changed class if it is inside the hypersphere whose center is at ub and radius is the distance between ub and uc else it is considered as unlabeled. The pattern set U is represented as $U = \{(\boldsymbol{u}(n), \boldsymbol{t}(n)), n = 1, 2, ..., N\}$, where $\boldsymbol{u}(n)$ is the nth input pattern vector and $\boldsymbol{t}(n)$ is the target vector of the corresponding input pattern. The target vector $\boldsymbol{t}(n)$ where $\boldsymbol{t}(n) = \{[t_1(n), t_2(n)] \mid t_i(n) \in (0, 1), \forall n\}$ represents the changed class when $\boldsymbol{t}(n) = [1, 0]$, unchanged class when $\boldsymbol{t}(n) = [0, 1]$ and unlabeled pattern when $\boldsymbol{t}(n) = [0, 0]$.

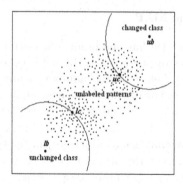

Fig. 1. Labeled and unlabeled patterns in two-dimensional feature space

2.4   Labeling Unknown Patterns

We train the network first using the labeled patterns that are automatically identified. The decision classes are considered as fuzzy sets [14] and we assume that the network's output values provide degree of membership to the fuzzy sets. Let us suppose the nth input pattern is presented to the network. The membership value $\mu_j(\boldsymbol{u}(n))$ of the nth input pattern $\boldsymbol{u}(n)$ to the jth fuzzy set is then given by the output $y_j^2(\boldsymbol{u}(n))$ of the jth $(j = 1, 2)$ output neuron. Contrast within the set of membership values $\mu_j(\boldsymbol{u}(n))$ is then increased [15] as follows:

$$\mu_j(\boldsymbol{u}(n)) = \begin{cases} 2[\mu_j(\boldsymbol{u}(n))]^2, & 0 \leq \mu_j(\boldsymbol{u}(n)) \leq 0.5 \\ 1 - 2[1 - \mu_j(\boldsymbol{u}(n))]^2, & 0.5 < \mu_j(\boldsymbol{u}(n)) \leq 1.0 . \end{cases} \tag{1}$$

Now we find out the k_{nn} nearest neighbors for each unlabeled pattern. Finding out the k_{nn} nearest neighbors considering all patterns is a time consuming task. To reduce this time complexity, instead of considering all the patterns we found out k_{nn} nearest neighbors for an unlabeled pattern corresponding to a pixel of the difference image by considering only the patterns generated by its surrounding pixels. Let M^n be the set of indices of the k_{nn} nearest neighbors of the unlabeled pattern $\boldsymbol{u}(n)$. Then the target vector $\boldsymbol{t}(n)$ for the unlabeled pattern $\boldsymbol{u}(n)$ is computed by

$$\boldsymbol{t}(n) = \left[\frac{\sum_{i \in M^n} t_1(i)}{k_{nn}}, \frac{\sum_{i \in M^n} t_2(i)}{k_{nn}} \right] \tag{2}$$

where for unlabeled pattern $t_j(n) = \mu_j(\boldsymbol{u}(n))$ and for labeled pattern $t_j(i) = 0 \ or \ 1$, $j = 1, 2$.

2.5   Learning Algorithm

The learning technique used here is inspired by the principle used in [15] and is given bellow.

Learning algorithm of the network

Step 1: Train the network using labeled patterns only.

Step 2: Assign soft class labels to each unlabeled or softly labeled pattern by passing it through the trained network.

Step 3: Re-estimate these soft class labels using Eq. (1) and Eq. (2).

Step 4: Train the network using both the labeled and the softly labeled patterns.

Step 5: If the sum of square error obtained from the network becomes constant or the number of iterations exceeds some given number then goto Step 6; else goto Step 2.

Step 6: Stop.

3   Description of the Data Sets

In our experiments we used two data sets representing areas of Mexico and Sardinia Island of Italy. For Mexico data multispectral images (of size 512×512) acquired by the Landsat Enhanced Thematic Mapper Plus (ETM+) sensor of the Landsat-7 satellite, in an area of Mexico on April 2000 and May 2002 were considered. Between the two acquisition dates, fire destroyed a large portion of the vegetation in the considered region. As band 4 is more effective to locate the burned area, we generated the difference image by considering only spectral band 4. Sardinia data is related to Mulargia lake of Sardinia Island. It consists of Landsat-5 TM images (of size 412×300) acquired on September 1995 and July 1996, respectively. Between the two acquisition dates the water level in the lake is changed. We applied the CVA technique [1] on spectral bands 1, 2, 4, and 5 of the two multispectral images, as preliminary experiments had demonstrated that the above set of channels contains useful information of the change in water level. The available ground truth concerning the location of the changes was used to prepare the "reference map" (Fig. 2) which was useful to assess change-detection errors.

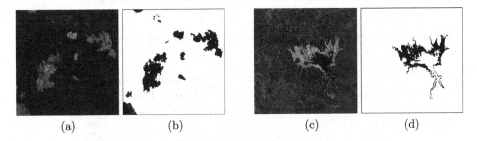

(a)   (b)   (c)   (d)

Fig. 2. Mexico data: (a) Difference image and (b) Corresponding reference map; Sardinia data: (c) Difference image and (d) Corresponding reference map

4   Experimental Results

In order to establish the effectiveness of the proposed technique, the present experiment compares the change-detection result provided by the proposed method with a context-insensitive Manual Trial and Error Thresholding (MTET) technique, the K-means clustering [13] technique and a context sensitive technique presented in [3] based on the combined use of the EM algorithm and Markov Random Fields (MRF) (we refer to it as EM+MRF technique). The MTET technique generates a minimum error change-detection map under the hypothesis of spatial independence among pixels by finding a minimum error decision threshold for the difference image. The minimum error decision threshold is obtained by computing change-detection errors (with the help of the reference map) for all values of the decision threshold. K-means clustering algorithm is applied on the generated patterns (as described in Sec. 2.1) with $K = 2$. Comparisons were carried out in terms of both overall change-detection error and number of false alarms (i.e., unchanged pixels identified as changed ones) and missed alarms (i.e., changed pixels categorized as unchanged ones).

In the present experiment the architecture of MLP is $9 : 8 : 2$ i.e., the network has 9 input neurons, 8 hidden neurons in a single hidden layer and 2 output neurons. To find out k_{nn} nearest neighbors for each input pattern we have taken 50×50 window and the value of k_{nn} is taken as 8.

4.1   Analysis of Results

Table 1 shows that the overall error produced (using both Mexico and Sardinia data) by the proposed technique is much smaller than that produced by the MTET technique. Figs. 3 and 4 depict the change-detection maps. A visual comparison points out that the proposed approach generates a more smooth change-detection map compared to the MTET procedure. From the tables one can also see that the proposed MLP based technique generates better change-detection results than the result produced by the K-means technique. The best result obtained by existing EM+MRF technique with a specific value of β of MRF is also close to the result obtained by the proposed technique.

Table 1. Overall error, missed alarms and false alarms resulting by MTET, K-means, EM+MRF, and the proposed technique

Techniques	Mexico data			Sardinia Island data		
	Missed alarms	False alarms	Overall error	Missed alarms	False alarms	Overall error
MTET	2404	2187	4591	1015	875	1890
K-mean	3108	665	3773	637	1881	2518
EM+MRF	946	2257	3203	592	1108	1700
Proposed	2602	703	3305	1294	303	1597

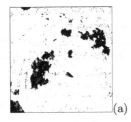 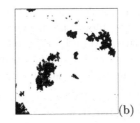

Fig. 3. Change-detection maps obtained for the data set related to the Mexico area using (a) MTET technique, (b) proposed technique

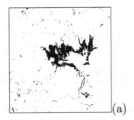 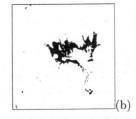

Fig. 4. Change-detection maps obtained for the data set related to the Sardinia area using (a) MTET technique, (b) proposed technique

5   Discussion and Conclusions

In this article, a semi-supervised and automatic context-sensitive technique for change detection in multitemporal images is proposed. The technique discriminates the changed and unchanged pixels in the difference image by using a multilayer perceptron. The number of neurons in the input layer is equal to the dimension of the input patterns and the number of neurons in the output layer is two. The input patterns are generated considering each pixel in the difference image along with its neighboring pixels, in order to take into account the spatial contextual information. On the basis of the characteristics of these input patterns, a heuristic technique is suggested to automatically identify a few input patterns that have very high probability to belong either to changed or to unchanged class. Depending on these labeled patterns and assuming that the less dense region can act as separator between the classes, a semi-supervised learning algorithm based on MLP is used to assign the soft class label for each unlabeled pattern.

The presented technique shows the following advantages: (i) it is distribution free, i.e., like EM+MRF model presented in [3] it does not require any explicit assumption on the statistical model of the distributions of classes of changed and unchanged pixels, (ii) it does not require human efforts to identify the labeled patterns, i.e., the labeled patterns are identified heuristically without using any ground truth information. Like other semi-supervised cases, the time requirement of this technique is little more. It is worth noting that for the considered kind of application it is not fundamental to produce results in real time.

Experimental results confirm the effectiveness of the proposed approach. The presented technique significantly outperforms the standard optimal-manual context-insensitive MTET technique and K-means technique. The proposed technique provides comparable overall change detection error to the best result achieved with the context-sensitive EM+MRF technique.

Acknowledgements

Authors like to thank the Department of Science and Technology, India and University of Trento, Italy, the sponsors of the ITPAR program and Prof. L. Bruzzone, the Italian collaborator of this project, for providing the data.

References

1. Singh, A.: Digital change detection techniques using remotely sensed data. Int. J. Remote Sensing 10, 989–1003 (1989)
2. Liu, X., Lathrop, R.G.: Urban change detection based on an artificial neural network. Int. J. Remote Sensing 23, 2513–2518 (2002)
3. Bruzzone, L., Prieto, D.F.: Automatic analysis of the difference image for unsupervised change detection. IEEE Trans. Geosci. Remote Sensing 38, 1171–1182 (2000)
4. Hame, T., Heiler, I., Miguel-Ayanz, J.S.: An unsupervised change detection and recognition system for forestry. Int. J. Remote Sensing 19, 1079–1099 (1998)
5. Chavez, P.S., MacKinnon, D.J.: Automatic detection of vegetation changes in the southwestern United States using remotely sensed images. Photogramm. Eng. Remote Sensing 60, 1285–1294 (1994)
6. Radke, R.J., Andra, S., Al-Kofahi, O., Roysam, B.: Image change detection algorithms: A systematic survey. IEEE Trans. Image Processing 14, 294–307 (2005)
7. Canty, M.J.: Image Analysis, Classification and Change Detection in Remote Sensing. CRC Press, Taylor & Francis (2006)
8. Kasetkasem, T., Varshney, P.K.: An image change-detection algorithm based on Markov random field models. IEEE Trans. Geosci. Remote Sensing 40, 1815–1823 (2002)
9. Ghosh, S., Bruzzone, L., Patra, S., Bovolo, F., Ghosh, A.: A context-sensitive technique for unsupervised change detection based on Hopfield-type neural networks. IEEE Trans. Geosci. Remote Sensing 45, 778–789 (2007)
10. Patra, S., Ghosh, S., Ghosh, A.: Unsupervised change detection in remote-sensing images using modified self-organizing feature map neural network. In: Int. Conf. on Computing: Theory and Applications (ICCTA-2007), pp. 716–720. IEEE Computer Society Press, Los Alamitos (2007)
11. Seeger, M.: Learning with labeled and unlabeled data. Technical report, University of Edinburgh (2001)
12. Haykin, S.: Neural Networks: A Comprehensive Foundation. Pearson Education, Fourth Indian Reprint (2003)
13. Duda, R.O., Hart, P.E., Stork, D.G.: Pattern Classification, 2nd edn. Wiley, Singapore (2001)
14. Klir, G.J., Yuan, B.: Fuzzy sets and Fuzzy Logic - Theory and Applications. Prentice Hall, New York (1995)
15. Verikas, A., Gelzinis, A., Malmqvist, K.: Using unlabelled data to train a multilayer perceptron. Neural Processing Letters 14, 179–201 (2001)

Automatic Guidance of a Tractor Using Computer Vision

Pedro Moreno Matías and Jaime Gómez Gil

University of Valladolid, Departament of Teory Signal, Comunications and Telematic
Engineering, de Teoría de la Señal, Comunicaciones e Ingeniería Telemática,
Campus Miguel Delibes, 47011, Valladolid,
Fax: 983423667
jgomez@tel.uva.es

Abstract. This paper presents a computer vision guidance system for
agricultural vehicles. This system is based on a segmentation algorithm that
uses an optimum threshold function in terms of minimum quadratic value over
a discriminant based on the Fisher lineal discriminant. This system has achieved
not only very interesting results in the sense of segmentation, but it has also
guided successfully a vehicle in a real world environment.

1 Introduction

This paper describes an automatic agricultural computer vision guided system. Precision
Agriculture is defined as the application of technical advances to agriculture; this field
has grown rapidly in recent years because it offers a cheaper and more effective way of
working the land. When using precision techniques it is more appropriate to use a local
rather than a global reference due to benefits as position independence, a priori terrain
knowledge independence and tolerance independence [1].

We have chosen a concrete local reference technique that has been rapidly
developed for agricultural purpose: Computer Vision. The first tentative trials were
direct applications of the traditional computer vision techniques like FFT convolution
filters, basic border detection or region division techniques [1], [2]. The Hayashi and
Fujii system based on border detection and Hough transform [3] and the system based
on texture and intensity from the group of Chateau [4] could be some examples.
Lately, some investigators like Reid and Searcy [5], Gerrish [6] and Klassen [7] have
developed quite effective systems which were never actually tested guiding a vehicle.
In recent years, some systems have achieved great results when guiding a vehicle in a
very specific task, like the Ollis and Stentz algorithm [8] for harvesting or the
Bulanon, Kataoka, Ota and Hiroma [9] guidance system to collect apples. However,
there is still no effective solution to guide an agricultural vehicle with independence
of the task.

2 Objectives

As mentioned above, the main goal for this project is to develop the most general-
purpose solution possible. Thus its success must be measured in terms of the number

A. Ghosh, R.K. De, and S.K. Pal (Eds.): PReMI 2007, LNCS 4815, pp. 169–176, 2007.

of agricultural tasks that can be guided by the system. The concrete effectiveness for each concrete task is less important.

The decision of whether a task can be reasonably done or not with the system will depend on the relative distance between the actual division line between areas and the approximation of the algorithm for the segmentation algorithm, and the real distance between rows in the guidance algorithm.

3 Development

As described, the system is based on a software application that decides the direction to be taken by the tractor and a hardware interface that communicates with the hardware guidance system. The most complex and interesting part of the system is the software application that implements the computer vision logic. The software is based on the algorithm of Ollis and Stentz [8], where the Fisher lineal discriminant [10], a mathematic tool, was introduced. This software is completed with a simple guidance algorithm.

3.1 Image Segmentation: The Ollis and Stentz Algorithm

The algorithm of Ollis and Stentz is row by row processing algorithm that is based on the following assumptions: the main idea is that the image is composed by a two well differentiated areas that have a frontier between the areas inside the image for every row. Along with this, the discriminant, the function that translates the 3D matrix that composes the image into a 1D matrix which is used to separate the two areas, is a bimodal function. This means that for any of the two differentiated areas in the image, there are presented values of the returning values of the discriminant around a medium value, m1 and m2, respectively.

Based on these suppositions, the algorithm is defined as a row by row processing algorithm because each border point can be calculated from the value of the discriminant of the points of the row. In Fig. 1, it is represented an example representation of the discriminant function against the position in the row.

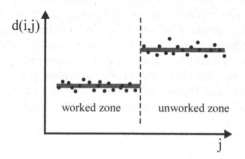

Fig. 1. Threshold example for a discriminant

With a bimodal discriminant function, the success of the algorithm depends on the correct election of the threshold function in term of MSE (Mean square error). Mathematically, the discriminant can be expressed as the following (1).

$$d(i, j) = \alpha \cdot r(i, j) + \beta \cdot g(i, j) + \gamma \cdot b(i, j) \qquad (1)$$

3.2 Discriminant Election: The Fisher Lineal Discriminant

Although the Ollis and Stentz algorithm was originally defined for a static discriminant, it can be variable in time, so it can be autooptimized as the Fisher lineal discriminant, which is the discriminant chosen for the system described in the paper.

Operatively, the Fisher discriminant is an adaptative stocastic tool obtained from the means and the variances of each area of the image. Concretely, if we analyze the discriminant as the geometric projection of the three colour spaces (one for each component RGB) over an hypothetic line in this 3D space, the Fisher lineal discriminant can be analyzed as the projection of the colour space over the line defined by the vector result of the difference of the two means of each zone. Projecting over this line, it minimizes the overlapping as shown in Fig. 2.

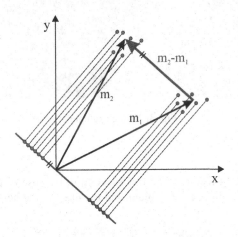

Fig. 2. Graphic example of a Fisher linear discriminant for a 2D colour space

3.3 Guiding Algorithm

We have chosen the simplest solution for the guiding algorithm: it consists on an average of the border points detected by the algorithm (which represents implicitly a low-pass filter that eliminates some noise). Although simple, this solution presents good results when the camera is positioned just over the border line and focusing the furthest possible.

For the hardware interface and initial calibration, a lineal approximation between the minimum, centred and maximum positions of the tractor direction and a parallel port communication were chosen. This lineal approximation revealed to be more efficient and simpler than the Ackermann algorithm.

4 Results

4.1 Tilling

In general, tilling is done in times when there is little vegetation, so this limited growth did not affect the results of the experiment; in addition, the conditions of illumination at this time of year were favourable for our trials. Consequently, the obtained segmentation line had a very small deviation from the real line. Nevertheless, the got results were worse when focusing on a point very near the vehicle because the granularity of the image is bigger and the algorithm can be fooled by clumps of earth and small shadows. Fig. 3 shows some examples.

Fig. 3. Tilling examples

4.2 Harvesting

We expected harvesting results to be less successful, because of the similarity of the two parts of the image in terms of colour and texture. As the height of the harvested

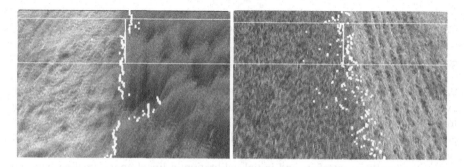

Fig. 4. Harvesting examples

and uncut plants are so different, the results could be affected by the shadows and light. However, the actual results were quite respectable as shown in Fig. 4.

4.3 Ploughing and Direct Sowing

Most of our experiments involved ploughing and direct sowing; therefore, we have a great deal of results. On the one hand, we achieved interesting results even in presumably uninteresting conditions, such as low illumination or low contrast between zones, as shown in Fig. 5.

Fig. 5. Correct sowing examples

On the other hand, poor results were also obtained due to shadows and irregular vegetation, where the system completely failed, as can be seen in Fig. 6.

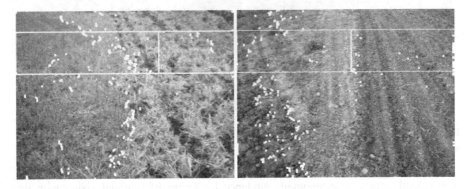

Fig. 6. Incorrect sowing examples

5 Autoguidance System

To get an autonomous guidance, it has been installed in a tractor a laptop where it has been connected a webcam (see Fig. 7).

Fig. 7. Installation of the webcam in the tractor

Adaptations in the tractor have had to be done to control the tractor by the notebook which processes the video. It has been installed a controller box which receives data from the notebook about the tractor direction and moves a DC motor which works thedirection. It has also been installed a strap which joins the motor with the steering wheel, a clutch that tightens a strap to pass from manual to autonomous driving and a potentiometer in the front wheel to know in every moment the steering state (see Fig. 8).

Fig. 8. Left: Adaptations for the movement of the engine. **Right**: Potentiometer in the front wheel.

Subsequently, tests of autonomous guidance by means of artificial vision with the system presented in this paper have been done. The control law represented in Fig. 9 left has been applied. When the vision system detected the treated area and the non-treated area, the guidance was made correctly. This happened when the tractor raised moist soil with a darker colour. Fig. 9 right presents a plot where the tractor guidance has been made without problems by means of artificial vision.

Fig. 9. Left: Graphic representation of the control law. **Right**: The dark area corresponds with the cultivated area by means of artificial vision. The light area is the non-cultivated area.

6 Conclusions

In summary, very positive results have been achieved for every task tested using the system under uniform conditions of illumination and terrain.

However, conditions of non-uniform illumination raise the rate of errors; therefore, the system gets confused by the irregularities shadows of the terrain or external factors.

It can also be observed that the effect of irregular vegetation make the system inefficient when the difference of texture between the processed and non-proccessed areas is not significant, as in tilling. This effect can be reduced with a correct location and focusing of the webcam.

An important result is the good behaviour of the guidance algorithm although it is quite simple. The system has been tested with straight rows but we are sure that with small modifications it would work correctly in curved trajectories.

References

1. Pajares, G., Cruz, D.l.: Visión por computador: Imágenes digitales y aplicaciones. Ed. Ra-Ma (2001)
2. Jahns, G.: Automatic Guidance in Agriculture: A Review. ASAE paper NCR, St. Joseph. MI, 83-404 (1983)
3. Hayashi, M., Fujii, Y.: Automatic Lawn Mower Guidance Using a Vision System. In: Proceedings of the USA-Japan Symposium on Flexible Automation, New York (1988)
4. Chateau, T.: An Original Correlation and Data Fusion Based Approach to Detect a Reap Limit into a Gray Level Image. In: Proceedings of the 1997 International Conference on Intelligent Robots and Systems, pp. 1258–1263 (1997)
5. Reid, J.F, Searcy, S.W.: Detecting Crop Rows Using the Hough Transform. Presented at the 1986 summer meeting of the ASAE, San Luis Obispo (1986)
6. Gerrish, J.: Path-finding by Image Processing in Agricultural Field Operations. Transactions of the Society of Automotive Engineers (1987).

7. Klassen: Agricultural Vehicle Guidance Sensor. Paper 93-1008 Presented at the 1993 International Summer Meeting of the ASAE, Spokane, WA (June 1993)
8. Ollis, M., Stentz, A.: First Results in Crop Line Tracking. In: Proceedings of IEEE Conference on Robotics and Automation (ICRA 1996), pp. 951–956. IEEE Computer Society Press, Los Alamitos (1996)
9. Bulanon, D.M., Kataoka, T., Ota, Y., Hiroma, T.: A Segmentation Algorithm for the Automatic Recognition of Fuji Apples at Harvest. Biosystems Engineering 83, 405–412 (2002)
10. Duda, R., Hart, P.: Pattern Classification and Scene Analysis, pp. 114–118. Wiley & Sons, Chichester (1973)

Multiscale Boundary Identification for Medical Images

ZuYing Wang, Mrinal Mandal, and Zoltan Koles

Department of Electrical and Computer Engineering,
University of Alberta, Edmonton, AB, Canada
{zywang, mandal}@ece.ualberta.ca, z.koles@ualberta.ca

Abstract. Boundary identification in medical images plays a crucial role in helping physicians in patient diagnosis. Manual identification of object boundaries is a time-consuming task and is subject to operator variability. Fully automatic procedures are still far from satisfactory in most real situations. In this paper, we propose a boundary identification method based on multiscale technique. Experimental results have shown that the proposed method provides superior performance in medical image segmentation.

Keywords: Boundary identification, multiscale edge detection.

1   Introduction

Various medical imaging modalities such as the radiograph, computed tomography (CT), and magnetic resonance (MR) imaging are widely used in routine clinical practice. Boundary identification (BI) of deformed tissue plays a crucial role in accurate patient diagnosis. For example, an MR brain image can be segmented into different tissue classes, such as gray matter, white matter, and cerebrospinal fluid. Unfortunately, manual BI methods are very time consuming and are often subjective in nature. Recently an edge-based MR image segmentation method, mtrack [1], has been proposed. This algorithm provides a reasonably good performance. However, the edge-based segmentation algorithm has some limitations, such as sensitivity to noise and presence of gaps in detected boundaries. Together, these effects degrade the quality of the detected boundaries. The discrete wavelet transform (DWT) has recently been shown to be a powerful tool in multiscale edge detection. Marr and Hildreth [2] introduced the concept of multiscale edge detection for detecting the boundaries of objects in an image. Mallat and Zhong [3] showed that multiscale edge detection can be implemented by smoothing a signal with a convolution kernel at various scales, and detecting sharp variation points as edges. Tang et al. [4] studied the characterization of edges with wavelet transform and Lipschitz exponents.

In this paper, we propose a multiscale boundary identification algorithm, based on DWT, for better edge-based segmentation of MR images. The remainder of this paper is organized as follows. Section 2 briefly reviews the related background work. Section 3 presents the proposed method developed based on

A. Ghosh, R.K. De, and S.K. Pal (Eds.): PReMI 2007, LNCS 4815, pp. 177–185, 2007.
© Springer-Verlag Berlin Heidelberg 2007

the existing theory for multiscale edge detection and boundary identification. Section 4 presents the performance evaluation of the proposed method. The conclusions are presented in Section 5.

2   Review of Related Work

In this section, we present a brief review of the related background work.

2.1   mtrack Algorithm

The mtrack method [1] has been developed at the University of Alberta for medical image segmentation, especially for MR images of the head. As shown in Fig. 1, it includes three major modules: edge detection, edge feature extraction, and edge tracing. The three modules are explained below.

Fig. 1. Block diagram of mtrack algorithm

1. *Edge detection*: In this module, the edge of an image is obtained using a Canny edge detector. The detector is applied for six operator directions. Subpixel edge resolution is obtained by linearly interpolating the zero crossings in the second derivative of a Gaussian filtered image.
2. *Edge Feature Extraction*: Edge features are extracted from the detected edges. For each edge point along the operator direction, the top and bottom intensity points of that edge slope are obtained by examining the 1st and 2nd derivatives curves. The coordinates of the subpixel edge points and the top and bottom intensity on each side of the edge points are combined as four-dimensional edge information.
3. *Edge Tracing*: A target tracking algorithm is used to link the edge points to form an object boundary. The edge information is assumed to be position information of a hypothetical object (target) that moves along the boundary. Target tracking starts from a starting point, follows the path of the target (edge) until no further edge point can be founded on the track or the starting point is revisited. Thus, the boundary can be drawn by linking all the edge points that are found in the path of the target, including the starting point.

The mtrack software can perform automatic tracking and can also manually cut and link tracks to form the desired contours. However, some problems remain: First, the tracking performance depends on the starting point. Secondly, it depends on the track direction. The tracking algorithm also involves the possibility of self-intersection, the selection of appropriate parameter values, and the possibility that a closed contour is not formed in all cases. The main limitation of this algorithm is that it is sensitive to noise.

2.2 Multiscale Edge Detection Technique

In this paper, we use the multiscale edge detection technique proposed in [3]. We present a brief review of Mallat's edge detection technique in this section. The schematic of multiscale edge detection is shown in Fig. 2. Consider a wavelet function $\Psi_s(x, y)$ at scale s. The 2-D wavelet-decomposed image $f(x, y)$ at scale s has two components: $W_s^1(x, y) = f * \Psi_s^1(x, y)$ and $W_s^2(x, y) = f * \Psi_s^2(x, y)$ where $\Psi_s^1(x, y)$ and $\Psi_s^2(x, y)$ are, respectively, horizontal and vertical filters. At each scale s, the modulus of the gradient vector is called the *wavelet transform modulus*, and is defined as

$$M_s(x, y) = \sqrt{|W_s^1(x, y)|^2 + |W_s^2(x, y)|^2}. \tag{1}$$

The angle of the wavelet gradient vector for each scale is given by

$$A_s(x, y) = \arctan(W_s^2(x, y)/W_s^1(x, y)). \tag{2}$$

Fig. 2. Block diagram of multiscale edge detection [3]

For a 2-D image, the wavelet transform modulus maxima (WTMM) are located at points where the wavelet transform modulus is larger than its two neighbors in one-dimensional neighborhoods along the angle direction. The local regularity of a signal is characterized by the decay of the wavelet transform across scales [3], and it can be measured in terms of Lipschitz exponent. For a signal $f(x)$, the Lipschitz exponent α satisfies the following:

$$|W_s(x)| \le As^\alpha \tag{3}$$

where $W_s(x)$ are wavelet coefficients at scale s, and A is a positive constant. Eq. (3) states that the modulus of the wavelet coefficients of a signal vary with the scale s according to the Lipschitz regularity of that signal. The characterization of edges by Lipschitz exponent and wavelet transform has been investigated by Mallat et al. [3]. It has been shown that positive Lipschitz exponent α corresponds to edge structure (i.e. step edge) and negative Lipschitz exponent α corresponds to Dirac structure (i.e. spike noise).

3   Proposed Method

In this section, we present the proposed multiscale BI method. The proposed method is developed based on the classical wavelet theory for multiscale edge detection and BI. It has been evaluated in the mtrack framework to see if the proposed method is useful for edge-based BI. The schematic of the integrated method is shown in Fig. 3. A comparison between Fig. 3 and Fig. 1 reveals that the proposed method replaces Canny edge detection with multiscale edge detection, and adds multiscale boundary identification before edge tracing. The two new modules in Fig. 3 are explained below.

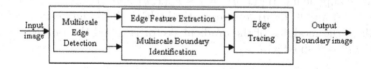

Fig. 3. Block diagram of the proposed boundary identification algorithm

3.1   Multiscale Edge Detection

The multiscale edge detection algorithm proposed by Mallat et al. [3] was reviewed in section 2.2. In this paper, we use this algorithm to calculate multiscale edges. The schematic of the proposed multiscale edge detection is shown in Fig. 4. A comparison between Fig. 4 and Fig. 2 reveals that the proposed multiscale edge detection adds two new modules to improve the edge detection performance after traditional edge detection technique. These new modules are explained in the following discussion.

Fig. 4. Block diagram of proposed multiscale edge detection

We can determine the WTMM for each scale of a MR image. The WTMM points at each decomposition level are shown in Fig. 5 as white pixels in the binary image. It is shown in Fig. 5 that when the scale increases, the details and the noise effect decrease quickly. As a result, the edge locations may change in higher scales. We use the WTMM of the first scale s_1 (i.e., the highest resolution) as the edge image. By looking at the WTMM at different scales, we find that edges of higher significance (i.e., stronger edges) are more likely to be kept by the wavelet transform across scales. Edges of lower significance (i.e., weaker edges) are more likely to disappear at higher scales.

In theory, WTMM can be found along the gradient angle. But practically, we found that only using one-dimensional neighborhood along the angle direction to decide an edge pixel is not sufficient, because a pixel can be part of one edge in one angle direction and be part of another edge in another angle direction. Therefore, we check all the eight pixels around one pixel in the centre for four different directions to decide whether the centre pixel is a WTMM in

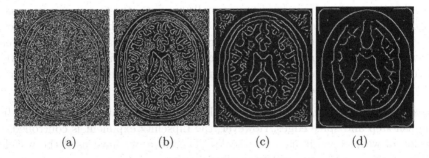

(a)   (b)   (c)   (d)

Fig. 5. WTMM at four scales: a) s_1; b) s_2; c) s_3; d) s_4

each direction. In order to achieve more accurate edge pixel location, subpixel edge coordinates need to be obtained. We can interpolate the edge pixel that we found in the WTMM process, and fit the pixel coordinates in a polynomial curve to get subpixel resolution. If the pixel is the local maximum, then we record it as an edge pixel in that direction.

3.2   Multiscale Boundary Identification

The schematic of the multiscale boundary identification is shown in Fig. 6. We present each module in the following sections.

Fig. 6. Block diagram of multiscale boundary identification

2-D WTMM Chain: After we obtain the WTMM for each scale, we need to chain those maxima across scales to indicate where the true edge is. If it is the true edge, it should not disappear during the wavelet decay. We start the WTMM chaining process from the highest scale down to the lowest scale. We need to link the maxima points between two adjacent scales and record the WTMM chain.

We do a four-level decomposition and have four scales with WTMM found for each scale. We start at the highest scale (i.e., scale s_4), and for each maxima pixel, we search in the previous scale (i.e., scale s_3) within a small region. Then a qualification test is performed to check if there is one maxima pixel among all these pixels in the window that has strongest connection to the considered pixel. We link these two maxima pixels together. It is possible that we may find two pixels in scale j that could connect to the same maxima pixel in scale $j - 1$. However, we would like to have a one-to-one match to the lower scale maxima pixel for this chaining process. In order to prevent this competing situation, an optimization method [5] is used to optimally perform one-to-one maxima pixel matching. A WTMM chaining record is maintained, in which a higher score is assigned to a stronger edge pixel that survives along the wavelet decay, whereas a lower score is assigned to a weaker edge pixel. We define our 2-D WTMM chaining record by 'Scale Depth', as it demonstrates how deep the WTMM pixel is connected through scales.

Lipschitz Regularity Analysis: The Lipschitz exponent can be used to distinguish step and spike edges. However, the Lipschitz exponent is controlled by the log of wavelet modulus across scales. Therefore, we have used the wavelet modulus across scales to identify the boundaries.

Edge Quality Analysis: In order to increase the robustness of the detected edges, we carry out an edge quality test. Here, the amplitude and slope information of the WTMM chain through scales are used to give the strength of each point. A fuzzy logic system is used to determine if a detected edge is a strong edge or a weak edge.

4   Experimental Setup and Results

The performance of the proposed technique has been evaluated using the MR images obtained from Montreal Neurological Institute (MNI) [6]. The data is obtained as 1mm slice thickness, with pixels per slice, and 8 bits per pixel resolution. The slices are oriented in the transverse plane. There are total 181 slices. We have used the MNI data with 1%, 3%, 5%, 7% and 9% noise, and 0%, 20% non-uniformity to evaluate performance. We use MNI slice 95 with 3% of noise and 20% of non-uniformity to show an example of evaluation. The comparisons of MR image edge tracing results are presented to demonstrate that the multiscale boundary identification method improves the edge tracing performance.

4.1   Evaluation Criteria

To evaluate the performance, we compare the detected boundary with the true boundary. Let (m, n) represent coordinate of a point, and $b(m, n)$ is the boundary image, i.e., $b(m, n) = 1$ if (m, n) is a true boundary point. Let the output tracked boundary image be denoted by $t(m, n)$. $t(m, n) = 1$ represents the detected boundary points. We define three criteria as follows:

(1) The *total number of track points* N_{TP} on the detected boundary image is obtained as: $N_{TP} = \sum_{m,n} t(m, n)$.

(2) The *total number of good points* N_{GP} on the detected boundary image is obtained as: $N_{GP} = \sum_{m,n} t(m, n).b(m, n)$.

(3) The *track point ratio* R is defined as: $R = N_{GP}/N_{TP}$.

Ideally, track point ratio R should be 1. However, because the edge tracing algorithm can trace on false boundary points, R is normally less than 1. The higher track point ratio corresponds to the better tracing result. Therefore, we

Table 1. Four evaluation methods. MED: Multiscale Edge Detection; ET: Edge Tracing; MBI: Multiscale Boundary Identification, EFE: Edge Feature Extraction.

Method 1	MED+ET
Method 2	MED+MBI+ET
Method 3	MED+EFE+ET
Method 4	MED+EFE/MBI+ET

use the track point ratio R to represent the performance of the output boundary image obtained. For each input image, four different testing methods for the edge tracing algorithm are used for the evaluation comparison. Table 1 shows the modules of the four methods. We have tested all four methods to evaluate the degree of improvement obtained using the multiscale boundary identification method. Note that Method 4 is the proposed BI method shown in Fig. 3.

4.2   Performance

Fig. 7 shows a gray/white boundary tracking result by four methods. The four methods use the same edge image obtained from multiscale edge detection, and the starting point chosen for edge tracing is in the same location of the edge image. It is observed that Methods 3 and 4 provide much better performance compared to Methods 1 and 2. Between Fig. 7 (c) and (d), we observe that the small boundaries tracked in the left-bottom quarter in Fig. 7 (d) are more accurate than in Fig. 7 (c). Therefore, Method 4 is better than Method 3.

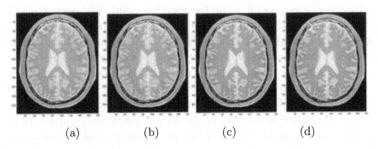

(a)   (b)   (c)   (d)

Fig. 7. Gray/white boundary tracks for test MR image. (a) Method 1; (b) Method 2; (c) Method 3; (d) Method 4.

Table 2 lists the statistical results of 30 tracks of test image for each method. The mean (μ), median (η), and standard deviation (σ) for N_{GP} , N_{TP} , and R are calculated. It is again observed that Method 4 provides the best edge tracing result than others. Also note that the standard deviation in the last row is the lowest among others (σ_R of Method 4 is five times less than Method 3). This means that the edge tracing using Method 4 is less dependent of the choice of the starting point. The performance evaluation is carried out with 1%, 3%, 5%, 7%

Table 2. Statistical results of 30 tracks

Method	N_{GP}			N_{TP}			R		
	η_{GP}	μ_{GP}	σ_{GP}	η_{TP}	μ_{TP}	σ_{TP}	η_R	μ_R	σ_R
1	201	209.4	132.8	2032.5	1813.2	821.4	0.126	0.155	0.141
2	158.5	137.9	68.6	590	533.2	217.7	0.255	0.283	0.127
3	1024	1007.8	449.1	1623	1352.4	580.4	0.745	0.704	0.103
4	1234.5	1144.6	249.3	1609.5	1500.4	308.9	0.765	0.759	0.025

Table 3. R of MNI data slice 95 with 20% non-uniformity

Noise Level	Method3			**Method4**		
	η_R	μ_R	σ_R	η_R	μ_R	σ_R
1%	0.836	0.823	0.029	**0.817**	**0.819**	**0.011**
3%	0.745	0.704	0.103	**0.765**	**0.759**	**0.025**
5%	0.356	0.360	0.144	**0.581**	**0.573**	**0.049**
7%	0.255	0.260	0.117	**0.430**	**0.419**	**0.047**
9%	0.238	0.224	0.095	**0.305**	**0.303**	**0.068**

and 9% of noise levels, and 0%, 20% non-uniformity levels in the synthetic MNI data. Table 3 lists the track point ratio of Method 3 and Method 4 on various noise levels (1%, 3%, 5%, 7% and 9%) with 20% non-uniformity. It is observed that as the noise level increases, the track point ratio of both methods decreases. At each noise level, Method 4 performs better than Method 3, as it has higher μ_R and η_R values and less σ_R value, which means the edge tracing by Method 4 is more accurate and less dependent on the starting point. The same trend has also been seen for 0% non-uniformity. As seen in the experiments, the edge tracing using Method 4 performs better than Method 3. This means that the multiscale boundary identification added into the edge tracing algorithm more accurately tracks the gray/white boundary in the synthetic MR images. Using edge quality from multiscale boundary identification, makes the edge tracing less sensitive to noise, reduces tracing of false edges, and most importantly, the edge tracing is less dependent on the choice of starting point. More simulation results with different data sets (i.e. real MR images) provide a similar trend in the performance [7].

5   Conclusion

In this paper, we propose a robust multiscale boundary identification technique for medical images. We enhance the traditional wavelet-based edge detection technique by obtaining directional edges with subpixel resolution. For multiscale boundary identification, we obtain the scale depth of the edge through WTMM chain. We further separate the true edge from noise by Lipschitz analysis, and we obtain the quality for each edge point by fuzzy logic. Experiments results show that the boundary identification preformed by multiscale edge quality analysis improves the accuracy of the edge tracing significantly, and the tracking is less dependent on the choice of starting point.

References

1. Withey, D.: Dynamic edge tracing: recursive methods for medical image segmentation, Ph.D. Thesis, University of Alberta, Canada (2006)
2. Marr, D., Hildreth, E.: Theory of edge detection. Proc. of the Royal Society of London 207, 187–217 (1980)
3. Mallat, S., Zhong, S.: Characterization of signals from multiscale edges. IEEE Trans. on PAMI 14(7), 710–732 (1992)
4. Tang, Y.Y., Yang, L., Feng, L.: Characterization and detection of edges by Lipschitz exponents and MASW wavelet transform. Proc. Intl. Conf. on Pattern Recognition 2, 1572–1574 (1998)
5. Bertsekas, D.P., Castanon, D.A.: A forward/reverse auction algorithm for asymmetric assignment problems. Computational Optimization and Applications 1, 277–297 (1992)
6. MNI data set (August 2005), http://www.bic.mni.mcgill.ca/brainweb/
7. Wang, Z.Y.: Multiscale Boundary Identification for Medical Images. M.Sc. Thesis, University of Alberta, Canada (2007)

FEM 2D Analysis of Mild Traumatic Brain Injury on a Child

Ernesto Ponce and Daniel Ponce

Escuela Universitaria de Ingeniería Mecánica,
Universidad de Tarapacá, Arica, Chile
eponce@uta.cl

Abstract. Traumatic brain injury is one of the most frequent causes of disability amongst children and adolescents. There are cognitive and neurological effects caused by repetitive head injuries. Learning deficiency is likely to be the result of early head injurie. This may impact the ability to control emotions and exhibit inappropriate behaviour. These children have trouble responding to subtle social cues and planning difficult tasks. Concussions can occur whenever there is a collision of the head against a hard object. The aim of this investigation is the modelling, by means of the two-dimensional Finite Element Method, of brain stress in children caused by head injuries. Three impact cases were analyzed: a concentrate left brain side blow, a diffused blow and a frontal head collision. The brain damage is determined by comparing the last resistance of the arteriole and neurone. The mathematical models can be used for protective design and demonstrating the brain damage.

Keywords: brain injury; computational neuroscience; bio-informatics; finite elements.

1 Introduction

A child's brain is extremely delicate as it is still developing and does not have the strength of an adult's. In young children the neck muscles are not fully developed, and therefore the head has little protection when exposed to sudden or violent swings. Moreover, as the skull is still developing, it provides little defence due to the low bone density. The head of a child is proportionately larger than his body, making accidents more likely to happen. The features of children's biological material are different to those of adults, as their tissues contain more water and are less resistant to stress. Children's skulls have a density of around 1,1 g/cm3, while adults reach 1,3 g/cm3, therefore the former skull mechanical resistance is less. Within a juvenile's skull there is a very delicate electrochemical mechanical balance that is evolving as the brain and its sensitivity grow. Physical impacts are likely to damage the neurones as they cut into the myelin causing its internal sodium ions to slowly escape and to be replaced by calcium ones.

In flexible isotropic materials, the magnitude of fault tensions caused by rupture is half of those of longitudinal tensions, making the brain tissue particularly vulnerable. In the long term, these fault tensions will cause the neurones

A. Ghosh, R.K. De, and S.K. Pal (Eds.): PReMI 2007, LNCS 4815, pp. 186–191, 2007.

to die. One potential cause of mild traumatic brain injury is the heading of soccer balls, head injury in a bike crash and playing others sports. Additional damage can occur due to communication interruption between the frontal lobe, and the amygdala, which resides in the centre of the brain controlling emotions. This type of damage may in the long term bring about aggressive behaviour. Research shows that physically abused children may become antisocial adults, Raine, [7] and The Franklin Institute on Line, [9]. There is little analysis, by means of finite elements, in relation to brain damage caused by light impact. The mild traumatic brain injury on a child is described and classified by Lee, [2] and Münsterberg, [4]. The images are based on the experience of Ourselin: [5], Pitiot: [6], The Virtual Human Brain [10] and Yong: [12].

2 Theory

The system of equations of finite elements shown below are of the elastic linear type, described by Zienkiewicz, [13]:

$$KD = F \tag{1}$$

Where:
K is the rigidity matrix.
D is the vector of nodal displacement.
F is the vector of applied loads.

In this work, four different materials have been used: Skin, skull bone, brain fluid and brain tissue. The first three materials are laid out in layers, the last one being the brain tissue. Every material has different properties and geometry. Each one will have a distinct matrix of rigidity. The equation (1) would therefore end up as:

$$(K_{skin} + K_{skull} + K_{fluid} + K_{brain})D = F \tag{2}$$

When modifications are made for viscoplastic material, such as skin, brain tissue and fluid there is a need to change each matrix of rigidity. Consequently, each of these matrices has to be split into two, one that contains all distortion terms (K_d) and the other that comprises the volumetric deformation (K_v), then:

$$K_{tissue} = (Ev/((1+v)(1-2v)))K_V + (E/(2(1-v)))K_d \tag{3}$$

Where:
E = bulk modulus of the material
v = Poissons ratio

In the theory of plasticity, there is conservation of the body volume under stress, where needs to be set to 0,5. However, when modelling for FEM, to avoid producing an indefinitely large end value in the volumetric effect and to eliminate the distortion effect, the module of Poisson has to be inferior to 0,5. This is an important data element that needs to be introduced in the software.

Horizontal slices at eye level are analysed. The FEM analysis are performed in two dimensions, with a layer thickness of 1 mm. For the purposes of symmetry, saving time and RAM memory, the work was carried out in a semi-piece in most cases, except for the diffuse impact. The biological materials used in the model correspond to a child's.

The Input data are the external charge and Table 1.

Some of the input are described by: Antish [1], Sklar [8], Torres [11] and Heys [3].

Table 1. Input data of the head materials of a child

	Mass Density	Modulus of Elasticity	Poisson's ratio	Shear Modulus
Units	$Ns^2/mm/mm^3$	N/mm^2		N/mm^2
Brain	0.00105	1935	0.4	0.00083
Brain Liquid	0.00102	2280	0.4	0.00002
Blood	0.00104	2250	0.4	0.00002
Bone	0.00112	7680	0.2	2800
Skin	0.00101	2000	0.4	0.002

3   Results

Three impact cases were analyzed:

1. A concentrate 90 kg blow on the left brain side. The results are the stresses and the strain. It is a punch of medium force, done by an adult.
2. A difussed 120 kg blow on the left brain side. The results are the nodal displacement. It is similar to a soccer ball impact, at short distance.
3. A frontal childs head collision at 5 m/s, like a cyclist accident. The results are the strain and nodal displacement.

Fig. 1. FEM brain structure showing different materials: skin, bone of the cranium, brain's fluid and brain tissue. It is a horizontal slice at eye level. For the purposes of symmetry is a semi-piece.

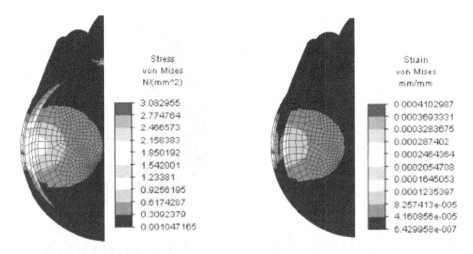

Fig. 2. FEM modeling showing stresses and strain produced by a concentrate 90 kg blow on the left brain side

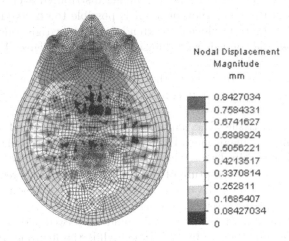

Fig. 3. FEM modeling. Nodal displacement. A head of a child under diffused 120 kg blow on the left brain side.

4 Conclusions

Figure 2 shows the analysis of distribution of tensions in the child's head caused by a non-traumatic impact focused on the bone. The impact is modelled on the left side of the brain. The pale colours indicate a greater concentration of stress just below the point of impact, but there are also some minor stresses towards the centre of the brain. The left hand side of figure 2 corresponds to the same case but in deformities. The prediction is that there would be local damage as well as internal damage to the brain.

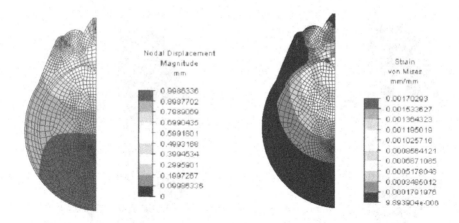

Fig. 4. Right side: FEM strain modeling. A frontal childs head collision at 5 m/s. The frontal lobe shows the biggest strain. Left side: Nodal displacement.

Figure 3 models the effect of a lateral impact distributed over a greater area. In this analysis of the nodal movements it is possible to observe movements of greater diffused magnitude which also extend towards the right side of the brain. This prediction would coincide with diffused traumas experienced by boxers, The Franklin Institute on Line, [9].

The right side of figure 4 shows the effect of an impact at 5 m/s, such as that happening to a cyclist. A flat impact on the child's forehead is modelled. The major deformations appearing on the frontal lobe would have a destructive result for the brain tissue. The damage appears beyond half of the brain, although diminishing in magnitude. The left hand side of figure 4 shows the nodal displacement of the brain. Due to the biggest displacement in the occipital lobe, it is possible damage by a rupture of the blood vessels and nerves of arachnoid net, between the skull and the brain. It is also possible a rebound against the back dome bone, getting a second internal impact.

The stress of rupture of a child's skull would be between 80 and 90 N/mm2, but the arterioles break up at 0,126 N/mm2 while the neurones do it at 0,133 N/mm2. The bone would not fracture with these kind of impact, but the soft tissue would suffer lesions. These mathematical models can be used for designing better protective equipment and for demonstrating the brain damage in children caused by non-traumatic impacts on the skull.

References

1. Antish, P., et al.: Measurements of mechanical properties of bone material in vitro by ultrasound reflection: methodology and comparison with ultrasound transmission. Journal of Bone Miner Res. 3, 417–426 (1991)
2. Lee, I.: Interpretación Clínica de la Escala de Inteligencia de Wechsler. TEA Ediciones S.A., Madrid (1989)

3. Heys, et al.: First-order system least-squares (FOSLS) for modelling blood flow. Medical Engineering and Phisics, Elsevier 28, 495–503 (2006)
4. Müsterberg, E.: Test Guestáltico Visomotor para Niños. Ed. Guadalupe. México (1999)
5. Ourselin, S.: Recalage d'images médicales par appariement de regions- Application á la construction d'atlas histologiques 3D., Tesis doctoral, Université de Nice (2002)
6. Pitiot, A.: Piecewise Affine Registration of Biological Images for Volume Reconstruction. Medical Image Analysis 3(3), 465–483 (2006)
7. Raine, A.: Prefrontal Damage in People with Antisocial Personality Disorder (Paper)(2000)
8. Sklar, F., Elashvili, I.: The pressure-volume function of brain elasticity. Physiological consideration and clinical applications. Journal of Neurosug 47(5), 670–679 (1977)
9. The Franklin Institute on Line, The human brain (2005), http://www.fi.edu.brain/index.htm#
10. The Virtual Human Brain on line. Actual human brain dissection images (2005), http://www.vh.org/adult/provider/anatomy/BrainAnatomy/ BrainAnatomy.html
11. Torres, H., Zamorano, M.: SAR Simulation of Chiral Waves in a Head model. Rev. Facing 1(9), 3–19 (2001)
12. Yong, F., Tianzi, J.: Volumetric Segmentation of Brain Images Using Paralell Genetic Algorithms. IEEE Transactions on Medical Imaging 21(8), 904–909 (2002)
13. Zienkiewicz, O.: El Método de los Elementos Finitos, Reverté, España (1980)

Discovery of Process Models from Data and Domain Knowledge: A Rough-Granular Approach

Andrzej Skowron

Institute of Mathematics,
Warsaw University
Banacha 2, 02-097 Warsaw, Poland
skowron@mimuw.edu.pl

Extended Abstract

The rapid expansion of the Internet has resulted not only in the ever growing amount of data therein stored, but also in the burgeoning complexity of the concepts and phenomena pertaining to those data. This issue has been vividly compared [14] by the renowned statistician, prof. Friedman of Stanford University, to the advances in human mobility from the period of walking afoot to the era of jet travel. These essential changes in data have brought new challenges to the development of new data mining methods, especially that the treatment of these data increasingly involves complex processes that elude classic modeling paradigms. "Hot" datasets like biomedical, financial or netuser behavior data are just a few examples. Mining such temporal or stream data is on the agenda of many research centers and companies worldwide (see, e.g., [31,1]). In the data mining community, there is a rapidly growing interest in developing methods for the discovery of structures of temporal processes from data. Works on discovering models for processes from data have recently been undertaken by many renowned centers worldwide (e.g., [34, 19, 36, 9], www.isle.org/~langley/, soc.web.cse.unsw.edu.au/bibliography/discovery/index.html).

We discuss a research direction for discovery od process models from data and domain knowledge within the program *Wisdom technology* (wistech) outlined recently in [15, 16].

Wisdom commonly means *rightly judging* based on available knowledge and interactions. This common notion can be refined. By *wisdom*, we understand an adaptive ability to make judgments correctly (in particular, correct decisions) to a satisfactory degree, having in mind real-life constraints. The intuitive nature of wisdom understood in this way can be metaphorically expressed by the so-called *wisdom equation* as shown in (1).

$$wisdom = adaptive\ judgment\ +\ knowledge\ +\ interaction. \qquad (1)$$

Wisdom could be treated as a certain type of knowledge. Especially, this type of knowledge is important at the highest level of hierarchy of meta-reasoning in intelligent agents.

A. Ghosh, R.K. De, and S.K. Pal (Eds.): PReMI 2007, LNCS 4815, pp. 192–197, 2007.
© Springer-Verlag Berlin Heidelberg 2007

Wistech is a collection of techniques aimed at the further advancement of technologies to acquire, represent, store, process, discover, communicate, and learn *wisdom* in designing and implementing intelligent systems. These techniques include approximate reasoning by agents or teams of agents about vague concepts concerning real-life, dynamically changing, usually distributed systems in which these agents are operating. Such systems consist of other autonomous agents operating in highly unpredictable environments and interacting with each others. Wistech can be treated as the successor of database technology, information management, and knowledge engineering technologies. Wistech is the combination of the technologies represented in equation (1) and offers an intuitive starting point for a variety of approaches to designing and implementing computational models for wistech in intelligent systems.

Knowledge technology in wistech is based on techniques for reasoning about knowledge, information, and data, techniques that enable to employ the current knowledge in problem solving. This includes, e.g., extracting relevant fragments of knowledge from knowledge networks for making decisions or reasoning by analogy.

Judgment technology in wistech is covering the representation of agent perception and adaptive judgment strategies based on results of perception of real life scenes in environments and their representations in the agent mind. The role of judgment is crucial, e.g., in adaptive planning relative to the Maslov Hierarchy of agents' needs or goals. Judgment also includes techniques used for perception, learning, analysis of perceived facts, and adaptive refinement of approximations of vague complex concepts (from different levels of concept hierarchies in real-life problem solving) applied in modeling interactions in dynamically changing environments (in which cooperating, communicating, and competing agents exist) under uncertain and insufficient knowledge or resources.

Interaction technology includes techniques for performing and monitoring actions by agents and environments. Techniques for planning and controlling actions are derived from a combination of judgment technology and interaction technology.

There are many ways to build foundations for wistech computational models. One of them is based on the *rough-granular computing* (RGC). Rough-granular computing (RGC) is an approach for constructive definition of computations over objects called granules, aiming at searching for solutions of problems which are specified using vague concepts. Granules are obtained in the process called granulation. Granulation can be viewed as a human way of achieving data compression and it plays a key role in implementing the divide-and-conquer strategy in human problem-solving [38]. The approach combines rough set methods with other soft computing methods, and methods based on granular computing (GC). RGC is used for developing one of the possible wistech foundations based on approximate reasoning about vague concepts.

As an opening point to the presentation of methods for discovery of process models from data we use the proposal by Zdzisław Pawlak. He proposed in 1992 [27] to use data tables (information systems) as specifications of concurrent systems. Since then, several methods for synthesis of concurrent systems from data have been developed (see, e.g., [32]).

Recently, it became apparent that rough set methods and information granulation have set out a promising perspective to the development of approximate reasoning methods in multi-agent systems. At the same time, it was shown that there exist significant limitations to prevalent methods of mining emerging very large datasets that involve complex vague concepts, phenomena or processes (see, e.g., [10, 30, 35]). One of the essential weaknesses of those methods is the lack of ability to effectively induce the approximation of complex concepts, the realization of which calls for the discovery of highly elaborated data patterns. Intuitively speaking, these complex target concepts are too far apart from available low-level sensor measurements. This results in huge dimensions of the search space for relevant patterns, which renders existing discovery methods and technologies virtually ineffective. In recent years, there emerged an increasingly popular view (see, e.g., [12, 18]) that one of the main challenges in data mining is to develop methods integrating the pattern and concept discovery with domain knowledge.

In this lecture, the dynamics of complex processes is specified by means of vague concepts, expressed in natural languages, and of relations between those concepts. Approximation of such concepts requires a hierarchical modeling and approximation of concepts on subsequent levels in the hierarchy provided along with domain knowledge. Because of the complexity of the concepts and processes on top levels in the hierarchy, one can not assume that fully automatic construction of their models, or the discovery of data patterns required to approximate their components, would be straightforward. We propose to use in discovery of process models and their components through an interaction with domain experts. This interaction allows steering the discovery process, therefore makes it computationally feasible. Thus, the proposed approach transforms a data mining system into an experimental laboratory, in which the software system, aided by human experts, will attempt to discover: (i) process models from data bounded by domain constraints, (ii) patterns relevant to user, e.g., required in the approximation of vague components of those processes.

This research direction has been pursued by our team, in particular, toward the construction of classifiers for complex concepts (see, e.g., [2–4, 6–8, 11, 20–23]) aided by domain knowledge integration. Advances in recent years indicate a possible expansion of so far conducted research into discovery of models for processes from temporal or spatio-temporal data involving complex objects.

We discuss the rough-granular modeling (see, e.g., [29]) as the basis for discovery of processes from data. We also outline some perspectives of the presented approach for application in areas such as prediction from temporal financial data, gene expression networks, web mining, identification of behavioral patterns, planning, learning interaction (e.g., cooperation protocols or coalition formation), autonomous prediction and control by UAV, summarization of situation, or discovery of language for communication.

The novelty of the proposed approach for the discovery of process models from data and domain knowledge lies in combining, on one side, a number of novel methods of granular computing for wistech developed using the rough set methods and other known approaches to the approximation of vague, complex concepts (see, e.g., [2–8, 17, 20–23, 25, 26, 28, 29, 37, 38]), with, on the other side, the discovery of process' structures from data through an interactive collaboration with domain experts(s) (see, e.g., [2–8, 17, 20–23, 29]).

Acknowledgments

The research has been supported by the grant from Ministry of Scientific Research and Information Technology of the Republic of Poland.

Many thanks to Mr Tuan Trung Nguyen for suggesting many helpful ways to improve this article.

References

1. Aggarwal, C. (ed.): Data Streams: Models and Algorithms. Springer, Berlin (2007)
2. Bazan, J., Peters, J.F., Skowron, A.: Behavioral pattern identification through rough set modelling. In: Ślęzak, D., et al. (eds.) pp. 688–697 [33] (2005)
3. Bazan, J., Skowron, A.: On-line elimination of non-relevant parts of complex objects in behavioral pattern identification. In: Pal, S.K., et al. (eds.) pp. 720–725 [24](2005)
4. Bazan, J., Skowron, A.: Classifiers based on approximate reasoning schemes. In: Dunin-Kęplicz, B., et al. (eds.) pp. 191–202 [13] (2005)
5. Bazan, J., Skowron, A., Swiniarski, R.: Rough sets and vague concept approximation: From sample approximation to adaptive learning. In: Peters, J.F., Skowron, A., Grzymała-Busse, J.W., Kostek, B., Świniarski, R.W., Szczuka, M. (eds.) Transactions on Rough Sets I. LNCS, vol. 3100, pp. 39–63. Springer, Heidelberg (2004)
6. Bazan, J., Kruczek, P., Bazan-Socha, S., Skowron, A., Pietrzyk, J.J.: Risk pattern identification in the treatment of infants with respiratory failure through rough set modeling. In: Proceedings of IPMU 2006, Paris, France, Paris, July 2-7, 2006, pp. 2650–2657. Éditions E.D.K (2006)
7. Bazan, J., Kruczek, P., Bazan-Socha, S., Skowron, A., Pietrzyk, J.J.: Automatic planning of treatment of infants with respiratory failure through rough set modeling. In: Greco, S., Hata, Y., Hirano, S., Inuiguchi, M., Miyamoto, S., Nguyen, H.S., Słowiński, R. (eds.) RSCTC 2006. LNCS (LNAI), vol. 4259, pp. 418–427. Springer, Heidelberg (2006)
8. Bazan, J.: Rough sets and granular computing in behavioral pattern identification and planning. In: Pedrycz, W., et al. (eds.) [29] (2007) (in press)
9. Borrett, S.R., Bridewell, W., Langely, P., Arrigo, K.R.: A method for representing and developing process models. Ecological Complexity 4(1-2), 1–12 (2007)
10. Breiman, L.: Statistical modeling: The two Cultures. Statistical Science 16(3), 199–231 (2001)
11. Doherty, P., Łukaszewicz, W., Skowron, A., Szałas, A.: Knowledge Representation Techniques: A Rough Set Approach. Studies in Fuzziness and Soft Computing 202. Springer, Heidelberg (2006)

12. Domingos, P.: Toward knowledge-rich data mining. Data Mining and Knowledge Discovery 15, 21–28 (2007)
13. Dunin-Kęplicz, B., Jankowski, A., Skowron, A., Szczuka, M.: Monitoring, Security, and Rescue Tasks in Multiagent Systems (MSRAS 2004). Series in Soft Computing. Springer, Heidelberg (2005)
14. Friedman, J.H.: Data mining and statistics. What's the connection? Keynote Address. In: Proceedings of the 29th Symposium on the Interface: Computing Science and Statistics, Houston, Texas (May 1997)
15. Jankowski, A., Skowron, A.: A wistech paradigm for intelligent systems. In: Transactions on Rough Sets VI: Journal Subline. LNCS, vol. 4374, pp. 94–132. Springer, Heidelberg (2006)
16. Jankowski, A., Skowron, A.: Logic for artificial intelligence: The Rasiowa - Pawlak school perspective. In: Ehrenfeucht, A., Marek, V., Srebrny, M. (eds.) Andrzej Mostowski: Reflections on the Polish School of Logic, IOS Press, Amsterdam (2007)
17. Jankowski, A., Skowron, A.: Wisdom Granular Computing. In: Pedrycz, W., et al. (eds.) (in press 2007)
18. Kriegel, H.-P., Borgwardt, K.M., Kröger, P., Pryakhin, A., Schubert, M., Zimek, A.: Future trends in data mining. Data Mining and Knowledge Discovery 15(1), 87–97 (2007)
19. de Medeiros, A.K.A., Weijters, A.J.M.M., van der Aalst, W.M.P.: Genetic process mining: an experimental evaluation. Data Mining and Knowledge Discovery 14, 245–304 (2007)
20. Nguyen, H.S., Bazan, J., Skowron, A., Nguyen, S.H.: Layered learning for concept synthesis. In: Peters, J.F., Skowron, A., Grzymała-Busse, J.W., Kostek, B., Świniarski, R.W., Szczuka, M. (eds.) Transactions on Rough Sets I. LNCS, vol. 3100, pp. 187–208. Springer, Heidelberg (2004)
21. Nguyen, T.T.: Eliciting domain knowledge in handwritten digit recognition. In: Pal, S., et al. (eds.) pp. 762–767 [24] (2005)
22. Nguyen, T.T.: Outlier and exception analysis in rough sets and granular computing. In: Pedrycz, W., et al. (eds.) [29] (in press 2007)
23. Nguyen, T.T., Willis, C.P., Paddon, D.J., Nguyen, S.H., Nguyen, H.S.: Learning Sunspot Classification. Fundamenta Informaticae 72(1-3), 295–309 (2006)
24. Pal, S.K., Bandyopadhyay, S., Biswas, S. (eds.): PReMI 2005. LNCS, vol. 3776, pp. 18–22. Springer, Heidelberg (2005)
25. Pawlak, Z.: Rough sets. International Journal of Computer and Information Sciences 11, 341–356 (1982)
26. Pawlak, Z.: Rough Sets: Theoretical Aspects of Reasoning about Data. System Theory, Knowledge Engineering and Problem Solving, vol. 9. Kluwer Academic Publishers, The Netherlands, Dordrecht (1991)
27. Pawlak, Z.: Concurrent versus sequential the rough sets perspective. Bulletin of the EATCS 48, 178–190 (1992)
28. Pawlak, Z., Skowron, A.: Rudiments of rough sets. Information Sciences 177(1): 3–27; Rough sets: Some extensions. Information Sciences 177(1): 28–40; Rough sets and boolean reasoning. Information Sciences 177(1): 41–73 (2007)
29. Pedrycz, W., Skowron, A., Kreinovich, V. (eds.): Handbook of Granular Computing. John Wiley & Sons, New York (in press)
30. Poggio, T., Smale, S.: The mathematics of learning: Dealing with data. Notices of the AMS 50(5), 537–544 (2003)

31. Roddick, J.F., Hornsby, K., Spiliopoulou, M.: An updated bibliography of temporal, spatial and spatio- temporal data mining research. In: Roddick, J.F., Hornsby, K. (eds.) TSDM 2000. LNCS (LNAI), vol. 2007, Springer, Heidelberg (2001)
32. Suraj, Z.: Rough set methods for the synthesis and analysis of concurrent processes. In: Polkowski, L., Lin, T.Y., Tsumoto, S. (eds.) Rough Set Methods and Applications: New Developments in Knowledge Discovery in Information Systems, Studies in Fuzziness and Soft Computing, vol. 56, pp. 379–488. Springer, Heidelberg (2000)
33. Ślęzak, D., Yao, J., Peters, J.F., Ziarko, W., Hu, X.(eds.): RSFDGrC 2005. LNCS (LNAI), vol. 3642. Springer, Heidelberg (2005)
34. Unnikrishnan, K.P., Ramakrishnan, N., Sastry, P.S., Uthurusamy, R.: 4th KDD Workshop on Temporal Data Mining: Network Reconstruction from Dynamic Data Aug 20, 2006, The Twelfth ACM SIGKDD International Conference on Knowledge Discovery and Data (KDD 2006) August 20 - 23, 2006 Philadelphia, USA (2006), http://people.cs.vt.edu/ ramakris/kddtdm06/cfp.html
35. Vapnik, V.: Statistical Learning Theory. John Wiley & Sons, New York (1998)
36. Wu, F.-X.: Inference of gene regulatory networks and its validation. Current Bioinformatics 2(2), 139–144 (2007)
37. Zadeh, L.A.: A new direction in AI-toward a computational theory of perceptions. AI Magazine 22(1), 73–84 (2001)
38. Zadeh, L.A.: Generalized theory of uncertainty (GTU)-principal concepts and ideas. Computational Statistics and Data Analysis 51, 15–46 (2006)

The Possibilities of Fuzzy Logic
in Image Processing

M. Nachtegael*, T. Mélange, and E.E. Kerre

Ghent University, Dept. of Applied Mathematics and Computer Science
Fuzziness and Uncertainty Modeling Research Unit
Krijgslaan 281 - S9, B-9000 Gent, Belgium
`mike.nachtegael@ugent.be`

Abstract. It is not a surprise that image processing is a growing research field. Vision in general and images in particular have always played an important and essential role in human life. Not only as a way to communicate, but also for commercial, scientific, industrial and military applications. Many techniques have been introduced and developed to deal with all the challenges involved with image processing. In this paper, we will focus on techniques that find their origin in fuzzy set theory and fuzzy logic. We will show the possibilities of fuzzy logic in applications such as image retrieval, morphology and noise reduction by discussing some examples. Combined with other state-of-the-art techniques they deliver a useful contribution to current research.

1 Introduction

Images are one of the most important tools to carry and transfer information. The research field of image processing not only includes technologies for the capture and transfer of images, but also techniques to analyse these images. Among the wide variety of objectives, we mention the extraction of additional information from an image and practical applications such as image retrieval, edge detection, segmentation, noise reduction and compression.

But how can image processing be linked to fuzzy set theory and fuzzy logic? A first link can be found in the modeling of images. On the one hand, an n-dimensional image can be represented as a mapping from a universe $\mathcal{X}$ (a finite subset of $\mathbb{R}^n$, usually a $M \times N$-grid of points which we call pixels) to a set of values. The set of possible pixel values depends on whether the image is binary, grayscale or color. Binary images take values in $\{0, 1\}$ (black $= 0$; white $= 1$), grayscale images in $[0, 1]$ (values correspond to a shade of gray); the representation of color images depends on the specific color model, e.g. RGB, HSV, La*b* [26]. On the other hand, a fuzzy set A in a universe $\mathcal{X}$ is characterized by a membership function that associates a degree of membership $A(x) \in [0, 1]$ with each element x of $\mathcal{X}$ [38]. In other words: a fuzzy set A can be represented as a $\mathcal{X} - [0, 1]$ mapping, just like grayscale images. Consequently, techniques from fuzzy set theory can be applied to grayscale images.

---

* Corresponding author.

A. Ghosh, R.K. De, and S.K. Pal (Eds.): PReMI 2007, LNCS 4815, pp. 198–208, 2007.

A second link can be found in the nature of (the information contained in) images. Image processing intrinsically encounters uncertainty and imprecision, e.g. to determine whether a pixel is an edge-pixel or not, to determine whether a pixel is contaminated with noise or not, or to express the degree to which two images are similar to each other. Fuzzy set theory and fuzzy logic are ideal tools to model this kind of imprecise information.

In this paper we will illustrate the possibilities of fuzzy logic in image processing applications such as image retrieval (Section 2), mathematical morphology (Section 3) and noise reduction (Section 4). We will focus on the first application, and briefly review the other two applications.

2 Similarity Measures and Image Retrieval

2.1 An Overview of Similarity Measures for Images

Objective quality measures or measures of comparison are of great importance in the field of image processing. Two applications immediately come to mind: (1) Image database retrieval: if you have a reference image, then you will need measures of comparison to select similar images from an image database, and (2) Algorithm comparison: similarity measures can serve as a tool to evaluate and to compare different algorithms designed to solve particular problems, such as noise reduction, deblurring, compression, and so on.

It is well-known that classical quality measures, such as the MSE, $RMSE$ (Root Mean Square Error) or the $PSNR$ (Peak Signal to Noise Ratio), do not always correspond to human visual observations. A first example is given in Figure 1. A second example is that it occurs that several distortions of a specific image (e.g. obtained by adding impulse noise, gaussian noise, enlightening, blurring, compression) have the same MSE (which would rank the images as "equally similar" w.r.t. the original image), while visual evaluation clearly shows that there is a distinction between the images.

To overcome these drawbacks, other measures have been proposed. Among the different proposals, we also encounter similarity measures that find their origin in fuzzy set theory. Extensive research w.r.t. the applicability of similarity measures for fuzzy sets in image processing has been performed [30,31]. This resulted in a set of 14 similarity measures, e.g. [5]:

$$M(A, B) = \frac{|A \cap B|}{|A \cup B|} = \frac{\sum\limits_{(i,j) \in \mathcal{X}} \min(A(i,j), B(i,j))}{\sum\limits_{(i,j) \in \mathcal{X}} \max(A(i,j), B(i,j))}.$$

Note that we use minimum and maximum to model the intersection and union of two fuzzy sets, and the sigma count to model the cardinality of a fuzzy set.

Unfortunately, similarity measures that are applied directly to grayscale images do not show a convincing perceptual behaviour. Therefore two types of extensions have been proposed:

Fig. 1. The MSE does not always correspond to human evaluation of image comparison: left = original cameraman image, middle = cameraman image with 35% salt & pepper noise (MSE w.r.t. left image = 7071), right = monkey image (MSE w.r.t. left image = 5061

(1) Neighbourhood based measures [33]: each image is divided into blocks, the similarity between the corresponding blocks of the images is calculated using the similarity measures discussed above and finally these values are combined in a weighthed sum, where the weights depend on the homogenity of the block images: the higher the homogenity, the higher the weight.

(2) Histogram based measures [32]: the similarity measures discussed above can also be applied on image histograms, which can be normalized and/or ordered. In this way, the human sensitivity to frequencies can be taken into account.

The results obtained using these extensions are already much better. In order to incorporate the image characteristics in an even more optimal way we proposed combined similarity measures. These measures are constructed by combining (multiplying) neighbourhood-based similarity measures and similarity measures which are applied to ordered histograms. Extensive and scientifically guided psycho-visual experiments confirm the good and human-like results of these measures [36,37], which is a very important feature (see Figure 1).

The next step is the construction of human-like similarity measures for color images. A first attempt has been carried out in [34,35]. The existing grayscale similarity measures have however also been applied successfully in the context of color image retrieval, as is outlined in the next section.

2.2 Application: Image Retrieval

Image retrieval applications are important because the increasing availability of images and the corresponding growth of image databases and users, makes it a challenge to create automated and reliable image retrieval systems [2,3,14,19,27,29]. We consider the situation in which a reference image is available, and that similar images from a database have to be retrieved. A main drawback of most existing systems is that the images are characterized by textual descriptors (describing features like color, texture, morphological properties, and so on), which usually have to be made by a person [4,8,28]. Using histogram based similarity measures, we have developed a retrieval system that does succeed in extracting images from a database of images automatically, based on the similarity values only.

The characteristics of the system can be summarized as follows [18]:

- We use the HSI color space to model colors. The hue component is enough to recognize the color, except when the color is very pale or very somber. In order to perform an extraction based on dominant colors, we limit ourselves to 8 fundamental colors, that are modelled with trapezoidal fuzzy numbers [3]. In those cases where there is nearly no color present in the image we will use the intensity component to identify the dominant "color". Also for this component we use a fuzzy partition.
- We calculate the membership degree of all the pixels in every image with respect to the fundamental colors modelled by the trapezoidal fuzzy numbers. In that way we obtain 8 new "images".
- We consider the histogram of each of these 8 images, and normalize them by dividing all the values by the maximum value. In that way we obtain for each image 8 fuzzy sets, representing the frequency distribution of the membership degrees with respect to the 8 fundamental colors.
- We use a specific histogram-based similarity measure [32] to calculate the similarity between these different histograms.
- We combine the similarity values corresponding to the 8 histograms or colors by using the standard average as aggregation operator.
- We perform the same procedure w.r.t. the intensity component.
- The overall similarity between two images is defined as the average of the similarity value w.r.t. the hue component and the similarity value w.r.t. the intensity component.

Experimental results confirm the ability of the proposed system to retrieve color images from a database in a fast and automated way. One example is illustrated in Figure 2. We used a database of over 500 natural images of animals, flowers, buildings, cars, texture images, ..., and we show the retrieval results (the 10 most similar images), together with the numerical results, using a flower image as query image. The results are quite good: the three most similar retrieved images are flowers in the same color as the one in the query image. The other retrieved images do not contain flowers but have a very similar layout, i.e., they all show an object with a natural background. This illustrates that the proposed approach has potential w.r.t. color image retrieval.

3 Mathematical Morphology

Mathematical morphology is a theory, based on topological and geometrical concepts, to process and analyze images. The basic morphological operators (dilation, erosion, opening and closing) resulted from this research, and form the fundamentals of this theory [25]. A morphological operator transforms an image into another image, using a structuring element which can be chosen by the user.

Query Image

| 0.6319 | 0.63162 | 0.62331 |

| 0.50014 | 0.49588 | 0.49582 |

| 0.4865 | 0.48566 | 0.4762 |

0.4742

Fig. 2. Retrieval result for the natural image experiment

The goal is to retrieve additional information from the image (e.g. edges) or to improve the quality of the image (e.g. noise reduction).

By the 1980's, binary morphology was extended to grayscale morphology (there are two classical approaches: the threshold approach [25] and the umbra approach [11]), and the lattice-based framework was shaped. As a scientific branch, mathematical morphology expanded worldwide during the 1990's. It is also during that period that different models based on fuzzy set theory were introduced. Today, mathematical morphology remains a challenging research field.

The introduction of "fuzzy" morphology is based on the fact that the basic morphological operators make use of crisp set operators, or equivalently, crisp logical operators. Such expressions can easily be extended to the context of fuzzy sets. In the binary case, where the image and the structuring element are represented as crisp subsets of $\mathbb{R}^n$, the dilation and erosion are defined as follows [25]:

$$D(A, B) = \{y \in \mathbb{R}^n | T_y(B) \cap A \neq \emptyset\},$$
$$E(A, B) = \{y \in \mathbb{R}^n | T_y(B) \subseteq A\},$$

with $T_y(B) = \{x \in \mathbb{R}^n | x - y \in B\}$ the translation of B by the point y.

The binary dilation and erosion have a very nice geometrical interpretation. For example, the dilation $D(A, B)$ consists of all points y in $\mathbb{R}^n$ such that the translation $T_y(B)$ of the structuring element has a non-empty intersection with the image A. Consequently, the dilation will typically extend the contours of objects in the image, and fill up small gaps and channels. Similar interpretations can be made for the erosion.

Fuzzy techniques can now be used to extend these operations to grayscale images and structuring elements. This can be realised by fuzzifying the underlying logical operators, using conjunctors and implicators. A $[0, 1]^2 - [0, 1]$ mapping $\mathcal{C}$ is called a conjunctor if $\mathcal{C}(0,0) = \mathcal{C}(1,0) = \mathcal{C}(0,1) = 0$, $\mathcal{C}(1,1) = 1$, and if it has increasing partial mappings. Popular examples are $T_M(a,b) = \min(a,b)$, $T_P(a,b) = a \cdot b$ and $T_W(a,b) = \max(0, a + b - 1)$, with $(a,b) \in [0,1]^2$. A $[0,1]^2 - [0,1]$ mapping $\mathcal{I}$ is called an implicator if $\mathcal{I}(0,0) = \mathcal{I}(0,1) = \mathcal{I}(1,1) = 1$, $\mathcal{I}(1,0) = 0$, and if it has decreasing first and increasing second partial mappings. Popular examples are $I_W(a,b) = \min(1, 1 - a + b)$, $I_{KD}(a,b) = \max(1 - a, b)$ and $I_R(a,b) = 1 - a + a \cdot b$, with $(a,b) \in [0,1]^2$. The resulting expressions for the fuzzy dilation and fuzzy erosion are as follows [7]:

$$D_{\mathcal{C}}(A, B)(y) = \sup_{x \in \mathbb{R}^n} \mathcal{C}(B(x - y), A(x)), \tag{1}$$

$$E_{\mathcal{I}}(A, B)(y) = \inf_{x \in \mathbb{R}^n} \mathcal{I}(B(x - y), A(x)), \tag{2}$$

with $A, B \in \mathcal{F}(\mathbb{R}^n)$ and with $\mathcal{C}$ a conjunctor and $\mathcal{I}$ an implicator.

Extensive studies [15] showed that the most general model of fuzzy mathematical morphology is the one based on the expressions above.

An application of fuzzy morphological operations is illustrated with the "cameraman" image in Figure 3. This figure shows the edge images (an edge image can be obtained by subtracting the eroded from the dilated image) for two different pairs of conjunctors and implicators. It also illustrates that the choice of

Fig. 3. Fuzzy morphological edge images for $(\mathcal{C},\mathcal{I}) = (T_M, I_{KD})$ (left) and $(\mathcal{C},\mathcal{I}) = (T_W, I_W)$ (right)

the fuzzy logical operators has a big influence on the result of the fuzzy morphological operators.

The next important step was the extension to and the use of fuzzy techniques for color morphology. Although a component-based approach, in which we apply grayscale operators on each color component separately, is straightforward, it also leads to disturbing artefacts because correlations between colors are not taken into account. Therefore, colors should be treated as vectors. A major problem is that one also has to define an ordering between the color vectors, in order to be able to extend concepts such as supremum and infimum; also the sum, difference and product of colors has to be defined if we want to extend fuzzy morphology to color images. Numerous orderings have been defined in different color spaces such as RGB, HSV and La*b*; our own ordering is discussed in [9]. This resulted in the setup of theoretical frameworks for (fuzzy) color morphology, in which pratical applications can now be studied [1,6,10,12,13].

4   Noise Reduction

Images can be contaminated with different types of noise. Among the most common types of noise we find impulse noise (e.g. salt & pepper noise), additive noise (e.g. gaussian noise) and multiplicative noise (e.g. speckle noise). It is a great challenge to develop algorithms that can remove noise from an image, without disturbing its content. The main disadvantage of classical filters is that they treat all the pixels in the same way, based on purely numerical information. This makes them incapable of taking into account any uncertainty and imprecision that usually occurs (e.g. not all the pixels will be contaminated with noise in the same way; one should be able to work with degrees of contamination).

The use of fuzzy techniques offers a solution. In general, a fuzzy filter for noise reduction uses both numerical information (just as classical filters) and linguistic information (modeled by fuzzy set theory; enabling us to work with e.g. "small" and "large" gradient values). This information is processed by a fuzzy rule base (approximate reasoning; enabling us to use rules such as "if most of the gradient

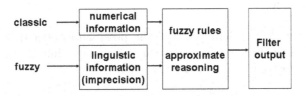

Fig. 4. Fuzzy filters use both numerical and linguistic information to process an image

values are large, then assume that the pixel is noisy"), resulting in a (defuzzified) filter output. The general scheme of fuzzy filters is shown in Figure 4.

Several proposals for noise reduction algorithms based on fuzzy set theory and fuzzy logic have been made. In [16,17] we have studied 38 different algorithms that were specifically designed for impulse noise and/or gaussian noise, and found out that the best performing filters always are based on fuzzy logic. The more recent construction of new fuzzy-logic-based filters resulted in new comparative studies, confirming that fuzzy logic is a very powerful tool for noise reduction [22], also for color images [20,23,24] and video sequences [39]. We illustrate the FIDRM filter, the Fuzzy Impulse Noise Detection and Reduction Method [21], which is designed for grayscale images, in Figure 5.

Fig. 5. The Lena image with 20% impulse noise (MSE = 3309,40), and the result of the FIDRM filter (MSE = 30,57)

In summary, fuzzy logic makes it possible to reason with uncertainty and imprecision (linguistic information), which is inherent to noise detection and reduction, which explains the very good results. On the other hand, quite a lot of research still has to be carried out: the continued development of new filters for noise reduction, for different types of noise (e.g. speckle noise gets less attention than impulse noise and gaussian noise), for different types of images (e.g. natural scenes, satellite images, medical images, ...), for different types of images (grayscale, color and video) and the necessary comparative studies to evaluate them w.r.t. each other, will require many future efforts.

5 Conclusion

In this paper, we have briefly outlined the possibilities offered by fuzzy logic in the field of image processing, in particular for image similarity and retrieval, mathematical morphology, and noise reduction

References

1. Angulo, J.: Unified morphological color processing framework in a lum/sat/hue representation. In: Proceedings of ISMM 2005, International Symposium on Mathematical Morphology, France, pp. 387–396 (2005)
2. Brunelli, R., Mich, O.: Histograms analysis for image retrieval. Pattern Recognition 34, 1625–1637 (2001)
3. Chamorro-Martínez, J., Medina, J.M., Barranco, C., Galán-Perales, E., Soto-Hidalgo, J.M.: An approach to image retrieval on fuzzy object-relational database using dominant color descriptors. In: Proceedings of the 4th Conference of the European Society for Fuzzy Logic and Technology, EUSFLAT, pp. 676–684 (2005)
4. Chang, S.: Content-based indexing and retrieval of visual information. IEEE Signal Processing Magazine 14(4), 45–48 (1997)
5. Chen, S.M., Yeh, M.S., Hsiao, P.Y.: A comparison of similarity measures of fuzzy values. Fuzzy Sets and Systems 72, 79–89 (1995)
6. Comer, M.L., Delp, E.J.: Morphological operations for color image processing. Journal of Electronic Imaging 8(3), 279–289 (1999)
7. De Baets, B.: Fuzzy morphology: a logical approach. In: Ayyub, B.M., Gupta, M.M. (eds.) Uncertainty Analysis in Engineering and Sciences: Fuzzy Logic, Statistics, and Neural Network Approach, pp. 53–67. Kluwer Academic Publishers, Boston (1997)
8. Del Bimbo, A.: Visual Information Retrieval. Morgan Kaufmann Publishers, San Francisco (2001)
9. De Witte, V., Schulte, S., Nachtegael, M., Van der Weken, D., Kerre, E.E.: Vector morphological operators for colour images. In: Kamel, M., Campilho, A. (eds.) ICIAR 2005. LNCS, vol. 3656, pp. 667–675. Springer, Heidelberg (2005)
10. Hanbury, A., Serra, J.: Mathematical morphology in the CIELAB space. Image Analysis and Stereology 21(3), 201–206 (2002)
11. Haralick, R.M., Sternberg, S.R., Zhuang, X.: Image analysis using mathematical morphology. IEEE Transactions on Pattern Analysis and Machine Intelligence 9(4), 532–550 (1987)
12. Li, J., Li, Y.: Multivariate mathematical morphology based on principal component analysis: initial results in building extraction. International Archives for Photogrammetry, Remote Sensing and Spatial Information Sciences 35(B7), 1168–1173 (2004)
13. Louverdis, G., Andreadis, I., Tsalides, P.: New fuzzy model for morphological color image processing. In: Proceedings of IEEE Vision, Image and Signal Processing, pp. 129–139 (2002)
14. Lu, G., Phillips, J.: Using perceptually weighted histograms for colour-based image retrieval. In: Proceedings of the 4th International Conference on Signal Processing, pp. 1150–1153 (1998)
15. Nachtegael, M., Kerre, E.E.: Connections between binary, gray-scale and fuzzy mathematical morphologies. Fuzzy Sets and Systems 124(1), 73–86 (2001)

16. Nachtegael, M., Schulte, S., Van der Weken, D., De Witte, V., Kerre, E.E.: Fuzzy filters for noise reduction: the case of impulse noise. In: Proceedings of SCIS-ISIS (2004)
17. Nachtegael, M., Schulte, S., Van der Weken, D., De Witte, V., Kerre, E.E.: Fuzzy Filters for noise reduction: the case of gaussian noise. In: Proceedings of FUZZ-IEEE (2005)
18. Nachtegael, M., Schulte, S., De Witte, V., Mélange, T., Kerre, E.E.: Color image retrieval using fuzzy similarity measures and fuzzy partitions. In: Proceedings of ICIP 2007, 14th International Conference on Image Processing, 7th edn., San Antonio, USA (2007)
19. Omhover, J.F., Detyniecki, M., Rifqi, M., Bouchon-Meunier, B.: Ranking invariance between fuzzy similarity measures applied to image retrieval. In: Proceedings of the 2004 IEEE International Conference on Fuzzy Systems, pp. 1367–1372. IEEE, Los Alamitos (2004)
20. Schulte, S., Nachtegael, M., De Witte, V., Van der Weken, D., Kerre, E.E.: A new two step color filter for impulse noise. In: Proceedings of the 11th Zittau Fuzzy Colloquium, pp. 185–192 (2004)
21. Schulte, S., Nachtegael, M., De Witte, V., Van der Weken, D., Kerre, E.E.: A fuzzy impulse noise detection and reduction method. IEEE Transactions on Image Processing 15(5), 1153–1162 (2006)
22. Schulte, S., De Witte, V., Nachtegael, M., Van der Weken, D., Kerre, E.E.: A new fuzzy filter for the reduction of randomly valued impulse noise. In: Proceedings of ICIP 2006, 13th International Conference on Image Processing, Atlanta, USA, pp. 1809–1812 (2006)
23. Schulte, S., Nachtegael, M., De Witte, V., Van der Weken, D., Kerre, E.E.: Fuzzy impulse noise reduction methods for color images. In: Proceedings of FUZZY DAYS 2006, International Conference on Computational Intelligence, Dortmund (Germany), pp. 711–720 (2006)
24. Schulte, S., De Witte, V., Nachtegael, M., Mélange, T., Kerre, E.E.: A new fuzzy additive noise reduction method. In: Image Analysis and Recognition - Proceedings of ICIAR 2007. LNCS, vol. 4633, pp. 12–23. Springer, Heidelberg (2007)
25. Serra, J.: Image analysis and mathematical morphology. Academic Press Inc, Londen (1982)
26. Sharma, G.: Digital Color Imaging Handbook. CRC Press, Boca Raton, USA (2003)
27. Smeulders, A.W.M., Worring, M., Santini, S., Gupta, A., Jain, R.: Content-based image retrieval at the end of the early years. IEEE Transactions on Pattern Analysis and Machine Intelligence 22(12), 1349–1379 (2000)
28. Stanchev, P., Green, D., Dimitrov, B.: High level color similarity retrieval. International Journal of Information Theories & Applications 10(3), 283–287 (2003)
29. Stanchev, P.: Using image mining for image retrieval. In: Proceedings of the IASTED International Conference on Computer Science and Technology, pp. 214–218 (2003)
30. Van der Weken, D., Nachtegael, M., Kerre, E.E.: The applicability of similarity measures in image processing. Intellectual Systems 6(1-4), 231–248 (2001)
31. Van der Weken, D., Nachtegael, M., Kerre, E.E.: An overview of similarity measures for images. In: Proceedings of ICASSP 2002, IEEE International Conference on Acoustics, Speech and Signal Processing, Orlando, USA, pp. 3317–3320 (2002)
32. Van der Weken, D., Nachtegael, M., Kerre, E.E.: Using similarity measures for histogram comparison. In: De Baets, B., Kaynak, O., Bilgiç, T. (eds.) IFSA 2003. LNCS, vol. 2715, pp. 396–403. Springer, Heidelberg (2003)

33. Van der Weken, D., Nachtegael, M., Kerre, E.E.: Using similarity measures and homogeneity for the comparison of images. Image and Vision Computing 22(9), 695–702 (2004)
34. Van der Weken, D., De Witte, V., Nachtegael, M., Schulte, S., Kerre, E.E.: A component-based and vector-based approach for the construction of quality measures for colour images. In: Proceedings of IPMU 2006, International Conference on Information Processing and Management of Uncertainty in Knowledge-Based Systems, Paris (France), pp. 1548–1555 (2006)
35. Van der Weken, D., De Witte, V., Nachtegael, M., Schulte, S., Kerre, E.E.: Fuzzy similarity measures for colour images. In: Proceedings of CIS-RAM 2006, IEEE International Conferences on Cybernetics & Intelligent Systems and Robotics, Automation & Mechatronics, Bangkok (Thailand), pp. 806–810 (2006)
36. Vansteenkiste, E., Van der Weken, D., Philips, W., Kerre, E.E.: Psycho-visual evaluation of fuzzy similarity measures. In: Proceedings of SPS-DARTS 2006, 2nd Annual IEEE BENELUX/DSP Valley Signal Processing Symposium, Antwerp (Belgium), pp. 127–130 (2006)
37. Vansteenkiste, E., Van der Weken, D., Philips, W., Kerre, E.E.: Evaluation of the perceptual performance of fuzzy image quality measures. In: Gabrys, B., Howlett, R.J., Jain, L.C. (eds.) KES 2006. LNCS (LNAI), vol. 4251, pp. 623–630. Springer, Heidelberg (2006)
38. Zadeh, L.: Fuzzy Sets. Information Control 8, 338–353 (1965)
39. Zlokolica, V., De Geyter, M., Schulte, S., Pizurica, A., Philips, W., Kerre, E.E.: Fuzzy logic recursive change detection for tracking and denoising of video sequences. In: Proceedings of IS&T/SPIE, 17th Annual Symposium Electronic Imaging Science and technology, Video Communications and Processing, vol. 5685, pp. 771–782 (2005)

Fuzzy Ordering Relation and Fuzzy Poset

Branimir Šešelja and Andreja Tepavčević*

Department of Mathematics and Informatics
Faculty of Sciences, University of Novi Sad
Trg D. Obradovića 4, 21000 Novi Sad, Serbia
seselja@im.ns.ac.yu, etepavce@eunet.yu

Abstract. Connections between (weakly) reflexive, antisymmetric and transitive lattice-valued fuzzy relations on a nonempty set X (fuzzy ordering relations on X) and fuzzy subsets of a crisp poset on X (fuzzy posets) are established and various properties of cuts of such structures are proved.

A representation of fuzzy sets by cuts corresponding to atoms in atomically generated lattices has also been given.

AMS Mathematics Subject Classification (1991): 04A72.

Keywords and Phrases: Lattice valued fuzzy ordering relation, fuzzy weak ordering relation, fuzzy poset, cutworthy approach.

1 Introduction

The present research belongs to the theory of fuzzy ordering relations. These are investigated as lattice valued structures. Therefore, the framework as well as the subject are within the order theory. Fuzzy structures presented here support cutworthy properties, that is, crisp fuzzified properties are preserved under cut structures. Therefore, the co-domains of all mappings (fuzzy sets) are complete lattices, without additional operations. Namely, such lattices support the transfer of crisp properties to cuts.

Due to the extensive research of fuzzy ordering relations, we mention only those authors and papers which are relevant to our approach. From the earlier period, we mention papers by Ovchinnikov ([9,10], and papers cited there); we use his definition of the ordering relation. In the recent period, there are papers by Bělohlávek (most of the relevant results are collected in his book [1]). He investigates also lattice-valued orders, the lattice being residuated. Recent important results concerning fuzzy orders and their representations are obtained by De Baets and Bodenhofer (see a state-of-the-art overview about weak fuzzy orders, [4]).

* This research was partially supported by Serbian Ministry of Science and Environment, Grant No. 144011 and by the Provincial Secretariat for Science and Technological Development, Autonomous Province of Vojvodina, grant "Lattice methods and applications".

A. Ghosh, R.K. De, and S.K. Pal (Eds.): PReMI 2007, LNCS 4815, pp. 209–216, 2007.

Most of the mentioned investigations of fuzzy orders are situated in the framework of T-norms, or more generally, in the framework of residuated lattices and corresponding operations.

As mentioned above, our approach is purely order-theoretic.

Motivated by the classical approach to partially ordered sets, we investigate fuzzy orders from two basic aspects. Firstly we consider fuzzy posets, i.e., fuzzy sets on a crisp ordered set. The other aspect is a fuzzification of an ordering relation. As the main result we prove that there is a kind of natural equivalence among these two structures. Namely, starting with a fuzzy poset, we prove that there is a fuzzy ordering relation on the same set with equal cuts. Vice versa, for every fuzzy ordering relation there is a fuzzy poset on the same underlying set, so that cut-posets of two structures coincide. We also present some properties of cuts, which give an additional description of fuzzy orders.

Our results are partially analogue to those in paper [14], in which similar problems are discussed for fuzzy lattices. We mention that we use fuzzy weak ordering relations (reflexivity is weakened), which differs from the notion of the weak order used in the paper [4].

2   Preliminaries

We use well known notions from the classical theory of ordered structures (see e.g., [6]). A **poset** $(P, \leq)$ is a nonempty set P endowed with an **ordering relation (order)** ρ, not necessarily linear. It is well known that an order on a given set by definition fulfills properties of reflexivity, antisymmetry and transitivity. We use also the notion of a **weak ordering relation** on a set P, which is an antisymmetric and transitive relation, satisfying also the **weak reflexivity** on P: for any $x, y \in P$

if $x\rho y$ then $x\rho x$ and $y\rho y$.

Throughout the paper L is a complete lattice with the top element (1) and the bottom element (0); sometimes these are denoted by 1_L and 0_L. An element $a \in L$ distinct from 0 is an **atom** in L if it covers the least element 0, i.e., if $0 \leq x \leq a$ implies $x = 0$ or $x = a$. If there is a single atom in L which is below all other non-zero elements in L, then it is called a **monolith**. A lattice L is **atomic** if for every non-zero element $x \in L$, there is an atom a such that $a \leq x$. Finally, a lattice L is **atomically generated** if every non-zero element is a join (supremum) of atoms.

Next we present notions connected with lattice valued fuzzy sets and structures. For more details we refer to the overview articles [11,12].

A **fuzzy set** on a nonempty set S is a mapping $\mu : S \to L$. A **cut set** (briefly **cut**) of μ is defined for every $p \in L$, as the subset μ_p of S, such that $x \in \mu_p$ if and only if $\mu(x) \geq p$ in L. A **strong cut** of μ is a subset $\mu_p^>$ of S such that for every $p \in L$, $x \in \mu_p^>$ if and only if $\mu(x) > p$ in L.

A property or a notion which is transferred or generalized to fuzzy structures is said to be **cutworthy** if it is preserved by all cuts of the corresponding fuzzy structure (see e.g., [8]).

If X is a nonempty set, any mapping $\rho : X^2 \to L$ is an L–**valued** (**lattice valued**) relation on X.

Since a fuzzy relation is a fuzzy subset of X^2, a cut relation is a cut of the corresponding fuzzy set: for $p \in L$, a p-**cut** of ρ is a subset ρ_p of X^2, such that $\rho_p = \{(x, y) \mid \rho(x, y) \geq p\}$.

Now, we recall fuzzy ordering relations and some of their main properties. Among many definitions of fuzzy orders, we use the following, which, in our opinion, is the most natural lattice-valued generalization of the crisp order. The reason for this opinion is that all non-trivial cuts of such fuzzy orders are crisp ordering relations, i.e., its importance should be understood in the cutworthy sense.

An L–valued relation is
- **reflexive** if

$$\rho(x, x) = 1, \text{ for every } x \in X;$$

- **weakly reflexive** if

$$\rho(x, x) \geq \rho(x, y) \text{ and } \rho(x, x) \geq \rho(y, x), \text{ for all } x, y \in X;$$

- **antisymmetric** if

$$\rho(x, y) \wedge \rho(y, x) = 0, \text{ for all } x, y \in X, x \neq y;$$

- **transitive** if

$$\rho(x, y) \wedge \rho(y, z) \leq \rho(x, z), \text{ for all } x, y, z \in X.$$

An L–valued relation ρ on X is an L–**fuzzy ordering relation** (**fuzzy order**) on X if it is reflexive, antisymmetric and transitive.

An L–valued relation ρ on X is a **weak L–fuzzy ordering relation** (**weak fuzzy order**) on X if it is weakly reflexive, antisymmetric and transitive.

Observe that, like in the crisp case, every fuzzy order is, at the same time, a fuzzy weak order on the same set, while the reverse need not be true.

Theorem 1. *A relation $\rho : S^2 \to L$ is an L-fuzzy ordering relation if and only if all cuts except 0-cut are ordering relations.*

Proof. Let $\rho : S^2 \to L$ be a fuzzy ordering relation, and let $p \in L$. If $p = 0$, then $\rho_0 = S^2$ and it is not ordering relation unless $\mid S \mid = 1$.

Let $p \neq 0$. Then, $(x, x) \in \rho_p$, by the reflexivity of ρ.

If $(x, y) \in \rho_p$ and $(y, x) \in \rho_p$, then $\rho(x, y) \geq p$ and $\rho(y, x) \geq p$ and hence $\rho(x, y) \wedge \rho(y, x) \geq p$, which is true only in case $x = y$.

If $(x, y) \in \rho_p$ and $(y, z) \in \rho_p$ then $\rho(x, y) \geq p$ and $\rho(y, z) \geq p$. Therefore $\rho(x, z) \geq \rho(x, y) \wedge \rho(y, z) \geq p$ and hence $(x, z) \in \rho_p$.

On the other hand, suppose that ρ is a fuzzy relation such that all cuts (except 0-cut) are ordering relations.

Since all p cuts are reflexive relations, $\rho(x, x) = \bigvee_{p \in L} \rho_p(x, x) = 1$.

Further, let $x \neq y$ and let $\rho(x,y) \wedge \rho(y,x) = p$. Then, $(x,y) \in \rho_p$ and $(y,x) \in \rho_p$, and by antisymmetry of ρ_p, we have that $p = 0$.

Similarly, we can prove the transitivity of ρ.

3   Fuzzy Posets and Fuzzy Orders

In paper [14], fuzzy lattices are considered from two aspects: fuzzy lattices as fuzzy algebraic structures and fuzzy lattices as particular fuzzy ordered sets. In the same paper, conditions for transferring of one concept to another are given.

A fuzzy lattice as a fuzzy algebraic structure is obtained by a fuzzification of the carrier, while a fuzzy lattice as a fuzzy ordered structure is obtained by a fuzzification of the ordering relation.

In this part we formulate analogue connections for fuzzy orders.

Fuzzy Poset

If $(P, \leq)$ is a poset and $(L, \wedge, \vee)$ a complete lattice, we consider a fuzzy poset as a fuzzy set $\mu : P \to L$ with the given crisp ordering relation $\leq$.

Cut sets are sub-posets of P, i.e., every cut is a subset of P endowed with the restriction of the relation $\leq$ to this subset. These restriction-relations can also be considered as weak orders on P. We denote these restriction-relations with the same symbol, $\leq$.

Fuzzy Weak Order

On the other hand, starting with the same poset $(P, \leq)$, we fuzzify a relation $\leq$ as a mapping from $P^2 \to \{0, 1\}$. Thus we obtain a fuzzy weak ordering relation on P, as a mapping from P^2 to L which is weakly reflexive, antisymmetric and transitive.

It is straightforward to check that the cut sets of fuzzy weak order on P are crisp weak orderings on P.

A connection between two fuzzifications (a fuzzy poset and a fuzzy weak order) is given in the next theorems.

Lemma 1. *Let $\rho : S^2 \to L$ be an L-fuzzy ordering relation, such that L is a complete lattice having a monolith a. Then, the strong cut $\rho_0^>$ is a crisp ordering relation on the set S.*

Proof. Since $\rho(x,x) = 1 > 0$, we have that $(x,x) \in \rho_0^>$.

If $\rho(x,y) > 0$ and $\rho(y,x) > 0$, we have that $\rho(x,y) \geq a$ and $\rho(x,y) \geq a$, hence $\rho(x,y) \wedge \rho(y,x) \geq a$ and thus $x = y$.

If $\rho(x,y) > 0$ and $\rho(y,z) > 0$, we have that $\rho(x,y) \wedge \rho(y,z) \geq a$ and hence by fuzzy transitivity, $\rho(x,z) \geq a$. Therefore $(x,z) \in \rho_0^>$.

Remark. The assumption that L has a monolith is here essential. Indeed, the strong 0 cut in a fuzzy ordered set with the co-domain being a complete lattice without the unique atom need not be an ordered set. Example 1 below illustrates this fact.

Theorem 2. *Let $\mu : P \to L$ be an L–fuzzy poset. Then, the mapping $\rho : P^2 \to L$ defined by*

$$\rho(x,y) = \begin{cases} 1, & \text{if } x = y \\ \mu(x) \wedge \mu(y), & \text{if } x < y \\ 0, & \text{otherwise.} \end{cases}$$

is an L–fuzzy ordering relation.

Proof. The mapping ρ is reflexive by the definition. To prove the antisymmetry, assume $x \neq y$. Then, by the antisymmetry of $<$, if $x < y$,then it is not $y < x$ and vice versa. Therefore, at least one of the values $\rho(x,y)$ or $\rho(y,x)$ is 0 and thus $\rho(x,y) \wedge \rho(y,x) = 0$.

We prove the transitivity. In case $\rho(x,y) \wedge \rho(y,z) = 0$, it is true that $\rho(x,y) \wedge \rho(y,z) \leq \rho(x,z)$. In case $\rho(x,y) \wedge \rho(y,z) > 0$, we have that either $x = y$ or $y = z$ or $x < y$ and $y < z$. In case $x = y$ (similarly for $y = z$), we have that $\rho(x,y) \wedge \rho(y,z) = 1 \wedge \rho(x,z) \leq \rho(x,z)$.

In case $x < y$ and $y < z$, we have $x < z$ and thus $\rho(x,y) \wedge \rho(y,z) = \mu(x) \wedge \mu(y) \wedge \mu(y) \wedge \mu(z) \leq \mu(x) \wedge \mu(z) = \rho(x,z)$.

In the following we use the notation $0 \oplus L$ for a lattice L' obtained by adding a new least element 0 to the lattice L. The least element in L becomes thus a monolith in L' and it is denoted by 0_L (see Figure 1).

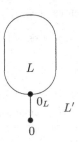

Fig. 1.

Theorem 3. *Let $\mu : P \to L$ be an L–fuzzy poset and let $L' = 0 \oplus L$. Then, the mapping $\rho : P^2 \to L'$ defined by*

$$\rho(x,y) = \begin{cases} \mu(x) \wedge \mu(y), & \text{if } x \leq y \\ 0, & \text{otherwise.} \end{cases}$$

is an L–fuzzy weak ordering relation. Moreover cuts μ_p and ρ_p, for $p \in L$ coincide as crisp sub-posets.

Proof. If $x \leq y$ then $\rho(x,x) = \mu(x) \geq \mu(x) \wedge \mu(y) = \rho(x,y)$. In case $x \nleq y$, we have $\rho(x,y) = 0$ and hence weak reflexivity is also satisfied.

The proof for antisymmetry is the same as in the previous theorem.

In the proof of transitivity, similarly as in the previous theorem, the only nontrivial case is when $\rho(x,y) \wedge \rho(y,z) > 0$. Thus, $x \leq y$ and $y \leq z$, and hence $x \leq z$. Therefore, $\rho(x,z) = \mu(x) \wedge \mu(z) \geq \mu(x) \wedge \mu(y) \wedge \mu(y) \wedge \mu(z) = \rho(x,y) \wedge \rho(y,z)$.

Let $p \in L$. We prove that ρ_p is a weak ordering relation on P, the same one which corresponds to μ_p. As already mentioned, a weak ordering relation on P is an ordering relation on a subset of P.

$(x, y) \in \rho_p$ if and only if $\rho(x, y) \geq p$ if and only if $\mu(x) \wedge \mu(y) \geq p$ if and only if $\mu(x) \geq p$ and $\mu(y) \geq p$ if and only if $x, y \in \mu_p$, completing the proof.

Next theorem gives the converse of the previous ones, and proves that every fuzzy ordering relation determines a fuzzy ordered set.

Theorem 4. *Let $\rho : P^2 \to L'$ be an L- fuzzy weak ordering relation, where $L' = 0 \oplus L$ is a complete lattice with a monolith 0_L, the top element 1_L and the bottom element 0. Let $S := \{x \mid \rho(x, x) > 0\}$. Then, the mapping $\mu : S \to L$ defined by:*

$$\mu(x) = \rho(x, x)$$

is an L–fuzzy poset on $(S, \leq)$, where the order is the strong 0-cut, i.e., $\leq = \rho_0^>$. In addition, for any $p \in L$, from $(x, y) \in \rho_p$ it follows that $x \in \mu_p$ and $y \in \mu_p$.

Proof. The relation $\rho_0^>$, by Lemma 1, is a crisp ordering relation, which is denoted by $\leq$. μ is a fuzzy set on P, so $(P, \leq)$ is an L-fuzzy ordered set.

Let $p \in L$. Then from $(x, y) \in \rho_p$ it follows that $\rho(x, y) \geq p$ and hence $\rho(x, x) \geq p$ and $\rho(y, y) \geq p$. Therefore, $x \in \mu_p$ and $y \in \mu_p$.

In the following we present some properties of fuzzy ordering relations connecting them to crisp cut-ordering relations.

Proposition 1. *Let $\rho : S^2 \to L$ be an L-fuzzy weak ordering relation, where L is an atomic lattice with the set of atoms A. Then, there is a family of posets $\{P_a \mid a \in A\}$ deduced from ρ such that each poset in the family is a maximum poset under inclusion, being a cut of fuzzy relation ρ. In addition, every cut poset is a sub-poset of a poset in this family.*

Proof. In Theorem 1 it is proved that any cut set, except 0-cut is a crisp weak ordering relation, or a crisp ordering relation on a subset of S. It is well known that from $p \leq q$ it follows that $\rho_q \subseteq \rho_p$. Since 0-cut is not an ordering relation in a general case (it is equal to S^2), the maximum posets being the cut sets are ρ_a for a atoms. Since L is atomic, by the same property we deduce that every cut except 0 is a subposet of a poset in this family.

ρ	a	b	c	d	e
a	p	p	0	0	0
b	0	1	0	0	r
c	p	1	1	0	r
d	0	q	q	q	0
e	p	p	0	0	1.

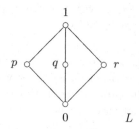

Fig. 2.

Example 1. Let ρ be a fuzzy weak ordering relation on set $S = \{a, b, c, d, e\}$, defined by the table, where the co-domain lattice is given in Figure 2.

Cuts of this fuzzy relation are as follows:

ρ_1	a b c d e
a	0 0 0 0 0
b	0 1 0 0 0
c	0 1 1 0 0
d	0 0 0 0 0
e	0 0 0 0 1

ρ_p	a b c d e
a	1 1 0 0 0
b	0 1 0 0 0
c	1 1 1 0 0
d	0 0 0 0 0
e	1 1 0 0 1

ρ_q	a b c d e
a	0 0 0 0 0
b	0 1 0 0 0
c	0 1 1 0 0
d	0 1 1 1 0
e	0 0 0 0 1

ρ_r	a b c d e
a	0 0 0 0 0
b	0 1 0 0 1
c	0 1 1 0 1
d	0 0 0 0 0
e	0 0 0 0 1.

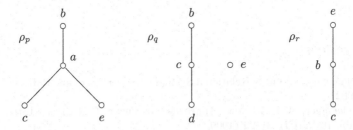

Fig. 3.

Considering three cuts that corresponds to atoms of L, we can see that posets corresponding to other cuts are sub-posets of some of these (in this case only ρ_1)(Figure 3).

Proposition 2. *Let $\rho : S^2 \to L$ be an L-fuzzy weak ordering relation, where L is an atomic lattice with the set of atoms A. Then $\rho(x, y) > 0$ if and only if there is an $a \in A$, such that $x \leq y$ in P_a, where $\{P_a \mid a \in A\}$ is a family of posets defined in the previous proposition.*

Proof. If $\rho(x, y) = p > 0$, for a $p \in L$, then there is an atom $a \in A$, such that $a \leq p$ and thus $x \leq y$ in P_a.

To prove the converse, if there is an $a \in A$ $(a \neq 0)$ such that $x \leq y$ in P_a, then $\rho(x, y) \geq a$ and hence $\rho(x, y) > 0$.

Proposition 3. *If $\rho : S^2 \to L$ is an L-fuzzy weak ordering relation, where L is an atomically generated lattice with the set of atoms A, then*

$$\rho(x, y) = \bigvee_{a \in A} a \cdot P_a(x, y),$$

where $P_a(x, y) = 1$ if $x \leq y$ in P_a and 0 otherwise, $a \cdot 1 = a$ and $a \cdot 0 = 0$.

Proof. Let $\rho(x, y) = p$ and let p be the supremum of some atoms $\{a_i \mid i \in I\}$. The proposition is a direct consequence of the fact that p is the supremum exactly of the atoms below it.

4 Conclusion

There are many possibilities to fuzzify and investigate fuzzy ordering relations and fuzzy posets. It turns out that the approach presented here is very convenient for the cutworthy approach. Indeed, we have established a connection among fuzzy posets as fuzzy subsets of crisp posets and fuzzy orderings, so that cut sets are preserved. Therefore, depending of the concrete situation in which fuzzy order appears, it is possible to use both, fuzzy poset and fuzzy ordering: cut sets would be the same.

In addition, last propositions point out the importance of atoms in atomically generated lattice in representation of values of fuzzy sets by cuts corresponding to atoms.

References

1. Bělohlávek, R.: Fuzzy Relational Systems. Kluwer Academic Publishers, Dordrecht (2002)
2. Bodenhofer, U.: A New Approach to Fuzzy Orderings. Tatra Mt. Math. Publ. 16(Part I), 21–29 (1999)
3. Bodenhofer, U.: Representations and constructions of similarity based fuzzy orderings. Fuzzy Sets and Systems 137, 113–136 (2003)
4. Bodenhofer, U., De Baets, B., Fodor, J.: A compendium of fuzzy weak orders: Representations and constructions. Fuzzy Sets and Systems 158, 811–829 (2007)
5. Gorjanac-Ranitovic, M., Tepavčević, A.: General form of lattice-valued fuzzy sets under the cutworthy approach. Fuzzy Sets and Systems 158, 1213–1216 (2007)
6. Davey, B.A., Pristley, H.A.: Introduction to Lattices and Order. Cambridge University Press, Cambridge (1992)
7. Janis, V., Šeselja, B., Tepavčević, A.: Non-standard cut classification of fuzzy sets. Information Sciences 177, 161–169 (2007)
8. Klir, G., Yuan, B.: Fuzzy sets and fuzzy logic. Prentice Hall P T R, New Jersey (1995)
9. Ovchinnikov, S.V.: Similarity relations, fuzzy partitions, and fuzzy orderings. Fuzzy Sets an Systems 40(1), 107–126 (1991)
10. Ovchinnikov, S.V.: Well-graded spaces of valued sets. Discrete Mathematics 245, 217–233 (2002)
11. Šeselja, B., Tepavčević, A.: Completion of Ordered Structures by Cuts of Fuzzy Sets, An Overview. Fuzzy Sets and Systems 136, 1–19 (2003)
12. Šeselja, B., Tepavčević, A.: Representing Ordered Structures by Fuzzy Sets. An Overview, Fuzzy Sets and Systems 136, 21–39 (2003)
13. Šeselja, B.: Fuzzy Covering Relation and Ordering: An Abstract Approach, Computational Intelligence. In: Reusch, B. (ed.) Theory and Applications, pp. 295–300. Springer, Heidelberg (2006)
14. Tepavčević, A., Trajkovski, G.: L-fuzzy lattices: an introduction. Fuzzy Sets and Systems 123, 209–216 (2001)

A Practical Fuzzy Logic Controller for Sumo Robot Competition

Hamit Erdem

Baskent University, 06530 Ankara, Turkey
herdem@baskent.edu.tr
http://www.baskent.edu.tr/~herdem/

Abstract. This paper describes the design of fuzzy logic based sumo wrestling robot. The designed robot has a simple control algorithm and single fuzzy microcontroller is used in hardware implementation. The designed robot meets the specifications needed to compete in a sumo robot competition. The main difference of the designed system with earlier sumo robots is in control algorithms. A simple fuzzy logic controller is developed for detection and tracking of opponent in competition ring. Three infrared (IR) sharp sensors are used for target detection. Fuzzy microcontroller fuses the sensor data's and provides the necessary control signal to motors for heading robot toward the opponent. The fuzzy rules were optimized for the best results possible in software which are loaded in fuzzy controller. The implemented control algorithm shows better performance and executes the opponent detection algorithm in less time in comparison with conventional sumo robot algorithm. Design procedure and experimental results are presented to show the performance of the intelligent controller in designed system.

1 Introduction

Mobile robots provide an attractive platform for combining mechanical, electronic, computer, control and communication systems to create an integrated system for education and research. Mobile robots are mechanical devices capable of moving in an environment with a certain degree of autonomy. Autonomous navigation is associated with the availability of external sensors that capture information from the environment through different sensors. Motion autonomy in robotics may be defined as the ability for a robot to perform a given movement without any external intervention. Performing the task in best way is depended on performance of sensors and control algorithm. Fuzzy logic controller (FL) as nonlinear control method is used for sensor outputs reasoning and control of robot navigation in many works. FL is used to overcome the difficulties of modeling the unstructured, dynamically changing environment, which is difficult to express using mathematical equations [1,2,3]. Fuzzy logic is simple sensor fusion method for combining sensor outputs. Sumo wrestling robot is one of important field to attract student attention for engineering science. The Sumo Robot competition has become more popular around the world [4,5]. Similar to traditional sumo wrestling, the main objective is to force the competitor out of the

A. Ghosh, R.K. De, and S.K. Pal (Eds.): PReMI 2007, LNCS 4815, pp. 217–225, 2007.

ring while staying in the ring. In sumo robot competition an autonomous mobile robot tries to find the opponent inside the wrestling board by using different sensors and force the opponent robot out of the ring. There are two main factors which effect wrestling result, faster opponent detection and applying more force to competitor.This work suggests FLC solution for first factor. This paper describes the design of a fuzzy logic controlled (FLC) sumo robot with considering the specifications needed to compete in a sumo robot competition. Fuzzy logic is used for opponent detection and tracking. This paper is organized as follows. Section II presents the Problem definition and discussion of sumo operation. Section III defines the developed control algorithm and FLC. Section IV describes the application of FLC to sumo robot and tracking of opponent problem. The performance of designed robot and applied control algorithm are discussed based on experimental results in last section.

2   Problem Definition and Discussion of Sumo Operation

Based on the specifications needed to compete in a sumo robot competition a simple mobile robot is designed. Generally there are four operation modes must be considered in design of sumo robot. These modes are;

Search or hunting mode: The robot moves around and its range sensors scans across the ring for sensing the opponent Opponent facing mode: The opponent has been sensed before and robot tries to face the opponent directly. Attack mode: Robot drive straight ahead at full power to push the opponent off the ring. Survive mode: The robot enters this state when it detects the ring border line. Its goal is to survive by not going off the competition ring.

In sumo robot competition, there are two main factor must be considered for wining. The first factor is the effective algorithm for facing and directing to target after sensing of opponent. The second factor is applying maximum power to motors during attack mode which dependent on motors power. This work suggests FLC solution for first factor. With consideration of above operation modes a sumo robot is designed. The main block diagram and structure of the robot are shown in Figure 1. Main parts of designed system are:

- Ring border line detection sensor (IR sensor)
- Opponent detection sensors (Sharp IR sensors: S1, S2, S3)
- Touch sensor (mechanical sensor)
- Right and left DC motors
- Main controller (ST52F513 fuzzy microcontroller)
- Motor driver
- Display (LCD)

Main controller receives inputs from touch, ring line border and target detection sensors and generate control signals for differential drive motion system. As shown in Figure 1, three Sharp GP2D12 infrared range sensors are used to sense the opponent. These sensors are mounted on the left, right and middle front of the robot. Target detection sensors generate an analog voltage. The voltage is

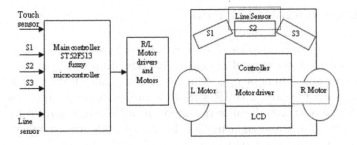

Fig. 1. Block diagram of designed robot and main structure of designed robot

higher the closer the sensor is to an object. Onboard ADC module of microcontroller is used for analog to digital conversion. Steering of robot is accomplished by rotating the motors at different angular velocities. When one motor rotates slower than the other, the robot will move in an arc curving to the side of the slower turning. By rotating the motors in the opposite direction of each other, the robot will rotate in place.

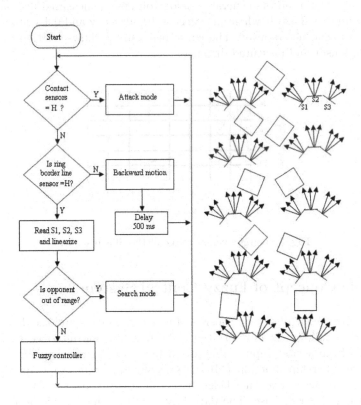

Fig. 2. Main control algorithm of sumo robot and some possible target positions

3 Main Control Algorithm

Based on the operation modes of sumo robot and applying FLC for directing to target a control algorithm is developed .This algorithm is presented in Figure 2. As shown in algorithms of operation, in search mode, Robot rotates around itself and scans the area with IR sensors to find opponent. Right after sensing the target, robot tries to localize the opponent and turns toward it by changing angular speed of left and right motors. In attack mode robot applies maximum power to motors to attack opponent and pushes it to out of ring. Attack mode finishes with sensing of ring border line. In FLC section, there is a simple sensor fusion by fuzzy logic. Due to outputs of S1, S2, S3 IR sensors, robot tries to localize the opponent and attack to it.

4 Opponent Detection Sensors

For localization and detecting of opponent three GP2D12 IR Sharp range sensors are used. This sensor is a compact, self-contained IR ranging system incorporating an IR transmitter, receiver, optics, filter, detection, and amplification circuitry [6]. These sensors generate analog voltage for measured distance. The output voltage is 2.5 volt when an object is 10 cm away and 0.4 volts when an object is 80 centimeters away. The curve in Figure 3 shows relation of output voltage of sensor and measured distance.

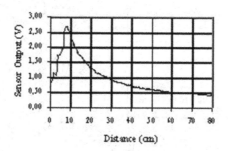

Fig. 3. Voltage versus range of the Sharp sensor

5 Basic Concept of Fuzzy Control System

A typical fuzzy control system consists of four components and the descriptions are stated as follows; 1) Fuzziffication: The fuzziffication interface performs a conversion from a crisp input value into a fuzzy set. In this application triangulation membership function (MF) is selected for input values fuzzification. 2) Knowledge Base: The knowledge base commonly consists of two sections: a database and a rule-base. The data base contains the MF of the fuzzy sets used in the fuzzy rules and the rule-base contains a number of fuzzy IF-THEN rules. 3) Inference Engine: The inference engine that performs the fuzzy reasoning upon the fuzzy control rules is the main component of the fuzzy controller. There are varieties of compositional methods in fuzzy inference, such as max-min compositional operation and max-product compositional operation etc. 4) Defuzzification: The defuzzification converts the fuzzy output of the rule-base into a non-fuzzy value. The center of area (COA) is the often used method in defuzzification. In this application COA method is used for determination of real value for directing of motors. For onboard application, the selected controller supports directly the fuzzy inference method.

5.1 Opponent Tracking with FLC

Based on sumo navigation and competition rules an algorithms is developed. There is a complete algorithm in Figure 2 for robot navigation in sumo competition. The algorithm takes advantage of the essential characteristics of a differential drive to direct the wrestling robot. There are three IR sensors for detection of opponent position. A simple MISO system is considered for opponent detection and tracking. As shown in main algorithm, the role of FLC begins when one of the sensors receives signals from opponent. A FLC with three input and single output is considered. The reason behind choosing three MF for each input sensor variable is to limit the number of fuzzy rules. This is one of the important considerations in designing a fuzzy logic controller. Fewer rules improve system response time especially in practical application.

5.2 Development of Fuzzy Rules

Fuzzy rules are intuitive rules that can be driven by considering all possible scenarios with input sensor values. For example, if S1 and S2 detects weak signal and S3 detects medium signal, the logical action of the robot is to take a small turn to right. Robot sensing area and possible aspects of the sumo robot when

Table 1. Fuzzy rules for sumo robot

Rule	S1	S2	S3	Motion	Rule	S1	S2	S3	Motion
Rule1	Low	Low	Medium	Small R.	Rule12	Medium	Medium	High	Center
Rule2	Low	Low	High	Full R.	Rule13	Medium	High	Low	Small L.
Rule3	Low	Medium	Low	Center	Rule14	Medium	High	Medium	Center
Rule4	Low	Medium	Medium	Small R.	Rule15	Medium	High	High	Center
Rule5	Low	Medium	High	Small R.	Rule16	High	Low	Low	Full L.
Rule6	Low	High	Low	Center	Rule17	High	Medium	Low	Small L.
Rule7	Low	High	Medium	Small R.	Rule18	High	Medium	Medium	Center
Rule8	Low	High	High	Center	Rule19	High	High	Low	Center
Rule9	Medium	Low	Low	Small L.	Rule20	High	High	Medium	Center
Rule10	Medium	Medium	Low	Small L.	Rule21	High	High	High	Center
Rule11	Medium	Medium	Medium	Center					

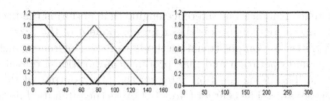

Fig. 4. Membership functions for input and output variables

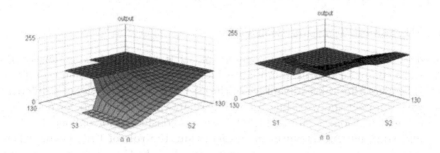

Fig. 5. Three dimensional plots for rule base (S2, S3 and output)(S1, S3 and output)

encountering an opponent are shown in Figure 2. Based on sumo behavior and possible aspects of the sumo robot when encountering an opponent, 21 rules are generated. The optimized rules for detection and tracking of target are shown in Table1.

For every input sensor three MF are selected, Low (L), Medium (M) and High (H) are fuzzy sets for input sensors. For output of fuzzy block singleton MF is selected. The selected membership functions for input and output variables are shown in Figure 4. IR sensors generate analog voltage approximately 0.42V to 2.66V. This range is divided into three sets for every input sensor. For the output of FLC which controls the angular speed of motors five MF are considered. The five singleton membership functions are Full Left (FL), Small Left (SL), Center (C), Small Right (SL) and Full Right (FR). Based on output of fuzzy controller, two different PWM value are sent to motors. For example generation of Center(C) value maintains the robot on a straight course while sending other values to controller will make robot turn to right or left.The result of fuzzy reasoning based on generated rules is evaluated and tested with using of Winfact 6 [7]. The Figure 5 shows three dimensional relations between sensors and fuzzy logic output.

5.3   Main Controller

In hardware application for onboard control, T52F513 fuzzy microcontroller is used [8]. T52F513 is a decision processor for the implementation of FL algorithms. This controller is a device of ST FIVE family of 8-bit Intelligent

Controller Units (ICU), which can perform, Both Boolean and fuzzy algorithms in an efficient manner. Sensor outputs are sent to analog inputs of controller. Target detection sensors outputs are sent to fuzzy block after A/D conversion. The result of fuzzy reasoning is sent to PWM module for motor control. Based on fuzzy reasoning two different PWM value are send to motor drive circuit.

6 Experimental Results

The first application is obtaining sensors perception area. Three sensors are used for sensing target. It is considered that sumo robot is in motionless condition and a cubic object (10x10 cm) is moving on horizontal direction from right to left with low speed. Due to the positions of Sharp sensors positions, firstly sensor S3 then sensor2 (S2) and finally sensor3 (S3) detects moving object nearly for 25 sampling time. The sampling time is 0.02s. Experiment condition and responses of sensors are shown in Figure 6. The sampling frequency of sensor values is 25Hz.

The second experiment is finding and pushing out the motionless object which is located at the back of robot. In start condition the head of the robot is against the object and sensors can not sense it. Robot starts in search mode and tries to find object. Firstly robot turns to left for a short time, the S1 detects the object. Based on control algorithm the rotating direction of motors changes until head of robot directed to the object, then it speeds up until hitting the object and pushes opponent out of the ring. For completely pushing out of ring, robot goes forward and backward due to sensor 2 response and hits the opponent several times. The steps of robot motion for detection of the opponent are shown in

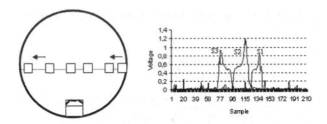

Fig. 6. Motionless robot and moving target from right to left

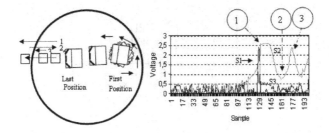

Fig. 7. Robot motions for attacking to target and sensors response

Figure 7. The motion steps 1 to 3 in Figure shows the backward and forward motions of robot after it touches the opponen.

During rotational motion, opponent is detected by S1 for a short time, then robot rotate to left and tries to center to opponent. As shown in figur 8, S2 has the maximum detection time. Considering the sampling time which is 0.02s. S2 senses the opponent nearly for 10 sampling time (0.2s). Sensing of target by middle sensor (S2) is the biggest time for sensor in this experiment. This long time shows directing of robot towards the target. Changing of S2 output shows forward and backward motion of robot for pushing out the opponent. After sensing ring border line backward motion begins. The third experiment is detection of two different targets which are located in right and left of the robot. In starting motion robot rotates to left and senses the object one. With applying heading algorithm and attack mode pushes out object one and go to search mode and start left hand rotation. With sensing the next object applies the same algorithm and pushes out the second object. Figure 8 shows robot motions and Figure 9 shows the sensors outputs during the third experiment

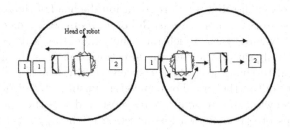

Fig. 8. Robot motions during attack to target 1 and 2

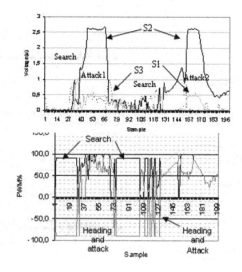

Fig. 9. Sensors responses in II experiment and PWM outputs for right and left motors

and the applied PWM duty cycles to motor control circuit. The positive and negative values of PWM represent turning of robot in search mode.

7 Conclusions and Results

In this work, an intelligent sumo wrestling robot is designed. The designed robot has a very simple and minimal hardware configuration and meets the specifications needed to compete in a sumo robot competition. Fuzzy logic controller is used as sub controller for tracking of the opponent in a competition. In sumo competitions, fast sensing, localization and maneuvering in right time are important factors to win. In comparison with traditional sumo robot control algorithm, the designed system takes minimum time for execution of control algorithms in hardware application and can attack to opponent with a minimal maneuvers. The performance of applied control method can be improved with adding some extera sensors for sensing attack of the opponent during competion.

References

1. Driankov, D., Saffiotti, A.: Fuzzy Logic for Autonomous Vehicle Navigation. Springer, New York (2001)
2. Seraji, H., Howard, A.: Behavior-based robot navigation on challenging terrain: A fuzzy logic approach. IEEE Trans on Robot Automat 18(3), 308–321 (2002)
3. Vadakkepat, P., Miin, O., Peng, X., Lee, T.: Fuzzy behavior based control of mobile robots. IEEE Trans on Fuzzy Syst 12(4), 559–565 (2004)
4. Liu, J., Pok, C.K., Keung.: Learning: coordinated maneuvers in complex environments: a sumo experiment. In: CEC 99 (1999)
5. Liu, J.M., Zhang, S.W.: Multi-phase sumo maneuver learning. ROBOTICA 22(2), 61–75 (2004)
6. Sharp Corporation: Distance Measuring Sensor GP2D120. Spec. ED-99170 (99)
7. Winfact 6: Simulation program, http://www.kahlert.com
8. ST Microelectronics: ST52F513 sheet's. FUZZYSTUDIOTM5 User Manual (2001)

Hierarchical Fuzzy Case Based Reasoning
with Multi-criteria Decision Making
for Financial Applications

Shanu Sushmita and Santanu Chaudhury

Electrical Engineering Department, Indian Institute of Technology Delhi,
New Delhi, India - 110016
{schaudhury,shanusushmita}@gmail.com

Abstract. This paper presents a framework for using a case-based reasoning system for stock analysis in financial market. The unique aspect of this paper is the use of a *hierarchical* structure for case representation. The system further incorporates a *multi-criteria* decision-making algorithm which furnishes the most suitable solution with respect to the current market scenario. Two important aspects of financial market are addressed in this paper: *stock evaluation* and *investment planning*. CBR and multi-criteria when used in conjunction offer an effective tool for evaluating goodness of a particular stock based on certain factors. The system also suggests a suitable investment plan based on the current assets of a particular investor. Stock evaluation maps to a flat case structure, but investment planning offers a scenario more suited for structuring the case into successive detailed layers of information related to different facets. This naturally leads to a hierarchical case structure.

1   Introduction

In this paper, we propose an application framework involving a fuzzy case based reasoning system with a hierarchical case structure and a multi-criteria decision making. So far, case structures used in majority of CBR applications have had a flat case structure which, to some extent, manages to incorporate example applications that can be mapped to flat case structures. Many of the complex real world problems require a framework capable of handling non-summarized information where one component's value is dependent on several other relevant factors. Investment planning is an example application having non-summarised information, which requires classification of case features, based on different aspects, into successive layers. This segregation of information into layers leads to a hierarchical case structure. Also, combining hierarchical CBR with multi-criteria is all together a new approach towards decision making systems tailored for applications having a dynamic and complex nature.

Past did witness some of the evolving research work, which looked into the financial areas and suggested how CBR can be efficiently used as a tool to produce a suitable decision making system. Some of the work like [1] addressed the movement of the stock market and its prediction. [2] proposed the daily financial condition indicator (DFCI) monitoring financial market built on CBR, [3], in their work proposed a new learning

A. Ghosh, R.K. De, and S.K. Pal (Eds.): PReMI 2007, LNCS 4815, pp. 226–234, 2007.

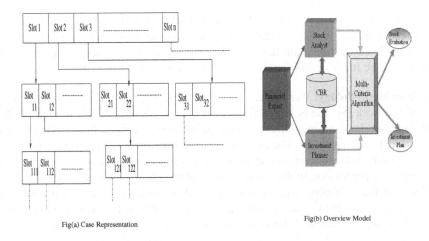

Fig(a) Case Representation

Fig(b) Overview Model

Fig. 1. Hierarchical Representation Of Case Structure and the System Model

technique which extracts new case vectors using Dynamic Adaptive Ensemble CBR, which again deals with the prediction of the overall stock market. Author [4] presents case-based reasoning approach for financial crises warning, which is again an approach that works on the broader level of financial issues. In other words, past research did manage to explore into the domain of financial market but, most of the research were confined to the working of the overall market situation and their prediction. Here, in this paper, we investigate use of CBR as an intelligent tool in considering and analysing an individuals stock with respect to current market scenario and investment planning of an individual. -Also, as stated before, CBR applications so far have always been mapped to flat case structures, while applications more suited to hierarchical case structures have not been explored. Although, work like [5] and [6] presented the concept of a hierarchical CBR, but the concept proposed involve reusing of multiple cases at various levels of abstraction. In [7] a CBR system with a two-level weight specification is developed. Our representation differs from the above mentioned approaches in terms of case representation style, where each case is a collection of independent sub cases.

The organisation of this paper is as follows: Section 2 gives a detailed explanation of the Case structure and working of the hierarchical case based reasoning system. Section 3 describes the implication of the proposed framework in financial market. In Section 4, we present the implementation details and the results of the two applications: Stock evaluation and Investment Planner.The last section presents the conclusion and future work.

2   Fuzzy Hierarchical Case Structure

Case - Structure: For our application, we have modified the classical flat case structure, such that each case is represented as a collection of sub-cases as shown in Fig. 1(a). Classical CBR uses symbolic and/or numeric attributes. However, many of the descriptors characterising the real world problems are associated with certain degree of fuzziness and uncertainty. Fuzzy CBR is a methodology which uses linguistic or re-alistic variables for case representation. It emulates human reasoning about similarity of real-world cases, which are fuzzy. Each sub case consists of its specific attributes, which could be fuzzy or non-fuzzy or complex (consisting of a sub case).

Working Methodology: Working methodology of the similarity computation is based on our past work [8]. The proposed framework operates in three different stages. In the first stage, similar cases are retrieved using the indexing attributes. In the second stage, we consider each retrieved case and compute a similarity score at every subsequent level for each child slot (referred also as a *node* in the context of a tree structure) of the respective parent slot. In third stage similarity scores are combined using weighted T-Norm for fuzzy attributes, and using Weighted similarity score for crisp attributes. Figure 2 represents the similarity computation of cases having hierarchical structure. A generalised representation of this computation is given in equations 1, 2 and 3.

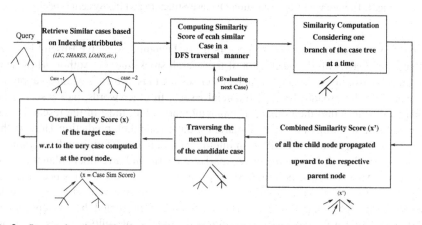

Fig. 2. Steps showing Similarity Score Computation of similar cases w.r.t the Query Case.(Similarity scores are combined using weighted T-Norm for the fuzzy child nodes and weighted summation for the crisp attributes.

Similarity Score Computation: For Crisp slots (non-fuzzy), similarity score of each node with respect to the query node is computed by exact matching, string comparison, and numerical comparison algorithms. The selection of the matching algorithm depends on the data type of the slot. At each parent slot, the aggregate similarity taking into account the non-fuzzy child slots is evaluated by a weighted summation of similarity of all the non-fuzzy child slots.

We show below the mathematical formulation for the evaluation of similarity score of a case with the hierarchical structure. Let $N_k^{\{l\}}$ denote the kth node at the level $\{l\}$. The node $N_k^{\{l\}}$ can have child nodes which could be fuzzy or non-fuzzy. The sets of fuzzy child nodes and crisp child nodes of $N_k^{\{l\}}$ are denoted as $fuzzyCh(N_k^{\{l\}})$ and $crispCh(N_k^{\{l\}})$ respectively. Let $s_k^{\{l\}}$ denote the aggregation of the similarity computed for the node $N_k^{\{l\}}$ and w be the weight. The similarity score at level $\{l\}$, that is $s_k^{\{l\}}$ has

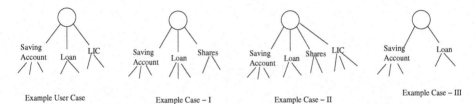

Fig. 3. The User case(query) represented above has three features (*Saving Account, Loan, LIC*. The database Case II has features (*Shares, Loan, LIC, Saving Account*), of which feature *Shares* is not there is the query case. This attribute(feature) and its detail expansion is not used in similarity computation. Situations where the number of similar nodes in the example cases become same, for example in case I and Case III, the computed similarity score would depend on the attribute value of the leaf nodes.

two components: fuzzy similarity score and crisp similarity score. The fuzzy component of the similarity score s_k is denoted as *fuzzySim*(s_k) and the crisp component is denoted as *crispSim*(s_k). We have:

$$fuzzySim(s_k^{\{l\}}) = \frac{\sum_i s_i^{\{l+1\}} w_i}{\sum_i w_i} \tag{1}$$

$$crispSim(s_k^{\{l\}}) = \frac{\sum_i s_i^{\{l+1\}} w_i}{\sum_i w_i} \tag{2}$$

where the summation index i varies for all the nodes $N_i^{\{l+1\}} \in crispCh(N_k^{\{l\}})$. We have:

$$s_k^{\{l\}} = \frac{fuzzySim(s_k^{\{l\}}) \times w_f + crispSim(s_k^{\{l\}}) \times w_c}{w_f + w_c} \tag{3}$$

where the weights w_f and w_c are sums of weights for fuzzy and crisp slots respectively.

Finally, we select the similar cases having a score above a threshold β. Thus we see that the similarity measure of a node is a weighted average of the similarity measure for its child nodes (both crisp and fuzzy).

Multi-criteria Decision Making Algorithm: Past researches have led to some useful multi-criteria decision making methods. ELECTRE [9], [10] and PRMETHEE [11] are some of the ranking function that could be used to outrank one similar case over the other based on certain parameters. In this paper, we have applied the ISD [10] method to furnish the most preferred solution, which considers the various criteria that affect the goodness of a particular stock or investment plan with respect to the current market situation. ISD(x_i, x_j) measures the degree of superiority or dominance of x_I over x_j in respect to all multiple attributes (or criteria) and all other cases.

3 Application Description

The application we propose primarily focuses on two important implications of financial market: stock evaluation and investment planning. Our system is a useful tool both

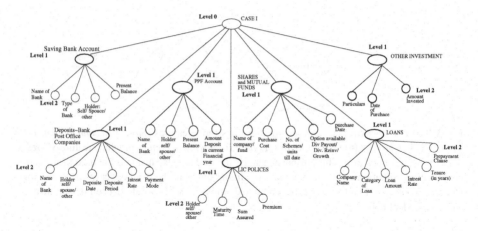

Fig. 4. Hierarchical Representation Of Case - Investment Planner: Example cases in the case base may have structure as shown above, or they may be an instance of the case structure having fewer branches and nodes, as shown in the figure 3 (Example Case -I) and (Example Case -II). Here, example Case-I possesses attributes, *Saving Account, Loan* and *Shares*. However, example Case -II consists of one more attribute (branch), *LIC Policy*.

for the people who are familiar to this field (stock analyst) as well as people who are actually not finance experts, but might be interested to buy a share or have an investment plan for themselves. The overview of the proposed methodology for this application is shown in Fig. 1(b).

Stock Analyst: Stock analyst works by taking values for parameters, which play an important role in determining the worth of any particular stock like: - Net sale, PAT, EPS, Growth percent, RoE, RoCE, etc. These values form the problem query of the case. If the most similar case suggests the query stock values to be good in past, do these values still continues to be good with respect to today's market status? To answer this question, the selected similar cases are assessed using multi-criteria algorithm, which makes use of certain criteria to determine their fitness in the present time.

Investment Planner: It works by taking inputs in two stages. In the first stage it takes in the financial goals of the person. In the second stage, it considers the worth of the current financial assets of that person.Once the information for both the stages are received, an investment plan is suggested which maps various assets to different financial goals. The Investment planner uses a hierarchical case structure as shown in Fig. 4. Initial retrieval, for the similar cases, is done by taking financial goals as indexing slots. Each retrieved cases, has a similarity score of one, since initially cases are retrieved by one to one mapping of each indexing slot. Overall similarity of the Case is done at each subsequent level by using equation 1, 2 and 3.

4   Implementation and Results

Visual C++ is used as the front end to create user interfaces and database handling. Microsoft Access servers as the back end to store cases in case base. When the stock analyst receives the query stock with its values for various parameters it compares the *query stock* against the stored stocks in the case base having similar values. Example stocks are selected based on qualitative or quantitative analysis, depending on the nature of the attribute. Finally the result, a suggestion regarding the query stock is given as a *"A Good Buy"* , or as a *"Not a Good buy"* or *"hold the stock"* if already possessing it.

Company name	Sector	Net Sales (Rs. m)	PAT (Rs. m)	EPS	EPS (Growth %)	P / E	RoE(%)	RoCE(%)	EV/ Sales	EV/ EBDIT
WIPRO	Information Technology	44910	10964.60	47.8	HIGH	37.9	HIGH	39.7	13.9	47

Fig. 5. Query Stock

Stock Analyst: In Response to a query stock, similar stocks are retrieved. These similar stocks are compared to obtain the stock having maximum preference, based on ranking criteria. How well these stocks fit in today's market was decided by considering the various criteria:

- Time: The current the better,
- Sector of the company: Should belong to same sector: IT, Fertilizer, etc,
- Percentage holding of the company: Comparing the market share,
- Balance sheet of the company: Comparing the assets and liabilities,
- Return over capital employed(RoCE%): Stocks having similar values are preferred,
- Public holding in the company,
- Value of Rupee against Dollar: The difference should be somewhat similar.

Company name	Sector	Net Sales (Rs. m)	PAT (Rs. m)	EPS	EPS (Growth %)	P / E	RoE(%)	RoCE(%)	EV/ Sales	EV/ EBDIT
WIPRO	Information Technology	44900	11964.52	47.8	HIGH	37.6	HIGH	38.5	13.9	45

(a) Retrieved Case 1

Company name	Sector	Net Sales (Rs. m)	PAT (Rs. m)	EPS	EPS (Growth %)	P / E	RoE(%)	RoCE(%)	EV/ Sales	EV/ EBDIT
TCS	Information Technology	45822	11000.20	47.8	36.7	37.9	37.9	39.7	13.9	40

(b) Retrieved Case 2

Company name	Sector	Net Sales (Rs. m)	PAT (Rs. m)	EPS	EPS (Growth %)	P / E	RoE(%)	RoCE(%)	EV/ Sales	EV/ EBDIT
POLARIS	Information Technology	43322	10500.34	47.0	35.4	36.98	37.9	40.0	14.01	42

(c) Retrieved Case 3

Fig. 6. Shows 3 retrieved stocks for the query of Fig. 5. Stock (a) was selected since it had the maximum Dominance

Goals	Purchase or construction of house	Major repairs or renovation	Your own Education	Education of children	Your own Marriage	Marriage of Children	Providding for medical or emergencies	Purchase of car or other vehicle	overseas Vacations	Retirments needs
Pririty	HIGH	LOW	LOW	LOW	HIGH	LOW	MEDIUM	HIGH	VERY HIGH	HIGH
Years to Go	20	5	–	–	2	25	ANY	2	1	30
Present Value of worth	40,00,000	5,00,000	––	–	2,00,000	10,00,000	50,000	8,00,000	5,00,000	20,00,000

Fig. 7. Financial Goals Described - Query(phase 1)

Short-term liquidity: Interest on deposits with banks is no more eligible for deduction from income taxes under section 80L of the IT Act. And considering our existing investments, we do not recommend you any fixed deposits.

Life Insurance: We recommend term insurance having coverage of Rs 20 Lacs coverage of Rs 20 Lacs with a term of 20 years for you.

Housing Loan: We recommend housing finance for 85% of the housing requirement i.e. 34 Lacs @ (Rate of Interest) 7.75% in year 2026 with a monthly EMI of Rs 40,804/-. However it needs to be reviewed periodically to ensure the rate of interest being charged is as per the market norms.

Car Loan: We recommend car loan for 100% of car requirement, i.e. 8 lacs @ 10% for 5 years in year 2008 with a monthly EMI of Rs 16,998/-. The same needs to be reviewed periodically to to ensure the rate of interest is charged as per market norms

Assumptions

1. Growth in Savings: Your savings are assumed to grow by 8%. Your current savings have been taken at Rs. 3 Lacs per annum approx. (excluding contribution to EPF), as discussed with you.

2. Inflation and interest rates: are projected to be 5% and 7% respectively in the long term.

3. Reinvestments: Investments made will not be withdrawn except for their earmarked purpose. Likewise, earnings on investments will remain reinvested till a goal is met.

4. Existing Portfolio: We have considered your existing portfolio of Rs. 4.21 Lacs (including your Bank Balances) during the construction of the plan.

Regular Invesment	Amount (Rs)
MIP Funds	200000
Insurance/Pension Premiums	136170
Equity/Balanced Mutual Funds	60000
Term Insurance	7000
Total (in Rs)	403170

Fig. 8. Recommended Investment Plan based on user profile and priorities

The aforementioned parameters are considered to evaluate the dominance of one retrieved stock over another, in response to the query stock. Selected Stock which got the highest rank amongst the similar stocks is shown in Fig. 6.

Investment Planner: The criteria used by the investment planner to rank the retrieved similar cases in response to a query case are as follows:

- Time: current the better,
- Risk assessment: Cases having similar risk tolerance are preferred.
- Priority of the goals,
- Age factor: Cases should belong to the recommended age group,
- Number of dependants,
- Profession: Self employed or salaried.

When the query (shown in Fig. 7(user's goals input) and 9(user's current assets worth) (A) is posed to the investment planner, the case (shown in Fig. 9) (B) having second highest score was selected.After applying the multi-criteria attributes on the selected cases,the case shown in figure 9 (B)dominated the other candidate cases in terms of its similarity as well as usefulness in present time. Profile and the priority(age=34, house plans (in 5 years) , retirement plans(high priority) and child education (two children)) of the query user had the maximum similarity with the goals and profile of the selected Example case (age=40, house plan (in 3 years), retirement plan(high priority), child education(two children)).The case shown in Fig. 9 suggested the Investment plan, shown in Fig. 8, for the posed query.

Having a case base size of 250 example case for Investment Planner and over 500 for Stock Analyst, we have achieved a reasonable performance of the proposed application. The results obtained were discussed with few expert stock analysts and they found them

(A) Current assets Described (Query Phase –II)

Holder: Self/Spouse	Time to Maturity	Sum Assured	Premium	Table Term
Self	20	1,00,000	4,422	107–20
Self	15	1,00,000	5,325	106–15

Name of Bank/Post Office	Holder: Self/Spouse	Deposite Period	Deposite Amount	Interest Rate	Payment mode: monthly/yearly
ICICI	Self	36	5000	3	Monthly

Name of Bank	Type of Account	Holder: Self/Spouse	Present Balance
ICICI	Savings	Self	30,000
HDFC	Savings	Self	2,50,000

Name of Company	Loan Category House/car/personal	Loan amount	Monthly Premium	Rate of Interest	Tenure (in years)	Prepayment clause
ICICI	House	1,200,000	13,208	10.75	15	Partial payment
ICICI	House	450,000	4198	7.5	20	Partial Prepayment

(B) Retrieved Inverstment Plan with Highest Ranking

Name of Company	category of Loan House/car/personal	Loan amount	Monthly Premium	Rate of Interest	Tenure (in years)	Prepayment clause
ICICI	House	1,000,000	13,208	10.75	17	Partial payment
ICICI	House	450,000	4198	7.5	20	Partial Prepayment
HDFC	Car	2,00,000	1,000	2.5	4	Zero Prepayment

Name of Bank/Post Office	Holder: Self/Spouse	Deposite Period	Deposite Amount	Interest Rate	Payment mode: monthly/yearly
SBI	Self	36	4500	3	Monthly

Name of Bank	Type of Account	Holder: Self/Spouse	Present Balance
ICICI	Current	Self	32,563
HDFC	Savings	Spouse	2,50,000

Fig. 9. Query Phase(II) and the selected Investment Plan with Highest Ranking

to be satisfactory 75% of the time. For certain set of test data the prediction made by the system were also compared with the predictions made by analysts themselves. We found the comparitive results to be quite satisfying. Although, some of the results were not justified, which we conclude were due to the fact that for such kind of application dealing with fluctuating market trend a more updated and large case base would be required. The proposed suggestion fits convincingly with going market trend. Also, the Investment Planner produces logically convincing Investment plan in response to the query.

5 Conclusion and Future Work

Having a hierarchical case structure offers a better representation of the complex details. This provides a clearer understanding of the application while taking comprehensive details into account. As a result the proposed solution suits more appropriately with the actual requirements of the user. As a future work, we propose to look at the next few other important applications of financial market, like mutual funds.

References

1. Kim, K.-J.: Toward Global Optimization of Case-Based Reasoning Systems for Financial Forecasting. Applied Intelligence 21, 239–249 (2004)
2. Kyong Joo Oh, T.Y.K.: Financial market monitoring by case-based reasoning. Expert Systems with Applications 32 (2007)
3. Chun, Y.J.P.S.H.: Dynamic adaptive ensemble case-based reasoning: application to stock market prediction. Expert Systems with Applications 28, 435–443 (2005)
4. Ying Wei, Y.W., Li, F., Li, F.: Case-Based Reasoning: An Intelligent Approach Applied for Financial Crises Warning. Springer, Heidelberg (2003)

5. Smyth, P.C.B., Keane, M.T.: Hierarchical case-based reasoning integrating case-based and decompositional problem-solving techniques for plant-control software design. Knowledge and Data Engineering (2001)
6. Tang, Y.: Time series extrapolation using hierarchical case-based reasoning. Signal Processing, Pattern Recognition, and Applications (2006)
7. Chi-I Hsua, P.-L.H., Chiub, C.: Predicting information systems outsourcing success using a hierarchical design of case-based reasoning (2004)
8. Singh, T., Goswami, S.C.P.S.: Distributed fuzzy case based reasoning. Elsevier:Journal of Applied Soft Computing 4, 323–343 (2004)
9. Roy, B.: The outranking approach and the foundations of ELECTRE methods ch in book: Readings in Multiple Criteria Decision Aid. In: Costa, C.A.B.e (ed.) Springer, Berlin (1990)
10. Pedro, J.S., Burstien, F.: A framework for case-based fuzzy multi-criteria decision support for tropical cyclone forecasting. In: Proceedings of the 36th Hawaii International Conference on System Sciences(HICSS), IEEE, Los Alamitos (2002)
11. Brans, J.P., Mareschal, B.: The PROMETHEE methods for MCDM; The PROMALC GAIA and BANKADVISER softwarech in book: Readings in Multiple Criteria Decision Aid. In: Costa, C.A.B.e. (ed.) Springer, Berlin (1990)

Comparison of Neural Network Boolean Factor Analysis Method with Some Other Dimension Reduction Methods on Bars Problem[*]

Dušan Húsek[1], Pavel Moravec[2], Václav Snášel[2], Alexander Frolov[3],
Hana Řezanková[4], and Pavel Polyakov[5]

[1] Institute of Computer Science, Dept. of Neural Networks,
Academy of Sciences of Czech Republic, Pod Vodárenskou věží 2,
182 07 Prague, Czech Republic
dusan@cs.cas.cz

[2] Department of Computer Science, FEECS, VŠB – Technical University of Ostrava,
17. listopadu 15, 708 33 Ostrava-Poruba, Czech Republic
{pavel.moravec,vaclav.snasel}@vsb.cz

[3] Institute of Higher Nervous Activity and Neurophysiology, Russian Academy of Sciences,
Butlerova 5a, 117 485 Moscow, Russia
aafrolov@mail.ru,

[4] Department of Statistics and Probability, University of Economics,
Prague, W. Churchill sq. 4, 130 67 Prague, Czech Republic
rezanka@vse.cz

[5] Institute of Optical Neural Technologies, Russian Academy of Sciences,
Vavilova 44, 119 333 Moscow, Russia
pavel.mipt@mail.ru

Abstract. In this paper, we compare performance of novel neural network based algorithm for Boolean factor analysis with several dimension reduction techniques as a tool for feature extraction. Compared are namely singular value decomposition, semi-discrete decomposition and non-negative matrix factorization algorithms, including some cluster analysis methods as well. Even if the mainly mentioned methods are linear, it is interesting to compare them with neural network based Boolean factor analysis, because they are well elaborated. Second reason for this is to show basic differences between Boolean and linear case. So called bars problem is used as the benchmark. Set of artificial signals generated as a Boolean sum of given number of bars is analyzed by these methods. Resulting images show that Boolean factor analysis is upmost suitable method for this kind of data.

1 Introduction

In order to perform object recognition (no mater which one) it is necessary to learn representations of the underlying characteristic components. Such components correspond

---

[*] The work was partly funded by the Centre of Applied Cybernetics 1M6840070004 and partly by the Institutional Research Plan AV0Z10300504 "Computer Science for the Information Society: Models, Algorithms, Appplications" and by the project 1ET100300419 of the Program Information Society of the Thematic Program II of the National Research Program of the Czech Republic.

A. Ghosh, R.K. De, and S.K. Pal (Eds.): PReMI 2007, LNCS 4815, pp. 235–243, 2007.

to objects, object-parts, or features. An image usually contains a number of different objects, parts, or features and these components can occur in different configurations to form many distinct images. Identifying the underlying components which are combined to form images is thus essential for learning the perceptual representations necessary for performing object recognition.

There exists many attempts that could be used for this reason. In this paper we concentrate on *neural network based algorithm for Boolean factor analysis* and compare it with some recently used dimension reduction techniques for automatic feature extraction. Here we empirically investigate their performance.

How does the brain form a useful representation of its environment? That is question that was behind the development of neural network based methods for dimension reduction. Among those neural network based Boolean factor analysis, see Frolov at al. [1,2], and Földiáks network [3] fall.

The comparison of linear approaches [4,5,6,7,8] capable to find an approximate solution of the Boolean factor analysis task with our attempt [2], if only for the reason they are fast and well elaborated, is very interesting. The most popular linear method is the *singular value decomposition* which was already successfully used for automatic feature extraction. However singular value decomposition is not suitable for huge data collections and is computationally expensive, so other methods of dimension reduction were proposed. Here we apply *semi-discrete decomposition* as example. As the data matrix has all elements non-negative, a new method called *non-negative matrix factorization* is applied as well.

However, for the sake of finding an hidden structure in the images data statistical methods used very often in the past are still good choice, mainly different algorithms for cluster analysis, perhaps in conjunction with a some data visualization methods.

The bars problem (and its variations) is a benchmark task for the learning of independent image features [3]. The collection, the *base vectors* can be interpreted as images, describing some common characteristics of several input signals. These base vectors are often called eigenfaces in the case of face recognition task, see M. Turk and A. Pentland [9].

The rest of this paper is organized as follows. The second section explains the dimension reduction method used in this study. Then in the section three we describe experimental results, and finally in the section four we made some conclusions.

2   Dimension Reduction

We used the three promising linear methods of dimension reduction for our comparison with Boolean factor analysis – singular value decomposition, semi-discrete decomposition, non-negative matrix factorization and two statistical clustering methods. All of them are briefly described bellow.

2.1   Neural Network Boolean Factor Analysis

Neural network Boolean Factor Analysis (*NBFA*) is a powerful method for revealing the information redundancy of high dimensional binary signals [2]. It allows to express every signal (vector of variables) from binary data matrix A of observations as super-position of binary factors:

$$A = \bigvee_{l=1}^{L} S_l f^l, \tag{1}$$

where S_l is a component of factor scores and $\mathbf{f}^l$ is a vector of factor loadings and $\vee$ denotes Boolean summation ($0 \vee 0 = 0, 1 \vee 0 = 1, 0 \vee 1 = 1, 1 \vee 1 = 1$). If we mark Boolean matrix multiplication by the symbol $\odot$, then we can express approximation of data matrix A in matrix notation

$$A_k \simeq F \odot S \tag{2}$$

where $\mathbf{S}$ is the matrix of factor scores and $\mathbf{F}$ is the matrix of factor loadings. The Boolean factor analysis implies that components of original signals, factor loadings and factor scores are binary values.

Optimal solution of $\mathbf{A_k}$ decomposition according to (2) by brute force search is NP-hard problem and as such is not suitable for high dimensional data. On other side the classical linear methods could not take into account non-linearity of Boolean summation and therefore are inadequate for this task.

2.2   Singular Value Decomposition

Singular Value Decomposition (*SVD*), see M. Berry, S. Dumais, and T. Letsche [5], is an algebraic extension of classical vector model. It is similar to the PCA method, which was originally used for the generation of eigenfaces. Informally, SVD discovers significant properties and represents the images as linear combinations of the base vectors. Moreover, the base vectors are ordered according to their significance for the reconstructed image, which allows us to consider only the first k base vectors as important (the remaining ones are interpreted as "noise" and discarded). Furthermore, SVD is often referred to as more successful in recall when compared to querying whole image vectors (M. Berry, S. Dumais, and T. Letsche [5]).

Formally, we decompose the matrix of images A by singular value decomposition, calculating singular values and singular vectors of A.

We have matrix A, which is an $n \times m$ rank-r matrix (where $m \geq n$ without loss of generality) and values $\sigma_1, \ldots, \sigma_r$ are calculated from eigenvalues of matrix AA^T as $\sigma_i = \sqrt{\lambda_i}$. Based on them, we can calculate column-orthonormal matrices $U = (u_1, \ldots, u_n)$ and $V = (v_1, \ldots, v_n)$, where $U^T U = I_n$ a $V^T V = I_m$, and a diagonal matrix $\Sigma = diag(\sigma_1, \ldots, \sigma_n)$, where $\sigma_i > 0$ for $i \leq r, \sigma_i \geq \sigma_{i+1}$ and $\sigma_{r+1} = \cdots = \sigma_n = 0$.

The decomposition

$$A = U \Sigma V^T \tag{3}$$

is called *singular decomposition* of matrix A and the numbers $\sigma_1, \ldots, \sigma_r$ are *singular values* of the matrix A. Columns of U (or V) are called *left* (or *right*) singular vectors of matrix A.

Now we have a decomposition of the original matrix of images A. We get r nonzero singular numbers, where r is the rank of the original matrix A. Because the singular values usually fall quickly, we can take only k greatest singular values with the corresponding singular vector coordinates and create a *k-reduced singular decomposition* of A.

Let us have k $(0 < k < r)$ and singular value decomposition of A

$$A = U\Sigma V^T \approx A_k = (U_k U_0) \begin{pmatrix} \Sigma_k & 0 \\ 0 & \Sigma_0 \end{pmatrix} \begin{pmatrix} V_k^T \\ V_0^T \end{pmatrix} \tag{4}$$

We call $A_k = U_k \Sigma_k V_k^T$ a k-reduced singular value decomposition (rank-k SVD).

Instead of the A_k matrix, a matrix of image vectors in reduced space $D_k = \Sigma_k V_k^T$ is used in SVD as the representation of image collection. The image vectors (columns in D_k) are now represented as points in k-dimensional space (the *feature-space*). represent the matrices U_k, Σ_k, V_k^T.

Rank-k SVD is the best rank-k approximation of the original matrix A. This means that any other decomposition will increase the approximation error, calculated as a sum of squares (*Frobenius norm*) of error matrix $B = A - A_k$. However, it does not implicate that we could not obtain better precision and recall values with a different approximation.

Once computed, SVD reflects only the decomposition of original matrix of images. If several hundreds of images have to be added to existing decomposition (*folding-in*), the decomposition may become inaccurate. Because the recalculation of SVD is expensive, so it is impossible to recalculate SVD every time images are inserted. The *SVD-Updating* is a partial solution, but since the error slightly increases with inserted images. If the updates happen frequently, the recalculation of SVD may be needed soon or later.

2.3 Semi-discrete Decomposition

Semi-Discrete Decomposition (*SDD*) is one of other SVD based methods, proposed recently for text retrieval in (T. G. Kolda and D. P. O'Leary [8]). As mentioned earlier, the rank-k SVD method (called *truncated SVD* by authors of semi-discrete decomposition) produces dense matrices U and V, so the resulting required storage may be even larger than the one needed by the original term-by-document matrix A.

To improve the required storage size and query time, the semi-discrete decomposition was defined as

$$A \approx A_k = X_k D_k Y_k^T, \tag{5}$$

where each coordinate of X_k and Y_k is constrained to have entries from the set $\varphi = \{-1, 0, 1\}$, and the matrix D_k is a diagonal matrix with positive coordinates.

The SDD does not reproduce A exactly, even if $k = n$, but it uses very little storage with respect to the observed accuracy of the approximation. A rank-k SDD (although from mathematical standpoint it is a sum on rank-1 matrices) requires the storage of $k(m + n)$ values from the set $\{-1, 0, 1\}$ and k scalars. The scalars need to be only single precision because the algorithm is self-correcting. The SDD approximation is formed iteratively.

The optimal choice of the triplets (x_i, d_i, y_i) for given k can be determined using greedy algorithm, based on the residual $R_k = A - A_{k-1}$ (where A_0 is a zero matrix).

2.4 Non-negative Matrix Factorization

The Non-negative Matrix Factorization (*NMF*)method calculates an approximation of the matrix A as a product of two matrices, W and H. The matrices are usually pre-filled with random values (or H is initialized to zero and W is randomly generated). During the calculation the values in W and H stay positive. The approximation of matrix A, matrix A_k, can be calculated as $A_k = WH$.

The original NMF method tries to minimize the Frobenius norm of the difference between A and A'_k using

$$\min_{W,H} ||V - WH||_F^2 \qquad (6)$$

as the criterion in the minimization problem.

Recently, a new method was proposed in (M. W. Spratling [10]), where the constrained least squares problem is solved

$$\min_{H_j}\{||V_j - WH_j||_2^2 - \lambda||H_j||_2^2\} \qquad (7)$$

as the criterion in the minimization problem. This approach gives yields better results for sparse matrices.

Unlike in SVD, the base vectors are not ordered from the most general one and we have to calculate the decomposition for each value of k separately.

2.5 Statistical Clustering Methods

To set our method in more global context we applied cluster analysis, too. The clustering methods help to reveal groups of similar features, it means typical parts of images. However, obtaining disjunctive clusters is a problem of the usage of traditional hard clustering. We have chosen two most promising techniques available in recent statistical packages – hierarchical agglomerative algorithm and two-step cluster analysis.

A hierarchical agglomerative algorithm (*HAA*) starts with each feature in a group of its own. Then it merges clusters until only one large cluster remains which includes all features. The user must choose dissimilarity or similarity measure and agglomerative procedure. At the first step, when each feature represents its own cluster, the dissimilarity between two features is defined by the chosen dissimilarity measure. However, once several features have been linked together, we need a linkage or amalgamation rule to determine when two clusters are sufficiently similar to be linked together. Several linkage rules have been proposed. For example, the distance between two different clusters can be determined by the greatest distance between two features in the clusters (complete linkage method – *CL*), or average distance between all pairs of objects in the two clusters (average linkage between groups – *ALBG*). Hierarchical clustering is based on the proximity matrix (dissimilarities for all pairs of features) and it is independent on the order of features.

For binary data (one for the black point and zero for the white point), we can choose for example Jaccard and Ochiai (cosine) similarity measures of two features. The former can be expressed as $S_J = \frac{a}{a+b+c}$ where a is the number of the common occurrences of ones and $b+c$ is the number of pairs in which one value is one and the second is zero.

The latter can be expressed as $S_O = \sqrt{\frac{a}{a+b} \cdot \frac{a}{a+c}}$. The further clustering techniques mentioned above are suitable for large data files but they are independent on the order of features.

In two-step cluster analysis (*TSCA*), the features are arranged into sub-clusters, known as cluster features (CF), first. These cluster features are then clustered into k groups, using a traditional hierarchical clustering procedure. A cluster feature represents a set of summary statistics on a subset of the data. The algorithm consists of two phases. In the first one, an initial CF tree is built (a multi-level compression of the data that tries to preserve the inherent clustering structure of the data). In the second one, an arbitrary clustering algorithm is used to cluster the leaf nodes of the CF tree. Advantage of this method is its ability to work with larger data sets; disadvantage then, is its sensitivity to the order of the objects (features in our case). In the implementation in the SPSS system, the log-likelihood distance is proposed.

This distance between clusters a and b is

$$d(a,b) = \zeta_a + \zeta_b - \zeta_{<a,b>}$$

where $< a, b >$ denotes a cluster created by joining objects from clusters a and b, and

$$\xi_g = -m_g \sum_{i=1}^{n} H_{gi} ,$$

where m_g is the number of features in the gth cluster, n is the number of images and H_{gi} is an entropy

$$H_{gi} = -\sum_{u=1}^{2} \frac{m_{giu}}{m_g} \ln \frac{m_{giu}}{m_g},$$

where m_{gi1} is the number of zeros and m_{gi2} is the number of ones.

3   Experimental Results

For testing of above mentioned methods, we used generic collection of 1600 32 × 32 black-and-white images containing different combinationas of horizontal and vertical lines (bars). The probabilities of bars to occur in images were the same and equal to 10/64, i.e. images contain 10 bars in average. An example of several images from generated collection is shown in Figure 1a.

For the first view on image structures, we applied traditional cluster analysis. We clustered 1024 (32 × 32) positions into 64 and 32 clusters. The problem of the use of traditional cluster analysis consists in that we obtain disjunctive clusters. So we can find only parts of horizontal or vertical bars respectively.

We applied HAA and TSCA algorithms in the SPSS system. The problem of the the latter consists in that it is dependent on the order of features. We used two different orders; in the second one the images from 1001 to 1009 were placed between the images 100 and 101 for example.

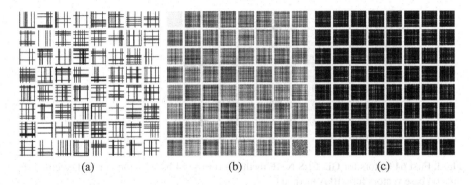

Fig. 1. Several bars from generated collection (a) First 64 base images of bars for SVD method (b) First 64 factors for original NMF method (c)

For hierarchical cluster analysis, we tried to use different similarity measures. We found that linkage methods have more influence to the results of clustering than similarity measures. We used Jaccard and Ochiai (cosine) similarity measures suitable for asymmetric binary attributes. We found both as suitable methods for the identification of the bars or their parts. For 64 clusters, the differences were only in a few assignments of positions by ALBG and CL methods with Jaccard and Ochiai measures.

The figures illustrate the application of some of these techniques for 64 and 32 clusters. Figure 2a,b show results of ALBG method with Jaccard measure. Figure 2a for 32 clusters and Figure 2b for 64 clusters. In the case of 32 clusters, we found 32 horizontal bars (see Figure 2c) by TSCA method for the second order of features.

Many of tested methods were able to generate a set of base images or factors, which should ideally record all possible bar positions. However, not all methods were truly successful in this.

With SVD, we obtain classic singular vectors, the most general being among the first. The first few are shown in Figure 1b. We can se, that the bars are not separated and different shades of gray appear.

The NMF methods yield different results. The original NMF method, based on the adjustment of random matrices W and H provides hardly-recognizable images even for

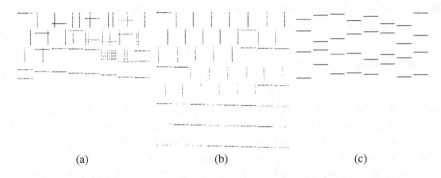

Fig. 2. ALBG Jaccard – 32 clusters (a) ALBG Jaccard – 64 clusters (b) TSCA – 32 clusters (c)

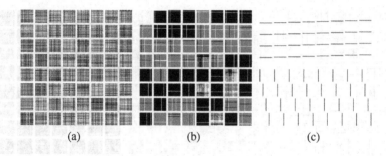

Fig. 3. First 64 factors for GD-CLS NMF method (a) First 64 base vectors for SDD method (b) First 64 base vectors for NBFA method (c)

$k = 100$ and 1000 iterations (we used 100 iterations for other experiments). Moreover, these base images still contain significant salt and pepper noise and have a bad contrast. The factors are shown in Figure 1c. We must also note, that the NMF decomposition will yield slightly different results each time it is run, because the matrix(es) are pre-filled with random values.

The GD-CLS modification of NMF method (proposed in [7]) tries to improve the decomposition by calculating the constrained least squares problem. This leads to a better overall quality, however, the decomposition really depends on the pre-filled random matrix H. The result is shown in Figure 3a.

The SDD method differs slightly from previous methods, since each factor contains only values $\{-1, 0, 1\}$. Gray in the factors shown in Figure 3b represents 0; -1 and 1 are represented with black and white respectively.

The base vectors in Figure 3b can be divided into three categories:

1. Base vectors containing only one bar.
2. Base vectors containing one horizontal and one vertical bar.
3. Other base vectors, containing several bars and in some cases even noise.

Finally, we made decomposition of images into binary vectors by Boolean factor analysis method proposed in [11]. The factor search was performed under assumption that the number of ones in factor is not less than 5 and not greater than 200. Since the images are obtained by Boolean summation of binary bars, it is not surprising, that NBFA is able to reconstruct all bars as base vectors, providing an ideal solution, as we can see in Figure 3c.

4   Conclusion

In this paper, we have compared several dimension reduction and clustering methods on bars collection. NBFA perfectly found basis (factors) from which the all learning pictures can be reconstructed. It is because factor analysis is looking for hidden dependencies among observed data. On the other side reason for 100% success resides also in this, that there was no noise superposed on the input data.

Cluster analysis is focused on finding original factors from which images were generated. Applied clustering methods were quite successful in finding these factors. The problem consists in created disjunctive clusters. So, only some bars or their parts were revealed. However, from the general view on 64 clusters, it is obvious, that images are compounded from vertical and horizontal bars (lines). By two-step cluster analysis 32 horizontal lines were revealed by clustering to 32 clusters.

The SVD method may also be used, but the results are not as good as in the case of NBFA. The SVD and NMF methods yield slightly worse results, since they are not focused on binary data.

References

1. Frolov, A.A., Sirota, A.M., Húsek, D., Muravjev, P.: Binary factorization in Hopfield-like neural networks: single-step approximation and computer simulations. In: Neural Networks World, pp. 139–152 (2004)
2. Frolov, A.A., Húsek, D., Muravjev, P., Polyakov, P.: Boolean Factor Analysis by Attractor Neural Network. Neural Networks, IEEE Transactions 18(3), 698–707 (2007)
3. Földiák, P.: Forming sparse representations by local anti-Hebbian learning. Biological cybernetics 64(22), 165–170 (1990)
4. Fierro, R.D., Jiang, E.P.: Lanczos and the Riemannian SVD in information retrieval applications. Numerical Algebra with Applications 3(1), 1–18 (2003)
5. Berry, M., Dumais, S., Letsche, T.: Computational Methods for Intelligent Information Access. In: Proceedings of the 1995 ACM/IEEE Supercomputing Conference, San Diego, California, USA (1995)
6. Berry, M.W., Fierro, R.D.: Low-Rank Orhogonal Decomposition for Information Retrieval Applications. Numerical Algebra with Applications 1(1), 1–27 (1996)
7. Shahnaz, F., Berry, M., Pauca, P., Plemmons, R.: Document clustering using nonnegative matrix factorization. Journal on Information Processing and Management 42, 373–386 (2006)
8. Kolda, T.G., O'Leary, D.P.: Computation and uses of the semidiscrete matrix decomposition. ACM Transactions on Information Processing (2000)
9. Turk, M., Pentland, A.: Eigenfaces for recognition. Journal of Cognitive Neuroscience 3(1), 71–86 (1991)
10. Spratling, M.W.: Learning Image Components for Object Recognition. Journal of Machine Learning Research 7, 793–815 (2006)
11. Frolov, A.A., Husek, D., Polyakov, P., Rezankova, H.: New Neural Network Based Approach Helps to Discover Hidden Russian Parliament Votting Paterns. In: Proceedings of International Joint Conference on Neural Networks, pp. 6518–6523. Omnipress, Madison, WI, USA (2006)

A GA-Based Pruning Strategy and Weight Update Algorithm for Efficient Nonlinear System Identification

G. Panda, Babita Majhi, D. Mohanty, and A.K. Sahoo

ECE department, National Institute of Technology Rourkela,
Rourkela - 769 008, Orissa, India

Abstract. Identification of nonlinear static and dynamic systems plays a key role in many engineering applications including communication, control and instrumentation. Various adaptive models have been suggested in the literature using ANN and Fuzzy logic based structures. In this paper we employ an efficient and low complexity Functional Link ANN (FLANN) model for identifying such nonlinear systems using GA based learning of connective weights. In addition, pruning of connecting paths is also simultaneously carried out using GA to reduce the network architecture. Computer simulations on various static and dynamic systems indicate that there is more than 50% reduction in original model FLANN structure with almost equivalent identification performance.

1 Introduction

The area of system identification is one of the most important areas in engineering because of its applicability to a wide range of problems. Recently, Artificial Neural Networks (ANN) has emerged as a powerful learning technique to perform complex tasks in highly nonlinear dynamic environments. At present, most of the work on system identification using neural networks are based on multilayer feedforward neural networks with backpropagation learning or more efficient variations of this algorithm [1,2]. On the other hand the Functional link ANN (FLANN) originally proposed by Pao [3] is a single layer structure with functionally mapped inputs. The performance of FLANN for system identification of nonlinear systems has been reported [4] in the literature. Patra and Kot [5] have used Chebyschev expansions for nonlinear system identification and have shown that the identification performance is better than that offered by the multilayer ANN (MLANN) model. Evolutionary computation has been applied to search optimal values of recursive least-square (RLS) algorithm used in the system identification model [6].

While constructing an artificial neural network the designer is often faced with the problem of choosing a network of the right size for the task to be carried out. The advantage of using a reduced neural network is less costly and faster in operation. However, a much reduced network cannot solve the required problem while a fully ANN may lead to accurate solution. Choosing an appropriate ANN

A. Ghosh, R.K. De, and S.K. Pal (Eds.): PReMI 2007, LNCS 4815, pp. 244–251, 2007.

architecture of a learning task is then an important issue in training neural networks. Giles and Omlin [7] have applied the pruning strategy for recurrent networks. In this paper we have considered an adequately expanded FLANN model for the identification of nonlinear plant and then used Genetic Algorithm (GA) to train the filter weights as well to obtain the pruned input paths based on their contributions. Procedure for simultaneous pruning and training of weights have been carried out in subsequent sections to obtain a low complexity reduced structure.

The rest of paper is organized as follows. In Section 2 the basics of adaptive system identification is presented. Section 3 illustrates the proposed GA based pruning and training method using FLANN structure. The performance of the proposed model obtained from computer simulations are presented in Section 4. We present some concluding remarks in Section 5.

2 Adaptive System Identification

The essential and principal property of an adaptive system is its time-varying, self-adjusting performance. System identification concerns with the determination of a system, on the basis of input output data samples. The identification task is to determine a suitable estimate of finite dimensional parameters which completely characterize the plant. Depending upon input-output relation, the identification of systems can have two groups. In static identification the output at any instant depends upon the input at that instant. The system is essentially a memoryless one and mathematically it is represented as $y(n) = f[x(n)]$ where $y(n)$ is the output at the nth instant corresponding to the input $x(n)$. In dynamic identification the output at any instant depends upon the input at that instant as well as the past inputs and outputs. These systems have memory to store past values and mathematically represented as $y(n) = f[x(n), x(n-1), x(n-2) \cdots y(n-1), y(n-2), \cdots]$ where $y(n)$ is the output at the nth instant corresponding to the input $x(n)$. A basic system identification structure is shown in figure 1. The impulse response of the linear segment of the plant is represented by $h(n)$ which is followed by nonlinearity (NL) associated with it. White Gaussian noise $q(n)$ is added with nonlinear output accounts for measurement noise. The desired output $d(n)$ is compared with the estimated output $y(n)$ of the identifier to generate the error $e(n)$ for updating the weights of the model. The training of the filter weights is continued until

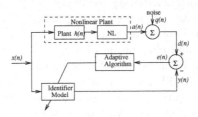

Fig. 1. Block diagram of system identification structure

the error becomes minimum and does not decrease further. At this stage the correlation between input signal and error signal is minimum.

3   Proposed GA Based Pruning and Training Using FLANN Structure

The FLANN based system identification is shown in figure 2. The FLANN is a single layer network in which the hidden nodes are absent. Here each input pattern is functionally expanded to generate a set of linearly independent functions. The functional expansion is achieved using trigonometric, polynomial or exponential functions. An N dimensional input pattern $X = \begin{bmatrix} x_1 & x_2 & \cdots & x_N \end{bmatrix}^T$. Thus the functionally expanded patterns becomes $X^* = \begin{bmatrix} 1 & x_1 & f_1(x_1) & \cdots & x_N & f_1(x_N) \end{bmatrix}$ where all the terms in the square bracket represents enhanced patterns. Then the improved patterns are used for pattern classification purpose.

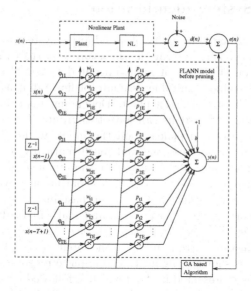

Fig. 2. FLANN based identification of nonlinear plants showing updating of weights and pruning path

In this Section a new algorithm for simultaneous training and pruning of weights using binary coded genetic algorithm (BGA) is proposed. Such a choice has led to effective pruning of branch and updating of weights. The pruning strategy is based on the idea of successive elimination of less productive paths (functional expansions) and elimination of weights from the FLANN architecture. As a result, many branches (functional expansions) are pruned and the overall architecture of the FLANN based model is reduced which in turn reduces the corresponding computational cost associated with the proposed model without sacrificing the performance. Various steps involved in this algorithm are dealt in this section.

Step 1-Initialization in GA: A population of M chromosomes is selected in GA in which each chromosome constitutes $TE(L + 1)$ number of random binary bits where the first TE number of bits are called Pruning bits (P) and the remaining bits represent the weights associated with various branches (functional expansions) of the FLANN model. Again $(T - 1)$ represents the order the filter and E represents the number of expansions specified for each input to the filter.

A pruning bit (p) from the set P indicates the presence or absence of expansion branch which ultimately signifies the usefulness of a feature extracted from the time series. In other words a binary 1 will indicate that the corresponding branch contributes and thus establishes a physical connection where as a 0-bit indicates that the effect of that path is insignificant and hence can be neglected. The remaining TEL bits represent the TE weight values of the model each containing L bits.

Step 2-Generation of Input Training Data: $K \geq (500)$ number of signal samples is generated. In the present case two different types of signals are generated to identify the static and feed forward dynamic plants.

  i. To identify a feed forward dynamic plant, a zero mean signal which is uniformly distributed between ± 0.5 is generated.
 ii. To identify a static system, a uniformly distributed signal is generated within ± 1.

Each of the input samples are passed through the unknown plant (static and feed forward dynamic plant) and K such outputs are obtained. The plant output is then added with the measurement noise (white uniform noise) of known strength, there by producing k number of desired signals. Thus the training data are produced to train the network.

Step 3-Decoding: Each chromosome in GA constitutes random binary bits. So these chromosomes need to be converted to decimal values lying between some ranges to compute the fitness function. The equation that converts the binary coded chromosome in to real numbers is given by

$$RV = R_{min} + \left\{ (R_{max} - R_{min})/(2^L - 1) \right\} \times DV \qquad (1)$$

Where R_{min}, R_{max}, RV and DV represent the minimum range, maximum range, decimal and decoded value of an L bit coding scheme representation. The first L number of bits is not decoded since they represent pruning bits.

Step 4-To Compute the Estimated Output: At nth instant the estimated output of the neuron can be computed as

$$y(n) = \sum_{i=1}^{T} \sum_{j=1}^{E} \varphi_{ij}(n) \times W_{ij}^m(n) \times P_{ij}^m(n) + b^m(n) \qquad (2)$$

where $\varphi_{ij}(n)$ represents j^{th} expansion of the ith signal sample at the nth instant. $W_{ij}^m(n)$ and $P_{ij}^m(n)$ represent the jth expansion weight and j^{th} pruning weight of the ith signal sample for m^{th} chromosome at k^{th} instant. Again $b^m(n)$ corresponds to the bias value fed to the neuron for m^{th} chromosome at nth instant.

Step 5-Calculation of Cost Function: Each of the desired output is compared with corresponding estimated output and K errors are produced. The Mean-square-error (MSE) corresponding to m^{th} chromosome is determined by using the relation (3). This is repeated for M times (i.e. for all the possible solutions).

$$MSE(m) = \sum_{k=1}^{K} e_k^2 / K \qquad (3)$$

Step 6-Operations of GA: Here the GA is used to minimize the MSE. The crossover, mutation and selection operators are carried out sequentially to select the best M individuals which will be treated as parents in the next generation.

Step 7-Stopping Criteria: The training procedure will be ceased when the MSE settles to a desirable level. At this moment all the chromosomes attain the same genes. Then each gene in the chromosome represents an estimated weight.

4   Simulation Study

Extensive simulation studies are carried out with several examples from static as well as feed forward dynamic systems. The performance of the proposed Pruned FLANN model is compared with that of basic FLANN structure. For this the algorithm used in Sections 3 is used in the simulation.

4.1   Static Systems

Here different nonlinear static systems are chosen to examine the approximation capabilities of the basic FLANN and proposed Pruned FLANN models. In all the simulation studies reported in this section a single layer FLANN structure having one input node and one neuron is considered. Each input pattern is expanded using trigonometric polynomials i.e. by using $\cos(n\pi u)$ and $\sin(n\pi u)$, for $n = 0, 1, 2, \cdots, 6$. In addition a bias is also fed to the output. In the simulation work the data used are $K = 500$, $M = 40$, $N = 15$, $L = 30$, probability of crossover $= 0.7$ and probability of mutation $= 0.1$. Besides that the R_{max} and R_{min} values are judiciously chosen to attain satisfactory results. Three nonlinear static plants considered for this study are as follows:

Example-1: $f_1(x) = x^3 + 0.3x^2 - 0.4x$

Example-2: $f_2(x) = 0.6\sin(\pi x) + 0.3\sin(3\pi x) + 0.1 sin(5\pi x)$

Example-3: $f_3(x) = (4x^3 - 1.2x^2 - 3x + 1.2)/(0.4x^5 + 0.8x^4 - 1.2x^3 + 0.2x^2 - 3)$

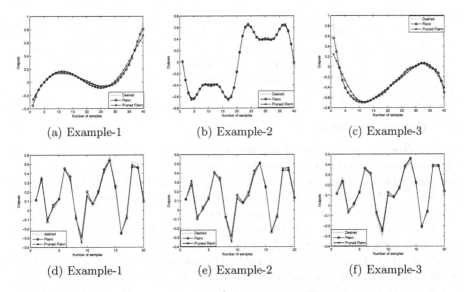

(a) Example-1   (b) Example-2   (c) Example-3

(d) Example-1   (e) Example-2   (f) Example-3

Fig. 3. Comparisons of output response; (a)–(c) for static systems of Example 1–3 and (d)–(f) for dynamic systems of Examples 1–3

At any n^{th} instant, the output of the ANN model $y(n)$ and the output of the system $d(n)$ is compared to produce error $e(n)$ which is then utilized to update the weights of the model. The LMS algorithm is used to adapt the weights of basic FLANN model where as a proposed GA based algorithm is employed for simultaneous adaptation of weights and pruning of the branches. The basic FLANN model is trained for 20000 iterations where as the proposed FLANN model is trained for only 60 generations. Finally the weights of the ANN are stored for testing purpose. The responses of both the networks are compared during testing operation and shown in figures 3(a), 3(b), and 3(c).

The results of identification of $f_1(\cdot)$, $f_2(\cdot)$ and $f_3(\cdot)$ are shown figures 3(a)–3(c). In the figures the actual system output, basic FLANN output and pruned FLANN output are marked as "Desired", "FLANN" and "Pruned FLANN" respectively. From these figures, it may be observed that the identification performance of the FLANN model with all the examples is quite satisfactory. For

Table 1. Comparison of computational complexities between a basic FLANN and a Pruned FLANN model for Static Systems

| Ex. No. | Number of operations | | | | Number of weights | |
| | Additions | | Multiplications | | | |
	FLANN	Prunned FLANN	FLANN	Prunned FLANN	FLANN	Prunned FLANN
Ex-1	14	3	14	3	15	4
Ex-2	14	2	14	3	15	3
Ex-3	14	5	14	5	15	6

the Pruned FLANN structure, quite close agreement between the system output and the model output is observed. In fact, the modeling error of the pruned FLANN structure is found to be comparable with that of the basic FLANN structure for all the three nonlinear static structures considered. Table 1 illustrates the total computational complexity involved in both the architectures to identify the same system.

4.2   Dynamic Systems

In the following the simulation studies of nonlinear dynamic feed forward systems has been carried out with the help of several examples. In each example, one particular model of the unknown system is considered. In this simulation a single layer FLANN structure having one input node and one neuron is considered. Each input pattern is expanded using the direct input as well as the trigonometric polynomials i.e. by using u, $\cos(n\pi u)$ and $\sin(n\pi u)$, for $n = 1$. In this case the bias is removed. In the simulation work we have considered $K = 500$, $M = 40$, $N = 9$, $L = 20$, probability of crossover $= 0.7$ and probability of mutation $= 0.03$. Besides that the R_{max} and R_{min} values are judiciously chosen to attain satisfactory results. The three nonlinear dynamic feed forward plants considered for this study are as follows:

Example-1:
i. Parameter of the linear system of the plant $\begin{bmatrix} 0.2600 \ 0.9300 \ 0.2600 \end{bmatrix}$
ii. Nonlinearity associated with the plant $y_n(k) = y_k + 0.2y_k^2 - 0.1y_k^3$

Example-2:
i. Parameter of the linear system of the plant $\begin{bmatrix} 0.3040 \ 0.9029 \ 0.3410 \end{bmatrix}$
ii. Nonlinearity associated with the plant $y_n(k) = \tanh(y_k)$

Example-3:
i. Parameter of the linear system of the plant $\begin{bmatrix} 0.3410 \ 0.8760 \ 0.3410 \end{bmatrix}$
ii. Nonlinearity associated with the plant $y_n(k) = y_k - 0.9y_k^3 + 0.2y_k^2 - 0.1y_k^3$

The basic FLANN model is trained for 2000 iterations where as the proposed FLANN is trained for only 60 generations. While training, a white uniform noise of strength $-30dB$ is added to actual system response to assess the performance of two different models under noisy condition. Then the weights of the ANN are stored for testing. Finally the testing of the networks model is undertaken by presenting a zero mean white random signal to the identified model. Performance comparison between the FLANN and pruned FLANN structure in terms of estimated output of the unknown plant has been carried out. The responses of both the networks are compared during testing operation and shown in figures 3(d), 3(e), and 3(f).

The results of identification of all the examples are shown in figures 3(d)–3(f). In the figures the actual system output, basic FLANN output and pruned FLANN output are marked as "Desired", "FLANN" and "Pruned FLANN" respectively. From the simulation results, it may be seen that the model output

Table 2. Comparison of computational complexities between a basic FLANN and a pruned FLANN model for Dynamic Systems

Ex. No.	Number of operations				Number of weights	
	Additions		Multiplications			
	FLANN	Prunned FLANN	FLANN	Prunned FLANN	FLANN	Prunned FLANN
Ex-1	8	3	9	4	9	4
Ex-2	8	2	9	3	9	3
Ex-3	8	2	9	3	9	3

responses closely agree with those of plant output for both the FLANN and the pruned FLANN based structures. Comparison of computational complexities between the conventional FLANN and the pruned FLANN is provided in Table 2. From Tables 1 and 2 it is evident that for all the identification performance cases studied, the computational load on the pruned FLANN is much lower than that of FLANN model.

5 Conclusions

The present paper has proposed simultaneous weight updating and pruning of FLANN identification models using GA. Computer simulation studies on static and dynamic nonlinear plants demonstrate that there is more than 50% active paths are pruned keeping response matching identical with those obtaining from conventional FLANN identification models.

References

1. Jagannathan, S., Lewis, F.L.: Identification of a class of nonlinear dynamical systems using multilayered neural networks. In: IEEE International Symposium on Intelligent Control, Columbus, Ohio, USA, pp. 345–351 (1994)
2. Narendra, K.S., Parthasarathy, K.: Identification and control of dynamical system using neural networks. IEEE Trans. Neural Networks 1(1), 4–26 (1990)
3. Pao, Y.H.: Adaptive Pattern Recognition and Neural Network. Addison Wesley, Reading, MA (1989)
4. Patra, J.C., Pal, R.N., Chatterji, B.N., Panda, G.: Identification of nonlinear dynamic systems using functional link artificial neural networks. IEEE Trans.on Systems, Man and Cybernetics-Part B 29(2), 254–262 (1999)
5. Patra, J.C., Kot, A.C.: Nonlinear dynamic system identification using chebyshev functional link artificial neural networks. IEEE Trans. on Systems Man and Cybernetics-Part B 32(4), 505–511 (2002)
6. Juang, J.G., Lin, B.S.: Nonlinear system identification by evolutionary computation and recursive estimation method. In: Proceedings of American Control Conference, pp. 5073–5078 (2005)
7. Giles, C.L., Omlin, C.W.: Pruning recurrent neural networks for improved generalization performance. IEEE Trans. on Neural Networks 5(5), 848–851 (1994)

Enhanced Quantum Evolutionary Algorithms for Difficult Knapsack Problems

C. Patvardhan[1], Apurva Narayan[1], and A. Srivastav

[1] Faculty of Engineering,
Dayalbagh Educational Institute,
Agra – 282005
cpatvardhan@hotmail.com, apurvanarayan@gmail.com
[2] Mathematisches Seminar, Christian-Albrechts-Universitat Zu Kiel, Kiel Germany
asr@numerik.uni-kiel.de

Abstract. Difficult knapsack problems are problems that are expressly designed to be difficult. In this paper, enhanced Quantum Evolutionary Algorithms are designed and their application is presented for the solution of the DKPs. The algorithms are general enough and can be used with advantage in other subset selection problems.

Keywords: Evolutionary Algorithms ,Quantum ,knapsack.

1 Introduction

The classical knapsack problem is defined as follows: Given a set of n items, each item j having an integer profit p_j and an integer weight w_j, the problem is to choose a subset of the items such that their overall profit is maximized, while the overall weight does not exceed a given capacity c. The problem may be formulated as the following integer programming model

$$(KP) \qquad \text{maximize} \quad \sum_{j=1}^{n} p_j x_j \tag{1}$$

$$\text{subject to} \quad \sum_{j=1}^{n} w_j x_j \le c, \tag{2}$$

$$x_j \in \{0,1\}, \quad j = 1,\ldots,n. \tag{3}$$

where the binary decision variables x_j are used to indicate whether item j is included in the knapsack or not. Without loss of generality it may be assumed that all profits and weights are positive, that all weights are smaller than the capacity c, and that the overall weight of the items exceeds c.

The standard knapsack problem (SKP) is *NP*-hard in the weak sense, meaning that it can be solved in pseudo-polynomial time through dynamic programming. The SKPs are quite easy to solve for the most recent algorithms [1-5]. Various branch-and-bound algorithms for SKPs have been presented. The more recent of these solve a core problem, i.e. an SKP defined on a subset of the items where there is a large probability of finding an optimal solution. The MT2 algorithm [1] is one of the most advanced of these algorithms. Realizing that the core size is difficult to estimate in

A. Ghosh, R.K. De, and S.K. Pal (Eds.): PReMI 2007, LNCS 4815, pp. 252–260, 2007.

advance, Pisinger [5] proposed to use an expanding core algorithm. Martello and Toth [3] proposed a special variant of the MT2 algorithm which was developed to deal with hard knapsack problems. One of the most successful algorithms for SKP was presented by Martello, Pisinger and Toth [7]. The algorithm can be seen as a combination of many different concepts and is hence called Combo.

The instances do not need to be changed much before the algorithms experience a significant difficulty. These instances of SKP are called the Difficult Knapsack Problems (DKP) [6]. These are specially constructed problems that are hard to solve using the standard methods that are employed for the SKPs. Pisinger [6] has provided explicit methods for constructing instances of such problems.

Heuristics which employ history of better solutions obtained in the search process, viz. GA, EA, have been proven to have better convergence and quality of solution for some "difficult" optimization problems. But, still, problems of slow/premature convergence remain and have to be tackled with suitable implementation for the particular problem at hand. Quantum Evolutionary Algorithms (QEA) is a recent branch of EAs. QEAs have proven to be effective for optimization of functions with binary parameters [8, 9].

Although the QEAs have been shown to be effective for SKPs in [8], their performance on the more difficult DKPs has not been investigated. Exact branch and bound based methods are not fast for DKPs. This provides the motivation for attempting to design newer Enhanced QEAs (EQEAs) with better search capability for the solution of the DKPs. In this paper, such EQEAs are presented. The QEAs are different from those in [8] as they include a variety of quantum operators for executing the search process. The computational performance of the EQEAs is tested on large instances of nine different varieties of the DKPs i.e. for problems up to the size of 10,000. The results obtained are compared with those obtained by the greedy heuristic. It is seen that the EQEAs are able to provide much improved results.

The rest of the paper is organized as follows. In section 2 we describe the various types of DKPs reported in the literature. A brief introduction to the QEAs and some preliminaries regarding the QEAs are provided in section 3. The EQEA is given in the form of pseudo-code in section 4. The computational results are provided section 5. Some conclusions are derived in section 6.

2 Difficult Knapsack Problems (DKPs)

Several groups of randomly generated instances of DKPs have been constructed to reflect special properties that may influence the solution process in [6]. In all instances the weights are uniformly distributed in a given interval with *data range with R* =1000. The profits are expressed as a function of the weights, yielding the specific properties of each group. Nine such groups are described below as instances of DKPs. The first six are instances with large coefficients for profit and the last three are instances of small coefficients for profits.

(i) Uncorrelated data instances: p_j and w_j are chosen randomly in $[1,R]$. In these instances there is no correlation between the profit and weight of an item.

(ii) Weakly correlated instances: weights w_j are chosen randomly in $[1,R]$ and the profits p_j in $[w_j - R/10 , w_j + R/10]$ such that $p_j >= 1$. Despite their name, weakly correlated instances have a very high correlation between the profit and weight of an item. Typically the profit differs from the weight by only a few percent.

(iii) Strongly correlated instances: weights w_j are distributed in [1,R] and $p_j = w_j + R/10$. Such instances correspond to a real-life situation where the return is proportional to the investment plus some fixed charge for each project.

(iv) Inverse strongly correlated instances: profits p_j are distributed in [1,R] and $w_j = p_j + R/10$. These instances are like strongly correlated instances, but the fixed charge is negative.

(v) Almost strongly correlated instances: weights w_j are distributed in [1,R] and the profits p_j in $[w_j + R/10 - R/500, w_j + R/10 + R/500]$. These are a kind of fixed-charge problems with some noise added. Thus they reflect the properties of both strongly and weakly correlated instances.

(vi) Uncorrelated instances with similar weights: weights w_j are distributed in [100000 , 100100] and the profits p_j in [1, 1000].

(vii) Spanner instances (v,m)**:** These instances are constructed such that all items are multiples of a quite small set of items — the so-called spanner set. The spanner instances span(v,m) are characterized by the following three parameters: v is the size of the spanner set, m is the multiplier limit, and any distribution (uncorrelated, weakly correlated, strongly correlated, etc.) of the items in the spanner set may be taken. More formally, the instances are generated as follows: A set of v items is generated with weights in the interval [1,R], and profits according to the distribution. The items (p_k, w_k) in the spanner set are normalized by dividing the profits and weights with $m+1$. The n items are then constructed, by repeatedly choosing an item $[p_k, w_k]$ from the spanner set, and a multiplier a randomly generated in the interval [1,m]. The constructed item has profit and weight $(a \cdot p_k, a.w_k)$. The multiplier limit was chosen as $m = 10$.

(viii) Multiple strongly correlated instances mstr(k_1,k_2,d)**:** These instances are constructed as a combination of two sets of strongly correlated instances. Both instances have profits $p_j := w_j + k_i$ where $k_i = 1$, 2 is different for the two instances. The multiple strongly correlated instances mstr(k_1,k_2,d) are generated as follows:

The weights of the n items are randomly distributed in [1,R]. If the weight w_j is divisible by d, then we set the profit $p_j := w_j + k_1$ otherwise set it to $p_j := w_j + k_2$. The weights w_j in the first group (i.e. where pj = wj + k_1) will all be multiples of d, so that using only these weights at most d $[c/d]$ of the capacity can be used. To obtain a completely filled knapsack some of the items from the second distribution need to be included.

(ix) Profit ceiling instances pceil(d)**:** These instances have the property that all profits are multiples of a given parameter d. The weights of the n items are randomly distributed in [1,R], and the profits are set to $p_j = d[w_j/d]$. The parameter d was chosen as $d=3$.

3   QEA Preliminaries

QEA is a population-based probabilistic Evolutionary Algorithm that integrates concepts from quantum computing for higher representation power and robust search.

Qubit
QEA uses qubits as the smallest unit of information for representing individuals. Each

qubit is represented as $q_i = \begin{bmatrix} \alpha_i \\ \beta_i \end{bmatrix}$. State of the qubit (q_i) is given by $\left| \Psi_i \right\rangle = \alpha_i \left| 0 \right\rangle + \beta_i \left| 1 \right\rangle$,

α_i and β_i are complex numbers representing probabilistic state of qubit, i.e. $|\alpha_i|^2$ is probability of the state being 0 and $|\beta_i|^2$ is the probability of the state being 1, such that $|\alpha_i|^2 + |\beta_i|^2 = 1$. For purposes of QEA, nothing is lost by regarding α_i and β_i to be real numbers. Thus, a qubit string with n bits represents a superposition of 2^n binary states and provides an extremely compact representation of the entire search space.

Observation
The process of generating binary strings from the qubit string, Q, is called Observation. To observe the qubit string (Q), a string consisting of same number of random numbers between 0 and 1 (R) is generated. The element P_i is set to 0 if R_i is less than square of Q_i and 1 otherwise. Table1 represents the observation process.

Table 1. Observation of qubit string

i	1	2	3	4	5	....	Ng
Q	0.17	0.78	0.72	0.41	0.89	....	0.36
R	0.24	0.07	0.68	0.92	0.15	....	0.79
P	1	0	0	1	0	....	1

Updating qubit string
In each of the iterations, several solution strings are generated from Q by observation as given above and their fitness values are computed. The solution with best fitness is identified. The updating process moves the elements of Q towards the best solution slightly such that there is a higher probability of generation of solution strings, which are similar to best solution, in subsequent iterations. A Quantum gate is utilized for this purpose so that qubits retain their properties [8]. One such gate is rotation gate, which updates the qubits as

$$\begin{bmatrix} \alpha_i^{t+1} \\ \beta_i^{t+1} \end{bmatrix} = \begin{bmatrix} \cos(\Delta\theta_i) & -\sin(\Delta\theta_i) \\ \sin(\Delta\theta_i) & \cos(\Delta\theta_i) \end{bmatrix} \begin{bmatrix} \alpha_i^t \\ \beta_i^t \end{bmatrix}$$

where, α_t^{t+1} and β_i^{t+1} denote probabilities for i^{th} qubit in $(t+1)^{th}$ iteration and $\Delta\theta_i$ is equivalent to the step size in typical iterative algorithms in the sense that it defines the rate of movement towards the currently perceived optimum.

The above description outlines the basic elements of QEA. The qubit string, Q, represents probabilistically the search space. Observing a qubit string 'n' times yields 'n' different solutions because of the probabilities involved. Fitness of these is computed and the qubit string, Q, is updated towards higher probability of producing strings similar to the one with highest fitness. The sequence of steps continues. The above ideas can be easily generalized to work with multiple qubit strings. Genetic operators like crossover and mutation can then be invoked to enhance the search power further.

4 Enhanced Quantum Evolutionary Algorithm

The algorithm is explained succinctly in the form of a pseudo-code below.

Notation

- Max = number of items in DKP, Cap = Capacity of knapsack
- Profit = Array of profit by selecting each item
- Weight = Array of weight by selecting each item
- Gsol = DKP Solution by Greedy heuristic, GProfit = Profit of items in Gsol
- P, ep ,rp : Strings of Quantum bits used in search
- NO1 = Operator used to evolve ep towards best solution found so far
- NO2 = Operator used to evove rp randomly
- NO3 = Rotation operator used to evolve p
- bestcep = String produced by operator NO1 giving best result on observation
- bestcrp = Quantum string produced by NO2 giving best result on observation
- Best = Best profit found by EQEA of solution maxsol

Algorithm EQEA

1	Initialize iteration number t=0, cap, profit, weight , max
2	Sort items in descending order of profit/weight
3	Find greedy solution G with profit Gcost.
4	Best = Gcost , Maxsol = G
5	Initialize for every 'k' If G[k] ==1 p[k] =ep[k]=rp[k]=0.8 else p[k] =ep[k]=rp[k]=0.2
6	Best Cap = ep , Best Cap = rp /*Initialization*/
7	Observe p[k] to get solution with cost tcost ; if(tcost > Best){Best = tcost; Maxsol = tsol;}
8	while (termination_criterion != TRUE) Do Steps 9-15

9	Apply NO1 to p to generate the current rp i.e crp;
	   for(i=0;i<num1;i++)
	{ Observe crp to obtain solution rsol with cost rcost;
	if(rcost > Best)
	{Crossover rsol with Maxsol to obtain tsol with cost tcost;
	If(tcost>rcost) {rsol = tsol;rcost = tcost}
	bestcrp = crp;}
	If(rcost >Best) {Best = rcost ; maxsol = rsol;}

10	Repeat step 9 with NO2 on p to obtain bestcep, ecost and esol;
11	Apply NO3 on p

12	For(i = 0 ; i < max; i++)
	{qmax[i] = findmax(bestcep[i], bestcrp[i],p[i]);
	qmin[i] = findmix(bestcep[i], bestcrp[i],p[i]);
	if(p[i] > 0.5)
	{{if((maxsol[i]==1)and (bestcep[i]==1)and (bestcrp[i]==1))
	p[i]=qmax[i];}
	   else {if((maxsol[i]==0)and (bestcep[i]==0)and (bestcrp[i]==0))
	p[i]=qmin[i];}}

13	Settled = number of p[i]'s with $\alpha_i > 0.98$ or $\alpha_i < 0.02$
14	$t = t + 1$
15	If((t > iter _count) or (settled > 0.98 * max)) termination _ criterion = TRUE

Algorithm EQEA starts with the greedy solution and initializes the qubit strings in accordance with the greedy solution as in step 4. Observe operation in step 5 is a modified form of observation described in section 3. In this, the solution string resulting from the observation is checked for violation of capacity constraint and repaired, if necessary, using a greedy approach i.e. selected items are deselected in increasing order of profit/weight till capacity constraint is satisfied. The crossover operator is a simple two-point crossover with greedy repair of constraint violations.

Evolving Qubits

In EQEA, the qubit is evolved state of the qubit, which is a superposition of state 0 and 1, is shifted to a new superposition state. This change in probability magnitudes $|\alpha|^2$ and $|\beta|^2$ with change in state is transformed into real valued parameters in the problem space by two neighborhood operators.

Neighborhood Operator 1 (NO1) generates a new qubit array crp from the rp. An array R is created with max elements generated at random such that every element in R is either +1 or -1. Let ρ_k be the kth element in R. Then θ_k^t is given by

$$\theta_k^t = \theta_k^{t-1} + \rho_k * \delta \qquad (4)$$

where, δ is alteration in angle and θ_k^t is the rotated angle given by arctan(β_k^t / α_k^t).

δ is randomly chosen in the range [0, θ_k^{t-1}] if $\rho_k = -1$ and in the range [θ_k^{t-1}, $\pi/2$] if $\rho_k = +1$.

The new probability amplitudes, α_k^t, β_k^t are calculated using rotation gate as

$$\begin{bmatrix} \alpha_{ijk}^t \\ \beta_{ijk}^t \end{bmatrix} = \begin{bmatrix} \cos\delta & \sin\delta \\ -\sin\delta & \cos\delta \end{bmatrix} \begin{bmatrix} \alpha_{ijk}^{t-1} \\ \beta_{ijk}^{t-1} \end{bmatrix} \qquad (5)$$

Neighborhood Operator 2 (NO2) NO2 works just as NO1 except that it generates a point between ep and BEST. It is primarily utilized for exploitation of search space.

The rationale for two neighborhood operators is as follows. NO1 has a greater tendency for exploration. NO2 has a greater tendency towards exploitation because, as the algorithm progresses, the values of ep converge towards BEST. Table 2 shows the frequency of use of NO1 and NO2.

Table 2. Frequency of use of NO1 and NO2

Stage of search	Proportion of NO1 (%)	Proportion of NO2 (%)
First one-fifth iterations	90	10
Second one-fifth iterations	70	30
Third one-fifth iterations	50	50
Forth one-fifth iterations	30	70
Last one-fifth iterations	10	90

The neighborhood operators thus evolve new quantum strings from the existing strings. This approach removes all disadvantages of binary representation of real numbers while, at the same time, balances exploration and exploitation in the sense that it adopts the "step-size" from large initially to progressively smaller size.

Updating Qubit String(NO3)

In the updating process, the individual states of all the qubits in p are modified so that probability of generating a solution string which is similar to the current best solution is increased in each of the subsequent iterations. Amount of change in these probabilities is decided by Learning Rate, $\Delta\theta$, and is taken as 0.001π. Updating process is done as explained in section 2. Table 3 presents the choice of $\Delta\theta$ for various conditions of objective function values and i^{th} element of p and BEST in t^{th} iteration. F(p) is the profit of the current solution observed from p. F(BEST) is the profit of maxsol.

Table 3. Calculation of $\Delta\theta$ for t^{th} iteration

Fitness	Elemental Values	$\Delta\theta$
X	$p_i = BEST i$	0
F(BEST)>F(p)	$p_i > BESTi$	0.001π
	$p_i < BESTi$	-0.001π
F(BEST)<F(p)	$p_i > BESTi$	-0.001π
	$p_i < BESTi$	0.001π
F(p)=F(BEST)	$p_i > BEST_i$	0
	$p_i < BEST_i$	0

The updating process is illustrated in figure 2 for k^{th} element for t^{th} iteration i.e. changes in state of kth qubit and corresponding change in probability amplitudes. Findmax finds the maximum of the three arguments whereas findmin finds the minimum of the three arguments.

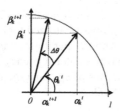

Fig. 2. Updating k^{th} element of qubit string Q

5 Results of Computational Experiments

The computational experiments have been performed on the above mentioned nine types of DKPs for the data range R = 1000 and the number of items ranging from 500 to 10000. For each instance type a series of $H = 9$, $c = \frac{h}{H+1}\sum_{j=1}^{n} w_j$, $h = 1,.., ...H$

instances is performed. The EQEA performed better in most of the cases where the problem did not get settled up with a definite structure. The table 4 shows computational results. AGP is the average profit in solutions obtained by the greedy algorithm whereas AQP is the average profit of solutions by the EQEA algorithm.

Table 4. Shows the relative values of solutions by Greedy and Quantum Evolutionary methods

DKP Grp.	Problem size = 500			Problem size = 5000			Problem size = 10,000		
	AGP	AQP	# EQEA better	AGP	AQP	# EQEA better	AGP	AQP	# EQEA better
1	48438	61018	7	483294	563412	8	984889	1158681	5
2	67582	69037	5	672893	688997	4	1363381	1395080	5
3	50076	61948	6	506102	620607	6	1015016	1246752	5
4	44036	44919	6	445676	451673	1	896304	906715	1
5	50792	62752	5	507447	621695	6	1021859	1251110	7
6	55633	57705	3	553570	584196	4	1127432	1177977	4
7	25255	31255	7	253710	310182	7	508955	622955	5
8	37893	43051	7	382858	430121	4	757420	856606	4
9	142162	153300	4	1437554	1541309	3	2897585	3106700	1

The Third column in each shows the number of problems (out of 9 tried for various capacities of the knapsack) in which EQEA gives better solution than Greedy method for that group. It is seen that the EQEA provides a considerable improvement in several instances.

6 Conclusions

An enhanced QEA is presented for Difficult knapsack problems. EQEA shows good performance even for large instances of the problems. EQEA can be used with advantage in any subset slection problems apart from the DKPs.

Acknowledgments

The authors are extremely grateful to the Revered Chairman, Advisory Committee on Education, Dayalbagh for constant guidance and encouragement.

References

[1] Martello, S., Toth, P.: A new algorithm for the 0-1 knapsack problem. Management Science 34, 633–644 (1988)
[2] Martello, S., Toth, P.: Knapsack Problems: Algorithms and Computer Implementations. J. Wiley, Chichester (1990)

[3] Martello, S., Toth, P.: Upper bounds and algorithms for hard 0-1 knapsack problems. Operations Research 45, 768–778 (1997)

[4] der Heide, F.M.a.: A polynomial linear search algorithm for the n-dimensional knapsack problem. Journal of the ACM 31, 668–676 (1984)

[5] Pisinger, D.: An expanding-core algorithm for the exact 0-1 knapsack problem. European Journal of Operational Research 87, 175–187 (1995)

[6] Pisinger, D.: Where are the Hard Knapsack Problems, Technical Report, 2003-8, DIKU, University of Copenhagen, Denmark

[7] Martello, S., Pisinger, D., Toth, P.: Dynamic programming and strong bounds for the 0-1 knapsack problem. Management Science 45, 414–424 (1999)

[8] Han, K.H., Kim, J.H.: Quantum-Inspired Evolutionary Algorithm for a Class of Combinatorial Optimization. IEEE Transactions on Evolutionary Computation 6(6), 580–593 (2002)

[9] Han, K.H., Kim, J.H.: Quantum-Inspired Evolutionary Algorithms with a New Termination Criterion, Hε Gate, and Two-Phase Scheme. IEEE Transactions on Evolutionary Computation 8(2), 156–169 (2002)

A Decomposition Approach for Combined Heuristic and Differential Evolution Method for the Reactive Power Problem

B. Bhattacharyya[1] and S.K. Goswami[2]

[1] Dept. of Electrical Engineering, NIT, Durgapur 713209, India
[2] Dept. of Electrical Engineering, Jadavpur University, Kolkata 700032, India
biplabrec@yahoo.com, skgoswami_ju@yahoo.co.in

Abstract. The author in the present paper has attempted a variable decomposition approach. All the variables of the reactive power planning optimization do not involve cost. Those involving costs are treated as planning variables and the variables having no cost involvement are treated separately as the dispatch variables. Solution approaches are also designed separately for the two types of variables and a mixed heuristic and evolutionary strategy has been developed. As the number of variables in the evolutionary technique thus decreases, the solution becomes faster.

Keywords: Heuristics, Evolutionary techniques, Planning, Dispatch.

1 Introduction

In this paper, the main concern is proper planning and co-ordination of control variables which are either transformer tap changers, shunt capacitors, Generators reactive VAr's in an interconnected power system such that real power loss becomes minimum. The problem of reactive power planning in a power system can be shown to be a combinatorial optimization problem and number of methods have been proposed to solve the problem Reactive power planning tool was described in [1]. Simulated Annealing technique [2,3] is applied for the capacitor placement problem. Fuzzy logic, found it's application for handling reactive power problem in [4-6]. Expert system [7], and AI method [8] is also used for solving the reactive power problem. Heuristics and approximate reasoning approach was used to get the solution of reactive power problem in [9-12]. Genetic Algorithm [13-20] is used as an optimization technique for capacitor placement and also for solving reactive power planning and dispatch problem. Evolutionary programming, Evolutionary strategies [21-27] were applied for reactive power problem. Particle Swarm optimization (PSO) [28] is used as a technique to deal reactive power problem. Differential Evolution and the Hybrid Evolutionary approaches [29-34] are recent trends to handle reactive power optimization problems. In the present paper, the authors propose a new approach to the solution of the reactive power problem based on heuristics for the capacitor placements and Differential Evolution (DE)

A. Ghosh, R.K. De, and S.K. Pal (Eds.): PReMI 2007, LNCS 4815, pp. 261–268, 2007.

method for controlling reactive generations of Generators and transformer tap
positions of an inter connected power system. Shunt capacitors which is treated as
planning variable are installed at candidate buses using heuristic approach considering loss sensitivity at that bus. Differential Evolution technique is then applied
for optimal setting of transformer tap positions and reactive Var's Generated by
the Generator's . The results obtained by each of this mixed heuristics and Differential Evolution technique is then compared with the results obtained by using
each of the Evolutionary (GA,PSO and DE) methods.

2 The Proposed Approach

The method has been developed for solving the reactive power planning problem
though it may be used for the dispatch problem as well. Installation of the generators and tap changing transformers in a power system require considerations of the
aspects beyond the reach of the reactive power optimization problem. On the other
hand, the decision of installing new capacitors is solely guided by the reactive
power considerations. As the capacitor installations involve costs, it is expected that
the decision for the installations of new capacitors will be taken after the best possible utilization of the existing sources only. This gives rise to the idea that capacitor
installation problem has to be solved after solving the reactive power dispatch problem for the generators and the tap changing transformers only. In such a case the
capacitor installation problem has to solved separately so as to minimize the additional cost to be incurred in the new capacitors. On the other hand, treatment of the
complete problem simultaneously, as in conventional way, will demand unnecessary searches requiring long time to generate the optimal solution. The above stated
idea motivated the development of a heuristic technique for the solution of the capacitor installation problem and an evolutionary approach for solving the generator
and tap changer dispatch problem.

2.1 The Heuristic Technique for the Capacitor Installation Problem

The installation of a new capacitor is justified by two related responses of the systems. The cost reduction due to reduced power loss and the improvement of the voltage profile.

Given the loss equation of a power system

$$P_{loss} = \sum_k v_i^2 + v_j^2 - 2v_i v_j \cos \delta_{ij} ,$$ and assuming that change in the Var generations

has negligible effect on the voltage phase angle such that voltage magnitudes only
will be changed, loss sensitivity with respect to node voltages may be used to judge
where new sources are to be introduced or old sources are to be made richer.

For changes in the bus voltages, incremental change in the power loss is given by

$$\Delta P_{loss} = \left[\frac{\partial P_{loss}}{\partial v_1} \frac{\partial P_{loss}}{\partial v_2} .... \frac{\partial P_{loss}}{\partial v_n} \right] \left[\Delta v_1 . \Delta v_2 ...... \Delta v_n \right]^T$$

The implication of the above relation is that loss reduction can be achieved by increasing the voltages at those nodes where loss sensitivity to voltage has a negative value and reducing the voltages of those nodes where loss sensitivity value is positive. Maximum reduction in the loss is possible where the product $\dfrac{\partial P_{loss}}{\partial v_i}.\Delta v_i$ is most negative and this node should first be selected for additional Var support. While $\dfrac{\partial P_{loss}}{\partial v_i}$ can be determined by putting the current values of the node voltages and phase angles in the relation $\dfrac{\partial P_{loss}}{\partial v_i} = \sum_k 2v_i - 2v_j \cos \delta_{ij}$, where k include all lines incident to node $- i$, problem arises in getting a value for Δv_i . One way to overcome the problem may be to know the maximum possible reduction in loss by adjusting the voltage of node i. Thus, in the present work Δv_i is computed as $\Delta v_i = v_i^{max} - v_i$ or, $\Delta v_i = v_i^{min} - v_i$ depending upon the sign of the loss sensitivity. v_i in the above is the known voltage at node i.

The above stated procedure for selecting the node for additional Var generation may work without any problem in case of a dispatch problem. In the planning problem, however, simply reducing the loss may not reduce the overall cost of the system as reduction has to be achieved by paying for additional capacitors to be installed. Moreover, operational constraint for system voltage also has to be maintained. To be on the safe side, therefore, it is decided to select that node as the first candidate for capacitor installation where the product of $\dfrac{\partial P_{loss}}{\partial v_i}.\Delta v_i$ is most negative and at the same time voltage is outside the permissible lower and upper limits. The amount of capacitor Var addition however has to be justified before taking an installation decision. The verification of cost reduction requires a load flow to be performed. A decoupled load flow is used here. A further gain in solution speed of the load flow is achieved by recognizing the fact that, addition of Var generation will not change the active power flow pattern. Thus, after the addition of the Var source only the Q-equation of the fast decoupled load flow has to be solved in order to assess the cost benefit. Still greater reduction in solution time is achieved by modeling the capacitor as a constant Var source. This eliminates the necessity of updating the B" matrix and simple forward and backward substitutions only serve the purpose.

The heuristic algorithm for capacitor location identification is as given below:

1. Determine the loss sensitivities at all the nodes of the system.
2. Determine maximum possible voltage adjustments
$$\Delta v_i = v^{limit} - v_i$$
3. Determine the product $\dfrac{\partial P_{loss}}{\partial v_i}.\Delta v_i$, i = 1,......n.

4. Identify the nodes having a negative value of the product.
5. Order the nodes with decreasing values of the product.
6. Select the first node of the list as the capacitor location.
7. Apply the minimum size of capacitor, at the selected location.
8. Perform only the Q-solution of the decoupled load flow.
9. Determine the loss reduction.
10. Compare with the capacitor cost.
11. If cost reduction is possible, install the capacitor and go to step –1. Otherwise go to the next step.
12. All nodes in the list exhausted? If yes go to step 14. Otherwise go to the next step.
13. Select the next node in the list as the capacitor location and go to Step 7.
14. End of the heuristic step 7.

2.2 The Evolutionary Solution

As already mentioned, the evolutionary approach is used in this algorithm to know the optimum dispatch only. For this, capacitors selected by the heuristic algorithm are assumed to be fixed in the evolutionary process and the generator reactive outputs and the tap changer ratios are used as variables. The evolutionary algorithm such as DE(Differential Evolution) then works in the usual manner. It is to be mentioned here that for the load flows to be performed during the execution of the evolutionary algorithms, capacitors need not be modeled as constant Q-source. Rather an admittance model this time is preferred to have a better result.

2.3 The Mixed Heuristic and Evolutionary Algorithm

Because of the sequential solutions of the two part of the problem, a one step solution will seldom be the optimum. An iterative approach is therefore necessary. The complete solution algorithm is as given in the flowchart below:

3 Application Results

The proposed mixed heuristic and differential evolution approach has been applied to solve the reactive optimization problems of IEEE 14 and 30 bus test systems.

In IEEE 14 bus system the first three elements of the string are for transformer tap position the next four positions are for Generator Var sources and as the weak buses are selected by heuristics, the string length has reduced to seven.

Transformer taps are in line 8, 9 and 11 of the IEEE 14 bus system, since Bus number 2, 3, 6, 8 are PV buses, generator vars are controlled at these four buses. Heuristically 10th , 13th and 14th buses are determined as weak buses. Hence Shunt capacitors are installed at these locations. Transformer taps are in line 11, 12, 15 and 36 of the IEEE 30 bus, hence 1st four positions are assigned for transformer taps in the string, 5th to 9th position are kept for generator reactive VAR sources as bus 2, 5, 8, 11 and 13 are the PV buses of the 30 bus system. So string length is nine and hence string length is reduced in comparison to that considered in chapter three. By Heuristic analysis 21st, 26th and 30th buses are found as candidate buses for installation of shunt capacitors. Table 1 – shows the result obtained by Heuristic – DE method and comparison of this method with that of GA, PSO and DE method for IEEE 14 and IEEE 30 bus.

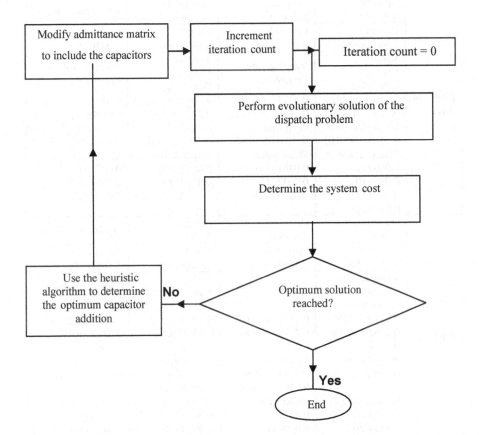

As already mentioned, the evolutionary approach is used in this algorithm to know the optimum dispatch only. For this, capacitors selected by the heuristic algorithm are assumed to be fixed in the evolutionary process and the generator reactive outputs and the tap changer ratios are used as variables. The evolutionary algorithm such as DE (Differential Evolution) then works in the usual manner. It is to be mentioned here that for the load flows to be performed during the execution of the evolutionary algorithms, capacitors need not be modeled as constant Q-source. Rather an admittance model this time is preferred to have a better result.

Table 1. Results of the Heuristic- DE & other Evolutionary algorithms : Planning Problem. (Note: Numbers within the parenthesis indicate the buses or branches where the devices are connected)

Test System	Heuristic – DE	GA	PSO	DE
IEEE 14 Bus	Optimum Solution	Optimum Solution	Optimum Solution	Optimum Solution
	Cost in $:- 6.9441×106	Cost in $:- 6.9493×106	Cost in $:- 6.9477×106	Cost in $:- 6.9470×106
	Variable Values:- Generator:- 0.3109 0.2514 0.2259 0.0421 tap position:- 0.9568 0.95 0.95 Shunt values:- 0.006 (7) 0.03 (10) 0.072 (13) 0.048 (14)	Variable values:- Generator:- 0.3221 0.2532 0.1650 0.1161 tap position:- 0.95 0.95 0.95 Shunt values:- 0.0461 (14) 0.0632 (13) 0.0203 (12)	Variable Values:- Generator:- 0.3082 0.2487 0.24 0.1014 tap position;_ 0.95 0.95 0.95 Shunt values:- 0.0476 (14) 0.0519 (13) 0 (12)	Variable Values:- Generator:- 0.3135 0.2499 0.2318 0.0907 tap position:- 0.95 0.95 0.95 Shunt values:- 0.0477 (14) 0.0513 (13) 0.0141 (12)
IEEE 30 Bus	Optimum Solution	Optimum Solution	Optimum Solution	Optimum Solution
	Cost in $:- 3.5852×106	Cost in $:- 3.5899×106	Cost in $:- 3.6029×106	Cost in $:- 3.5877×106
	Variable Values:- Generator:- 0.1513 0.2561 0.22629 0.0382 0.2514 tap position:- 0.9 0.9 0.9 0.9 Shunt values:- 0.036 (21) 0.036 (26) 0.048 30)	Variable Values:- Generator:- 0.1371 0.2603 0.2575 0.0887 0.2061 tap position:- 0.9084 0.9007 0.9101 0.9001 Shunt values:- 0.0136 (29) 0.0394 (20) 0.0232 (30) 0.0245 (14)	Variable Values:- Generator:- 0.1818 0.2721 0.3466 0.1 0.2639 tap position;_ 0.9 0.9104 0.9090 0.9 Shunt values:- 0.0204 (29) 0.0636 (20) 0.0169 (30) 0 (14)	Variable Values:- Generator:- 0.1572 0.2523 0.2495 0.432 0.2366 tap position:- 0.9 0.9 0.9 0.9 Shunt values:- 0.0140 (29) 0.0455 (20) 0.0307 (30) 0.0169 (14)

4 Conclusion

Here, it is observed that if heuristic approach is used for shunt capacitor placement purpose and then differential evolution technique is continued for solving reactive power planning problem, considerable improvement is noticed. So this approach could be a new method for solving reactive power optimization problem. This approach is used for finding the solution of the reactive power problem by using variable decomposition technique in the sense that some of the power system control variable is treated as planning variable and the rest of the variables those does not incorporate cost in the objective function are treated as dispatch variables.

References

1. Losi, A., Rossi, F., Russo, M., Verde, P.: A new tool for reactive power planning. IEE Proc, Genr, Transm, Distrib. 1140(4), 256–262 (1993)
2. Chiang, H.D., Wang, J.C., Cockings, O., Shin, H.D.: Optimal Capacitor Placements in Distribution Systems: Part 1: A new Formulation and the Overall Problem. IEEE Trans, power Delivery 5(2), 634–642 (1990)
3. Chiang, H.D., Wang, J.C., Cockings, O., Shin, H.D.: Optimal Capacitor Placements in Distribution Systems: Part 2: A new Formulation and the Overall Problem. IEEE Trans, power Delivery 5(2), 643–649 (1990)
4. Rahaman, K.H.A., Shahidehpour, S.M.: A Fuzzy Based Optimal Reactive Power Control. IEEE Trans, Power Syst. 8(2), 66–668 (1993)
5. Rahaman, K.H.A., Shahidehpour, S.M.: Reactive Power Optimization Using Fuzzy Load Representation. IEEE Trans, Power Syst. 9(2), 898–905 (1994)
6. Chen, Y.L.: An interactive Fuzzy – Norm satisfying method for Multiobjective Reactive power sources planning. IEEE Trans on Power Syst. 15(3), 1154–1160 (2000)
7. Jwo, W.S., Liu, C.W., Liu, C.C., Hsiao, Y.T.: Hybrid expert and simulated annealing approach to optimal reactive power planning. IEE Proc, Genr, Transm, Distrib. 142(4), 381–385 (1995)
8. Rahaman, K.H.A., Shahidehpour, S.M., Daneshdoost, M.: AI Approach to Optimal Var control with Fuzzy Reactive Load. IEEE Trans, Power Syst. 10(1), 88–97 (1995)
9. Yokoyama, R., Niimura, T., Nakanishi, Y.: A coordinated Control Of Voltage And Reactive Power By Heuristic Modelling And Approximate Reasoning. IEEE Trans, Power Syst. 8(2), 636–670 (1993)
10. Mantovani, J.R.S., Garcia, A.V.: A heuristic method for reactive power planning. IEEE Trans on Power Syst. 11(1), 68–74 (1996)
11. Chis, M., Salama, M.M.A., Jayaram, S.: Capacitor Placement In Distribution Systems Using Heuristic Search Strategies. IEE Proc. – C. 97, 225–230
12. Hong, Y., Liu, C.C.: A Heuristic and Algorithmic Approach to Var Planning. IEEE Trans on Power Syst. 7(2), 505–512 (1992)
13. Iba, K.: Reactive Power Optimization by Genetic Algorithm. IEEE Trans, Power Syst. 9(2), 685–692 (1994)
14. Lee, k.y., Bai, X., Park, Y.M.: Optimization method for reactive power planning by using a modified simple Genetic Algorithm. IEEE Trans on Power Syst. 10(4), 1843–1850 (1995)
15. Miu, K.N., Chiang, H.D., Darling, G.: Capacitor placement, replacement and control in large scale distribution systems by a GA- based two stage Algorithm. IEEE Trans on Power Syst 12(3) (1997)
16. Abdullah, W.N.W., Saibon, H., Zaia, A.A.M., Lo, K.L.: Genetic Algorithm for reactive power Dispatch. IEEE Catalogue No 98EX137, 160–163 (1998)
17. Xiangping, M., Zhishan, L., Huaguang, Z.: Fast synthetic Genetic Algorithm and its application to optimal control of reactive power flow. IEEE 0-7803-4754-4/98, 1454–1458 (1998)

18. Ghose, T., Goswami, S.K., Basu, S.K.: Solving Capacitor placement problems in Distribution systems using Genetic Algorithms. Electric Machines and Power systems. 27, 429–441 (1999)
19. Mantovani, J.R.S., Modesto, S.A.G., Garcia, A.V.: Var Planning Using Genetic Algorithm and linear programming. IEE Proc, Gener. Transm. Distrib. 148(3), 257–262 (2001)
20. Bakirtzis, A.G., Biskas, P.N., Zoumas, C.E., Petridis, V.: Optimal Power flow by Enhanced Genetic Algorithm. IEEE Trans on Power Syst. 17(2), 229–236 (2002)
21. Wu, O.H., Ma, J.T.: Power System Reactive Power Dispatch Using Evolutionary Programming. IEEE Trans, Power Syst. 10(3), 1243–1248 (1995)
22. Ma, J.T., Lai, L.L.: Evolutionary programming approach to reactive power planning. IEE Proc, Gener. Transm. Distrib. 143(4), 365–370 (1996)
23. Yeh, E.C., Venkata, S.S., Sumic, Z.: Improved Distribution system planning using computational evolution. IEEE Trans on Power Syst. 11(2), 668–674 (1996)
24. Lai, L.L., Ma, J.T.: Application of Evolutionary programming to reactive power planning-comparision with nonlinear approach. IEEE Trans, Power Syst. 12(1), 198–206 (1997)
25. Lee, K.Y., Yang, F.F.: Optimal Reactive Power Planning using Evolutionary Algorithms: A Comparative Study for Evolutionary Programming, Evolutionary strategy, Genetic Algorithm and Linear Programming. IEEE Trans, Power Syst. 13(1), 101–108 (1998)
26. Wong, K.P., Yuryevich, J.: Evolutionary-programming-based Algorithm for environmentally-constrained economic dispatch. 13(2), 301–306 (1998)
27. Gomes, J.R., Saavedra, O.R.: Optimal Reactive Power Dispatch Using Evolutionary Computation: Extended algorithms. IEE Proc, Gener. Transm. Distrib. 146(6), 586–592 (1999)
28. Yoshida, H., Kawata, K., Fukuyama, Y., Takayama, S., Nakanishi, Y.: A Particle Swarm Optimization for Reactive Power and Voltage Control Considering Voltage Security Assessment. IEEE Trans, Power Syst. 15(4), 1232–1239 (2000)
29. Storn, R., Price, K.: Minimizing the real functions of the ICEC' 96 contest by differential evolution. IEEE 0-7803-2902-3/9, 842–844 (1996)
30. Cheng, S.L., Hwang, C.: Optimal Approximations of Linear systems by a Differential Evolution Algorithm. IEEE Trans on Systems, Man and Cybernetics, Part A: Systems and Humans 31(6), 698–707 (2001)
31. Lampinen, J.: A Constraint Handling approach for the Differential Evolution Algorithm. IEEE 0-7803-7282-4/02, 1468–1473 (2002)
32. Su, C.T., Lee, C.S.: Modified Differential Evolution method for capacitor placement of Distribution systems. IEEE 03-7525-4/02, 208–213 (2002)
33. Zhang, W.J., Xie, X.F.: DEPSO: Hybrid particle swarm with differential operator. IEEE 07803-7952-7/03-2003, 3816–3821 (2003)
34. Yan, W., Lu, S., Yu, D.C.: A Novel Optimal Reactive Power Dispatch Method Based on Improved Hybrid Evolutionary Programming Technique. IEEE Trans, Power Syst. 19(2), 913–918 (2004)

Cunning Ant System for Quadratic Assignment Problem with Local Search and Parallelization

Shigeyoshi Tsutsui

Hannan University, Matsubara, Osaka 580-8502 Japan
tsutsui@hannan-u.ac.jp

Abstract. The previously proposed *cunning* ant system (*c*AS), a variant of the ACO algorithm, worked well on the TSP and the results showed that the *c*AS could be one of the most promising ACO algorithms. In this paper, we apply *c*AS to solving QAP. We focus our main attention on the effects of applying local search and parallelization of the *c*AS. Results show promising performance of *c*AS on QAP.

1 Introduction

In a previous paper [1,2], we have proposed a variant of the ACO algorithm called the *cunning* Ant System (*c*AS) and evaluated it using TSP which is a typical NP-hard optimization problem. The results showed that the *c*AS could be one of the most promising ACO algorithms. In this paper, we apply *c*AS to solving the quadratic assignment problem (QAP). The QAP is also an NP-hard optimization problem and it is considered one of the hardest optimization problems [3,4]. The QAP is also a good set of problems for testing the capabilities of solving combinatorial optimization problems.

There are many studies on solving QAP with ACO showing better results than with other meta-heuristics. These studies are summarized in [5]. Typical examples of ACO algorithms for the QAP are AS-QAP, MMAS-QAP, and ANTS-QAP. Among these, it is reported that MMAS-QAP [3] is the best performing algorithm [5].

We performed a preliminary study which applied *c*AS to solving QAP in [6]. In this paper, we apply *c*AS to solving QAP and compare the performance with the performance of MMAS [3]. We also discuss an approach for parallelization of the *c*AS for QAP.

In the remainder of this paper, Section 2 gives a brief overview of *c*AS when it is applied in TSP. Then, Section 3 describes how the solutions with *c*AS for the QAP are constructed. In Section 4, we provide an empirical analysis of the *c*AS and compare the results with MMAS. In Section 5, we study the use of a kind of parallelization of *c*AS, with the aim of achieving faster execution of the algorithm in a network environment. Finally, Section 6 concludes this paper.

A. Ghosh, R.K. De, and S.K. Pal (Eds.): PReMI 2007, LNCS 4815, pp. 269–278, 2007.

2   A Brief Overview of cAS

cAS [1,2] introduced two important schemes. One is a scheme to use partial solutions which we call *cunning*. The other is to use the colony model, dividing colonies into units. Using partial solutions to seed solution construction in the ACO can be found in [7,8,9] with other frameworks. The agent introduced in cAS is called *cunning ant* (*c-ant*). The c-ant differs from traditional ants in its manner of solution construction. It constructs a solution by borrowing a part of existing solutions. The remainder of the solution is constructed based on $\tau_{ij}(t)$ probabilistically as usual. In a sense, since this agent in part appropriates the work of others to construct a solution, we named the agent *c-ant* after the metaphor of its cunning behavior. An agent from whom a partial solution has been borrowed by a *c-ant* is called a *donor ant* (*d-ant*).

We use a colony model which consists of m units [1,2]. Each unit consists of only one $ant^*_{k,t}$ ($k = 1, 2, \ldots, m$). At iteration t in unit k, a new $c\text{-}ant_{k,t+1}$ creates a solution with the existing ant in the unit (i.e., $ant^*_{k,t}$) as the $d\text{-}ant_{k,t}$. Then, the newly generated $c\text{-}ant_{k,t+1}$ and $d\text{-}ant_{k,t}$ are compared, and the better one becomes the next $ant^*_{k,t+1}$ of the unit. Thus, in this colony model, $ant^*_{k,t}$, the best individual of unit k, is always reserved.

Pheromone density $\tau_{ij}(t)$ is then updated with $ant^*_{k,t}$ ($k=1, 2, \ldots, m$) and $\tau_{ij}(t+1)$ is obtained as:

$$\tau_{ij}(t+1) = \rho \cdot \tau_{ij} + \sum_{k=1}^{m} \Delta^* \tau_{ij}^k(t), \tag{1}$$

$$\Delta^* \tau_{ij}^k(t) = 1/C^*_{k,t} : \text{ if } (i,j) \in ant^*_{k,t}, \ 0 : \text{ otherwise}, \tag{2}$$

where the parameter ρ ($0 \leq \rho < 1$) is the trail persistence (thus, $1-\rho$ models the evaporation), $\Delta^* \tau_{ij}^k(t)$ is the amount of pheromone $ant^*_{k,t}$ puts on the edge it has used in its tour, and $C^*_{k,t}$ is the fitness of $ant^*_{k,t}$.

In cAS, pheromone update is performed with m $ant^*_{k,t}$ ($k=1,2,\ldots,m$) by Eq. 3 within $[\tau_{min}, \tau_{max}]$ as in MMAS [3]. Here, τ_{max} and τ_{min} for cAS is defined as

$$\tau_{\max}(t) = \frac{1}{1-\rho} \times \sum_{k=1}^{m} \frac{1}{C^*_{k,t}}, \tag{3}$$

$$\tau_{\min}(t) = \frac{\tau_{\max} \cdot (1 - \sqrt[n]{p_{best}})}{(n/2 - 1) \cdot \sqrt[n]{p_{best}}}, \tag{4}$$

where p_{best} is a control parameter introduced in MMAS [3].

3   Cunning Ant System for QAP

The QAP is a problem in which a set of facilities or units are assigned to a set of locations and can be stated as a problem to find permutations which minimize

$$f(\phi) = \sum_{i=0}^{n-1} \sum_{j=0}^{n-1} a_{ij} b_{\phi(i)\phi(j)}, \tag{5}$$

where $A = (a_{ij})$ and $B = (b_{ij})$ are two $n \times n$ matrices and ϕ is a permutation of $\{0, 1, \ldots, n-1\}$. Matrix A is a distance matrix between locations i and j, and B is the flow between facilities r and s. Thus, the goal of the QAP is to place the facilities on locations in such a way that the sum of the products between flows and distances are minimized.

3.1 The c-ant for QAP

The c-ant in QAP acts in a slightly differ- ent manner than a c-ant in TSP. In TSP, pheromone trails $\tau_{ij}(t)$ are defined on each edge between city i and j. On the other hand, the pheromone trails $\tau_{ij}(t)$ in the QAP ap- plication correspond to the desirability of as- signing a facility i to a location j [3]. In this paper, we use this approach for cAS on QAP. Fig. 1 shows how the c-ant acts in QAP.

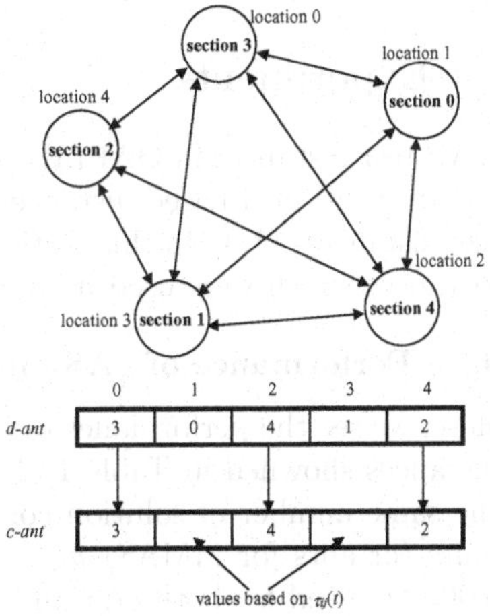

In this example, the c-ant borrows part of the node values at location 0, 2, and 4. The c-ant constructs the remainder of the node values for location 1 and 3 according to the following probability:

$$p_{ij}(t) = \frac{\tau_{ij}(t)}{\sum_{k \in N(i)} \tau_{ik}}, \qquad (6)$$

Fig. 1. c-ant and d-ant in QAP

where $N(i)$ is the set of still unassigned facilities. Using c-ant in this way, we can prevent premature stagnation the of search, because only a part of the nodes in a string are newly generated, and this can prevent over exploitation caused by strong positive feedback to $\tau_{ij}(t)$ as we observed in cAS in [1,2]. The colony model of cAS for QAP is the same as was used in [1,2] for TSP.

3.2 Sampling Methods

Let us represent the number of nodes that are constructed based on $\tau_{ij}(t)$, by l_s. Then, l_c, the number of nodes of partial solution, which c-ant borrows from d-ant, is $l_c = n - l_s$. Following cAS in TSP, we use the control parameter γ which define $E(l_s)$ (the average of l_s) by $E(l_s) = n \times \gamma$ and use the following probability density function $f_s(l)$ used in [1,2] as

$$f_s(l) = \begin{cases} \frac{1-\gamma}{n\gamma} \left(1 - \frac{l}{n}\right)^{\frac{1-2\gamma}{\gamma}} & \text{for } 0 < \gamma \le 0.5, \\ \frac{\gamma}{n(1-\gamma)} \left(\frac{l}{n}\right)^{\frac{2\gamma-1}{1-\gamma}} & \text{for } 0.5 < \gamma < 1. \end{cases} \qquad (7)$$

In cAS for TSP, nodes in continuous positions of d-ant are copied to c-ant, because the partial solutions of d-ant are represented by nodes in continuous positions. However, in QAP there is no such constraint and it is not necessary for nodes, which are copied from *d-ant* or sampled according to $\tau_{ij}(t)$, to be in continuous positions. Thus, in creating a new *c-ant* in QAP, nodes at some positions are copied and others are sampled with a random sequence of positions as follows: The number of nodes to be sampled l_s is generated by Eq. 7 with a given γ value. Then we copy the number of nodes, $l_c = n-l_s$, from *d-ant* at random positions and sample the number of remaining nodes, l_s, according to Eq. 6 with random sequence.

4   Experiments

QAP test instances in QAPLIB [10] can be classified into i) randomly generated instances, ii) grid-based distance matrix, iii) real-life instances, and iv) real-life-like instances [11,3]. In this section, we evaluate *c*AS on the QAP using QAPLIB instances which were used in [3] and compare the performance with MMAS.

4.1   Performance of *c*AS on QAP Without Local Search

Here, we see the performance of *c*AS without local search using relativlely small instances showned in Table 1. The comparison with MMAS was performed on the same number of solution constructions $E_{max} = n \times 800,000$. For number of units (or ants for MMAS) $m = n \times 4$ is used. ρ value of 0.9 and p_{best} value of 0.005 are used for both *c*AS and MMAS. 25 runs were performed.

Table 1 summarizes the results. The values in the table represent the deviation from the optimum value by *Error* (%) $((f(\phi) - best)/best \times 100)$. The results of *c*AS are with γ value of 0.3. The code for MMAS is implemented by us and tuned for the appropriate use of *global best* and *iteration best* in the pheromone update so as to get the smallest values of *Error*. We got the smallest value when the *global best* was applied every 5 iterations to the pheromone update in MMAS. The *pts* strategy [3] in MMAS was also tuned.

The values in bold-face show the best performance for each instance. From this table, we can see that *c*AS has good performance.

4.2   The Effect of γ Values

Table 1 shows *Error* for $\gamma = 0.3$. Fig. 2 shows the variations of *Error* for various γ values on tai30aCnug30Ckra30b, and tai30b. Here, γ values were varied starting from 0.1 to 0.9 with step 0.1. From this figure, we can see the effectiveness of us-

Table 1. Results without local search. *Error*(%) is average over 25 independent runs.

QAP class	QAP instance	*c*AS (γ=0.3)	MMAS	
			MMAS+pts	MMAS
i	tai20a	**1.006**	2.996	3.140
	tai25a	**1.566**	3.217	3.380
	tai30a	**1.843**	3.004	2.758
	tai35a	**2.194**	3.690	3.600
ii	nug30	**0.455**	1.675	1.962
iii	kra30a	**1.147**	3.963	4.014
	kra30b	**0.447**	2.736	2.836
iv	tai20b	**0.000**	0.348	0.388
	tai25b	**0.003**	1.753	2.385
	tai30b	**0.066**	2.274	2.279
	tai35b	**0.252**	2.453	2.472

ing *c-ant*; i.e., with the smaller values of γ (in the range of $[0.1, 0.5]$), the better values in *Error* are observed as was the case with *c*AS on TSP in [1,2].

4.3   Analysis of the Convergence Process of cAS

As we discussed in Section 2 and Subsection 4.2, the cunning action can be expected to prevent premature stagnation the of search, because only a part of the nodes in a solution are newly generated, and this prevent over exploitation caused by strong positive feedback to $\tau_{ij}(t)$. In this subsection, we analyze the convergence process using *Entropy* of pheromone density $\tau_{ij}(t)$ to measure the diversity of the system.

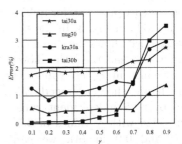

Fig. 2. Change of $\overset{\gamma}{Error}$ of cAS without local search for various γ

Definition of entropy of pheromone density. We define $I(t)$, entropy of pheromone density $\tau_{ij}(t)$, as follows:

$$I(t) = -\frac{1}{n}\sum_{i=0}^{n-1}\sum_{j=0}^{n-1} p_{ij}(t) \log p_{ij}(t), \tag{8}$$

where $p_{ij}(t)$ is defined as

$$p_{ij}(t) = \frac{\tau_{ij}(t)}{\sum\limits_{j=0}^{n-1} \tau_{ij}(t)}. \tag{9}$$

The upper bound of $I(t)$ is obtained when all elements of $\tau_{ij}(t)$ have the same values as found during the initialization stage ($t=0$). This value is calculated as

$$\overline{I} = \log(n). \tag{10}$$

To calculate the lower bound of $I(t)$, let's consider an extreme case in which all strings have the same set of node values and pheromones have been distributed across the set. If this iteration continues for a long time, all elements of $\tau_{ij}(t)$ converge to τ_{min} or τ_{max}. The lower bound of the *entropy* $I(t)$ is obtained in these situations and can be calculated as Eq. 11 as follows:

$$\underline{I} = \log(r + n - 1) - \frac{r \log(r)}{r + n - 1} \tag{11}$$

where $r = \tau_{max}/\tau_{min}$. In the following analysis, we use the normalized entropy $I_N(t)$ which is defined with $I(t)$, $\overline{I}$, and $\underline{I}$ as

$$I_N(t) = \frac{I(t) - \underline{I}}{\overline{I} - \underline{I}} \tag{12}$$

Then, $I_N(t)$ takes values in $[0.0, 1.0]$.

Analysis of the convergence process. Here we show the convergence processes for tai25b and tai30b in Fig. 3. In the figure, the left shows the change in *Error* (%) and the right shows the change in $I_N(t)$. Values in the figure show averaged values over 25 independent runs. On tai25 with γ values of 0.5, 0.7, and 0.9 in (a), we can see that I_N converges around 80000, 40000, and 20000 iterations, respectively. These iterations coincide with the iterations where stagnations in *Error* occur. With γ value of 0.3, the value of I_N gradually decreases and the search continues with less stagnation. With γ value of 0.1, I_N keeps larger values until the end of run, resulting in slow convergence in *Error*. Sim-

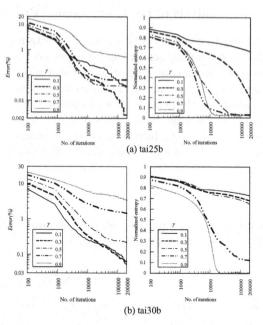

(a) tai25b

(b) tai30b

Fig. 3. Convergence processes of tai25b and tai30b without local search

ilar results for tai30b are observed, although their values in detail are different from tai25b.

From this convergence process analysis using the entropy measure, we can see the effectiveness of the cunning scheme with smaller values of γ. That is, on average, taking the rate of $(1-\gamma)$ partial solution from existing solutions, and having the rate of γ partial solution being generated anew from the pheromone density can maintain diversity of the system, resulting in good balance between exploration and exploitation in the search. However, with extreme smaller values of γ, i.e., $\gamma \leq 0.1$, the search processes becomes much slower, though the diversity of pheromone density can be maintained.

4.4   Performance of *c*AS with Local Search

Here we study *c*AS with a local search on QAP. In [3], MMAS is combined with two local searches, i.e., Robust Taboo search algorithm (Ro-TS) developed by Taillard [11] and 2OPT. In this paper, we combined *c*AS with Ro-TS (*c*AS-TS) and compare the results with results described in [3].

Parameter settings and the methods of applying *c*AS with Ro-TS (*c*As-TS) are the same as were used for MMAS with To-TS (MMAS-TS) in [3] as follows: m value of 5, ρ value of 0.8, and p_{best} value of 0.005. 250 times, short Ro-TS runs of length $4n$ were applied. This setting was designed in [3] so that the computational time is the same as the Ro-TS carried out alone in which $1000 \times n$ iterations was allowed. We used the Ro-TS code which is available at [12], though the code, which is originally written in C, was rewritten in Java since our cAS code is written in Java.

Table 2 summarizes the results. For comparison, we show results of other algorithms, i.e., MMAS-TS, MMAS-2OPT, GH (Gentic Hybrid), HAS (Hybrid Ant System), and Ro-TS. These are taken from [3]. Results of cAS-TS is for γ=0.4 and 0.8. First, we compare cAS-TS with MMAS-TS. In this comparison, we showed the better values in bold-face, and showed the best performing values in bold-face with an under line.

For instances in class i), cAS-TS performed better than MMAS-TS and and was the best performer. Here note that cAS-TS with $\gamma = 0.4$ showed better performance than cAS-TS with $\gamma = 0.8$. For instances in class ii), cAS-TS showed bet-

Table 2. Results of cAS in *Error* (%) with local search. Results except for cAS are average over 10 runs.

QAP class	QAP	cAS-TS		MMAS-TS	MMAS-2OPT	HAS-QAP	GH	Ro-TS
		(γ=0.4)	(γ=0.8)					
i	tai35a	**0.582**	**__0.572__**	0.715	1.128	1.762	0.698	0.589
	tai40a	**__0.726__**	0.793	0.794	1.509	1.989	0.884	0.990
	tai50a	**__1.051__**	1.190	1.060	1.795	2.800	1.049	1.125
	tai60a	**__1.059__**	1.289	1.137	1.882	3.070	1.159	1.203
	tai80a	**__0.740__**	1.029	0.836	1.402	2.689	0.796	0.900
ii	sko42	**0.005**	0.008	0.032	0.051	0.076	__0.003__	0.025
	sko49	**0.044**	0.062	0.068	0.115	0.141	__0.040__	0.076
	sko56	**__0.055__**	0.065	0.075	0.098	0.101	0.060	0.088
	sko64	0.077	**__0.035__**	0.071	0.099	0.129	0.092	0.071
	sko72	0.126	0.091	**__0.090__**	0.172	0.277	0.143	0.146
	sko81	0.105	0.105	**__0.062__**	0.124	0.144	0.136	0.136
	sko90	**__0.104__**	0.168	0.114	0.140	0.231	0.196	0.128
iii	ste36a	0.139	0.118	**__0.061__**	0.126	n.a.	n.a.	0.155
	ste36b	**__0__**	**__0__**	**__0__**	**0**	n.a.	n.a.	0.081
iv	tai35b	0.098	**__0__**	0.051	**0**	0.026	0.107	0.064
	tai40b	0.403	**__0__**	0.402	**0**	0.000	0.211	0.531
	tai50b	0.183	**0.113**	0.172	**__0.009__**	0.192	0.214	0.342
	tai60b	0.280	0.091	**__0.005__**	**__0.005__**	0.048	0.291	0.417
	tai80b	0.716	**0.445**	0.591	**__0.266__**	0.667	0.829	1.031
	tai100b	0.263	**0.155**	0.230	**__0.114__**	n.a.	n.a.	0.512

ter performance than MMAS-TS except for with sko72 and sko81. Note here that GH showed the best values among all algorithms, but the performance differences between GH and cAS-TS were very small.

For instances in class iii), MMAS-TS performe better than cAS-TS on ste36a, and *Error*=0 on ste36b for both cAS-TS and MMAS-TS. For instances in class iv), with all instances except tai60b, cAS-TS showed better *Error* values than MMAS-TS. Note here that for all instances in this class, MMAS-2OPT has the best *Error* values among the algorithms.

In comparison between cAS-TS and MMAS-TS, cAS-TS outperforms MMAS-TS on 15 instances and MMAS-TS outperforms cAS-TS on 4 instances. Thus, cAS-TS has relatively better performance than MMAS-TS. However, the performance differences between cAS-TS and MMAS-TS were not as large as those we saw in Table 1 where no local search is applied. This is because the effect of local searches become more dominant.

5 Parallelization of cAS (p-cAS)

Parallelization of evolutionary algorithms including ACO is a well-known and popular approach [13,14,15]. There are two main reasons for using parallelization: (i) given a fixed time to search, to increase the quality of the solutions found within that time; (ii) given a fixed solution quality, to reduce the time to find a solution not worse than that quality [15]. In this study, we ran a *parallel cAS* (*p-cAS*) exploiting the second reason, i.e., aiming to reduce the time to find solutions which are of the same quality as those found with a single processor.

All of our code for evolutionary computation research are written in Java, including ACO algorithms. Java has many *classes* of programming for network environments. Typical examples are RMI and Jini [16]. In our implementation of *p-cAS*, we used *Java applet* scheme to send the program to client computers. This enables us to use all computers in the network which have a web browser with Java runtime. Communication between the server and clients is performed by exchanging *objects* with the *Serializable interface*. The server program runs as a Java application.

5.1 Load Sharing *p-cAS* on QAP

In the ACO framework, the most popular parallel architecture is the *island model* in which multiple sub-colonies are run in parallel on distributed computers exchanging information among them periodically [15]. The main priority of the *island model* is placed on improving the solution quality. In contrast, our main priority is to reduce computational time using the *load sharing model* (*load sharing p-cAS*).

Table 3. Comptational time of *cAS* with Ro-TS in millisecond

QAP	sampling with τ_{ij}	applying Ro-TS	updating of τ_{ij}	other	TOTAL (ms)
tia40b	3.8	6833.8	1.3	12.3	6851.2
(%)	0.1%	99.7%	0.0%	0.2%	100.0%
tia50b	4.9	13465.8	2.3	13.3	13486.2
(%)	0.0%	99.8%	0.0%	0.1%	100.0%
tia60b	7.0	23448.7	3.0	15.7	23474.4
(%)	0.0%	99.9%	0.0%	0.1%	100.0%
tia80b	10.6	56688.7	5.9	19.9	56725.0
(%)	0.0%	99.9%	0.0%	0.0%	100.0%
tia100b	14.8	114411.7	9.0	21.4	114456.9
(%)	0.0%	100.0%	0.0%	0.0%	100.0%

Table 3 shows the computation times on QAP in *cAS* with Ro-TS which were performed in Section 4.4. The machine we used had two Opteron 280 (2.4GHz, Socket940) processors with 2GB main memory. The OS was 32-bit WindowsXP. Java2 (j2sdk1.4.2_13) was installed. From this table, we can see that more than 99% of computation time is used for Ro-TS. Therefore we distribute the calculation for local search over computers in the network. Fig. 4 shows the functions of the server and clients. When we use m ants, local searches for m ants are distributed over the server and $m-1$ clients.

Experimental conditions for the *load sharing p-cAS* in this research are as follows: We used two Opteron-based machines, say machine A and machine B, each which has the structure described above and thus each machine has 4 processing units. The machines A and B are connected via a 1000BASE-T switching hub. We assigned server functions to machine A and client functions to machine B. In machine A, we installed an Apache [17] http server. We assigned four clients so that the logical experimental conditions are the same as the experiments in Section 5 with a single machine. To do so, we ran 4 independent browser processes to access the server. The experiments were performed with 25 independent runs and the time to complete the computation was measured. We used γ value of 0.8. With this scheme, among 5 ants, local search for one ant is performed by server machine A, and local searches for the other 4 ants are performed by client machine B.

Fig. 5 summarizes the results on the computation time. The results are averaged over 25 runs. Here, *gain* indicates (run time of cAS) / (run time of p-cAS). If there is no communication overhead, gain should be five. However, as we can see in Fig. 5 there is a communication overhead between the server and clients. Due to this overhead, the gain is smaller than 1 for tai40b and tai50b,

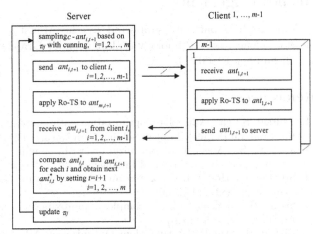

Fig. 4. Structure of load sharing p-cAS

and is 1 for tai60b. On the other hand, gain is 1.8, 2.9, and 4.1 for tai80b, tai100b, and tai150b, respectively. This is because the communication overhead for larger problems becomes relatively smaller compared with the time required for the local search. To illustrate this, we also showed $T_{comm.}$, the total time used by the server for communication between the server and the client.

6   Conclusions

In this paper, we applied cAS to solving the QAP and compared it agaist MMAS. The results showed cAS has promising performance. We analyzed the convergence process and the results showed that the cunning scheme is effective in maintaining diversity of pheromone density. An implementation for a simple load sharing parallel cAS (p-cAS) is also shown and a

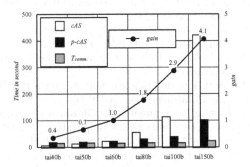

Fig. 5. Structure of load sharing p-cAS

meaningful speedup of computation in the network environment was observed. However, the following study subjects remain for future work: combining cAS with other local search, such as 2OPT; study on other types of p-cAS such as the island model; improving the server and client programs to reduce communication overhead.

Acknowledgements

This research is partially supported by the Ministry of Education, Culture, Sports, Science and Technology of Japan under Grant-in-Aid for Scientific Research number 19500199.

References

1. Tsutsui, S.: cas: Ant colony optimization with cunning ants. In: Proc. of the 9th Int. Conf. on Parallel Problem Solving from Nature (PPSN IX), pp. 162–171 (2006)
2. Tsutsui, S.: Ant colony optimization with cunning ant. Transactions of the Japanese Society for Artificial Intelligence 22(1), 29–36 (2007)
3. Stützle, T., Hoos, H.: Max-min ant system. Future Generation Computer Systems 16(9), 889–914 (2000)
4. Sahni, S., Gonzalez, T.: P-complete approximation problems. Journal of the ACM 23, 555–565 (1976)
5. Dorigo, M., Stützle, T.: Ant Colony Optimization. MIT Press, Massachusetts (2004)
6. Tsutsui, S., Liu, L.: Solving quadratic assignment problems with the cunning ant system. In: Proc. of the 2007 CEC (to appear)
7. Acan, A.: An external memory implementation in ant colony optimization. In: Dorigo, M., Birattari, M., Blum, C., Gambardella, L.M., Mondada, F., Stützle, T. (eds.) ANTS 2004. LNCS, vol. 3172, pp. 73–84. Springer, Heidelberg (2004)
8. Acan, A.: An external partial permutations memory for ant colony optimization. In: Proc. of the 5th European Conf. on Evolutionary Computation in Combinatorial Optimization, pp. 1–11 (2005)
9. Wiesemann, W., Stützle, T.: An experimental study for the the quadratic assignment problem. In: Dorigo, M., Gambardella, L.M., Birattari, M., Martinoli, A., Poli, R., Stützle, T. (eds.) ANTS 2006. LNCS, vol. 4150, pp. 179–190. Springer, Heidelberg (2006)
10. QAPLIB-A Quadratic Assignment Problem Library, http://www.opt.math.tu-graz.ac.at/qaplib/
11. Taillard, É.D.: Robust taboo search for the quadratic assignment problem. Parallel Computing 17, 443–455 (1991)
12. Taillard, E.: Robust tabu search implementation, http://mistic.heig-vd.ch/taillard/
13. Cantu-Paz, E.: Efficient and Accurate Parallel Genetic Algorithm. Kluwer Academic Publishers, Boston (2000)
14. Tsutsui, S., Fujimoto, Y., Ghosh, A.: Forking gas: Gas with search space division schemes. Evolutionary Computation 5(1), 61–80 (1997)
15. Manfrin, M., Birattari, M., Stützle, T., Dorigo, M.: Parallel ant colony optimization for the traveling salesman problems. In: Dorigo, M., Gambardella, L.M., Birattari, M., Martinoli, A., Poli, R., Stützle, T. (eds.) ANTS 2006. LNCS, vol. 4150, pp. 224–234. Springer, Heidelberg (2006)
16. Sun Microsystems, Inc.: Java 2 Platform, Standard Edition, v1.4.2 at API Specification, http://java.sun.com/j2se/1.4.2/docs/api/
17. Apache Software Foundation: Apache HTTP server project, http://httpd.apache.org/

Use of Ant Colony Optimization for Finding Neighborhoods in Image Non-stationary Markov Random Field Classification

Sylvie Le Hégarat-Mascle[1], Abdelaziz Kallel[2], and Xavier Descombes[3]

[1] IEF/AXIS, Université de Paris-Sud 91405, Orsay Cedex, France
[2] CETP/IPSL, 10, 12 avenue de l'Europe 78140, Vélizy, France
[3] CNRS/INRIA/UNSA, INRIA, 06902 Sophia Antipolis, Cedex, France

Abstract. In global classifications using Markov Random Field (MRF) modeling, the neighborhood form is generally considered as independent of its location in the image. Such an approach may lead to classification errors for pixels located at the segment borders. The solution proposed here consists in relaxing the assumption of fixed-form neighborhood. Here we propose to use the Ant Colony Optimization (ACO) and to exploit its ability of self-organization. Modeling upon the behavior of social insects for computing strategies, the ACO ants collect information through the image, from one pixel to the others. The choice of the path is a function of the pixel label, favoring paths within a same image segment. We show that this corresponds to an automatic adaptation of the neighborhood to the segment form. Performance of this new approach is illustrated on a simulated image and on actual remote sensing images SPOT4/HRV.

1 Introduction

Classification processes were among the first attempts to interpret image quantitatively. Global approaches, such as Maximum A Posteriori (MAP), have been possible using Markov Random Fields (MRF) modeling [1]. For a given neighborhood system with clique potential functions, a global energy term is defined that should be minimized. Now, if the neighborhood form is considered as stationary within the image (that is generally the case), classification errors may occur on pixels having neighbors belonging to different classes. To overcome this problem, alternative approaches have been proposed, such as a line process [2] or specific potentials [3,4]. The solution proposed here is to relax the assumption of neighborhood stationary. Ant Colony Optimization [5] (ACO) that belongs to evolutionary computation [6,7,8,9] algorithms has been successfully applied to routing in telecommunication networks (e.g. [10,11]), quadratic assignment problem [12], graph coloring problem [13], traveling salesman problem [14]. The information gathered by simple autonomous mobile agents, called 'ants' (ACO derives from the behavior of social insects), is shared and exploited to solve a problem. Generally, each ant constructs its own solution and a trace of the best solution is kept (though the pheromone deposit technique) during the iterative

A. Ghosh, R.K. De, and S.K. Pal (Eds.): PReMI 2007, LNCS 4815, pp. 279–286, 2007.
© Springer-Verlag Berlin Heidelberg 2007

construction of the final solution. At each iteration, to produce the solution which will be memorized, several new solutions (as many as ants) are constructed: some of them taking into account a combination of several previous solutions (their number depending on the pheromone evaporation rate), and some of them being more or less randomly constructed. Like for SA or genetic algorithms, the introduction of randomness in the search procedure permits to escape from local minima. In this study, we propose to use the ACO by exploiting its ability of self-organization. Considering a MRF modeling with a non-stationary neighborhood, the ACO scheme jointly estimates the regularized classification map and the optimal non-stationary neighborhood. Section 2 deals with the image modeling; in section 3, the proposed method based on ACO heuristics is presented; section 4 shows some results, firstly on a simulated image, and secondly, on a SPOT/HRV image; finally, section 5 gathers our conclusions.

2   Classification Problem Assuming Non-stationary Neighborhood

The problem of image classification is to determine the realization of L, the label image, knowing those of X, the 'radiometric' image. Ω being the set of the pixel locations (image lattice), X is a random field that takes values in $\mathbb{R}^{|\Omega|}$, and L is a random field that takes values in $\Lambda^{|\Omega|}$, where $\Lambda = \{1,\ldots,c\}$ and c is the class number. The cardinal of Ω is $|\Omega| = N_l \times N_c$, N_l being the image dimension in lines and N_c its dimension in columns. According to the *Maximum A Posteriori* criterion, the optimum solution maximizes $p(X/L).p(L)$. Assuming that the distribution law of the pixel values conditionally to their class l_s is an independent Gaussian of mean μ_{l_s} and standard deviation σ_{l_s},

$$p(X/L) = \left(\frac{1}{\sqrt{2\pi}}\right)^{|\Omega|}.\exp\left\{-\frac{1}{2}\sum_{s\in\Omega}\left[\left(\frac{x_s - \mu_{l_s}}{\sigma_{l_s}}\right)^2 + \log(\sigma_{l_s}^2)\right]\right\}. \tag{1}$$

The prior model $p(L)$ is defined assuming a MRF modeling. Neighborhood system is such that, if pixel s' belongs to pixel s neighborhood $N(s)$, then s is a neighbor of s':

$$s' \in N(s) \Leftrightarrow s \in N(s'). \tag{2}$$

According to the Hammerley-Clifford theorem, $p(L)$ follows a Gibbs distribution:

$$p(L) = \frac{1}{Z}.\exp\left\{-\sum_{\gamma\in\Gamma}\mathcal{V}_\gamma(l_s, s \in \gamma)\right\}, \tag{3}$$

where Γ is the set of the image cliques γ that describe the interactions between pixels, Z is a normalization constant, and $\mathcal{V}_\gamma$ is the γ potential. Generally, the clique potentials are defined such that a pixel and its neighbors have a high probability to share the same label. Finally, the MAP criterion leads to the minimization of

$$E = \frac{1}{2}\sum_{s\in\Omega}\left[\left(\frac{x_s - \mu_{l_s}}{\sigma_{l_s}}\right)^2 + \log(\sigma_{l_s}^2)\right] + \sum_{\gamma\in\Gamma}\mathcal{V}_\gamma(l_s, s \in \gamma). \tag{4}$$

The minimization of (4) is performed using the fact that the global energy difference between two label image configurations only differing by one pixel label only depends on this pixel s and its neighborhood. Under the assumption of pixel neighborhood having the same geometry at each pixel location s, this calculation can be performed handling a reasonable number of terms. However, the diversity of the areas or object shapes makes questionable the stationarity assumption for neighborhood form. Therefore, here we propose to relax this assumption and to adopt an approach where the neighborhood form is automatically adjusted. At each location s, the neighborhood is constructed based on three criteria:

- each pixel has the same number of neighbors,
- the neighbor pixels are connected,
- the neighbor pixels have a high probability to share the same label.

Now, for the classification problem, (5) is still valid except that the cliques are now defined over non stationary neighborhoods. In the simplest case, only cliques of cardinal 2 are considered. Assuming the Potts model for the potentials, for a given neighborhood system, the function to minimize is given by

$$E = \sum_{s \in \Omega} \left[\frac{(x_s - \mu_{l_s})^2}{2\sigma_{l_s}^2} + \log(\sigma_{l_s}) + \beta \frac{|\{r \in N(s); l_r \neq l_s\}|}{2} \right], \qquad (5)$$

where $|\{r \in N(s); l_r \neq l_s\}|$ is the number of neighbors having a different label than s, and β is a positive parameter weighting the relative importance of the 'data attach' term and the neighborhood one. The factor $1/2$ is due to the fact that, for cliques of order 2, their potentials are counted 2 times when the sum is done over the pixels rather than over the image clique set.

Relaxing the neighborhood stationarity assumption, the s optimal neighborhood now depends on the label l_s. Therefore, we cannot directly obtain an expression of the energy difference between two label image configurations. Indeed, now the cliques involving s are not the same ones when $l_s = l_s^{(1)}$ and when $l_s = l_s^{(2)}$ since, they depend on the neighborhood geometry which varies with l_s. Moreover, due to the constraint of constant neighborhood cardinal, when s neighborhood is changed, the neighborhoods of some other pixels are also changed: the pixel s previous neighbors, that have lost one neighbor (namely s), have to find another pixel neighbor replacing it, and the new s neighbors, that have gain one neighbor (s), have to get rid of another pixel neighbor, and so on. We showed [15] that it corresponds to assume the existence of another random field H that corresponds to the definition of the non-stationary neighborhood in every pixel, and that the couple (L, H) is Markovian.

At each pixel location s, the optimal neighborhood is researched with the constraint of its cardinality (third criterion for neighborhood construction), and such that the symmetry property (2) is satisfied: once a pixel has constructed its neighborhood (h_s), the neighborhoods of some other pixels (those belonging to hs) are already partially constructed, and subsequent neighborhood constructions should take into account already achieved neighborhood constructions.

Then, one must use a metaheuristic to find the 'optimal' neighborhood config-uration defined (i) knowing the image observation, i.e. X realization, and (ii) assuming the image label, i.e. L realization.

3   Application of the ACO Meta-heuristic

Ant Colony Optimization mimics to the way social insects are able to solve some optimization problems. The ant problem is to find the shortest path between their nest and food. While searching for food, ants deposit trails of a chemical substance called pheromones to which other ants are attracted. As shorter paths to food will be traversed more quickly, they have a better chance of being sought out and reinforced by other ants before the volatile pheromones evaporate.

More conceptually, the problem of the ant colony is the following: given a function to minimize, different solutions are examined (randomly in a small percentage of cases), and each of them is memorized (thanks to the pheromone deposit) depending on its quality. In our case, we use ACO because of the analogy that can be done between the research of an optimal path by ants and the research of an optimal set of connected pixels from a given 'originate' pixel, and the required interaction (obtained using pheromone deposit) between local neighborhood solutions for (2) ascertaining. We now explain the way ACO is used for neighborhood construction. Assuming that the order of neighbor selection is without importance, in 8-connectivity, during the neighborhood construction, the following neighbor can be selected among any of the pixels located in a range of $[1, +1]$ lines and column of an already selected neighbor.

According to the third neighborhood criterion, the cardinal of the neighbor-hood of pixel s, $|N(s)|$, is assumed to be constant. In the following, it is noted N_n. Denoting $\delta(.,.)$ the Kroenecker function: $\delta(i,j) = 1$ if $i = j$, $\delta(i,j) = 0$ otherwise, and $h_s = N(s)$ the s neighborhood $|\{r \in N(s); l_r \neq l_s\}|$ writes $\sum_{r \in h_s} 1 - \delta(l_s, l_r)$. Then, using (2), (5) is:

$$E = \sum_{s \in \Omega} \left\{ \sum_{r \in h_s} \left[\frac{\beta}{2}(1 - \delta(l_s, l_r)) + \frac{1}{N_n}\left(\frac{(x_r - \mu_{l_r})^2}{2\sigma_{l_r}^2} + \log(\sigma_{l_r})\right) \right] \right\}. \quad (6)$$

Now, according to the Markovian property of (L, H), we are able to compare the energy of two configurations only differing by the label l_s of pixel s and the neighborhood realizations h_t of the pixels t included in $W_{N_n}(s)$, the $(2N_n + 1) \times (2N_n + 1)$ sized neighborhood around s. Practically, this means that, for any pixel t of $W_{N_n}(s)$ having n_{out} 'active' neighbor(s) outside of $W_{N_n}(s)(n_{out}[0, N_n])$, these n_{out} neighbors are fixed, and only the other $(N_n - n_{out} = n_{in})$ 'active' neighbors are 'free' and can be changed with other possible neighbors belonging to $W_{N_n}(s)$. This is the choice of these 'free' neighbors that will be optimized by the ants. The algorithm is as follows [15].

The pixels t emit ants that gather the information about neighbor label, along some paths of connected pixels including t. During their paths, neighborhood constructions, the ants select the following neighbor from 'routing indicators'.

These latter are based on previously deposed pheromones (either by ants emitted by t or by ants emitted by other pixels having chosen t as neighbor) and the 'energy hop' $E_{t \to r}$ defined as:

$$E_{t \to r} = \beta(1 - \delta(l_t, l_r)) + \frac{1}{N_n} \left(\frac{(x_t - \mu_{l_t})^2}{2\sigma_{l_t}^2} + \log(\sigma_{l_t}) \frac{(x_r - \mu_{l_r})^2}{2\sigma_{l_r}^2} + \log(\sigma_{l_r}) \right). \quad (7)$$

(7) shows that the following neighbor r is chosen considering its label l_r (relative to the emitting pixel label l_t) and also its 'data attach' energy, which gives a hint about the l_r confidence. According to the ACO procedure, the next neighbor is chosen either randomly according to the probability of random exploration, or minimizing a function of $E_{t \to r}$ and pheromone deposition. Practically, to simulate the pheromone deposition we define, for each pixel s a 'neighborhood' matrix of size $(2N_n + 1) \times (2N_n + 1)$ representing all the possible neighbor pixel locations from s. The 'neighborhood' matrix values are real, with the matrix norm equal to 1, i.e. this matrix is somewhat a fuzzy representation of the neighbor feature of the pixels around s. Denoting $[N_s](r)$ the value of the neighborhood matrix of s in r ($r \in W_{N_n}(s)$, $[N_s](r) = [N_r](s)$), we define the cost of the choice of r as t following neighbor as $c_{t \to r} = \beta(1 - [N_t](r)) + E_{t \to r}$.

Arriving at a selected neighbor pixel, an ant waits a time proportional to the cost $c_{t \to r}$, before selecting the next neighbor. Each ant has to find a number of neighbors equal to the number of 'free active neighbors' of its emitting pixel. Then, it stops and on its return deposits pheromones on the visited neighbor pixels. Practically, for each pixel r visited by an ant emitted by pixel t, $[N_t](r)$ is increased by the quantity of pheromone deposit q. Due to (2), pheromones are also deposited on the 'neighborhood' matrices of the r selected neighbors, on the pixel corresponding to t in these 'neighborhood' matrices, $[N_r](t)$. Due to the waiting time, pixels on 'good' paths are visited frequently by ants, thus increasing neighborhood matrix values for pixels contained in those paths and diminishing other ones.

Then, the classification global algorithm is as follows. First are set the image parameters: neighborhood size N_n and weight β, and the ACO parameters: pheromone deposit quantity q, random exploration probability p_e, and experience duration T_e. For initialization, a blind classification of X is performed. As long as the stopping criterion is not verified, the pixels s are considered successively. For each s, the local energy E_1 term is computed according to the current label $l_s^{(1)}$ of s and the current neighborhoods h_t of the pixels t included in $W_{N_n}(s)$. For each new label to test $l_s(2)$, reconstruct the neighborhoods h_t of the pixels t included in $W_{N_n}(s)$ using ACO: $W_{N_n}(s)$ pixels t emit ants, which explore neighborhood solutions and actualize neighborhood matrices as described previously; each time an ant is back to its source pixel t, if the 'experience duration' T_e is not expired, t generates a new ant. At the end of the ACO experience, the new the local energy E_2 term is computed. Then, the decision to change the s label from $l_s^{(1)}$ to $l_s^{(2)}$ is taken if the local energy has decreased, i.e. if $\Delta E = E_1 - E_2$ is positive. With this algorithm, the convergence is ensured by

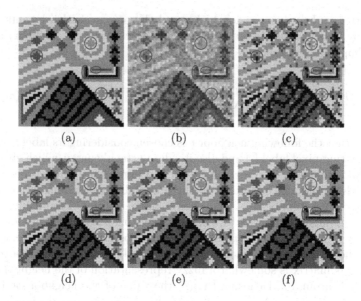

Fig. 1. Simulated data (noise $\sigma = 40$): (a) 'true' label image, (b) data image, and obtained classifications: (c) blind result, (d) ACO result, (e) 8-connectivity stationary neighborhood result using SA algorithm, (f) line process [1] result

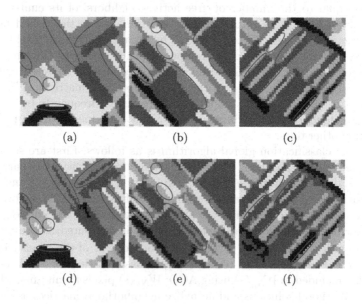

Fig. 2. Classification results: isotropic neighborhood (first line), and adaptive neighborhood (second line)- 50×50 pixel subparts of an actual SPOT image

the fact that the global energy is decreased at each step, just as for the Iterative Conditional Modes (ICM, [16]).

4   Results

We first consider simulated data with Gaussian $\mathcal{N}(\mu_i, \sigma_i)$ conditional distribution. Fig.1a shows the 4-class label image, and Fig.1b the data image when $\mu_i \in \{100, 200, 300, 400\}$, and σ_i equal to 40 for all i. Classical classification corresponding to (4) minimization with stationary neighborhood, either for null clique potential functions (blind classification, Fig.1c), or for clique potential functions corresponding to 8-connectivity Potts model was first performed. The MAP is obtained using simulated annealing process, testing different β parameter values and keeping the best result (Fig.1e). The blind result is very noisy. With the classical Potts stationary neighborhood, most of the 'isolated' errors have been corrected. To correct the packet errors the β parameter must be increased, however doing that some fine structures are lost. To overcome this limitation in performance, one has to change the image model revisiting the assumption of stationary neighborhood. Fig.1f is the result provided by the line process [1], much more sophisticated (and complex) than the isotropic neighborhood model, and *a priori* able to preserve image fine structures while regularizing the configuration. The ACO result is presented on Fig.1d. Among the considered approaches, ACO leads to the best result even if some errors remain due to the high level of noise. Some examples illustrating the interest of the proposed approach are pointed on Fig.1.

We now consider actual data acquired by the SPOT4/HRVIR sensor, having pixel size equal to $20 \times 20m^2$ and from whose measurements a 'vegetation index' can be derived for the study of vegetation areas (e.g. agricultural areas). Fig.2 shows the results of classification considering five main classes of vegetation densities (corresponding to different growing stages), and pointed areas where the great complexity of the landscape, and the thinness some fields illustrates the interest of adaptive neighborhood approach (relative to isotropic one) both allowing better preservation of fine structures (areas 1, 2, 4, 5, 11, 13, 14, 16) and a removal of blind classification errors (3, 12) even in the presence of mixed pixels (7, 9, 16, 17, 18, 19).

5   Conclusion

In this study, we present a method to estimate non-stationary neighborhood shape in the framework of MRF Bayesian classification. It uses the meta-heuristic ACO, based on the behavior of social insects (ants) and their ability to find optimal solution thanks to the deposit of pheromones for their communications. Applied, as we propose, to the construction of pixel neighborhoods in global classification problems, it yields a performance superior to that of classical fixed form neighborhoods. The advantage of having a neighborhood shape, which automatically adapts to the image segment, clearly appears in the case of images containing fine elements. The proposed method shows a stable performance relative to image parameter β, and is relatively robust to the exact fitting of the ACO parameters that can be calibrated to the proposed default values.

References

1. Geman, S., Geman, D.: Stochastic relaxation, gibbs distribution and bayesian restoration of images. IEEE Trans. on PAMI 6, 721–741 (1984)
2. Geman, D., Geman, S., Graffigne, C., Dong, P.: Boundary detection by constrained optimization. IEEE Trans. on PAMI 12, 609–628 (1990)
3. Geman, S., Reynolds, G.: Constrained restoration and recovery of discontinuities. IEEE Trans. on PAMI 14, 367–383 (1992)
4. Descombes, X., Kruggel, F., von Cramon, Y.: Spatio-temporal fmri analysis using markov random fields. IEEE Trans. on Medical Imaging 17(6), 1028–1039 (1998)
5. Dorigo, M., Maniezzo, V., Colorni, A.: The ant system: optimization by a colony of cooperating agents. IEEE Trans. on Systems, Man, and Cyber. 26, 29–41 (1996)
6. Yao, X.: Evolutionary Computation: Theory and Applications. World Scientific, Singapore (1999)
7. Tan, K., Lim, M., Yao, X., L., P., W. (eds.): Recent Advances in simulated Evolution And Learning. World Scientific, Singapore (2004)
8. Rodríguez-Vázquez, K., Fonseca, C.M., Fleming, P.J.: Identifying the structure of nonlinear dynamic systems using multiobjective genetic programming. IEEE Trans. on Sys., Man, and Cyber. 34(4), 531–545 (2004)
9. Yeun, Y.S., Ruy, W.S., Yang, Y.S., Kim, N.J.: Implementing linear models in genetic programming. IEEE Trans. Evol. Comp. 8(6), 542–566 (2004)
10. di Caro, G., Dorigo, M.: Antnet: distributed stigmeric control for communications networks. J. of Artificial Intelligence Research 9, 317–365 (1998)
11. Sigel, E., Denby, B., Le Hégarat-Mascle, S.: Application of ant colony optimization to adaptive routing in a leo telecommunications satellite network. Ann. of Telecommunications 57, 520–539 (2002)
12. Colorni, A., Dorigo, M., Maffioli, F., Maniezzo, V., Righini, G., Trubian, M.: Heuristics from nature for hard combinatorial problems. Int. Trans. in Operat. Research 3, 1–21 (1996)
13. Costa, D., Hertz, A.: Ants can colour graphs. J. of the Operat. Resea. Soc. 48, 295–305 (1997)
14. Dorigo, M., Gambardella, L.: Ant colony system: A cooperative learning approach to the traveling salesman problem. IEEE Trans. on Evol. Comp. 1, 53–66 (1997)
15. Le Hégarat-Mascle, S., Kallel, A., Descombes, X.: Ant colony optimization for image regularization based on a non-stationary markov modeling. IEEE Trans. on Image Processing 16(3), 865–878 (2007)
16. Besag, J.: On the statistical analysis of dirty pictures. J. of the Royal Statistical Society, Series B 3(48), 259–302 (1986)

A Neuro-Fuzzy Scheme for Integrated Input Fuzzy Set Selection and Optimal Fuzzy Rule Generation for Classification

Santanu Sen[1] and Tandra Pal[2]

[1] Tejas Networks India Ltd., Bangalore - 560078, India
santanusen_82@yahoo.co.in
[2] Department of Computer Science and Engineering,
National Institute of Technology, Durgapur 713209, India
tandranit@yahoo.com

Abstract. This paper proposes a scheme for designing a classier along with fuzzy set selection. It detects discontinuities in the domain of input and prunes the corresponding fuzzy sets using a neuro-fuzzy architecture. This reduces the size of the network and so the number of rules. The network is trained in three phases. In the rst phase, the network learns the important fuzzy sets. In the subsequent phases, the network is pruned to produce optimal rule set by pruning conicting and less significant rules. We use a four-layered feed-forward network and error back propagation learning. The second layer of the network learns a modulator function for each input fuzzy set that identies the unnecessary fuzzy sets. In this paper, we also introduced the notion of utility factors for fuzzy rules. Rules with small utility factors are less signicant or less used rules and can be eliminated. They are learned and detected by the third layer. After training in phase 3 in its reduced structure, the system retains almost the same level of performance. The proposed system has been tested on synthetic data set and found to perform well.

1 Introduction

A classifier assigns a class label to an object taking feature vector of the object as input. Fuzzy systems implemented with neural networks are known as neural fuzzy systems [1]-[4]. Fuzzy rule based systems have been successfully applied to various classication tasks [4],[6],[8]. The neuro-fuzzy systems proposed so far for function approximation [1]-[3] or classication [4] are capable of learning a mapping from inputs to their corresponding outputs and/or feature selection. But they are not capable of analyzing the input data to nd discontinuities and utilizing this information to nd unnecessary fuzzy sets in reducing the size of the system. In [7], Krishnapuram and Lee uses fuzzy aggregation function to develop a neural network for classication. Feature selection is also done in their network in certain conditions.

Shann and Fu [2] proposed a fuzzy neural network (FNN) for selection of rules in a fuzzy controller. Initially, the network contained all possible fuzzy rules. The

A. Ghosh, R.K. De, and S.K. Pal (Eds.): PReMI 2007, LNCS 4815, pp. 287–294, 2007.

redundant rules were pruned from the system to get a concise rule base after EBP learning. Pal and Pal [1] discussed various limitations of the network proposed by Shann and Fu and provided a better scheme to maintain nonnegative certainty factors of rules for learning and rule pruning. Chakraborty and Pal [3] proposed a scheme for simultaneous feature selection with system identication in neuro-fuzzy paradigm. In their network they used a special modulator function for each input feature to identify the redundant features. In another scheme for classication, they [4] used the same strategy for feature selection. Certainty factors are a measure of the condence in a rule, but they have used it as means of identifying a rule as a less used one. We train the network in three phases. In phase 1, the network is trained with all the nodes and links. After learning, the redundant fuzzy sets are discarded from the network. In phase 2, the system is retrained keeping only the impor- tant fuzzy sets. After training, the inconsistent (rules with the same antecedent but dierent consequences) and less signicant (rules with small utility factors) rules are pruned. The network is retrained to adjust its weights in its reduced architecture In phase 3. The paper is organized as follows. In section 2, we present the network structure of the proposed system. In section 3, the learning scheme used by the network for fuzzy set selection and also for rule generation is discussed. Section 4 gives the simulation results. We conclude in section 5.

2   Network Structure

2.1   Network Structure for System Identification and Pattern Classification

Let there be s input features $(x_1, x_2, ..., x_s)$ and c classes $(t_1, t_2, ..., t_c)$. For a given sample x in the s dimentional input feature space, the proposed network can deal with fuzzy rules of the form.

R_i: If x_1 is A_{1i} and x_2 is A_{2i} ... and x_s is A_{si} then x belongs to class t_l with a certainty d_l, $(1 \leq l \leq c)$. Here, A_{ji} is the ith fuzzy set defined on the domain of x_j.

The neural fuzzy system is realized using a four-layered feed forward network which is quite similar to the network structure used in [4]. The functions of the nodes with their inputs and outputs are discussed below layer by layer. We use suffixes p, n, m and l to denote the suffixes of the nodes corresponding to the layers 1 through 4 respectively. The output of each node is denoted by z. The total number of nodes in layer i is denoted by N^i.

Each node in layer 1 represents an input linguistic variable and works as a buffer and transmits the input to the next layer. Hence, if x_p is the input to a node of this layer, the output of the layer is

$$z_p = x_p. \tag{1}$$

Primarily, the nodes in layer 2 act as fuzzifiers. Hence, if there are N_i fuzzy sets associated with the ith input feature and there are s input features then the

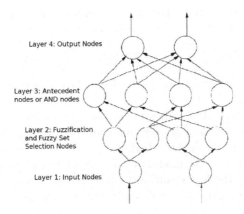

Fig. 1. The network structure for classification

number of nodes in layer 2 will be $N^2 = \sum_{i=1}^{s} N_i$. This layer is also responsible for analysis of input domain for discontinuity. In our simulation we have used the *bell-shaped Gaussian function*. All connections between nodes in layer 1 and layer 2 is unity and the output of a node in layer 2 is

$$\bar{z}_n = exp\{-\frac{(z_p-\mu_n)^2}{\sigma_n^2}\}. \tag{2}$$

The subscript n denotes the nth fuzzy set of the linguistic variable x_p. μ_n and σ_n are the mean and spread of the Gaussian membership function representing a linguistic term of x_p associated with the node n of this layer.

For identification of null fuzzy sets we use a different modulator function for each fuzzy set (linguistic term) of each feature. The idea is that, the null linguistic terms should always produce a membership value of 0. Moreover, this will lead the rules involving null fuzzy sets to be fired with strength 0. This is realized by modulating the outputs of the nodes in the following way

$$z_n = \bar{z}_n \times (1 - e^{-\nu_n^2}) \tag{3}$$

The parameter ν_n can be learnt using error back-propagation. We can see that when ν_n^2 takes a large value then z_n tends to $\bar{z}_n$. On the other hand, for small values of ν_n^2, z_n tends to 0. Therefore, after learning, ν_n^2 should obtain large values for non-null fuzzy sets whereas, it should take small values for null fuzzy sets.

Layer 3 is responsible for performing intersections of the input fuzzy sets of the fuzzy rules. Product operation cannot be chosen for this purpose, because it will produce a very small number as the firing strength of the rules. This does not allow the output nodes to be fired with the desired strength [4]. Hence, we used a soft and differentiable version of *min* regarded as *softmin* as in [4]:

$$softmin(x_1, x_2, ..., x_s, q) = (\frac{x_1^q + x_2^q + ... + x_s^q}{s})^{\frac{1}{q}}.$$

As $q \to -\infty$, *softmin* tends to the minimum of all x_is, $i = 1, 2, ..., s$. For all the simulations we have used $q = -12$. The output of the mth node of this layer is

$$\bar{z}_m = (\frac{\sum_{n \in P_m} z_n^q}{|P_m|})^{\frac{1}{q}} \tag{4}$$

where P_m is the set of indices of the nodes in layer 2 connected to node m of layer 3.

This layer also learns the utility factors corresponding to a fuzzy rule. The utility factor of the mth node of this rule is denoted by U_m. To keep the utility factors non-negative, we calculate them as $U_m = u_m^2$ as negative utility factors are difficult to interpret. We modulate the output of the nodes in this layer in such a manner that the less significant rules should always produce an output close to zero. This is realized by calculating the output of the mth node of this layer as

$$z_m = \bar{z}_m u_m^2 \tag{5}$$

The utility factor u_m is initialized with a small value (say, 0.0001) for each node in this layer.

Layer 4 is the output layer of the network. Each node of this layer represents the consequent of the fuzzy rules. Each node of this layer picks up only one antecedent based on the maximum agreement with facts in terms of product of firing strength and certainty factor. This represents a fuzzy union or OR operation implemented by the max operator. The output of node l of layer 4 is given by

$$z_l = \max_{m \in P_l}(z_m g_{lm}^2) \tag{6}$$

where P_l represents the set of indices of nodes in layer 3 connected to node l in layer 4.

3   Learning Rules and Fuzzy Set Modulators

The training phases are designed to minimize the error function

$$e = \frac{1}{2}\sum_{i=1}^{N} E_i = \frac{1}{2}\sum_{i=1}^{N}\sum_{l=1}^{c}(y_{il} - z_{il})^2 \tag{7}$$

where c is the number of nodes in layer 4 and y_{il} and z_{il} are the target and actual outputs of node l in layer 4 for input sample $x_i; i = 1, 2, ..., N$. We can drop the subscript i without loss.

The delta for layer 4 is given by

$$\delta_l = -(y_l - z_l) \tag{8}$$

Then delta for layer 3 becomes

$$\begin{aligned}\delta_m &= \sum_{l \in Q_m} \delta_l g_{lm}^2, \quad \textit{if } z_m g_{lm}^2 = \max_{m'}\{z_{m'} g_{lm'}^2\} \\ &= 0, \qquad\qquad \textit{otherwise}\end{aligned} \tag{9}$$

where, Q_m is the set of indices of the nodes in layer 4 connected to node m of layer 3.

Similarly, the delta for layer 2 is calculated as

$$\delta_n = \sum_{m \epsilon R_n} \delta_m u_m^2 \left(\frac{z_m z_n^{q-1}}{\sum_{n \epsilon P_m} z_n^q}\right). \tag{10}$$

where, R_n is the set of indices of nodes in layer 3 connected to node n in layer 2.

Now, having calculated the δ for each layer, we can calculate the equations for updating weights g_{lm}, utility factors u_m and modulators for input fuzzy sets ν_n. For updating weights we calculate,

$$\begin{aligned}\frac{\partial E}{\partial g_{lm}} &= \sum_{l \epsilon Q_m} 2\delta_l z_m g_{lm}, \quad if \ z_m g_{lm}^2 = \max_{m'}\{z_{m'} g_{lm'}^2\} \\ &= 0, \quad\quad\quad\quad\quad\quad otherwise\end{aligned} \tag{11}$$

Where, Q_m is the set of indices of nodes in layer 4 connected to node m in layer 3.

For updation of utility factors,

$$\frac{\partial E}{\partial u_m} = 2\partial_m u_m \left(\frac{\sum_{n \epsilon P_m} z_n^q}{|P_m|}\right)^{\frac{1}{q}} \tag{12}$$

Where, P_m is the set of indices of nodes in layer 2 connected to node m in layer 3.

Similarly, for updating ν_n we calculate,

$$\frac{\partial E}{\partial \nu_n} = 2\delta_n \nu_n \exp\{-((\tfrac{z_p - \mu_n}{\sigma_n})^2 + \nu_n^2))\} \tag{13}$$

where, the node n in layer 2 is connected to node p in layer 1.

The weights are updated during learning. Initially the weights are assigned some random values in the range $[0, 1]$. As weights are updated the weights connecting the actcedtents to its proper consequent get larger values and the rest smaller values resulting in minimization of total system error.

Initially all the ν_n and u_m are assigned a very small positive value (say 0.0001) so that all the nodes in layer 2 and layer 3 produce an output close to 0. Thus, all the fuzzy sets are initially considered to be null sets and all rules as less significant ones. As learning proceeds, the fuzzy sets necessary to reduce the system error, i.e, the non-null fuzzy sets, have their ν_ns raised to larger values and the rules necessary to reduce the system error have their u_m increased to higher values. Hence, only the non-null fuzzy sets and significant fuzzy rules are allowed to be passed throgh the network.

4   Results

The network for pattern classification was tested on many artificially generated data sets. Here we have shown only the results of XOR data. There are two

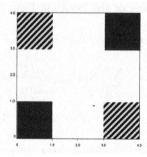

Fig. 2. Plot of XOR data

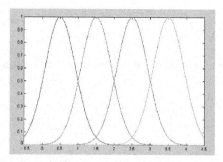

Fig. 3. Plot of fuzzy sets of input variables x_1 and x_2 for XOR data

Table 1. Initial architeture of the neuro-fuzzy system for XOR

Layer no	no of nodes
1	2
2	8
3	16
4	2

input features x_1, x_2 and two different classes c_1, c_2. The plot of the XOR data is shown in figure 2. We take four fuzzy sets for both x_1 and x_2 as shown in figure 3. The initial network architecture is shown in table 1.

In phase 1 the network was trained for 300 epochs with $\eta = 0.1$, $\theta = 0.01$ and $\chi = 0.01$. The values for modulators for the input fuzzy sets thus obtained are shown in table 2. Accordingly fuzzy sets x_{12}, x_{13}, x_{22} and x_{23} are removed from the system. In Phase 2 learning is performed with $\eta = 0.1$ and $\theta = 0.01$ for 300 epochs. The utility factors obtained after this phase are shown in table 3. Clearly, all the four rules are significant and are to be kept.

Hence, from an intial value $4 \times 4 \times 2 = 32$ the number of rules was ultimately reduced to $2 \times 2 \times 1 = 4$. Thus an 87.5% reduction in the number of rules was

Table 2. Values of ν_n for different linguistic terms of the input features for XOR

	x_{11}	x_{12}	x_{13}	x_{14}	x_{21}	x_{22}	x_{23}	x_{24}
ν_n	1.39	0.18	0.24	1.98	1.16	0.23	0.20	1.22
$1 - e^{-\nu_n^2}$	0.86	0.03	0.06	0.98	0.74	0.06	0.04	0.77

Table 3. Values of utility factors for the fuzzy rules for XOR

	x_{11}	x_{14}
x_{21}	1.91	2.14
x_{24}	1.94	1.97

achieved. Moreover, the system could recognize the points not lying to any class even though it was not trained for that. The system was tested with randomly chosen 1000 input points and there were 6 missclassifications with the threshold for belonging to a class being 0.8 resulting the percentage of error 0.6%. There were no missclassifications when the threshold was reduced to 0.6.

5   Conclusion

We demonstrated the methodologies to remove the null fuzzy sets and less significant rules from the system after identifying them apart from revisiting the methodologies for pruning the incompatible rules. The system (for pattern classification) is also capable of saying "Don't know" as output when the corresponding sample does not lie in any of the classes the system was trained for [5]. We did not tune the parameters of the fuzzy membership functions which can improve the system performance further. No guideline is also provided to analyze the domains of the output features. We are unable to find a single modulator function which can do both fuzzy set selection as well as feature selection by eliminating all the fuzzy sets corresponding to a redundant feature.

References

1. Pal, N.R., Pal, T.: On rule pruning using fuzzy neural networks. Fuzzy Sets Syst. 106, 335–347 (1999)
2. Shann, J.J., Fu, H.C.: A fuzzy neural network for rule acquiring on fuzzy control systems. Fuzzy Sets Syst. 71, 345–357 (1995)
3. Chakraborty, D., Pal, N.R.: Integrated Feature Analysis and Fuzzy Rule-Based System Identification in a Neuro-Fuzzy Paradigm. IEEE Trans. Systems Man Cybernetics 31, 391–400 (2001)
4. Chakraborty, D., Pal, N.R.: A Neuro-Fuzzy Scheme for Simultaneous Feature Selection and Fuzzy Rule-Based Classification. IEEE Trans. on Neural Networks 15(1), 110–123 (2004)

5. Chakraborty, D., Pal, N.R.: Making a Multilayer Perceptron Network Say- "Don't Know" When It Should. In: ICONIP, Orchid Country Club, Singapore (November 18-22, 2002)
6. Ishibuchi, H., Nakashima, T., Morisawa, T.: Voting in fuzzy rule-based systems for pattern classification problem. Fuzzy Sets Syst. 103, 223–238 (1999)
7. Krishnapuram, R., Lee, J.: Propagation of uncertainty in neural networks. In: Proc. SPIE Conf. Robot, Computer Vision, Bellingham, WA, pp.377—383 (1988)
8. Nozaki, K., Isibuchi, H., Tanaka, H.: Adaptive fuzzy rule-based classification systems. IEEE Trans. Fuzzy Syst. 4, 238–250 (1996)

Rough Set Theory of Pattern Classification in the Brain

Andrzej W. Przybyszewski

Department of Psychology, McGill University, Montreal, Canada
Dept of Neurology, University of Massachusetts Medical Center, Worcester, MA, USA
przy@ego.psych.mcgill.ca

Abstract. Humans effortlessly classify and recognize complex patterns even if their attributes are imprecise and often inconsistent. It is not clear how the brain processes uncertain visual information. We have recorded single cell responses to various visual stimuli in area V4 of the monkey's visual cortex. Different visual patterns are described by their attributes (condition attributes) and placed, together with the decision attributes, in a decision table. Decision attributes are divided into several classes determined by the strength of the neural responses. Small cell responses are classified as *class 0*, medium to strong responses are classified as *classes 1 to n-1* (*min(n)=3*), and the strongest cell responses are classified as class *n*. The higher the class of the decision attribute the more preferred is the stimulus. Therefore each cell divides stimuli into its own family of equivalent objects.

By comparing responses of different cells we have found related concept classes. However, many different cells show inconsistency between their decision rules, which may suggest that parallel different decision logics may be implemented in the brain.

Keywords: visual brain, imprecise computation, bottom-up, top-down processes, neuronal activity.

1 Introduction

We define after Pawlak [1] an information system as $S = (U, A)$, where U, A are nonempty finite sets called the *universe of objects* and the *set of attributes*, respectively. If $a \in A$ *and* $u \in U$, the value $a(u)$ is a unique element of V (where V is a value set). The *indiscernibility relation* of any subset B of A or $I(B)$, is defined [3] as follows: $(x, y) \in I(B)$ or $xI(B)y$ if and only if $a(x) = a(y)$ for every a $\in B$, where $a(x) \in V$. $I(B)$ is an equivalence relation, and $[u]_B$ is the equivalence class of u, or a *B-elementary granule*. The family of all equivalence classes of $I(B)$ will be denoted $U/I(B)$ or U/B. The block of the partition U/B containing u will be denoted by $B(u)$. The concept $X \subseteq U$ is *B-definable* if for each $u \in U$ either $[u]_B \subseteq X$ or $[u]_B \subseteq U\backslash X$. $\underline{B} X = \{u \in U: [u]_B \subseteq X \}$ is a lower approximation of X. The concept $X \subseteq U$ is B *indefinable* if there exists $u \in U$ such that $[u]_B \cap X \neq \phi\}$. $\overline{B} X = \{u \in U: [u]_B \cap X \neq \phi\}$ is an *upper approximation* of X. The set $BN_B (X) = \overline{B} X - \underline{B} X$ will be referred to as the *B-boundary region* of. X If the boundary region of X is the empty set

A. Ghosh, R.K. De, and S.K. Pal (Eds.): PReMI 2007, LNCS 4815, pp. 295–303, 2007.
© Springer-Verlag Berlin Heidelberg 2007

than X is *exact* (*crisp*) with respect to B; otherwise if $BN_B(X) \neq \phi$ X is not *exact* (i.e., it is *rough*) with respect to B. We say that the B-lower approximation of a given set A is the set of union of all B-granules that are included in the set A, and the B-upper approximation of A is a set of the union of all B-granules that have nonempty intersection with A. We will distinguish in the information system two disjoint classes of attributes: condition and decision attributes. The system S will be called a decision table $S = (U, C, D)$ where C and D are condition and decision attributes.

In this paper the universe U will be assumed to be all visual patterns that are characterized by their attributes C. The purpose of our research is to find how these objects are classified in the brain. Therefore we are looking to determine D on the basis of a single neuron recording from the visual area in the brain.

Imprecise reasoning is a characteristic of natural languages and is related to human decision-making effectiveness [2]. The brain, in contrast to the computer, is constantly integrating many asynchronous parallel streams of information [3], which help in its adaptation to the environment. Most of our knowledge about the function of the brain is based on electrophysiological recordings from single neurons. In this paper we will describe properties of cells from the visual area V4. This intermediate area of the ventral stream mediates shape perception, but different laboratories propose different often-contradictory hypotheses about properties of V4 cells. We propose the use of rough set theory (Pawlak, [1]) to classify concepts as related to different stimuli attributes. We will show several examples of our method.

2 Method

Results of electrophysiological experiments are placed into the following decision table. Neurons are identified using numbers related to a collection of figures in [4]. Different measurements of the same cell are denoted by additional letters (*a, b, ...*) and placed in the first column adjacent to the cell number. The next columns of the table describe stimulus attributes and their values. Stimulus attributes are as follows:

1. orientation in degrees appears in the column labeled *o*, and orientation bandwidth is labeled by *ob*.
2. spatial frequency is denoted as *sf*, and spatial frequency bandwidth is *sfb*
3. *x*-axis position is denoted by *xp* and the range of x-positions is *xpr*
4. *y*-axis position is denoted by *yp* and the range of y-positions is *ypr*
5. *x*-axis stimulus size is denoted by *xs*
6. *y*-axis stimulus size is denoted by *ys*
7. stimulus shape is denoted by *s*, with values of *s* are defined as follows: for grating *s=1*, for vertical bar *s= 2*, for horizontal bar *s= 3*, for disc *s= 4*, for annulus *s=5*.

Thus the full set of stimulus attributes is expressed as B = {*o, ob, sf, sfb, xp, xpr, yp, ypr, xs, ys, s*}. The cell's responses *r* are divided into several classes are placed in the last column of the table.

3   Results

We have analyzed the experimental data from several neurons recorded in the monkey's V4 [4]. Below we show a modified figure from the above work (Fig.1), along with the associated decision table (table 1).

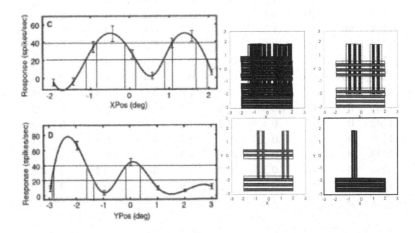

Fig. 1. Curves represent approximated responses of a cell from area V4 to vertical (C), and horizontal (D) bars. Bars change their position along the x-axis (Xpos) or along the y-axis (Ypos). Responses of the cell are measured in spikes/sec. Mean cell responses $\pm$ SE are marked in C and D plots. Cell responses are divided into 5 ranges (classes) but for simplicity only two horizontal lines are plotted. On the right are schematic representations of cell response on the basis of Table 1. Vertical and horizontal bars in certain x- and y-positions give significant responses: class 1 - upper left schematic, class 2 – upper right, class 3 - lower left, and class 4 – lower right schematic. These schematics represent decision rules for each response class.

On the basis of the decision table we have made a schematic of the optimal stimulus for this cell (Fig. 1, right side). Fig. 1 (left side) shows the cell's responses to the stimulus, which was a long narrow bar with vertical (Fig.1 C) or horizontal (Fig.1 D) orientation. The cell's responses are divided into strength classes (horizontal lines in plots of Fig. 1) with stimuli attributes placed in the decision table (Table 1). This table is converted into a schematic (right side of Fig. 1), which can be read as the decision rules related to four classes of cell responses. On the basis of this schematic the receptive field can be divided into smaller areas with different preferences, and these subfields can be stimulated independently as is shown in Fig. 2. Table 2 divides data from Fig. 2 and from Fig. 5 in [4] into decision classes, which determine equivalent classes of stimuli as shown on the schematic in lower part of Fig. 2.

Table 1. Decision table for the cell shown in Fig. 1. Attributes *ob, sf, sfb* were constant and are not presented in the table. Cell responses *r* below 10 spikes/s were defined *class 0*, above 10 spikes/s is defined as *class 1*, above 20 sp/s – *class 2*, above 40 sp/s - *class 3*, and above 50 sp/s – *class 4*.

Cell	o	xp	xpr	yp	ypr	xs	ys	s	r
12c	90	-0.6	1.4	0	0	0.4	4	2	1
12c1	90	-0.6	1.2	0	0	0.4	4	2	2
12c2	90	-0.6	0.6	0	0	0.4	4	2	3
12c3	90	1.35	1.3	0	0	0.4	4	2	1
12c4	90	1.3	1	0	0	0.4	4	2	2
12c5	90	1.3	0.5	0	0	0.4	4	2	3
12c6	90	-0.6	0	0	0	0.4	4	2	4
12d	0	0	0	-2	1.8	4	0.4	3	1
12d1	0	0	0	-2.2	1.6	4	0.4	3	2
12d2	0	0	0	-2.2	1.2	4	0.4	3	3
12d3	0	0	0	0.1	1.8	4	0.4	3	1
12d3	0	0	0	0.15	1.3	4	0.4	3	2
12d4	0	0	0	0.15	0.7	4	0.4	3	3
12d5	0	0	0	-2.2	0.9	4	0.4	3	4

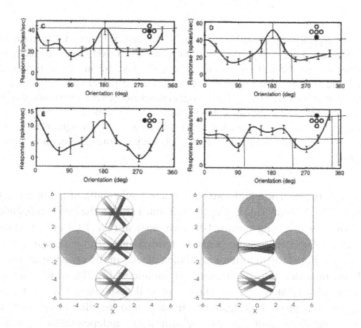

Fig. 2 Modified plots on the basis of [4] (upper plots), and their representation on the basis of table 2 (lower plots). C-F Curves represent responses to different orientations of one V4 cell when its subfields (their positions are shown in plots) are covered with 2 degree grating discs 2 degrees apart in a 6 degree receptive field. Lower plots: Gray circles indicate cell response below 10 spikes/s. Plots on the left are related to class 1, in the middle – class 2, and plots on the right are related to responses of class 3.

Table 2. Decision table for one cell shown in Fig. 2 (Figs. 3, 5 in [4]). Attributes xpr, ypr, s are constant and are not presented in the table. Cell # 3* from Fig. 3, cell# 5* from Fig. 5 cell #3₅* combined Figs. 3 and 5. Cell responses r below 10 spikes/s were defined as *class 0*, 10 - 20 spikes/s is defined as *class 1*, 20 - 40 sp/s – *class 2*, above 40 sp/s - *class 3*.

Cell	o	ob	sf	sfb	xp	yp	r
3₅c0	180	180	2.5	1.5	0	0	1
3c	172	105	2	0	0	0	2
3c1	10	140	2	0	0	0	2
3c2	180	20	2	0	0	0	3
3₅d0	180	180	2.5	1.5	0	0	1
3d	172	105	2	0	0	-2	2
3d1	5	100	2	0	0	-2	2
3d2	180	50	2	0	0	-2	3
3₅e	180	10	2	0	-2	0	1
3₅f0	180	180	2.5	1.5	0	2	1
3f	170	100	2	0	0	2	2
3f1	10	140	2	0	0	2	2
3f2	333	16	2	0	0	2	3
5a	180	0	2.3	2.6	0	-2	2
5b	180	0	2.5	3	0	2	2
5c	180	0	2.45	2.9	0	0	1
5c1	180	0	2.3	1.8	0	0	2

In order to find general decision rules (decision table reduct) we introduce a tolerance on a certain attribute values (discretization problem): all *ob* values with $0 < ob < 60$ we denote as ob_n (narrow orientation bandwidth), $ob > 100$ we denote as ob_w (wide orientation bandwidth), we write sfb_n if $0 < sfb < 1$ (small bandwidth), and sfb_w if $sfb > 1$. The **Decision rules** are as follows:

DR1: $ob_n \wedge xp_0 \rightarrow r_3$, **DR2**: $ob_n \wedge xp_{-2} \rightarrow r_1$,
DR3: $ob_w \wedge sfb_n \rightarrow r_2$, **DR4**: $ob_w \wedge sfb_w \rightarrow r_1$,
Notice that Figs 1 and 3 show possible configurations of the optimal stimulus. However, they do not take into account interactions between several stimuli, when more than one subfield is stimulated. In addition there are **Subfield Interaction Rules:**
SIR1: facilitation when stimulus consists of multiple bars with small distances (0.5-1 deg) between them, and inhibition when distance between bars is 1.5 -2 deg.
SIR2: inhibition when stimulus consists of multiple similar discs with distance between them ranging from 0 deg (touching) to 3 deg.
SIR3: Center-surround interaction, which is described below in detail.

The next part is related to the center-surround interaction **SIR3**. The decision table (Table 3) shows responses of 8 different cells stimulated with discs or annuli (Fig. 10 in [4]). In order to compare different cells, we have normalized their optimal orientation and denoted it as 1, and removed them from the table. We have introduced

Table 3. Decision table for eight cells comparing the center-surround interaction. All stimuli were concentric discs or annuli with xo – outer diameter, xi – inner diameter. All stimuli were localized around the middle of the receptive field, so that ob = xp = yp = xpr = ypr = 0 were fixed and we did not put them in the table. Cell responses r below 20 spikes/s were defined as *class 0*, 20 – 40 sp/s is defined as *class 1*, 40 – 100 sp/s – *class 2*, above 100 sp/s - *class 3*.

Cell	sf	sfb	xo	xi	s	r
101	0.5	0	7	0	4	0
101a	0.5	0	7	2	5	1
102	0.5	0	8	0	4	0
102a	0.5	0	8	3	5	0
103	0.5	0	6	0	4	0
103a	0.5	0	6	2	5	1
104	0.5	0	8	0	4	0
104a	0.5	0	8	3	5	2
105	0.5	0	7	0	4	0
105a	0.5	0	7	2	5	1
106	0.5	0	6	0	4	1
106a	0.5	0	6	3	5	2
107	0.5	0.25	6	0	4	2
107a	0.9	0.65	6	3	5	2
107b	3.8	0.2	6	3	5	2
107c	2.3	0.7	6	3	5	3
107d	2	0	6	2	5	2
107e	2	0	4	0	4	1
108	0.5	0	6	0	4	1
108a	1.95	0.65	4	0	4	2
108b	5.65	4.35	6	2	5	2
108c	0.65	0.6	6	2	5	3

a tolerance on values of sf. We have denoted as sf_{low} (low spatial frequency) all $sf < 1$, as sf_m for $1.7 < sf < 3.5$, and sf_h for $4 < sf$. We have calculated if the stimulus contains sf_{low}, sf_m or sf_h by taking from the table $sf \pm sfb$ and skipping sfb. For example, in the case of 108b the stimulus has sf: 8.6 ± 7.3 c/deg, which means that sf_m or sf_h are values of the stimulus attributes. We can also skip s, which is determined by values of xo and xi.

Stimuli used in these experiments can be placed in the following categories:

$Y_o = |sf_{low}\ xo_7\ xi_0| = \{101, 105\}$; $Y_1 = |\ sf_{low}\ xo_7\ xi_2| = \{101a, 105a\}$; $Y_2 = |sf_{low}\ xo_8\ xi_0| = \{102, 104\}$; $Y_3 = |\ sf_{low}\ xo_8\ xi_3| = \{102a, 104a\}$; $Y_4 = |\ sf_{low}\ xo_6\ xi_0| = \{103, 106, 107, 108, 20a\}$; $Y_5 = |sf_{low}\ xo_6\ xi_2| = \{103a, 106a, 107a, 108b, 20b\}$; $Y_6 = |sf_{low}\ xo_8\ xi_0| = \{104, 108a\}$; $Y_7 = |sf_{low}\ xo_8\ xi_3| = \{104a, 108a\}$; $Y_6 = |sf_{low}\ xo_4\ xi_0| = \{107e, 108a\}$.

These are equivalence classes for stimulus attributes, which means that in each class they are indiscernible $IND(B)$. We have normalized orientation bandwidth to 0 in $\{20a, 20b\}$ and spatial frequency bandwidth to 0 in cases $\{107, 107a, 108a, 108b\}$.

There are four classes of responses, denoted as r_0, r_1, r_2, r_3. Therefore the expert's knowledge involves the following four concepts:

$\mid r_o \mid = \{101, 102, 102a, 103, 104, 105\}$, $\mid r_1 \mid = \{101a, 103a, 105a, 106, 107b, 108\}$
$\mid r_2 \mid = \{104a, 106a, 107, 107a, 107b, 107d, 108a, 108b\}$, $\mid r_3 \mid = \{107c, 108c\}$, which are denoted as X_o, X_1, X_2, X_3.

We want to find out whether equivalence classes of the relation $IND\{r\}$ form the union of some equivalence relation $IND(B)$, or whether $B \Rightarrow \{r\}$.

We will calculate the lower and upper approximation [1] of the basic concepts in terms of stimulus basic categories:

$\underline{B} X_o = Y_o = \{101, 105\}$, $\overline{B} X_o = Y_o \cup Y_2 \cup Y_3 \cup Y_4 = \{101, 105, 102, 104, 102a,$
$104a, 103, 106, 107, 108\}$,

$\underline{B} X_1 = Y_1 \cup Y_5 = \{101a, 105a, 103a\}$, $\overline{B} X_1 = Y_1 \cup Y_5 \cup Y_6 \cup Y_4 = \{101a,$
$105a, 103a, 106, 108, 107e\}$,

$\underline{B} X_2 = 0$, $\overline{B} X_2 = Y_3 \cup Y_4 \cup Y_5 \cup Y_6 = \{102a, 104a, 103a, 107a, 108b, 106a,$
$20b, 103, 107, 106, 108, 20a, 107b, 108a\}$

Concept 0 and concept 1 are roughly *B-definable*, which means that only with some approximation can we say that stimulus Y_0 did not evoke a response (concept 0) in cells 101, 105. Other stimuli Y_2, Y_3 evoked no response (*concept 0*) or weak (*concept 1*) or strong (*concept 2*) response. This is similar for concept 1. However, concept 2 is internally *B-undefinable*. Stimulus attributes related to this concept should give us information about cell characteristics, but data from table 3 cannot do it. We can find quality [1] of our experiments by comparing properly classified stimuli $POS_B(r)=\{101, 101a, 105, 105a\}$ to all stimuli and responses:

$$\gamma\{r\} = \mid\{101,101a,105,105a\}\mid/\mid\{101,101a, ...,20a,20b\}\mid = 0.2.$$

We can also ask what percentage of cells we have fully classified. We obtain consistent responses from 2 of 9 cells, which means that $\gamma\{cells\} = 0.22$. This is related to the fact that for some cells we have tested more than two stimuli. What is also important from an electrophysiological point of view is there are negative cases. There are many negative instances for the concept 0, which means that in many cases this brain area responds to our stimuli; however it seems that our concepts are still only roughly defined. We have the following decision rules:

DR5: $x_{o7} x_{i2} s_5 \rightarrow r_1$; **DR6**: $x_{o7} x_{i0} s_4 \rightarrow r_0$, **DR7**: $x_{o8} x_{i0} s_4 \rightarrow r_0$.
They can be interpreted as the statement that a large annulus (s5) evokes a weak response, but a large disc (s4) evokes no response. However, for certain stimuli there is inconsistency in responses of different cells (Table 3): 103: $x_{o6} x_{i0} s_4 \rightarrow r_0$, 106: $x_{o6} x_{i0} s_4 \rightarrow r_1$.

4   Discussion

The purpose of our study has been to determine how different categories of stimuli and particular concepts are related to the responses of a single cell. We test our theory

Fig. 3. In their paper David et al. [5] stimulated V4 neurons (medium size of their receptive fields was 10.2 deg) with natural images. Several examples of their images are shown above. We have divided responses of cells into three concept categories. The two images on the left represent cells, which give strong responses, related to our expertise *concept 2*. The two images in the middle evoke medium strength responses and they are related to *concept 1*. The two images on the right gave very weak responses; they are related to *concept 0*.

on a set of data from David et al. [5], shown in Fig. 3. We assume that the stimulus configuration in the first image on the left is similar to that proposed in Fig. 2; therefore it should give a strong response. The second image from the left can be divided into central and surround parts. The stimulus in the central disc is similar to that from Fig. 2 (*DR1*). Stimuli on the upper and right parts of the surround have a common orientation and a larger orientation bandwidth ob_w in comparison with the center (Fig. 2). These differences make for weak interactions between discs as in *SIR2* or between center-surround as in *SIR3*. This means that these images will be related to *concept 2*. Two middle images show smaller differences between their center and surround. Assuming that the center and surround are tuned to a feature of the object in the images, we believe that these images would also give significant responses. However, in the left image in the middle part of Fig. 3, stimuli in the surround consist of many orientations (ob_w) and many spatial frequencies (sfb_w); therefore medium class response is expected *DR4* (*concept 1*). The right middle image shows an interesting stimulus but also with a wide range of orientations and spatial frequencies *DR4*. There are small but significant differences between center and surround parts of the image. Similar rules as to the previous image can be applied. In consequence brain responses to both images are related to *concept 1*. In the two images on the right there is no significant difference between stimulus in the center and the surround. Therefore the response will be similar to that obtained when a single disc covers the whole receptive field: *DR6, DR7*. In most cells such a stimulus is classified as *concept 0*.

In summary, we have showed that using rough set theory we can divide stimulus attributes in relationships to neuronal responses into different concepts. Even if most of our concepts were very rough, they determine rules on whose basis we can predict neural responses to new, natural images.

Acknowledgement. I thank M. Kon for useful discussion and comments.

References

1. Pawlak, Z.: Rough Sets - Theoretical Aspects of Reasoning about Data. Kluwer Academic Publishers, Boston, London, Dordrecht (1991)
2. Zadeh, L.A.: Toward a perception-based theory of probabilistic reasoning with imprecise probabilities. Journal of Statistical Planning and Inference 105, 233 (2002)

3. Przybyszewski, A.W., Linsay, P.S., Gaudiano, P., Wilson, C.: Basic Difference Between Brain and Computer: Integration of Asynchronous Processes Implemented as Hardware Model of the Retina. 18, 70–85 (2007)
4. Pollen, D.A., Przybyszewski, A.W., Rubin, M.A., Foote, W.: Spatial receptive field organization of macaque V4 neurons. Cereb Cortex 12, 601–616 (2002)
5. David, S.V., Hayden, B.Y., Gallant, J.L.: Spectral receptive field properties explain shape selectivity in area V4. J. Neurophysiol. 96, 3492–3505 (2006)

Rough Core Vector Clustering

CMB Seshikanth Varma, S. Asharaf, and M. Narasimha Murty*

Computer Science and Automation,
Indian Institute of Science, Bangalore-560012
Tel.: +91-80-22932779
mnm@csa.iisc.ernet.in

Abstract. Support Vector Clustering has gained reasonable attention from the researchers in exploratory data analysis due to firm theoretical foundation in statistical learning theory. Hard Partitioning of the data set achieved by support vector clustering may not be acceptable in real world scenarios. Rough Support Vector Clustering is an extension of Support Vector Clustering to attain a soft partitioning of the data set. But the Quadratic Programming Problem involved in Rough Support Vector Clustering makes it computationally expensive to handle large datasets. In this paper, we propose Rough Core Vector Clustering algorithm which is a computationally efficient realization of Rough Support Vector Clustering. Here Rough Support Vector Clustering problem is formulated using an approximate Minimum Enclosing Ball problem and is solved using an approximate Minimum Enclosing Ball finding algorithm. Experiments done with several Large Multi class datasets such as Forest cover type, and other Multi class datasets taken from LIBSVM page shows that the proposed strategy is efficient, finds meaningful soft cluster abstractions which provide a superior generalization performance than the SVM classifier.

1 Introduction

Several domains that employ cluster analysis deal with massive collections of data and hence demands algorithms that are scalable both in terms of time and space. Some of the algorithms that have been proposed in the literature for clustering large data sets are DB-SCAN[1], CURE[2], BIRCH[3], etc. Another major concern in data clustering is whether the clustering scheme should produce a hard partitioning (crisp sets of patterns representing the clusters) of the data set or not. Yet another concern is that the clusters that may exist in any data set may be of arbitrary shapes. We can say that there is an intrinsic softness involved in cluster analysis.

Several Rough Set[4] based algorithms like Rough Kmeans[5], Rough Support Vector Clustering (RSVC)[6] etc have been proposed for achieving soft clustering. RSVC is a natural fusion of Rough sets and Support Vector Clustering[7] paradigm. But the Quadratic Programming(QP) Problem involved in RSVC

---

* Corresponding author.

A. Ghosh, R.K. De, and S.K. Pal (Eds.): PReMI 2007, LNCS 4815, pp. 304–310, 2007.

makes it computationally expensive(at least quadratic in terms of the number of training points) to handle large datasets. In this paper we propose Rough Core Vector Clustering (RCVC) which is computationally efficient for achieving Rough Support Vector Clustering. The rest of the paper is organized as follows. RSVC is discussed in Section 2. In Section 3 the RCVC technique is introduced. Empirical results are given in Section 4 and Section 5 deals with Conclusions.

2   Rough Support Vector Clustering (RSVC)

RSVC uses a non linear transformation ϕ from data space to some high dimensional feature space and looks for the smallest enclosing rough sphere of inner radius R and outer radius T. Now the primal problem can be stated as

$$\min \quad R^2 + T^2 + \frac{1}{vm}\sum_{i=1}^{m}\xi_i + \frac{\delta}{vm}\sum_{i=1}^{m}\xi_i'$$

$$\text{s.t.} \quad \|\phi(x_i) - \mu\|^2 \leq R^2 + \xi_i + \xi_i'$$

$$0 \leq \xi_i \leq T^2 - R^2 \quad \xi_i' \geq 0 \quad \forall i \tag{1}$$

The Wolfe Dual[8] of this problem can be written as

$$\min \quad \sum_{i,j}^{m}\alpha_i\alpha_j K(x_i,x_j) - \sum_{i=1}^{m}\alpha_i K(x_i,x_i) \text{ s.t. } 0 \leq \alpha_i \leq \frac{\delta}{vm} \ \forall i, \quad \sum_{i=1}^{m}\alpha_i = 2 \tag{2}$$

It can be observed that for RSVC, $\delta > 1$ and it reduces to the original SVC formulation for $\delta = 1$. It may also be observed that the images of points with s $\alpha_i = 0$ lie in lower approximation (hard points), $0 < \alpha_i < \frac{1}{vm}$ form the Hard Support Vectors, (Support Vectors which mark the boundary of lower approximation), $\alpha_i = \frac{1}{vm}$ lie in the boundary region (patterns that may be shared by more than one cluster, a soft point). $\frac{1}{vm} < \alpha_i < \frac{\delta}{vm}$ form the Soft Support Vectors (Support Vectors which mark the boundary of upper approximation) and $\alpha_i = \frac{\delta}{vm}$ lie outside the sphere (Bounded Support Vectors or outlier points). From $\sum_{i=1}^{m}\alpha_i = 2$ and $0 \leq \alpha_i \leq \frac{1}{vm}$ we can see that the number of BSVs $n_{bsv} < 2(\frac{vm}{\delta})$ for $\delta = 1$ $n_{bsv} < 2(vm) = v'm$ This corresponds to all the patterns x_i with $\| \phi(x_i) - \mu \|^2 > R^2$. Since $\delta > 1$ for RSVC, we can say that $\frac{v'}{\delta}$ is the fraction of points permitted to lie outside T and v' is the fraction of points permitted to lie outside R. Hence v and δ together give us control over the width of boundary region and the number of BSVs. The algorithm to find clusters in RSVC is given in Algorithm 1.

3   Rough Core Vector Clustering

Consider a situation where $k(x,x) = \kappa$, a constant. This is true for kernels like Gaussian given by $k(x_i,x_j) = e^{-g\|x_i-x_j\|^2}$, where $\| \cdot \|$ represents the L_2 norm

Data: Input Data
Result: Cluster Labels for the Input Data

- Find adjacency matrix M as $M[i,j] = \begin{cases} 1 \text{ if } G(y) \leq R \quad \forall y \in [x_i, x_j] \\ 0 \text{ otherwise} \end{cases}$
- Find connected components for the graph represented by M.
  This gives the Lower Approximation of each cluster.
- Now find the Boundary Regions as $x_i \in L\_A(C_i)$ and
  pattern $x_k \notin L\_A(C_j)$ for any cluster j,
  if $G(y) \leq T \; \forall y \in [x_i, x_k]$ then $x_k \in B\_R(C_i)$

Algorithm 1: Algorithm to find clusters

and g is a user given parameter. The dot product kernel like polynomial kernel given by $k(x_i, x_j) = (< x_i, x_j > +1)^\lambda$ with normalized inputs x_i and x_j also satisfy the above condition(λ is a non-negative integer).Now the dual of the RSVC problem given in equation (2) reduces to the form

$$\min \sum_{i,j}^{m} \alpha_i \alpha_j K(x_i, x_j) \;\; s.t. \;\; 0 \leq \alpha_i \leq \frac{\delta}{vm} \;\; \forall i \;\; \sum_{i=1}^{m} \alpha_i = 2 \qquad (3)$$

The above equation can be solved by an adapted version of the Minimum Enclosing Ball(MEB) finding algorithm used in CVM[9]. Let us define a Soft Minimum Enclosing Ball(SMEB) as an MEB of the data allowing $\frac{v'}{\delta}$ fraction of the points to lie outside T(see the discussion in Section 2). To solve the above problem we use an approximate SMEB algorithm given in next Section.

3.1   Approximate Soft MEB Finding Algorithm

The traditional algorithms for finding MEBs are not efficient for $d > 30$ and hence as in CVM[9], the RCVC technique adopts a variant of the faster approximation algorithm introduced by Badou and Clarkson[10]. It returns a solution within a multiplicative factor of $(1 + \epsilon)$ to the optimal value, where ϵ is a small positive number.

The $(1 + \epsilon)$ approximation of the SMEB problem is obtained by solving the problem on a subset of the data set called *Core Set*. Let $B_S(a, T)$ be the exact SMEB with center a and outer radius T for the data set S and $B_Q(\tilde{a}, \tilde{T})$ be the SMEB with center $\tilde{a}$ and radius $\tilde{T}$ found by solving the SMEB problem on a subset of S called Core Set(Q). Given an $\epsilon > 0$, a SMEB $B_Q(\tilde{a}, (1 + \epsilon)\tilde{T})$ is an $(1 + \epsilon)$-approximation of $SMEB(S) = B_S(a, T)$ if $B_Q(\tilde{a}, (1 + \epsilon)\tilde{T})$ forms the $SMEB(S)$ and $\tilde{T} \leq T$.

Formally, a subset $Q \subseteq S$ is a core set of S if an expansion by a factor $(1+\epsilon)$ of its SMEB forms the $SMEB(S)$. The approximate MEB finding algorithm uses a simple iterative scheme: At the tth iteration, the current estimate $B_Q(\tilde{a}_t, \tilde{T}_t)$ is expanded incrementally by including that data point in S that is farthest from the center $\tilde{a}$ and falls outside the $(1 + \epsilon)$-ball $B_Q(\tilde{a}_t, (1 + \epsilon)\tilde{T}_t)$. To speed up the process, CVM uses a probabilistic method. Here a random sample S' having 59 points is taken from the points that fall outside the $(1 + \epsilon)$-ball $B_Q(\tilde{a}_t, (1 + \epsilon)\tilde{T}_t)$. Then the point in S' that is farthest from the center $\tilde{a}_t$ is taken as the approximate farthest point from S. The iterative strategy to include the farthest point in the MEB is repeated until only $\frac{v'}{\delta}$ fraction of the points to lie outside T. The set of all such points that got added forms the core set of the data set.

3.2   The Rough Core Vector Clustering Technique

The cluster labeling procedure explained in RSVC can now be used to find the soft clusters in Q. Now the question is how can we cluster the rest of the points in the data set S. For that RCVC technique employs the Multi class SVM algorithm discussed below.

A Multi class SVM(MSVM). MSVM is formulated here as an SVM[11] with vector output. This idea comes from a simple reinterpretation of the normal vector of the separating hyperplane. This vector can be viewed as a projection operator of the feature vectors into a one dimensional subspace. An extension of the range of this projection into multi-dimensional subspace gives the solution for vector labeled training of SVM.

Let the training data set be $S = \{(x_i, y_i)\}_{i=1}^{m}$ where $x_i \in R^d$, $y_i \in R^T$ for some integers $d, T > 0$. i.e. we have m training points whose labels are vector valued. For a given training task having T classes, these label vectors are chosen out of the finite set of vectors $\{y_1, y_2, ...y_T\}$. The primal is given as

$$\min_{W, \rho, \xi_i} \quad \frac{1}{2}trace(W^T W) - \rho + \frac{1}{\nu_2 m}\sum_{i=1}^{m}\xi_i$$

$$\text{s.t.} \quad y_i^T(W\phi(x_i)) \geq \rho - \xi_i \tag{4}$$

$$\xi_i \geq 0 \quad \forall i \tag{5}$$

The corresponding Wolfe Dual[8] is

$$\min_{\alpha_i} \quad \sum_{i,j=1}^{m}\alpha_i\alpha_j(\frac{1}{2} < y_i, y_j > k(x_i, x_j)) \text{ s.t.} \sum_{i=1}^{m}\alpha_i = 1 \quad 0 \leq \alpha_i \leq \frac{1}{\nu_2 m} \quad \forall i \tag{6}$$

The decision function predicting one of the labels from $1...T$ is

$$\arg\max_{t=1...T} < y_t, (W\phi(x_j)) >= \arg\max_{t=1...T}\left(\sum_{i=1}^{m}(\alpha_i < y_i, y_t > (k(x_i, x_j)))\right) \tag{7}$$

Let $y_i(t)$ denote the tth element of the label vector y_i corresponding to the pattern x_i. One of the convenient ways is to choose the label vectors is as

$$y_i(t) = \begin{cases} \sqrt{\dfrac{(T-1)}{T}} & \text{if item } i \text{ belongs to category } t \\[3mm] \sqrt{\dfrac{1}{T(T-1)}} & \text{otherwise} \end{cases}$$

It may be observed that this kind of an assignment is suitable for any $T > 2$. Now it may be observed that the MSVM formulation given by equation (6) will become an MEB problem with the modified kernel $\tilde{k}(z_i, z_j) = <y_i, y_j> k(x_i, x_j)$ given $k(x, x) = \kappa$. So we can use the approximate SMEB finding algorithm discussed above to train the MSVM.

3.3 Cluster Labeling in RCVC

Once the RSVC clustering is done for the core set, we train the MSVM(used as classifier) on the coreset using the cluster labels. Now any point $x_i \in (S - Q)$ or the test data is clustered as follows. If it is a hard point, take the output cluster label predicted by MSVM (*Hard decision*) else if it is a soft point or outlier point, we take a soft labeling procedure where the point is allowed to belong to more than one cluster. Here the number of allowed overlaps *limit* is taken as a user defined parameter and the point is allowed to belong to any of the top ranked *limit* number of clusters found by the MSVM(*Soft decision*).

Table 1. Details of the data sets(x/y) x and y denote Train and Test sizes respectively, NC denote Number of Classes, and Parameter values g, δ, ν, ϵ are used for RCVC and ν_2 used for MSVM and the description of these parameters are same as explained in section 2 and 3

Dataset(x/y)/parameter	dim	NC	g	δ	ν	ϵ	ν_2
combined(78823/19705)	100	3	9.70375	10.922222	0.09	0.0922	0.06
acoustic(78823/19705)	100	3	177.70375	11.07	0.09	0.0922	0.06
seismic(78823/19705)	100	3	177	7.956	0.125	0.0922	0.061
shuttle(43500/14500)	9	7	777	5.6	0.125	0.0922	0.06
forest(526681/56994)	54	7	77.70375	10.666667	0.09	0.0922	0.06

4 Experimental Results

Experiments are done with five real world data sets viz; the shuttle, acoustic, seismic, combined from the LIBSVM[12] page available at *"http://www.csie.ntu. edu.tw/~cjlin/libsvmtools/datasets/multiclass"* and Forest covertype data set from the UCI KDD archive available at *"http://kdd.ics.uci.edu/databases/cover type/covertype.html"*. Ten percent of this forest data set is sampled uniformly keeping the class distribution and is taken as the test data set. The rest is taken as the train data set. For all the data sets, a comparison with the one-against-one

Table 2. Results on **One Vs One SVM** and **RCVC**. The abbreviations used are - #SV : No. of Support Vectors, NC : No. of clusters, CA : Classification Accuracy on test data in percentage. Train and Test time are given in seconds.

Algo./Data		combined	acoustic	seismic	shuttle	forest
	#SV	34875	52593	45802	301	339586
1Vs1 SVM	Train Time	3897	5183	3370	170	67328
	Test time	750	686	565	7	6321
	CA(%)	**81.91**	**66.9728**	**72.5704**	**99.924**	**72.3409**
RCVC	coreset size	142	140	126	161	145
	#SV	142	140	126	159	145
	NC	69	105	54	91	43
	train time	131	141	81	38	370
	test time	22	27	13	8	25
	CA(%)	**92.971327**	**91.727988**	**87.561533**	**99.965517**	**88.08822**

SVM(1Vs1 SVM)[12] is done. In all the experiments we used Gaussian kernel function. All the experiments were done on a Intel Xeon(TM) 3.06GHz machine with 2GB RAM. The details of the data sets along with the parameters used in the experiments and the results obtained are given in Table 1 and 2 respectively.

5   Conclusion

RCVC is computationally efficient than RSVC. Soft clustering scheme used in RCVC enables us to identify ambiguous regions (having overlap between clusters of heterogeneous nature) in the data set. RCVC allows soft clusters of any arbitrary shape and the extent of softness can be varied by controlling the width of boundary region and number of outliers done by changing the user defined parameters. So, this method is scalable and involves less computation when compared to RSVC.

References

1. Ester, M., Kriegel, H.P., Sander, J., Xu, X.: A density-based algorithm for discovering clusters in large spatial databases with noise. In: KDD, pp. 226–231 (1996)
2. Guha, Rastogi, Shim: CURE: An efficient clustering algorithm for large databases. SIGMODREC: ACM SIGMOD Record 27 (1998)
3. Zhang, T., Ramakrishnan, R., Livny, M.: BIRCH: an efficient data clustering method for very large databases. In: Proceedings of ACM-SIGMOD International Conference of Management of Data, pp. 103–114 (1996)
4. Pawlak, Z.: Rough sets. International Journal of Computer and Information Sciences 11, 341–356 (1982)
5. Lingras, P., West, C.: Interval set clustering of web users with rough K -means. J. Intell. Inf. Syst 23, 5–16 (2004)

6. Asharaf, S., Shevade, S.K., Murty, N.M.: Rough support vector clustering (2005)
7. Ben-Hur, A., Horn, D., Siegelmann, H.T., Vapnik, V.: Support vector clustering. Journal of Machine Learning Research 2, 125–137 (2001)
8. Fletcher: Practical Methods of Optimization. Wiley, Chichester (1987)
9. Tsang, I.W., Kwok, J.T., Cheung, P.M.: Core vector machines: Fast SVM training on very large data sets. Journal of Machine Learning Research 6, 363–392 (2005)
10. Badoiu, M., Clarkson, K.L.: Smaller core-sets for balls. In: SODA, pp. 801–802 (2003)
11. Burges, C.J.C.: A tutorial on support vector machines for pattern recognition. Knowledge Discovery and Data Mining 2, 121–167 (1998)
12. Chang, C.C., Lin, C.J.: LIBSVM: a library for support vector machines. Online (2001)

Towards Human Level Web Intelligence:
A Brain Informatics Perspective

Ning Zhong

Department of Life Science and Informatics
Maebashi Institute of Technology, Japan &
The International WIC Institute
Beijing University of Technology, China
`zhong@maebashi-it.ac.jp`

In this talk, we outline a vision of Web Intelligence (WI) research from the viewpoint of Brain Informatics (BI), a new interdisciplinary field that studies the mechanisms of human information processing from both the macro and micro viewpoints by combining experimental cognitive neuroscience with advanced information technology. BI can be regarded as brain sciences in WI centric IT age and emphasizes on a systematic approach for investigating human information processing mechanism. Advances in instruments like fMRI and information technologies offer more opportunities for research in both Web intelligence and brain sciences. Further understanding of human intelligence through brain sciences fosters innovative Web intelligence research and development. Web intelligence portal techniques provide a powerful new platform for brain sciences. The synergy between WI and BI advances our ways of analyzing and understanding of data, knowledge, intelligence and wisdom, as well as their relationship, organization and creation process. Web intelligence is becoming a central field that changes information technologies, in general, and artificial intelligence, in particular, towards human-level Web intelligence.

A. Ghosh, R.K. De, and S.K. Pal (Eds.): PReMI 2007, LNCS 4815, p. 311, 2007.
© Springer-Verlag Berlin Heidelberg 2007

Quick Knowledge Reduction Based on Divide and Conquer Method in Huge Data Sets*

Guoyin Wang[1,2] and Feng Hu[1,2]

[1] Institute of Computer Science and Technology,
Chongqing University of Posts and Telecommunications,
Chongqing 400065, P.R.China
[2] School of Information Science and Technology,
Southwest Jiaotong University,
Chengdu 600031, P.R.China
{wanggy,hufeng}@cqupt.edu.cn

Abstract. The problem of processing huge data sets has been studied for many years. Many valid technologies and methods for dealing with this problem have been developed. Random sampling [1] was proposed by Carlett to solve this problem in 1991, but it cannot work when the number of samples is over 32,000. K. C. Philip divided a big data set into some subsets which fit in memory at first, and then developed a classifier for each subset in parallel [2]. However, its accuracy is less than those processing a data set as a whole. SLIQ [3] and SPRINT [4], developed by IBM Almaden Research Center in 1996, are two important algorithms with the ability of dealing with disk-resident data directly. Their performance is equivalent to that of classical decision tree algorithms. Many other improved algorithms, such as CLOUDS [5] and ScalParC [6], are developed later. RainForest [7] is a framework for fast decision tree construction for large datasets. Its speed and effect are better than SPRINT in some cases. L. A. Ren, Q. He and Z. Z. Shi used hyper surface separation and HSC classification method to classify huge data sets and achieved a good performance [8, 9].

Rough Set (RS) [10] is a valid mathematical theory to deal with imprecise, uncertain, and vague information. It has been applied in such fields as machine learning, data mining, intelligent data analyzing and control algorithm acquiring, etc, successfully since it was proposed by Professor Z. Pawlak in 1982. Attribute reduction is a key issue of rough set based knowledge acquisition. Many researchers proposed some algorithms for attribution reduction. These reduction algorithms can be classified into two categories: reduction without attribute order and reduction with attribute order. In 1992, A. Skowron proposed an algorithm for attribute reduction based on discernibility matrix. It's time complexity is $t = O(2^m \times n^2)$, and space complexity is $s = O(n^2 \times m)$ (m is the

---

* This work is partially supported by National Natural Science Foundation of China under Grant No.60573068, Program for New Century Excellent Talents in University (NCET), Natural Science Foundation of Chongqing under Grant No.2005BA2003, Science & Technology Research Program of Chongqing Education Commission under Grant No.KJ060517.

number of attributes, n is the number of objects) [11]. In 1995, X. H. Hu improved Skowron's algorithm and proposed a new algorithm for attribute reduction with complexities of $t = O(n^2 \times m^3)$ and $s = O(n \times m)$ [12]. In 1996, H. S. Nguyen proposed an algorithm for attribute reduction by sorting decision table. It's complexities are $t = O(m^2 \times n \times \log n)$ and $s = O(n + m)$ [13]. In 2002, G. Y. Wang proposed an algorithm for attribute reduction based on information entropy. It's complexities are $t = O(m^2 \times n \times \log n)$ and $s = O(n \times m)$ [14]. In 2003, S. H. Liu proposed an algorithm for attribute reduction by sorting and partitioning universe. It's complexities are $O(m^2 \times n \times \log n)$ and $s = O(n \times m)$ [15]. In 2001, using Skowron's discernibility matrix, J. Wang proposed an algorithm for attribute reduction based on attribute order. Its complexities are $t = O(m \times n^2)$ and $s = O(m \times n^2)$ [16]. In 2004, M. Zhao and J. Wang proposed an algorithm for attribute reduction with tree structure based on attribute order. Its complexities are $t = O(m^2 \times n)$ and $s = O(m \times n)$ [17].

However, the efficiency of these reduction algorithms in dealing with huge data sets is not high enough. They are not good enough for application in industry. There are two reasons: one is the time complexity, and the other is the space complexity. Therefore, it is still needed to develop higher efficient algorithm for knowledge reduction.

Quick sort for a two dimension table is an important basic operation in data mining. In huge data processing based on rough set theory, it is a basic operation to divide a decision table into indiscernible classes. Many researchers deal with this problem using the quick sort method. Suppose that the data of a two dimension table is in uniform distribution, many researchers think that the average time complexity of quick sort for a two dimension table with m attributes and n objects is $O(n \times \log n \times m)$. Thus, the average time complexity for computing the positive region of a decision table will be no less than $O(n \times \log n \times m)$ since the time complexity of quick sort for a one dimension data with n elements is $O(n \times \log n)$. However, we find that the average time complexity of sorting a two dimension table is only $O(n \times (\log n + m))$ [18]. When $m > \log n$, $O(n \times (\log n + m))$ will be $O(n \times m)$ approximately.

Divide and conquer method divides a complicated problem into simpler sub-problems with same structures iteratively and at last the sizes of the sub-problems will become small enough to be processed directly. Since the time complexity of sorting a two dimension table is just $O(n \times (m + \log n))$, and quick sort is a classic divide and conquer method, we may improve reduction methods of rough set theory using divide and conquer method.

Based on this idea, we have a research plan for quick knowledge reduction based on divide and conquer method. We have two research frameworks: one is attribute order based, and the other one is without attribute order.

(1) Quick knowledge reduction based on attribute order.

In some huge databases, the number of attributes is small and it is easy for domain experts to provide an attribute order. In this case, attribute reduction algorithms based on attribute order will be preferable.

Combining the divide and conquer method, a quick attribute reduction algorithm based on attribute order is developed. Its time complexity is $O(m \times n \times (m + \log n))$, and its space complexity is $O(n + m)$ [19].

(2) Quick knowledge reduction without attribute order.

In some huge databases, the number of attributes is big, even over 1000. It is very difficult for domain experts to provide an attribute order. In this case, attribute reduction algorithm without attribute order will be needed. Although many algorithms are proposed for such applications, their complexities are too high to be put into industry application. Combining the divide and conquer method, we will develop new knowledge reduction algorithms without attribute order. The aim of complexities of such algorithms will be: $t = O(m \times n \times (m + \log n))$ and $s = O(n + m)$. In this research framework, we have had some good results on attribute core calculation, its complexities are $t = O(m \times n)$, $s = O(n + m)$ [20].

Keywords: Huge data sets processing, rough set, quick sort, divide and conquer, attribute reduction, attribute order.

References

1. Catlett, J.: Megainduction: Machine Learning on Very Large Databases. PhD thesis, Basser Department of Computer Science, University of Sydney, Sydney, Australia (1991)
2. Chan, P.: An Extensible Meta-learning Approach for Scalable and Accurate Inductive Learning. PhD thesis, Columbia University, New York, USA (1996)
3. Mehta, M., Agrawal, R., Rissanen, J.: SLIQ: A fast scalable classifier for data mining. In: Apers, P.M.G., Bouzeghoub, M., Gardarin, G. (eds.) EDBT 1996. LNCS, vol. 1057, pp. 18–32. Springer, Heidelberg (1996)
4. Shafer, J., Agrawal, R., Mehta, M.: SPRINT: A scalable parallel classifier for data mining. In: VLDB. Proceedings of 22nd International Conference on Very Large Databases, pp. 544–555. Morgan Kaufmann, San Francisco (1996)
5. Alsabti, K., Ranka, S., Singh, V.: CLOUDS: A Decision Tree Classifier for Large Datasets. In: KDD 1998. Proceedings of the Fourth International Conference on Knowledge Discovery and Data Mining, New York, USA, pp. 2–8 (1998)
6. Joshi, M., Karypis, G., Kumar, V.: ScalParC: A new scalable and efficient parallel classification algorithm for mining large datasets. In: Proceedings of the 12th International Parallel Processing Symposium (IPPS/SPDP 1998), Orlando, Florida, USA, pp. 573–580 (1998)
7. Gehrke, J., Ramakrishnan, R., Ganti, V.: RainForest: A Framework for Fast Decision Tree Constructionof Large Datasets. In: Proceedings of the 24th International Conference on Very Large Databases (VLDB), New York, USA, pp. 416–427 (1998)
8. Ren, L.A., He, Q., Shi, Z.Z.: A Novel Classification Method in Large Data (in Chinese). Computer Engineering and Applications 38(14), 58–60 (2002)
9. Ren, L.A., He, Q., Shi, Z.Z.: HSC Classification Method and Its Applications in Massive Data Classifying (in Chinese). Chinese Journal of Electronics, China 30(12), 1870–1872 (2002)
10. Pawlak, Z.: Rough sets. International Journal of Computer and Information Sciences 11, 341–356 (1982)

11. Skowron, A., Rauszer, C.: The discernibility functions matrics and functions in information systems. In: Slowinski, R. (ed.) Intelligent Decision Support - Handbook of Applications and Advances of the Rough Sets Theory, pp. 331–362. Kluwer Academic Publisher, Dordrecht (1992)
12. Hu, X.H., Cercone, N.: Learning in relational database: A rough set approach. International Journal of Computional Intelligence 11(2), 323–338 (1995)
13. Nguyen, H.S, Nguyen, S.H.: Some efficient algorithms for rough set methods. In: Proceedings of the Sixth International Conference, Information Procesing and Management of Uncertainty in Knowledge-Based Systems(IPMU 1996), Granada, Spain, July 1-5, vol. 2, pp. 1451–1456 (1996)
14. Wang, G.Y., Yu, H., Yang, D.C.: Decision table reduction based on conditional information entropy (in Chinese). Chinese Journal of computers 25(7), 759–766 (2002)
15. Liu, S.H., Cheng, Q.J., Shi, Z.Z.: A new method for fast computing positve region(in Chinese). Journal of Computer Research and Development 40(5), 637–642 (2003)
16. Wang, J., Wang, J.: Reduction algorithms based on discernibility matrix: the ordered attributed method. Journal of Computer Science and Technology 11(6), 489–504 (2001)
17. Zhao, M., Wang J.: The Data Description Based on Reduct PhD thesis, Institute of Automation, Chinese Academy of Sciences, Beijing, China (in Chinese) (2004)
18. Hu, F., Wang, G.Y.: Analysis of the Complexity of Quick Sort for Two Dimension Table (in Chinese). Chinese Journal of computers 30(6), 963–968 (2007)
19. Hu, F., Wang, G.Y.: Quick Reduction Algorithm Based on Attribute Order(in Chinese). Chinese Journal of computers 30(8) (to appear, 2007)
20. Hu, F., Wang, G.Y, Xia, Y.: Attribute Core Computation Based on Divide and Conquer Method. In: Kryszkiewicz, M., et al. (eds.) RSEISP 2007. Lecture Note in Artificial Intelligence, Warsaw, Poland, vol. 4585, pp. 310–319 (2007)

Prefix-Suffix Trees: A Novel Scheme for Compact Representation of Large Datasets

Radhika M. Pai[1] and V.S Ananthanarayana[2]

[1] Manipal Institute of Technology, Manipal
[2] National Institute of Technology Karnataka, Surathkal

Abstract. An important goal in data mining is to generate an abstraction of the data. Such an abstraction helps in reducing the time and space requirements of the overall decision making process. It is also important that the abstraction be generated from the data in small number of scans. In this paper we propose a novel scheme called Prefix-Suffix trees for compact storage of patterns in data mining, which forms an abstraction of the patterns, and which is generated from the data in a single scan. This abstraction takes less amount of space and hence forms a compact storage of patterns. Further, we propose a clustering algorithm based on this storage and prove experimentally that this type of storage reduces the space and time. This has been established by considering large data sets of handwritten numerals namely the OCR data, the MNIST data and the USPS data. The proposed algorithm is compared with other similar algorithms and the efficacy of our scheme is thus established.

Keywords: Data mining, Incremental mining, Clustering, Pattern-Count(PC) tree, Abstraction, Prefix-Suffix Trees.

1   Introduction

In today's technologically developing world, there is an increase in both collection and storage of data. With the increase in the amounts of data, the methods to handle them should be efficient in terms of computational resources like memory and processing time. In addition to this the database is dynamic. The state of database changes due to either addition/deletion of tuples to /from the database. So, there is a need for handling this situation incrementally, without accessing original data more than once for mining applications. Clustering is one such data mining activity which deals with large amounts of data. Clustering has been widely applied in many areas such as pattern recognition and image processing, information processing, medicine, geographical data processing, and so on. Most of these domains deal with massive collections of data. In data mining applications, both the number of patterns and features are typically large. They cannot be stored in main memory and hence needs to be transferred from secondary storage as and when required. This takes a lot of time. In order to reduce the time, it is necessary to devise efficient algorithms to minimize the disk I/O operations. Several algorithms have been proposed in the literature for

A. Ghosh, R.K. De, and S.K. Pal (Eds.): PReMI 2007, LNCS 4815, pp. 316–323, 2007.

clustering large data sets[2,3,4]. Most of these algorithms need more than one scan of the database. To reduce the number of scans and hence the time, the data from the secondary storage are stored in main memory using abstractions and the algorithms access these data abstractions and hence reduce the disk scans. Some abstractions to mention are the CF-tree, FP-tree[4], PC-tree[8], PPC-tree[6], kd-trees[5], AD-trees[1], the PP-structure[13].The CF-tree[4]is the cluster feature vector tree which stores information about cluster descriptions at each node. This tree is used for clustering. The construction of this tree requires a single scan provided the two factors B and T are chosen properly. The FP-tree[4] is used for association rule mining and stores crucial and quantitative information about frequent sets. The construction of this tree requires two database scans and the tree is dependent on the order of the transactions. The kd-tree[5] and the AD-trees[1] reduce the storage space of the transactions by storing only the prototypes in the main memory. These structures are well suited for the applications for which they have been developed, but the use of these structures is limited as it is not possible to get back the original transactions. i.e. the structure is not a complete representation. PC-tree[8] is one such structure which is order independent , complete and compact. By using this structure, it is possible to retrieve back the transactions. The PC-tree[8] is a compact structure and the compactness is achieved by sharing the branches of the tree for the transactions having the same prefix. The tree generates new branches if the prefixes are different. One more abstraction called PPC-tree[6] is similar to PC-tree but it partitions the database vertically and constructs the PC-tree for each partition and for each class separately. The drawback of this structure is that, it is not possible to retrieve back the original transactions and the use of this structure in clustering is very much dependent on the partitioning criteria. The advantage of this structure is that it generates some synthetic patterns useful for supervised clustering. Both these abstractions need only a single database scan. The problem with the PPC-tree is that, by looking at the data set, it is difficult to predict the number of partitions in advance. The PP-structure[13] is similar to the PC-tree in the sense that it shares the branches for the transactions having the same prefix. But it differs from the PC-tree, in that it also shares the branches for the transactions having the same suffix. This structure also is a compact representation, but the construction of the structure is very complex as it searches the already constructed structure to check whether the current pattern's postfix is already present. In this paper, we propose a scheme called Prefix-Suffix Trees which is a variant of the PC-tree and which in principle is similar to the Prefix-Postfix structure in that the transactions having the same prefix or the same postfix have their branches being shared. The construction is simple compared to the Prefix-Postfix structure.The advantage of the Prefix-Suffix tree is that this also generates some synthetic patterns not present in the database which aids in clustering. The advantage of this scheme over the PPC-tree is that it is possible to get back the original transactions by storing a little extra information. In case of PPC-tree, if the number of parts is not selected properly, it results in overtraining and hence decrease in accuracy.

2   The Prefix-Suffix Trees Based Scheme

The Prefix-Suffix Trees based scheme which we propose is an abstract and compact representation of the transaction database. It is also a complete representation, order independent and incremental. The Prefix-Suffix Trees stores all transactions of the database in a compact way.

The Prefix-Suffix Trees are made up of nodes forming trees. Each node consists of four fields.

They are 'Feature' specifies the feature value of a pattern. The feature field of the last node indicates the transaction-id of the transaction which helps in retrieving the original transactions. 'Count' . The count value specifies the number of patterns represented by a portion of the path reaching this node. 'Child-pointer' represents the pointer to the following path. 'Sibling-pointer' points to the node which indicates the subsequent other paths from the node under consideration.

Fig.1 shows the node structure of Prefix-Suffix trees.

Feature	Count	Sibling-ptr	child-ptr

Fig. 1. Node Structure of the Prefix-Suffix Trees

2.1   Construction of the Prefix-Suffix Trees

The algorithm for the construction of the Prefix-Suffix Trees is as follows. Let T_r be the transaction database.

Partition the transaction database T_r into 2 equal parts. Let the 2 parts be T_{r1} and T_{r2} respectively.

For T_{r1} construct the tree as follows which is called Prefix Tree.

Let the root of the Prefix-tree be TR1.

For each pattern, $t_i \in T_{r1}$ Let m_i be the set of positions of non-zero values in t_i.

If no sub-pattern starting from TR1 exists corresponding to m_i,
THEN
    Create a new branch with nodes having 'Feature' fields as values of mi and 'Count' fields with values set to 1.
    ELSE
    Put values of m_i in an existing branch e_b by incrementing the corresponding 'count' field values of mi by appending additional nodes with 'count' field values set to 1 to the branch e_b.

For T_{r2}, reverse T_{r2} to get T_{r2r}.

Construct the tree for T_{r2r} as described earlier which is called the Suffix Tree.

Let TR1 be the root of Prefix Tree corresponding to T_{r1} and let TR2R be the root of the Suffix Tree corresponding to T_{r2r}.

2.2   Clustering Algorithm

In order to cluster the test pattern, the test pattern is also partitioned into 2-partitions using the same partitioning criteria as used for the training patterns. Let c be the number of classes, k, the number of nearest neighbours. The algorithm proceeds as follows.

For each branch b_l in the Prefix-Tree TR1.

Find the matches between the test pattern and the branch b_l. let it be C_1^l. Find k largest counts in decreasing order. Let them be $C_1^l, C_1^2, \ldots, C_1^k$ .Let the corresponding labels be $O_1^l, O_1^2, \ldots, O_1^k$.

Similarly, for each branch b_m in Suffix-Tree TR2R,

Find the matches between the test pattern and the branch b_m. let it be C_2^l. Find k largest counts in decreasing order. Let them be $C_2^l, C_2^2, \ldots, C_2^k$ and the corresponding labels be $O_2^l, O_2^2, \ldots, O_2^k$

For i= 1 to k

  For j= 1 to k

    Find $C_p = C_i + C_j$ if $O_i == O_j$ where $0 \le p \le k-1$.

Find k largest counts in decreasing order among all C_p where $0 \le p \le k$.

Compute the weight , $W_p = 1 - (C_k - C_p)/(C_p - C_1)$

For n = 1 to c

  $Sum_n = \sum 1m[W_m]$ where $(O_m == n)$

Output (label = O_x) for which Sum_x is maximum for $x \in 1, 2, \ldots, c$

2.3   Comparision of the Prefix-Suffix Trees and the PC-Tree

The PC-tree[8] compacts the database by merging the nodes of the patterns having the same prefix. But in Prefix-Suffix Trees, still compaction is achieved by merging the branches of the trees having the same suffix also and so the number of nodes is reduced thus saving considerable space. An example-Consider the following set of transactions as shown in Fig.2A. The first column gives the transaction number, the second column gives the set of features and the last column gives the label. The partitioned set of transactions Part1 and Part2 are given in Fig.2B and Fig.2C and the set of transactions of Part2 in reverse order are given in Fig.2D respectively. The Prefix- tree , The Suffix-tree and the PC-tree for the set of transactions is given in Fig.3A, Fig.3B and Fig.3C respectively. In the figures, the nodes are indicated by circles, the right arrow is the child pointer , the downward pointer is the sibling pointer. The first number inside the circle is the feature value and the number after the colon is the count. The last node in all branches is the label node.

3   Experiment and Results

To evaluate the performance of our algorithm, the following three real world datasets were considered.

Tr.No.   Features   Label   Tr.No.   Features   Label
1 1,2,3,4,5,8,9,10,11,12,14,15,16   0   1   1,2,3,4,5,8   0
2 1,2,3,4,7,10,11,12,14,15,16   0   2   1,2,3,4,7   0
3 2,3,4,5,6,12,14,15,16   0   3   2,3,4,5,6   0
4 2,4,5,7,9,12,13,14   3   4   2,4,5,7   3
5 2,4,5,6,8,12,13,14   3   5   2,4,5,6,8   3

   A. Sample set of transactions   B.Set of Transactions in Part1

Tr.No.   Features   Label   Tr.No.   Features   Label
1   9,10,11,12,14,15,16   0   1   16,15,14,12,11,10,9   0
2   10,11,12,14,15,16   0   2   16,15,14,12,11,10   0
3   12,14,15,16   0   3   16,15,14,12   0
4   12,13,14   3   4   14,13,12   3
5   12,13,14   3   5   14,13,12   3

   C.Set of Transactions in Part2   D.Set of Transactions of Part2 in
                                             reverse order

Fig. 2. Sample set of Transactions

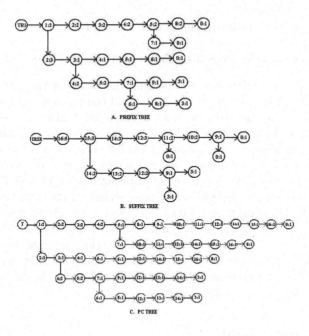

A. PREFIX TREE

B. SUFFIX TREE

C. PC TREE

Fig. 3. Prefix Tree, Suffix Tree and PC Tree for the transactions in Fig.2

Table 1. Comparison of the Prefix-Suffix Trees, PP-structure, PC-Tree, PPC-Tree and K-NNC based algorithms

Expt-No	Dataset	Algorithm	Storage space in Bytes	Time in secs training+test time	Accuracy
1	OCR data (2000 patterns)	Prefix-Suffix Trees based alg.	1106880	120	91.45
		PP-structure based alg.	1123696	34	89.56
		PC-tree based alg.	1406528	153	89.56
		PPC-tree based alg. (with 4 parts)	1119280	131	41.58
		k-NNC	1544000	18	89.44
2	OCR data (4000 patterns)	Prefix-Suffix Trees based alg.	2046928	230	93.07
		PP-structure based alg.	2070112	110	92.04
		PC-tree based alg.	2717792	245	91.96
		PPC-tree based alg. (with 4 parts)	1970304	198	41.01
		k-NNC	3088000	31	91.9
3	OCR data (6670 patterns)	Prefix-Suffix Trees based alg.	3202576	315	94.12
		PP-structure based alg.	3216400	863	93.76
		PC-tree based alg.	4408256	386	93.61
		PPC-tree based alg. (with 4 parts)	3812320	314	64.3
		k-NNC	5149240	47	93.55
4	USPS data	Prefix-Suffix Trees based alg.	7877088	435	93.32
		PP-structure based alg.	7991504	554	93.27
		PC-tree based alg.	10030656	537	92.68
		PPC-tree based alg. (with 4 parts)	6490336	337	92.23
		k-NNC	7495148	39	93.27
5	MNIST data	Prefix-Suffix Trees based alg.	107430848	25317	96.7
		PP-structure based alg.	108528480	5996	96.5
		PC-tree based alg.	126535296	28451	96.5
		PPC-tree based alg. (with 4 parts)	77355312	17182	68.3
		k-NNC	188400000	cannot be stored in main mem.	cannot be implemented as it can't be stored

3.1   Dataset1: OCR Data

This is a handwritten digit dataset. There are 6670 patterns in the training set, 3333 patterns in the test set and 10 classes. Each class has approximately 670 training patterns and 333 test patterns. We conducted experiments separately with 2000(200 patterns from each class), 4000(400 patterns from each class) and 6670 training patterns(667 patterns from each class).

3.2   Dataset2: USPS Data

This is a dataset which is a collection of handwritten digits scanned from the U.S. postal services. There are 7291 patterns in the training set and 2007 patterns in the test set and 10 classes. Each pattern represents a digit and has 256 features.

3.3   Dataset3: MNIST Data

This is a data which is a mixture of the NIST(National Institute of standards and technology) special database 3 and 1. This is collection of handwritten digits written by census bureau employees and high school students. This is a large data set having 60000 patterns in the training set and 10000 patterns in the test set and 10 classes. Each pattern has 784 features. We have compared our algorithm with the PC-tree, PPC-tree, the Prefix-postfix structure and the k-NNC algorithms for all the above datasets and the results are tabulated in Table 1. From the table, we observe that the Prefix-Suffix trees based algorithm consumes less space than the PC-tree, the Prefix-Postfix structure and the k-NNC algorithm without sacrificing for the classification accuracy. For all the datasets, dataset 1 , 2 and 3 we observe that the accuracy is increased by a certain order. All the experiments were executed on Xeon processor based Dell precision 670 workstation having a clock frequency of 3.2 GHZ and 1 GB RAM.

4   Conclusion

In this paper, a novel scheme called Prefix-Suffix Trees for compact storage of patterns is proposed which stores the transactions of a transaction database in a compact way. This scheme is complete, order independent and incremental. The use of this scheme in clustering is given and the effectiveness of the algorithm is established by comparing the scheme with the PC-tree, PPC-tree, Prefix-Postfix structure based algorithms and the benchmark algorithm k-NNC. The new scheme is found to be more compact than the PC-tree without sacrificing for the accuracy as shown. The performance of the algorithm is evaluated by testing the algorithm with 3 different datasets of handwritten digits and the effectiveness of our algorithm is thus established.

References

1. Moore, A., Lee, M.S.: Cached Sufficient statistics for efficient machine learning with large datasets. Journal of Artificial Intelligence Research 8, 67–91 (1998)
2. Jain, A.K., Dubes, R.C.: Algorithms for Clustering Data. Prentice-Hall, Englewood Cliffs (Advanced Reference Series)

3. Jain, A.K., Murty, M.N., Flynn, P.J.: Data Clustering: A Review. ACM Computing Surveys 31(3), 264–323 (1999)
4. Pujari, A.K.: Data Mining techniques. University Press, New Haven (2001)
5. Friedman, J.H., Bentley, J.L., Finkel, R.A.: An algorithm for finding best matches in logarithmic expected time. ACM trans. Math software 3(3), 209–226 (1997)
6. Viswanath, P., Murthy, M.N.: An incremental mining algorithm for compact realization of prototypes. Technical Report, IISC, Bangalore (2002)
7. Prakash, M., Murthy, M.N.: Growing subspace pattern recognition methods and their neural network models. IEEE trans. Neural Networks 8(1), 161–168 (1997)
8. Ananthanarayana, V.S., NarasimhaMurty, M., Subramanian, D.K.: Tree structure for efficient data mining using rough sets. Pattern Recognition Letters 24, 851–886 (2003)
9. http://www.cs.cmu.edu/15781/web/digits.html
10. http://wwwi6.informatik.rwthaachen.de/~keysers/usps.html
11. Duda, R.O., Hart, P.E.: Pattern Classification and Scene Analysis. Wiley, New York (1973)
12. Ravindra, T., Murthy, M.N.: Comparison of Genetic Algorithms based prototype selection scheme. Pattern Recognition 34, 523–525 (2001)
13. Pai, R.M., Ananthanarayana, V.S.: A novel data structure for efficient representation of large datasets in Data Mining. In: Proceedings of the 14th international Conference on Advanced Computing and Communications, pp. 547–552 (2006)

Efficient Multi-method Rule Learning for Pattern Classification Machine Learning and Data Mining

Chinmay Maiti[1] and Somnath Pal[2]

[1] Dept. of Info. Tech.
Jadavpur University
chinmay@it.jusl.ac.in
[2] Dept. of Comp. Sc. & Tech.
Bengal Engg. & Sc. University
sp@cs.becs.ac.in

Abstract. The work presented here focuses on combining multiple classifiers to form single classifier for pattern classification, machine learning for expert system, and data mining tasks. The basis of the combination is that efficient concept learning is possible in many cases when the concepts learned from different approaches are combined to a more efficient concept. The experimental result of the algorithm, EMRL in a representative collection of different domain shows that it performs significantly better than the several state-of-the-art individual classifier, in case of 11 domains out of 25 data sets whereas the state-of-the-art individual classifier performs significantly better than EMRL only in 5 cases.

Keywords: Machine learning, Multiple Classifiers, Missing values, Discretization, Classification.

1 Introduction

The task of constructing ensembles of classifiers[7] can be categorized into subtasks . One approach is to generate classifiers by applying different learning algorithm with heterogeneous[12] model representations to a single data set.As per our knowledge, such approach was introduced for decision tree learning in[15] and was subsequently improved in[16]. Another approach is to apply a single learning algorithm with different parameters settings to a single data set. Then, methods like bagging[2] and boosting[9] generate multiple classifiers by applying a single learning algorithm to different versions of a given data set. Bagging produces replicate training sets by sampling with replacement from the training instances. Boosting considers all instances at each repetition, but maintains a weight for every instance in the training set that reflects its importance, re-weighting of the misclassified instances leads to different classifiers. In both cases, voting to form a composite classifier combines the multiple classifiers. In bagging, each component classifier has the equal vote, whereas in boosting uses different voting strengths to each component classifiers on the basis of their accuracy.

A. Ghosh, R.K. De, and S.K. Pal (Eds.): PReMI 2007, LNCS 4815, pp. 324–331, 2007.

This paper here focuses on the combining multiple classifiers using multi-method approach for pattern classification, machine learning for expert system and data mining task. This method is to apply different learning algorithms with homogeneous model representation on a single data set and the combinational algorithm is called EMRL(Efficient Multi-method Rule Learning). The algorithm EMRL heuristically combines the outputs of different state-of-the-art rule inducing algorithms, namely C4.5rules[18],CN2[5][6],RISE[8] and PRISM[4] and produces efficient composite classifier. The following section briefly describes EMRL algorithm. The experimental evaluation of EMRL algorithm for comparison of performances with the state-of-the-art individual classifier is described in section 3. Finally, the conclusion is summarized in the last section.

2   EMRL Algorithm

2.1   The Basis of EMRL Algorithm

Fundamentally, EMRL(Efficient Multi-method Rule Learning) provides a framework to combine the outputs of various base classifiers of empirical inductive generalization, namely C4.5rules, CN2, RISE and PRISM. This is done by heuristically combining outputs of different classifiers in the rule plane. This approach is called as multi-method learning in[15] for decision tree plane.

The heuristic that is the basis of this paper is that efficient concept learning is possible in many cases when the concepts learned from different approaches are combined to a more efficient concept. EMRL algorithm performs in two steps. It generates sets of rules for a given instance set by using different rule inducing algorithms, and then combines the rule set to get an efficient composite classifier.

2.2   The EMRL Algorithm

The Efficient Multimethod Rule Learning(EMRL) algorithm is given below. The objective of EMRL algorithm is to select a resultant set of rules which is able enough to classify the example set completely and successfully. An EMRL algorithm is based on k-fold cross-validation of the training data sets to select the best possible set of rules generated by k numbers of (k-1) parts forming temporary training sets with k numbers of one part formimg temporary test set. It then generates the set of rulesets obtained from m ($m \geq 2$) rule inducing algorithm like C4.5rules, CN2, RISE, PRISM for each temporary training set. Next, it computes the union of the rulesets and then calculates the weighted vote for each rule using the corresponding temporary training data set. First, the weighted vote is initialized to zero for each rule in the unified set. Then weighted vote is computed for each rule one by one. The weighted vote of a rule = the number of instances in the temporary training set correctly classified by the rule. Next, the weighted vote for each rule is modified using the instances in the temporary test set.

Modification of vote for each rule is done whether the rule classifies an instance correctly or misclassifies. If an instance is classified correctly by the rule then weighted vote of that rule is increased by adding RV(reward vote) to the existing vote. Again, if an instance is misclassified by it then decrease the vote by subtracting PV(penalty vote)from the existing vote. If an instance is not covered by the rule[1] then no changes to the vote.In EMRL algorithm both reward and penalty vote are totally experimental. After modification of weighted vote for rules in all rule sets $R_j j = 1, ..., m(m \geq 2)$ compute the average weight and the final average weighted vote is determined by using the following formula Final Average Weighted Vote = (average weighted vote × APW) / 100.

The APW(Acceptable Percentage of Weight)value is also experimental and varying from 1 to 100(in steps of ϵ). Here, APW acts as a factor for varying the threshold value of avreage weighted vote which is used for final selection of rules in the accetable rule set. Now, the computed final weighted vote is used to select the best possible set of rules. The rule with weighted vote greater than or equal to the final weighted vote is considered in the final set.

EMRL Algorithm

Step1. Divide the training set into n mutually exclusive parts and prepare temporary training data set (L_i) with (n-1) parts and call the rest one part as temporary test data set (T_i). Generate L_i and $T_i, i = 1, 2, ..., n$.

Step2. For each L_i generate Rule sets $R_j, j = 1, ..., m$ $(m \geq 2)$ by applying m rule including algorithms.

Step3. Construct union of all input rule sets $R_j, j = 1, ..., m$.

Step4. For all rules initialize weighted vote as zero.

Step5. Calculate the weighted vote for each rule using temporary training data set L_i, i.e., weighed vote of a rule is equal to the Number of instances correctly classified by it.

Step6. Modify the weighted vote for each rule of the union set using temporary test data set T_i

  a) If an instance is correctly classified by a rule then increase the weighted vote of that rule by RV amount (user defined reward vote).

  b) If an instance is not covered by a rule then do nothing to the weighted vote of that rule.

  c) If an instance is misclassified by a rule then decrease the weighted-vote of that rule by PV amount (user defined penalty vote).

Step7. Calculate the average weighted vote and modify it by the formula Final average weighted vote= (average weighted vote × APW)/100 (APW means Acceptable Percentage of Weight varying from 1 to 100).

Step8. Composite efficient rule set is then formed by selecting the rules from union of all R_j's and having weighted vote greater than or equal to the modified weighted vote.

---

[1] A rule in a rule set may classify or misclassify an instance or may not be applicable to the instance at all. The rule that is not applicable to an instance is considered as not covering the instance.

2.3   Time Complexity

Worst case time complexity of the algorithm EMRL is $O(mr^2) + O(mer)$, where m is the number of terms in each rule, e is the total number of instances in the training set and r is the number of unified rules.

3   Empirical Evaluation

An empirical evaluation is carried out with the goal of testing whether at all improvement is possible in the multi-method framework proposed in EMRL algorithm.

3.1   Data Sets

Empirical calculations of the performance of EMRL algorithm are carried out on 25 real world data sets drawn from the University of California at Irvin data repository [14]. The data sets are chosen vary across a number of dimensions including the type of application area, the type of features in the data sets, the number of output classes and the number of instances. These are data sets with both discrete and continuous data.

3.2   Treatment of Missing Values

Real-world data is often incomplete and often has missing values. Several authors [13][18] have developed different answers to this problem, usually based either on filling in missing attributes values with most probable value or a value determined by exploiting interrelation-ship among the values of different attributes, or on looking at the probability distribution of values of attributes. In this work [17], a different approach is followed in the treatment of missing values. The missing values in attribute means that a particular test is not performed (and hence not required) to classify the particular instance. This itself is information. To keep this information in resulting rules the missing values of an attribute are assigned a separate value is chosen as the highest value of the attribute which would not be the case with other methods. Such treatment of missing values as a separate value is well-known in machine learning literature [10].

3.3   Discretization

Efficient discretization of continuous attributes is an important because it has effects on speed, accuracy and understandability of the induction models. Furthermore, many machine learning alogorithms such as PRISM [4] can not handle continuous features. All data sets are discretized using global discretization based on successive pseudo deletion of instances [17] to reduce the conflicting instances, i.e., by successive reduction of noises in the database. Pre-discretization of data is also a step towards learning itself and can be considered as an elementary inductive method that predicts the local majority class in each learned interval [24] and therefore, adds informations unlike loss of information to the model.

3.4   Experimental Design

We have considered C4.5rules, CN2, RISE and PRISM rule inducing algorithms as base learners for EMRL algorithm. The EMRL, being a general framework, can combine output rules of any two or more base rule inducing algorithms. All possible combinations of C4.5rules, CN2, RISE and PRISM have tried using EMRL for all 25 data sets. In our experimental design, we set reward vote as one and penalty vote as two. There is a basis that why the penalty vote is greater than the reward vote. The task of EMRL algorithm is to find the best possible set of rules which are efficient in future for prediction. So, EMRL is trying to severely penalize the rules which are less significant to classify the instances and penalize them by decreasing the vote by twice the amount than for successful classification when it misclassifies an instance. Althoug we have conducted our experiment with RV=1 and PV=1, but the chosen values with RV=1 and PV=2 were formed to result in slightly improved accuracy. Also, in our experimental design to determine the value of APW, we make several runs of the EMRL algorithm with interval 5 (ϵ=5) of the APW value starting from 20 to 70(instead of 1 to 100 because the performance of the final acceptable rule set degrades drastically beyond APW value of 20 and 70) and finally set the value as 45 where it provides maximum accuracy for most data sets. We have considered the combination with highest predictive accuracy and compare with the best of single state-of-the-art base algorithms and that is reported here. The accuracies have calculated using k-fold cross-validation method. Here, we set the value of k equals to 10. Such 10-fold cross-validation experiments are repeated 5 times for each of the data sets and each of the possible combination of algorithms. Thus, accuracy calculations are averaged over 50 values for each data set and each combination.

3.5   Experimental Results

The Table 1 shows empirical results in terms of average accuracies for both EMRL (combination with highest accuracies) and best of single state-of-the-art base algorithms. Superscripts indicates confidence level for difference between the EMRL and state-of-the-art base algorithm using one tailed paired t-test. The results of Table 1 are interesting in themselves; however, they are summarized in a few global measures of comparative performance in Table 2.

4   Related Work

The main idea of ensemble methodology is to combine a set of models, each of which solves the same original task, in order to obtain a better composite global model, with more accurate and reliable estimates or decisions than can be obtained from using a single model.The strategy of combining the classifiers generated by an induction algorithm [19]. The simplest combiner determines the output solely from the outputs of the individual inducers. Ali and Pazzani [1]

Table 1. Empirical Result: acc.=average accuracy and s.d.=standard deviation. Superscripts denote confidence level :1 is 99.5%, 2 is 99%, 3 is 97.5%, 4 is 95%, 5 is 90%, and 6 is below 90%. Symbols denote the algorithms : A1 is C4.5rules, A2 is CN2, A3 is RISE, A4 is PRISM.

Data Sets	Individual Classifiers		EMRL	
	$acc. \pm s.d.$	Algorithm	$acc. \pm s.d.$	Algorithms Combined
heart hungary	76.39 ± 8.32	A4	67.02 ± 9.45^1	A3,A4
heart switzerland	45.46 ± 10.56	A2	46.47 ± 10.47^6	A1, A3,A4
heart cleveland	58.86 ± 7.28	A2	61.42 ± 7.51^3	A3,A4
primary-tumar	42.52 ± 7.15	A2	48.42 ± 8.44^1	A1,A2
iris	96.65 ± 3.35	A1	95.33 ± 3.06^2	A1,A2/A2,A3
glass	72.86 ± 7.36	A3	75.19 ± 7.41^4	A1,A2,A3,A4
pima-indians	74.47 ± 6.02	A1	76.56 ± 5.92^3	A1,A2
liver disorder	67.30 ± 7.22	A3	71.64 ± 6.96^1	A2,A3
hepatitis	85.29 ± 8.06	A3	85.87 ± 6.28^6	A2,A3
cpu-performance	59.14 ± 14.99	A3	58.24 ± 13.96^6	A1,A2,A3,A4
dermatology	94.85 ± 5.60	A3	80.30 ± 10.78^1	A2,A3
hdva	31.50 ± 13.24	A4	34.00 ± 8.31^5	A2,A4
credit	86.09 ± 4.31	A3	87.68 ± 3.12^2	A1,A2,A3
ecoli	81.26 ± 4.71	A3	82.74 ± 6.45^4	A2,A3
hypothyroid	99.20 ± 0.39	A1	99.30 ± 0.31^4	A1,A2,A3
wisconsin	95.29 ± 2.57	A2	97.28 ± 1.35^1	A1,A2
wine	98.33 ± 3.56	A3	98.33 ± 3.56^6	A3,A4
echo cardiogram	65.99 ± 11.69	A3	66.65 ± 8.55^6	A1,A3
breast cancer	72.71 ± 7.29	A3	74.50 ± 6.04^4	A2,A3
voting	95.39 ± 3.09	A1	96.10 ± 3.67^5	A2,A3
zoo	97.00 ± 6.40	A2	94.00 ± 6.63^2	A1,A2,A3 / A2,A3,A4
lymphography	82.34 ± 9.30	A2	85.10 ± 11.14^4	A2,A3
new-thyroid	99.09 ± 2.73	A2/A3	99.09 ± 2.73^6	A2,A3
annealing	99.50 ± 0.83	A3	98.75 ± 0.97^1	A1,A2
imports-85	79.62 ± 8.44	A2	78.67 ± 9.54^6	A1,A2,A4

Table 2. Summary of The Results

Algorithm → Measure ↓	State-of-the-art Individual Classifier	EMRL (C4.5rules,CN2,RISE,PRISM)
Average Accuracy Over 25 data sets	77.92	78.34
Number of wins in Accuracy	7	16
Number of significant wins in Accuracy	5	11
Wilcoxson sign rank test in Accuracy	— —	90.56%

have compared several combination methods: uniform voting, Bayesian combination, distribution summation and likelihood combination. Moreover, theoretical analysis have been developed for estimating classification improvement [21]. Along with simple combiners there are other more sophisticated method, such as stacking [23]. On the otherhand, there is the success of bagging [2] and boosting [9] and similar other re-sampling techniques. These have the disadvantage of generating incomprehensible concept descriptions. MDTL2 [15], MDTL3.1 [16] was a valuable tool for combining large rule sets into simpler comprehensible decision tree. EMRL algorithm presented here combines rulesets generated by different heterogeneous classifiers into a single efficient set of rules. The difference of MDTL and EMRL type of combination of classifiers with other ensemble and multiple classifier systems is that the former type use the concept description of an ensemble of classifiers (instead of their output) for combination whereas the latter type use the outputs of emsemble of classifiers for the classification tasks. Although, the resultant classifier from MDTL and EMRL type of algorithm can be used for bagging, boosting, voting or any other classifier combination method like Data Envelop Analysis Weighting method [20], logarithmic Opinion Pool [11], Order Statistics [22], to produce still improved results.

5   Conclusion and Future Research

Recently, combining classifiers to achieve greater accuracy has become an increasingly important research topic in machine learning. In this paper, we describes an algorithm EMRL that can combine various base rule inducing algorithm into an efficient composite classifier. The empirical evaluation of the algorithm shows that combination of state-of-the-art base classifier using EMRL produced significantly better accuracies than best of single state-of-the-art base classifier. Further, performance improvement can possibly be expected if EMRL is incorporated as a base classifier in different ensemble methods such as bagging, boosting, voting(heterogeneous) etc methods on which our current work is going on. Other direction of future research include applicability of EMRL to large databases.

References

1. Ali, K.M., Pazzani, M.J.: Error Reduction through Learning Multiple Descriptions. Machine Learning 24(3), 173–202 (1996)
2. Breiman, L.: Bagging Predictors. Machine Learning 24(2), 123–140 (1996)
3. Bruzzon, L., Cossu, R.: A Robust Multiple Classifier System for a Partially Unsupervised Updating of Land-cover Maps. In: Kittler, J., Roli, F. (eds.) MCS 2001. LNCS, vol. 2096, pp. 259–268. Springer, Heidelberg (2001)
4. Candrowska, J.: PRISM: An algorithm for inducing modular rules. International Journal for Man-Machine Studies 27, 349–370 (1987)
5. Clark, P., Boswell, R.: Rule Induction with CN2: Some recent improvements. In: Proc. of the 5th ECML, Berlin, pp. 151–163 (1991)

6. Clark, P., Niblett, T.: The CN2 Induction Algorithm. ML 3, 261–283 (1989)
7. Dietterich, T.G.: Machine Learning Research: Four Current Directions. AI Magazine 18(4), 97–136 (1997)
8. Domings, P.: Unification Instance-based and Rule-base Induction. Machine Learning 3, 139–168 (1996)
9. Freund, Y., Schapire, R.E.: Experiments with a New Boosting Algorithm. In: Proc. of the 13th International Conference on Machine Learning, pp. 146–148. Morgan Kaufmann, San Francisco (1996)
10. Grzymala-Busse, J.W., Grzymala-Busse, W.J.: Handling Missing Attributes Values. In: Maimon, O., Rokach, L. (eds.) Data Mining & Knowledge Discovery Hand Book, pp. 35–57. Springer, Heidelberg (2005)
11. Hansen, J.: Combining Predictors Meta Machine Learning Methods abd Bias Variance & Ambiguity Decompositions Ph. D. dissertation, Aurhns University (2000)
12. Merz, C.J.: Using Correspondence Analysis to Combine Classifiers. Machine Learning 36(1/2), 33–58 (1999)
13. Mingers, J.: An Empirical Comparison of Selection Measure for Decision-tree Induction. Machine Learning 3, 319–342 (1989)
14. Murphy, P. M.: UCI Repository of Machine Learning Databases (Machine-readable data repository). Department of Information and Computer Science, Irvine, CA: University of California (1999)
15. Pal, S., Saha, T.K., Pal, G.K., Maiti, C., Ghosh, A.M.: Multi-method Decision Tree Learning for Data Mining. In: IEEE ACE 2002, pp. 238–242 (2002)
16. Pal, S., Maiti, C., Mazumdar, A.G., Ghosh, A.M.: Multi-Method Combination For Learning Search-Efficient And Accurate Decision Tree. In: 5-th Intenational conference on Advances in Pattern Recognition, ISI, Kolkata, pp. 141–147 (2003)
17. Pal, S., Maiti, C., Debnath, K., Ghosh, A.M.: SPID3: Discretization Using Pseudo Deletion. Journal of The Institute of Engineers 87, 25–31 (2006)
18. Quinlan, J.R.: Program for Machine Learning. Morgan Kaufmann, CA (1993)
19. Rokach, L.: Ensemble Methods for Classifiers. In: Maimon, O., Rokach, L. (eds.) The Data Mining and Knowledge Discovery Hand Book, pp. 956–980. Springer, Heidelberg (2005)
20. Sohn, S.Y., Choi, H.: Ensemble based on Data Envelopment Analysis. In: ECML Meta Learning Workshop (September 2004)
21. Tumer, K., Ghosh, J.: Linear and Order Statistics Combiners for Pattern Classification. In: Sharkey, A. (ed.) Combining Artificial Neural Nets, pp. 127–162. Springer, Heidelberg (1999)
22. Tumer, K., Ghosh, J.: Robust Order Statistics based Ensemble for Distributed Data Mining. In: Kargupto, H., Chan, P. (eds.) Advances in Distributed and Parallel Knowledge Discovery, pp. 185–210. AAAI / MIT Pess (2000)
23. Wolpert, D.: Stacked Generalization. Neural Networks 5(2), 241–260 (1992)
24. Zighed, D.A., Rabaseda, S., Rakotomalala, R., Feschet, F.: Discretization methods in supervised learning. In: Ency. of Comp. Sc.& Tech. vol. 44, pp. 35–50. Marcel Dekker Inc (1999)

High Confidence Association Mining Without Support Pruning

Ramkishore Bhattacharyya[1] and Balaram Bhattacharyya[2]

[1] Microsoft India (R&D) Pvt. Ltd., Gachibowli, Hyderabad – 500 032, India
[2] Dept of Computer and System Sciences, Visva-Bharati Universty,
Santiniketan – 731235, India
rk_ju@yahoo.com, balaramb@gmail.com

Abstract. High confidence associations are of utter importance in knowledge discovery in various domains and possibly exist even at lower threshold support. Established methods for generating such rules depend on mining itemsets that are frequent with respect to a pre-defined cut-off value of support, called support threshold. Such a framework, however, discards all itemsets below the threshold and thus, existence of confident rules below the cut-off is out of its purview. But, infrequent itemsets can potentially generate confident rules. In the present work we introduce a concept of *cohesion* among items and obtain a methodology for mining high confidence association rules from itemsets irrespective of their support. Experiments with real and synthetic datasets corroborate the concept of cohesion.

Keywords: Association rule, high-confidence, support threshold, cohesion.

1 Introduction

Association Rule Discovery (ARD) [1] is a foundational technique in exploring relationships between attributes. An association rule is an expression $A \xrightarrow{\sigma, \mu} C$, where A and C are itemsets and $A \cap C = \varnothing$, A and C are called antecedent and consequent respectively. Support σ of the rule is given by the percentage of transaction records containing the itemset $A \cup C$. Confidence μ of the rule is the conditional probability that a record, containing A, also contains C and is given as $\mu = \sigma(A \cup C)/\sigma(A)$. Classical ARD task concerns the discovery of every such rule with both support and confidence greater than or equal to pre-defined thresholds.

However, numerous applications such as medical diagnosis, bio-informatics, image categorization, weather forecasting, web mining, etc often contain potentially large number of low-support high-confidence rules, possibly with notable significance, but owing to the support constraint classical ARD misses them. Consider the rule {Down's syndrome} → { trisomy 21} (a chromosomal defect). Patients, with trisomy 21, suffer from Down's syndrome with almost 100% confidence. But Down's syndrome is found in about one in thousand cases. So a moderately low support threshold may miss it. Successively lowering the threshold cannot be a solution as the process is to capture unknown knowledge.

A. Ghosh, R.K. De, and S.K. Pal (Eds.): PReMI 2007, LNCS 4815, pp. 332–340, 2007.
© Springer-Verlag Berlin Heidelberg 2007

Looking differently, high confidence of a rule indicates high probability of coexistence of the consequent with the antecedent. That is, we must look for a new property of an itemset that can reveal information on tendency of coexistence of items. We term this property as *cohesion* of an itemset.

The essence of mining support independent high-confidence rules has long been realized in the backdrop of classical Apriori algorithm [2]. Literature [4] employs dataset partitioning technique and border-based algorithm in mining high confidence rules. However, such an algorithm is restricted to discover all top rules (confidence = 100%) with the consequent being specified. Literature [3] proposes a measure called *similarity*, as a substitute for support, to explore low-support high confidence rules. However, the phenomenon has been studied only for two-itemsets for its computationally prohibitive nature in higher order itemsets. Similarity measure has been generalized for k-itemset in literature [5] with different nomenclature *togetherness*. The authors have discussed some positive sides of using togetherness in mining high-confidence rules. However, due to a less efficient Apriori-like algorithm, experimental opportunity with several real datasets has been missed. In fact, both the later two literatures do not establish the fact that due to some natural phenomenon of application domain, a set of items appear collectively which similarity or togetherness reflects. Therefore, we use the more spontaneous terminology cohesion. With the notion of cohesion of an itemset, the present work formalizes the problem of mining confident rules irrespective of their support information. We introduce a new algorithm CARDIAC[1] for mining cohesive itemsets which is the gateway of association mining. Theoretical as well as experimental studies prove its superiority over support threshold framework in mining confident associations. Since the problem is NP-Complete, we will not go into complexity analysis of our algorithm.

2 Formalism

Let $I = \{1, 2, ..., m\}$ be a set of items and $T = \{1, 2, ..., n\}$ be a set of transactions identifiers or tids. Assume $t(X)$ the set of all tids containing the itemset X as a subset.

Definition 1. (Cohesion) Cohesion of an itemset X is the ratio of cardinalities of two sets viz. the set all of tids containing X as a subset and the set of all tids containing any non-empty subset of X as a subset and is given by,

$$\xi(X) = \left| \bigcap_{x \in X} t(x) \right| / \left| \bigcup_{x \in X} t(x) \right| . \tag{1}$$

For the sake of clarity, we choose $\lambda(X) = \left| \bigcap_{x \in X} t(x) \right|$ and $\rho(X) = \left| \bigcup_{x \in X} t(x) \right|$.

Definition 2. (Cohesive itemset) An itemset X is called a cohesive, if $\xi(X)$ is no less than pre-specified threshold.

As the name suggests, expression (1) gives the measure of a set of items to stick together. Clearly, every item of the dataset is cohesive (cohesion is unity). It is one of the major advantages achieved by shifting the domain from support to cohesion. So, cohesion of an itemset X needs both $\lambda(X)$ and $\rho(X)$ to be computed.

---

[1] CARDIAC stands for Confident Association Rule Discovery using Itemset Cohesion; 'A' is gratuitous.

Lemma 1. Cohesive itemsets retain downward closure property [2].

Proof. For all itemsets Y and X such that $Y \supseteq X$, $\lambda(Y) \leq \lambda(X)$ and $\rho(Y) \geq \rho(X)$. So, $\xi(Y) = \lambda(Y) / \rho(Y) \leq \lambda(X) / \rho(Y) \leq \lambda(X) / \rho(X) = \xi(X)$. Hence, all subsets of a cohesive itemset must also be cohesive.

3 Algorithm

The methodology, we develop for mining cohesive itemsets, incorporates the concept of vertical mining in [6]. Let $X = \{x_1, x_2, \ldots, x_n\}$ be an itemset. We follow the notation X' for $\{x_1', x_2', \ldots, x_n'\}$. If $t(X)$ is the set of tids containing X as a subset, $t(X')$ is the set of tids containing X' as a subset i.e. the set of tids containing none of $x \in X$ as a subset. So for an itemset X,

$$\bigcup_{x \in X} t(x) = T \setminus \bigcap_{x' \in X'} t(x') . \tag{2}$$

$$\left| \bigcup_{x \in X} t(x) \right| = n - \left| \bigcap_{x' \in X'} t(x') \right|, \, n = |T| . \tag{3}$$

Proof. Equation (2) follows directly from definition.
As $\bigcap_{x' \in X'} t(x') \subseteq T$, $T \setminus \bigcup_{x' \in X'} t(x') = |T| - \left| \bigcap_{x' \in X'} t(x') \right| = n - \left| \bigcap_{x' \in X'} t(x') \right|$.
Hence is the result.

Lemma 2. For all Y such that $Y \supseteq X$, $\bigcap_{y' \in Y'} t(y') \subseteq \bigcap_{x' \in X'} t(x')$.

Proof. For all Y such that $Y \supseteq X$, $\bigcup_{y \in Y} t(y) \supseteq \bigcup_{x \in X} t(x)$. So, $T \setminus \bigcup_{y \in Y} t(y) \subseteq T \setminus \bigcup_{x \in X} t(x)$. Hence, $\bigcap_{y' \in Y'} t(y') \subseteq \bigcap_{x' \in X'} t(x')$.

So, instead of maintaining the set $\bigcup_{x \in X} t(x)$ of monotonic increasing cardinality, it is wise to maintain $\bigcap_{x' \in X'} t(x')$ which is of monotonic decreasing cardinality.

```
CARDIAC([P],mincohesion)
//[P] is the set of cohesive prefixes
for all X_i ∈[P]do
  S = ∅;
  for all X_j ∈[P]| j>i do
    X = X_i∪X_j;
    t(X)= t(X_i)∩t(X_j);λ(X)= |t(X)|;
    t(X')= t(X_i')∩t(X_j');ρ(X)= n - |t(X')|;
    ξ(X)= λ(X)/ρ(X);
    if ξ(X) ≥ mincohesion
        S = S∪X;
  if S ≠ ∅ then call CARDIAC(S,mincohesion)
```

Fig. 1. CARDIAC algorithm for cohesive itemset generation

Lemma 3. Under the same threshold for cohesion and support, number of cohesive itemsets is greater or equal to that of frequent itemsets.

Proof. For an itemset X, $\xi(X) = \lambda(X) / \rho(X) \geq \lambda(X) / |T| = \sigma(X)$ as $\rho(X)$ is upper bounded by $|T|$. So, under the same threshold for both cohesion and support, all frequent itemsets are cohesive. But there possibly exists cohesive itemsets that are not frequent. For a dataset, count of such itemsets can be large enough as will be shown in experimental section.

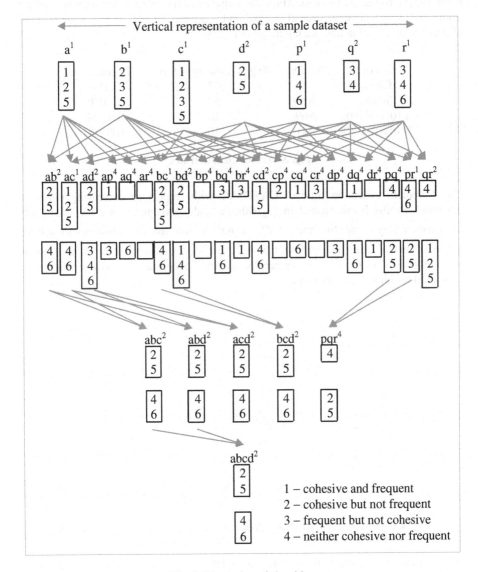

Fig. 2. Illustration of algorithm

Example. Fig. 2 illustrates the algorithm with a vertical sample dataset in vertical representation, threshold for cohesion or support being set to 50%. The superscript for itemsets indicates its status which is defined at right-bottom corner inside Fig. 2.

4 Experiment

All experiments are conducted on a 2GHz AMD Athlon PC with 1GB primary memory. Algorithm CARDIAC (Fig. 1) is implemented in C++ and run on Debian Linux platform. Characteristics of the real and synthetic datasets, used for the purpose of performance evaluation, are tabulated in Fig. 3.

Dataset	# Items	Avg. Transaction Size	# Transactions
Chess	76	37	3,196
Pumsb*	2087	50	49,046
T10I4D100K	869	10	101,290
T40I10D100K	941	40	100,000

Fig. 3. Dataset Characteristics

Fig. 4 depicts execution time of CARDIAC with respect to decreasing cohesion threshold, results being plotted in logarithmic scale. Execution time has drastically been reduced by computing the set $\cap_{x' \in X'} t(x')$ which quickly converges to null set but still equally serves the purpose. Datasets, that are sparse with fewer items, contain less number of tidsets with lesser cardinality and hence require less execution time compared to other dense datasets.

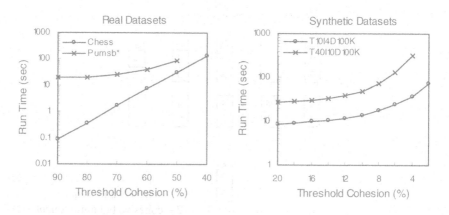

Fig. 4. Performance of CARDIAC algorithm with decreasing threshold cohesion

Fig. 5 presents a comparative study on count of frequent and cohesive itemsets under fixed thresholds. The chess dataset contains only 76 items with average transaction length 37. Hence it exhibits minimum variation from one record to other. So number of cohesive itemsets is slightly more than the frequent itemsets. Remaining datasets, however, contain lots of items and their variations in record to record. Here, number of cohesive itemsets is orders of magnitude higher than corresponding frequent itemsets. These itemsets actually contribute to the low support high confidence rules.

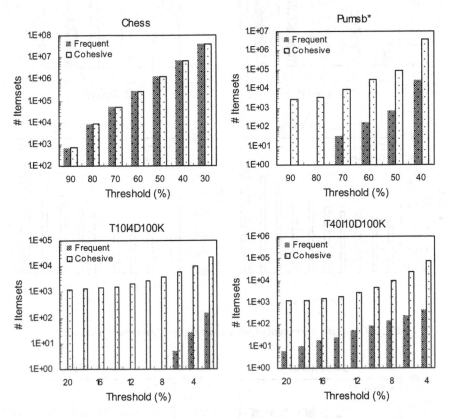

Fig. 5. Comparison on count of frequent and cohesive itemsets

Fig. 6 shows a comparative study on count of association rules under support and cohesion measure, threshold confidence being set to 90% in each. Again, in chess dataset, there is no much difference in the count of rules generated by both the measures. But for other datasets, count of rules is incomparably greater for cohesion as expected. The excess rules obtained are actually the low support high confidence ones which classical ARD misses. If we stick to the support measure and try to mine those rules, threshold support has to be set to such a low value that mining would rather be intractable. In addition, it would incur the cost of pruning huge low confidence rules. A higher choice for cohesion successfully mines the high confidence rules with lesser number of rules to be pruned.

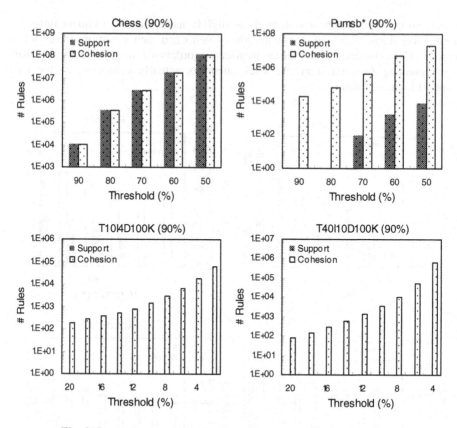

Fig. 6. Comparison on count of rules under support and cohesion measure

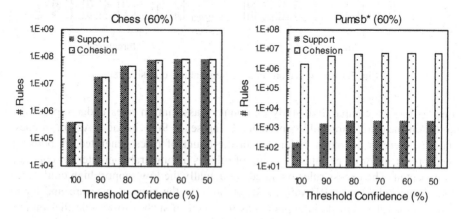

Fig. 7. (*continued*)

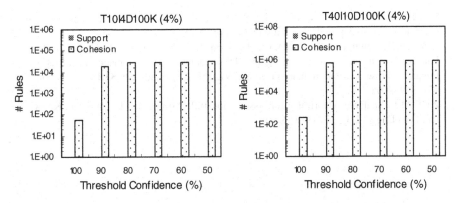

Fig. 7. Comparison on count of rules with decreasing threshold confidence

Fig. 7 shows a comparative study on count of rules versus confidence under fixed support and cohesion measures. In all datasets, except chess, cohesion mines significant number of confident rules which support measure fails to achieve. For the two synthetic datasets, count of rules under respective threshold support is nil down to 50% threshold of confidence. But cohesion successfully mines them.

5 Conclusion

Associations, explored using cohesion, are indicative of natural properties prevalent in the application domain, no matter whatever are their support counts. Mining such rules using support constraint is well-nigh infeasible as it would incur huge cost of dealing with intractable number of itemsets. Additionally, cohesion explores comparatively lesser number of itemsets than support to discover same number of rules, reducing burden on the rule miner process.

Acknowledgment. The authors wish to acknowledge Dr. Raghunath Bhattacharyya for his valuable suggestions on applicability of cohesion in medical domain.

References

1. Aggarwal, R., Imielinski, T., Swami, A.: Mining Association Rules between Sets of Items in Large Databases. In: Proceedings of ACM SIGMOD 1993, Washington DC, pp. 207–216 (1993)
2. Aggarwal, R., Srikant, R.: Fast Algorithm for Mining Association Rules. In: VLDB. Proceedings of 20th Very Large Database Conf., pp. 487–499 (1994)
3. Cohen, E., Datar, M., Fujiwara, S., Gionis, A., Indyk, P., Motwani, R., Ullman, J.D., Yang, C.: Finding Interesting Associations without Support Pruning. In: Proceedings of IEEE ICDE, pp. 489–500 (2000)

4. Jinyan, L., Xiuzhen, Z., Guozhu, D., Kotagiri, R., Qun, S.: Efficient Mining of High Confidence Association Rules without Support Thresholds. In: Żytkow, J.M., Rauch, J. (eds.) PKDD 1999. LNCS (LNAI), vol. 1704, pp. 406–411. Springer, Heidelberg (1999)
5. Pal, S., Bagchi, A.: Association against Dissociation: some pragmatic considerations for Frequent Itemset generation under Fixed and Variable Thresholds. SIGKDD Explorations 7(2), 151–159 (2005)
6. Zaki, M.J.: Scalable Algorithms for Association Rule Mining. IEEE Trans. On Knowledge and Data Engg 12(3), 372–390 (2000)

An Unbalanced Data Classification Model Using Hybrid Sampling Technique for Fraud Detection

T. Maruthi Padmaja[1], Narendra Dhulipalla[1], P. Radha Krishna[1],
Raju S. Bapi[2], and A. Laha[1]

[1] Institute for Development and Research in Banking Technology (IDRBT),
Hyderabad, India
{tmpadmaja,dnarendra,prkrishna,alaha}@idrbt.ac.in
[2] Dept of Computer and Information Sciences,
University of Hyderabad, India – 500046
bapics@uohyd.ernet.in

Abstract. Detecting fraud is a challenging task as fraud coexists with the latest in technology. The problem to detect the fraud is that the dataset is unbalanced where non-fraudulent class heavily dominates the fraudulent class. In this work, we considered the fraud detection problem as unbalanced data classification problem and proposed a model based on hybrid sampling technique, which is a combination of random under-sampling and over-sampling using SMOTE. Here, SMOTE is used to widen the data region corresponding to minority samples and random under-sampling of majority class is used for balancing the class distribution. The value difference metric (VDM) is used as distance measure while doing SMOTE. We conducted the experiments with classifiers namely k-NN, Radial Basis Function networks, C4.5 and Naive Bayes with varied levels of SMOTE on insurance fraud dataset. For evaluating the learned classifiers, we have chosen fraud catching rate, non-fraud catching rate in addition to overall accuracy of the classifier as performance measures. Results indicate that our approach produces high predictions against fraud and non-fraud classes.

Keywords: Fraud detection, SMOTE, VDM, Hybrid Sampling and Data Mining.

1 Introduction

Fraud is defined as 'criminal deception, the use of false representations to gain an unjust advantage'. It is a costly problem for many industries such as banking and insurance and co-exist with the latest in technology. Fraud detection is a continuously evolving discipline and requires a tool that is intelligent enough to adapt to criminals' strategies and ever changing tactics to commit fraud.

We can reduce some of these losses by using data mining techniques and collecting the customer data from the organizations. In general, fraud detection is

A. Ghosh, R.K. De, and S.K. Pal (Eds.): PReMI 2007, LNCS 4815, pp. 341–348, 2007.

a binary classification problem. However, in reality it is an n-class problem as fraud can be done in various ways and each *modus operandi* is different from the other. Compared to non-fraud data distribution, the fraud data is sparsely distributed. So it is extremely difficult to extract the fraud patterns in this scenario. In this work, we consider fraud detection as an unbalanced data classification problem where the majority samples (non-fraud samples) outnumber the minority samples (fraud samples). Usually, the classification algorithms exhibit poor performance while dealing with unbalanced datasets and results are biased towards the majority class. Hence, an appropriate model is needed to classify unbalanced data, especially for the fraud detection problem. For these type of problems, we cannot rely upon the accuracy of the classifier because the cost associated with fraud sample being predicted as a non-fraud sample is very high. The performance measures that can be used here are cost based metrics and ROC analysis.

In this paper, we proposed a model for fraud detection that uses hybrid sampling technique, which is a combination of random under-sampling and Synthetic Minority Over-sampling Technique (SMOTE [2]) to oversample the data. Here we have used Value Difference Metric (VDM [10]) as a distance measure in SMOTE. We identified optimal classifier based on its True Positive (TP) rate, True Negative (TN) rate and consider accuracy as one of the measures.

The remainder of this paper is organized as follows. In Section 2 we present the brief literature review of the problem. Section 3 describes the proposed model and experimental setup. Section 4 provides the results and discussion. In section 5 we conclude the paper.

2   Related Work

In this section, we describe the related work from two perspectives: (i) The techniques available to handle the unbalanced data, and (ii) Available methods for fraud detection.

Some of the techniques available to handle the unbalanced data are: (a) one-class classification, (b) sampling techniques (c) ensembling of classifiers and (d) rule based approaches. Foster [7] presented a review on the issues related to unbalanced data classification. From the experiments on 25 unbalanced data sets at different unbalanced levels (20 of them are from UCI), Weiss and provost [14] concluded that the natural distribution is not usually the best distribution for learning. Kubat and Matwin [9] did selective under-sampling of majority class by keeping minority classes fixed. They categorized the minority samples into some noise overlapping, the positive class decision region, borderline samples, redundant samples and safe samples. By using Tomek links concept, which is a type of data cleaning procedure used for under-sampling, they deleted the borderline majority samples. Kubat [8] proposed SHRINK system which searches for "best positive regions" among the overlapping regions of the majority and minority classes. Domingos [5] compared 'Metacost', a general method for making classifiers cost sensitive, with under-sampling and over-sampling techniques. He identified 'Metacost' outperform both the sampling techniques and he also observed that majority under-sampling is preferable than minority over-sampling technique. Chawla et al [2] proposed Synthetic Minority Over-sampling technique (SMOTE). It is an over-sampling approach in which the minority sample is over-sampled by creating synthetic (or artificial) samples rather than by over-sampling with replacement. The minority class is over-sampled by taking each minority class sample and introducing synthetic samples along the line segments joining any/all of the k minority class' nearest neighbors. Depending upon the amount of over-sampling required, neighbors from the k nearest neighbors are randomly chosen. This approach effectively forces the decision region of the minority class to become more general. Study of "whether over-sampling is more effective than under-sampling" and "which over-sampling or under-sampling rate should be used" was done by Estabrooks et al [6],who concluded that combining different expressions of the resampling approach is an effective solution.

An excellent survey of work on fraud detection has recently been done by Clifton [4]. This work defines the professional fraudster, formalizes the main types and subtypes of known fraud and presents the nature of data evidence collected within affected industries. Stolfo et al [11] outlined a metaclassifier system for detecting credit card fraud by merging the results obtained from the local fraud detection tools at different corporate sites to yield a more accurate global tool. Similar kind of work has been carried out by Stolfo et al [12] and elaborated by Chan et al [1] and Stolfo et al [13]. Their work described a more realistic cost model to accompany the different classification outcomes. Wheeler and Aitken [15] have also explored the combination of multiple classification rules. Clifton et al [3] proposed a fraud detection method, which uses stacking-bagging approach to improve cost savings.

3   Proposed Model

Figure 1 depicts our proposed hybrid sampling model for fraud detection. The model is a combination of random under-sampling and over-sampling. It mainly works based on determining how much percentage of minority samples (original minority + artificial minority samples) and majority samples to add to the training set such that a classifier can achieve best TP and TN rates. Here, TP is the number of fraud samples correctly classified and TN is the number of non-fraud samples correctly classified. The uniqueness of our model lies in generating synthetic minority samples using SMOTE, doing different levels of SMOTE, varying original fraud samples to be added to the training set and validating the classifier models on entire dataset. The majority data is randomly under-sampled and the minority data is over-sampled using SMOTE to emphasize the minority data regions, which uses distance metric called VDM.

Here, our idea is to do guided search for identifying the optimal SMOTE factor in the data space. This optimal SMOTE factor is dataset dependent. First we select some initial SMOTE factor for the data space and apply this SMOTE factor on varying minority original fraud percentages, continue this search by

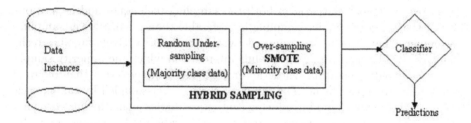

Fig. 1. Hybrid sampling model for fraud detection

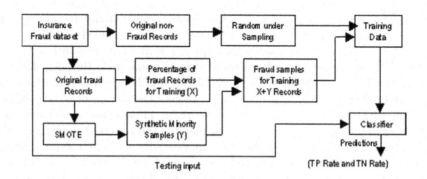

Fig. 2. Generation of samples for classification

systematically increasing SMOTE factor from initial value to the optimal SMOTE factor. At the optimal SMOTE point the underlying classifier exhibits best *TP* and *TN* rates. As the dataset we are experimenting contains nominal attributes, we used VDM distance measure, suitable for nominal attributes in finding the nearest neighbors while doing SMOTE. Using this distance measure two values are considered to be closer if they have more similar classifications (i.e., more similar correlations with the output classes), regardless of what order the values may be given in.

We varied the number of synthetic minority samples (generated using SMOTE) to be given to the classifier by doing different levels of SMOTE and compared the effect of this on various classifiers. This is to know, how classifiers behave with respect to increase in number of minority samples added to the training set, i.e., by widening the boundary across the class region. In the presence of unbalanced datasets with unequal error costs, it is appropriate to use *TP* rate and *TN* rate than accuracy alone.

Figure 2 shows the process of generating training samples for training the classifier. Initially, the fraud samples and non-fraud samples are separated into two different data sets and applied SMOTE on fraud samples for the given level of SMOTE factor. For example, if we specify SMOTE factor as 5 and input fraud samples are x, then artificial fraud samples generated after SMOTE are 5x. Generally the choice of optimal SMOTE factor is data dependent. For the dataset under consideration, the class distribution of non-fraud (or majority samples) and fraud (or minority samples) samples is 94:6. So for experiments we considered SMOTE factors of 5,7,9,11 and 15. The training dataset is an amalgamation of artificially generated fraud samples, original fraud samples and non-fraud samples. We varied the Original Fraud Data Percentage (OFD percentage or rate), i.e., number of original fraud samples to be added to training, ranging from 0 to 75. Similarly non-fraud samples were randomly under sampled from non-fraud data set in such a way that class distribution for training becomes 50:50. In this work, we have chosen classifiers namely k-NN, Radial Basis Function networks, C4.5 and Naive Bayes. Total experiments conducted are 100; 25 for each classifier by doing different levels of SMOTE and varying OFD rate.

4   Experimental Results and Discussion

Experiments are conducted on an insurance dataset used in [3]. The data set pertains to automobile insurance and it contains 15421 samples, out of which 11338 samples are from January-1994 to December-1995 and remaining 4083 samples are form January-1996 to December -1996. There are 30 independent attributes and one dependent attribute (class label). Here, six are numerical attributes and remaining are categorical attributes. The class distribution of non-fraud and fraud is 94:6 which indicates that the data is highly unbalanced. We discarded the attribute *PolicyType*, because it is an amalgamation of existing attributes *VehicleCategory* and *BasePolicy*. Further, we created three attributes namely weeks-past, *is-holidayweek-claim* and *age-price-wsum* to improve the predictive accuracy of classifiers as suggested in [3]. So the total numbers of attributes used are 33. All the numerical attributes are discrete in nature and thus converted into categorical attributes in order to compute the distance between the samples using VDM.

We implemented SMOTE in MATLAB 7.0 and used the WEKA3.4 toolkit for experimenting with four classification algorithms selected. We computed classification accuracy, *TP* rate and *TN* rate by varying OFD rate on each SMOTE factor. We have set k,i.e., the number of nearest neighbors to be selected while doing SMOTE, as 7,9,11,13 and 17 for SMOTE factors 5,7,9,11 and 15 respectively. Tables 1, 2, 3 and 4 show the *TP* rate and *TN* rate for k-NN, RBF neural network, C4.5 and Naive Bayes classifiers respectively. It was observed from the experiments that the accuracy is more than 90% in more than 40 cases out of 100 experiments. Figures 3 to 6 shows how classifiers behave when we change OFD rate and SMOTE factor. The X-axis represents the OFD rate and Y-axis represents the classification accuracy.

The observations from experiments conducted are as follows. K-NN has recorded slight decrease in accuracy with increase in OFD rate for fixed SMOTE factor. Accuracy of the classifier is improved when SMOTE factor is increased, but with increase in the OFD rate, it has decreased slightly. This is because non-fraud catching rate has decreased with increase in OFD rate and non-fraud samples account for more percentage in the test set. It has shown a best fraud

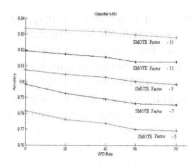

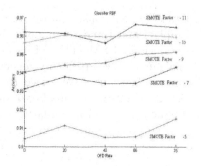

Fig. 3. Plot of OFD rate Vs Accuracy for classifier k-NN (k=10)

Fig. 4. Plot of OFD rate Vs Accuracy for classifier RBF

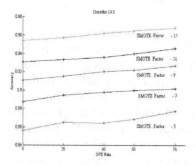

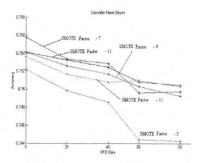

Fig. 5. Plot of OFD rate VS Accuracy for classifier C4.5

Fig. 6. Plot of OFD rate Vs Accuracy for classifier Naive Bayes

catching rate of 99.9%. In all the 25 experiments conducted on this classifier, the fraud catching rate of more than 95% (Table 1). This may be due to inherent use of k-NN in generating artificial minority samples by SMOTE. RBF has got accuracy and non-fraud catching rate of more than 90% in all the 25 cases. Fraud catching rate is also equally good which is between 86.2% and 96.6% (Table 2). With respect to increase in OFD rate, C4.5 has recorded a slight increase in accuracy for all SMOTE Levels. It has shown good improvement in accuracy from 0.856 to 0.967 by increasing SMOTE factor ranging from 5 to 15. In the case of Naive Bayes classifier, accuracy has reduced slightly by increase in OFD rate for each SMOTE factor. This classifier has not shown any improvement in accuracy for varied SMOTE factor and OFD rate (Table 4). For all classifiers at all SMOTE factors, with increase in OFD rate, *TP* rate has increased slightly but there is not much impact on *TN* rate.

We also ranked the classifiers according to accuracy, fraud catching rate and non-fraud catching rate (results are not shown due to paucity of space). Classifiers C4.5 and RBF have performed very well compared to the remaining two classifiers. Though, Naive Bayes classifier works well in general, in our experiments it has shown poor performance on the data set under consideration com-

Table 1. *TP* and *TN* values for k-NN

SMOTE factor		OFD rate				
		0	20	40	55	75
5	*TP*	.94	.95	.95	.95	.96
	TN	.77	.76	.76	.75	.75
7	*TP*	.97	.97	.97	.97	.98
	TN	.78	.78	.77	.77	.77
9	*TP*	.98	.99	.99	.99	.99
	TN	.79	.79	.79	.78	.78
11	*TP*	.99	.99	.99	.99	.99
	TN	.80	.80	.80	.80	.80
15	*TP*	.99	.99	.99	.99	.99
	TN	.82	.82	.82	.81	.81

Table 2. *TP* and *TN* values for RBF

SMOTE factor		OFD rate				
		0	20	40	55	75
5	*TP*	.86	.88	.90	.92	.96
	TN	.90	.91	.90	.90	.91
7	*TP*	.88	.90	.94	.94	.96
	TN	.93	.94	.93	.93	.94
9	*TP*	.88	.90	.92	.93	.96
	TN	.94	.94	.94	.95	.95
11	*TP*	.91	.94	.93	.95	.96
	TN	.96	.96	.95	.96	.96
15	*TP*	.90	.92	.92	.94	.93
	TN	.96	.96	.96	.96	.96

Table 3. *TP* and *TN* values for C4.5

SMOTE factor		OFD rate				
		0	20	40	55	75
5	*TP*	.67	.72	.77	.83	.86
	TN	.86	.87	.87	.87	.87
7	*TP*	.70	.76	.80	.83	.88
	TN	.89	.90	.90	.90	.90
9	*TP*	.71	.77	.81	.83	.87
	TN	.92	.92	.92	.92	.93
11	*TP*	.72	.77	.80	.84	.89
	TN	.94	.94	.94	.94	.94
15	*TP*	.73	.78	.82	.85	.88
	TN	.96	.96	.97	.97	.97

Table 4. *TP* and *TN* values for Naive Bayes

SMOTE factor		OFD rate				
		0	20	40	55	75
5	*TP*	.59	.60	.61	.62	.65
	TN	.76	.75	.75	.75	.75
7	*TP*	.59	.60	.60	.61	.63
	TN	.76	.76	.76	.75	.75
9	*TP*	.59	.60	.60	.61	.62
	TN	.76	.76	.76	.75	.75
11	*TP*	.59	.60	.60	.60	.61
	TN	.76	.76	.76	.76	.75
15	*TP*	.60	.60	.60	.60	.61
	TN	.76	.76	.76	.76	.75

pared to other classifiers. Generally, if we consider only accuracy as performance measure for fraud detection, classifiers applied on original dataset will give more accuracy by predicting all fraud samples as non-fraud samples. But high classification accuracy is achieved using proposed model by increasing fraud catching rate, besides maintaining high non-fraud catching rate.

5   Conclusion

In this paper, we explored the existing methods for fraud detection and proposed a new model based on hybrid sampling approach. Our model uses superior approach SMOTE to oversample the fraud data. We observed that by using SMOTE, the fraud regions emphasized well so that the classifiers can able to learn from the fraud class. Experiments are conducted on insurance data set for four classifier namely k-NN, RBF neural networks, C4.5 and Naive Bayes. Results demonstrated that our model is efficient for fraud detection . For instance, when we applied RBF by taking SMOTE factor as 11 and kept OFD rate as 55, we got non-fraud catching rate of 96.7% (*TN* rate), fraud catching rate of 95.9% (*TP* rate) and the accuracy of 96.6%. Thus, intelligent use of SMOTE for generating artificial fraud samples resulted in improving the accuracy, non-fraud catching rate and fraud catching rate. Though the proposed model implemented for insurance domain, it can be applicable to other domains as well for fraud detection.

References

1. Chan, P., Fan, W., Prodromidis, A., Stolfo, S.: Distributed Data Mining in Credit Card Fraud Detection. IEEE Intelligent Systems 14, 67–74 (1999)
2. Chawla, N.V., Bowyer, K.W., Hall, L.O., Kegelmeyer, W.P.: SMOTE: Synthetic Minority Over-sampling Technique. JAIR 16, 324–357 (2004)
3. Phua, C., Damminda, A., Lee, V.: Minority Report in Fraud Detection: Classification of Skewed Data. Sigkdd Explorations 6(1) (2004)
4. Clifton phua Lee, V., Smith, K., Gayler, R.: A Comprehensive Survey of Data Mining-based Fraud Detection Research. In: Artificial Intelligence review (2005)
5. Domingos, P.: Metacost: A General Method for Making Classifiers Cost-sensitive. In: Proceedings of the Fifth ACM SIGKDD International Conference on Knowledge Discovery and Data Mining, pp. 155–164. ACM Press, New York (1999)
6. Estabrooks, A., Jo, T., Japkowicz, N.: A Multiple Resampling Method for Learning from Imbalances Data Sets. Computational Intelligence 20(1) (2004)
7. Foster, P.: Machine learning from imbalanced data sets. Invited paper for the AAAI 2000 Workshop on Imbalanced Data Sets (2000)
8. Kubat, M., Holte, R., Matwin, S.: Machine Learning for the Detection of Oil Spills in Satellite Radar Images. Machine Learning 30, 195–215 (1998)
9. Kubat, M., Matwin, S.: Addressing the Curse of Imbalanced Training Sets: One Sided Selection. In: Proceedings of the Fourteenth International Conference on Machine Learning, Nashville, Tennesse, pp. 179–186. Morgan Kaufmann, San Francisco (1997)
10. Wilson, R., Tony,: Improved Heterogeneous Distance Functions. JAIR 6, 1–34 (1997)
11. Stolfo, J., Fan, D.W., Lee, W., Prodromidis, A.L.: Credit card fraud detection using meta-learning: Issues and initial results. In: AAAI Workshop on AI Approaches to Fraud Detection and Risk Management, pp. 83–90. AAAI Press, Menlo Park, CA (1997)
12. Stolfo, S., Andreas, L.P., Tselepis, S., Lee, W., Fan, D.W.: JAM: Java agents for meta-learning over distributed databases. In: AAAI Workshop on AI Approaches to Fraud Detection and Risk Management, pp. 91–98. AAAI Press, Menlo Park, CA (1997)
13. Wei Fan, S., Lee, W., Prodromidis, A., Chan, P.: Cost-based modeling for fraud and intrusion detection: Results from the JAM Project. In: Proceedings of the DARPA Information Survivability Conference and Exposition 2, pp. 130–144. IEEE Computer Press, New York (1999)
14. Weiss, G., Provost, F.: The Effect of Class Distribution on Classifier Learning. Technical Report ML-TR-43, Department of Computer Science, Rutgers University (January 2001)
15. Wheeler, R., Aitken, S.: Multiple algorithms for fraud detection. Knowledge-Based Systems 13(2/3), 93–99 (2000)

Automatic Reference Tracking with On-Demand Relevance Filtering Based on User's Interest

G.S. Mahalakshmi, S. Sendhilkumar, and P.Karthik

Department of Computer Science and Engineering,
Anna University, Chennai, Tamilnadu, India
mahalakshmi@cs.annauniv.edu, thamaraikumar@cs.annauniv.edu

Abstract. Automatic tracking of references involves aggregating and synthesizing references through World Wide Web, thereby introducing greater efficiency and granularity to the task of finding publication information. This paper discusses the design and implementation of crawler-based reference tracking system, which has the advantage of online reference filtering. The system automatically analyses the semantic relevance of the reference article by harvesting keywords and their meanings, from title and abstract of the respective article. Indirectly this attempts to improve the performance of the reference database by reducing the articles that are actually being downloaded thereby improving the performance of the system. The number of levels for recursive downloads of reference articles are specified by the user. According to user's interest the system tracks up the references required for the understanding of the seed article, stores them in the databases and projects the information by threshold based view filtering.

Keywords: Clustering, Information filtering, Internet search, Relevance feedback, Retrieval models, Reference tracking, Citations.

1 Introduction

Everyday numerous publications are deposited in a highly distributed fashion across various sites. WWW's current navigation model of browsing from site to site does not facilitate retrieving and integrating data from multiple sites. Often budding researchers find themselves misled amidst conceptually ambiguous references while exploring a particular seed scholarly literature either aiming at a more clear understanding of the seed document or trying to find the crux of the concept behind the journal or conference article. Integration of bibliographical information of various publications available on the web by interconnecting the references of the seed literature is very much essential in order to satisfy the requirements of the research scholar.

An automatic online reference mining system [7] involves tracking down the reference section of the input seed publication thereby pulling down each reference entry and parsing them out to get metadata [2,6]. This metadata is populated into the database to promote further search in a recursive fashion. The representation of stored articles framed year wise, author wise or relevance wise, provides complete information about the references for the journal cited. The proposed system searches the

A. Ghosh, R.K. De, and S.K. Pal (Eds.): PReMI 2007, LNCS 4815, pp. 349–356, 2007.
© Springer-Verlag Berlin Heidelberg 2007

entire web for referring to the scholarly literatures and retrieves the semantically relevant reference articles anywhere from the web even though the authorized copy available online is secured. More often, upon submitting the research findings to journals, the authors generally attach the draft article into their home page and of course, without violation of copyrights of the publisher involved. Our proposed system fetches these articles and lists to the user's purview thereby satisfying the only objective of aiding a researcher during the online survey of literatures. By this, we mean, there is no security breach over the automatic fetching of online articles and further, upon acquiring the key for protected sites, a thorough search and reference retrieval shall be achieved.

Though we take utmost care in filtering the cross-reference articles, the recursion can still grow infinite. Also, the harvested articles may be far from understanding the concept of seed paper. Therefore, filtering the reference articles by semantic relevance would be of much use. But filtering the harvested articles ends up with discarding the harvested articles at the cost of system performance. This paper discusses the crawler-based reference tracking system, which analyses the semantic relevance of the referred scholarly literature on demand i.e. at the time of downloading the reference article. Relevance analysis is done by extracting the semantic information from the abstract and keywords section of the particular article and by comparing it with that of the seed reference article thereby distinguishing closely relevant scholarly literatures. The relevant articles that qualify for downloading are submitted to any commercial search engine and the results are retrieved. The retrieved reference articles are filtered for cross references, and are stored in a database.

2 Related Work

Citation systems like CiteSeer [1] do not retrieve the entire contents of the file referred to and instead, only the hyperlink to the file in their respective database storage is displayed at the output. Therefore, such systems are not reliable since the successful retrieval is highly dependent on the cached scholarly articles in their own databases (the display 'Document not in database' in CiteSeer). The Google Scholar [4] also performs the search for a given query over scholarly articles but the search results are more unrealistic and deviating like CiteSeer when compared to the results of our system.

3 Background

3.1 Automatic Reference Tracker (ART)

The simple reference tracking system has two sections, namely, reference document retrieval and reference document representation. Of this, reference document representation is merely a hierarchical representation of the collected documents based on indexes like, author, title, and year of publication or relevance of key information. Reference document retrieval includes a) Reference Extraction b) Reference Parsing c) Reference filtering based on keyword relevance d) Link Extraction and Data retrieval.

TrackID	Start_rec	End_rec
1	1	44
2	45	100
3	101	150

Fig. 1. Track details

The entire ART framework [6] is designed such that each reference document retrieved is stored as unique paper and can be tracked from the seed document. Assigning a unique paper id for each reference document facilitates the accessing of track details. Each document consists of a back reference id which points to the base document *pid*, which refers the document. This helps to track down the hierarchy of retrieved documents (refer Fig 1).

Reference documents are retrieved by a recursive algorithm, which inputs the seed scholarly literature and returns a hierarchy of reference documents that are relevant to the seed scholarly literature. The main tracker framework inputs the seed document and preprocess the seed document. Preprocessing involves converting the document file format to an appropriate one in order to extract text from the document.

From the preprocessed text of the seed article, the references are extracted; each individual references are parsed. All the references listed out in a publication's reference section may not be in the same reference format. Reference formatting for journals, books and conference proceedings (called as reference sub-formats) vary in their structure and style, and hence, identification of their individual formats and subformats is essential to parse them accordingly. We have limited our parsing to IEEE format. After eliminating duplicating and irrelevant links by comparing the cross references, the reference link extraction is initiated [3].

The references are first segregated into linkable and non-linkable formats. The linkable ones (http references) are extracted directly from the WWW. In non-linkable references, key words are identified and provided to a search engine (Yahoo, for our work) for locating the hyperlink of the reference document. The locations obtained through the search results are analyzed and the particular reference publication is downloaded. These documents retrieved are again pre-processed to sort out their content reference sections and they are again used for tracking up to a certain extent based on their relevance.

Reference Filtering. In reference filtering, initially keywords are harvested from the seed document. Two fundamental approaches of relevance calculation discussed in [6] are: title based analysis and abstract based analysis. Since, the title of every reference article is obtained at the stage of reference parsing, decisions regarding whether to include the reference article for reference extraction or not shall be made before submitting the query to the search engine. Therefore, title based keyword analysis is followed for filtering the references.

The metadata of each reference is compared with the set of keywords to obtain relevance count, a measure of relevance to the seed document. From the relevance count, the user can segregate documents based on a threshold limit. After the articles are retrieved, the relevance of every tracked reference article with respect to the abstract of the seed article is calculated at the time of projecting the results to the user.

Abstract based reference filtering is used as a second filter before the data is actually projected to the user. However, the tracked articles were found to be deviating from the seed scholarly literature on practical reference tracking scenario, which needed a serious analysis along the terms of semantics of the retrieved articles.

Data Representation. The representation can be channelised so that, using color legends and other layout schemes, the user is able to identify the exact publication that aids his/her literature search during research. The representation can be author based, relevance based or year based [6].

3.2 Need for Semantically Enhanced Reference Filtering

Automatic extraction of bibliographic data and reference linking information from the online literature has certain limitations. The primary limitation [6] was the performance bottleneck caused by recursively retrieving online reference documents. However, other publications of authors of seed reference article were maintained assuming that such articles will express the continuity of the author's previous work. Therefore, fixing a threshold in terms of relevance of documents retrieved for a seed document was an obvious solution [6], but poor selection of threshold resulted in reduction of quality of retrieved documents [7]. The reason is that, the number of scholarly articles retrieved will be actually more with a more popular research area. This happened to be the motivation for providing an alternate means of fixing dynamic thresholds by sense-based keyword analysis [5] to prioritize the reference entries in the track database, thereby improving the retrieval performance and clarity. Therefore, application of semantic based relevance-filtering algorithm over the track database of online reference tracker [6] would be of much use.

3.3 Semantically Enhanced Reference Tracking

In [6], seed document's title-based keyword analysis resulted in prioritizing the reference entries in a document so that least relevant reference entries may be discarded, resulting in limited articles getting projected to the user. Since the fixing of threshold in projecting the contents of the track database highly depends on reference filtering based on relevancy, new measures of relevance calculation and threshold fixing were explored. The modification of relevance based reference filtering is two-fold: 1. analysing the keywords collected not only from the title but also from the abstract of the seed document (Modified relevance based reference filtering) 2. analysing the semantic relevancy between keywords of the seed document and the retrieved reference document (Sense based reference filtering). In semantically enhanced reference tracking [7], the parsing of reference format has been extended to html and doc types of scholarly articles. After retrieval, the metadata of relevant references is stored in database (refer Figure 2. for details).

Modified relevance based reference filtering. In plain keyword based relevance filtering [6], elimination of reference articles was based only on the no. of keyword matches against the title of the reference article with that of the seed literature. Here,

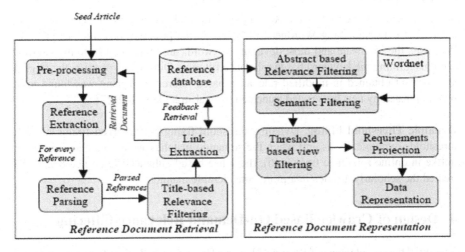

Fig. 2. Architecture of Semantically Enhanced Reference Tracker

that has been extended to keyword based relevance filtering of scholarly articles to involve extraction of words from the title, abstract and highlights in the seed article and the retrieved articles. Based on the frequency of word matches, the threshold to eliminate or include the article into the track database is calculated (refer Figure 3.)

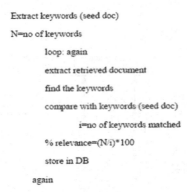

Fig. 3. Algorithm for Modified Relevance based Reference Filtering

The database column "percentage" indicates the percentage of keyword matches across the scholarly literatures. The above solution seemed to be obvious since the keyword region has been extended to abstract and all through the text of the reference article. However, the tracked articles were found to be deviating from the seed scholarly literature on practical reference tracking scenario, which needed a serious analysis along the terms of semantics of the retrieved articles.

Sense based Enhanced Reference Filtering This involved filtering relevant documents based on the semantics of the keywords obtained. Here, the keywords from the seed document are first extracted and then for each keyword, the synonymous words are further extracted using Wordnet [8]. Based on the priorities of senses as given by WordNet, the irrelevant and most deviating keywords are eliminated from the keyword set, thus refining the keyword based relevance filtering. During this process, the original keyword set obtained in modified relevance based reference filtering was found to be both populated by new senses and truncated from abstract keywords [7].

Dynamic Threshold Fixing. Dynamic threshold fixing [7] involved fixing adaptive thresholds dynamically as and when the semantic matches were obtained. Therefore, unlike in online reference tracker [6], the sense-based enhanced reference tracker [7] treated the scholarly articles in a more considerable manner.

4 Design of Crawler Based On-Demand Relevance Filtering

Semantic based relevance filtering [7] was efficient in reducing the articles in the reference projection. However, the performance of the reference database was still poor [6,7] due to populating varied reference articles, which actually have no meaning to lie only in the database storage without actually being projected to the user. Crawler based online relevance filtering involves the title-based, abstract-based and sense-based filtering of references in the pre-fetching phase of the crawler before every article is actually downloaded and stored into the reference database.

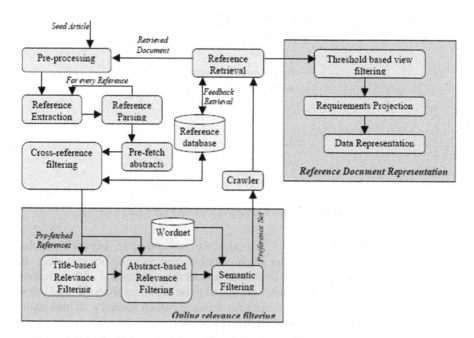

Fig. 4. Automatic Reference Tracking with On-demand Relevance Filtering

The architecture of online relevance filtering based reference tracker is shown in figure 4. The duplication of references are avoided during pre-fetching of articles by cross-reference filtering with respect to the reference database. Only after eliminating all cross-references, the relevance of the pre-fetched article is checked though online relevance filtering. Online relevance filtering basically performs the decision of inclusion / omission of the reference article for downloading which is actually carried out on–demand over the pre-fetched reference abstracts which results in a preference set. The links in the preference set form the seed urls of the crawler using which the reference articles are downloaded.

5   Results of On-Demand Reference Tracking

For implementation purposes, we have assumed the semantic relevance for any reference article as 50% of that of the seed document. The recursion levels are assumed as 2. We have conducted the experiment for 3 different seed research papers seed 1: Probabilistic models for focused web crawling, seed 2: An efficient adaptive focused crawler based on ontology learning, seed 3: Dominos: A new web crawlers' design. The observations are given below in figure 5. There are many reasons for discarding the retrieval of reference articles. The situations encountered in our retrieval process are tabulated in table 1.

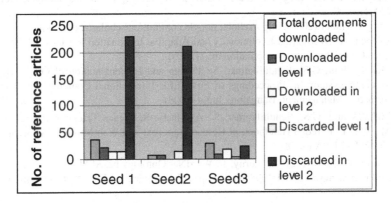

Fig. 5. Result Summary for selected seed research articles

Table 1. On-demand Reference Tracking

Levels	1			2		
Reasons for discarding	Seed 1	Seed 2	Seed 3	Seed 1	Seed 2	Seed 3
Database miss	8	0	1	30	20	5
Query miss	6	10	0	50	40	3
URL not found	0	1	0	0	0	0
Relevance miss (<50%)	1	4	2	150	150	17

6 Conclusion

This paper has explicitly proposed the integrated framework for on-demand retrieving and representation of online documents and listed the issues involved. The most relevant documents are filtered using separate semantic analysis. The system still has a serious performance bottleneck caused by fixing the semantic relevance and number of recursion levels. Since this scenario highly depends on reference filtering based on relevancy, new and adaptive measures of relevance calculation and threshold fixing should be explored. Sense based filtering based on WordNet is found to be more time consuming. Other techniques like ontology based semantic analysis shall be considered to improve the reference filtering of scholarly articles, which are the future directions of this work.

References

[1] Citeseer.: Scientific Literature Digital Library (2006), Retrieved from http://citeseer.ist.psu.edu/
[2] Day, M.Y., Tsai, T.-H., Sung, C.L., Lee, C.W., Wu, S.H., Ong, C.S., Hsu, W.L.: A Knowledge – based Approach to Citation Extraction. In: Proceedings of IEEE IRI-2005 (2005)
[3] Bergmark, D., Lagoze, C.: An Architecture for Automatic Reference linking. In: Constantopoulos, P., Sølvberg, I.T. (eds.) ECDL 2001. LNCS, vol. 2163, Springer, Heidelberg (2001)
[4] Google scholar (2006), http://scholar.google.com/intl/en/scholar/about.html
[5] Kushchu, I.: Web-based Evolutionary and Adaptive Information Retrieval. IEEE Transactions On Evolutionary Computation 9(2) (2005)
[6] Mahalakshmi, G.S., Sendhilkumar, S.: Design and Implementation of Online Reference Tracking System. In: Proceedings of First IEEE International Conference on Digital Information Management, Bangalore, India (2006)
[7] Mahalakshmi, G.S., Sendhilkumar, S.: Automatic Reference Tracking System. In: Song, M., Wu, Y.-F. (eds.) Handbook of Research on Text and Web Mining Technologies, Idea Group Inc, USA (2008)
[8] Snášel, V., Moravec, P., Pokorný, J.: WordNet Ontology based Model for Web Retrieval. In: Proceedings of WIRI 2005 Workshop, Tokyo, Japan, IEEE Press, Los Alamitos (2005)

User's Search Behavior Graph for Aiding Personalized Web Search

Sendhilkumar .S[1] and Geetha .T.V[2]

[1, 2] Department of Computer Science & Engineering
College of Engineering, Anna University
Chennai – 600 025, India
{thamaraikumar, tvgeedir}@cs.annauniv.edu

Abstract. The www is a dynamic environment and it is difficult to capture the user preferences and interests without interfering with the normal activity of the user. Hence to improve user searching in the World Wide Web, we have proposed a personalized search system that supports user searches by learning about user preferences and by observing responses to prior search experiences aided by a new index called the User Conceptual Index (UCI). This paper models every user's search behavior as a User's Search Behavior (USB) Graph. The main focus of this paper is the analysis of the USB graph and the redesign of the UCI using the results arrived from the analysis.

Keywords: Personalization, Web Search, Web Information Retrieval, User Conceptual Index, User's Search Behavior Graph.

1 Introduction

The Web is a large collection of semi-structured and structured information sources and web users often suffer from information overload. General web search is performed predominantly through text queries to search engines. Because of the enormous size of the web, text alone is usually not selective enough to limit the number of query results to a manageable size. To alleviate this problem, personalization becomes a popular remedy to customize the Web environment for users and help them search their information need easily. Personalized search can be of two types: context oriented and individual oriented. By context oriented it means the interrelated conditions that occur within an activity. Individual oriented search means the totality of characteristics that distinguishes an individual. Context includes factors like the nature of information available, the information currently being examined, the applications in use, when, and so on. Individual oriented search encompasses elements like the user's goals, prior and tacit knowledge, past information seeking behaviors, among others.

The context of a search can be derived from the terms used in a search query. Likewise the individual search behaviors like, the links which the user clicks, the way in which the user moves from one page to the other and the actions like saving, printing or copy full or part of the page's content that were performed on the page viewed can be used to confirm the context of search derived form the search query. This insight has led S. Sendhilkumar and T. V. Geetha to design a search-aiding index called

A. Ghosh, R.K. De, and S.K. Pal (Eds.): PReMI 2007, LNCS 4815, pp. 357–364, 2007.

the UCI [2] that makes a search to be both context as well as individual oriented. The developed UCI utilizes the relationships between the search query and the pages visited by the user and thus it provides a ranking of the result pages based on the individual's context of search.

The main focus of this paper is towards the redesigning of the UCI based on the individual search behavior. A new graph called the User's Search Behavior (USB) graph has been discussed in this paper to model the user's browsing behavior. Also this paper explores the various analyses that can be performed on the USB graph and how these results can be of use for redesigning the UCI. Thus the redesigned UCI which provides both context as well as individual oriented search can improve the performance of search in the web by recommending pages that might be the direct answers to the user's information need.

2   The User Conceptual Index (UCI) and Its Significance

The proposed User conceptual index or User co-ordination index (UCI) is mathematically defined as in definition 1.

Definition 1. Given a set of search queries SQ and a set of relevant pages represented by their index words IW, User Conceptual Index can be defined as conceptual relation between the SQ and the relevant pages, represented by a weighted function as given in (1).

$$\text{UCI} = f(W_{SKW \rightarrow IW}) \tag{1}$$

where, SKW is the search query, IW is the index word(s) for a page and W is the sum of the weight of any SQ and their relevant IWs.

In equation 1 W is the sum of the weight of any SQ and their relevant $IW(s)$. This conceptual relation is an index for expanding the search query and for further page ranking modification. The proposed personalized search system identifies a matching (SQ, IW) pair that has the highest UCI value and these pages will be recommended first to the users. These recommended combinations give the pages, which are direct answers to the user's information need.

While computing the average weight for SQ factors like the frequency of terms used in the query, frequency of the query and the query usage time were considered. Similarly for computing the average weight for IWs points like the page hits and the page view time were used.

Sample searches were conducted through Yahoo! Search engine to demonstrate the effectiveness of the proposed UCI based recommendation scheme [3].

The traditional performance evaluation measures for any IR based like precision and recall were used to evaluate [4] the search based on UCI. Recall measures how well a search system finds what the user wants and precision measures how well it weeds out the user don't wants and from the experimental results it was observed that there is a significant improvement of 9% in the overall efficiency of the UCI based search than the normal search [4].

Though it was a tedious process to compute the UCI values for all the result pages, the results proved the search to be more context oriented because the computed UCI contains information contributed directly by the user (i.e., the user queries) and also inferred by the system (like the hit count and page view times).

But still the UCI based search can be made more realistic and more relevant to the user's context of search by analyzing the individual's search behavior. From the user's browsing behavior useful patterns like user's interested pages, links that give the relationship between pages, pages that are indirect answers to the query, etc. can be derived. The following sections discuss about the link analyses technique and how it can be utilized to improve the UCI based search.

3 Link Analyses and Its Usage in Tracing User Behavior

Link analysis is a method that is used to determine what pages in the collection are important to users. Analysis of how pages link to each other in the web has led to significant improvements in web information retrieval [5]. Link analysis technology has been widely used to analyze the pages' importance, such as Hits [6] and Page Rank [7]. The goal of web search is to find all documents relevant for a user query in a collection of distributed web documents. Most search engines and related tools continue to ignore an essential part of the web – the links and continue making improvements in the information retrieval algorithms.

With the aim of utilizing the benefits of link analysis, this work represents the users click data (the links in a page which the user clicks while searching) as a graph. The user navigation through the web while searching for his/her information need can be represented as a graph, which can be used to analyze the user behavior, and hence the interests of the user from link analysis. The following sections describe how every user transaction and sessions as modeled mathematically for further analysis.

3.1 Modeling User Search Behaviors

From the client side user data collected two sets namely: the transaction set and the sessions set can be defined as in definition 2 and 3.

Definition 2. Given a search query SQ_i and the initial set of pages (technically represented by their IWs) retrieved by the search engine, a transaction T is defined as a set of pages visited by the user by clicking the links of the pages in the initial set of search results retrieved and is represented as $T = \{P_1, P_2, \cdots, P_n\}$, where $P_i \in T$ is represented as a set of n page views, $P = \{p_1, p_2, \cdots, p_n\}$.

Page views are semantically meaningful entities which can be used for extracting useful information like users' interested web pages, relation between various pages, user interests, etc. Frequently occurring words in a web page are used as the IW.

Definition 3. Given a set of search queries Q, where $SQ_i \in Q$, a session is defined as a set of transactions, which can be represented as $S = \{T_1, T_2, \cdots, T_m\}$, where every T_i represents one unique SQ_i.

A session S_i with a set of search queries Q will directly represent the user's information need in that session. Every user search session can be modeled as a graph called the Search Flow Graph and the next session discusses about the modeling of the SFG.

3.2   The User's Search Behavior (USB) Graph and Link analysis

A USB graph is a graph that represents a user's search transaction T showing how the user moves from one page to the other while searching for their information need, hence called as a flow graph. Mathematically a USB graph can be defined as given in definition 4.

Definition 4. A User's Search Behavior (USB) graph is defined as directed cyclic graph representing each user transaction $T = \{P_1, P_2, \cdots, P_n\}$, where $P_i \in T$ is represented as a set of n pageviews, $P_i = \{p_{i1}, p_{i2}, \cdots, p_{in}\}$.

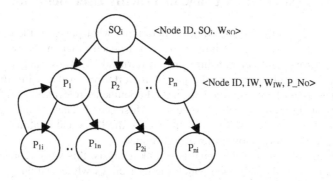

Fig. 1. User's Search Behavior (USB) graph

In figure 1 the root node represents the search query and the descendents of the root represents the pages viewed by the user. The backward arrow shows the user's movement from the current page to the previously viewed page. The root node in the SFG is represented using a 3-tuple <Node ID, SQi, W_{SQ}>, where SQ_i is the Search Query, W_{SQ} is the average SQ_i weight , Node ID comprises of Depth, Offset, where Depth represents the level and Offset is the node's position in the ith level and it is {0,0} for the root node. All other nodes are represented using a 4-tuple <Node ID, IW, W_{IW}, P_No>, where IW is the index word(s) for a page, W_{IW} is the average weight for the index words, P_No is the {Depth, Offset} of parent node.

Table 2 presents the graph constructed in a tabular format for the sample query "IICAI 07" represented by the Node ID 00. The weight in the first row of table 2 represents the W_{SQ} and all other weights indicate the W_{IW}. The weights are computed by taking an average of the time and frequency of usage of a SQ/IW through the various search sessions.

It can be seen from table 2 that the pages with node ID 20 and 30 represent the same page, but first the page with node id 20 is viewed by clicking the link from the page with node ID 10. So the parent page ID for the page 20 is 10. And once again the same page was viewed by clicking a link present in another page (represented by the node ID 21) and so the parent page ID for page 30 is 21. Also the table contains the ID of those pages that were viewed by clicking the back button. For example, the page with node ID 30 was visited by the user well before by

Table 2. Sample USB graph constructed for the query IICAI 07 in tabular format

Node ID <Depth, Off-set>	Visited Page's URL	Index Words (IW)	Weights (W_{SQ} /W_{IW})	Parent Page ID	Back Track ID
00	www.google.com		0.498705	-	
10	http://www.iiconference.org/	{conference, AI}	0.780034	00	
11	http://www.iiconference.org/iicai07 /	{index}	0.008243	00	
12	http://www.cl.cam.ac.uk/~jac22/cfp /msg00248.html	{iicai-07}	0.012534	00	
20	http://www.iiconference.org/session s.html	{AI}	0.267801	10	10
21	http://www.iiconference.org/dates.h tml	{2007}	0.253788	10	10
22	http://www.iiconference.org/submis sion.html	{paper}	0.076756	10	10
30	http://www.iiconference.org/session s.html	{AI}	0.006329	21	21
31	http://www.iiconference.org/iicai07 /swir.html	{semantic, web}	0.849056	20	20

clicking the link available in page 10. But the user on seeing the same page by clicking a link from page 21 goes back to the previous page from where he/she selected the link. This is indicated by the back track ID in table 2. Such back track-ing will help us to identify the pages that provide the user with useful and non-redundant links. Using this detail limited set of useful pages without redundant links can be recommended to the user.

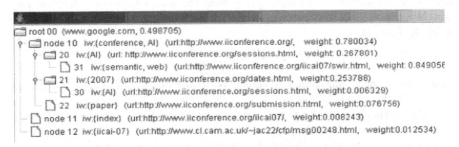

Fig. 2. USB Graph constructed for the data in table 2

The graph constructed for the data given in table 2 is shown in figure 2. In figure 2 the node ID, URL, index words and the index word weight, represents each node. Every node in the graph is an expandable node and by clicking the expandable button we can view the links visited from a specific page.

Thus the main advantage of the USB graph is that it includes only those pages viewed by the user and maintains the order in which the pages were viewed. Hence only a reduced set of pages can be used for the computation of the UCI. Using the page weight and view time it can be concluded whether the transaction is useful for the user's search or not.

4 USB Graph Analyses and Modification of the UCI

The USB graph thus constructed includes both the spatial features as well as the temporal feature. The spatial temporal features like the hits and the page view time respectively are considered for index weight calculation. Hence Link analysis can give significant contributions to web page retrieval from search engines, to web community discovery, and to the measurement of web page influence. It can help to rank results and find high-quality *index/hub/link* pages that contain links to the best sites on the user's topic of interest.

Also from the sub-trees of the USB graph, page view levels (the depth to which a user might move to find their information needs) can be identified and such information depth levels can be used as an indication for the user to stop searching further. Also from the experiments conducted and from the USB graph it was found that the maximum level of depth that was traversed by the user was 12 and it was for the query "News NDTV". For other queries like "NLP", "java", "data mining" and "Artificial Intelligence Conferences" the maximum depth was 9. For the queries "IICAI 07", "deadlock" and "Anna University exam results" the depth of search was 3(from table 2) and 2 respectively.

4.1 Redesign of the UCI

Each page is assigned a weight based on its relevance to the search word, where the relevancy is to what extent the search keyword *SKW* matches with the index/feature words *IW* of the pages returned as a result of the user search. The page weight w is a defined as a function $f(W_{SKW \rightarrow IW})$ called the search keyword – relevant page weight function [10]. The average weight for each transaction can be calculated by using the formula:

$$W_{ti} = \Sigma W_{Pi} + (-\Sigma W_{Pj}) \qquad (2)$$

where, W_{ti} is the ith transaction's weight, W_{Pi} is the weight of ith page P_i. The negative sign in the second half of the above formula signifies back tracking. Every time the user backtracks from the jth page to the ith page we increase the weight of the ith page and reduce weight of the jth page.

In any user transaction if the user has already visited a page, say *P1* and during the process of searching the user finds that the same page *P1* getting referred or linked in another page, then we consider that the page *P1* as an important/relevant page for the given search keyword *SKW* and hence the weight for the page *P*i can be increased. Now for each search keyword we have n number of transactions and the weights of respective transactions W_{ti} can be used in identifying the pages of users interest. In the previous work of Sendhilkumar .S and Geetha .T.V [10] search keyword-relevant page weight function $f(W_{SKWi \rightarrow Pi})$ and the page relation weight function $f(W_{Pi \rightarrow Pj})$ from the data collected was calculated. The page relation weight is assigned based on the relevancy between any two pages, i.e., to what extent the index/feature words IW_i

of the ith page P_i matches with that of the index/feature words IW_j of the jth page P_j. These two weighted functions can be used to represent user's current task.

From the transaction weights W_{ti} user-interested transactions can be identified and with the aid of such interest-oriented transactions filtering of unwanted transactions can be done and thus user's information needs can be reached more easily. These user-interested transactions will contain pages relevant to the given user's search keyword SKW. Therefore from the link analysis can be identified the most relevant pages to a particular SKW. The percentage of relevancy between the pages and the search keyword can be confirmed from the user's current search behavior, which is represented as a graph in our work. Hence we increase the weight of a search keyword and its relevant page in the search keyword-relevant page weight function $f(W_{SKWi \to Pi})$ by adding the new search keyword-relevant page weight derived from the transaction link analysis to the previously calculated search keyword-relevant page weight. Similarly the related pages can be identified from the user-interested transactions and hence i the weight of related pages can be increased in the page relation weight function $f(W_{Pi \to Pj})$. Thus the two weighted functions: the search keyword-relevant page weight function and the page relation weight function get modified from the results of transaction link analysis.

Thus the final weight for a search keyword $w(SKW_i)$ for which a user is searching in a session is given in (3).

$$w(SKW_i) = \Sigma(W_{SKWi \to Pi}) . \Sigma (W_{Pi \to Pj}) + W_{Ti} \qquad (3)$$

Thus a new modified weight is calculated for the search keywords by taking into consideration the user's current search behavior.

Finally the new search keyword weight $\mathbf{w(SKW_i)}$ can be added to the previously calculated UCI as in (4).

$$UCI = f(W_{SKWi \to Pi}) + \mathbf{w(SKW_i)} \qquad (4)$$

This newly modified UCI weight that takes care of both current and previous interests of users can be used in the recommendation process such that if the weight is below a threshold then those combinations can be rejected and others can be recommended. These recommended combinations give the pages, which are direct answers to the user's information need. Filtering systems determine which documents in the result sets are relevant and which are not. Good filters remove many non-relevant documents and preserve the relevant ones in the results set. Now these filtered pages can be re-ranked using any ranking algorithm and presented to the user.

5 Conclusions

In summary this paper has incorporated a relation between the web content and the search key words by means of a new index called the modified User Conceptual Index. This paper models every user search behavior as a User's Search Behavior (USB) graph. The modified User Conceptual Index takes into account about the user's interested transaction and sessions that contain relevant pages, which match their information need. This refined User Conceptual Index can be used in any recommendation process at any stage of the search. We are still in the process of refining the UCI so that it also takes into account the user's shift in interests and the rate of decay of user's interest on some concepts. Further clustering techniques have been planned for clustering the user groups from the data collected by the proposed agent program and to include dimensionality reduction techniques to make use of a reduced set of data for future analysis.

References

[1] Page, L., Brin, S., Motwani, R., Winograd, T.: The PageRank citation ranking: Bringing order to the Web. Technical report, Stanford University Database Group (1998), http://citeseer.nj.nec.com/368196.html
[2] Sendhilkumar, S., Geetha, T.V.: Web Search using personalized user conceptual index. In: Proceedings of 2nd Indian Interantional Conference in Artificial Intelligence, pp. 1719–1728 (December 2005)
[3] Sendhilkumar, S., Geetha, T.V.: Personalized Web Search Using Enhanced Probabilistic User Conceptual Index. Special Issue of the Journal of Intelligent systems (in Press)
[4] Sendhilkumar, S., Geetha, T.V.: An Evaluation of Personalized Web Search for an Individual User. In: the International conference in Artificial Intelligence and Pattern Recognization (AIPR 2007), Orlando, USA, pp. 484–490 (July 2007)
[5] Henzinger, M.: Link Analysis in Web Information Retrieval. IEEE Data Engineering 23(3), 3–8 (2000)
[6] Kleinberg, J.M.: Authoritative sources in a hyperlinked environment. Journal of the ACM 46(5), 604–632 (1999)
[7] Brin S., Page, L.: The anatomy of a large-scale hypertextual web search engine. In: Proc. of WWW 2007 Brisbane, Australia, pp. 107–117 (April 1998)

Semantic Integration of Information Through Relation Mining - Application to Bio-medical Text Processing

Lipika Dey,[1] Muhammad Abulaish,[2] Rohit Goyel[3], and Jahiruddin[4]

[1] Innovation Labs, Tata Consultancy Services, New Delhi, India
lipika.dey@tcs.com
[2] Department of Mathematics, Jamia Millia Islamia, New Delhi, India
abulaish@ieee.org
[3] Department of Mathematics, Indian Institute of Technology, New Delhi, India
[4] Department of Computer Science, Jamia Millia Islamia, New Delhi, India

Abstract. Semantic frameworks can be used to improve the accuracy and expressiveness of natural language processing for the purpose of extracting meaning from text documents. Such a framework represents knowledge using semantic networks and can be generated using information mined from text documents. The key issue however is to identify relevant concepts and their inter-relationships. In this paper, we have presented a scheme for semantic integration of information extracted from text documents. The extraction principle is based on linguistic and semantic analysis of text. Entities and relations are extracted using Natural Language Processing techniques. A method for collating information extracted from multiple sources to generate the semantic net is also presented. The efficacy of the proposed semantic framework is established through experiments carried out for visualizing information embedded in biomedical texts extracted from PubMed database.

Keywords: Relation extraction, Semantic net, Knowledge visualization, NLP.

1 Introduction

While Search Engines provide an efficient way of accessing relevant information, the sheer volume of the information repository on the Web makes assimilation of this information a potential bottleneck in the way its consumption. One approach to overcome this difficulty could be to use intelligent techniques to collate the information extracted from various sources into a semantically related structure which the user can aid the visualization of the content at multiple levels of complexity. Such a visualiser provides a semantically integrated view of the underlying text repository in the form of a consolidated view of the concepts that are present in the collection, and their inter-relationships as derived from the collection along with their sources. The semantic net thus built can be presented to the user at arbitrary levels of depth as desired. It may be noted that

A. Ghosh, R.K. De, and S.K. Pal (Eds.): PReMI 2007, LNCS 4815, pp. 365–372, 2007.

the proposed semantic net is different from a domain ontology [4] which is a more rigid structure aimed at presenting a shared conceptualization of a domain, usually representing knowledge that is ratified by domain experts. The semantic net based visualization system proposed in this paper is more of an aid towards assimilating knowledge from a large collection. However, the semantic net thus built can be used to build ontology, as proposed in [1], when used with restraint over a focused and authentic text corpus, armed with appropriate feasibility analysis mechanisms.

In this paper, we have presented a scheme for semantic integration of information extracted from text documents. The information components extracted from text are either concepts or inter-concept relations. The extraction principle is based on semantic analysis of text from which entities and relations are extracted using Natural Language Processing (NLP) techniques. Though there has been a lot of work on entity extraction from text documents, relation mining has received far less attention. We have shown that relation mining can yield significant information components from text whose information content is much more than entities. We have also proposed a method for collating information extracted from multiple sources and present them in an integrated fashion. The system functions are designed to work in a domain-independent way though here we predominantly present results from the biological domain. This has been chosen since the growth of articles in this area over the last decade has necessitated development of dedicated search engines for locating relevant documents to enable scientists and researchers assimilate information about ongoing research. The proposed system has been shown to be capable of collating and presenting information from multiple scientific abstracts to present a global view of the collection. It is possible to slice and dice or aggregate to get more detailed or more consolidated view as desired.

The remaining paper is structured as follows. Section 2 elaborates on the text mining approach to extract relations and their arguments from text documents. Section 3 presents the semantic net generation algorithm. Section 4 presents the experimental results. Section 5 is a review of the related works and finally section 6 concludes the paper with future directions.

2   Relation Extraction Through Text Mining

Over the last decade, a lot of work has been done on locating and recognizing biological entities in documents. The proposed approach to building a semantic net of information explores *the roles of biological entities* in a collection of document and integrates the information components thus extracted into a cohesive structure. *Roles* of entities are characterized by relations expressed in a sentence in which these entities occur. These relations can be identified through semantic and linguistic analysis [2]. The relation mining framework presented in this paper uses NLP tools to identify entities and relations in a document. It is domain-agnostic in nature. Entities within a document can be identified as Noun Phrases. For bio-medical documents one can additionally use biological entity extractors. Relation extraction from text is a two step process which is explained in the following sub-sections.

2.1   Document Pre-processing and Parsing

The purpose of this step is to expedite the parsing process and facilitate the extraction of information components from text documents. While working with biological abstracts extracted through PubMed, we have eliminated author's names, their affiliations etc. The text documents are parsed whereby each word is assigned a Parts-Of-Speech (POS) tag and each sentences is converted into a dependency tree. We have used a statistical parser (Stanford Parser[1]) that has been developed by the standard natural language processing group, Stanford University. Two sample sentences, their tagged forms and corresponding dependency trees created by the Stanford parser are shown in Table 1. The dependency tree extracts linguistic relationships like *subject, object, possession, conjunction* etc. among words in a sentence.

2.2   Relation Extraction

The proposed approach to relation extraction traverses the dependency tree and analyzes the linguistic dependencies in order to trace biologically significant relations. A biological relation is usually manifested in a document as a relational verb. All relational verbs however do not represent biological relations and only those which are located in the proximity of biological entities are considered. In order to identify valid biological relations we apply a pattern-mining based technique. A biological relation along with the associated biological entities is termed as relation triplets (RT). A biological relation is characterized by verb and may occur in a sentence in its root form or as a variant of it. Different classes of variants of a relational verb are recognized by our system. *Morphological variants* of a root verb consist of self-modifications. For example, the root verb *"activate"*, has three inflectional verb forms: *"activates"*, *"activated"* and *"activating"*. In the context of biological relations, we also observe that the occurrence of a verb in conjunction with a preposition very often changes the nature of the verb. For example, the relation *"activates in"* denotes a significant class of biological reactions. Thus, we also consider a second category of biological relations, which are combinations of *root verbs* or their *morphological variants*, and *prepositions* that follow these. Typical examples of biological relations identified in this category include *"activated in"*, *"binds to"*, *"stimulated with"* etc.

RT extraction process is implemented as a rule-based system. Dependencies output by the Parser are analyzed to identify *subject, object, verb, preposition,* and various other relationships among elements in a sentence. Some sample rules are presented below to highlight the functioning of the system.

Rule 1: If there exist two dependencies involving two different entities E_i and E_j associated with single verb V satisfying the condition $[Subj(V, E_i) \land Obj(V, E_j)]$,

---

[1]   http://nlp.stanford.edu/downloads/lex-parser.shtml

Table 1. Sample sentences, their tagged form and corresponding dependency tree generated by the Stanford Parser

---

Sentence No. 1. [PMID: 17446028]
Alzheimer's disease (AD) is the commonest form of degenerative dementia and is characterised by progressive cognitive decline.

Taged Sentence:
Alzheimer/NNP 's/POS disease/NN -LRB-/-LRB- AD/NNP -RRB-/-RRB- is/VBZ the/DT common-est/JJ form/NN of/IN degenerative/JJ dementia/NN and/CC is/VBZ characterised/VBN by/IN progres-sive/JJ cognitive/JJ decline/NN ./.

Dependency Tree:
poss(disease-3, Alzheimer-1), nsubj(is-7, disease-3), dep(disease-3, AD-5), det(form-10, the-8), amod(form-10, commonest-9), dobj(is-7, form-10), amod(dementia-13, degenerative-12), of(form-10, dementia-13), dep(characterised-16, is-15), and(is-7, characterised-16), amod(decline-20, progressive-18), amod(decline-20, cognitive-19), by(characterised-16, decline-20)

Sentence No. 2. [PMID: 17445916]
Alzheimer's disease is characterised by both cognitive deterioration and the development of a wide range of neuropsychiatric disturbances...

Taged Sentence:
Alzheimer/NNP 's/POS disease/NN is/VBZ characterised/VBN by/IN both/DT cognitive/JJ deteriora-tion/NN and/CC the/DT development/NN of/IN a/DT wide/JJ range/NN of/IN neuropsychiatric/JJ disturbances/NNS ...

Dependency Tree:
poss(disease-3, Alzheimer-1)
nsubjpass(characterised-5, disease-3), aux(characterised-5, is-4), det(deterioration-9, both-7), amod(deterioration-9, cognitive-8), by(characterised-5, deterioration-9), det(development-12, the-11), and(deterioration-9, development-12), det(range-16, a-14), amod(range-16, wide-15), of(development-12, range-16), ...

---

then V is identified as a relational verb between the two entities E_i and E_j. It is characterized as an instance of binary relation represented by $E_i \rightarrow V \leftarrow E_j$. During RT extraction, E_i is treated as head noun and along with other related words in its proximity forms the subject of the sentence. Similarly, the head noun E_j along with related words in its proximity forms the object of the sentence. By applying this rule, the two relation triplets identified from the first sentence in table 1 is ⟨*Alzheimer's disease (AD) → is ← the commonest form of degenerative dementia*⟩ and ⟨*Alzheimer's disease (AD) → characterised by ← progressive cognitive decline*⟩.

Rule 2: If there exist two dependencies involving two different entities E_i and E_j associated with single verb V satisfying the condition $[Subj(V, E_i) \wedge P(V, E_j)]$, where P is a prepositional word, then the verb V along with the prepositional word P is identified as a relational verb between the entities E_i and

E_j. It can be characterized as an instance of a binary relation represented by $E_i \rightarrow V - P \leftarrow E_j$.

Rule 2 presents another kind of rule in which the object component is not explicitly marked by the parser in the dependency tree. Rule 2 helps in identifying the relation triplet $\langle$ *Alzheimer's disease* $\rightarrow$ *characterised by* $\leftarrow$ *both cognitive deterioration and the development of a wide range of neuropsychiatric disturbances* $\rangle$ from sentence 2 shown in table 1.

Table 2 shows a partial list of relation triplets $\langle$*subject, relation, object*$\rangle$, extracted by using rules 1 and 2 from a collection of abstracts which were returned by PubMed for the query term *"Alzhemer's Disease"*. Relations thus extracted are used to generate a semantic net as explained in the next section.

Table 2. A partial list of relation triplets extracted by using rules 1 and 2 from a collection of text documents describing Alzheimer's disease

Subject	Relation	Object
BACE1	is	the *protease* responsible for the production of *amyloid-beta peptides* that accumulate in the brain of *Alzheimer 's disease (AD) patients*
Alzheimer's disease	is	the commonest form of *degenerative dementia*
Alphav integrins	be	important mediators of *synaptic dysfunction* prior to *neurodegeneration* in *Alzheimer 's disease*
Interaction between the *ADAM12* and *SH3MD1* genes	confer	*Late-onset Alzheimer 's disease*
Alzheimer 's disease (AD)	characterised by	Both *cognitive deterioration* and the development a wide range of *neuropsychiatric disturbances*
Alzheimer's disease	characterised by	progressive *cognitive decline*

3   Semantic Net Generation

The major idea of generating a semantic net is to highlight the role of a concept in a text corpus by eliciting its relationship to other concepts. The nodes in a semantic net represent entities/concepts and links indicate relationships. While concept ontologies are specialized types of semantic net, which also highlight the taxonomical and partonimical relations among concepts, the proposed semantic net is designed only to represent the biological relations mined from the text corpus. For an extracted relation triplet $\langle E_i, R_a, E_j \rangle$, the entities E_i and E_j are used to define classes and R_a is used to define relationships. Biological entities can be *complex* in nature which includes relations among *atomic* entities. For example, in the current scenario a Noun Phrase *"Interaction between the ADAM12 and SH3MD1 genes"* represents a complex entity, which contains atomic entities like *ADAM12* and *SH3MD1*. Linguistic analysis rules based on *preposition* and *conjunctive* analysis are recursively applied over complex entities to identify atomic entities and their relationships as well. A relation of the type ¡Ec1, Rc, Ec2¿ where Ec1 (or Ec2) is a complex entity ¡E1, Ri, E2¿ is represented as an n-ary relation whose arguments include the atomic entities and relations extracted from the complex entities. A formal algorithm for semantic net generation is given below:

Algorithm: Semantic Net Generation
Input: *The set of relation triplets (R) mined from text documents*
Output: *Semantic net - a directed graph (G)*
Steps:
1. Initialize G with ϕ
2. For every relation triplet $\langle E_i, R_a, E_j \rangle \in R$ do
3. If $E_i, E_j \notin G$ then
   a. Create separate nodes for both left and right entities
   b. Draw a directed edge from E_i to E_j and label with R_a
4. If E_i (or E_j) is a complex entity then
   a. Break E_i (or E_j) into a set $S = \{E_{i1}, E_{i2}, ..., E_{in}\}$ of atomic entities on the basis of connectors and prepositions
   b. Create a connected sub-graph G_s as follows:

   – Create a separate node for each atomic entity $E_{ik} \in S$ if $E_{ik} \notin G$, $1 < k < n$
   – Draw a directed edge from left entity to right entity - with edge marked by the connector between the entities

   c. Replace node E_i (or E_j) of G with sub-graph G_s
5. Stop

4   Results

We elucidate the proposed approach through results generated from querying the PubMed database for the query term *"Alzheimer's disease"*. A total of 100 documents consisting 1047 sentences were filtered out of 38135 sentences as likely to conatin valid relations triplets. Using *typedDependenciesCollapsed* option, the documents were parsed by Stanford Parser and a total of 751 triplets were extracted from them by applying the RT extraction rules mentioned in section 2.2. A partial list of such triplets is shown in table 2. To initiate the generation of semantic net, we first identified those triplets that contained the query string within the left or right entities as the candidate concept nodes. After applying the linguistic rules over these entity sets the list of concepts is extended to contain the atomic entities also. The semantic net generation algorithm is applied over these elements. Due to limitation of space, a partial view of the generated semantic net is shown in figure 1.

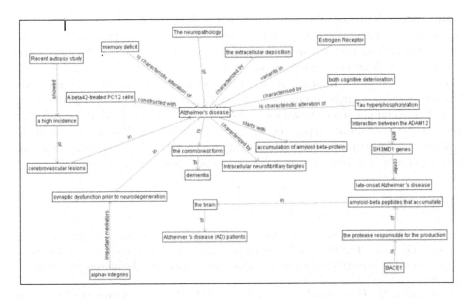

Fig. 1. Semantic Net generated around the biological concept *"Alzheimer's disease"*

5   Related Work

Visualization is a key element for effective consumption of information. Semantic
Nets provide a consolidated view of domain concepts and can aid in the process.
Wagner *et al.* [6] have suggested building a semantic net using the Wiki technology for making e-governance easier through easy visualization of information. In
[5], similar approaches have also been proposed for integrating and annotating
multimedia information. In [3], a soft-computing based technique is proposed
to integrate information mined from biological text documents with the help
of biological databases. [1] proposes building a semantic net for visualization of
relevant information with respect to use cases like the *nutrigenomics* use case,
wherein the relevant entities around which the semantic net is built are predefined.

The proposed method differs from all these approaches predominantly in its
use of pure linguistic techniques rather than using any pre-existing collection
of entities. Though biological relation mining [2] have gained attention of researchers for unraveling the mysteries of biological reactions, their use in biological information visualization is still limited.

6   Conclusion and Future Work

In this paper, we have presented a scheme for extracting relevant information
from text documents and their semantic integration. The extraction principle is
based on semantic analysis of text from which entities and relations are extracted

using Natural Language Processing techniques. We have also proposed a method for collating information extracted from multiple sources and present them in an integrated fashion with the help of semantic net. The system is being integrated to work as a front-end visualizer for a search engine which can enable quick comprehension of information. As the graph shows, the semantic net highlights the role of a single entity in various contexts which are useful both for a researcher as well as a layman. The limitations of the currently used graph drawing software restrict the appropriate representation of n-ary relations. We are also working towards a proper graph-based visualizer in which we shall also add a method to point to the original documents where concepts occur.

References

1. Castro, A.G., Rocca-Serra1, P., Stevens, R., Taylor1, C., Nashar, K., Ragan, M.A., Sansone, S.-A.: The use of concept maps during knowledge elicitation in ontology development processes - the nutrigenomics use case, BMC Bioinformatics (May 25, 2006)
2. Ciaramita, M., Gangemi, A., Ratsch, E., Saric, J., Rojas, I.: Unsupervised Learning of Semantic Relations between Concepts of a Molecular Biology On-tology. In: Proceedings of the 19th International Joint Conference on Artificial Intelligence (IJCAI 2005), pp. 659–664 (2005)
3. Cox, E.: A Hybrid Technology Approach to Free-Form Text Data Mining, http://scianta.com/pubs/AR-PA-007.htm
4. Fensel, D., Horrocks, I., van Harmelen, F., McGuinness, D.L., Patel-Schneider, P.: OIL: Ontology Infrastructure to Enable the Semantic Web. IEEE Intelligent Systems 16(2), 38–45 (2001)
5. García, R., Celma, O.: Semantic Integration and Retrieval of Multimedia Metadata. In: Knowledge Mark-up and Semantic Annotation Workshop, Semannot 2005. CEUR (2005)
6. Wagner, C., Cheung, K.S.K., Rachael, K.F.: Building Semantic Webs for e-government with Wiki technology. Electronic Government 3(1) (2006)

Data Analysis and Bioinformatics

Vito Di Gesù

C.I.T.C., Università di Palermo, Italy

Abstract. Data analysis methods and techniques are revisited in the case of biological data sets. Particular emphasis is given to clustering and mining issues. Clustering is still a subject of active research in several fields such as statistics, pattern recognition, and machine learning. Data mining adds to clustering the complications of very large data-sets with many attributes of different types. And this is a typical situation in biology. Some cases studies are also described.

Keywords: Clustering, data mining, bio-informatics, Kernel methods, Hidden Markov Models, Multi-Layers Model.

1 Introduction

Bio-informatics is a new discipline devoted to the solution of biological problems, usually on the molecular level, by the use of techniques including applied mathematics, statistics, computer science, and artificial intelligence. Major research efforts regard sequence alignment [1], gene finding [2], genome assembly, protein structure alignment [3] and prediction [4], prediction of gene expression, protein-protein interactions, and the modeling of evolution [5].

Mining in structured data is particularly relevant for bio-informatics applications, since the majority of biological data is not kept in databases consisting of a single, flat table [6]. In fact, bio-informatics databases, BDB, are structured and linked *objects*, connected by relations representing a rich internal structure. Examples of BDB are databases of proteins [7], of small molecules [8], of metabolic and regulatory networks [9]. Moreover, biological data representations are structured and heterogeneous; they consist of large sequences (e.g. 10^6 gene sequences), $2D$ large structures (e.g. $10^5 \sim 10^6$ spots on DNA chips), $3D$ structures (e.d. DNA phosphate model, Figure 1a), graphs, networks, expression profiles, and phylogenetic trees (Figure 1b). Several issues are dealing with mining biological data, among them there are *kernel methods* for classification of microarray time series data [10]. This classification of gene expression time series has many potential applications in medicine and pharmacogenomics, such as disease diagnosis, drug response prediction or disease outcome prognosis, contributing to individualized medical treatment. Graph kernels representations of proteins have been designed to retrieve structure and bio-chemical information and protein function prediction. Feature graphs are considered to represent potential docking sites and retrieve activity maps $3D$ protein databases.

A. Ghosh, R.K. De, and S.K. Pal (Eds.): PReMI 2007, LNCS 4815, pp. 373–388, 2007.

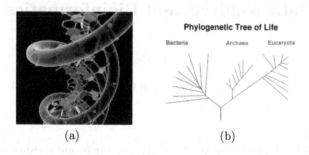

(a)   (b)

Fig. 1. (a) 3D structure of the DNA phosphate model; (b) an example of phylogenetic tree

Concept of similarity play a relevant role in search both $2D$ and $3D$ shape matching in bio-molecular databases. For example, similar $3D$ shape can be retrieved by using a similarity model based on $3D$ shape histograms, $3D$ surface segments, and parametric surface functions including paraboloid and trigonometric polynomials that approximate surface segments.

Finally, methods for finding all subspaces of high-dimensional data containing density-based clusters are necessary because finding clusters in high-dimensional data is usually futile. Moreover, high-dimensional data may be clustered differently in varying subspaces of the feature space. Subspace clustering aims at finding all subspaces of high-dimensional data in which clusters exist.

Specific topics include: preprocessing tasks such as data cleaning and data integration as applied to biological data; classification and clustering techniques for microarrays; comparison of RNA structures based on string properties and energetics; discovery of the sequence characteristics of different parts of the genome; mining of haplotype to find disease markers; sequencing of events leading to the folding of a protein; inference of the subcellular location of protein activity; classification of chemical compounds based on structure; special purpose metrics and index structures for phylogenetic applications; query languages for protein searching based on the shape of proteins, and very fast indexing schemes for sequences and pathways.

The paper is structured as follows: Section 2 outlines both the descriptive and the predictive mining in databases; Section 3 reviews recent clustering algorithms for biological data; in Section 4 two cases studies are described; Section 5 provides final remarks and new perspectives in mining biological data.

2   Mining in Biological Database

Data mining techniques are classified in *descriptive* and *predictive*. In descriptive mining, local structures are searched to discover pattern embedded in data. In predictive mining, models are designed to make predictions for new, unseen cases.

2.1   Descriptive Data Mining

The problem of finding patterns of interest in a data-set is a typical pattern recognition problem, that depends on the nature of problem. For example, the problem of finding frequent item-sets in coregulated genes by estimating number of motif instances has been considered in [11]. Authors ground their analysis on TOUCAN system [12] and Hidden Markov Model inference nets [13]. The prediction of Cis-Regulatory Elements is analyzed in [14] combining different algorithms (Clover, Cluster-Buster, sequence identity, and ITB-algorithm).

Search technique have been developed to solve pattern matching problems in other domains, such as approximate string matching on large DNA sequences [15,16]. These methods include star alignments and tree alignments, which are usually based on dynamic programming. In [17] a polynomial-time dynamic programming algorithm for solving the maximum common subtree of two trees is considered to implement an accurate and efficient tool for finding and aligning maximally matching glycan trees (see Figure 2). For a review on trees matching see [18].

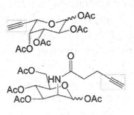

sugar analogies

Fig. 2. (a) An example of biological tree-structure

Structure similarity of two proteins from the matching of pairs of secondary structure elements. In [19] the matching is performed using a fast bipartite graph-matching algorithm that avoids the computational complexity of searching for the full subgraph isomorphism between the two sets of interactions. More information on graph matching in biology can be found in [20].

Matching algorithms are interesting to find all sub-patterns occurring with a minimum frequency in a database of patterns (strings, trees, graphs). The problem can be extended in finding all patterns with a minimum frequency in one data-set and a maximum frequency in another. This is a question relevant for the analysis of differentially expressed genes with applications to protein structure folding prediction and drug discovering, both of them are characterized by 3D structures (see Figures 3a,b).

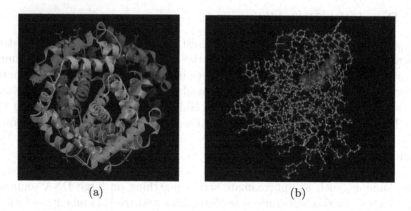

(a)   (b)

Fig. 3. (a) 3D structure of a protein molecule; (b) an example of binding drug molecule into a protein molecule

2.2   Predictive Mining

Data Mining is an analytic process designed to explore a large amount of data to find consistent patterns and/or systematic relationships between variables, and then to validate the findings by applying the detected patterns to new subsets of data. *Predictive Data Mining (PDM)* is usually applied to identify a statistical models that can be used to predict some response of interest [21]. For example, a *PDM* may be more exploratory in nature to identify cluster or to aggregate or amalgamate the information in very large data sets into useful and manageable chunks. *PDM* techniques can be very useful in the case of inductive databases. The process of data mining consists of three stages:

1) *The initial exploration* usually starts with data preprocessing followed by data transformations to select subsets of records. In case of data sets with large numbers of variables ("fields"), preliminary feature selection operations are performed to lower the number of variables to a manageable range. Depending on the nature of the analytic problem, data mining may involve a simple choice of straightforward predictors for a regression model, and a wide variety of graphical and statistical methods in order to identify the most relevant variables and determine the complexity and/or the general nature of models that can be taken into account in the next stage.

2) *Model building and validation* considers and evaluates various models to chose the best one based on their predictive performance. This may be a very elaborate process. There are a variety of techniques developed to achieve this goal - many of which are based on the so-called "competitive evaluation of models", that is, applying different models to the same data set and then comparing their performance to choose the best. These techniques - which are often considered the core of predictive data mining - include: Bagging (Voting, Averaging), Boosting, Stacking (Stacked Generalizations), and Meta-Learning.

3) *Deployment* uses the model selected as best and applies it to new data in order to generate predictions or estimates of the expected outcome (i.e., the application of the model to new data in order to generate predictions).

Traditionally *PDM* has been applied to business or market related. Recently, *PDM*, due to the enormous growing of biological repositories of gene expression data generated by DNA microarray experiments, is providing a new tool for both medical diagnosis and genomic studies. In [22] a systematic approach for learning and extracting rule-based knowledge from gene expression data is presented. A class of predictive self-organizing network, known as *Adaptive Resonance Associative Map* (ARAM), is used for modeling gene expression data, whose learned knowledge can be transformed into a set of symbolic *IF-THEN* rules for interpretation.

3 Survey on Clustering Methods in Bioinformatics

Clustering is the process of grouping data objects into a set of disjoint classes so that objects within a class have high *similarity* to each other, while objects in separate classes are more *dissimilar*. Clustering is part of exploratory data analysis, where *rules* are *eventually* found as a *creative* induction scheme that implies the need for experimental and theoretical models validations.

Currently, typical microarray experiments may contain 10^6 genes. One of the characteristics of gene expression data is that it is meaningful to cluster both genes and samples [23,24]. Here, genes are treated as elements, while samples are features. On the other hand, samples can be partitioned into homogeneous groups that may correspond to some particular macroscopic phenotype. The distinction of gene-based clustering and sample-based clustering is grounded on different characteristics of clustering tasks for gene expression data. Some clustering algorithms, such as K-means and hierarchical approaches, can be used both to group genes and to partition samples. In the following a list of most used clustering techniques to analyze biological data is listed.

K-means [25] is a partition-based clustering method. Given a pre-specified number K, K-means partitions the data set into K disjoint clusters such that the sum of the squared distances of elements from their cluster centers is minimized.

K-means has been applied on gene expression data [26], finding clusters that contain a significant portion of genes with similar functions. Moreover, upstream sequences of DNA-genes within the same cluster allowed to extract 18 motifs, which are promising candidates for novel cis-regulatory elements. The K-means algorithm has some drawbacks (setting of number of clusters, it produces a large number of outliers). To overcome them, several algorithms have been proposed. For example, the K-medoids algorithm uses an element closest to the center of a cluster as the representative (medoid) such that the total distance between the K selected medoids and the other elements is minimized. This algorithm is more robust to the outliers than K-means. Another group of algorithms use some thresholds to control the coherence of clusters. For example, the maximal similarity between two separate cluster centroids and the minimal similarity between an element and its cluster centroid, [27]. In [28], clusters are constrained to have a diameter no larger than d (compact clusters); in [29] a more efficient algorithm (*Adapt_Cluster*) is proposed. Here, an element will be assigned to a given cluster if the assignment has a *higher probability* than a given threshold. It turns out that only clusters with qualified coherence from the data set are extracted. Therefore, users do not need to input the number, K of clusters.

In any case, K-means algorithm and its derivatives require either the number of clusters or some coherence threshold. The clustering process is like a *black box*. Therefore, they are not flexible to the local structures of the data set, and can hardly support interactive exploration for coherent expression patterns.

SOM (Self-Organizing Maps) were developed on the basis of a single layered neural network [30]. Elements, usually of high dimensionality, are mapped onto a set of neurons organized with low dimensional structures, e.g., a two dimensional $p \times q$ grid. Each neuron is associated with a reference vector, and each element is mapped to the neuron with the closest reference vector. During the clustering process, each data object acts as a training sample that directs the movement of the reference vectors towards the denser areas of the input vector space, so that those reference vectors are trained to fit the distributions of the input data set. When the training is complete, clusters are identified by mapping all data points to the output neurons.

One of the remarkable features of SOM is that it allows one to impose partial structure on the clusters, and arranges similar patterns as neighbors in the output neuron map. This feature facilitates an easy visualization and interpretation of clusters, partly supporting the explorative analysis of gene expression patterns. However, similar to the K-means algorithm, SOM requires the number of clusters, which is typically unknown in advance for gene expression data.

In [31] the SOM algorithm is applied to study hematopoietic differentiation. The expression patterns of 1,036 human genes are mapped to a 6 × 4 SOM. The SOM organizes genes into biologically relevant clusters that suggest novel hypotheses about hematopoietic differentiation and this provides interesting insights into the mechanism of differentiation.

Recently, the Department of Information Technology (National University of Ireland) has developed the system SOMBRERO (Self-Organizing Map for Biological Regulatory Element Recognition and Ordering). SOMBRERO finds regulatory binding sites by using SOM to find over-represented motifs in a set of DNA sequences [32,33]. It includes prior knowledge in the initialization phase that significantly improves accuracy when known motifs are present in the input data, while accuracy is not negatively affected for the discovery of novel motifs.

MBA (Model Based Algorithms) [34] provides a statistical framework to model the cluster structure of gene expression data. The data set is assumed to come from a mixture of underlying probability distributions, with each component corresponding to a different cluster. The goal is to estimate the parameters $\Theta = \{\theta_i | 1 \leq i \leq k\}$ and $\Gamma = \{\gamma_r^i | 1 \leq i \leq k; 1 \leq r \leq n\}$ that maximize the likelihood $L_{mix}(\Theta, \Gamma) = \sum_{i=1}^{k} \gamma_r^i f_i(x_r, \theta_i)$, where n is the number of elements, k is the number of components, x_r is a data object (i.e., a gene expression profile),

$f_i(x_r, \theta_i)$ is the density function of x_r in component C_i with some unknown set of parameters θ_i; γ_r^i represents the probability that x_r belongs to C_i.

An important advantage of MBA approaches is that they provide an estimated probability that an elements belongs to a given cluster. The probabilistic feature of MBA is particularly suitable for gene expression data because it is typical for a gene to participate multiple cellular processes, so that a single gene may have a high correlation with two different clusters. Moreover, MBA does not need to define a distance (or similarity) between two gene profiles. Instead, the measure of coherence is inherently embedded in the statistical framework.

However, MBA assumes that the data set fits a specific distribution. This may not be true and there is currently no well-established general model for gene expression data. Several MBA approaches claim a multivariate Gaussian distribution. Although the Gaussian model works well for gene-sample data where the expression levels of genes are measured under a collection of samples, it may not be effective for time-series data (the expression levels of genes are monitored during a continuous series of time points).

To better describe the gene expression dynamics in time-series data, several new models have been introduced. For example, each gene expression profile can be modeled as a cubic spline so that each time point influences the overall smooth expression curve [35]. In addition, time-series may follow an autoregressive model, where the value of the series at time t is a linear function of the values at several previous time points [36].

Time series data are often treated with *Hidden Markov model* (HMM) as an extension of a Markov model, in which a state has a probability of emitting some output. Formally, an HMM is a finite state machine with probabilities for each transition, that is, a probability of the next state is given by the current state. The states are not directly observable; instead, each state produces one of the observable outputs with a certain probability.

HMM's can be used to represent the alignment of multiple sequences or sequence segments by attempting to capture common patterns of residue conversion. They are widely used in the analysis of biological sequences to take in account for the dependencies in time-series bio-data [37].

GBA(Graph Based Algorithm) models a gene expression data set as a undirected weighted graph $G(V, E, W)$, where each gene is represented by a vertex $v \in V$, an arc $(x, y) \in E$ connects a pair of genes $x, y \in V$ with a weight, $W(x, y)$, based on the similarity between the expression patterns of x and y. The similarity is often normalized to $[0, 1]$, where 0 imply the non existence of an arc, and 1 the perfect fit of two genes. The problem of clustering a set of genes is then isomorph to some classical graph-theoretical problems, such as searching for the *minimum cut* [38], the *minimum spanning tree* [39], or the *maximum cliques* [23] in graph G. Other algorithms recursively split G into a set of *Highly Connected Components* (HCC) along the minimum cut, and each HCC is considered as a cluster. For example, the algorithm CLICK (CLuster Identification via Connectivity Kernels) sets up a statistic framework to measure the coherence within a subset of genes and determine the criterion to stop the recursive splitting process.

The graph-based algorithms stem from some classical graph theoretical problems. Although with solid mathematical ground, they may not be suitable for gene expression data without adaption. For example, in gene expression data, groups of co-expressed genes may be highly connected by a large amount of *intermediate* genes. In this case, the approaches based on minimum spanning tree and minimum cut may lead to clusters including genes with incoherent profiles but highly connected by a series of *intermediate* genes [40].

HA's (Hierarchical Algorithms) fall in two categories:

Agglomerative (i.e., bottom-up approach) that initially regards each data object as an individual cluster. Agglomerative approaches merge, at each step, the closest pair of clusters until all the groups are merged into one cluster.

Divisive (i.e., top-down approach) that starts with one cluster containing all the data objects. Divisive approaches iteratively split clusters until each cluster contains only one data object or certain stop criterion is met. For divisive approaches, the essential problem is to decide how to split clusters at each step.

An example of agglomerative hierarchical clustering is proposed in [41]; it combines tree-structured vector quantization and partitive K-means clustering. This hybrid technique reveals clinically relevant clusters in large publicly available data sets. The system is less sensitive to data preprocessing and data normalization. Moreover, results obtained have strong similarities with those obtained by self-organizing maps.

A clique graph is an undirected graph that is the union of disjoint complete graphs. In [23] the idea of a corrupted clique graph data model is introduced. Clustering a dataset is equivalent to identifying the original clique graph from the corrupted version with as few errors as possible. CAST Algorithm is an example of graph theoretic approach that relies on the concept of a clique graph and uses a divisive clustering approach. Thus, the model assumes that there is a *true biological partition of the genes into disjoint clusters bases on the functionality of the genes*. In [42] an enhanced version of CAST, called E-CAST, is described. The main difference with CAST is the use of a dynamic threshold is introduced. The threshold value is computed at the beginning of each new cluster.

PBCA (Pattern-based Clustering Algorithms) have been proposed to capture coherence exhibited by a subset of genes on a subset of attributes. This approach takes in account the fact that in molecular biology any cellular process may take place only in a subset of the attributes (samples or time points). For example, in [43] the concept of *bicluster* to measure the coherence between genes and attributes is introduced. *Biclustering* was first introduced in [44,45] , it finds a partition of the vectors and a subset of the dimensions such that the projections along those directions of the vectors in each cluster are close to one another. Then the problem requires to cluster vectors and dimensions simultaneously, thus the name biclustering.

The complexity of the biclustering problem depends on the exact problem formulation, and particularly on the merit function used to evaluate the quality of a given bicluster. The exact solution of biclustering is NP-complete so that clever heuristics are considered to solve it with small lossy of information. For example, in [46] a stochastic algorithm based on Simulated Annealing [47] is presented and validated on a variety of data-sets showing that Simulated Annealing find significant biclusters in many cases. Evolutionary algorithms have been used in [48] to implement biclustering clustering of gene expression on yeast and lymphoma data-sets.

A multi-objective evolutionary clustering for gene expression data is described in [49]. Here, a set of solutions, which are all optimal and involving trade-offs between conflicting objectives, are considered. Unlike single-objective optimization problems, the multiple-objective approach tries to optimize $m \geq 2$ conflicting solutions evaluated by fitness functions. Validation was carried out on microarray data consisting of a benchmark gene expression dataset, viz., Yeast.

ECA (Evolutive Clustering Algorithms) have been recently proposed to analyze biological data to overcome both the computational complexity of greedy algorithms and to improve the space solution scan.

In [50] the *GenClust* (Genetic Clustering) algorithm has been introduced for clustering of gene expression data. GenClust has two key features: (a) a novel coding of the search space that is simple, compact and easy to update; (b) it can be naturally used in conjunction with *data driven* internal validation methods.

In [51] a new classifier, based on fuzzy-integration schemes, is introduced. Schemes are controlled by a genetic optimization procedure. Two versions of integration are proposed and validated by experiments on real data representing: (a) biological cellsBreast cancer databases from the University of Wisconsin and Waveform ((ftp://ftp.ics.uci.edu/pub/machine-learning-databases)); (b) Urine analysis cells database kindly provided by IRIS Diagnostic, CA, USA. Comparison with feed-forward neural network and Support Vector Machine classifiers have been considered for comparison. Results show the good performance and robustness of the integrated classifier.

In [52] an incremental Genetic K-means Algorithm (IGKA) is presented. IGKA is an extension of previously proposed genetic algorithm to improve the computation of K-means algorithm. The main idea of IGKA is to calculate the objective value *Total Within-Cluster Variation* and to cluster centroids incrementally whenever the mutation probability is small. IGKA always converges to the global optimum. Experiments indicate that IGKA algorithm has a better time performance when the mutation probability decreases to some point.

4   Case Studies

4.1   An Example of Evolutive Algorithm: GenClust

GenClust [50] is one of the most recent evolutive clustering algorithm, that can be seen as a genetic variant of ISODATA and it is an incarnation of the technique devised in [53] for clustering based on Genetic Algorithms. The main difference between Genetic algorithms for clustering already present in the literature [54] and *GenClust* consists in the generated solution space. GenClust codes a solution (label) for each element instead of coding the whole partition of the data set.

Such much simpler coding technique allows a very efficient update of the state of the algorithm and also guarantees a more efficient search of the solution space.

The general idea behind *GenClust* is quite simple. The algorithm proceeds in stages; at each stage, t, a partition $\mathcal{P}_t$ of $X \subset \mathbb{R}^d$ into K classes $C_1, C_2, \cdots, C_K$ is generated. The initial partition, $\mathcal{P}_0$, is obtained by a random assignment of elements to classes or by the computation of the partition through another clustering algorithm. Based on $\mathcal{P}_t$ and using genetic operators (cross-over and mutation) and a suitable fitness function, the algorithm computes $\mathcal{P}_{t+1}$. Note that, there is no guarantee that the new partition is such that $VAR(\mathcal{P}_{t+1}) \leq VAR(\mathcal{P}_t)$. Where, $VAR(\mathcal{P})$ denoted the internal partition variance.

Each element $x \in X$ is coded via a 32 bit string α_x (referred to as *chromosome*). The chromosome encodes the class that x belongs to in a partition using the 8 least significant bits. We refer to it as the label λ_x. The remaining 24 bits give the position of x within its cluster, referred to as pos_x. The chosen coding is compact and easy to handle and allows to represents up to 256 classes and data sets of size up to 1.6793.604 elements. These values are adequate for real applications. The genetic operators of one point crossover and mutation are applied to each chromosome with probability 0.9 and 0.1, respectively. The fitness function of individual (x, λ) in partition $\mathcal{P}$ is given by:

$$f((x, \lambda)) = \sqrt{\frac{1}{d} \sum_{j=1}^{d} \frac{(x_j - \mu_j^\lambda)^2}{\max(x_j, \mu_j^\lambda))^2}} \tag{1}$$

where, μ^λ is centroid of the cluster λ in $\mathcal{P}$.

GenClust has been validated using the FOM methodology, conceived for gene expression data [58]. GenClust has been validated on several set of biological data; among them the Rat Central Nervous System data set [59], Yeast Cell Cycle [60], Reduced Yeast Cell Cycle [61], and Peripheral Blood Monocytes [62].

4.2   Analysis of Genes Expression Patterns

Analyzing coherent gene expression patterns is an important task in bioinformatics research and biomedical applications. This issue is important because co-expressed genes may belong to the same or similar functional categories and indicate co-regulated families, while coherent patterns may characterize important cellular processes and suggest the regulating mechanism in the cells. Examples of co-expressed gene groups are shown in Figure 4. adapted or proposed to identify clusters of co-expressed genes and recognize coherent expression patterns as the centroids of the clusters. However, the interpretation of co-expressed genes and coherent patterns mainly depends on the knowledge domain, which presents several challenges for coherent pattern mining. In such cases, the design of interactive clustering systems may be useful. An examples of interactive exploration system is GeneX (Gene eXplorer) for mining coherent expression patterns [63]. GeneX is composed of a preprocessing module to perform to estimate missing values, logarithmic transformation and standardization of each

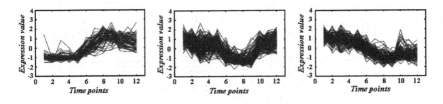

Fig. 4. Examples of co-expressed gene groups.

gene expression profile. A pattern manager module allows users to explore coherent patterns in the data set and save/load the coherent patterns. A working zone integrates parallel coordinates, coherent pattern index graphs (pulses of coherent pattern index graphs indicate the potential existence of a coherent pattern), and tree views. For example, users can select a node in the tree view, then the working zone will display the corresponding expression profiles and coherent pattern index graph.

4.3 Study of Proteins Sequences

The analysis of stochastic signals aims to both extract *significant* patterns from noisy background and to study their spatial relations (periodicity, long term variation, burst, etc.). Examples of such kind of data are protein-sequences in molecular biology where protein folding are studied [64] and the positioning of nucleosomes along chromatin [55]. The analysis carried out in both cases has been tackled by using probabilistic networks (e.g., Hidden Markov Models [13], Bayesian networks,...). However, probabilistic networks may suffer of high computational complexity, and results can be biased from locality that depends on the memory steps they use [56]. In [57] a Multi-Layers Model (MLM) is proposed that is computational efficient, providing a better structural view of the input data. The MLM consists in the generation of several sub-samples from the input signal. For example, in the case of input signal fragment, representing the *Saccharomyces cerevisiae* microarray data, each value in the x axes represents a spot on the microarray and its intensity is the log ratio Green/Red (see Figure 5a). The problem is the identification of particular patterns in the DNA called *nucleosome* and *linker* regions. Nucleosomes correspond to peaks of about 140 base pairs long, or six to eight microarray spots (black circle in Figure 5a), surrounded by lower ratio values corresponding to linker regions (marked by dashed circles). The multi-layer view is obtained by intersecting the signal with horizontal lines, each one representing a threshold value t_k (see Figure 5b). The persistence of the signal at increasing threshold values together with its width and power is considered to perform the discrimination of linkers from nucleosomes. From the biological point of view, the accurate positioning of nucleosomes provides useful information regarding the regulation of gene expression

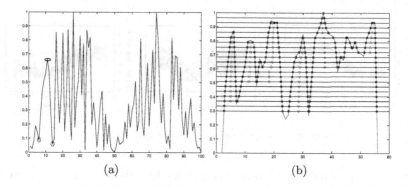

(a)   (b)

Fig. 5. An example of analysis by *MLM*: (a) Input Signal; (b) Pattern identification and extraction

in eukaryotic cells. In fact, how eukaryotic DNA is packaged into a highly compact and dynamic structure called chromatin may provide information about a variety of diseases, including cancer.

5 Final Remarks and Perspectives

A survey of current data analysis methods in bioinformatics has been provided. Main emphasis has been given to clustering techniques because of their impact in many biological applications, as the mining of biological data. The topic is so wide that several aspects have been omitted or not fully developed. The aim was to introduce main ideas and to stimulate new research directions. Challenges with bioinformatics are the need to deal with interdisciplinary directions, the difficulties in the validation of development data analysis methods, and, more important to be addressing important biological problems.

References

1. Brudno, M., Malde, S., Poliakov, A.: Glocal alignment: finding rearrangements during alignment. Bioinformatics 19(1), 54–62 (2003)
2. Rogic, S.: The role of pre-mRNA secondary structure in gene splicing in Saccharomyces cerevisiae, PhD Dissertation, University of British Columbia (2006)
3. Bourne, P.E., Shindyalov, I.N.: Structure Comparison and Alignment. In: Bourne, P.E., Weissig, H. (eds.) Structural Bioinformatics, Wiley-Liss, Hoboken, NJ (2003)
4. Zhang, Y., Skolnick, J.: The protein structure prediction problem could be solved using the current PDB library. Proc. Natl. Acad. Sci. USA 102(4), 1029–1034 (2005)
5. Gould, S.J.: The Structure of Evolutionary Theory. Belknap Press (2002)
6. Matsuda, T., Motoda, H., Yoshida, T., Washio, T.: Mining Patterns from Structured Data by Beam-wise Graph-Based Induction. In: Lange, S., Satoh, K., Smith, C.H. (eds.) DS 2002. LNCS, vol. 2534, pp. 422–429. Springer, Heidelberg (2002)

7. Schaffer, A.A., Aravind, L., Madden, T.L., Shavirin, S., Spouge, J.L., Wolf, Y.I., Koonin, E.V., Altschul, S.F.: Improving the accuracy of PSI-BLAST protein database searches with composition-based statistics and other refinements. Nucleic Acids Res. 29(14), 2994–3005 (2001)
8. http://www.netsci.org/Resources/Web/small.html
9. Karp, P.D., Riley, M., Saier, M., Paulsen, I.T., Paley, S.M., Pellegrini-Toole, A.: The EcoCyc and MetaCyc databases. Nucleic Acids Research 28, 56–59 (2000)
10. Vert, J.-P.: Support Vector Machine Prediction of Signal Peptide Cleavage Site Using a New Class of Kernels for Strings. In: Proceedings of the Pacific Symposium on Biocomputing, vol. 7, pp. 649–660 (2002)
11. Aerts, S., Thijs, G., Coessens, B., Staes, M., Moreau, Y., De Moor, B.: Toucan: deciphering the cis-regulatory logic of coregulated genes. Nucleic Acids Research 31(6), 1753–1764 (2003)
12. http://homes.esat.kuleuven.be/~saerts/software/toucan.php
13. Cappé, O., Moulines, E., Rydén, T.: Inference in Hidden Markov Models. Springer, Heidelberg (2005)
14. Kielbasa, S.M., Blüthgen, N., Sers, C., Schäfer, R., Herze, H.: Prediction of Cis-Regulatory Elements of Coregulated Genes Szymon. Genome Informatics 15(1), 117–124 (2004)
15. Cheng Cheung, L.-L., Siu-Ming Yiu, D.W.: Approximate string matching in DNA sequences. In: Proceedings DASFAA 2003, pp. 303–310 (2003)
16. Myers, G.: A fast bit-vector algorithm for approximate string matching based on dynamic programming. Journal of the ACM 46(3), 395–415 (1999)
17. Aoki, K.F., Yamaguchi, A., Okuno, Y.: Effcient Tree-Matching Methods for Accurate Carbohydrate Database Queries. Genome Informatics 14, 134–143 (2003)
18. Gusfield, D.: Algorithms on Strings, Trees and Sequences: Computer Science and Computational Biology, The Press Syndacate of the University of Cambridge, UK (1999)
19. Taylor, W.R.: Protein Structure Comparison Using Bipartite Graph Matching and Its Application to Protein Structure Classification. Molecular & Cellular Proteomics 1(4), 334–339 (2002)
20. Yang, Q., Sze, S.-H.: Path Matching and Graph Matching in Biological Networks. Journal of Computational Biology 14(1), 56–67 (2007)
21. Sholom, M.W., Indurkhya, N.: Predictive Data-Mining: A Practical Guide. Morgan Kaufmann, San Francisco (1998)
22. Tana, A.H., Panb, H.: Predictive neural networks for gene expression data analysis. Neural Networks 18, 297–306 (2005)
23. Ben-Dor, A., Shamir, R., Yakhini, Z.: Clustering gene expression patterns. Journal of Computational Biology 6(3/4), 281–297 (1999)
24. Eisen, M.B., Spellman, P.T., Brown, P.O., Botstein, D.: Cluster analysis and display of genome-wide expression patterns. Proc. Natl. Acad. Sci. USA 95(25), 14863–14868 (1998)
25. MacQueen, J.B.: Some Methods for classification and Analysis of Multivariate Observations. In: Proceedings of 5-th Berkeley Symposium on Mathematical Statistics and Probability, Berkeley, vol. 1, pp. 281–297. University of California Press (1967)
26. Tavazoie, S., Hughes, J.D., Campbell, M.J., Cho, R.J., Church, G.H.: Systematic determination of genetic network architecture. Nature Genet. 22(3), 281–285 (1999)
27. Herwig, R., Poustka, A.J., Muller, C., Bull, C., Lehrach, H., O'Brien, J.: Large-Scale Clustering of cDNA Fingerprinting Data. Genome Research 9(11), 1093–1105 (1999)

28. Heyer, L.J., Kruglyak, S., Yooseph, S.: Exploring expression data: identification and analysis of coexpressed genes. Genome Research 9(11), 1106–1115 (1999)
29. De Smet, F., Mathys, J., Marchal, K., Thijs, G., De Moor, B., Moreau, Y.: Adaptive quality-based clustering of gene expression profiles. Bioinformatics 18, 735–746 (2002)
30. Kohonen, T.: Self-Organization and Associative Memory. Springer, Berlin (1984)
31. Tamayo, P., Slonim, D., Mesirov, J., Zhu, Q., Kitareewan, S., Dmitrovsky, E., Lander, E.S., Golub, T.R.: Interpreting patterns of gene expression with self-organizing maps: Methods and application to hematopoietic differentiation. Proc. Natl. Acad. Sci. USA 96(6), 2907–2912 (1999)
32. http://bioinf.nuigalway.ie/sombrero/
33. Mahony, S., Golden, A., Smith, T.J., Benos, P.V.: Improved detection of DNA motifs using a self-organized clustering of familial binding profiles. Bioinformatics 21(Suppl 1), 283–291 (2005)
34. Yeung, K.Y., Fraley, C., Mura, A., Raftery, A.E., Ruzzo, W.L.: Model-based clustering and data transformations for gene expression data. Bioinformatics 17, 977–987 (2001)
35. Yeang, C.-H., Jaakkola, T.: Time Series Analysis of Gene Expression and Location Data. In: Proceedings of the Third IEEE Symposium on BioInformatics and BioEngineering (BIBE 2003), pp. 1–8 (2003)
36. Ramoni, M.F., Sebastiani, P., Kohane, I.S.: Cluster analysis of gene expression dynamics. Proc. Natl. Acad. Sci. USA 99(14), 9121–9126 (2002)
37. Koski, T.T.: Hidden Markov Models for Bioinformatics. Series: Computational Biology, vol. 2. Springer, Heidelberg (2002)
38. Hartuv, E., Shamir, R.: A clustering algorithm based on graph connectivity. Information Processing Letters 76(4/6), 175–181 (2000)
39. Xu, Y., Olman, V., Xu, D.: Clustering gene expression data using a graph-theoretic approach: an application of minimum spanning trees. Bioinformatics 18, 536–545 (2002)
40. Jiang, D., Pei, J., Zhang, A.: Interactive Exploration of Coherent Patterns in Time-Series Gene Expression Data. In: Proceedings of the Ninth ACM SIGKDD International Conference on Knowledge Discovery and Data Mining (KDD 2003), Washington, DC, USA, pp. 24–27 (2003)
41. Sultan, M., Wigle, D.A., Cumbaa, C.A., Marziar, M., Glasgow, J., Tsao, M.S., Jurisca, J.: Binary tree-structured vector quantization approach to clustering and visualizing microarray data. Bioinformatics 18(1), 111–119 (2002)
42. Bellaachia, A., Portnoy, D., Chen, Y., Elkahloun, A.G.: E-CAST: a data mining algorithm for gene expression data. In: Proceedings of the ACM SIGKDD Workshop on Data Mining in Bioinformatics (BIOKDD 2002), pp. 49–54 (2002)
43. Cheng, Y., Church, G.M.: Biclustering of expression data. In: Proceedings of the Eighth International Conference on Intelligent Systems for Molecular Biology (ISMB), vol. 8, pp. 93–103 (2000)
44. Mirkin, B.: Mathematical Classification and Clustering. Kluwer Academic Publishers, Dordrecht (1996)
45. Van Mechelen, I., Bock, H.H., De Boeck, P.: Two-mode clustering methods:a structured overview. Statistical Methods in Medical Research 13(5), 363–394 (2004)
46. Bryan, K., Cunningham, P., Bolshakova, N.: Biclustering of Expression Data Using Simulated Annealing. In: 18th IEEE Symposium on Computer-Baseds Medical Systems (CBMS 2005), pp. 383–388 (2005)

47. Kirkpatrick, S., Gelatt, C.D., Vecchi, M.P.: Optimization by Simulated Annealing. Science 220(4598), 671–680 (1983)

48. Chakraborty, A., Maka, H.: Biclustering of Gene Expression Data Using Genetic Algorithm. In: IEEE Symposium on Computational Intelligence in Bioinformatics and Computational Biology (CIBCB 2005), vol. 14(15), pp. 1–8 (2005)

49. Sushmita, M., Haider, B.: Multi-objective evolutionary biclustering of gene expression data. Pattern Recognition 39(12), 2464–2477 (2006)

50. Di Gesù, V., Giancarlo, R., Lo Bosco, G., Raimondi, A., Scaturro, D.: GenClust: A Genetic Algorithm for Clustering Gene Expression Data. BMC Bioinformatics 6(289) (2005)

51. Di Gesù, V., Lo Bosco, G.: A genetic integrated fuzzy classifier. Pattern Recognition Letters 26(4), 411–420 (2005)

52. Lu, Y., Lu, S., Fotouhi, F., Deng, Y., Brown, S.J.: Incremental genetic K-means algorithm and its application in gene expression data analysis. BMC Bioinformatics 5(172) (2004)

53. Di Gesù, V., Lo Bosco, G.: GenClust: a Genetic Algorithm for Cluster Analysis. In: Proc. ADA III, pp. 12–18 (2004)

54. Jain, A.K., Murty, M.N., Flynn, P.J.: Data Clustering: A Review. ACM Computing Surveys 31(3), 264–323 (1999)

55. Yuan, G.C., Liu, Y.J., Dion, M.F., Slack, M.D., Wu, L.F., Altschuler, S.J., Rando, O.J.: Genome-Scale Identification of Nucleosome Positions in S. cerevisiae. Science 309, 626–630 (2005)

56. Delcher, A.L., Kasif, S., Goldberg, H.R., Hsu, W.H.: Protein secondary structure modelling with probabilistic networks. In: Proc. of Int. Conf. on Intelligent Systems and Molecular Biology, pp. 109–117 (1993)

57. Corona, D., Di Gesù, V., Lo Bosco, G., Pinello, L., Yuan, G.-C.: A new Multi-Layers Method to Analyze Gene Expression. In: Proc. KES 2007. LNCS, Springer, Heidelberg (in press, 2007)

58. Yeung, K.Y., Haynor, D.R., Ruzzo, W.L.: Validating clustering for gene expression data. Bioinformatics 17, 309–318 (2001)

59. Somogyi, R., Wen, X., Ma, W., Barker, J.L.: Developmental kinetic of GLAD family mRNAs parallel neurogenesis in the rat Spinal Cord. Journal Neurosciences 15, 2575–2591 (1995)

60. Spellman, P., Sherlock, G., Zhang, M., et al.: Comprehensive identification of cell cycle regulated genes of the yeast Saccharomyces Cerevisiae by microarray hybridization. Journal of Mol. Biol. Cell 9, 3273–3297 (1998)

61. Cho, R.J., et al.: A genome-wide transcriptional analysis of the mitotic cell cycle. Journal of Molecular Cell 2, 65–73 (1998)

62. Hartuv, E., Schmitt, A., Lange, J., et al.: An Algorithm for Clustering of cDNAs for Gene Expression Analysis Using Short Oligonucleotide Fingerprints. Journal Genomics 66, 249–256 (2000)

63. Jiang, D., Pei, J., Zhang, A.: Towards Interactive Exploration of Gene Expression Patterns. SIGKDD Explorations 5(2), 79–90 (2003)

64. Delcher, A.L., Kasif, S., Goldberg, H.R., Hsu, W.H.: Protein secondary structure modelling with probabilistic networks. In: Proc. of Int. Conf. on Intelligent Systems and Molecular Biology, pp. 109–117 (1993)

65. Yuan, G.C., Liu, Y.J., Dion, M.F., Slack, M.D., Wu, L.F., Altschuler, S.J., Rando, O.J.: Genome-Scale Identification of Nucleosome Positions in S. cerevisiae. Science 309, 626–630 (2005)
66. Delcher, A.L., Kasif, S., Goldberg, H.R., Hsu, W.H.: Protein secondary structure modelling with probabilistic networks. In: Proc. of Int. Conf. on Intelligent Systems and Molecular Biology, pp. 109–117 (1993)
67. Corona, D., Di Gesù, V., Lo Bosco, G., Pinello, L., Yuan, G.-C.: A new Multi-Layers Method to Analyze Gene Expression. In: Proc. KES 2007 11th International Conference on Knowledge-Based and Intelligent Information & Engineering Systems. LNCS, Springer, Heidelberg (in press, 2007)

Discovering Patterns of DNA Methylation: Rule Mining with Rough Sets and Decision Trees, and Comethylation Analysis

Niu Ben[1], Qiang Yang[1], Jinyan Li[2], Shiu Chi-keung[3], and Sankar Pal[4]

[1] Department of Computer Science and Engineering, Hong Kong University of Science & Technology, Hong Kong, China
{csniuben, qyang}@cse.ust.hk
[2] Institute for Infocomm Research, Singapore
JYLi@ntu.edu.sg
[3] Department of Computing, Hong Kong Polytechnic University, Hong Kong, China
csckshiu@comp.polyu.edu.hk
[4] Indian Statistical Institute, Kolkata, India
sankar@isical.ac.in

Abstract. DNA methylation regulates the transcription of genes without changing their coding sequences. It plays a vital role in the process of embryogenesis and tumorgenesis. To gain more insights into how such epigenetic mechanism works in the human cells, we apply the two popular data mining techniques, i.e., Rough Sets, and Decision Trees, to uncover the logical rules of DNA methylation. Our results show that the Rough Sets method can generate and utilize fewer rules to fully separate the methylation dataset, whereas Decision Trees method relies on more rules but involves fewer decision variables to do the same task. We also find that some of the gene promoters are highly comethylated, demonstrating the evidence that genes are highly interactive epigenetically in human cells.

1 Introduction

DNA methylation is the epigenetic modification of eukaryotic DNA involved in various biological activities including gene silencing, X chromosome inactivation, gene imprinting, and genome defense [1]. Gene silencing mediated by DNA methylation has tight relation to the tumorgenesis and the embryogenesis in human cells. In tumor progression, the aberrant methylation of the normally unmethylated promoter CGI has been found to be associated with the transcriptional inactivation of over half of the classic tumor suppressor genes. In stem cell development, the methylation of the promoter CGI is highly involved in the maintenance of the embryonic stem cell pluripotency, and the orderly differentiation of these cells into many other cell types. Analysis of the DNA methylation patterns thus has its clinical significance to the treatment of cancers and cell therapy.

Recently, DNA methylation has aroused considerable interests in the area of computational biology and bioinformatics. The classification, clustering, and feature

A. Ghosh, R.K. De, and S.K. Pal (Eds.): PReMI 2007, LNCS 4815, pp. 389–397, 2007.
© Springer-Verlag Berlin Heidelberg 2007

selection methods in machine learning have been successfully applied to the DNA methylation data [2-6]. In this paper, we employ two data mining techniques, Rough Sets and Decision Trees to learn the logical rules of DNA methylation. In doing so, we are able to advance our knowledge on the epigenetic mechanism of human caner cells, embryonic stem cells, and normally differentiated cells.

2 Method

2.1 Data Set

We use the recently published data from Bibikova et al. [3] in our study. Three types of cells, namely, the human embryonic stem cell, the cancer cell and the normally differentiated cell, are collected and investigated, as shown in Table I,

Table I. Sample cells used in experiment

Cell type	Name of sample cells
ES cells	BG01, BG01V, BG02, BG03; ES02, ES03; HUES7; NTERA-2; Relicell hES1; SA01, SA02, SA02.5; TE04, TE06; WA01, WA07, WA09.
Cancer cell	A431; C33A; EC109; Fet; HCE4; HCT116, HT29; LNCaP; LS174; MCF7; MDA_MB_435, MDA_MB_468; NCI_H1299, NCI_H1395, NCI_H2126, NCI_H358, NCI_H526, NCI_H69; PC3; SW480; T47D, TE3, TT, TTn.
Normal cell	Breast; Colon; Lung; Ovary; Prostate; NA06999, NA07033, NA10923, NA10924.

Totally, there are 37, 24, and 9 sample cell lines collected for each type of the cells, and 1536 CpG sites are selected from the 5' regulatory regions of 371 genes. The selected genes are chosen on the basis of their importance to the cellular behaviors, including the tumor suppressor genes, the oncogenes, and the genes that are responsible for the cell growth, the apoptosis, the DNA damage repair and the oxidative metabolism. The profiling of DNA methylation consists of three major steps, the extraction of DNA, the Bisulfite conversion of the CpG sites, and the GoldenGate assay of the methylation levels. For a specific CpG site its methylation level $\alpha \in [0, 1]$ is defined as the ratio of the intensities of the fluorescent signals from the methelated (m) and the unmethylated (u) alleles,

$$\alpha = \frac{m}{m + u}. \tag{1}$$

Finally, we have 70 entries of the different types of cells. Each of them contains 1536 attributes of the CpG sites with the numerical attribute values in [0, 1].

2.2 Rough Sets Method

Rough Sets theory was first developed by Pawlak in the early 1980s [7]. It is an effective machine learning method that can be utilized to represent and reason about the imprecise and the uncertain data. We can apply Rough Sets to extract the DNA methylation rules directly from the training data set. Let A represent the set of the CpG sites. u_i^1 and u_j^2 denotes the i-th and the j-th cell sample in class C_1 and C_2, respectively. Define $f_{ij} \subseteq A$ as the set of the CpG sites whose methylation intensities in the cell sample u_i^1 and u_j^2 are not identical, i.e.,

$$f_{ij} = \left\{ a \in A \mid a(u_i^1) \neq a\left(u_j^2\right), i = 1,..., N_1, j = 1,..., N_2 \right\}, \tag{2}$$

where $a(u)$ is the function that returns the value of the attribute 'a' of sample 'u'. N_1 and N_2 is the number of samples of the two classes, C_1 and C_1. The methylation rules can be generated by evaluating the Bool function in (3).

$$f = \wedge\left(\vee f_{ij}^k\right), \tag{3}$$

where f_{ij}^k is the $k-th$ element in f_{ij}, $k = 1,..., |f_{ij}|$. $\wedge$ and $\vee$ are the Bool operators. Given some new input the methylation rules can then be used for classification.

2.3 Decision Trees

Decision Trees method, introduced by Quinlan [8], is popular for rule induction due to its mathematical simplicity and the Bayesian optimality. Let A be the set of all the CpG sites and S is the sample set of two classes C_1 and C_1. For the CpG site, 'a', $a \in A$, let Value(a) represent the set of its possible values and S_v is the subset of the samples whose attribute 'a' takes the value of v, $v \in$ Value (a). The construction of the decision tree is based on the computation of Information Gain by (4),

$$Gain(S, a) = Entropy(S) - \sum_{v \in Value(a)} \frac{|S_v|}{|S|} Entropy(S_v), \tag{4}$$

where

$$Entropy(S) = \sum_i - p_i \log_2 p_i \tag{5}$$

and p_i, $i = 1, 2$, is the conditional probability of the class C_i in the sample set S. We choose the attribute with the maximum information gain to separate the samples. Recursively, in doing so, we can derive a decision tree where all the samples are classified in the leaf node. The association rules can thus be obtained by traversing from the root node to the leaf node of the tree, and then truncating the intermediate nodes with the Bool 'AND' operators.

3   Results

To obtain the rules that can be easily interpreted and understood by human scientists, we first discrete the data by performing clustering with the k-Means method. Five clusters are generated for each attribute to describe the different levels of methylation intensity, i.e., {Very Low, Low, Middle, High, Very High}, or {VL, L, M, H, VH} in short. Fig. 1 illustrates the result of clustering for the gene CpG site APBA2-1274.

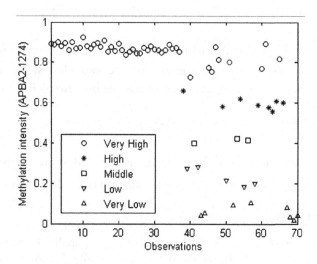

Fig. 1. Clusters generated for five language variables

3.1   Learn DNA Methylation Rules with Rough Sets

In our experiment on Rough Sets, we choose the Rosetta software [9] for implementation. We obtained three sets of methylation rules that can be utilized to

Table II. Methylation rules of embryonic stem cells

No.	Rules	Matched
1	(SMARCA3-1167=VH)&(EPO-1269=VH)&(ASC-350=VH)& (CCND2-596=VH)&(CDH3-152=VH)&(CFTR-1051=VH)	14
2	(ARHI-521=VH)&(SLC6A8-519=VH)	8
3	(ASCL2-1038=H)&(CRIP1-1227=H)	5
4	(ASCL2-1143=M)&(MOS-1474=M)&(TSC2-307=VL)	5
5	(RARRES1-893=M)&(HIC1-1081=M)&(CAPG-337=VH)	4
6	(ASCL2-1048=L)&(ASCL2-856=L)&(IRF5-1259=M)& (ASC-1350=M)	3

fully separate the different types of cells. These rules are presented in TABLE II, III and IV, respectively. They can explain the dependency relations among the genes that contribute to the differentiation of the human embryonic stem cells, the growth and migration of the cancer cells, and the normal functions of the differentiated cells.

Table III. Methylation rules of cancer cells

No.	Rules	Matched
1	(PGR-1223=VH)&(CDH3-152=VH)	10
2	(ABL1-217=VH)&(ASC-1416=VH)&(PTPRO-1357=H)	6
3	(CFTR-1097=H)&(CRIP1-1227=VH)&(CFTR-1051=VH)&(ASCL2-1038=VH)	3
4	(ASCL2-1038=M)&(HTR1B-573=M)&(ASCL2-1143=VH)&(CFTR-1097=H)	3
5	(APBA2-537=M)&(ASC-1335=M)&(ASCL2-1048=VL)&(ASCL2-856=VL)&(ATP10A-344=M)	2

Table IV. Methylation rules of normally differentiated cells

No.	Rules	Matched
1	(ASCL2-1339=VH)&(CCND2-596=VH)&(CRIP1-1227=VH)&(CYP1A1-330=VH)&(DBC1-1053=VH)&(EDNRB-1255=VH)&(EPO-1186=VH)&(EPO-1269=VH)&(GABRA5-535=VH)&(GDF10-1382=VH)&(GSTM2-1323=VH)&(HBII-52-1450=VH)&(HLA-DRA-1353=VH)	6
2	(DLC1-1012=VL)&(EPM2A-666=VL)&(F2R-473=VH)&(IL13-298=VL)	3

3.2 Induce Methylation Rules with Decision Tree

In learning the rules from Decision Trees, we use the SPASS Clementine package [10] for our implementation. The rules obtained are listed in Table V, VI, and VII, respectively.

Table V. Methylation rules of embryonic stem cells

No.	Rules	Matched
1	(PTPNS1-765=VH)&(GABRG3-1299=VH)	13
2	(PTPNS1-765=VL)&(CHGA-1371=VL)	10
3	(PTPNS1-765=M)&(HOXA11-558=M)	7
4	(PTPNS1-765=H)&(IL13-55=VH)	2
5	(PTPNS1-765=L)&(MSF-1020=M)	2
6	(PTPNS1-765=L)&(MSF-1020=VH)&(ABCB1-562=VH)	2
7	(PTPNS1-765=M)&(HOXA11-558=VH)&(PAX6-1337=VL)	1

Table VI. Methylation rules of cancer cells

No.	Rules	Matched
1	(PTPNS1-765=H)&(IL13-55=H)	7
2	(PTPNS1-765=L)&(MSF-1020=L)	6
3	(PTPNS1-765=M)&(HOXA11-558=VH)&(PAX6-1337=VH)	5
4	(PTPNS1-765=VL)&(CHGA-1371=M)	3
5	(PTPNS1-765=L)&(MSF-1020=VH)&(ABCB1-562=H)	1
6	(PTPNS1-765=M)&(HOXA11-558=VH)&(PAX6-1337=M)	1
7	(PTPNS1-765=M)&(HOXA11-558=VL)	1

Table VII. Methylation rules of normally differentiated cells

No.	Rules	Matched
1	(PTPNS1-765=VH)&(GABRG3-1299=M)	4
2	(PTPNS1-765=L)&(MSF-1020=VL)	2
3	(PTPNS1-765=VH)&(GABRG3-1299=H)	1
4	(PTPNS1-765=VL)&(CHGA-1371=L)	1
5	(PTPNS1-765=VL)&(CHGA-1371=VH)	1

In Table VIII, for comparison purpose, we show the statistics on the rules generated using the Rough Sets and the Decision Tree methods.

Table VIII. Statistics on the results of three methods

Method	Num. rules	Aver. length of rules	Num. CpGs	Num. genes
Decision Tree	19	2.1	8	8
Rough Sets	13	4.2	43	35

Decision Trees method results into 19 rules that fully summarize the methylation data, where 8 distinct CpG sites are used in building these rules. The rules have the average length of 2.1 attributes in their conditional part. The rough sets method, on the other hand, obtains 13 DNA methylation rules, six rules fewer than the decision tree method. It thus generates a more concise description of the methylation profile. But the average length of the rough sets rules is 4.2, larger than that of Decision Tree. It employs 43 distinct CpG sites of 35 genes in formulating these rules, compared with the 8 CpG sites in Decision Trees.

3.3 Co-methylation Analysis

To investigate the epigenetic interactions of genes, we compute the correlation score of the methylated CpG sites. We discover that CpG sites in the embryonic stem cells, the cancer cells and the normally differentiated cells can b highly comethylated, indicating that they are modulated by the same epigenetic process for cell function. In

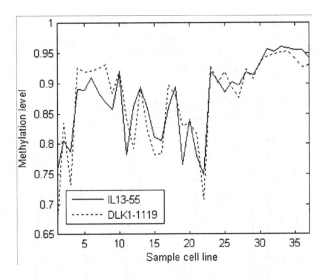

Fig. 2. Comethylation of gene IL13-55 with DLK1-1119 (embryonic stem cell)

Fig. 2, the DLK1 and the IL13 gene promoters have the comethylation score 0.88. DLK1 encodes the proteins mediating the differentiation of the B cells, while IL13 produces their growth factors. They coordinate to mobilize the immune system.

For cancer cells (see Fig. 3), the gene MYC-813 and PRKAR1A-1317 are highly comethylated with the high correlation coefficient of 0.974. The MYC gene is known as a very strong oncongene upregulated in many types of cancers. Surprisingly, the gene PRKAR1A, whose activation leads to the increased apoptosis of human B Lymphocytes, is comethylated with MYC. How this comethylation process occurs is still to be clarified.

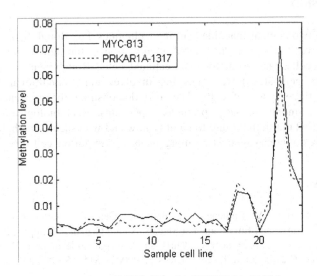

Fig. 3. Comethylation of MYC-813 with PRKAR1A-1317 (cancer cells)

Fig. 4 depicts the relation between G6PD and ELK1 in human embryonic stem cells. The comethylation score is 0.997. The G6PD produces the pentose sugars for nucleic acid synthesis, while the ELK1 regulates the nucleic acid metabolism. These two genes are involved in assembling and de-assembling the structure of the nucleic acid.

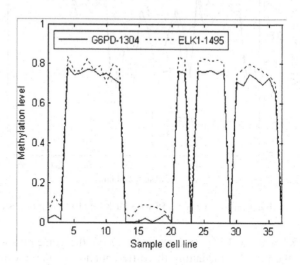

Fig. 4. Comethylation of G6PD-1304 with ELK1-1495 (embryonic stem cells)

We have performed clustering analysis on the methylation profile of all the three types of cells. The names of the genes with the comethylation coefficients higher than 0.80 are listed in the supplemental files.

4   Conclusion

We apply the two popular machine learning techniques, i.e., Rough Sets and Decision Tree, to uncover the logical rules DNA methylation in the human embryonic stem cells, cancer cells, and normally differentiated cells. Rough Sets, compared with Decision tree, generates fewer rules but involves more conditional variables to separate the three types of cells. We also demonstrate the existence of strong comethylation among the gene promoter CpG sites. Real biological experiments should be carried out in the future to identify how and why such comethylation occurs in the processes of embryogenesis, tumorgenesis, and in normal cell functions.

References

1. Jaenisch, R., Bird, A.: Epigenetic Regulation of Gene Expression: How the Genome Integrates Intrinsic and Environmental Signals. Nature Genetics 33 Suppl., 245–254 (2003)
2. Fabian, M., Peter, A., Alexander, O., Christian, P.: Feature Selection for DNA Methylation based Cancer Classification. Bioinformatics 17(90001), S157–S164 (2001)

3. Bibikova, M., et al.: Human Embryonic Stem Cells Have a Unique Epigenetic Signature., Genome Research, online article (August 2006)
4. Bhasin, M., Zhang, H., Reinherz, E., Reche, P.A.: Prediction of Methylated CpGs in DNA Sequences Using a Support Vector Machine. FEBS Letters 579, 4302–4308 (2005)
5. Marjoram, P., Chang, J., Laird, P.W., Siegmund, K.D.: Cluster Analysis for DNA Methylation Profiles Having a Detection Threshold. BMC Bioinformatics 7, 361 (2006)
6. Das, R., et al.: Computational Prediction of Methylation Status in Human Genomic Sequences. PNAS 103(28), 10713–10716 (2006)
7. Pawlak, Z., Wong, S.K.M., Ziarko, W.: Rough sets: Probabilistic versus Deterministic Approach. International Journal of Man-Machine Studies 29, 81–95 (1988)
8. Quinlan, J.R.: Induction of Decision Trees. Machine Learning 1, 81–106 (1986)
9. Rosetta software, http://rosetta.lcb.uu.se/
10. SPASS Clementine software, http://www.spss.com/clementine/

Parallel Construction of Conflict Graph for Phylogenetic Network Problem

D.S. Rao[1], G.N. Kumar[1], Dheeresh K. Mallick[2], and Prasanta K. Jana[1]

[1] Department of Computer Science and Engineering,
Indian School of Mines University, Dhanbad - 826 004, India
prasantajana@yahoo.com
[2] Department of Computer Science and Engineering,
Birla Institute of Technology, Mesra, Ranchi – 835 215, India
dkmallick@gmail.com

Abstract. Conflict graph is used as the major tool in various algorithms [14] - [18] for solving the phylogenetic network problem. The over all time complexity of these algorithms mainly depends on the construction of the conflict graph. In this paper, we present a parallel algorithm for building a conflict graph. Given a set of n binary sequences, each of size m, our algorithm is mapped on a triangular array in $O(n)$ time using $O(m^2)$ processors.

Keywords: Phylogenetic network, galled tree, conflict graph, parallel algorithm.

1 Introduction

A large number of databases are now available along with a massive volume of sequence data. In order to explore and analyze this massive data for many biological applications like phylogenetic inference, the use of high performance computing systems is inevitable. A phylogenetic tree that represents the evolutionary history of organisms has the drawback that it lacks the consideration of some important events such as genetic recombination, hybrid specifications, homoplasy and horizontal gene transfer. A perfect phylogeny problem, which states that each site mutates at most once can be solved in linear time for binary sequences under the infinite-sites assumption [1]. However, recombination is very important as it may provide some valuable clues such as locating the origin of the gene causing genetic diseases. The effects of avoiding recombination have been shown in [2] - [4]. The recombination event results from two individuals that lead a non-tree like structure. Therefore, a more biologically complete evolutionary model is a general network in which both the major evolutionary phenomena, i.e., mutation and recombination are taken care. This network is commonly known as phylogenetic network or ancestral recombination graph (ARG) in the population genetics literature. Hein [5], [6] introduced the problem of phylogenetic network with recombination. Wang et al. [11] showed that computing the minimum number of recombination is NP-hard. Other papers dealing with the lower bound computation on this number are due to Hudson and Kaplan [12], Myers and Griffiths

A. Ghosh, R.K. De, and S.K. Pal (Eds.): PReMI 2007, LNCS 4815, pp. 398–405, 2007.

[13], Song and Hein [7], [8]. Their algorithms are based on combinatorial methods and have exponential time complexity for the worst-case.

There have been many algorithms developed for reconstructing a phylogenetic network for a set of binary sequences. A comprehensive survey can be found in [9], [10]. Given a set of n binary sequences, each of size m, Wang et al. [11] gave an $O(nm + n^4)$ time algorithm to solve a special case of phylogenetic network problem called galled tree problem. In the recent years, several other algorithms have been developed for the same in which the structural properties of a conflict graph are used as the main tool. Gusfield et al. [14], [15] proposed an $O(nm + n^3)$ time algorithm with all-zero ancestral sequence. Later Gusfield [16] extended the results to the case when the ancestral sequence is not known in advance. They also reported a faster algorithm in [17] for site arrangement in gall that runs in $O(n^2)$ time. Bafna et al. [18] used the combinatorial properties of a conflict graph and developed an algorithm in $O(nm^2)$ time. However, the construction of the conflict graph is the principal computation that dominates the over all time complexity of the above algorithms. In this paper, we present a parallel algorithm for building the conflict graph with the motivation that the over all time complexity for solving the galled tree problem will be reduced. Our proposed algorithm requires $O(n)$ time on a triangular array using $O(m^2)$ processors. To the best of our knowledge no parallel algorithm has been developed for the same.

The rest of the paper is organized as follows. We describe the galled tree problem along with some basic terminologies in section 2. The proposed parallel algorithm is given in section 3 followed by the conclusions in section 4.

2   Basic Terminologies and Problem Definition

The following preliminaries will help in understanding our algorithm.

Definition 2.1 (*Phylogenetic network*). An (n, m) phylogenetic network (see Fig. 1) is a directed acyclic graph N that has exactly one root node, a set of internal nodes having indegree 1 or 2 and exactly n leaf nodes. A node with two incoming edges is called a recombination node. Each node is labeled with a binary sequence of length m starting with the root node labeled with all zero-sequence. Each edge except those entering into a recombination node is also assigned an integer (called site or column) within the range 1 to m. If e is an edge coming into a non-recombination node say, u, then the binary sequence (i.e., label) of u is obtained from u's parent by changing from 0 to 1 at position i where i is the integer assigned to the edge e. This corresponds to a mutation at site i on the edge e. Each recombination node w is associated with an integer r_w, $2 \leq r_w \leq m$. We call r_w as the recombination point for w. One of the two sequences labeling the parents of w is designated as P (to mean prefix) and the other as S (suffix). Then the sequence labeling w is formed by concatenating the first $r_w -1$ bits of P and the last $m - r_w + 1$ bits of S.

Given a set of binary sequence M (matrix of size $n \times m$), a phylogenetic network N derives M if and only if each sequence in M labels exactly one of the leaves of N. As an example, a phylogenetic network deriving a binary matrix is shown in Fig. 1 for n = 7 and m = 5. The interpretation of binary sequences and its motivation is discussed in [15].

Definition 2.2 (*Galled tree*). In a phylogenetic network two paths out of a node meeting at a recombination node can form a cycle called recombination cycle and whenever a recombination cycle has no node common with any other recombination cycles it is called a gall. A phylogenetic network where every recombination cycle is a gall is called a galled tree.

Definition 2.3 (*Galled tree problem*). Given a set of binary sequences M, it is to determine whether there exists a galled tree that derives M and if it does, construct it.

Definition 2.4 (*Conflict*). We say that two columns in M conflict each other if and only if they have three rows with the combinations (0,1), (1,0) and (1,1) and a column is called conflicted if it has conflict with at least one other column; otherwise it called unconflicted.

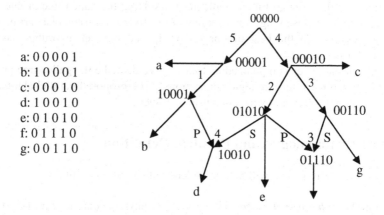

Fig. 1. A phylogenetic network deriving a set of binary sequences (shown in left)

Definition 2.5 (*Conflict graph*). A conflict graph G is formed with all the sites in M where each node is labeled by a distinct site and there exist an undirected edge $<\alpha, \beta>$ if and only if the sites α and β conflict (see Fig. 2).

A connected component in G is the maximal subgraph of G such that for any pair of nodes in G there is at least one path between those nodes in G. A trivial connected component has only one node and has no edge and the site associated with that node is unconflicted. Note that the conflict graph shown in Fig. 2(b) has a single nontrivial connected component that consists of all the sites and there is no trivial connected component. In other words, there is no unconflicted site.

We now state some established theorems with the context of conflict graphs. A phylogenetic network is called perfect if it has no recombination node. The following theorem gives the necessary and sufficient conditions for the existence of a perfect phylogenetic network.

$$M = \begin{bmatrix} 1 & 0 & 1 & 0 \\ 0 & 1 & 0 & 0 \\ 1 & 0 & 1 & 1 \\ 0 & 1 & 1 & 0 \\ 0 & 0 & 1 & 1 \\ 1 & 0 & 0 & 1 \end{bmatrix}$$

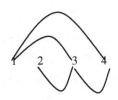

(a) A set of binary sequences   (b) Conflict graph

Fig. 2. Conflict graph corresponding to the set of binary sequences M

Theorem 2.1. There exists a perfect phylogenetic network that derives M if and only if there is no conflicted site in M. Moreover, if there is a perfect phylogenetic network and all columns are distinct, then there is a unique phylogenetic network for M and each edge is also uniquely labeled. If there are identical columns then the perfect phylogeny is unique up to any ordering given to multiple sites that label the same edge [1], [2].

The following theorems show the one-one correspondences between the nontrivial connected components of a conflict graph and the galls.

Theorem 2.2. Each gall in a phylogenetic network with conflicted sites contains all the sites of one non-trivial connected component but no sites from a different non-trivial component [15].

Theorem 2.3. If there is a galled tree for M, then every non-trivial connected components of the conflict graph must be bipartite, and the bipartition is unique, i.e., the sites on one side of the bipartite graph must be strictly smaller than the sites on the other side [15].

Thus the above theorems imply that the construction of the conflict graph is the major computation towards the solution of the phylogenetic netowrk/ galled tree problem.

3 Proposed Algorithms

Let us represent the binary matrix M as follows

$$M = \begin{bmatrix} a_{11} & a_{12} & \cdots & a_{1m} \\ a_{21} & a_{22} & \cdots & a_{2m} \\ \vdots & \vdots & & \vdots \\ a_{n1} & a_{n2} & \cdots & a_{nm} \end{bmatrix}$$

In the construction of the conflict graph, we need to compare each column with all its subsequent columns. As the size of the matrix is $n \times m$, it requires $n\left[\dfrac{m(m-1)}{2}\right]$, i.e., $O(nm^2)$ time. It can be noted that conflict checking between any pair of columns say, i and j is same as that of between j and i. Further, we do not require conflict checking of i with itself. Therefore, a triangular array of $\dfrac{m(m-1)}{2}$ processors will suffice the conflict calculation. Such a triangular array for $n = 6$ and $m = 4$ is shown in Fig. 3 in which a single '*' indicates one unit delay. The columns of the matrix are fed accordingly. Note that we start inputting from column 1 row wise while we start from column 2 column wise in the triangular array. We label the processor with a pair of indices (i, j) according to the columns i and j of the matrix M that are fed to the processor. Assume that each processor has some local registers. We use three flag registers $C1$, $C2$ and $C3$ to indicate the pattern 01, 10 and 11 respectively. At the end of the algorithm, the result of conflict is stored in the status register S, which is basically the adjacency matrix representation of the conflict graph.

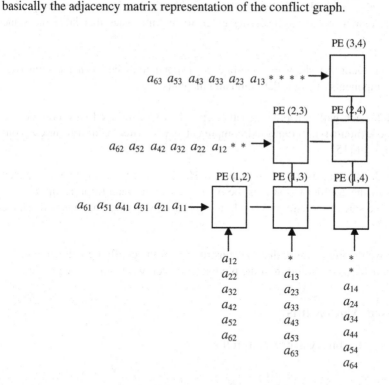

Fig. 3. Triangular array for computing the conflict graph of 6×4 matrix

The basic idea of our algorithm is as follows. Initially all the flag registers are set to 0. The columns of the matrix M are fed through the boundary processors. On receiving two inputs, the content of $C1$ / $C2$ / $C3$ is OR'ed (logical OR) with '1' depending on the input patterns 01, 10 or 11 respectively. It then sends the inputs to the corresponding next row / column processor. This method is continued until the processing is reached to the processor $PE(m-1, m)$. Finally we obtain the conflict result between the columns i and j by AND operation on the contents of $C1$, $C2$ and $C3$ registers. The algorithm is given stepwise as follows.

Step 1. /* Initialization of the flag registers */
> *for all* PE(i, j), $1 \leq i < j \leq m$ *do in parallel*
> $C1(i, j):= 0$;
> $C2(i, j):= 0$;
> $C3(i, j):= 0$
> *end forall*

Step 2. /* Conflict calculations */
> *for all* PE(i, j), $1 \leq i < j \leq m$ *do in parallel*
> *while* PE(i, j) receives two inputs a (from a row)and a'(from a column) *do*
>> (i)   *if* $(a = \text{'0' AND } a' = \text{'1'})$ *then*
>>> $C1(i, j):= C1(i, j) \text{ OR '1'}$;
>>> *else if* $(a = \text{'1' AND } a' = \text{'0'})$ *then*
>>> $C2(i, j):= C2(i, j) \text{ OR '1'}$;
>>> *else if* $(a = \text{'1' AND } a' = \text{'1'})$ *then*
>>> $C3(i, j):= C3(i, j) \text{ OR '1'}$;
>>
>> (ii)   *if* $(i < m)$ *then*
>>> *send* a' *to* PE$(i + 1, j)$
>>> *if* $(j < m)$ *then*
>>> *send* a *to* PE$(i, j + 1)$
>
> *end while*
> *end forall*

Step 3. /* Output */
> *for all* PE(i, j), $1 \leq i < j \leq m$ *do in parallel*
> $S(i, j):= C1(i, j) \text{ AND } C2(i, j) \text{ AND } C3(i, j)$
> *if* $S(i, j):= 1$ *then* write ("There is a conflict between i and j ")
> *end forall*

Step 4. Stop

The above algorithm is illustrated in Fig. 4 for the matrix given in Fig. 2. The contents of the flag registers after 1st, 2nd and final clock are shown in Fig. 4(a), 4(b) and Fig. (c) respectively. At the end of final clock (shown in Fig. 4 d), '1' is stored in the status registers $S(1, 3)$, $S(1, 4)$, $S(2,3)$ and $S(3,4)$ which indicates the conflict between the pair of columns (1, 3), (1,4), (2,3) and (3,4) respectively.

Complexity: Step 1 and Step 3 require constant time. In step 2, the elements $a_{n\ m-1}$ and $a_{n\ m}$ take $2m + n - 4$ communication time from the beginning of the computation to its termination reaching the last processor PE$(m-1, m)$. Since m can be at most $2n$ if there exists a galled tree [11], the time complexity of the above algorithm is $O(n)$.

It is important to note that the above algorithm can work independent of the value of n, i.e., the number of rows (binary sequences)

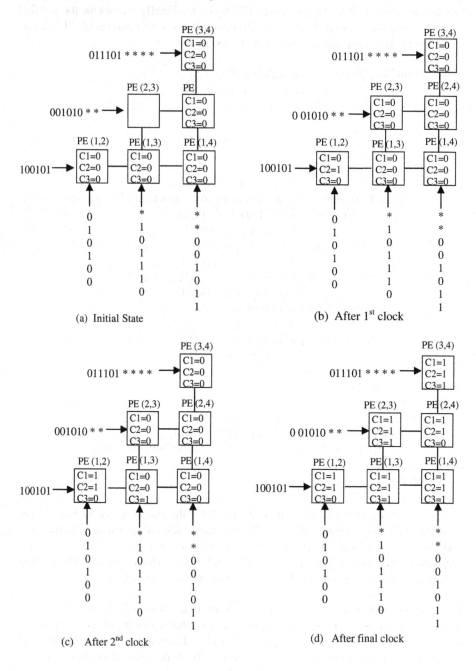

Fig. 4. An example of conflict computation

4   Conclusions

We have presented a parallel algorithm for the construction of a conflict graph, which is used as a major tool to solve a galled tree problem. Given set of n binary sequences, each of size m, our algorithm is mapped on a triangular array in $O(n)$ time using $O(m^2)$ processors and shown to be scalable with respect to n.

References

[1] Gusfield, D.: Efficient algorithms for inferring evolutionary trees. Networks 21, 19–28 (1991)
[2] Schierup, M.H., Hein, J.: Consequences of recombination on traditional phylogenetic analysis. Genetics 156, 879–891 (2000)
[3] Schierup, M.H., Hein, J.: Recombination and the molecular clock. Mol. Biol. Evol. 17, 1578–1579 (2000)
[4] Posada, D., Crandall, K.: The effect of recombination on the accuracy of phylogeny estimation. Journal of Molecular Evolution 54, 396–402 (2002)
[5] Hein, J.: Reconstructing evolution of sequences subject to recombination using parsimony. Math. Biosci. 98, 185–200 (1990)
[6] Hein, J.: A heuristic method to reconstruct the history of sequences subject to recombination. J. Mol. Evol. 36, 396–405 (1993)
[7] Song, Y., Hein, J.: On the minimum number of recombination events in the evolutionary history of DNA sequences. Journal of Mathematical Biology 48, 160–186 (2003)
[8] Song, Y., Hein, J.: Parsimonious reconstruction of sequence evolution and haplotype blocks: Finding the minimum number of recombination events. In: Proc. of 2003 Workshop on Algorithms in Bioinformatics, Berlin, Germany (2003)
[9] Linder, C.R., Moret, B.M.E., Nakhleh, L., Warnow, T.: Reconstructing networks part II: computational aspects. In: the ninth pacific symposium on Biocomputing (2004)
[10] Zahid, M.A.H., Mittal, A., Joshi, R.C.: Use of phylogenetic networks and its reconstruction algorithms. Journal of Bioinformatics India 4, 47–58 (2005)
[11] Wang, L., Zhang, K., Zhang, L.: Perfect phylogenetic networks with recombination. Journal of Computational Biology 8, 69–78 (2001)
[12] Hudson, R.R., Kaplan, N.L.: Statistical properties of the number of recombination events in the History of a sample of DNA sequences. Genetics 111, 147–164 (1985)
[13] Myers, S.R., Griffths, R.C.: Bounds on the minimum number of recombination events in a sample history. Genetics 163, 375–394 (2003)
[14] Gusfield, D., Satish, E., Langley, C.: Efficient reconstruction of phylogenetic networks (of SNPs) with constrained recombination. In: Proceedings of 2nd CSB Bioinformatics Conference, Los Alamitos, CA (2003)
[15] Gusfield, D., Satish, E., Langley, C.: Optimal efficient reconstruction of phylogenetic networks with constrained recombination. Journal of Bioinformatics and Computational Biology 2, 173–213 (2004)
[16] Gusfield, D.: Optimal, efficient reconstruction of root-unknown phylogenetic networks with constrained recombination, Technical Report, Department of Computer Sc., University of California, Davis, CA
[17] Gusfield, D., Satish, E., Langley, C.: The fine structure of galls in phylogenetic networks. Informs Journal on Computing 16, 459–469 (2004)
[18] Bafna, V., Bansal, V.: The number of recombination events in a sample history conflict graph and lower bounds. IEEE Ttrans. on Computational Biology and Bioinformatics 1, 78–90 (2004)

Granular Support Vector Machine Based Method for Prediction of Solubility of Proteins on Overexpression in Escherichia Coli

Pankaj Kumar[2], V.K. Jayaraman[1,*], and B.D. Kulkarni[1,*]

[1] Chemical Engineering Division, National Chemical Laboratory, Pune-411008, India
vk.jayaraman@ncl.res.in, bd.kulkarni@ncl.res.in
[2] Department of Chemical Engineering, Indian Institute of Technology,
Kharagpur-721302, India

Abstract. We employed a granular support vector Machines(GSVM) for prediction of soluble proteins on over expression in *Escherichia coli* . Granular computing splits the feature space into a set of subspaces (or information granules) such as classes, subsets, clusters and intervals [14]. By the principle of divide and conquer it decomposes a bigger complex problem into smaller and computationally simpler problems. Each of the granules is then solved independently and all the results are aggregated to form the final solution. For the purpose of granulation association rules was employed. The results indicate that a difficult imbalanced classification problem can be successfully solved by employing GSVM.

1 Introduction

The enteric bacterium *Escherichia coli* is the most commonly used organism for the production of recombinant proteins. *E coli* has been preferred over other expression hosts because it is well characterized, is easy to handle and manipulate genetically, and has a relatively high growth and production rate [3]. However only some proteins are soluble upon overexpression in *E coli* and others are generally expressed as insoluble aggregate folding intermediates known as inclusion bodies [3]. It has been observed that primary sequence of the protein is the most important determinant of the solubility status of the overexpressed protein [12]. Protein sequences are difficult to understand and model because of their random length; furthermore solubility of protein on overexpression in *E coli* is manifestation of the net effect of several sequence dependent and sequence independent factors [11]. Wilkinson and Harrison [18] observed that inclusion body formation is correlated, in descending order, to charge average, turn forming residue fraction, cysteine fraction, proline fraction, hydrophobicity and molecular weight. But later it was found by Davis *et al*, 1999 that only the first two features are critical in distinguishing soluble and insoluble proteins. This problem was further investigated by several authors [4,7,15] etc.

---

* Corresponding author.

A. Ghosh, R.K. De, and S.K. Pal (Eds.): PReMI 2007, LNCS 4815, pp. 406–415, 2007.

Aliphatic index, the frequency of occurrence of Asn, Thr and Tyr and the dipeptide and tripeptide-composition were found to be the most informative features by Idiculla Thomas and Balaji, 2005. Recently, Idicula-Thomas *et al*, 2006 employed Support Vector Machines(SVM) to predict solubility on overexpression. As the data was unbalanced they had employed the weighted version of the SVM which yielded an accuracy of 72% with a specificity of 76% and sensitivity of 55 %. The algorithm could satisfactorily predict the change in solubility for most of the point mutations reported in literature. Due to the immense importance of this classification problem, it would be highly desirable to increase the prediction accuracy and the sensitivity. In the present work a granular computing based machine learning approach has been employed with a view to improve the prediction performance. Granular computing, unlike traditional, computing, is knowledge oriented. In this work association rules have been used to make granules while SVM is employed as a classifying method. Our result shows superiority of the proposed method over building a single contiguous hyperplane to classify the data using SVM.

2   System and Methods

In this section a brief introduction of the principle of granular computing and support vector machine is presented. Subsequently the methodology employed in building Granular Support vector machines (GSVM) is explained. GSVM combines statistical machine learning algorithm with knowledge-based classification to build a robust model.

2.1   Granular Computing

Granular computing('GrC') was first introduced by T.Y.Lin, in 1997. Since then it has been successfully employed in various fields like diakoptics, divide and conquer structured programming, interval computing, cluster analysis, fuzzy and rough set theories, neutrosophic computing, quotient space theory, belief functions, machine learning, databases, and many others. [9,14,19].

Granular computing splits the feature space into a set of subspaces (or information granules) such as classes, subsets, clusters and intervals [14]. By the principle of divide and conquer it decomposes a bigger complex problem into smaller and computationally simpler problems. Each of the granules is then solved independently and all the results are aggregated to form the final solution. Proper granulation is capable of removing some redundant and irrelevant information and at the same time facilitates getting rid of overfitting problem [16]. Thus granulation helps in building a computationally more efficient model for a complex problem.

In granular computing information granules are first constructed and computations are subsequently carried out with the granules [19] In the literature several methods have been used for granulation like clustering [21], fuzzy sets [20], decision tree and association rules. In this work association rules [16] have been used for the purpose of granulation.

2.2   Association Rules

Association rules tend to capture the underlying hidden patterns in datasets [1]. It provides information in the form of "IF - THEN" statements. In the most general form an association rule has the form IF C_1 THEN C_2 where C_1 and C_2 are conjunctions of condition and each condition is of form either $A_i = V_i$ or $A_i \in (L_i, U_i)$. For e.g., IF *frequency of occurrence of Cysteine* (Cys) lies between 0.024 and 0.3279 THEN *that protein belongs to class -1* (inclusion bodies) The antecedent part ('IF part') can have one or more than one condition joined by an operator *and*. It must, however, be beneficial to form association rules with short IF-parts to avoid overfitting, yielding better generalization. To estimate the quality of a rule formed the confidence and support parameters are used:

Confidence. Confidence is defined as the fraction of instances that are correctly classified by the rule among the instances for which it makes any prediction. Thus if the confidence of a rule is one, we can say that all the data in the training sample that satisfies the rule are correctly classified. Mathematically confidence for an association rule can be represented as,

$$con = \frac{count\_then}{count\_if}$$

Where, *count_then*, is the number of sample that satisfies the $THEN$ part of the rule, and *count_if*, is the number of samples that satisfy only the IF part of association rule.

Support. Support is defined as the ratio of the training data that are correctly classified by the rule to the size of training data with same class label as then part. Hence a support indicates the fraction of data in a class correctly classified by the rule:

$$sup = \frac{count\_then}{size(class_{then})}$$

Where, $size(class_{then})$ is the size of training data with same class label as rule consequent.

While making a rule a threshold for support and confidence is used to prune out all the rules that will have support and confidence below the user defined threshold [2]. This is done so as to obtain a set of association rules that will enable efficient and reliable classification of unseen test instances. If the threshold confidence of the rule is kept high then less number of association rules will be mined but their prediction accuracy will be quite high. Similarly if support is kept very high generalization will be more but number of rules obtained will be very few while if support is too low then rule obtained will tend to overfit the training sample.

2.3   Support Vector Machines(SVM)

SVM are a machine learning algorithm introduced by Vapnik [17].It classifies a nonlinearly separable problem by building a linear separating hyperplane in a high dimensional feature space. The general hyperplane equation is of the form $w^T.x + b = 0$ where w is the weight vector, b is the bias and (.) denotes the dot product. Here w and b are selected so as to maximize the margin, $\frac{1}{\|w\|}$ between the hyperplane and the closest data points belonging to the different classes. The computational intractability problem introduced due to the high dimensionality is over come by defining an appropriate Kernel function.

Given a training dataset of the form (x_i, y_i), $i = 1, 2, ..., N$ where $x_i \in \mathbb{R}^m$ and $y_i \in \{-1, 1\}$, the final SVM classifier function can be given as:

$$f(x) = \sum_{i=1}^{m} y_i \alpha_i K(x_i, x) + b \qquad (1)$$

Here x_i represent i^{th} vector of input pattern and y_i is the target output corresponding to the ith vector and , K is the kernel matrix and $m(N)$ is the number of input pattern having non zero Langrangian multipliers(α_i); these are also called support vectors. The Langrangian multiplier is found by solving the following dual form of quadratic programming problem,

$$w(\alpha) = \sum_{i=1}^{N} \alpha_i - \frac{1}{2} \sum_{i,j=1}^{N} \alpha_i \alpha_j y_i y_j K(x_i, x_j) \qquad (2)$$

Subject to the constraint

$$0 \leq \alpha_i \leq C, i = 1, 2......N$$

and

$$\sum_{i=1}^{N} \alpha_i y_i = 0$$

Where C is the regularization parameter known as cost function that determines the tradeoff between the model complexity and the misclassification. For imbalanced classification problem SVM uses different error cost for the positive C^+ and negative C^- classes. Here the langrangian equation is modified to

$$L_p = \frac{\|w\|^2}{2} + C^+ \sum_{i|y_i=+!}^{n_+} \xi_i + C^- \sum_{j|y_j=-!}^{n_-} \xi_j - \sum_{i=1}^{n} \alpha_i \left[y_i (w.x_i + b) - 1 + \xi_i \right] - \sum_{i=1}^{n} r_i.\xi_i$$

Subject to the constraint

$$0 \leq \alpha_i \leq C^+ \text{ify}_i = +1, \text{ andify}_i = -1$$

After the optimal value of α^i is found the decision function is based on the sign of $f(x)$ as given by eq.(1).

Different types of kernel function (Burges 1998) are used for transformation of input space to a higher dimension feature space .Most commonly used kernel function are

Linear: $K(x_i, x) = x_i^T x_j$

Polynomial: $K(x_i, x) = (\gamma x_i^T x_j + r)^d$

Radial bias function: $K(x_i, x) = \exp(-\gamma \|x_i - x_j\|^2)$

In the present work RBF kernel was found to provide the best possible results.

2.4   Granular Support Vector Machine (GSVM)

For many complex problems it can never be guaranteed that a contiguous hyperplane as discussed above will be able to classify the data correctly. Better classification performance can be achieved by judicious granulation of the feature space. For example, consider the traditional XOR problem; as such without any transformation it is non-linearly separable but if we divide the whole feature into two equal halves then each half becomes linearly separable. Even in the case where a single linear hyperplane is available the use of granulation will help in maximizing the margin between the hyperplane and the closest data points belonging to the different classes.

Furthermore for the case of imbalanced data where the number of instances in one class is far more than the number of instances in other, the separating hyperplane tends to shifts towards the minority class so SVM misclassifies most of the instances into the majority class, thus giving higher accuracy for the majority class but poor predictivity for minority class. In such cases granulation may become an effective means to handle data imbalance. In GSVM the whole feature space is first divided into granules, *viz.*, pure (where almost all the instances belong to one class) and mixed granules (where instances from both classes are present). After separating out pure granules instances present in the mixed granule may become more balanced and hence the probability of prediction accuracy of SVM can be expected to be higher.

3   Modeling Method

3.1   Association Rules Formation as Granulation Methodology

In this work we employed the association rules methodology of Tang *et al., 2005* for the purpose of granulation. As explained in section (2.2) association rules with optimal support and confidence were mined. Among all the association rules formed with different attributes the one with highest confidence was added to the set called *selected_rule_set*. After a rule was chosen, all instances classified by the rule were removed and the rule formation process was repeated until no further rule with support and confidence greater than the predetermined threshold is formed or all the instances has already been classified. Care was taken to apply them in the order in which they are discovered. Table 1 shows the *selected_rule_set* for dataset with all 446 features.

3.2 GSVM Modeling

After all the association rules were obtained GSVM model was built by iteratively combining the association rules from *selected_rule_set* to find the optimal
granules which were both pure and significant. Thus using association rules the
complete feature space can be divided into three different granules[16]:

Positive pure granule (PPG) in which almost all the data belong to positive
class.

Negative pure granule (NPG) in which almost all the instances belong to
negative class.

Mixed zone (MG) or mixed granule which contains instances belonging to
both the classes

To begin with, over the complete feature space, cross validation performance
of SVM in the training dataset was obtained and was taken as baseline accuracy.
The subsequent algorithmic steps in the GSVM model are:

1. A rule from *selected_rule_set* was taken

2. All the instances that satisfy the antecedent (*If part*) part of rule was
removed as pure granules and was assigned the class label as predicted by consequent part of rule.

3. the remaining instances that do not satisfy the rule, form the mixed granule.
SVM model was built with the instances in the mixed granule by tuning the
algorithm parameters to obtain the best accuracy.

4. If the considered rule was added to set called *final_rule_set* if it was found
to improve the classification performance. The improved accuracy was now considered to be the new baseline accuracy.

5. Otherwise the next rule in the list from *selected_rule_set* was taken and
steps 3 to 5 were repeated.

6. The above steps were continued until the entire set of rules in the list had
been processed.

Table 2 shows the *final_rule_set* for the dataset comprising of all 446 features.
When unseen test instances are to be classified, they are first checked by the
formed association rules in *final_rule_set*. All the instances that satisfy the
antecedent of rule are assigned the class predicted by the rule and the class label
for the remaining instances (not predicted by rule) were predicted by the SVM
model built on mixed granule.

4 Experimental Evaluation

4.1 Data Description

The Dataset of Idicula-Thomas et al. [12] were employed for the GSVM experiments. This dataset consist of 192 protein sequences, 62 of which are soluble
on overexpression in *E.Coli* and the remaining 130 sequences form inclusion
body. The 446 features extracted by them include i) six physiochemical prop-
erties(Attribute nos. 1-6), *viz.*, aliphatic index, instability index of the entire
protein, instability index of the N-terminus and net charge. ii) twenty single

aminoacid residues(Attribute nos. 7-26) arranged in alphabetical order (A,C,D) followed by 20 reduced alphabets(attribur nos. 27-46). The reduced alphabets employed includes 7 reduced class of conformational similarity,8 reduced class of BLOSUM50 substitution matrix and 5 reduced class of hydrophobicity [12]. Finally the features in the list includes 400 attributes(attribute nos. 47-446) comprising of the dipeptide compositions.

4.2   Model Building

The instances(each comprising of 446 features)were randomly divided into training and test sets keeping the inclusion body forming and the soluble proteins approximately in ratio of 2:1. The training dataset comprised 128 sequences, 87 inclusion body-forming and 41 soluble proteins. The test dataset comprised 64 sequences, 43 inclusion body forming and 21 soluble proteins. [12] The modeling process was initiated by first forming association rules with the instances in the training dataset. As explained in section (3.1), only single feature association rules with substantial support and confidence were mined to form selected_rule_set. Table 1 shows the mined set of association rules in the form:
. IF $X_0 \leq attribute_i \leq X_1$ THEN $class = y$

Table 1. Mined association rule on original unscaled training data

S.No.	X_0	X_1	Attribute Number	Confidence	Support	Class
1	0.0051	0.0242	443	1	0.1954	-1
2	0.0059	Inf	435	1	0.1609	-1
3	0.0078	0.0127	296	1	0.1609	-1
4	0.0084	0.0141	330	0.9231	0.2927	1

GSVM model was built employing these rules for pure zones and SVM classification for the mixed zone. However before applying SVM, as a preprocessing step all the features were scaled by making their mean zero and standard deviation one. SVM experiments done in this work were performed using an implementation of LIBSVM (chang Lin, 2001). As our data was imbalanced weighted SVM was used. The SVM parameters C,γ and weights were tuned by grid search. Table 2 shows the final set of rule selected by GSVM algorithm to make a model. Out of the 4 rules shown in table 1 only the rules shown in table 2 were found to increase the cross-validation performance over training data, so only those two rules were selected.

The algorithm performance was subsequently tested on unseen test dataset using the same test measure as used by Idicula-Thomas and Kulkarni et al. [12]. 50 random splits of the dataset were taken (with the same ratio of nearly 1:2 between the two classes of proteins), and their average performance was measured. Table 3 shows the comparison of results obtained by using GSVM and SVM (as reported by [12]). These results shows that the GSVM is capable

Table 2. Final set of rules selected by GSVM algorithm

S.No.	X_0	X_1	Attribute Number	Confidence	Support	Class
1	0.0078	0.127	296	1	0.1609	-1
2	0.0084	0.141	330	0.9231	0.2927	1

Table 3. Classification result on test dataset averaged over 50 random splits

Number of features	Algorithm	ROC	Accuracy (%)	Specificity(%)	Sensitivity(%)
446	SVM	0.5316	72	76	55
446	GSVM	0.7227	75.41	81.40	63.14
27	GSVM	0.7635	79.22	84.70	68

of capturing inherent data distribution more accurately as compared to a single SVM build over complete feature space.

As the number of proteins forming inclusion bodies is far more than number of soluble proteins(nearly 2 times,) our dataset is imbalanced. So accuracy alone does not give the correct measure of performance. For an imbalanced data, receiver operation characteristic (ROC) curve is generally used as test measure. Our result shows a marked increase in the value of ROC from 0.5316 using SVM over complete feature to 0.72227 using GSVM for the *bestclassifier* reported by Idicula-Thomas A.J.Kulkarni *et al.*, 2006 using 446 features. The value of sensitivity and specificity has also gone up which has increased the overall accuracy to 75.41%. The increased ROC shows that our model is not biased towards majority class and is capable of predicting the minority class (soluble proteins) as well with equally good accuracy.

We also tried feature selection in the mixed granule with the original 446 features to find the most informative subset. After feature selection only 27 features were found critical for predicting the solubility. The selected features were aliphatic index, frequency of occurrence of residues Cysteine (Cys), Glutanic acid (Glu), Asparagine (Asn) and Tyrosine (Tyr). Among the reduced alphabets, only the reduced class [CMQLEKRA] was selected from the seven reduced classes of conformational similarity. Similarly from the five reduced classes of hydrophobicity originally reported, only [CFILMVW] and [NQSTY] were selected. And from the eight reduced classes of BLOSUM50 substitution matrix the only reduced class selected was [CILMV]. The 18 dipeptide whose composition were found to significant. These include [VC], [AE], [VE], [WF], [YF], [AG], [FG], [WG], [HH], [MI], [HK], [KN], [KP], [ER], [YS], [RV], [KY], and [TY]. A new GSVM model was built with these most informative features.

In this case we didn't get any positive rule, which satisfied our minimum support and confidence threshold condition (kept as 0.18 and 0.85 respectively for positive rule while for negative rule the values are 0.15 and 0.95 respectively). After applying GSVM all 3 rules were selected in the final model. So our final

Table 4. Mined association rule on original unscaled training data after feature selection

S.No.	X_0	X_1	Attribute Number	Confidence	Support	Class
1	0.0059	Inf	26	1	0.1954	-1
2	0.0240	0.3279	2	1	0.1609	-1
3	0.0027	0.3279	12	1	0.1954	-1

model with 27 selected features comprised of association rules shown in table 4 and SVM parameters C=32, $\gamma = 0.0039$ and W=1.3. Our result (Table 3) shows that performance was further improved by feature selection.

5   Conclusion

In this work Granular Support Vector Machines(GSVM) was successfully employed for classification of soluble and insoluble proteins. GSVM systematically combines statistical learning theory with granular computing to build a hybrid system exhibiting superior performance with the inherently unbalanced data set. By splitting the feature space into granules it reduces the complexity of problem thereby improving the classification efficiency. The significant increase in ROC values as compared to that obtained using SVM alone bears testimony to the excellent generalization capability of the hybrid model.

Acknowledgement

The authors are grateful to Department of Biotechnology (DBT) for providing financial assistance.

References

1. Agrawal, et al.: Mining association rules between sets of items in large databases. In: Proceedings of the ACM SIGMOD Conference on Management of Data, Washington, D.C., pp. 207–216 (May 1993)
2. Agrawal, R., Ramakrishnan, S.: Fast algorithms for mining association rules. In: Proc. 20th Int. Conf. Very Large Data Bases, VLDB, pp. 12–15. Morgan Kaufmann, San Francisco (1994)
3. Baneyx, F.: Recombinant protein expression in Escherichia coli. Curr. Opin. Biotechnol. 10, 411–421 (1999)
4. Bertone, P., et al.: SPINE: an integrated tracking database and data mining approach for identifying feasible targets in high-throughput structural proteomics. Nucleic Acids Res. 29, 2884–2898 (2001)
5. Burges, C.J.C.: A tutorial on support vector machines for pattern recognition. Data Mining Knowledge Disc 2(2), 121–167 (1998)
6. Davis, G.D., Elisee, C., Newham, D.M., Harrison, R.G.: New Fusion Protein Systems Designed to Give Soluble Expression in Escherichia coli. Biotechnol. Bioeng. 65, 382–388 (1999)

7. Goh, C.S., et al.: Mining the structural genomics pipeline: identification of protein properties that affect high-throughput experimental analysis. J. Mol. Biol. 336, 115–130 (2004)
8. Harrison, R.G.: Expression of soluble heterologous proteins via fusion with NusA protein. inNovations 11, 4–7 (2000)
9. Hirota, K., Pedrycz, W.: Fuzzy computing for data mining. Proceedings of the IEEE 87, 1575–1600 (1999)
10. Witten, I.H., Frank, E.: Data Mining: Practical machine learning tools and techniques, 2nd edn. Morgan Kaufmann, San Francisco (2005)
11. Idicula-Thomas, S., Balaji, P.V.: Understanding the relationship between the primary structure of proteins and its propensity to be soluble on overexpression in Escherichia coli. emphProtein Sci. 14, 582–592 (2005)
12. Idicula-Thomas, S., Kulkarni, A.J., Kulkarni, B.D., Jayaraman, V.K., Balaji, P.V.: A support vector machine-based method for predicting the propensity of a protein to be soluble or to form inclusion body on overexpression in Escherichia coli. Bioinformatics 22, 278–284 (2006)
13. Keerthi, S.S., Lin, C.-J.: Asymptotic behaviors of support vector machines with Gaussian kernel. Neural Computation 15(7), 1667–1689 (2003)
14. Lin, T.Y.: Granular computing, Announcement of the BISC Special Interest Group on Granular Computing (1997)
15. Luan, C.H., et al.: High-throughput expression of C. elegans proteins. Genome Res. 14, 2102–2110 (2004)
16. Yuchun, T., Bo, J., Zhang, Y.-Q.: Granular support vector machines with association rules mining for protein homology prediction, Artificial Intelligence in Medicine. Computational Intelligence Techniques in Bioinformatics 35(1-2), 121–134 (2005)
17. Vapnik, V.: The Nature of Statistical Learning Theory. Springer, New York (1995)
18. Wilkinson, D.L., Harrison, R.G.: Predicting the solubility of recombinant proteins in Escherichia coli. Biotechnology 9, 443–448 (1991)
19. Yao, Y.Y.: Granular computing: basic issues and possible solutions. In: Wang, P.P. (ed.) Proceedings of the 5th Joint Conference on Information Sciences, Atlantic City, New Jersey, USA. Association for Intelligent Machinery, vol. I, pp. 186–189 (2000)
20. Zadeh, L.A.: Toward a theory of fuzzy information granulation and its centrality in human reasoning and fuzzy logic. Fuzzy Sets and Systems 90(2), 111–127 (1997)
21. Zhong, W., He, J., Harrison, R., Tai, P.C., Pan, Y.: Clustering support vector machines for protein local structure prediction. Expert Systems with Applications 32(2), 518–526 (2007)

Evolutionary Biclustering with Correlation for Gene Interaction Networks

Ranajit Das[1], Sushmita Mitra[1], Haider Banka[2], and Subhasis Mukhopadhyay[3]

[1] Machine Intelligence Unit, Indian Statistical Institute, Kolkata 700 108, India
{ranajit_r,sushmita}@isical.ac.in
[2] Center for Soft Computing Research, Indian Statistical Institute,
Kolkata 700 108, India
hbanka_r@isical.ac.in
[3] Bioinformatics Center, Department of Bio-Physics, Molecular Biology and Genetics,
Calcutta University, Kolkata 700 009, India
smbmbg@caluniv.ac.in

Abstract. In this study, a novel rank correlation-based multiobjective evolutionary biclustering method is proposed to extract simple gene interaction networks from microarray data. Preprocessing helps to preserve those gene interaction pairs which are strongly correlated. Experimental results on time series gene expression data from Yeast are biologically validated based on standard databases and information from literature.

Keywords: Bioinformatics, transcriptional regulatory network extraction, gene expression profile, gene interaction network.

1 Introduction

Advent of DNA microarray technology, leading to complete-genome expression profiling, coupled with various analytical methods provide a lot of information about cellular function. This forms an indispensable tool for exploring transcriptional regulatory networks from the system level and is useful when one dwells into the cellular environment to investigate various complex interactions [1]. Biological pathways can be represented as networks and broadly classified as *metabolic pathways*, *signal transduction pathways* and *gene regulatory networks or gene interaction networks*. Biological networks connect genes, gene products (in the form of protein complexes) to one another. A network of co-regulated genes may form gene clusters that can encode proteins, which interact amongst themselves and take part in common biological processes. Clustering of gene expression profiles have been employed to identify co-expressed groups of genes [2] as well as to extract gene interaction/gene regulatory networks [3].

Sharing of the regulatory mechanism amongst genes, in an organism, is predominantly responsible for their co-expression. Genes with similar expression profiles are very likely to be regulators of one another or be regulated by some other common parent gene [4]. Often, it is noted that during few conditions a small set of genes are co-regulated and co-expressed, their behavior being almost

A. Ghosh, R.K. De, and S.K. Pal (Eds.): PReMI 2007, LNCS 4815, pp. 416–424, 2007.
© Springer-Verlag Berlin Heidelberg 2007

independent for rest of the conditions. The genes share local rather than global similar patterns in their gene expression profiles. Generally, group of genes are identified in the form of biclusters using continuous columns biclustering because biological processes start and terminate over a continuous interval of time [5], [6]. The aim of biclustering is to bring out such local structure inherent in the gene expression data matrix. It refers to the clustering of both rows (genes) and columns (conditions) of a data matrix (gene expression matrix), simultaneously, during knowledge discovery about local patterns from microarray data [7].

In this paper we use continuous column multiobjective evolutionary biclustering to propose the extraction of rank correlated gene pairs for generating gene interaction subnetworks based on regulatory information among genes. Preprocessing, involving the discretization of the rank correlation matrix (using quantile partitioning) and subsequent elimination of weak correlation links, is employed to retain strongly rank correlated (positive or negative) gene interaction pairs. An adjacency matrix is formed from the resulting correlation matrix, based on which the network is generated and biologically validated. The usefulness of the model is demonstrated, using time-series gene expression data from *Yeast*.

2   Extraction of Gene Interaction Network

Biological networks involving gene pairs which demonstrate transcription factor (TF)-target relationship, is an important research problem. A gene interaction network is a complex structure comprising various gene products activating or repressing other gene products. A gene that regulates other genes is termed the transcription factor, while the gene being regulated is called its target. The presence of a TF, can alternatively switch "ON" some genes in the network while others remain "OFF", orchestrating many genes simultaneously. The proper understanding of gene interaction networks is essential for the understanding of fundamental cellular processes involving growth and decay, development, secretion of hormones, etc. During transcription of gene expression specific groups of genes may be made active by certain signals which on activation, may regulate similar biological processes. The genes may also be regulators of each others transcription. Target genes sharing common TFs demonstrate similar gene expression patterns along time [8], [9]. Analysis of similar expression profiles brings out several complex relationships between co-regulated gene pairs, including co-expression, time shifted, and inverted relationships [10].

In this paper an attempt has been made to model the relationship between a transcription factor and its target's expression level variation over time in the framework of the generated biclusters. The extraction of the relationship between the gene pair is biologically more meaningful and computationally less expensive as a bicluster is a subset of highly correlated genes and conditions. Rank correlation provides a similarity measure, which retains the relevant information necessary for computing pairwise correlation between gene pairs. The relationship is presented in terms of rules, where a TF is connected to its regulated target gene. These rules are subsequently mapped to generate parts of the entire regulatory network. It may be noted that intra-pathway gene interactions, responsible for a particular biological function and possibly within a bicluster, are generally stronger than any inter-pathway interactions.

2.1   Multi-objective Evolutionary Biclustering

Biclustering refers to the simultaneous clustering and redundant feature reduction involving both attributes and samples. This results in the extraction of biologically more meaningful, less sparse partitions from high-dimensional data, and exhibit similar characteristics. The partitions are known as biclusters. Biclustering has been applied to gene expressions from cancerous tissues [11], mainly for identifying co-regulated genes, gene functional annotation, and sample classification.

A bicluster can be defined as a pair (g, c), where $g \subseteq \{1, \ldots, m\}$ represents a subset of genes and $c \subseteq \{1, \ldots, n\}$ represents a subset of conditions (or time points). The optimization task [7] involves finding the maximum-sized bicluster not exceeding a certain homogeneity constraint mentioned below. The size (or volume) $f(g, c)$ of a bicluster is defined as the number of cells in the gene expression matrix E (with values e_{ij}) that are covered by it. The homogeneity $\mathcal{G}(g, c)$ is expressed as a mean squared residue score. We maximize

$$f(g, c) = |g| \times |c|, \tag{1}$$

subject to a low $\mathcal{G}(g, c) \leq \delta$ for $(g, c) \in X$, with $X = 2^{\{1, \ldots, m\}} \times 2^{\{1, \ldots, n\}}$ being the set of all biclusters, where

$$\mathcal{G}(g, c) = \frac{1}{|g| \times |c|} \sum_{i \in g, j \in c} (e_{ij} - e_{ic} - e_{gj} + e_{gc})^2. \tag{2}$$

Here $e_{ic} = \frac{1}{|c|} \sum_{j \in c} e_{ij}$ and $e_{gj} = \frac{1}{|g|} \sum_{i \in g} e_{ij}$ are the mean row and column expression values for (g, c), and $e_{gc} = \frac{1}{|g| \times |c|} \sum_{i \in g, j \in c} e_{ij}$ is the mean expression value over all cells contained in the bicluster (g, c). The threshold δ, a user-defined quantity, represents the maximum allowable dissimilarity within the bicluster. A good bicluster is one for which $\mathcal{G}(g, c) < \delta$ for some $\delta \geq 0$.

Multi-objective evolutionary algorithm, a global search heuristic, has been used for biclustering [12]. The maximal set of genes and conditions were generated keeping the "homogeneity" criteria of the biclusters intact. Since these two characteristics of biclusters are conflicting to each other, multi-objective optimization may be employed to model them. To optimize this conflicting pair, the fitness function f_1 is always maximized while function f_2 is maximized as long as the residue is below the threshold δ. The formulation is as follows:

$$f_1 = \frac{g \times c}{|G| \times |C|}, \tag{3}$$

$$f_2 = \begin{cases} \frac{\mathcal{G}(g,c)}{\delta} & \text{if} \quad \mathcal{G}(g, c) \leq \delta \\ 0 & \text{otherwise,} \end{cases} \tag{4}$$

where g and c represent, respectively, the number of ones in the genes and conditions within the bicluster, $\mathcal{G}(g, c)$, δ are as defined earlier, and G and C are the total number of genes and conditions of the initial gene expression array. The Multi-objective GA (NSGA II), in association with the local search procedure discussed in [12], were used for the generation of the set of biclusters. The algorithm followed is discussed in details in [12].

2.2 Correlation Between Gene Pairs

In this paper we propose a rank correlation-based approach for the extraction of gene interaction networks. A small number of genes participate in a cellular process of interest, being expressed over few conditions. Co-regulated genes are often found to have similar patterns in their gene expression profiles locally, rather than globally. The genes share similar sub-profiles, over a few time points, instead of the complete gene expression profiles. Thus, considering the global correlation amongst genes, *i.e.*, computation of correlation amongst genes employing the complete gene expression data matrix, would not reveal proper relationship between two of them. The *Spearman rank correlation* provides such a local similarity measure between the two time-series curves, since it is shape-based. The expression profile e of a gene may be represented over a series of n time points. Since the genes in a bicluster are co-expressed, the concept of correlation have been used to quantify their similarity. Instead of the commonly used similarity measures like the Euclidean distance or the Pearson correlation the *Spearman rank correlation* (RC) have been employed due to its robustness towards outliers and measurement errors [13]. Moreover, RC does not assume a Gaussian distribution of points. $RC(e_1, e_2)$ between gene expression profile pair e_1 and e_2 provides a shape-based similarity measure between the two time-series curves, sampled at e_{1i} and e_{2i} over n time intervals. This is expressed as

$$RC(e_1, e_2) = 1 - \frac{6}{n(n^2 - 1)} \sum [r_{e_1}(e_{1i}) - r_{e_2}(e_{2i})]^2, \tag{5}$$

where $r_{e_1}(e_{1i})$ is the rank of e_{1i}. Here an extended version of the RC has been used which takes into account the resolving of ties, *i.e.* $e_{1j} = e_{1i}$ for $i \neq j$. The RC satisfies $-1 \leq RC(e_1, e_2) \leq 1$ for all e_1, e_2.

The first preprocessing step is to filter correlation coefficients which contribute minimally towards regulation. This is because often an exhaustive search of the possible interactions between genes is intractable. Next those coefficients are selected whose absolute values are above a detection threshold, suggesting greater correlation amongst the gene pairs. In this way we focus on a few highly connected genes, that possibly link the remaining sparsely connected genes. The correlation range $[RC_{\max}, RC_{\min}]$ is divided into three partitions each, using *quantiles* or *partition values*[1] [14] so that the influence of extreme values or noisy patterns are lessened. It has to be noted that negative correlation between

---

[1] Quantiles or partition values are the values of a variate which divide the total frequency into a number of equal parts.

two gene profiles is essentially not zero correlation between them. Furthermore, any correlation coefficient (i) having value above Q_2^+ (below Q_2^-) indicates high positive (negative) correlation, and (ii) having value in $[Q_1^+, Q_2^+)$ $([Q_2^-, Q_1^-))$ indicates moderate positive (negative) correlation.

An adjacency matrix is computed as

$$A(i,j) = \begin{cases} -1 \text{ if } & RC \leq Q_2^- \\ +1 \text{ if } & RC \geq Q_2^+ \\ 0 & \text{otherwise,} \end{cases} \tag{6}$$

where we assume absence of self correlations among the genes. Thereafter, a network connecting the various genes is generated.

3 Experimental Results

3.1 Network Extraction

Yeast cell-cycle CDC28 data [15] is a collection of 6178 genes (attributes) for 17 conditions (time points), taken at 10-minute time intervals covering nearly two cycles. The missing values present in the data set are imputed according to the methodology provided in [16][2]. At first pairwise rank correlation coefficients between gene pairs are computed by eqn. (5) to generate the network architecture from the extracted biclusters. Quantile partitioning is employed next, to choose the strong positive as well as negative correlation links. In this way, the top $\frac{1}{3}$ of the positive and negative links are chosen to be connected in a network. A sample network consisting of three biclusters of sizes 7, 10, and 14, respectively, are shown in Fig. 1. A transcription factor is connected to its target gene by an arrow when such a TF-Target pair is found to exist within any of the biclusters. Gene pairs connected by solid lines depict positive correlation, while those connected by dashed lines are negatively correlated. TFs external to the network, but having targets within the network, are connected to their corresponding targets by dotted arrows. As an example, the TF $YHR084W$ (encircled with solid lines) is a member of the network of 10 genes and has targets in all the three networks. An external TF $YJL056C$ (encircled with dotted lines) has targets in networks of 7 and 10 genes. The biclusters were biologically validated from gene ontology study, based on the statistically significant GO annotation database[3].

Fig. 2 illustrates the expression profiles of the six external TFs $GCR1/ YPL075W$, $RPN4/YDL020C$, $YAP1/YML007W$, $SIN3/YOL004W$, $ZAP1/ YJL056C$ and $CYC8/YBR112C$ along with the target gene $SSC1/YJR045C$, corresponding to the 7-node bicluster network (generated in terms of pairwise rank correlation values) in Fig. 1. The symbols *viz.* •, ∘, △, □, ◇, ▽, connected by the solid lines, are the expression profiles for the TFs, respectively, while the dashed line connecting the symbol ⋆ is that corresponding to the regulated gene.

---

[2] LSimpute: accurate estimation of missing values in microarray data with least squares methods.

[3] http://db.yeastgenome.org/cgi-bin/GO/goTermFinder

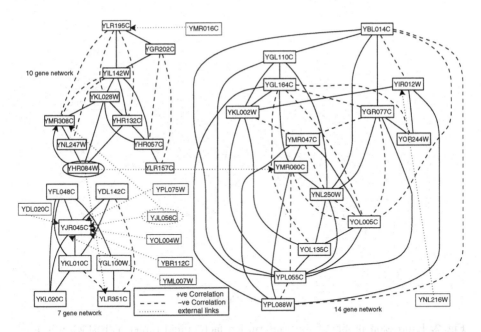

Fig. 1. Network (bicluster) of 10 genes connected by transcription factor $YHR084W$ to networks (biclusters) of 7 and 14 genes

The behavior of $GCR1/YPL075W$ with the target is considered as an example. Here the two genes are found to behave almost simultaneously over conditions 0-40 minutes, while the same two demonstrate a marked inverted relationship during conditions 40-80 minutes. Again the time span of 110-130 minutes displays a shifted response of the target as compared to the TF. However, the expression profiles of $CYC8/YBR112C$ and the target appear to share a strong simultaneous relationship almost along the entire time span. So the genes are found to behave similarly in their gene expression sub-profiles.

3.2 Biological Validation

During the prediction of regulatory networks [17] the genes $YHR084W$-$YLR351C$ were reported to form a TF-Target pair. We also obtained the summary of the TF-Target pair $YHR084W$-$YLR351C$ (Fig. 1) in terms of *Molecular Function*, *Biological Process* and *Cellular Component* from the Saccharomyces Genome Database (SGD)[4]. From our calculations we have also confirmed that an interaction exists between the target and its TF. It is reported in the database that the biological process involving protein $YLR351C$ is not fully understood as yet and $YHR084W$ has transcription factor activity. It becomes more difficult when one attempts to extract some biologically meaningful information

---

[4] A scientific database of the molecular biology and genetics of the yeast *Saccharomyces cerevisiae* - http://db.yeastgenome.org/

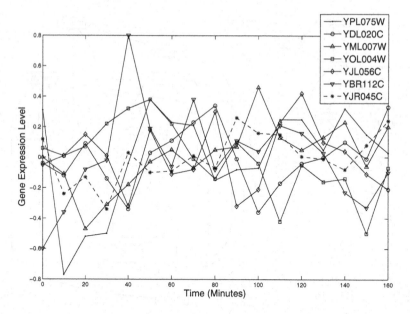

Fig. 2. Expression profile of six transcription factors and target YJR045C, in the 7-node network

involving these two entities. From such scanty information our method has been able to identify that there exists a link between a TF and its target. From their cellular components we model, as an efficacy of the biclustering, the transcription of $YLR351C$ by $YHR084W$ occurring inside the nucleus, and then the regular translation mechanism follows. In like manner for the TF-Target pair of $YPL075W$ and $YJR045C$ (Fig. 1) reported in [17], we obtained their summary from SGD and found $YPL075W$ to be transcriptional activator of genes involved in glycolysis while $YJR045C$ has ATPase, enzyme regulator and protein transporter activity. Again we were able to predict that $YPL075W$ is involved in the transcription of $YJR045C$ and would go into the glycolysis process.

One can arrive at similar kind of conclusions, for the rest TF-Target pairs, with a certain definite degree of confidence. As relevant literature in this area are really very sparse a large number negative results is only expected. Our algorithm has not yet detected any false positive or false negative TF-Target pairs, which is consistent with the information available either in the literature or in the databases.

4   Conclusions and Discussion

In this paper we have introduced an approach based on multiobjective evolutionary biclustering and subsequent extraction of rank correlated gene pairs for the extraction of gene interaction networks. Biologically relevant small biclusters were obtained, using time-series gene expression data from *Yeast*. These were validated using the statistically significant GO annotation database. The pairwise rank correlation coefficients among gene pairs were computed by eqn. (5) followed by the quantile partitioning to select the strong positive as well as negative correlation links. The strongly correlated genes were then chosen to be connected in a network. *TF*-Target gene pairs in the network, shown in Fig. 1, were found to exhibit strong correlations. We tried to model the interaction among them from information available in the literature/databases *viz.*, SGD. We have also analyzed the expression profiles of the regulator and the regulated genes which reveals several complex (time shifted, inverted, simultaneous, etc.) relationships between them. The sparse nature of gene regulatory networks was reflected well on choosing Spearman rank correlation as the similarity measure.

References

1. Mitra, S., Pedrycz, W.: Special Issue on Bioinformatics. Pattern Recognition 39 (2006)
2. Eisen, M.B., Spellman, P.T., Brown, P.O., Botstein, D.: Cluster analysis and display of genome-wide expression patterns. Proceedings of National Academy of Sciences USA 95, 14863–14868 (1998)
3. Tavazoie, S., Hughes, J.D., Campbell, M.J., Cho, R.J., Church, G.M.: Systematic determination of genetic network architecture. Nature Genetics 22, 281–285 (1999)
4. Gasch, A.P., Eisen, M.B.: Exploring the conditional coregulation of yeast gene expression through fuzzy *k*-means clustering. Genome Biology 3, research0059.1–0059.22 (2002)
5. Ji, L., Tan, K.L.: Identifying time-lagged gene clusters using gene expression data. Bioinformatics 21, 509–516 (2005)
6. Madeira, S.C., Oliveira, A.L.: A Linear Time Biclustering Algorithm for Time Series Gene Expression Data. In: Casadio, R., Myers, G. (eds.) WABI 2005. LNCS (LNBI), vol. 3692, pp. 39–52. Springer, Heidelberg (2005)
7. Cheng, Y., Church, G.M.: Biclustering of gene expression data. In: Proceedings of ISMB 2000, pp. 93–103 (2000)
8. Bansal, M., Belcastro, V., Ambesi-Impiombato, A., di Bernardo, D.: How to infer gene networks from expression profiles. Molecular Systems Biology 3, 1–10 (2007)
9. Jong, H.D.: Modeling and simulation of genetic regulatory systems: A literature review. Journal of Computational Biology 9, 67–103 (2002)
10. Zhang, Y., Zha, H., Chu, C.H.: A time-series biclustering algorithm for revealing co-rregulated genes. In: Proceedings of the International Conference on Information Technology: Coding and Computing (ITCC 2005), pp. 1–6 (2005)
11. Madeira, S.C., Oliveira, A.L.: Biclustering Algorithms for Biological Data Analysis: A Survey. IEEE Transactions on Computational Biology and Bioinformatics 1, 24–45 (2004)
12. Mitra, S., Banka, H.: Multi-objective evolutionary biclustering of gene expression data. Pattern Recognition 39, 2464–2477 (2006)
13. Balasubramaniyan, R., Hllermeier, E., Weskamp, N., Kamper, J.: Clustering of gene expression data using a local shape-based similarity measure. Bioinformatics 21, 1069–1077 (2005)
14. Davies, G.R., Yoder, D.: Business Statistics. John Wiley & Sons, London (1937)

15. Cho, R.J., Campbell, M.J., Winzeler, L.A., Steinmetz, L., Conway, A., Wodicka, L., Wolfsberg, T.G., Gabrielian, A.E., Landsman, D., Lockhart, D.J., Davis, R.W.: A genome-wide transcriptional analysis of the mitotic cell cycle. Molecular Cell 2, 65–73 (1998)
16. Bo, T.H., Dysvik, B., Jonassen, I.: Lsimpute: accurate estimation of missing values in microarray data with least squares methods. Nucleic Acids Research 32, 1–8 (2004)
17. Qian, J., Lin, J., Luscombe, N.M., Yu, H., Gerstein, M.: Prediction of regulatory networks: genome-wide identification of transcription factor targets from gene expression data. Bioinformatics 19, 1917–1926 (2003)

Identification of Gene Regulatory Pathways: A Regularization Method

Mouli Das[1], Rajat K. De[1], and Subhasis Mukhopadhyay[2]

[1] [1]Machine Intelligence Unit, Indian Statistical Institute, Kolkata 700 108, India
{mouli_r,rajat}@isical.ac.in
[2] Bioinformatics Center, Department of Bio-Physics, Molecular Biology and Genetics,
Calcutta University, Kolkata 700 009, India
smbmbg@caluniv.ac.in

Abstract. Network based pathways are emerging as an important paradigm for analysis of biological systems. In the present article, we introduce a new method for identifying a set of extreme regulatory pathways by using structural equations as a tool for modeling genetic networks. The method, first of all, generates data on reaction flows in a pathway. A set of constraints is formulated incorporating weighting coefficients. The effectiveness of the present method is demonstrated on two genetic networks existing in the literature. A comparative study with the existing extreme pathway analysis also forms a part of this investigation.

Keywords: flux balance analysis, gene regulatory networks, apoptosis, incidence matrix, yeast cell cycle.

1 Introduction

The abundance of genomic data currently available has led to the construction of genome-scale models of metabolism [1]. Recently several mathematical and computational approaches for defining the functions of networks have emerged. These properties analyze the systems properties of networks and move beyond the traditional pathway definitions present in biochemistry networks. A genetic network being a dynamic system provides important information on how a biological network changes from one state to another. But they need extensive quantitative information which is difficult to obtain [2]. The development of dynamic models of genetic networks is severely hampered due to the lack of experimental procedures to measure the dynamic quantities.

Due to the recent advances in genomics, the reconstruction of the networks of microorganisms has become feasible by using biochemical knowledge and information from genetic databases. However the analysis of such large scale systems remains a major challenge in computational biology. To study complex biological networks that are assumed to operate in the steady state it is necessary to develop a mathematical framework [3]. The stationary state condition allows for detecting routes in the system which are coupled with the stoichiometric coefficients. Constraint based approaches have become a major tool to analyze

A. Ghosh, R.K. De, and S.K. Pal (Eds.): PReMI 2007, LNCS 4815, pp. 425–432, 2007.

the network of microorganisms [4]. Pathway analysis is becoming increasingly important for assessing inherent network properties of biochemical reaction networks [5]. Of the two most promising concepts for pathway analysis, one relies on elementary flux modes [6] and the other on extreme pathways popularly known as flux balance analysis. Flux balance analysis [7] is based on the fundamental law of mass conservation and the application of optimization principles to determine the optimal distribution of resources within a network.

Here we develop a method for identification of extreme regulatory pathways in genetic networks. The method, first of all, generates the possible flow vectors in the pathway. We consider only those flow vectors which, by taking convex combination of the basis vectors spanning the null space of the given node-edge incidence matrix, satisfy the quasi-steady state condition. Then a set of weighting coefficients is incorporated. A set of constraints incorporating these weighting coefficients is formulated. An objective function, in terms of these weighting coefficients, is formed, and then minimized under regularization method. The weighting coefficients corresponding to a minimum value of the objective function represent the extreme regulatory pathway. The effectiveness of the present method is demonstrated on two genetic systems designed in [8]. The method is compared with the existing extreme pathway analysis [9].

2   Genetic Network Model

Most of the structural and regulatory analyses consider the networks as unweighted (directed or undirected) graphs, with the genes as nodes and the interactions among them as edges [10]. The limitations of these methods is that the strengths of the interactions, the actual magnitude of flow through individual genes, the concentration of intermediates, and the allosteric interactions known to be crucial to the regulation of intracellular biochemistry are not considered. Gene regulatory networks [11] is based on a map of allosteric interactions, and are composed of individual elements that interact with each other in a complex fashion to regulate and control the production of proteins necessary for cell function. There are two important aspects of every genetic network that have to be modeled and analyzed. The first is the topology (connectivity structure) and the second is the set of interactions between the elements, i.e. determining the dynamical behavior of the system. A genetic network can be represented as a directed graph where the nodes represent genes and the directed edge represents the regulatory relationship between two connected genes. Let g_i be the expected level of gene i associated with node i in the graph. There is a flow, associated with each directed edge (i, j) from node i to node j, which measures the amount of expression of gene i transported through the edge (i, j). The regulatory coefficient measures the regulatory strength between two connected genes. A system boundary can be drawn around a network which consists of both internal and exchange flows. There are n_I number of internal flows and n_E number of exchange flows. The k-th internal flow is denoted by V_k and the l-th exchange flow is denoted by b_l. The internal flows are constrained to be positive and the exchange flows are either positive, negative or either positive or negative depending on the flow of the gene across the network.

2.1   Linear Structural Equation Model

The structure and dynamics of biochemical reaction networks, at the lowest level of detail, we distinguish the stoichiometric structure of a biochemical reaction network. It is a description of all biochemical conversions that take place in the network (e.g., of catalysis, transport, and binding). It represents the topology of mass flow through the network. It does not incorporate inhibitory and activatory effects of allosteric effectors. Linear structural equations can be used for constructing a first order approximation model of a genetic network using steady state gene expression measurements [3]. Let $\mathbf{g}$ denote the expression levels of the genes in the network and $\mathbf{f}$ denote the vector of non-linear functions. Rate equations indicating the expression levels of the genes in the network are given in a simplified form as [12].

$$dg/dt = f(g, u) \tag{1}$$

where $\mathbf{u}$ is the set of transcriptional perturbations. When the system reaches a steady state which is equivalent to setting the time derivative of $\mathbf{g}$ to zero, the system can be approximated by a linear set of equations

$$dg/dt = AX \tag{2}$$

where $X = [\mathbf{g}^T, \mathbf{u}^T]^T$. The above equation can be mathematically formulated

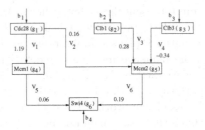

Fig. 1. Path diagram for a genetic network reconstructed from yeast cell cycle data

by a node-edge incidence matrix, B where the column in the matrix associated with edge (i, j) contains a '-1' in row i, a '+1' in row j and zeros elsewhere. The rank of the matrix B is equal to the number of genes in the network. So we decompose the matrix A into $A = BY$. As the number of columns in B is quite large than the number of rows the above decomposition may not be unique. Let $V = YX$. V is the vector of flows consisting of both internal and exchange flows. Thus at steady state, we get

$$BV = 0 \tag{3}$$

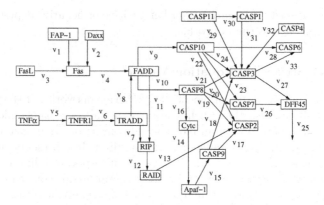

Fig. 2. Path diagram for apoptotic genetic network

which indicates the flow balance equations for the network. We model genetic networks by introducing two path diagrams (Fig. 1) and (Fig. 2) [8] which is a directed graph representing a system of structural equations. The path diagram consists of nodes represented by letters and edges represented by lines. The directed edges between the nodes denote the direction of the regulatory relationship between the nodes connected by the edges and this indicates a directed regulatory influence of one gene on another. The directed edges can represent either activation (positive control) or inhibition (negative control). The genetic network in the Fig. 1 reconstructed from the yeast cell cycle data consists of 6 genes from which the node-edge incidence matrix B can be constructed. Here for convenience of presentation we arrange the matrix B so that the first series of columns represent the internal flows and the remaining columns represent the exchange flows. All internal flows are positive yielding, $v_i \geq 0$, $i \epsilon n_I$. Like the stoichiometric matrix S in metabolic networks, the node-edge incidence matrix B plays a similar role in genetic networks. Here the b_{ij} element of the node-edge incidence matrix is the coefficient of the i-th gene in the regulatory process j. Here we further give an explanation for interpreting extreme directions as extreme pathways and develop a regularization approach for generating these pathways for genetic networks. Any cycle or a path having a starting point with entering exchange flow and an ending point with exiting exchange flow is an extreme direction, and, is referred to as an extreme regulatory pathway. The genes $Cdc28$ and $Clb1$ have an entering exchange flow. So $b1$ and $b2$ are entering roots. As the regulatory coefficient of the gene $Clb3$ on the gene $Mcm2$ is negative, the actual flow of the gene is from gene $Mcm2$ to the gene $Clb3$ which implies exchange flow $b3$ is negative and hence it is an exiting root. To balance the flow at the gene $Swi4$, the exchange flow $b4$ is negative and hence $b4$ is an exiting root.

3   Proposed Method

We consider the genetic network in (Fig. 1) with starting genes as $Cdc28$ and $Clb1$ and the target gene as $Swi4$ [12]. The target gene can be reached through

s different paths. That is, there are s flows/paths $V_1, V_2, \ldots, V_s$ in the network involving $Swi4$. We take the algebraic sum of the weighted flows of reactions $V_1, V_2, \ldots, V_s$ to reach the target $Swi4$ and it is given by

$$z = \sum_{k=1}^{s} c_k v_k \qquad (4)$$

Let us also consider that there are n flows comprising of both internal and exchange flows and m genes in the network. Here v_k is the gene flow involving genes $Swi4$ and $Clb3$. The term c_k denotes the weighting factor corresponding to the flow V_k. Here we have considered the genetic network where there is no feedback loop. The role of c_i in extreme pathway analysis is different from our method. In the earlier case, $\mathbf{c}$ is a *unit vector*, along a particular flow of the gene, whereas in the present method, $\mathbf{c}$ indicates the connection of other transcription factors (not shown in the diagram). The same procedure is applied for the genetic network in (Fig. 2) [8] where the starting genes are $FasL$ and $TNF\alpha$ and the target gene is $DFF45$.

3.1   Generation of Gene Flow Vectors

We require the values of the gene flow vectors $\mathbf{v} = [v_1, v_2, \ldots, v_n]^T$. We propose a method for generating flow vectors that approximately satisfies the quasi-steady state condition. That is, we generate those $\mathbf{v}$ which satisfies

$$\mathbf{B.v} \approx \mathbf{0} \qquad (5)$$

where $\mathbf{B}$ is the $m \times n$ node-edge incidence matrix that describes the relationship between genes and their regulatory interactions. $\mathbf{B}$ is computed from the diagram. As $m > n$, equation (5) is under determined. So we proceed in the following ways:
Step I: Generate basis vectors $\mathbf{v}_b$ that form the null space of the node-edge incidence matrix $\mathbf{B}$. Let the number of such basis vectors be l. (This is done by standard functions available in MATLAB).
Step II: Generate l number of random numbers $a_p, p = 1, 2, \ldots, l$. Then generate a vector $\mathbf{v}$ as a linear combination of the basis vectors using a_p.

3.2   Formulation of a New Constraint

As the genes are not expressed at the required level there comes further restrictions on the system, and we define a new constraint as

$$\mathbf{B.(C.v)} = \mathbf{0} \qquad (6)$$

$\mathbf{C}$ is an $n \times n$ diagonal matrix whose diagonal elements are the components of the vector $\mathbf{c}$. That is, if $\mathbf{C} = [\gamma_{ij}]_{n \times n}$, then $\gamma_{ij} = \delta_{ij} c_i$, where δ_{ij} is the Kronecker delta. Thus the problem of determining the extreme regulatory pathways starting from the genes $Cdc28$ and $Clb1$ to the target genes $Swi4$ and $Clb3$ boils down to an optimization problem, where z has to be optimized with respect to $\mathbf{c}$, subject to the inequality constraints and the new constraint.

3.3 Estimation of Weighting Coefficients c_i

Combining equations (4) and (6), we can reformulate the objective function as

$$y = 1/z + \boldsymbol{\Lambda}^T.(\mathbf{B}.(\mathbf{C}.\mathbf{v})) \tag{7}$$

that needs to be minimized with respect to the weighting factors c_i for all i. The term $\boldsymbol{\Lambda} = [\Lambda_1, \Lambda_2, \ldots, \Lambda_m]^T$ is the regularizing parameter. For the sake of simplicity, we have considered here $\Lambda_1 = \ldots = \Lambda_m = \Lambda$ (say). Initially, a set of random values in $[0, 1]$ corresponding to c_i's are generated. Then c_i's are modified iteratively using gradient descent technique, where the amount of modification for c_i in each iteration is defined as

$$\Delta c_i = -\eta \frac{\partial y}{\partial c_i} \tag{8}$$

The term η is a small positive quantity indicating the rate of modification. Thus the modified value of c_i is $c_i(t+1) = c_i(t) + \Delta c_i$, $\forall i$, $t = 0, 1, 2, \ldots c_i(t+1)$ is the value of c_i at iteration $(t+1)$, which is computed based on the c_i-value at the iteration t. We now analyze the results in Section 4.

4 Results

Following the method described in Section 3.1 we have generated a set of flow vectors. Then the objective function y (Equation(7)) is minimized, where the expression for z is defined as $z = c_5 v_5 + c_6 v_6 + c_{10} v_{10}$ for the genetic network in the (Fig. 1) and for the genetic network in the (Fig. 2) the expression for z is defined as $z = c_{26} v_{26} + c_{27} v_{27} - c_{25} v_{25}$. We vary the value of λ from 0.1 to 1.0. Initially λ should be kept small. For each value of λ, we minimize y, and consider that set of c_i-values corresponding to λ as the final solution, for which y becomes minimum. The genetic network in the (Fig. 1) reconstructed from the yeast cell cycle data consists of 6 genes where we have obtained the 5 extreme regulatory pathways as $p_1 : g_1 \rightarrow g_4 \rightarrow g_6, p_2 : g_1 \rightarrow g_5 \rightarrow g_6, p_3 : g_2 \rightarrow g_5 \rightarrow g_6, p_4 : g_1 \rightarrow g_5 \rightarrow g_3, p_5 : g_2 \rightarrow g_5 \rightarrow g_3$. The genetic network in the (Fig. 2) reconstructed from the yeast cell cycle data consists of 23 genes where we have obtained the 2 extreme regulatory pathways as $p_1 : v_3 \rightarrow v_4 \rightarrow v_{10} \rightarrow v_{20} \rightarrow v_{26}, p_2 : v_5 \rightarrow v_6 \rightarrow v_8 \rightarrow v_{10} \rightarrow v_{16} \rightarrow v_{14} \rightarrow v_{15} \rightarrow g_{21} \rightarrow v_{18} \rightarrow v_{27}$.

These are the two major experimentally confirmed pathways (extrinsic and intrinsic apoptosis pathways) [13]. The pathway p_1 involves response to binding of death ligands to their receptor FasL which triggers apoptosis via activating FADD and CASP8. Activation of complex of ligand receptor, FADD and CASP8 leads to the formation of a death inducing signaling complex (DISC), which in turn activates downstream effecters CASP7 and DFF45, resulting in DNA fragmentation. The seond apoptotic initiator pathway p_2 is induced by the formation of cytochrome c (Cytc) released from mitochondria with the adaptor Apaf-1, which in turn activates CASP9 and CASP3. The gene CASP3 will again trigger DNA fragmentation factor DFF45 and lead to DNA fragmentation.

A structure is the most essential feature of the networks. It provides information to assess the function of the gene. The extreme regulatory pathways helps to identify key components of the network structure and evaluate the relative importance of the gene in the network. These regulatory flows play an important role in apoptosis. The flows from the gene Fas to the gene FADD, and from the gene TRADD to the gene FADD involve responses to DISC [13]. Binding of death ligands to their receptor FADD activates the genes CASP8 and CASP10 and initiates a major extrinsic pathway. The genes CASP3, FADD, CASP8 and CASP10 are essential in apoptosis. A large number of extreme pathways are lost in apoptosis network due to the deletion of the above mentioned genes. The gene CASP3 is a major effecter gene and carries out the majority of substrate proteolysis during apoptosis [13]. FADD, CASP8 and CASP10 participate in DISC formation and play important roles in apoptosis initiation.

These pathways determines the gene regulatory route leading from the transcription of a given gene to the transcription of another gene. Genes communicate (interact) via the proteins they encode and protein production (transcription and translation) is controlled by a series of biochemical reactions, which are in turn influenced by many factors, both internal and external to the cell. In the case of metabolic networks, we can directly find a link between the concentration of the starting metabolite with that of the target metabolite, but it is not the same in case of genetic networks. Here the amount of mRNA produced by transcription and hence the amount of protein synthesized by translation by a starting gene will affect the protein synthesis of the target gene but they have no direct link between them. The gene flows at the steady state are combinations of the gene flows of the set of extreme regulatory pathways. We also use the algorithms developed in [9] for the genetic networks in (Figs. 1, 2) for generating extreme regulatory pathways. The extreme pathways generated by these two methods are the same for the genetic network in (Fig. 1). The extreme regulatory pathways generated by [9] and our regularization method is the same for the genetic network in (Fig. 2).

5   Conclusions and Discussions

The extreme regulatory pathways represent the regulatory capabilities i.e. the structural and functional properties of the genetic network [14]. These pathways determine the route starting from a particular gene to a given target gene. It is the set of interactions occurring between a group of genes which depend on each other's individual functions in order to make the aggregate function of the network available to the cell. The decrease in network functionality due to deletion of the genes and identifying the most important genes in a network can be done as these pathways indicate the flexibility and robustness of the networks. Selecting a model for a genetic network can have a great impact on how well the model follows the underlying dynamics of the actual genetic network. The balance between the available data, estimation techniques and the model complexity determines the usefulness of a given model.

Here we have given a brief overview of the methods and modeling descriptions available in computational systems biology. With the ability to reconstruct genetic networks on a large scale, the need to develop network-based pathway definitions and pathway analysis procedures has grown. We need to bring such pathway definitions into biological reality and use them productively to enhance our understanding of the systemic functions of real reconstructed networks.

References

1. Papin, J.A., Price, N.D., Wiback, S.J., Fell, D.A., Palsson, B.O.: Metabolic pathways in the post-genome era. Trends in Biochemical Sciences 28, 250–258 (2003)
2. Gardner, T.S., Bernardo, D., Lorenz, D., Collins, J.J.: Inferring genetic networks and identifying compound mode of action via expression profiling. Science 301, 102–105 (2003)
3. Datta, S.: Exploring relationships: a partial least square approach. Gene Exp. 9, 257–264 (2001)
4. Urbanczik, R., Wagner, C.: An improved algorithm for stoichiometric network analysis: theory and applications. Bioinformatics 21(7), 1203–1210 (2005)
5. Schilling, C.H., Edwards, J.S., Letscher, D.: Combining pathway analysis with flux balance analysis fot the comprehensive study of metabolic systems. Biotechnology and Bioengineering 71(4), 286–306 (2000)
6. Klamt, S., J., G., Kamp, A.V.: Algorithmic approaches for computing elementary modes in large biochemical reaction networks. IEEE Proc.- Syst. Biol. 152(4), 249–255 (2005)
7. Klamt, S., Stelling, J.: Two approaches for metabolic pathway analysis? Trends in Biotechnology 212(2), 64–69 (2003)
8. Xiong, M., Zhao, J., Xiong, H.: Network-based regulatory pathways analysis. Bioinformatics 20, 2056–2066 (2004)
9. Schilling, C.H., Letscher, D., Palsson, B.O.: Theory for the systemic defnition of metabolic pathways and their use in interpreting metabolic function from a pathway-oriented perspective. J. theor. Biol. 203, 229–248 (2000)
10. Kriete, A., Elis, R. (eds.): Computational systems Biology. Elsevier, San Diego, California, USA (2006)
11. Shen-Orr, S.S., Milo, R., Mangan, S., Alon, U.: Network motifs in the transcriptional regulation network of escherichia coli. Nat. Genet. 31, 64–68 (2002)
12. Xiong, M.M., Li, J., Fang, X.Z.: Identification of genetic networks. Genetics 166, 1037–1052 (2004)
13. Shivapurkar, N., Reddy, J., Chaudhary, P.M., Gazdar, A.F.: Apoptosis and lung cancer: A review. J. Cell. Biochem. 88, 885–898 (2003)
14. Covert, M.W., Schilling, C.H., Palsson, B.O.: Regulation of gene expression in flux balance models of metabolism. Journal of theoretical biology 223, 73–88 (2001)

New Results on Energy Balance Analysis of Metabolic Networks

Qinghua Zhou[1], Simon C.K. Shiu[2], Sankar K. Pal[3], and Yan Li[1]

[1] College of Mathematics and Computer Science, Hebei University,
Baoding City 071002, Hebei Province, China
[2] Department of Computing, The Hong Kong Polytechnic University,
HungHom, Kowloon, Hong Kong
[3] Indian Statistical Institute, Kolkata 700 108, India
{qinghua.zhou@gmail.com,csckshiu@comp.polyu.edu.hk
sankar@isical.ac.in, hbuliyan@gmail.com}

Abstract. A central and long-standing objective of cellular physiology has been to understand the metabolic capabilities of living cells. In this paper, we perform optimization of metabolic networks in E.Coli under both flux and energy balance constraints. The impact of the energy balance is investigated on the behavior of metabolic networks. Different from the existing work, the preliminary results showed that the impact of imposing energy balance constraints on the biological metabolic systems is very limited.

Keywords: Metabolic networks, Flux Balance Analysis, Energy Balance Analysis, Constraint optimization.

1 Introduction

Cellular metabolism, the integrated interconversion of thousands of metabolic substrates through enzyme-catalysed biochemical reactions, is the most investigated complex intracelluar web of molecular interactions. In particular, the ability to quantitatively describe metabolic fluxes through metabolic networks has been long desired(Reich & Sel'kov [1]). However, this endeavor has been hampered by the need for extensive kinetic information describing enzyme catalysis within the living cell. Detailed information about all the enzymes in a specific metabolic network is not available. This dilemma has recently been partially resolved by the development of flux balance-based metabolic models(Savinell & Palsson [2], Varma & Palsson [3]).

Flux Balance Analysis(FBA) assumes that metabolic networks will reach a steady state constrained by the stoichiometry. It only requires information regarding the stoichiometry of metabolic pathways along with known metabolic requirements for growth to describe metabolic flux distributions and cell growth. The stoichiometric constraints lead to an underdetermined system and the network's behavior can be studied by optimizing the steady-state behavior with respect to some objective function(Segre et al. [4]). FBA has been shown to provide meaningful predictions in Escherichia Coli(Varma & Palsson [5], Edwards et al. [6]).

A. Ghosh, R.K. De, and S.K. Pal (Eds.): PReMI 2007, LNCS 4815, pp. 433–438, 2007.

In 2002, Daniel et al. [7] introduced Energy Balance Analysis(EBA)-the theory and methodology for enforcing the laws of thermodynamics-on the base of constraints for FBA. They want to get more physically realistic results by considering of the energy balance and thermodynamics of the network reactions. Actually, the authors provided very promising results. In this paper, we test the FBA and EBA methods to analyze their difference and quality. Our results showed that the impact of imposing energy balance constraints on the biological metabolic systems is very limited. There is not only obvious improvement but the cpu running time extended from seconds to several hours for the large-scale metabolic networks. Therefore, there is no need to simply consider energy balance constraints when to deal with the FBA models of biologically metabolic networks.

The paper is organized as follows: In section 2, we introduce some basic concepts of FBA and EBA models. After that we give the detailed computational results in section 3. Finally, we give our conclusions in section 4.

2   Basis Concepts

FBA models assumed that there is a metabolic steady state, in which the metabolic pathway flux leading to the formation of a metabolite and that leading to the degradation of the metabolite must balance, which generates the flux balance equation (Savinell et al. [8])

$$S \cdot v = b \tag{2.1}$$

where S is a matrix comprising the stoichiometry of the catabolic reactions, v is a vector of metabolic fluxes, and b is a vector containing the net metabolic uptake by the cell. Typically, equation (2.1) is underdetermined, since the number of fluxes normally exceeds the number of metabolites. Then a particular solution may be found using linear optimization by stating an objective and seeking its maximal or minimal value within the stoichiometrically defined domain. In special, the ability to meet growth requirements of the cell is of central importance. It is represented by a single reaction(Varma & Palsson[5]):

$$\sum_{all\ M} d_M \cdot M \xrightarrow{V_{gro}} Biomass \tag{2.2}$$

where d_M represents the requirements in millimoles per gram of biomass of the M biosynthetic precursors and cofactors for biomass production. V_{gro} is the growth flux(grams of biomass produced), which with the basis of 1g(dry weight) per h reduces to the growth rate(grams of biomass produced per gram per hour). Minimizing the objective function $-V_{gro}$, then we get the particular solution.

What FBA lacks is the explicit consideration of the energy balance and thermodynamics of the network reactions(Daniel et al. [7]). So, they consider the following constraints in addition to the ones used in FBA models:

$$K \cdot \Delta\mu = 0 \tag{2.3}$$

$$v_i \cdot \Delta\mu_i < 0 \tag{2.4}$$

where K is a matrix which stores the null space vectors of S', and S' is a matrix converted from S by combining redundant fluxes and removing the columns that correspond to boundary columns. Where $\Delta\mu$ is the vector of chemical potential differences associated with the reaction fluxes.

3 Computational Results

We tested FBA and EBA models on examples selected from Daniel et al. [7] and [9], Also on a real biological metabolic systems of E.Coli. Details are as follows:

Example 1
Reaction network:

$$v_1 : A + 2B \leftrightarrow C$$

$$v_2 : C + D \leftrightarrow 2A + 2B$$

$$v_3 : A + B \leftrightarrow 2D$$

$$v_4 : A + C \leftrightarrow B + 3D$$

$$v_5 : B \leftrightarrow D$$

which contains 5 reactions acting on 4 metabolites. And where $v_1, \cdots, v_5$ represents the flux through the relative reaction. Following the ideas of Daniel et al. [7], we consider the same problem: Determine the maximum steady-state production of reactant D, for a given set of maximal input fluxes of reactants A,B, and C. (Three related problems are considered below. The first problem assumes that reactant A is the only available input substrate. For this case, the maximal input fluxes of A,B and C are set to 1, 0, and 0 mmol sec^{-1}, respectively. The remaining two cases use B only and C only as input substrates.)

Optimal fluxes are obtained by using the MATLAB software package. In particular, for FBA we use the procedure *linprog* in the optimization toolbox of MATLAB. For EBA, we choose the procedure *fmincon*. Further, for all the cases tested, we select the zeros vector as initial point needed by the procedure *fmincon*. Details are reported in Table 1. Where the variable A_{in} means the input flux of metabolite A, D_{out} means the output flux of metabolite D. Where the symbol FBA means flux balance constraints only and EBA means both flux and energy balance constraints.

From Table 1, we can see that for all the three cases, the computational results by FBA and EBA procedures are similar. That is to say, the impact of imposing energy balance constraints on Flux balance equation is very limited and we did not get more biologically meaningful results.(It is different with the ones in Daniel et al. [7]).

Table 1. Computational results of Example 1

	A input		B input		C input	
	FBA	EBA	FBA	EBA	FBA	EBA
v_1	-0.0284	0	0.1477	0.1250	-0.1660	-0.125
v_2	-0.3440	-0.3333	0.2134	0.2083	0.4300	0.4583
v_3	0.0249	0	0.3449	0.3750	0.6219	0.6250
v_4	0.3156	0.3333	-0.657	-0.0833	0.4041	0.4167
v_5	-0.3405	-0.3333	0.7208	0.7083	0.9741	0.9583
A_{in}	1	1	0	0	0	0
B_{in}	0	0	1	1	0	0
C_{in}	0	0	0	0	1	1
D_{out}	1	1	1	1	3	3

Example 2

Reaction network:

$$v_1 : A \leftrightarrow B$$

$$v_2 : B \leftrightarrow C$$

$$v_3 : C \leftrightarrow A$$

$$v_4 : C \leftrightarrow D$$

$$v_5 : D \leftrightarrow B$$

Which is similar with the ones in Daniel et al. [9] except every reaction is reversible for convenience. Further, we perform the same procedure as in Example 1 and the results are shown in Table 2.

As we can see, the results between FBA and EBA is quite similar and we got the same outcomes in all the cases. It has proved that there is no actual differences when consider the energy balance constraints in addition to the flux balance ones.

Table 2. Computational results of Example 2

	A input		B input		C input	
	FBA	EBA	FBA	EBA	FBA	EBA
v_1	0.5	0.5	-0.1296	-0.1250	0.1296	0.1250
v_2	0	0	0.2592	0.2500	-0.2592	-0.2500
v_3	-0.5	-0.5	-0.1296	-0.1250	0.1296	0.1250
v_4	0.5	0.5	0.3887	0.3750	0.6113	0.6250
v_5	-0.5	-0.5	-0.6113	-0.6250	-0.3887	-0.3750
A_{in}	1	1	0	0	0	0
B_{in}	0	0	1	1	0	0
C_{in}	0	0	0	0	1	1
D_{out}	1	1	1	1	1	1

Then, we tested the above ideas on the E.Coli metabolism systems(Edwards & Palsson [10]). The reaction network contains 953 fluxes acting on 536 metabolites(Detailed information can be found in [11]). Within the 953 fluxes, there have 214 boundary fluxes and 114 redundant fluxes, which results a 645×1578 coefficients matrix representing the scale of linear constraints, also with 625 nonlinear constraints. We reproduced the above programming and optimized the production of biomass with glucose and oxygen uptake constrained to be less than or equal to 10 and 15 mmol g/DW· h, respectively. For the flux of representing the non-growth related maintenance, we let it to be 7.6 mmol g/DW· h. For FBA algorithm, the resulting flux produces a growth rate of 0.78 mmol g/DW· h on glucose minimal media. And it only takes for several minutes. While for EBA algorithm, it stops after more than 9 hours and does not give optimal solution.

4 Conclusions

In this paper, we have tested FBA and EBA models respectively on two examples of little-scale metabolic networks and a real large-scale ones of E.Coli. Our initial results showed the limited impact of imposing energy balance constraints on flux balance constraints. There is no obvious improvement and the results of them are very similar for the two little-scale examples. As for the metabolic networks of E.Coli, the number of constraints nearly double the original size and the cpu running time increases from seconds to several hours. Therefore, we think there is no need to simply add the energy balance constraints to the flux balance constraints when considering the function of metabolic networks of real livings.

Acknowledgments. We would like to thank Dr. Jennie Reed and Dr. Palsson of UCSD for their kindness and patience to our questions. Also we were very grateful to the referees for their valuable comments and suggestions. And this work is supported by Doctoral Foundation of Hebei University(Grant No.Y2006084).

References

1. Reich, J.G., Sel'kov, E.E.: Energy metabolism of the cell. Academic Press, New York
2. Savinell, J.M., Palsson, B.O.: Network analysis of intermediary metabolism using linear optimization. II. Interpretation of hybridoma cell metabolism. J. Theor. Biol. 154, 455–473 (1992)
3. Varma, A., Palsson, B.O.: Metabolic capabilities of Escherichia coli. I. Synthesis of biosynthetic precursors and cofactors. J. Theor. Biol. 165, 477–502 (1993)
4. Segre, D., Vitkup, D., Church, G.M.: Analysis of optimality in natural and perturbed metabolic networks. Proc. Natl. Acad. Sci. USA 99, 15112–15117 (2002)
5. Varma, A., Palsson, B.O.: Stoichiometric flux balance models quantitatively predict growth and metabolic by-product secretion in wild-type Escherichia coli W3110. Appl. Environ. Microbiol. 60, 3724–3731 (1994)

6. Edwards, J.S., Ibarra, R.U., Palsson, B.O.: In silico predictions of Escherichia coli metabolic capabilities are consistent with experimental data. Nature Biotechnol. 19, 125–130 (2001)
7. Beard, D.A., Liang, S.-d., Qian, H.: Energy balance for analysis of complex metabolic networks. Biophysical Journal 83, 79–86 (2002)
8. Savinell, J.M., Palsson, B.O.: Network analysis of intermediary metabolism using linear optimization. I. Development of mathematical formalism. J. Theor. Biol. 154, 421–454 (1992)
9. Beard, D.A., Babson, E., Curtis, E., Qian, H.: Thermodynamics constraints for biochemical networks. J. Theor. Biol. 228, 327–333 (2004)
10. Edwards, J.S., Palsson, B.O.: Metabolic flux balance analysis and the in silico analysis of Escherichia coli K-12 gene deletions. BMC Bioinformatics 1, 1 (2000)
11. http://gcrg.ucsd.edu/supplementary_data/deletionanalysis/main.htm

Computation of QRS Vector of ECG Signal for Observation of It's Clinical Significance

S. Mitra[1], M. Mitra[2], and B.B. Chaudhuri[1]

[1] Computer Vision and Pattern Recognition Unit,Indian Statistical Institute,
Kolkata, India
[2] Department of Applied Physics, Faculty of Technology, University of Calcutta,
Kolkata, India
{sucharita_r,bbc}@isical.ac.in,
{mmaphy}@caluniv.ac.in

Abstract. An automated approach for computation of the frontal plane QRS vector and an important observation of its clinical significance is described in this paper. Frontal plane QRS vector is computed from the six frontal plane leads (Standard leads I, II, III , AVR, AVL and AVF). The R-R interval of each ECG wave is detected by square derivative technique. The baseline or isoelectric level of every ECG wave is determined. After that the net positive or net negative deflection (NQD) of QRS complex is detected. Net positive or net negative deflection in any lead is obtained by subtracting the smaller deflection (+ve or −ve) from the larger deflection (-ve or +ve). An algorithm is developed for computation of the exact angle,amplitude and direction of the frontal plane QRS vector from maximum and minimum NQD. In the present work, the PTB diagnostic ECG database of normal and Myocardial Infarction (MI) subjects is used for computation of the QRS vector. An interesting clinical observation that, the rotation of QRS axis for MI data may significantly detect the region of the infarcted cardiac wall, is reported in this paper.

Keywords: ECG, software, QRS vector, NQD, baseline.

1 Introduction

The electrocardiogram (ECG) represents a graph of variation of electric potential generated by the heart and recorded at the body surface. Clinically, the frontal plane QRS vector represents the average direction of ventricular activation in the frontal plane. So, magnitude and direction of QRS vector is an important aid for accurate and deductive evaluation of the Electrocardiogram. It is important to realize that two components are important in determining the appearance of the QRS complex in any given lead: the direction of electrical forces relative to the lead, and the magnitude of forces. If the cardiac vector is parallel to a particular lead, it will make greatest impression on that lead. So electrocardiograph will record maximum deflection. If the cardiac vector is directed towards the positive pole of that lead, the deflection will be positive.

A. Ghosh, R.K. De, and S.K. Pal (Eds.): PReMI 2007, LNCS 4815, pp. 439–446, 2007.

Minimal deflections will be formed when cardiac vector is perpendicular to the lead. Cardiac vector is used for identifying myocardial ischaemia during carotid endarterectomy[3] and also to locate major anatomical position of the injured region in the heart ventricles for various localizations of the lesion[4]. On the other hand, detection of R-R interval is essential for the task of ECG analysis. A great amount of research effort has been devoted in the development and evaluation of automated QRS detectors [5]. Two types of QRS detectors have been developed. There are two parts in these detectors. One is the processor and the other is the decision rule[6]. The processor performs linear filtering and non-linear transformation. It produces a noise-suppressed signal in which the QRS complexes are enhanced. The decision rule detects or classifies whether the enhanced parts are QRS complexes or not. Hidden Markov models [7] and other pattern recognition approaches are also used for detection of QRS complexes [8, 9]. In another scheme, simple morphological operator is used for detection of the same complexes[10]. In this approach the operator (openings and closings) works as peak-valley extractor. Different wavelet functions are also used for QRS complex detection in ECG. It has been noted that wavelet functions that support symmetry and compactness provide better results [11]. In another scheme a QRS complex detector based on the dyadic wavelet transform (DWT), which is robust to time varying QRS complex morphology and to noise, has also been developed [12]. Another approach for QRS detection is an adaptive matched filtering algorithm based upon an artificial neural network (ANN). An ANN adaptive whitening filter is used to model the lower frequencies of the ECG signals which are inherently non linear and non-stationary. The residual signal which contained mostly high frequency QRS complex energy was then passed through a linear matched filter to detect the location of the QRS complex [13]. In the present work, a software-based approach for detection of frontal plane QRS vector for both normal and pathological subjects is proposed. Some interesting clinical properties are observed, which are stated in the paper.

2   Methodology

For computation of frontal plane cardiac axis the following steps are involved:

2.1   Determination of R-R Interval of ECG Wave

Proper detection of the R-R interval between two consecutive ECG waves is very important for various applications. In the present work,differentiation technique is used to get second order derivative of ECG wave by using 5-point Lagrangian interpolation formula for differentiation [1]. The formula is given below:

$$f_0' = \frac{1}{12h}(f_{-2} - 8f_{-1} + 8f_1 - f_2) + \frac{h^4}{30}f^v(\xi) \tag{1}$$

Here ξ lies between the extreme values of the abscissas involved in the formula. After squaring the values of second order derivative, a square-derivative curve

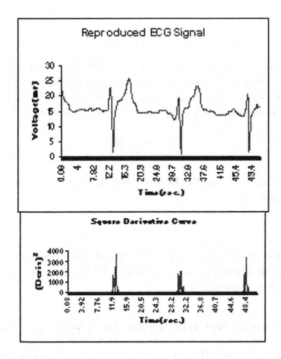

Fig. 1. R-R Interval Detection

having only high positive peaks of small width at the QRS complex region can be obtained (fig. 1). A small window of length (say W) was taken to detect the area of this curve and we obtained maximum area at those peak regions. The local maxima of these peak regions are considered as R-peak. An experiment with considerable number of test samples was done for choosing of appropriate value of W and was set at ~0.07sec. The system was tested for both noise free and noisy signals. The levels of all types of noise were increased from 10% to 30% and we achieved 99.8% to 98.1% accuracy in the detection of QRS complexes. The performance comparison of this method with other existing methods is given in [15].

2.2 Detection of Base Line or Isoline

In order to accurately detect the net QRS deflection, isoelectric line must be correctly identified. Most methods are based upon the assumption that the isoelectric level of the signal lies in the region ~80 ms left of the R-peak, where the first derivative becomes minimum.

In particular, let y_1, y_2, ..., y_n be the samples of a beat,

y'_1, y'_2, ..., y'_{n-1} be their first differences and y_r the sample where the R-peak occurs. The isoelectric level samples y_b are then defined if either of the two following criteria is satisfied:

$$|y'_{r-j-int(0.08f)}| = 0, j = 1, 2, \ldots., 0.01f \ or$$
$$|y'_{r-j-int(0.08f)}| \le |y'_{r-i-int(0.08f)}|, i, j = 1, 2, \ldots., 0.02f \tag{2}$$

where f is the sampling frequency. After the isoelectric level is found, and after comparing the current beat with the previous corrected one ($y^p_{b_p}$), it is easy to align the current beat with the previous one, using the declination of the line connecting the isoelectric levels of the two beats. If we define,

$$\gamma = \frac{y_b - y^p_{b_p}}{n_b} \tag{3}$$

where n_b is the number of samples between the two baseline points, the alignment procedure becomes:

$$y_b \quad \rightarrow \quad y^{p_{b_p}} + \gamma \ \ y^p_{b_p} \tag{4}$$

Other baseline correction techniques rely on adaptive filtering of the ECG beat and provide reliable results as well, however the QRS wave shape is corrupted by the use of such techniques [2].

2.3   Computation of Net QRS Deflection (NQD) in a Lead

After detection of baseline, the location of Q wave is determined by searching the first transition from inclination to declination or vice versa from the R point. After getting q point the net QRS deflection can be computed from the following equation:

$$NQD = (Qpt - avg\_bpt) + (Rpt - avg\_bpt)(4) \tag{5}$$

Where, Qpt = Q point, Rpt = R point, avg-bpt = Average baseline point

2.4   Detection of Frontal Plane Cardiac Axis

An algorithm is developed for detection of cardiac axis on the basis of the following rules.

(1) Look for a limb lead with NQD close to zero. This could be either a lead with very little recognizable deflection in either direction, or equally large positive and negative deflections, such that the sum of positive and negative deflections is close to zero. The mean electrical axis will then be either perpendicular or 270^0 out of phase to this lead depending on the sign (+ve or −ve) of NQD at that lead. For example, if lead I has close to zero net amplitude, the axis would be either +90°or -90°(fig. 2). Then, look at any of the inferior leads: II, III, or AVF. If these leads register a predominantly positive deflection, the QRS axis will be closer to +90°. If the inferior leads (II, III, AV_F) are predominantly negative, the axis is closer to -90°.

(2) Sometimes, no ECG lead can be identified which appears to have zero netamplitude i.e, NQD $\neq 0$ in every lead. In this case, decide which lead(s)have

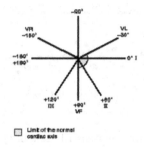

Fig. 2. Angle of different limb leads from the frontal plane

the largest positive or negative deflections. The QRS axis will be close to the direction of this lead. The algorithm for detection of the angle and the amplitude of the QRS vector is as follows:

1. Find the maximum and minimum NQD for all the six limb leads. Let them be MAX and MIN, respectively.
2. Calculate the angle of the lead having maximum NQD from the lead I. Let the angle be $\theta1$.
3. Calculate $\theta2 = |tan^{-1}(MIN/MAX)|$, where $\theta2$ is the angle of the QRS vector from the lead containing minimum NQD.
4. Calculate the angle of the lead having minimum NQD from the lead I. Let the angle be $\theta3$.
5. If $\theta3$ is clockwise rotated from $\theta1$ then $\theta = \theta1 + \theta2$, where θ is the angle of QRS axis from lead I.
6. Otherwise, if $\theta3$ is anti-clockwise rotated from $\theta1$ then $\theta = \theta1 - \theta2$.

3   Result

Our observation is being continued on 50 normal and 50 diseased subjects having myocardial infraction (MI) of PTB diagnostic ECG database available in

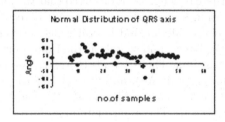

Fig. 3. Normal distribution of QRS axis

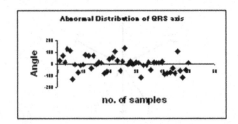

Fig. 4. Abnormal distribution of QRS axis

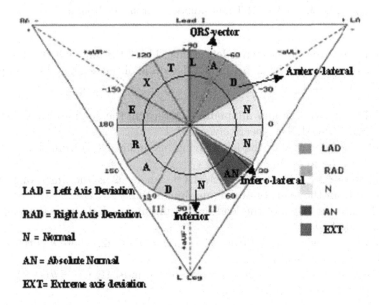

Fig. 5. The separation of Myocardial wall along with the electrical axis of heart

physionet. The net QRS height along with the angle of direction of QRS vector for all the six limb leads for normal and MI subjects are given in table 1 and 2. Fig. 3 shows that the normal distribution of QRS axis ranges between 30^0 –60^0 in nearly 84% cases whereas in fig. 4 it is observed that the abnormal distribution of QRS axis varies within a wider range and also it detects the location of the infarcted regions specially inferior and lateral walls between the 4 Myocardial walls (anterior, inferior, lateral and septal) as the frontal plane leads cover these two walls of the myocardium. These 4 walls there are also divided into another 4 regions named antero-septal (AS), antero-lateral (AL), infero-septal (IS)and infero-lateral (IL) and infarction at AL and IL can also be detected from the rotation of the QRS vectors. The antero-lateral surface of the left ventricle is oriented towards lead AVL and positive pole of standard lead I. On the other hand, according to the basic principle, the initial QRS vector is directed away

Table 1. A portion of detected NQD and angle of QRS vector for Normal Subjects

Patient ID.	Net QRS Deflection (NQD)						θ1	θ2	θ3	θ
	L1	L2	L3	AV$_R$	AV$_L$	AV$_F$				
s0461_re	0.73990	0.398598	-0.25546	-0.5354	0.567818	-0.09567	-360	7.36712	-90	-7.3671
s0462_re	0.38046	0.9155	0.7445	-0.68522	0.134846	0.682308	-300	8.37895	-30	51.6210
s0463_re	0.39290	0.882308	0.547	-0.68593	0.216833	0.709231	-300	13.80716	-30	46.1928
s0464_re	0.451	0.937385	0.545833	-0.69027	0.221727	0.689615	-300	13.30799	-30	46.6920
s0466_re	0.72590	0.675692	0.151286	-0.7732	0.4868	0.4558	-330	11.07073	-240	41.0707
s0467_re	0.42207	0.282833	-0.34071	-0.29127	0.4014	0.156636	-360	20.36023	-270	20.3602
s0469_re	0.8292	1.66623	1.1145	-1.1342	0.210001	1.371286	-300	7.18330	-30	52.8167
S0470_re	0.8815	1.705454	1.047833	-1.170154	0.238909	1.40409	-300	7.97438	-30	52.0256
s0474_re	0.65472	0.835545	0.291833	-0.7794	0.228455	0.511833	-300	15.29203	-30	44.7079
s0478_re	0.5584	1.254	0.806143	-0.906334	0.230167	0.9052	-300	10.40063	-30	49.5993

Table 2. A portion of detected NQD and angle of QRS vector for MI Subjects

Patient ID.	Net QRS Deflection (NQD)						θ1	θ2	θ3	θ
	L1	L2	L3	AV$_R$	AV$_L$	AV$_F$				
s0015_lre	-0.0411	0.3284	0.3421	-0.314636	-0.1552	0.275143	-240	6.8628	-180	126.8628
s0016_lre	0.0611	0.3284	0.3421	-0.267	-0.1552	0.275143	-240	10.1262	-360	109.8737
s0017_lre	0.3046	-1.2153	-1.1342	0.568545	0.773	-1.006727	-120	14.06988	-360	-105.93
s0054_lre	0.50175	-0.3998	-0.6950	0.556091	0.942833	-0.479889	-30	22.97888	-120	-52.9788
s0050_lre	0.4056	-0.2722	-0.408	-0.3715	0.304615	-0.476143	-90	29.75553	-120	-119.755
s0059_lre	0.556	-0.3438	-0.6003	0.500455	0.58	-0.352091	-60	29.8013	-120	-89.8013
s0060_lre	0.556	-0.3438	-0.6183	0.500455	0.612	-0.300364	-60	25.90875	-90	-85.908
s0039_lre	0.2884	-0.0905	-0.536	0.011667	0.320364	-0.428273	-60	1.246945	-150	-61.2469
s0037_lre	-0.2466	0.61388	0.77	-0.2406	-0.474833	0.668	-240	17.35228	-330	102.6477
s0043_lre	-0.1474	0.6128	0.6434	-0.306455	-0.351834	0.614154	-240	12.90350	-180	132.9035

from the necrosis within the infarcted region. So, for the inferior wall, it is directed upwards and to the left i.e. towards lead AVL or positive pole of lead I [16]. Following these principles the rules are developed to detect the infarcted regions accordingto the MIN, MAX and θ.

RULE 1: For Antero-lateral MI θ will be in RAD region [fig. 5].

RULE 2: For Inferior / Infero-lateral MI θ will be in LAD or EXT region [fig. 5].

4   Conclusion

A computerized approach for accurate computation of QRS vector of ECG signals is described in this paper. For this purpose the accurate detection of both baseline and NQD is necessary. Hence, modules for computation of those are been developed. Finally, depending on the rules of cardiac axis determination by searching the minimum and maximum NQD, an algorithm is developed for computation of the angle, amplitude and direction of the frontal plane QRS vector for both normal and diseased subjects to find out the clinical significance of this vector in cardiac disease identification. Still we use two types of data set (normal and MI) and achieved interesting result and also expect that the resultant cardiac vector of frontal and horizontal plane axis will play significant role in disease identification.

Acknowledgments. The part of the work is funded by Department of Biotechnology (DBT),Govt. of India.

References

1. Hildebrand, F.B.: Introduction To Numerical Analysis, TMH edn., pp. 82–84. Tata Mcgraw-Hill Publishing Company Ltd
2. Maglaveras, N., Stamkopoulos, T., Diamantaras, K., Pappas, C., Strintzis, M.: ECG pattern recognition and classification using non-linear transformations and neural networks: a review. International Journal of Medical Informatics 82, 191–208 (1998)
3. Kawahito, S., Kitahata, H., Tanaka, K., Nozaki, J., Oshita, S.: Dynamic QRS-complex and ST-segment monitoring by continuous vectorcardiography during carotid endarterectomy. British Journal of Anaesthesia 90(2), 142–147 (2003)
4. Aidu, E.A.I., Trunov, V.G., Titomir, L.I., Szathmary, V., Tyšler, M.: Noninvasive Location of Acute Ischemic Lesion in the Heart Ventricles Using a Few-lead System: Study on a Realistic Mathematical Model. Measurement Science Rreview 3(2), 33–36 (2003)
5. Friensen, G.M., Jannett, T.C., Jadallah, M.A., Yates, S.L,, Quint, S.R, Nagle, H.T: A comparison of the noise sensitivity of nine QRS detection algorithms. IEEE Trans. Biomed. Eng. BME-37, 85–89 (1990)
6. Pahlm, O., Sommo, L.: Software QRS detection in ambulatory monitoring – A review. Med. Biomed. Eng. Comput., 22, 289–297 (1984)
7. Coast, D.A., Stem, R.M., Cano, G.G., Briller, S.A.: An approach to cardiac arrhythmia analysis using hidden Markov models. IEEE Trans. Biomed. Eng. BME-37, 826–836 (1990)
8. Trahanias, P.E., Skordalakis, E.: Bottom-up approach to ECG pattern recognition problem. Med. Biomed. Eng. Comput. 27, 221–229 (1989)
9. Trahanias, P.E., Skordalakis, E.: Syntactic pattern recognition of the ECG. IEEE Trans. Pattern Anal. Mach. Intell. PAMI-12, 648–657 (1990)
10. Trahanias, P.E.: An approach to QRS complex detection using mathematical morphology. IEEE Trans. Biomed. Eng. BME-40, 201–205 (1993)
11. Dinh, H.A., Kumar, D.K., Pah, N.D., Burton, P.: Wavelets for QRS detection. Australas Phys. Eng. Sci. Med. 24, 207–211 (2001)
12. Kadamb, S., Murray, R., Boudreaux-Bartels, G.F.: Wavelet transform-based QRS complex detector. IEEE Trans Biomed Eng. 46, 838–848 (1999)
13. Yu, X., Xu, X.: QRS detection based on neural-network. Journal of Biomed. Eng. 17, 59–62 (2000)
14. Chung, E.K.: Pocket Guide to ECG Diagonosis, pp. 17–21. Oxford University Press, Mumbai (1997)
15. Mitra, S., Mitra, M.: Detection of QRS Complex of ECG Signals from Square-Derivative Curve. AMSE journal (Advances in Modeling), France, modeling C 65(2), 15–28 (2004)
16. Schamroth, L.: An Introduction to Electrocardiography, 7th edn., pp. 142–155. Blackwell Science

Signal Resampling Technique Combining Level Crossing and Auditory Features

Nagesha* and G Hemantha Kumar

Dept of Studies in Computer Science, University of Mysore,
Mysore - 570 006, India
shan_bk@yahoo.com

Abstract. Level crossing based sampling might be used as an alternative to Nyquist theory based sampling of a signal. Level crossing based approach take advantage of statistical properties of the signal, providing cues to efficient nonuniform sampling. This paper presents new threshold level allocation schemes for level crossing based nonuniform sampling. Intuitively, it is more reasonable if the information rich regions of the signal are sampled finer and those with sparse information are sampled coarser. To achieve this objective, we proposed non-linear quantization functions which dynamically assign the number of quantization levels depending on the importance of the given amplitude range. Various aspects of proposed techniques are discussed and experimentally validated. Its efficacy is investigated by comparison with Nyquist based sampling.

1   Introduction

The concept of level crossing sampling scheme has been proposed for analog to digital conversion [3]. Several case studies in analog to digital converters shows that level crossing based sampling technique can be more effective than existing synchronous Nyquist analog to digital converters (ADC). The level crossing sampling scheme has also demonstrated for speech applications using CMOS technology and a voltage mode approach for the analog parts of the converter [4]. Electrical simulations proved that the Figure of Merit of asynchronous level crossing converters increased compared to synchronous Nyquist ADCs. Level crossing sampling scheme has been suggested in literature for bursty signals, non-bandlimited signal[5] and band limited Gaussian random processes[6].

Conventional Nyquist based signal sampling is with uniform time-step variable amplitude. The new class of nonuniform sampling approach has been used to improve the performance of Nyquist ADCs exploiting the inner statistical properties of the signal. Several signals has interesting statistical properties, but Nyquist based sampling do not take the advantage of them. Signals such as electro cardiograms, speech signals, temperature sensors, pressure sensors are almost always constant and may vary significantly during brief moments. Due to Nyquist theory, which is to ensure the sampling to be at least twice that of the

---

* Corresponding author: 0821-2510789.

A. Ghosh, R.K. De, and S.K. Pal (Eds.): PReMI 2007, LNCS 4815, pp. 447–454, 2007.
© Springer-Verlag Berlin Heidelberg 2007

input signal frequency bandwidth, it is obvious that, in the time domain, this condition may result in a large number of samples with redundant information. It has been proved in [1,4] that level crossing sampling approach can lead to reduction in number of samples.

Level crossing scheme sends a pulse whenever the source signal crosses a threshold level. The threshold levels are uniformly distributed in the amplitude range [1,3,4,5]. The uniform threshold step size used in the level crossing regardless of the signal amplitude characteristics. The linear threshold allocation will result in a higher SNR at the region of higher amplitude than the region of lower amplitude. Hence, increased SNR at the higher signal amplitude does not increase the perceived audio quality because humans are most sensitive to lower amplitude components. To exploit this factor, a nonuniform threshold level allocation schemes based on Incomplete Beta Function (IBF), logarithmic and linear function are proposed in this paper. The parameters of the IBF are chosen in such a way that it focuses on sensitive lower amplitude regions.

The paper is organized as follows. Section 2 describes the level crossing based sampling approach with the proposed nonuniform threshold allocation scheme. Section 3 discusses the incorporation of IBF, logarithmic and linear functions to formulate a rule for allocation of nonuniform threshold levels in multilevel crossing. Section 4 explains the experimentation setup for testing the proposed approach and gives the results and analysis. Section 5 is devoted to major conclusions as well as gives directions for future lines of work.

2   Irregular Sampling Model

Level Crossing Analysis represents an approach to interpretation and characterization of time signals by relating frequency and amplitude information. Measurement of level crossing of a signal is defined as the crossings of a threshold level l by consecutive samples.

The level sampling with a level allocation function $f(s)$ is the mapping $L_{f(s)}$: $R \rightarrow f(s)Z : L_{f(s)} = f(s) \lfloor s/f(f(s)) \rfloor$ where $L = (l_1, l_2, \cdots, l_N)$ set of nonuniform spaced levels.

Since the quantization levels are irregularly spaced across the amplitude range of the signal, it increases the efficiency of bit usage. The spacing of the levels is decided by the importance of the amplitude segments which is discussed in section 3. A sample is recorded when the input signal crosses one of the nonuniformly spaced levels. The precession of time of the recorded sample is decided by the local timer T.

Let $L = (l_1, l_2, \ldots, l_N)$ be the set of nonuniform spaced levels and $2^b - 1 = N$ quantization levels with b bit resolution.

Definition 2.1: The level crossing of the threshold level $l \in L$ by signal $s(t)$ with period T is given by

$$L_{f(s)}(s, I_{ni}) = l \qquad iff \; s\left[\frac{i-1}{n}T\right]\left[s\left[\frac{i-1}{n}T\right] - l\right] < 0 \qquad (1)$$

This completes the security definition as far as Alice is concerned. From the point of view of the verifier, the goal of a malicious environment in which the verifier operates is to provide him with a ciphertext that encrypts a witness for a public relation that does not open to a witness even if all the group members apply their decryption function to it. Immunity to this attack, which we call soundness, guarantees that at least one group key will open to a valid witness.

Formulation of the Soundness Property. A soundness attack proceeds as follows: the adversary will create adaptively the group of recipients communicating with the GM. In this attack game, the adversary wins if, while playing the role of Alice, she convinces the verifier that a ciphertext is valid with respect to a public-relation $\mathcal{R}$ of the adversary's choice, but it holds that either (1) if the opening authority applies $\mathsf{sk_{OA}}$ to the ciphertext the result is a value that is not equal to a public-key of any group member, or (2) the revealed key satisfies $\mathsf{pk} \notin \mathcal{L}_{pk}^{\mathsf{param}}$. To formalize soundness we introduce the following group registration oracle:

$\mathsf{REG(sik, \cdot)}$: this is an oracle that simulates $\mathsf{J_{GM}}$, i.e., it is given $\mathsf{sk_{GM}}$ and registers users in the group; the oracle has access to a string **database** that stores the public-keys and their certificates.

Definition 4. *A GE scheme satisfies* soundness *if the following "soundness game", when instantiated with any PPT adversary $\mathcal{A}$, the probability it returns 1 is negligible.*

1. $\mathsf{param} \leftarrow \mathsf{SETUP_{init}}(1^\nu)$; $\langle \mathsf{pk_{OA}}, \mathsf{sk_{OA}} \rangle \leftarrow \mathsf{SETUP_{OA}}(\mathsf{param})$;
2. $\langle \mathsf{pk_{GM}}, \mathsf{sk_{GM}} \rangle \leftarrow \mathsf{SETUP_{GM}}(\mathsf{param})$;
3. $\langle \mathsf{pk_{\mathcal{R}}}, x, \psi, L, \mathsf{aux} \rangle \leftarrow \mathcal{A}^{\mathsf{REG(sk_{GM}, \cdot)}}(\mathsf{param}, \mathsf{pk_{GM}}, \mathsf{pk_{OA}}, \mathsf{sk_{OA}})$;
4. $\langle \mathsf{aux}, \mathsf{out} \rangle \leftarrow \langle \mathcal{A}(\mathsf{aux}), \mathcal{V} \rangle (\mathsf{param}, \mathsf{pk_{GM}}, \mathsf{pk_{OA}}, \mathsf{pk_{\mathcal{R}}}, x, \psi, L)$;
5. $\mathsf{pk} \leftarrow \mathsf{OPEN}(\mathsf{sk_{OA}}, [\psi]_{\mathsf{oa}}, L)$;
6. if $\mathsf{pk} \notin \mathsf{database}$ or $\mathsf{pk} \notin \mathcal{L}_{pk}^{\mathsf{param}}$ or $\psi \notin \mathcal{L}_{ciphertext}^{x, L, \mathsf{pk}_{\mathcal{R}}, \mathsf{pk_{GM}}, \mathsf{pk_{OA}}, \mathsf{pk}}$ then ret. 1 else 0;

Note that $\mathcal{L}_{ciphertext}^{x, L, \mathsf{pk}_{\mathcal{R}}, \mathsf{pk_{GM}}, \mathsf{pk_{OA}}, \mathsf{pk}} = \{\mathsf{ENC}(\mathsf{pk_{GM}}, \mathsf{pk_{OA}}, \mathsf{pk}, \mathsf{cert}, w, L) \mid w : (x, w) \in \mathcal{R}, \langle \mathsf{pk}, \mathsf{cert} \rangle \in \mathsf{Valid}\}$. This means that the soundness adversary wins if the key obtained by OA after opening is either not in the database, or is invalid, or the ciphertext ψ is not a valid ciphertext under pk encrypting a witness for x under $\mathcal{R}$.

A GE scheme should satisfy correctness, security, anonymity and soundness. Note that: (1) By defining the oracles USER and REG one can allow concurrent attacks or force sequential execution of the group registration process. (2) CPA variants of the security and anonymity definition w.r.t. either group members or the OA can be obtained by dropping the corresponding DEC oracles. (3) Soundness and security assume a trusted setup; extension to malicious setup can be done by enforcing trustworthy initialization by standard methods (e.g. threshold cryptography or ZK proofs).

3   Necessary and Sufficient Conditions for GE Schemes

Given that a GE scheme is a complex primitive it would be helpful to break down its construction to more basic primitives and provide a general methodology for constructing GE schemes. The necessary components for building a GE scheme will be the following:

1. Adaptively Chosen Message Secure Digital Signature. It will be used to generate the public-key certificates by the GM during the JOIN procedure.

2. Public-key Encryption with CCA2 Security and Key-Privacy. We will employ an encryption scheme $\langle \mathcal{G}_e, \mathcal{E}, \mathcal{D} \rangle$ that satisfies (1) CCA2-security and (2) CCA2-Key-privacy. We note that in public-key encryption with key-privacy the key-generation has two components, one called $\mathcal{Z}_e$ that produces public-parameters shared by all key-holders and the key-generation $\mathcal{G}_e$ that given the public-parameter of the system produces a public/secret-key pair. Note that using $\mathcal{Z}_e$ is mandatory since some agreement between the receivers is necessary for key-privacy (at minimum all users should employ public-keys of the same length).

3. Proofs of Knowledge. Such protocols in the zero-knowledge setting satisfy three properties: completeness, soundness with knowledge extraction and zero-knowledge. These proofs exist for any NP language assuming one-way functions by reduction, e.g., to the graph 3-colorability proof of knowledge [29]. In certain settings, zero-knowledge proofs can be constructed more efficiently by starting with a honest-verifier zero-knowledge (HVZK) proof of language membership protocol (i.e., a protocol that requires no knowledge extraction and it is only zero-knowledge against honest verifiers) and then coupling such protocol with an extractable commitment scheme (to achieve knowledge extraction) and with an equivocal commitment (to enforce zero-knowledge against dishonest verifiers, cf. [25]).

Modular Design of GE schemes.   Consider an arbitrary relation $\mathcal{R}$ that has an associated paramter generation procedure $\mathcal{G}_r$ and a witness sampler $\mathsf{sample}_{\mathcal{R}}$. In the modular construction we will employ: (1) a digital signature scheme $\langle \mathcal{G}_s, \mathcal{S}, \mathcal{V}_s \rangle$ that is adaptively chosen message secure; (2) a public-key encryption scheme $\langle \mathcal{Z}_e, \mathcal{G}_e, \mathcal{E}, \mathcal{D} \rangle$ that satisfies CCA2 security and Key-privacy; (3) two zero-knowledge proofs of language membership (defined below); to facilitate knowledge extraction we will employ also an extractable commitment scheme $\langle \mathcal{Z}_{c,1}, \mathcal{C}_1, \mathcal{T}_1 \rangle$. Without loss of generality we will assume that all employed primitives operate over bitstrings. The construction of a GE scheme $\langle \mathsf{SETUP}, \mathsf{JOIN}, \langle \mathcal{G}_r, \mathcal{R}, \mathsf{sample}_{\mathcal{R}} \rangle, \mathsf{ENC}, \mathsf{DEC}, \mathsf{OPEN}, \langle \mathcal{P}, \mathcal{V} \rangle, \mathsf{recon} \rangle$ is as follows:

SETUP. The SETUP_{init} procedure will select the parameters param by performing a sequential execution of $\mathcal{Z}_e, \mathcal{Z}_{c,1}$. The SETUP_{GM} procedure will be the signature-setup $\mathcal{G}_s$ and the SETUP_{OA} will be the encryption-setup $\mathcal{G}_e$.

JOIN. Each prospective user will execute $\mathcal{G}_e$ to obtain pk, sk and then engage in a protocol $\langle \mathcal{P}_{pk}, \mathcal{V}_{pk} \rangle$ which is proof of language membership with the GM for the language $\mathcal{L}_{pk}^{\mathsf{param}} = \{\mathsf{pk} \mid \exists \mathsf{sk}, \rho : \langle \mathsf{pk}, \mathsf{sk} \rangle \leftarrow \mathcal{G}_e(\mathsf{param}; \rho)\}$. The GM will respond with the signature cert $\leftarrow \mathcal{S}(\mathsf{sk}_{GM}, \mathsf{pk})$.

ENC. The procedure ENC, given a witness w for a value x such that $(x, w) \in \mathcal{R}$ and a label L, it will return the pair $\psi =_{\mathsf{df}} \langle \psi_1, \psi_2, \psi_3, \psi_4 \rangle$ where $\psi_1 \leftarrow \mathcal{E}(\mathsf{pk}, w, L_1)$, $\psi_2 \leftarrow \mathcal{E}(\mathsf{pk_{OA}}, \mathsf{pk}, L_2)$, $\psi_3 \leftarrow \mathcal{C}_1(\mathsf{cpk}, \mathsf{pk})$ $\psi_4 \leftarrow \mathcal{C}_1(\mathsf{cpk}, \mathsf{cert})$ where $L_1 = \psi_2 \| \psi_3 \| \psi_4 \| L$ and $L_2 = \psi_3 \| \psi_4 \| L$.

DEC. Given sk, a ciphertext $\langle \psi_1, \psi_2, \psi_3, \psi_4 \rangle$ and a label L, it will return $\mathcal{D}(\mathsf{sk}, \psi_1, \psi_2 \| \psi_3 \| \psi_4 \| L)$.

OPEN. Given $\mathsf{sk_{OA}}$, a ciphertext $\langle \psi_2, \psi_3, \psi_4 \rangle =_{\mathsf{df}} [\psi]_{\mathsf{oa}}$ and a label L it will return $\mathcal{D}(\mathsf{sk_{OA}}, \psi_2, \psi_3 \| \psi_4 \| L)$.

Finally, the protocol $\langle \mathcal{P}, \mathcal{V} \rangle$ is a zero-knowledge proof of language membership for the language:

$$\{ \langle \mathsf{param}, \mathsf{pk_{GM}}, \mathsf{pk_{OA}}, \mathsf{pk_{\mathcal{R}}}, x, \psi_1, \psi_2, \psi_3, \psi_4, L \rangle \mid \exists \, (coins_{\psi_1}, coins_{\psi_2},$$

$$coins_{\psi_3}, coins_{\psi_4}, \mathsf{pk}, \mathsf{cert}, w):$$

$$\wedge (\mathcal{C}_1(\mathsf{cpk}, \mathsf{pk}; coins_{\psi_3}) = \psi_3) \wedge (\mathcal{C}_1(\mathsf{cpk}, \mathsf{cert}; coins_{\psi_4}) = \psi_4) \wedge (\mathcal{V}_{\mathsf{s}}(\mathsf{pk}, \mathsf{cert}) = \mathsf{true})$$

$$\wedge (\mathcal{E}(\mathsf{pk}, w, (\psi_2 \| \psi_3 \| \psi_4 \| L); coins_{\psi_1}) = \psi_1)$$

$$\wedge (\mathcal{E}(\mathsf{pk_{OA}}, \mathsf{pk}, (\psi_3 \| \psi_4 \| L); coins_{\psi_2}) = \psi_2) \wedge ((x, w) \in \mathcal{R})$$

Note that the reconstruction procedure recon will be set to simply the identity function.

Theorem 1. *The* GE *scheme above satisfies (i)* Correctness, *given that all involved primitives are correct and* $\langle \mathcal{P}_{pk}, \mathcal{V}_{pk} \rangle$, $\langle \mathcal{P}, \mathcal{V} \rangle$ *satisfy completeness. (ii)* Anonymity, *given that the encryption scheme for users satisfies CCA2-key-privacy, the encryption scheme for OA satisfies CCA2-security, the commitment scheme* $\mathcal{C}_1$ *is hiding and* $\langle \mathcal{P}_{pk}, \mathcal{V}_{pk} \rangle$ *and* $\langle \mathcal{P}, \mathcal{V} \rangle$ *are zero-knowledge. (iii)* Security, *given that the employed encryption scheme for users satisfies CCA2-security, the commitment scheme* $\mathcal{C}_1$ *is hiding and* $\langle \mathcal{P}_{pk}, \mathcal{V}_{pk} \rangle$, $\langle \mathcal{P}, \mathcal{V} \rangle$ *are zero-knowledge. (iv)* Soundness, *given that the employed digital signature scheme satisfies adaptive chosen message security, the commitment scheme* $\mathcal{C}_1$ *is binding and extractable and* $\langle \mathcal{P}_{pk}, \mathcal{V}_{pk} \rangle$ *and* $\langle \mathcal{P}, \mathcal{V} \rangle$ *satisfy soundness.*

Necessity of the basic primitives. We consider the reverse of the above results: the existence of GE would imply public-key encryption that is CCA2 secure and private as well as digital signature and zero-knowledge proofs for any NP-language. More details are given in the full version [32].

4   Efficient GE of Discrete-Logarithms

In this section we will consider the discrete-logarithm relation $\langle \mathcal{G}_{\mathsf{dl}}, \mathcal{R}_{\mathsf{dl}}, \mathsf{sample}_{\mathsf{dl}} \rangle$: $\mathcal{G}_{\mathsf{dl}}$ given 1^ν samples a description of a cyclic group of ν-bits order and a generator γ of that group; $\mathcal{R}$ contains pairs of the form (x, w) where $x = \gamma^w$; note that $\mathsf{pk_{\mathcal{R}}} = \langle \mathsf{desc}(G), \gamma \rangle$ and $\mathsf{sk_{\mathcal{R}}}$ is empty. Finally $\mathsf{sample}_{\mathsf{dl}}$ on input $\mathsf{pk_{\mathcal{R}}}$ selects a witness w and returns the pair $(x = \gamma^w, w)$. In this section we will present a GE scheme for the above relation. Note that the results of this section can be easily extended to other relations based on discrete-logs such as a commitment to w.

Design of a public-key encryption for discrete-logarithms with key-privacy and security.   One of the hurdles in designing a GE for discrete-logarithms is finding a suitable encryption scheme for the group members. In this section we will present a public-key encryption scheme that is suitable for verifiable encryption of discrete-logarithms while it satisfies CCA2-key-privacy and CCA2-security. The scheme is related to previous public-key encryption schemes of [24,40,28,19,10] and it is the first Paillier-based public-key encryption that is proven to satisfy key-privacy and security against chosen ciphertext attacks. Below we give a detailed description of our public-key encryption $\langle \mathcal{Z}_e, \mathcal{G}_e, \mathcal{E}, \mathcal{D} \rangle$ and of the accompanying intractability assumptions that ensure its properties.

Public-parameters. The parameter selection function $\mathcal{Z}_e$, given 1^ν selects a composite modulus $n = pq$ so that n is a ν-bit number, $p = 2p' + 1, q = 2q' + 1$ and p, p', q, q' are all prime numbers with p, q of equal size at least $\lfloor \nu/2 \rfloor + 1$. Then it samples $g \leftarrow \mathbb{Z}_{n^2}^*$ and computes $g_1 \leftarrow g^{2n} (\mathrm{mod}\, n^2)$. Observe that $\langle g_1 \rangle$ with very high probability is a subgroup of order $p'q'$ within $\mathbb{Z}_{n^2}^*$. In such case $\langle g_1 \rangle$ is a group that contains all square n-th residues of $\mathbb{Z}_{n^2}^*$ and we will call this group $\mathcal{X}_{n^2}$. We note further that all elements of $\mathbb{Z}_{n^2}^*$ can be written in a unique way in the form $g_1^r (1 + n)^v (-1)^\alpha (p_2 p - q_2 q)^\beta$ where $r \in [p'q'], v \in [n], \alpha, \beta \in \{0, 1\}$ (in this decomposition, p_2, q_2 are integers that satisfy $p_2 p^2 \equiv_{q^2} 1, q_2 q^2 \equiv_{p^2} 1$). We will denote by $\mathcal{Q}_{n^2}$ the subgroup of quadratic residues modulo n^2 which can be easily seen to contain all elements of the form $g_1^r (1 + n)^v$ with $r \in \mathbb{Z}_{p'q'}$ and $v \in \mathbb{Z}_n$ and has order $np'q'$ (precisely one fourth of $\mathbb{Z}_{n^2}^*$ and is generated by $g_1(1 + n)$). Note that we will use the notation $h =_{\mathrm{df}} 1 + n$. Finally, a second value g_2 is selected as follows: w is sampled at random from $[\frac{n}{4}] =_{\mathrm{df}} \{0, \ldots, \lfloor \frac{n}{4} \rfloor\}$ and we set $g_2 \leftarrow g_1^w$. A random member $\mathcal{H}$ of a universal one-way hash function family UOWHF is selected [39]; the range of $\mathcal{H}$ is assumed to be $[0, 2^{\nu/2-2})$. The global parameters of the cryptosystem that will be shared by all recipients are equal to $\mathsf{param} = \langle n, g_1, g_2, \mathrm{desc}\mathcal{H} \rangle$, where $\mathrm{desc}\mathcal{H}$ is the description of $\mathcal{H}$.

Key-Generation. The key-generation algorithm $\mathcal{G}_e$ receives the parameters $\langle n, g_1, g_2, \mathrm{desc}\mathcal{H} \rangle$, samples $x_1, x_2, y_1, y_2 \leftarrow_R [\frac{n^2}{4}]$ and sets $\mathsf{pk} = \langle c, d, y \rangle$ where $c = g_1^{x_1} g_2^{x_2}$, $d = g_1^{y_1} g_2^{y_2}$ and $y = g_1^z$; the secret-key is $\mathsf{sk} = \langle x_1, x_2, y_1, y_2, z \rangle$. Note that below we may include the string param as part of the pk and sk strings to avoid repeating it, nevertheless it should be recalled in all cases that $n, g_1, g_2, \mathrm{desc}\mathcal{H}$ are global parameters that are available to all parties.

Encryption. The encryption function $\mathcal{E}$ operates as follows: given the pk, a message w and a label L it samples $r \leftarrow_R [\frac{n}{4}]$ and outputs the triple $\langle u_1, u_2, e, v \rangle$ computed as follows: $u_1 \leftarrow g_1^r \bmod n^2$, $u_2 \leftarrow g_2^r \bmod n^2$, $e \leftarrow y^r(1+n)^w \bmod n^2$, $v \leftarrow ||c^r d^{r\mathcal{H}(u_1, u_2, e, L)} \bmod n^2||$ where $|| \cdot || : \mathbb{Z}_{n^2}^* \to \mathbb{Z}_{n^2}^*$ is defined as follows $||x|| = x$ if $x \leq n^2/2$ and $||x|| = -x$ if $x > n^2/2$. We note that the "absolute value" function $|| \cdot ||$ is used to disallow the malleability of a ciphertext with where n subintervals are defined by $I_{ni} = \left[\frac{i-1}{n}T, \frac{i}{n}T\right], i = 1, 2, \cdots, n$. If a sample is recorded and transmitted every time a level crossing occurs, the encoding procedure is called asynchronous delta modulation[2].

3   Rules for Irregular Sampling

Determining the threshold levels is very important as it has a huge impact on the performance of coding. Unfortunately there is no theory available to determine those values. The uniform threshold levels are not the efficient coding of the levels because they do not take advantage of the statistical properties of the signal. The present study considers nonuniform threshold level allocation scheme depending on the signal amplitude characteristics and rules which controls the importance of amplitude regions. As a result, signals with lesser activity in higher amplitude regions compared to the lower amplitude regions, will have less number of levels at higher amplitude region. In this section, IBF, logarithmic and linear function rules are analyzed.

3.1   Incomplete Beta Function

The lack of data to decide the exact number of levels for a given amplitude range, creates problems concerning the selection of number of levels. In such cases, an expert will have to assume the levels. For this reason, the flexible incomplete beta distribution, capable of attaining a variety of shapes could be used in Level Crossing applications. Because of its extreme flexibility, the distribution appears ideally suited for the computation of number of levels for a specific amplitude region of a signal. The Incomplete beta function $I(z, \alpha, \beta)$ is defined by [7]:

$$I(z, \alpha, \beta) \equiv \frac{1}{B(\alpha, \beta)} \int_0^z u^{\alpha-1}(1-u)^{\beta-1}du \quad \alpha > 0, \beta > 0, 0 \leq z \leq 1 \quad (2)$$

Eq.2 has the limiting values $I_0(\alpha, \beta) = 0, I_1(\alpha, \beta) = 1$. The shape of the incomplete beta function obtained from Eq.2 depends on the choice of its two parameters α and β. Estimating these parameters is a challenge since these parameters control the number of levels for a given amplitude range of a speech signal along with the signal probability density function (PDF). We have empirically chosen the values of α and β such that IBF gives more weight to amplitude regions of importance.

3.2   Linear Function

Although human auditory perception is certainly does not use a linear function, this group of mapping methods renders acceptable results for a wide range of applications. Its strength is its simplicity and speed. The linear function is defined by

$$linear(n) = n \quad (3)$$

However, computing the importance of amplitude regions in linear scale is not merely a mater of mathematical convenience. There is more compelling, physical consideration to be taken into account, related to the importance of amplitude regions. Natural worst case representation for characterization of physical systems is linear. This means that all amplitude regions become equally important.

3.3 Logarithmic Function

A logarithmic of a number x in base b is a number n such that $x = b^n$, where the value b must be neither 0 nor a root of 1. It is usually written as

$$log_b(x) = n \qquad (4)$$

When x and b are further restricted to positive real numbers, the logarithm is a unique real number.

Representation of importance of amplitude on a logarithmic scale can be helpful when the importance of regions varies slowly. Logarithmic rule assigns less number of levels to the corner amplitude regions and more levels are assigned logarithmically in important amplitude regions. The center amplitude regions (near zero amplitude regions) are considered to be important amplitude regions. This issue, however is not whether to accept or reject logarithmic rule but to appreciate where it fits in, and where it does not.

3.4 Level Estimation

In a deterministic environment, the accuracy of the signal reconstruction depends on several parameters such as positioning of the levels, total number of levels, statistical properties of the signal etc. Specifically, for a given signal we analyze the structural behavior by estimating its PDF. The signal histogram is approximated to obtain the signal PDF $p(x)$.

Now consider a signal with PDF $p(x)$ and level allocation rule $R(x)$. Let N be the total number of levels. The locations of N levels are estimated by

$$L(x) = p(x) \otimes R(x) \qquad (5)$$

The $\otimes$ symbol represents linear filtering. $L(x)$ gives the probability of distribution of levels and it guides the distribution of N levels over the amplitude range. As expected, the spacing of N levels are not uniform they are nonuniformly spaced over the amplitude range. Each level can be represented with $log_2(N)$ bits. Hence only amplitude regions with high activity will be allocated more number of levels using the rule $R(x)$ and signal amplitude PDF.

4   Experimental Evaluation

In this section, the performance of the proposed approach is evaluated for speech signals. We have run simulations for the level crossing based sampling of speech

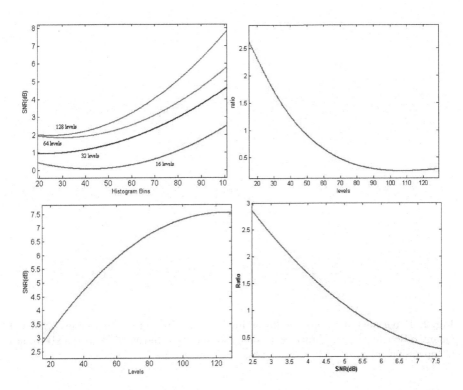

Fig. 1. Experimental results for IBF (a) Histogram bin versus SNR. (b) Quantization level versus ratio (c) Quantization level versus SNR.(d) SNR versus Compression ratio.

signals from TIMIT database. The TIMIT speech signals are sampled by 16 KHz sampling rate and each sample size is 16 bit. The PDF of the speech signal is estimated by computing the amplitude histogram of the signal with 100 bins. The total numbers of quantization levels needed to sample the given signal are set to 16, 32, 64 and 128. The accuracy of distribution of the levels computed from Eq.5, also depends on the number of bins used to compute convoluted PDF of the signal. The levels are estimated for 20, 40, 60, 80, 100 bins for comparison and analysis. We evaluated the system with proposed rules IBF, logarithmic and linear functions.

By comparing IBF, logarithmic and linear rule results, we analyze the performances. SNR of the resampled signal generally improves as the bins increase for all the levels. The IBF rule gives high SNR consistently compared to the logarithmic and linear rule at all bins. The performance of logarithmic rule is slightly less than that of IBF rule for all the levels (Fig.1(a) and Fig.3(a)). The best performance is observed for IBF rule with 128 levels. In case of linear rule, 1 dB drop in SNR is observed compared to IBF for 128 levels. Similarly, higher SNR is achieved for IBF and logarithmic rule compared to linear rule at all levels (Fig.1(a), Fig.2(a) and Fig.3(a)).

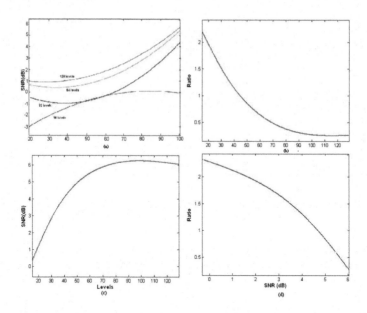

Fig. 2. Experimental results for linear function (a) Histogram bin versus SNR. (b) Quantization level versus ratio (c) Quantization level versus SNR. (d) SNR versus Compression ratio.

The comparison of compression ratio at various levels for the three rules is shown in Fig.1(b), Fig.2(b) and Fig.3(b). We observe that logarithmic rule slightly outperforms IBF rule. The logarithmic rule gives higher SNR for lesser levels and the ratio decreases as the levels are increased. Linear rule results in low compression ratio for all levels. For higher levels all the rules give similar results. The results of SNR versus levels (Fig.1(c), Fig.2(c) and Fig.3(c)) show that IBF and logarithmic rule performance is superior to linear rule at all levels. It is observed that IBF and logarithmic rules produce almost similar performance results. Minimum SNR for 16 levels is near 3 dB in IBF and logarithmic rule whereas minimum SNR in linear rule for 16 levels is 0.4 dB. Fig.1(d), Fig.2(d) and Fig.3(d) shows the plot of SNR versus compression ratio. The characteristic curve appears to be concave for IBF rule, linear for logarithmic rule and convex for linear rule. Performance of IBF and logarithmic rules are considerably better than linear rule, with higher SNR for higher compression ratio, which is nonetheless better performance. Comparison of IBF, logarithmic and linear rule shows that, the IBF rule and logarithmic outperforms linear rule. Also, performances of IBF slightly outperforms logarithmic rule in compression ratio and SNR.

The behavioral patterns of IBF, logarithmic and linear rule appear to be similar except in SNR versus ratio analysis. IBF rule is based on the auditory properties of the humans. IBF rule distributes more levels in the critical amplitude regions. Similar to IBF, logarithmic rule also considers that near zero amplitude regions are important than the corner amplitude regions. The priority

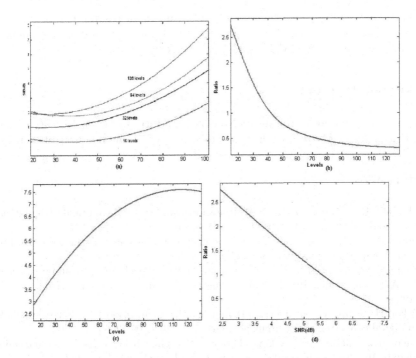

Fig. 3. Experimental results for logarithmic function (a) Histogram bin versus SNR. (b) Quantization level versus ratio (c) Quantization level versus SNR. (d) SNR versus Compression ratio.

of the amplitude regions varies logarithmically from corner amplitude regions to near zero value amplitude regions. Linear rule considers that each amplitude region is equally important. Hence, the SNR of the resampled signal remains consistently superior to linear rule. Lack of levels at critical amplitude regions of the signal decreases the SNR of the resampled signal. The performance of the proposed approaches is fairly consistent with that of Sayiner[3]. This experimental analysis illustrates that signal with special statistical behavior such as speech, medical signals are unsuitable for uniform sampling. These types of signals can be more efficiently sampled using a level crossing scheme.

5   Conclusion

In summary, this paper presents new threshold level allocation schemes for level crossing based nonuniform sampling which dynamically assigns the number of quantization levels depending on the importance of the given amplitude range of the input signal. Proposed methods take the advantage of statistical properties of the signal and allocate the nonuniformly spaced quantization levels across the amplitude range. The proposed level allocation scheme for nonuniform sampling based on level crossing may motivate directed attempts to augment traditional methods that will improve their ability. Overall, these results motivate continued work on level crossing based nonuniform sampling for improving sampling performance and analysis the signals as a whole.

Acknowledgement

The authors profoundly thank Mr.Noushath and Mr.Aradhya, Research Scholars, University of Mysore, for helping them to typeset the manuscript in LATEX.

References

1. Mitchell, R.J., Gonzalez, R.C.: Multilevel crossing rates for automated signal classification. In: Proc. of ICASSP 1978, vol. 3, pp. 218–222 (April 1978)
2. Inose, H., Aoki, T., Wantanable, K.: Asynchronous delta modulation systems, Electron. Commun., Japan, pp. 34–42 (March 1966)
3. Sayiner, N., Sorensen, H.V., Viswanathan, T.R.: A level crossing sampling scheme for A/D conversion. IEEE Transactions on Circuits and Systems II 43, 335–339 (1996)
4. Allier, E., Sicard, G., Fesquet, L., Renaudin, M.: Asynchronous level crossing analog to digital converters. Measurement Journal 37, 296–309 (2005)
5. Guan, K., Singer, A.C.: A Level Crossing Sampling Scheme for Non-Bandlimited Signals. In: Proc. of ICASSP 2006, vol. 3, pp. III-381-383 (May 2006)
6. Miskowicz, M.: Efficiency of Level-Crossing Sampling for Bandlimited Gaussian Random Processes. In: Proc. of IEEE International Workshop on Factory Communication Systems-2006, pp. 137–142 (June 2006)
7. Press, W.H., Teukolsky, S.A., William, T.V., Brian, P.F.: Numerical Recipes in C++, 2nd edn. Cambridge university press, Cambridge (2002)

Cepstral Domain Teager Energy for Identifying Perceptually Similar Languages

Hemant A. Patil[1] and T.K. Basu[2]

[1] Dhirubhai Ambani Institute of Information and Communication Technology,
DA-IICT , Gandhinagar, Gujarat, India
hemant_patil@daiict.ac.in
[2] Department of Electrical Engineering, Indian Institute of Technology, IIT
Kharagpur, West Bengal, India
tkb@ee.iitkgp.ernet.in

Abstract. Language Identification (LID) refers to the task of identifying an unknown language from the test utterances. In this paper, a new feature set, *viz.,*T-MFCC by amalgamating Teager Energy Operator (TEO) and well-known Mel frequency cepstral coefficients (MFCC) is developed. The effectiveness of the newly derived feature set is demonstrated for identifying perceptually similar Indian languages such as Hindi and Urdu. The modified structure of polynomial classifier of 2nd and 3rd order approximation has been used for the LID problem. The results have been compared with state-of-the art feature set, *viz.,*MFCC and found to be effective (an average jump 21.66%) in majority of the cases. This may be due to the fact that the T-MFCC represents the combined effect of airflow properties in the vocal tract (which are known to be language and speaker dependent) and human perception process for hearing.

1 Introduction

Language Identification (LID) refers to the task of identifying an unknown language from the test utterances. LID applications fall into two main categories: pre-processing for machine understanding systems and preprocessing for human listeners. Alternatively, an LID system could be run in advance of the speech recognizer. Alternatively, LID might be used to route an incoming telephone call to a human switchboard operator fluent in the corresponding language [6]. Several techniques such spectral, prosody, phoneme, word-level, etc. have been proposed in the literature for LID problem. In this paper, we adopt spectral-based approach [5] and show the effectiveness of the newly derived feature set,*viz.,*Teager Energy based Mel Frequency Cepstral Coefficients (T-MFCC) for identification of perceptually similar Indian languages, *viz.,*Hindi and Urdu.

2 Data Collection and Corpus Design

Database of 180 speakers (60 in each of Marathi, Hindi and Urdu) is created from the different states of India, *viz.,*Maharashtra, Uttar Pradesh and West Bengal

A. Ghosh, R.K. De, and S.K. Pal (Eds.): PReMI 2007, LNCS 4815, pp. 455–462, 2007.

with the help of a voice activated tape recorder (Sanyo model no. M-1110C & Aiwa model no. JS299) with microphone input, a close talking microphone (*viz.*,Frontech and Intex). During recording of the contextual speech, the interviewer asked some questions to speaker in order to motivate him or her to speak on his or her chosen topic. Other details of the experimental setup and data collection are given in [7].

Table 1. Database Description for LID system

Item	Details
No. of speakers	180 (60 in each of Marathi, Hindi and Urdu)
No. of sessions	1
Data type	Speech
Sampling rate	22,050 Hz
Sampling format	1-channel, 16-bit resolution
Type of speech	Read sentences, isolated words and digits, combination-lock phrases, questions, contextual speech of considerable duration
Application	Text-independent language identification (LID) system
Training language	Marathi, Hindi, Urdu.
Testing language	Marathi, Hindi, Urdu.
No. of repetitions	10 except for contextual speech.
Training segments	30 s, 60 s, 90 s, 120 s.
Test segments	1 s, 3 s, 5 s, 7 s, 10 s, 12 s, 15 s.
Microphone	Close talking microphone
Recording Equipment	Sanyo Voice Activated System (VAS), Aiwa, Panasonic magnetic tape recorders
Magnetic tape	Sony High-Fidelity (HF) voice and music recording cassettes
Channels	EP to EP Wire
Acoustic environment	Home/slums/college/remote villages/roads

3   The Teager Energy Operator (TEO)

Features derived from a linear speech production models assume that airflow propagates in the vocal tract as a linear plane wave. This pulsatile flow is considered the source of sound production [9]. According to Teager [8], this assumption may not hold since the flow is actually separate and concomitant vortices are distributed throughout the vocal tract. He suggested that the true source of sound production is actually the vortex-flow interactions, which are non-linear and a non-linear model has been suggested based on the energy of airflow. Fig.1 shows Teager's original investigations about distinct flow pattern of vowel '*i*' at top and bottom rear of the front oral cavity (due to the non-linear airflow)[8].

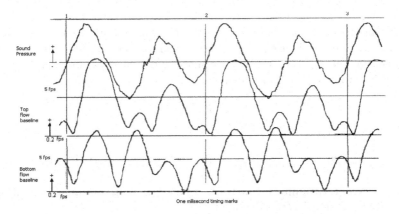

Fig. 1. Representative simultaneous normalized sound and air flow for the vowel 'i'. Top trace: sound pressure. Middle trace: airflow velocity measured by anemometers at the top rear of the front oral cavity. Bottom trace: air flow velocity measured at the bottom rear of the front oral cavity. (After Teager [8]).

There are two broad ways to model the human speech production process. One approach is to model the vocal tract structure using a source-filter model. This approach assumes that the underlying source of speaker's identity is coming from the vocal tract configuration of the articulators (i.e., size and shape of the vocal tract) and the manner in which speaker uses his articulators in sound production [4]. An alternative way to characterize speech production is to model the airflow pattern in the vocal tract. The underlying concept here, is that while the vocal tract articulators do move to configure the vocal tract shape (making cues for speaker's identity [4]), it is the resulting airflow properties which serve to excite those models which a listener will perceive for a particular speaker's voice [8],[9]. Modeling the time-varying vortex flow is a formidable task and Teager devised a simple algorithm which uses a non-linear energy-tracking operator called as Teager Energy Operator (TEO) (in discrete-time) for signal analysis with the supporting observation that hearing is the process of detecting energy. The concept was further extended to continuous-domain by Kaiser [3]. According to Kaiser, energy in a speech frame is a function of amplitude and frequency as well. Let us now discuss this point in brief.

The dynamics and solution (which is a S.H.M.) of mass-spring system are described as

$$\frac{d^2x}{dt^2} + \frac{k}{m}x = 0 \Rightarrow x(t) = A\cos(\Omega t + \phi)$$

and the energy is given by

$$E = \frac{1}{2}m\Omega^2 A^2 \Rightarrow E \propto (A\Omega)^2 \qquad (1)$$

From (1), it is clear that the energy of the S.H.M. of displacement signal $x(t)$ is directly proportional not only to the square of the amplitude of the signal but also to the square of the frequency of the signal. Kaiser and Teager proposed the algorithm to calculate the running estimate of the energy content in the signal. (1) can be expressed in discrete-time domain as

$$x(n) = A\cos(\omega n + \phi)$$

By trigonometry,

$$x^2(n) - x(n+1)x(n-1) = A^2\sin^2\omega \approx A^2\omega^2 \approx E_n$$

where E_n gives the running estimate of signal's energy. In continuous and discrete-time, TEO of a signal $x(t)$ is defined by

$$\Psi_c[x(t)] = \left[\frac{dx}{dt}\right]^2 - x(t)\frac{d^2x}{dt^2} \mapsto \Psi_d[x(n)] = x^2(n) - x(n+1)x(n-1) \qquad (2)$$

It is a well known fact that the speech can be modeled as a linear combination of AM-FM signals in some cases [7],[9]. Each resonance or formant is represented by an AM-FM signal of the form

$$x(t) = a(t)\cos(\phi(t)) = a(t)\cos[\int_0^t \omega_i(\tau)d\tau + \phi_0] \Rightarrow \Psi_c[x(t)] \approx \left(a\frac{d\phi}{dt}\right)^2 \qquad (3)$$

where $a(t)$ is a a time varying amplitude signal and $\omega_i(t)$ is the instantaneous frequency given by $\omega_i(t) = d\phi/dt$. This model allows the amplitude and formant frequency (resonance) to vary instantaneously within one pitch period. It is known that TEO can track the modulation energy and identify the instantaneous amplitude and frequency. Motivated by this fact, in this paper a new feature set based on nonlinear model of (3) is developed using the TEO. The idea of using TEO instead of the commonly used instantaneous energy is to take advantage of the modulation energy tracking capability of the TEO. This leads to a better representation of *formant information* (which is speaker and possibly language specific) in the feature vector than MFCC [7]. In the next section, we will discuss the details of T-MFCC.

4   Teager Energy Based MFCC (T-MFCC)

For a particular speech sound in a *language* , the *human perception* process responds with better frequency resolution to lower frequency range and relatively low fre-quency resolution in high frequency range with the help of human ear. To mimic this process MFCC is developed. For computing MFCC, we warp the speech spectrum into Mel frequency scale. This Mel frequency warping is done by multiplying the magnitude of speech spectrum for a preprocessed frame by magnitude of triangular filters in Mel filterbank followed by log-compression of sub-band energies and finally DCT. Davis and Mermelstein proposed one such filterbank to simulate this in 1980 for speech recognition application [2]. *Thus, MFCC can be a potential feature to identify perceptually distinct languages (because for perceptually similar languages there will be confusion in MFCC due to its dependence of human perception process for hearing).* Traditional MFCC-based feature extraction involves preprocessing; Mel-spectrum of preprocessed speech, followed by log-compression of subband energies and finally DCT is taken to get MFCC per frame [2]. In our approach, we employ TEO for calculating the energy of speech signal. Now, one may apply TEO in frequency domain, i.e., TEO of each subband at the output of Mel-filterbank, but there is difficulty from implementation point of view. Let us discuss this point in detail. In frequency-domain, (2) for pre-processed speech $x_p(n)$ implies,

$$F\left\{\Psi_c[x_p(t)]\right\} \mapsto F\left\{x_p^2(n) - x_p(n+1)x_p(n-1)\right\}$$

$$F\left\{\Psi_c[x_p(t)]\right\} = F\left\{x_p^2(n)\right\} - F\left\{x_p(n+1)x_p(n-1)\right\} \tag{4}$$

Using shifting and multiplication property of Fourier transform, we have

$$F\left\{x_p(n+1)x_p(n-1)\right\} = \frac{1}{2\pi}\int_0^{2\pi} X_{1p}(\theta)X_{2p}(\omega - \theta)d\theta$$

where $X_{1p}(\omega) = e^{-j\omega}X_p(\omega)$ and $X_{2p}(\omega) = e^{j\omega}X_p(\omega)$. Hence (4) becomes

$$F\left\{\Psi_c[x_p(t)]\right\} = \frac{1}{2\pi}\left\{\int_0^{2\pi}\left(1 - e^{j\omega}e^{-2\theta}\right)X_p(\theta)X_p\left(\omega - \theta\right)d\theta\right\} \tag{5}$$

Thus (5) is difficult to implement in discrete-time and also time-consuming. So we have applied TEO in the time-domain. Let us now see the computational details of T-MFCC.

Speech signal $x(n)$ is first passed through pre-processing stage to give pre-processed speech signal $x_p(n)$. Next we calculate the Teager energy of $x_p(n)$:

$$\Psi_d[x_p(n)] = x_p^2(n) - x_p(n+1)x_p(n-1) = \psi_1(n)(say)$$

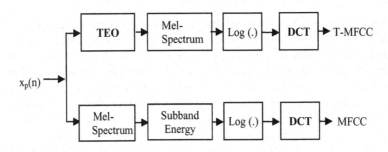

Fig. 2. Block diagram for T-MFCC and MFCC

The magnitude spectrum of the TEO output is computed and warped to Mel frequency scale followed by usual log and DCT computation (of MFCC) to obtain T-MFCC.

$$T - MFCC = \sum_{l=1}^{L} \log \left[\Psi_1(l) \right] \cos \left(\frac{k(l - 0.5)}{L} \pi \right), k = 1, 2, ..., Nc$$

where $\Psi_1(l)$ is the filterbank output of $F\{\psi_1(n)\}$ and $\log[\Psi(l)]$ is the log-filterbank output and T-MFCC(k) is the kth feature. T-MFCC differs from the traditional MFCC in the definition of *energy measure*, i.e., MFCC employs L^2 energy in frequency domain (due to Parseval's equivalence) at each subband whereas T-MFCC employs Teager energy in time domain. Fig. 2 shows the functional block diagram of MFCC and T-MFCC.

5   Experimental Results

In this paper, modified polynomial classifier of 2nd and 3rd order approximations is used as the basis for all the experiments [1]. The detailed discussion on modified classifier structure is beyond the scope of the paper and is given in [7]. Feature analysis was performed using 23.2 ms frame with an overlap of 50% and feature dimension is kept as 12. Each frame was pre-emphasized with the filter $1 - 0.97z^{-1}$, followed by Hamming windowing and then. We have taken 2 samples more to com-pute T-MFCC than that for MFCC because of TEO processing. The experiments are performed for different testing speech durations (i.e., 1 s, 3 s, 5 s, 7 s, 10 s, 12 s and 15 s) and training speech durations (i.e., 30 s, 60 s, 90 s, and 120 s). The results are shown as average success rates (over testing speech durations) in Table 2 (for Hindi and Urdu) and Table 3 (for Marathi and Hindi). In addition to this, the results are shown as overall success rates (computed as average over testing speech durations followed by average over training speech durations) in Tables 4 and 5 for polynomial classifiers of 2nd and 3rd order polynomial approximation. Finally, Tables 6-7 show confusion matrices (diagonal elements indicate % correct identification in a particular linguistic group and off-diagonal elements show the misidentification) for Hindi and Urdu with MFCC and T-MFCC, respectively.

Some of the observations from the results are as follows:

– Average success rates increase with the increase in training speech durations.
– For both 2nd order and 3rd order polynomial approximation and identifi-cation of perceptually similar languages (i.e., Hindi and Urdu), T-MFCC outperformed MFCC in all the cases of training speech durations. This may be due to the fact that MFCC is known to be developed to mimic human perception process and since the present problem deals with identification of perceptually similar languages (i.e., confusion in perception of phonemes of two languages, viz., Hindi and Urdu), MFCC gets confused in discriminating the language-specific features. On the other hand, T-MFCC represents the combined effect of airflow properties in the vocal tract (which are known to

Table 2. Average Success Rates for Hindi & Urdu with 2nd Order Approximation

TRFS	30s	60s	90s	120s
MFCC	21.42	22.97	23.57	23.69
T-MFCC	41.42	42.26	42.73	42.14

Table 3. Average Success Rates for Marathi & Hindi with 2nd Order Approximation

TR FS	30s	60s	90s	120s
MFCC	62.97	67.02	68.09	67.97
T-MFCC	55.83	57.85	58.21	56.42

Table 4. Overall Average Success Rates for Hindi and Urdu

OrderFS	2	3
MFCC	22.91	19.46
T-MFCC	42.14	43.56

Table 5. Overall Average Success Rates for Marathi and Hindi

Order FS	2	3
MFCC	66.51	62.22
T-MFCC	57.07	56.09

Table 6. Confusion Matrix with 2nd order approximation for MFCC (TR=120 s and TE=15 s)with Hindi(H) & Urdu(U)

Ident. Act.	H	U
H	85.55	14.44
U	76.66	23.33

Table 7. Confusion Matrix with 2nd order approximation for T-MFCC (TR=120 s and TE=15 s) with Hindi(H) & Urdu(U)

Ident. Act.	H	U
H	71.11	28.88
U	5.55	94.44

be language and speaker dependent [7]) and human perception process. So, T-MFCC is able to capture the speaker and language -specific information better than MFCC.

- On the other hand, for both 2nd order and 3rd order polynomial approximation and identification of perceptually *distinct* languages (i.e., Marathi and Hindi), MFCC outperformed T-MFCC.
- There is a significant improvement in the performance of T-MFCC for 3rd order approximation as compared to the 2nd order approximation. This is quite expected for a classifier of higher order polynomial approximation.
- Confusion matrix for T-MFCC performed better than MFCC. This shows that T-MFCC has better *class discrimination* power than MFCC for distinguishing perceptually similar languages.

6   Conclusion

In this paper, Teager Energy based MFCC (T-MFCC) features are proposed for identifying perceptually similar Indian languages, *viz.*, Hindi and Urdu. The performance of newly proposed feature set was compared with MFCC and found to be effective. This research work can be readily extended to identifying other perceptually similar Asian or European languages.

References

1. Campbell, W.M., Assaleh, K.T., Broun, C.C.: Speaker recognition with polynomial classifiers. IEEE Trans. on Speech and Audio Processing 10, 205–212 (2002)
2. Davis, S.B., Mermelstein, P.: Comparison of parametric representations for monosyllabic word recognition in continuously spoken sentences. IEEE Trans. Acoust., Speech and Signal Processing 28, 357–366 (1980)
3. Kaiser, J.F.: On a simple algorithm to calculate the 'energy' of a signal. Proc. of Int. Conf. on Acoustic, Speech and Signal Processing 1, 381–384 (1990)
4. Kersta, L.G.: Voiceprint Identification. Nature 196, 1253–1257 (1962)
5. Mary, L., Yegnanarayana, B.: Autoassociative neural network models for language identification. In: Int. Conf. on Intelligent Sensing and Information Processing, ICISIP, pp. 317–320 (2004)
6. Muthusamy, Y.K., Barnard, E., Cole, R.A.: Reviewing automatic language identification. IEEE Signal Processing Mag. 11, 3341 (1994)
7. Patil, H.A.: Speaker Recognition in Indian languages: A feature based approach. Ph.D. Thesis, Department of Electrical Engineering, IIT Kharagpur, India (July 2005)
8. Teager, H.M.: Some observations on oral air flow during phonation. IEEE Trans. Acoust., Speech, Signal Process. 28, 599–601 (1980)
9. Zhau, G., Hansen, J.H.L., Kaiser, J.F.: Non-linear feature based classification of speech under stress. IEEE Trans. on Speech and Audio Processing 9, 201–216 (2001)

Spoken Language Identification for Indian Languages Using Split and Merge EM Algorithm

Naresh Manwani, Suman K. Mitra, and M.V. Joshi

Dhirubhai Ambani Institute of Information and Communication Technology,
Gandhinagar, India

Abstract. Performance of Language Identification (LID) System using Gaussian Mixture Models (GMM) is limited by the convergence of Expectation Maximization (EM) algorithm to local maxima. In this paper an LID system is described using Gaussian Mixture Models for the extracted features which are then trained using *Split and Merge Expectation Maximization Algorithm* that improves the global convergence of EM algorithm. It improves the learning of mixture models which in turn gives better LID performance. A maximum likelihood classifier is used for classification or identifying a language. The superiority of the proposed method is tested for four languages

1 Introduction

Language Identification (LID), as the name suggests is an issue of identifying the language of any utterance irrespective of its length (duration of speech), context (topic and emotions) and speaker (gender, age and demographic region). "Humans have the best capability to identify the language" [1]. Due to the increasing demand of global communications, it is required to break the boundaries of languages. This gives new challenges to machine translation system of languages and speech recognition system also. For that the first step is identifying the language of the speech. Once a particular language has been identified, a translation or a recognition system can be trained to solve the problem based on the identified language.

LID based on language independent phoneme recognition followed by language modeling (PRLM) [2] needs phoneme recognizer. LID based language dependent parallel phoneme recognition (PPR) [2] requires labeled speech. It needs language dependent phoneme recognizer for each language. Both PRLM and PPR perform very well but are computationally very expensive. Alternate methods which do not require labeled speech have also been proposed but their reliability depends on the speech quality and the parameterization technique.

Parallel syllable like unit recognizers [3] can also be used in place of parallel phoneme recognizer for LID. This approach does not require annotated corpora. But its performance depends on how efficiently speech is segmented into syllables like sounds. Recently Auto Associative Neural Network (AANN) [4] are also used for LID. Which does not require transcribed database, butuses heuristics for modelling. Gaussian Mixture Models (GMM) are also used for LID [2]. Although

A. Ghosh, R.K. De, and S.K. Pal (Eds.): PReMI 2007, LNCS 4815, pp. 463–468, 2007.

performance of this approach is comparable to other approaches, it still suffers from the problem of its convergence to local maxima.

Feature extraction methods play important role in language discrimination. Mel Frequency Cepstral Coefficients (MFCC), Perceptual Linear Predictive (PLP) coefficients, Linear Prediction Coefficients (LPC) etc are some of the most commonly used feature extraction methods in speech applications. Recently new feature extraction techniques such as Modified Group Delay Feature (MGDF)[5], Time Frequency Principal Component (TFPC) [6] are explored.

In this paper, we first extract the MFCC and their delta as well as delta-delta coefficients as the features for the speech utterences. These features are then modelled as GMM and a split and merge EM(SMEM) algorithm is used to obtain the model parameters. The use of SMEM overcomes the difficulty of local maxima dur to EM. We show that the accuracy of the system can be improved by using split and merge EM algorithm.

The rest of the paper is organized as follows: section 2 discusses in brief about the GMM, their learning using EM algorithm and its limitations. In section 3 the split and merge is described which is used to overcome the limitation of EM algorithm. Section 4 shows the experiments and performance results of LID system using SMEM.

2   Gaussian Mixture Models and Expectation Maximization Algorithm

Gaussian mixture models (GMMs) play a very important role in pattern recognition. GMMs are used to approximate the distribution of the data as weighted sum of the multivariate Gaussian probability density function (pdf).

Efficient computation of the maximum likelihood parameter estimates of the GMM can be done with the EM(expectation maximization) algorithm. It optimizes the likelihood that the given data points (feature points as used in this study) are generated by a mixture of Gaussian probability density function [7].

In EM algorithm two steps are repeated iteratively. The first step also called E-step is used to calculate the expected data log-likelihood function. In the second step called M-step estimates of new parameter are obtained by maximizing the log-likelihood function. Finally, these two steps give estimated parameters.

1. EM algorithm breaks down when any Gaussian component has its covariance matrix singular. It happens when clusters contain insufficient observations or too many components are used to fit the data set where there are actually fewer clusters[9].
2. Another limitation of EM algorithm is it does not give the global maximum of the log-likelihood of the data, instated it gives us the local maxima.

3   Split and Merge Expectation Maximization Algorithm

SMEM algorithm was basically proposed by Ueda et al.[8]. It overcomes the problem of local maxima in parameter estimation of mixture models using EM algorithm. The main idea behind SMEM algorithm is that after usual convergence of EM algorithm split and merge operations are performed to update the parameters of some mixture components. Then again EM is performed. This process is repeated iteratively until log-likelihood is increased. The number of components are kept constant. This process improves the global convergence of the EM algorithm. This make GMMs to learn the languages better and the result is better LID performance. Split and merge criterion are described as below.

3.1 Split Criterion

For splitting, a local Kullback divergence can be defined as [8]:

$$J_{split}(m;\Theta) = \int f_m(x;\Theta) log \frac{f_m(x;\Theta)}{p_m(x;\theta_m)} dx, \quad (1)$$

which is a splitting measure for the mth component of mixture model, $\forall\ \theta$ is the model parameter vector. The above equation actually represents the distance between two distributions: the local data density $f_m(x)$ around the mth model and the density of the mth model specified by the current parameter estimate Θ [8]. The local data density is written as:

$$f_m(x;\Theta) = \frac{\sum_{n=1}^{N} \delta(x - x_n)p(m|x_n;\Theta)}{\sum_{n=1}^{N} p(m|x_n;\Theta)}. \quad (2)$$

The expression given in Eqn. (2) is a modified empirical distribution weighted by the posterior probability so that the data around the mth model is focused on. When the weights are equal, $i.e., p(m|x_n;\Theta) = 1/M$, then $f_m(x;\Theta)$ becomes $p_m(x;\Theta)$ where:

$$p_m(x;\Theta) = \frac{1}{N} \sum_{n=1}^{N} \delta(x - x_n). \quad (3)$$

The splitting measure $J_{split}(m;\Theta)$ is calculated for all components in the mixture model and the component corresponding to the maximum value of $J_{split}(m;\Theta)$ has the worst estimate of the local density and this is the best candidate for split.

3.2 Merge Criterion

If there are two mixture components such that the posterior probabilities of several data points belonging to these two components are same, then the two components should get merged. To calculate a suitable measure of this, a merge criterion is defined as follows:

$$J_{merge}(i,j;\Theta) = \frac{p_i(\Theta)^t p_j(\Theta)}{||p_i(\Theta)||\ ||p_j(\Theta)||}, \quad (4)$$

where $p_i(\Theta) = (p(i|x_1;\Theta), p(i|x_2;\Theta), ..., p(i|x_N;\Theta))^t \in \mathcal{R}^N$ is an N-dimensional vector consisting of the posterior probabilities for data points to belong component $\imath$. t denotes the transpose operation and $||.||$ denotes the Euclidean vector norm. Two components $\imath$ and $\jmath$ with large value of $J_{merge}(i, j; \Theta)$ are supposed to be good candidates for merge.

To get the parameters of the components after split and merge operation a method proposed by Zhang et al. [11] is used.

4   Experimental Results

Testing of thealgorithm has been done on four language viz. English, HIndi, Gujarati and telegu. For English language IViE corpus is used. The statistics of speech samples that are used for training and testing of different languages are shown in Table (1) and (2) correspondingly.

Table 1. Statistics of training data

Language	Speakers	Lengths of Sentences	Total Duration of Training Samples	No. of Training sentences
Hindi	27 speakers, 23 male and 4 female	2-5 sec	440 sec	135
Telugu	24 speakers 20 male and 4 female	3-8 sec	440 sec	98
Gujarati	22 speakers 18 male and 4 female	2-7 sec	472 sec	132
English	25 speakers 24 male and 11 female	2-9 sec	420 sec	138

First of all, the speech files are hand-segmented to remove silence regions. with the help of WAVE-PAD software. Then speech is segmented into frames of length 23 msec (256 samples) and the overlapping between two frame was taken half of the frame length which is 11.5 msec (128 samples). Hamming window is used for smoothing. Then 12-dimensional Mel Frequency Cepstral Coefficients (MFCC) are extracted for each frame and were augmented in their time context. After taking MFCC its Delta and Delta-Delta Cepstral Coefficients are also extracted. The window length for delta and Delta-Delta Coefficients is K=9 and K=5 respectively. Cepstral Mean Subtraction (CMS) is applied to remove the effect of convoluting noises.

Separate GMM is used for each of the coefficient stream(MFCC, its Delta and Delta-Delta) for each language. Number of components in each GMM are kept 40. Now, for every language there are three GMMs, one each corresponding to different feature stream.

Table 2. Statistics of test data

Language	Speakers	Lengths of Sentences	No. of Test utterances
Hindi	35 speakers, 31 male and 4 female	2-5 sec	105
Telugu	22 speakers 18 male and 4 female	3-9 sec	62
Gujarati	22 speakers 18 male and 4 female	2-10 sec	88
English	28 speakers 14 male and 14 female	2-10 sec	91

Table 3. Performance comparisons for LID using simple EM and SMEM

Languages taken	Simple EM	SMEM	Efficiency gained
Hindi, English, Gujarati, Telugu	81.20 %	82.65 %	1.45 %
Hindi, English, Gujarati	85.21 %	85.21 %	0.00 %
Hindi, English, Telugu	84.70 %	86.20 %	1.50 %
Hindi, Telugu, Gujarati	80.78 %	81.96 %	1.18 %
English, Telugu, Gujarati	87.96 %	90.87 %	2.91 %
Hindi, English	91.26 %	92.72 %	1.46 %
Hindi, Gujarati	87.05 %	85.50 %	-1.55 %
Hindi, Telugu	91.02 %	91.62 %	0.60 %
English, Gujarati	93.85 %	94.41 %	0.56 %
English, Telugu	98.69 %	98.69 %	0.00 %
Telugu, Gujarati	86.67 %	91.33 %	4.66 %

In the first experiment all GMMs are trained using EM algorithm. Next we apply the split and merge algorithm and perform the usual EM iteratively until the log-likelihood is increasing. The log-likelihood is given by

$$\mathcal{L}(\{x_n, y_n, z_n\}|\Theta_l^x, \Theta_l^y, \Theta_l^z) = \sum_1^N [a * logp(x_n|\Theta_l^x)$$
$$+ b * logp(y_n|\Theta_l^y) + c * logp(z_n|\Theta_l^z)], \tag{5}$$

where Θ_l^x are the parameters of GMM modeled using MFCC for language l, Θ_l^y are the parameters of GMM using delta Cepstral coefficients for language l and Θ_l^z are the parameters of GMM using delta-delta Cepstral coefficients for language l. x_n, y_n, z_n are Cepstral coefficients, delta Cepstral coefficients and delta-delta Cepstral coefficients correspondingly. It is assumed that these three streams are jointly statistically independent of each other. The maximum likelihood classifier hypothesizes i as the language of the unknown utterance, where

$$i = argmax_l \big[\mathcal{L}(\{x_n, y_n, z_n\} | \Theta_l^x, \Theta_l^y, \Theta_l^z) \big] \qquad (6)$$

Table(3) shows the LID performance for both using simple EM and split and merge EM. The test is performed for values $a = 0.6, b = 10, c = 10$. These values of a, b and c are approximated by experiments for which the performance is better. From the comparison results shown in Table(3) it is clear that SMEM outperform simple EM algorithm and gives better performance for LID.

5   Conclusions

A Split and Merge EM algorithm based approach is proposed to solve the language identification problem by using Gaussian mixture models. The problem of local maxima occurs in a mixture model is avoided by this split and merge EM (SMEM) algorithm. SMEM algorithm changes the parameters of some GMM components by split and merge operations. It improves the distribution of Gaussian components in the space which in-turn increases the log-likelihood of observing the data. This makes GMMs to learn the languages better in comparison to using the simple EM algorithm.

References

1. Muthusamy, Y.K.: A Segmental Approach to Automatic Language Identification. PhD thesis, Oregon Graduate Institute (1993)
2. Zissman, M.A., Singer, E.: Automatic language identification of telephone speech messages using phoneme recognition and N-GRAM modeling. In: Proc. ICASSP 1994, Adelaide, Austrailia (1994)
3. Nagrajan, T., Murthy, H.A.: Language identification using parallel syllable like unit recognition. In: Proc. ICASSP (2004)
4. Mary, L., Yegnanarayana, B.: Autoassociative Neural Network Models for Language Identification. In: Proc. IClSIP (2004)
5. Hegde, R.M., Murthy, H.A.: Automatic Language Identification and Discrimination Using the Modified Group Delay Feature. In: Proc. ICISIP (2005)
6. Bimbot, F.E., Magrin-chagnolleau, I., Dutat, M.: Language recognition using time-frequency principal component analysis and acoustic modeling (2000)
7. Bilmes, J.A.: A gentle tutorial of the EM algorithm and its application to parameter estimation for gaussian mixture and hidden markov models. Technical Report tr-97-021, International Computer Science Institute, Berkeley, California, USA (1997)
8. Ueda, N., Nakano, R., Ghabramani, Z., Hinton, G.E.: SMEM algorithm for mixture models. Neural Computation (2000)
9. Cheng, S.S., Wang, H.M., Fu, H.C.: A model-selection-based self-splitting gaussian mixture learning with application to speaker identification. In: EURASIP (2004)
10. Ormoneit, D., Tresp, V.: Improved gaussian mixture density estimates using bayesian penalty terms and network averaging. In: NIPS (1995)
11. Zhang, Z., Chibiao Chen, J.S., Chan, K.L.: EM algorithms for gaussian mixtures with split-and-merge operation. Pattern Recognition (2003)

Audio Visual Speaker Verification Based on Hybrid Fusion of Cross Modal Features

Girija Chetty and Michael Wagner

School of Information Sciences and Engineering
University of Canberra, Australia
girija.chetty@canberra.edu.au
michael.wagner@canberra.edu.au

Abstract. In this paper, we propose hybrid fusion of audio and explicit correlation features for speaker identity verification applications. Experiments were performed with the GMM based speaker models with a hybrid fusion technique involving late fusion of explicit cross-modal fusion features, with implicit eigen lip and audio MFCC features. An evaluation of the system performance with different gender specific datasets from controlled VidTIMIT data base and opportunistic UCBN database shows a significant performance improvement.

Keywords: Audio-visual, speaker identity verification, liveness checking, cross modal correlations.

1 Introduction

The performance of a speaker verification system can be enhanced by including visual information from the lip region, as it would be more difficult for an impostor to imitate both audio and dynamical visual information simultaneously [1] and [4]. Some of the recent findings in psychophysical analysis of visual speech by Kuratate, Munhall et.al [2], and Shinji Maeda [3] suggest that a speaking face is a kinematic-acoustic system in motion, and the shape, the texture, and the acoustic features during speech production are correlated in a complex way, with a single neuromotor source controlling the vocal tract behavior, and being responsible for both the acoustic and the visible attributes of speech production. Hence, the speaker models built with explicit audio-lip correlation features can allow better modeling of intrinsic temporal correlations between acoustic-labial articulators and vocal tract dynamics during speech production, enhancing the performance of speaker identity verification systems. Further, it would also allow liveness checks to be performed as it would be extremely difficult for an impostor to manufacture the complex intrinsic temporal correlations and make fraudulent replay attacks on speaker verification system.

In this paper, we propose several such explicit correlation features to model the intrinsic temporal correlations that exist in visual speech. We first perform a cross correlation analysis on the audio and lip modalities to extract the correlated part of the information, and then employ an hybrid fusion approach based on the optimal combination of feature-level and late fusion techniques to fuse the correlated and the

A. Ghosh, R.K. De, and S.K. Pal (Eds.): PReMI 2007, LNCS 4815, pp. 469–478, 2007.
© Springer-Verlag Berlin Heidelberg 2007

mutually independent components. Further, an automatic weight selection technique which automatically adapts the fusion weights to the audio noise conditions is proposed. We propose three different types of cross-modal association approaches under the linear correlation model: the Latent Semantic Analysis (LSA), the Cross-modal Factor Analysis (CFA), and the Canonical Correlation Analysis (CCA).

Experiments performed with the hybrid fusion based on the optimal combination of feature-level fused explicit cross-modal features (LSA, CFA and CCA), and the late fusion of lip features (eigenlip) features and audio MFCC features allow a considerable improvement in EER performance for both speaker identity verification scenarios. To improve the EER performance for noisy audio SIV scenarios, the late fusion weights were determined automatically, by performing a mapping between an audio reliability estimate and the modality weightings. It was found that the hybrid fusion with automatic weight adaptation improves the EER performance for different SIV scenarios including both the clean and the noisy audio conditions. This paper is organised as follows.

In the next section, the proposed explicit cross modal features (LSA, CFA and CCA), are described. Section 3 describes scheme for hybrid fusion of the explicit correlation features with audio and lip features for modeling the correlated and uncorrelated components. The automatic weight adaptation scheme is described in the Section 4. The details of the experimental results for the proposed approach are described in Section 5. Section 6 summarizes the findings on cross-modal features and describes details of the next stage of investigations.

2 Cross-Modal Association

In this section we describe the details of extracting explicit correlation features based on cross modal association (CMA) techniques which allow the modelling of the correlated components in audio and lip modalities.

2.1 Latent Semantic Analysis

Latent semantic analysis (LSA) is used as a powerful tool in text information retrieval to discover underlying semantic relationship between different textual units (.e.g. keywords and paragraphs) [5]. It is possible to detect the semantic correlation between visual faces and its associated speech based on LSA technique. The method consists of three major steps: the construction of a joint multimodal feature space, the normalization, the singular value decomposition (SVD), and the semantic association measurement.

Given n visual features and m audio features at each of the t video frames, the joint feature space can be expressed as:

$$X = [V_1, \dots, V_i, \dots, V_n, A_1, \dots, A_i, \dots A_m], \quad \text{where} \tag{1}$$

$$V_i = (v_i(1), v_i(2), \dots, v_i(t))^T, \quad \text{and} \tag{2}$$

$$A_i = (a_i(1), a_i(2), \dots, a_i(t))^T \tag{3}$$

Various visual and audio features can have quite different variations. Normalization of each feature in the joint space according to its maximum elements (or certain other statistical measurements) is thus needed and can be expressed as:

$$\hat{X}_{ij} = \frac{X_{ij}}{\max(\ abs\ (X_{ij})}\quad \forall j \tag{4}$$

After normalization all elements in normalized matrix $\hat{X}$ have values between -1 and 1. SVD can then be performed as follows:

$$\hat{X} = S . V . D^T \tag{5}$$

where S and D are matrices composing of left and right singular vectors and V is diagonal matrix of singular values in descending order. Keeping only the first and most important k singular vectors in S and D, we can derive an optimal approximation of $\hat{x}$ with reduced feature dimensions, where semantic (correlation) information between visual and audio features is mostly preserved.

2.2 Cross-Modal Factor Analysis

LSA does not distinguish features from different modalities in the joint space. The optimal solution based on overall distribution which LSA models, may not best represent semantic relationships between features of different modalities, since distribution patterns among features from the same modality will also greatly impact LSA's results. A solution to the above problem is to treat the features from different modalities as two separate subsets and focus only on the semantic patterns between these two subsets. Under the linear correlation model, the problem now is to find the optimal transformations that can best represent (or identify) the coupled patterns between the features of the two different subsets. We adopt the following optimization criterion to obtain the optimal transformations:

Given two mean centered matrices X and Y, which compose of row-by-row coupled samples from two subsets of features, we want orthogonal transformation matrices A and B that can minimize the expression:

$$\left\| XA - YB \right\|_F^2 ,\ \text{where} \tag{6}$$

$$A^T A = I\ \text{and}\ B^T B = I$$

$\left\| M \right\|_F$ denotes the Frobenius norm of the matrix M and can be expressed as:

$$\left\| M \right\|_F = \left(\sum_i \sum_j \left| m_{ij} \right|^2 \right)^{1/2} \tag{7}$$

In other words, A and B define two orthogonal transformation spaces where coupled data in X and Y can be projected as close to each other as possible.
Since we have:

$$\| XA - YB \|_F^2 = trace \ ((XA - YB).(XA - YB)^T)$$
$$= trace \ (XAA^T X^T + YBB^T Y^T - XAB^T Y^T - YBA^T X^T)$$
$$= trace \ (XX^T) + trace \ (YY^T) - 2.trace \ (XAB^T Y^T) \tag{8}$$

where trace of a matrix is defined to be the sum of the diagonal elements. We can easily see from above that matrices A and B which maximize *trace* $(XAB^T Y^T)$ will minimize Eqn. 3. It can be shown that such matrices are given by:

$$\begin{cases} A = S_{xy} \\ B = D_{xy} \end{cases} \quad \text{where}$$

$$X^T Y = S_{xy}.V_{xy}.D_{xy} \tag{9}$$

With the optimal transformation matrices A and B, we can calculate the transformed version of X and Y as follows:

$$\begin{cases} \tilde{X} = X.A \\ \tilde{Y} = Y.B \end{cases} \tag{10}$$

Corresponding vectors in $\tilde{X}$ and $\tilde{Y}$ are thus optimized to represent the coupled relationships between the two feature subsets without being affected by distribution patterns within each subset.

2.3 Canonical Correlation Analysis

Following the development of the previous section, we can adopt a different optimization criterion: Instead of minimizing the projected distance, we attempt to find transformation matrices A and B that maximize the correlation between X.A and Y.B. Given two mean centered matrices X and Y as defined in the previous section, we seek matrices A and B such that

$$correlatio \ n(XA, XB) = correlatio \ n(\tilde{X},\tilde{Y}) = diag \ (\sigma_1, \cdots \sigma_i, \cdots, \sigma_l) \tag{11}$$

where $\tilde{X} = Y.B$, and $1 \geq \sigma_1 \geq, \cdots, \sigma_i, \cdots, \geq \sigma_l \geq 0$. σ_i represents the largest possible correlation between the ith translated features in $\tilde{X}$ and $\tilde{Y}$. The CCA analysis is described in further detail in [6].

3   Hybrid Audio-Visual Fusion

In this section, we describe proposed hybrid fusion scheme for combining the audio-lip cross-correlation features extracted in Section 2, with the mutually independent audio and lip region features. The algorithm for audio-visual correlated component extraction is described now.

3.1 Feature Fusion of Correlated Components

Let f_A and f_L represent the audio MFCC and lip-region eigenlip features respectively. A and B represent the cross modal transformation matrices. One can apply LSA, CCA or CFA to find two new feature sets $f_A' = A^T f_A$ and $f_L' = B^T f_L$ such that the between-class cross modal association coefficient matrix of f_A' and f_L' is diagonal with maximised diagonal terms. However, maximised diagonal terms do not necessarily mean that all the diagonal terms exhibit strong cross-modal association. Hence, one can pick the maximally correlated components that are above a certain correlation threshold θ. Let us denote the projection vector that corresponds to the diagonal terms larger than the threshold θ by $\tilde{w}_A$ and $\tilde{w}_L$. Then the corresponding projections of f_A and f_L are given as:

$$\tilde{f}_A = \tilde{w}_A^T \cdot f_A \tag{12}$$

$$\tilde{f}_A = \tilde{w}_L^T \cdot f_A \tag{13}$$

Here $\tilde{f}_A$ and $\tilde{f}_L$ are the correlated components that are embedded in f_A and f_L. By performing feature fusion of correlated audio and lip components, we obtain the CFA optimized feature fused audio-lip feature vector:

$$\tilde{f}_{AL} = \begin{bmatrix} \tilde{f}_A & \tilde{f}_L \end{bmatrix} \tag{14}$$

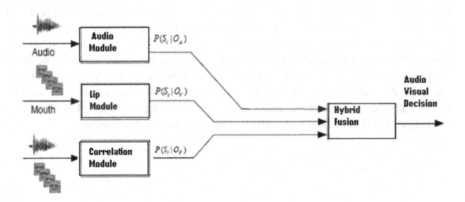

Fig. 1. System Overview of Hybrid Fusion Method

3.2 Late Fusion of Mutually Independent Components

In the Bayesian framework, late fusion can be performed using the product rule assuming statistically independent modalities. Various methods have been proposed in the literature [2] [3] and [6] as an alternative to the product rule such as the max rule, the min rule and the reliability-based weighted summation rule. We can compute joint or scores as a weighted summation:

$$\rho(\lambda_r) = \sum_{n=1}^{N} w_n \log P(f_n|\lambda_r) \, for \, \, r = 1,2,\dots,R \tag{15}$$

Where $\rho_n(\lambda_r)$ is the logarithm of the class-conditional probability $P(f_n|\lambda_r)$ for the nth modality, with class λ_r, and w_n denotes the weighting coefficient for modality n, such that $\Sigma_n w_n = 1$. Note that when $w_n = \frac{1}{N} \, \forall n$, Eqn. 15 is equivalent to the product rule. Since the w_n values can be regarded as the reliability values of the classifiers, this combination method is also referred to as RWS (Reliability Weighted Summation) rule [4]. The statistical and the numerical range of these likelihood scores vary from one classifier to another. Thus using sigmoid and variance normalization as described in [4], the likelihood scores can be normalized to be within the (0, 1) interval before the fusion process. The hybrid audio visual fusion vector is finally obtained by late fusion of feature fused correlated components ($\tilde{f}_{AL}$) with uncorrelated and mutually independent eigenlip features, and audio features with weights selected using RWS rule. An overview of the fusion method described is given in Figure 1.

4 Automatic Weight Adaptation

For the RWS rule, the fusion weights are chosen empirically, whereas for the automatic weight adaptation, a mapping needs to be developed between an audio reliability estimate and the modality weightings. The late fusion scores can be fused via addition or multiplication as shown in Eqn. 16 and 17. Both methods were investigated and it was found that the results achieved for both were similar (based on empirically chosen weights). However, additive fusion has been shown to be more robust to classifier errors [4], and should perform better when the fusion weights are automatically, rather than empirically determined. Hence the results for additive fusion only, are presented in this chapter. Prior to late fusion, all scores were normalized to fall into the range of [0, 1], using *min-max* normalisation.

$$P(S_i|x_A,x_T) = \alpha P(S_i|x_A) + \beta P(S_i|x_T), \quad \text{(a)}$$
$$P(S_i|x_A,x_T) = (P(S_i|x_A))^\alpha \times (P(S_i|x_T))^\beta. \quad \text{(b)}$$

$$\tag{16}$$

where

$$\alpha = \begin{cases} 0, & c \leq -1, \\ 1+c, & -1 < c < 0, \\ 1, & c \geq 0, \end{cases}$$

$$\beta = \begin{cases} 1, & c \leq 0, \\ 1-c, & 0 < c < 1, \\ 0, & c \geq 1, \end{cases}$$

$$\tag{17}$$

where x_A and x_V refer to the audio test utterance and visual test sequence/image respectively.

To carry out automatic fusion, that adapts to varying acoustic SNR conditions, a single parameter c, the *fusion parameter*, is used to define the weightings; the audio weight α and the visual weight β, i.e., both α and β dependent on c. Fig. 2 and Eqn. 17 show how the fusion weights, α and β, depend on the fusion parameter c. Higher values of c (>0) place more emphasis on the audio module whereas lower values (<0) place more emphasis on the visual module. For $c \geq 1$, $\alpha = 1$ and $\beta = 0$, hence the audio-visual fused decision is based entirely on the audio likelihood score, whereas, for $c \leq -1$, $\alpha = 0$ and $\beta = 1$, the decision is based entirely on the visual score. So in order to account for varying acoustic conditions, only c has to be adapted.

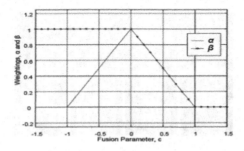

Fig. 2. The module weightings versus the fusion parameter "c"

The reliability measure was the audio likelihood score $\rho_n(\lambda_r)$. As the audio SNR decreases, this reliability measure decreases in absolute, and becomes closer to threshold for client likelihoods. Under clean test conditions, this reliability measure increases in absolute value because the client model yields a more distinct score. So, a mapping between ρ and c can automatically vary α and β and hence place more/less emphasis on the audio scores. To determine the mapping function $c(\rho)$, the values of c which provided for optimum fusion, copt, were found by exhaustive search for the N tests at each SNR levels. This was done by varying c from -1 to $+1$, in steps of 0.01, in order to find out which c value yielded the best performance. The corresponding average reliability measures were calculated, ρmean, across the N test utterances at each SNR level.

$$c(\rho) = c_{o_\epsilon} + \frac{h}{1 + \exp[d.(\rho + \rho_{o_\epsilon})]}.$$

(18)

The described method can be employed to combine any two modules. It can also be adapted to include a third module. We assume here that only the audio signal is degraded when testing, and that the video signal is of fixed quality. The third module we use here is an audio-lip correlation module, which involves a cross modal transformation of feature fused audio-lip features based on CCA, CFA or LSA cross modal analysis described in Sections 2 and 3.

5   Experimental Setup

The audio visual data from two different data corpora, VidTIMIT and UCBN was used for evaluating the performance of explicit cross modal features and hybrid fusion approach. The VidTIMIT multimodal person authentication database [4], consists of video and corresponding audio recordings of 43 people (19 female and 24 male). The mean duration of each sentence is around 4 seconds, or approximately 100 video frames.

Fig. 3. VidTIMIT and UCBN databases

The second type of data used is the UCBN database, a free to air broadcast news database. The broadcast news is a continuous source of video sequences, which can be easily obtained or recorded, and has optimal illumination, colour, and sound recording conditions. The database consists of 20 - 40 second video clips for anchor persons and newsreaders with frontal/near-frontal shots of 10 different faces (5 female and 5 male). Figure 3 shows some sample images from the VidTIMIT database (first two rows) and UCBN database (last two rows).

6   Results and Discussion

Different sets of experiments were conducted to evaluate the performance of the explicit cross modal features and hybrid fusion features in terms of DET curves and equal error rates (EER). As can be seen in % EER table in Table 1, the performance of ordinary features fusion of audio lip features $f_{mfcc-eigLip}$, can be improved by cross modal analysis. For the feature fusion of the correlated components $\tilde{f}_{mfcc-eigLip}$, the EER improves from 7.2 % to 4.7 % for CFA analysis for VidTIMIT male subset. Since each modality also carries mutually independent information, e.g. the texture of the lip region, possibly containing the information about the identity of a speaker, the overall performance can be enhanced with hybrid fusion, with an optimal

Table 1. EER (%) performance with late fusion of correlated ($\tilde{f}_{mfcc-eigLip}$) components with mutually independent (f_{eigLip} & f_{mfcc}) components: (+ represents RWS rule for late fusion, - represents feature level fusion)

Dataset	VidTIMIT male subset			UCBN female subset		
Cross Modal	CFA	CCA	LSA	CFA	CCA	LSA
Features	EER	EER	EER	EER	EER	EER
f_{mfcc}	4.88	4.88	4.88	5.7	5.7	5.7
f_{eigLip}	6.2	6.2	6.2	7.64	7.64	7.64
$f_{mfcc-eigLip}$	7.2	7.87	12.47	8.9	9.54	17.36
$\tilde{f}_{mfcc-eigLip}$	4.7	5.18	8.09	5.81	6.28	9.74
$f_{mfcc} + f_{mfcc-eigLip}$	1.03	1.03	1.03	1.12	1.12	1.12
$f_{mfcc} + \tilde{f}_{mfcc-eigLip}$	0.68	0.86	1.29	0.79	1.17	1.34
$f_{mfcc} + f_{eigLip} + f_{mfcc-eigLip}$	1.26	1.26	1.26	1.46	1.46	1.46
$f_{mfcc} + f_{eigLip} + \tilde{f}_{mfcc-eigLip}$	1.06	1.85	2.22	1.23	2.31	2.46

combination of the feature-level and the late fusion techniques combining lip, audio and correlated audio-lip feature vectors.

Also, for the VidTIMIT male subset, the hybrid fusion involving late fusion of audio features with feature-level fusion of correlated audio–lip features based on CFA analysis $f_{mfcc} + \tilde{f}_{mfcc-eigLip}$, yields a best EER of 0.68 %. Similar performance can be observed for different combinations of correlated component and independent component fusion for UCBN female dataset. For both data sets, around 22% improvement in EER is achieved with correlated component hybrid fusion ($f_{mfcc} + f_{eigLip} + \tilde{f}_{mfcc-eigLip}$) as compared to uncorrelated component hybrid fusion ($f_{mfcc} + f_{eigLip} + f_{mfcc-eigLip}$).It can also be noted that all the hybrid fusion modes (last four rows in Table 1) resulted in synergistic fusion, with the EER performance better than baseline audio only and visual only EERs of 4.88% and 6.2% for VidTIMIT male subset and 5.7 % and 7.64 % for the UCBN female subset.

7 Conclusions

In this paper, the performance evaluation of a novel hybrid fusion approach involving the correlated and the independent audio and lip modalities is proposed. The proposed cross modal factor analysis technique allows the extraction of the optimal correlated audio-lip features. An EER of less than 2 % was achieved for hybrid fusion with correlated features, and an EER improvement of around 22% is achieved with correlated component hybrid fusion when compared to uncorrelated component fusion. The EER performance for UCBN female subset for all fusion experiments was quite close to VidTIMIT male subset, even with poor quality of the visual data in UCBN dataset, with low resolution, small facial images, and presence of mostly irrelevant background information in the image sequences. Nevertheless, the performance for proposed technique with UCBN dataset from an opportunistic database depicts a more realistic speaker identity verification scenario.

References

[1] Brunelli, R., Falavigna, D.: Person Identification Using Multiple Cues. IEEE Transactions on Pattern Analysis and Machine Intelligence 17, 955–966 (1995)
[2] Kuratate, T., Munhall, K.G., Rubin, P.E., Vatikiotis-Bateson, E., Yehia, H.: Audio-visual synthesis of talking faces from speech production correlates. In: Proc. EuroSpeech 1999, ESCA (1999)
[3] Maeda, S.: A face model derived from a guided PCA of motion capture data and McGurk effects. In: Proceedings of the ATR symposium on Cross-modal Processing of Faces and Voices, pp. 63–64 (January 2005)
[4] Sanderson, C., Paliwal, K.K.: Fast features for face authentication under illumination direction changes. Pattern Recognition Letters 24, 2409–2419 (2003)
[5] Deerwester, S., Dumais, S.T., Furnas, G.W., Landauer, T.K., Harshman, R.: Indexing by latent semantic analysis. Journal of the American Society for Information Science 41(6), 391–407 (1990)
[6] Borga, M., Knutsson, H.: Finding Efficient Nonlinear Visual Operators using Canonical Correlation Analysis. In: Proc. of SSAB 2000, Halmstad, pp. 13–16

Voice Transformation by Mapping the Features at Syllable Level

K. Sreenivasa Rao[1], R.H. Laskar[2], and Shashidhar G. Koolagudi[1]

[1] School of Information Technology, IIT Kharagpur, Kharagpur-721302,
West Bengal, India
ksrao@iitkgp.ac.in, koolagudi@yahoo.com
[2] Department of Electrical Engineering, NIT Silchar, Silchar, Assam, India
laskar_r@nits.ac.in

Abstract. Voice transformation involves modifying the source speaker voice to target speaker voice. Voice characteristics of a speaker depends on the shape of the glottal pulse (source characteristics), shape of the vocal tract system (system characteristics) and the long term features (prosody or supra-segmental) of the speech signal produced by the speaker. In this paper we proposed the mapping functions to transform the vocal tract characteristics and intonation characteristics from source speaker to target speaker. Mapping functions are developed by the features extracted from syllable level. The shape of the vocal tract system is characterized by linear prediction coefficients, and the mapping function is realized by a five layer feedforward neural network. Mapping of the intonation characteristics (pitch contour) is provided by associating the code books derived from the pitch contours of the source and target speakers. The proposed mapping functions are used in voice transformation task. The target speaker's speech is synthesized and evaluated using listening tests. The results of the listening tests indicate that the proposed voice transformation provides better mapping of the voice characteristics compared to the earlier method proposed by the author. The original and the synthesized speech signals obtained using mapping functions are available for listening at *http://shilloi.iitg.ernet.in/~ksrao/result.html*

1 Introduction

Voice transformation is generally performed in two steps. In the first step, the training stage, a set of speech feature parameters of both the source and target speakers are extracted and appropriate mapping rules that transform the parameters of the source speaker on to those of the target speaker are generated. In the second step, the transformation stage, the features of the source signal are transformed using mapping rules developed in the training stage so that the synthesized speech possesses the personality of the target speaker [2007a].

To implement voice transformation, two problems need to be considered: what features are extracted from the underlying speech signals, and how to modify these features in such a way so that the transformed speech signals mimic target

A. Ghosh, R.K. De, and S.K. Pal (Eds.): PReMI 2007, LNCS 4815, pp. 479–486, 2007.

speakers voices. The first problem can be solved by identifying the speaker-specific features from the given speech signals. It is known that the shape of the glottal pulse, vocal tract transfer function and the prosodic features are uniquely characterize the speaker [2001a]. Feature parameters representing the vocal tract transfer function have been widely used in voice transformation. They include formant frequencies, Linear Prediction Coeffients (LPC), cepstrum and Line Spectrum Pair (LSP) coefficients [1995, 1999a, 1996]. In our previous work, we carried out the voice conversion by modifying the formant frequencies, pitch contour, duration patterns and energy profile by fixed scale factors [2006a]. As a result, the desired speaker characteristics are not much perceived in the synthe-sized speech. Therefore in this paper we are proposing the mapping functions, which accurately model the relationships between source and target speaker voices.

For mapping the speaker-specific features between source and target speakers, various models have been explored in the literature. These models are specific to the kind of features used for mapping. For instance, Gaussian Mixture Model (GMM) and Vector Quantization (VQ) are widely used for mapping the vocal tract characteristics [2001b, 1998a]. Scatter plots, GMMs and linear models are used for mapping the prosodic features [2003, 1998b]. In this work, we used neural network model for mapping the vocal tract characteristics, and code books for mapping the intonation characteristics. The main reason for exploring neural network models for mapping the vocal tract characteristics is that, it captures the nonlinear relations present in the patterns. The changes in the vocal tract shape corresponds to different speakers is highly nonlinear, therefore to model these nonlinearities, we have chosen neural network model in this work.

For mapping the intonation patterns, there exists different methods with vary-ing complexity. Mapping functions derived by linear, cubic and GMM models fail to predict the local variations at syllable and word levels [2003]. By using the code books consisting of intonation patterns at the utterance level can provide mapping to some extent at global level [2003]. But there is an error between the predicted pitch contour and the target contour with respect to local rise-falls in the intona-tion patterns. These rise-falls characterize the stress patterns present in the ut-terance, which are basically depend on the nature of the syllable (i.e., the basic constituents (consonants and vowels) of the syllable), and the linguistic context (nature of the preceding and the following syllables) associated to the syllable.

For Indian context, syllables are the most suitable basic units for the analy-sis and synthesis of speech. Syllables implicitly capture the duration patterns, shape of the vocal tract and coarticulation effects [2007b]. Eventhough the global characteristics of the intonation patterns depend on the nature of the utterance, but the local variations (rise-falls) depend on the nature of the individual syl-lables. Therefore for mapping the intonation patterns from source speaker to target speaker, the segments of the pitch contour derived from the syllables are used in this paper. Similarly for mapping the shape of the vocal tract system between the speakers the time-aligned linear prediction coefficients derived from the syllables are used.

The rest of the paper is organized as follows: In section 2, we discuss about the mapping of vocal tract system characteristics using feedforward neural network. Mapping of intonation characteristics between source and target speakers by using code books is discussed in section 3. Synthesis and evaluation of target speaker's voice is given in section 4. The final section contains the summary of the paper, conclusions derived from the work, and some future extensions to this work.

2 Mapping of the Vocal Tract System Characteristics

The basic shape of the vocal tract can be characterized by the gross envelope of the Linear Prediction (LP) spectrum. LP spectrum can be roughly represented by a set of resonant frequencies (formants) and their bandwidths. In our previous work, we used the formants and their bandwidths for characterizing the shape of the vocal tract, and further derived a gross relationship between formant frequencies and average pitch. This method provides the poor estimate of the shape of the vocal tract system for the desired speaker. Therefore in this paper we explored feedforward neural network (FFNN) model for capturing the relationship between the vocal tract system characteristics of the source and target speakers. Here LPCs are used to represent the shape of the vocal tract system. The LPCs are the design parameters of the LP filter, which models the vocal tract system accurately.

For deriving the mapping function using neural network, the network has to be trained with the LPCs derived from the spoken utterances of source and target speakers. For this study, we prepared the text transcription for 100 English sentences. These 100 sentences are recorded by 2 male and 2 female speakers. The duration of the sentence is varying between 3-5 secs. Each sentence has roughly about 15-20 syllables. The recorded speech files are segmented into syllables.

To capture the relationship between the vocal tract shapes of source and target speakers, we need to feed the time-aligned vocal tract features of source and target speakers at the input and output of the neural network respectively. In this work, we used Dynamic Time Warping (DTW) to derive the time-aligned vocal tract features. The procedure for deriving the time-aligned LPCs is given in Table 1. The neural network model used in this work is a five layer feedforward neural network, and it is shown in the Fig. 1.

Here the FFNN model is expected to capture the functional relationship between the input and output feature vectors of the given training data. The mapping function is between the 10-dimensional input vector and the 10-dimensional output. The 10-dimensional input and output vectors correspond to the time-aligned frame LPCs. Several network structures are explored in this study. The (empirically arrived) final structure of the network is $10L$ $20N$ $5N$ $20N$ $10L$, where L denotes a linear unit, and N denotes a nonlinear unit. The integer value indicates the number of units used in that layer. The nonlinear units use $tanh(s)$ as the activation function, where s is the activation value of that unit. The back-propagation learning algorithm is used for adjusting the weights of the network to minimize the mean squared error for each pair of time aligned LPCs [1999b].

Table 1. Steps for deriving the time-aligned LPCs

1. Derive the syllable pair from the utterance pair spoken by the pair of speakers.
2. Preemphasize the speech segments corresponding to the syllable pair.
3. Compute LPCs with 10th order LP analysis, with a frame size of 20 ms and a frame shift of 5 ms.
4. Apply DTW on the syllable pair (with a frame size of 20 ms and a frame shift of 5 ms) and derive the matching frames.
5. Select the LPCs corresponding to the matching frames of the syllable pair and they turned to be pairs of time-aligned LPCs.
6. Same process (steps 1-5) is repeated for other syllable pairs.

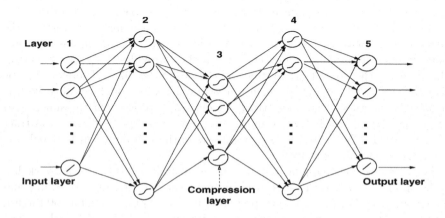

Fig. 1. A five layer FFNN model

A separate model is developed for each pair of the speakers. There are about 76000 matched frame LPCs present in the database, 80% of them are used for training the models. After the training phase, the weights in the network represents the mapping function between input and output. The performance of the model can be evaluated by both subjective and objective measures. Subjective evaluation consists of synthesizing the speech for the desired speaker using the LPCs derived from the neural network model, and conducting the listening tests to asses the desired speaker characteristics present in the synthesized speech. Objective measures consists of spectral distance between the predicted LPCs and actual LPCs. In this paper subjective evaluation is performed, and it is illustrated in Section 4.

3   Mapping of the Intonation Characteristics

For mapping the intonation characteristics between the source and target speakers, code books are prepared using the pitch contours derived from the syllables. For preparing the code books, all the feature vectors should have the same length.

Since the syllables in the database have varying durations, we used resampling technique to obtain the fixed length pitch contour for each syllable. In this study, we explored different lengths (6, 8, 10, 12 and 14), finally the optimum value is found to be 10. The database consists of 1783 syllables, and 90% of them are used for preparing the code books. The code books are prepared separately for source and target speakers using vector quantization [1980]. The entries of the code book represents the mean vectors (centroids) of the clusters formed in the 10-dimensional space. Size of the code book indicates the number of clusters considered for the analysis. In this study we explored different code book sizes (8, 16, 32 and 64), among them 32-size code book shows the better representation. Validation of the code books is performed using the test data.

3.1 Mapping of Code Book Entries

After preparing the code books for source and target speakers, the next step is to derive the mapping function between the entries of the code books. This will be carried out as follows:

1. The resized (10-dimensional) pitch contours of the source speaker which are used for preparing the code book are partitioned according to the entries of the source speaker code book (mean vectors or the cluster centroids). That is the syllable based pitch contours of the source speaker are divided into 32 groups corresponds to the entries of the source speaker code book.

2. For each group of pitch contours of the source speaker, determine the corresponding pitch contours of the target speaker and label them with the same identity as that of the source speaker.

3. The pitch contours of the target speaker belongs to group 1 are projected onto the entries of the target speaker code book, which corresponds to the mean vectors or centroids of the target vector space. The histogram distribution of the projected vectors with respect to the code book entries of the target speaker is determined, and is used to derive the sequence of weights for the sequence of mean vectors present in the target speaker code book.

Now it provides the mapping for the first entry of the source code book as the summation of the weighted code book entries of the target speaker, where the corresponding weights are derived using step 3. Similarly for other entries of the source code book, the corresponding weight vectors are determined. Each weight vector indicates the sequence of weights corresponds to the sequence of code book entries of the target speaker. The overall procedure for the transformation of source speaker pitch contour to target speaker pitch contour is given in Table 2.

4 Synthesis and Analysis of Target Speaker Voice

In the previous sections we described the methods for mapping the vocal tract characteristics and intonation patterns between source and target speakers. In this section, we discuss about the synthesis of target speaker's speech from source speaker's speech by using mapping functions, and then evaluating the presence

Table 2. Steps for transforming the pitch contour

1. Derive the pitch contour from the utterance spoken by the source speaker.
2. Segment the pitch contour with respect to the syllables present in the utterance.
3. For each segment of the pitch contour, determine the closest code book
   entry of the source speaker.
4. By using the weight vector (sequence of weights corresponds to the sequence
   of code book entries of the target speaker) corresponding to the code book
   entry of the source speaker, generate the pitch contour by the summation of
   the weighted code book entries of the target speaker.
5. Pitch contour for the target speaker at the utterance level is derived
   by concatenating the syllable level pitch contours derived from step 4.

of desired speaker characteristics in the synthesized speech. The transformation of source speaker speech to target speaker speech is performed as follows:

1. The LPCs representing the vocal tract shape and the pitch contour representing the intonation pattern of the source speaker are derived.

2. The LPCs corresponding to the target speaker are derived from the output of the 5-layer FFNN model, by giving the LPCs of the source speaker speech utterance as input to the FFNN model.

3. The pitch contour of the target speaker is obtained by concatenating the syllable level pitch contours, which are in turn derived from the syllable level pitch contours of the source speaker by using the code books.

Once the pitch contour for the target speaker is available, the pitch contour of the source speaker's speech utterance is replaced by the target speaker's pitch contour. In this work the desired pitch modification is carried out in linear prediction residual domain. The basic reason for choosing the residual domain for modification is that the successive samples in the LP residual are less correlated compared to the samples in the speech signal [2006b]. Therefore the residual manipulation is likely to introduce less distortion in the speech signal synthesized by using the modified LP residual. The details of the pitch modification method are discussed in our previous work [2006a, 2006b]. Finally, target speaker speech is synthesized by exciting the time varying filter representing the target speaker LPCs with the modified LP residual according to the target speaker pitch contour.

The performance of the mapping functions can be evaluated by using subjective and objective measures. In this work the basic goal is the voice transformation, therefore the mapping functions are evaluated by using perceptual tests (i.e., by conducting listening tests). In this work four separate mapping functions are developed for transforming the speaker voices: (1) male to female (M1-F1), (2) female to male (F2-M2), (3) male to male (M1-M2) and (4) female to female (F1-F2). Here M1, M2, F1 and F2 represents male and female speaker voices present in the database. For each case five utterances are synthesized using their associated mapping functions. Listening tests are conducted to assess the desired (target) speaker characteristics present in the synthesized speech. The recorded speech utterances of the target speaker correspond to the synthesized speech utterances are made available to the listeners to judge the relative performance. Twenty students are participated in conducting these tests. Each of the synthesized speech utterance is played to the listener, after playing its original recorded utterance. The subjects were asked to give their opinion score on a 5-point scale. The rating 5 indicates the excellent match between the original target speaker speech and the synthesized speech (i.e., synthesized speech is close to the original speech of the target speaker). The rating 1 indicates very poor match between the original and synthesized utterances, and the other ratings indicate different levels of deviation between 1 and 5. Each listener has to give the opinion scores for each of the five utterances in all the four cases (altogether 20 scores) mentioned above. The mean opinion scores (MOS) for male to female (M1-F1), female to male (F2-M2), male to male (M1-M2) and female to female (F1-F2) transformations are found to be 3.61, 3.94, 3.12 and 2.93, respectively. The obtained MOS indicate that the transformation is effective, if the source and target speakers are from different genders. The basic reason for this variation in MOS with respect to gender is that, the variation in the shapes of the vocal tract and intonation patterns may be large in the case of source and target speakers belongs to different genders. Since the listener is exposed to source and target speakers voices (original) as well as transformed voice with respect to target speaker, he or she may observe wide transformation in the case of male to female or female to male voice conversions compared to the other cases (male to male or female to female). Hence their feeling is reflected in the judgement. While comparing the performance of the voice transformation by the proposed method with the previous work (using formant modification and modification of prosody by a fixed factor), the MOS shows the superiority of the present method. The synthesized speech utterances of the target speaker derived from the proposed method and from the previous work are available for listening at *http://shilloi.iitg.ernet.in/~ksrao/result.html*

5   Summary and Conclusions

In this paper, a five layer FFNN model was proposed for mapping the vocal tract characteristics between source and target speakers. In the present work LPCs were used for representing the shape of the vocal tract. The final structure of the FFNN model was arrived at empirically. A mapping function between source and target pitch contour code books was derived in the form of weight vectors for transforming the intonation patterns between source and target speakers. The target speaker's speech was synthesized by using the parameters derived from the mapping functions correspond to vocal tract system characteristics and intonation characteristics. Subjective tests were conducted to evaluate the target speaker characteristics present in the synthesized speech. From the perceptual tests, it was found that the voice transformation is more effective, if the source and target speakers belongs to different genders. Subjective evaluation also indicated that the developed voice conversion system has improved, compared to its earlier version proposed by the author.

The performance of each mapping function has to be evaluated separately for analyzing the results of the subjective tests. Investigating the performance of mapping functions using objective measures may give some directions for further improvement in the present system. In this paper, we have not included the mapping functions correspond to source characteristics (shape of the glottal pulse), duration patterns and energy patterns (intensity profile) between souce and target speakers. The overall performance of the voice conversion system may be improved by including the above mapping functions, which are not used in the present study.

References

1. Lee, K.-S.: Statistical approach for voice personality transformation. IEEE Trans. Audio, Speech, and Language processing 15, 641–651 (2007)
2. Yegnanarayana, B., Reddy, K.S., Kishore, S.P.: Source and system features for speaker recognition using AANN models. In: Proc. ICASSP, Salt lake city, Utah, USA, pp. 409–412 (May 2001)
3. Narendranadh, M., Murthy, H.A., Rajendran, S., Yegnanarayana, B.: Transformation of formants for voice conversion using artificial neural networks. Speech Communication 16, 206–216 (1995)
4. Arslan, L.M.: Speaker transformation algorithm using segmental code books (STASC). Speech Communication 28, 211–226 (1999)
5. Lee, K.S., Youn, D.H., Cha, I.W.: A new voice personality transformation based on both linear and non-linear prediction analysis. In: Proc. ICSLP, pp. 1401–1404 (1996)
6. Rao, K.S., Yegnanarayana, B.: Voice conversion by prosody and vocal tract modification. In: Proc. Int. Conf. Information Technology, pp. 111–116 (December 2006)
7. Toda, T., Saruwatari, H., Shikano, K.: Voice conversion algorithm based on Gaussian mixture model with dynamic frequency warping of STRAIGHT spectrum. In: Proc. ICASSP, vol. 2, pp. 841–844 (May 2001)
8. Abe, M., Nakanura, S., Shikano, K., Kuwabara, H.: Voice conversion through vector quantization. In: Proc. ICASSP, pp. 655–658 (May 1998)
9. Inanoglu, Z.: Transforming pitch in a voice conversion framework, M.Phil thesis, St.Edmund's College University of Cambridge (July 2003)
10. Stylianou, Y., Cappe, Y., Moulines, E.: Continuous probabilistic transform for voice conversion. IEEE Trans. Speech and Audio Processing 6, 131–142 (1998)
11. Rao, K.S., Yegnanarayana, B.: Modeling durations of syllables using neural networks, Computer Speech and Language, pp. 282–295 (April 2007)
12. Haykin, S.: Neural Networks: A Comprehensive Foundation. Prentice-Hall Inc., New Jersey (1999)
13. Linde, Y., Buzo, A., Gray, R.M.: An algorithm for vector quantizer design. IEEE Trans. Commn. 28(1), 84–95 (1980)
14. Rao, K.S., Yegnanarayana, B.: Prosody modification using instants of significant excitation. IEEE Trans. Audio, Speech and Language Processing 14, 972–980 (2006)

Language Independent Skew Estimation Technique Based on Gaussian Mixture Models: A Case Study on South Indian Scripts

V.N. Manjunath Aradhya[1], Ashok Rao[2], and G. Hemantha Kumar[1]

[1] Dept of Studies in Computer Science,University of Mysore,
Mysore - 570 006, India
mukesh_mysore@rediffmail.com
[2] Dept of Electronics and Communication, S.J. College of Engineering
Mysore - India
ashokrao.mys@gmail.com

Abstract. During document scanning, skew is inevitably introduced into the incoming document image. Presence of additional modified characters, which get plugged in as extensions and remain as disjointed protrusions of a main character is really challenging in estimating inclination in skewed documents made up of texts in south Indian languages (Kannada, Telugu, Tamil and Malayalam). In this paper, we present a novel script independent (for south Indian) skew estimation technique based on Gaussian Mixture Models (GMM). The Expectation-Maximization (EM) algorithm is used to learn the mixture of Gaussians. Subsequently the cluster means are subjected to moments to estimate the skew angle. Experiments on printed and handwritten documents corrupted by noise is done. Our method shows significantly improved performance as compared to other existing methods.

1   Introduction

The volume of paper based documents continue to grow at a rapid rate in spite of the use of electronic version. As a result, both the transformation of a paper document to its electronic version, and its subsequent image processing and understanding have become an important application domain in computer vision and pattern recognition research. Document analysis and character recognition are usually performed through several phases: scanning, image enhancement, skew estimation and correction, segmentation, and character classification. The skew estimation of document images is particularly crucial among the document processing operations as it affects the subsequent understanding of the document. Several attempts have been made for skew detection and the methods can be mainly categorized into five groups: Hough transform, Cross Correlation, Projection profile, Fourier transformation and K nearest neighbor (K-NN) clustering.

Hough transform based technique for skew detection is presented in [11]. To reduce the computational burden, the bottom pixels of the candidate objects

A. Ghosh, R.K. De, and S.K. Pal (Eds.): PReMI 2007, LNCS 4815, pp. 487–494, 2007.
© Springer-Verlag Berlin Heidelberg 2007

within a selected region are subjected to Hough transformation [10]. The hierarchical Hough transformation technique was also adapted for skew estimation[16]. The main idea of the these methods is to reduce the amount of input data which in turn reduces their computational complexities. An improved method to overcome the drawback of the method proposed in [10] is presented in [13]. The cross-correlation method proposed in [4] is based on the correlation between two vertical lines in a document image. The horizontal projection profile (HPP), proposed in [9], is a histogram of the number of black pixels along the horizontal lines of a scanned document. The method works based on text line profile peaks and troughs to estimate the skew angle. However the method works only for ideal cases and it is known to be time consuming. To alleviate this, modifications are done to this iterative approach for quick convergence [8]. An approach for skew estimation based on HPP is described in [12]. Here HPP's are calculated for each strip and from the correlation of the profiles of the neighboring strips the skew angle is determined. Although the proposed method is computationally inexpensive, it cannot not work well if the document is skewed beyond $\pm10^0$. The Fourier transform based algorithm for skew estimation is presented in [14]. According to the method, skew angle of a document image corresponds to the direction where the density of Fourier space becomes the largest. However its computational complexity is very high. Nearest neighbor chain based approach for skew estimation in document images is proposed in [6]. Cao et al [15] proposed skew detection and correction in document images based on straight-line fitting. A skew detection and correction technique using Radon transform projection profile technique is described in [5]. An algorithmic technique that performs skew angle correction to handwritten Bengali text is reported in [1].

Aforementioned methods are script dependent and also they perform poorly if documents contain noise, degraded texts and varying font size of texts. More importantly, these methods may not obtain accurate results for south Indian scripts. This is due to additional modifying characters that remain as disconnected protrusions of a main character, which is one of the dominant feature of south Indian language particularly Kannada and Telugu. Hence, in this paper we present an improved technique of skew estimation for documents containing south Indian scripts. In addition, the proposed technique handles handwritten documents.

The organization of the paper is as follows: Proposed skew estimation technique is presented section 2. In section 3, Experiment and Comparative study are reported. Discussion and conclusion are drawn in section 4.

2   Proposed Methodology

This section presents the proposed methodology that is based on GMM and moments. The proposed method first extracts individual text lines present in the document image using the method described in [7]. This technique is based on boundary growing algorithm, which helps us in extracting text line present in document page. The resultant text line obtained from this algorithm is then passed on to GMM process to extract mean vector points. The resultant mean vector points are then used to estimate skew angle using moments. To implement this, we first explain the concept of GMM described in [2].

2.1   Gaussian Mixture Models (GMM)

GMM is a simple linear superposition of Gaussian components, aimed at providing a rich class of density than a single Gaussian. The formulation of Gaussian mixtures will provide us with a deeper insight into this important distribution and will serve to motivate the expectation-maximization algorithm. A distribution can be written as a linear superposition[1] of Gaussian in the form:

$$P(x) = \sum_{k=1}^{K} \pi_k \eta(x/\mu_k, \Sigma_k) \tag{1}$$

which is called a *mixture-of-Gaussians*. Where $\eta(x/\mu_k, \Sigma_k)$ is the multivariate Gaussian distribution of the form:

$$\eta(x/\mu_k, \Sigma_k) = \frac{1}{(2\pi)^{D/2} |\Sigma|^{1/2}} e^{-\frac{1}{2}(x-\mu_k)^T \Sigma_k^{-1}(x-\mu_k)} \tag{2}$$

Each Gaussian density $\eta(x/\mu_k, \Sigma_k)$ is called a component of the mixture and has its own mean μ_k and covariance Σ_k. The parameter π_k in Eq.(1) is called mixing coefficient. If we integrate both sides of Eq.(1) w.r.t x, both p(x) and the indidividual Gaussian components are normalized, we obtain $\sum_{k=1}^{K} \pi_k = 1$. Also, the requirement that $p(x) \geq 0$, together with $\eta(x/\mu_k, \Sigma_k) \geq 0$, implies $\pi_k \geq 0 \ \forall k$. Combining this with Eq.(1) we obtain $0 \leq \pi_k \leq 1$.

From the sum and product rules, the marginal density is given by

$$p(x) = \sum_{k=1}^{K} p(K)p(x/K) \tag{3}$$

which is equivalent to Eq.(1) in which we can view $\pi_k = p(k)$ as the prior probability of picking the kth component, and the density $\eta(x/\mu_k, \sum_k) = p(x/K)$ as the probability of x conditioned on k.

From Baye's theorem the posterior probabilities $p(K/x)$, which is also known as responsibilities , are given by:

$$\gamma(z_k) \equiv \quad p(K/x) \tag{4}$$

$$= \frac{p(K)p(x/K)}{\sum_l p(l)p(x/l)} \tag{5}$$

$$= \frac{\pi_k \eta(x/\mu_k, \Sigma_k)}{\sum_l \eta(x/\mu_l, \Sigma_l)} \tag{6}$$

The form of the Gaussian mixture distribution is governed by the parameters π, μ and Σ, where we have used the notation $\pi \equiv \pi_1, \pi_2, \ldots, \pi_K$, $\mu \equiv \mu_1, \mu_2, \ldots, \mu_K$ and $\Sigma \equiv \Sigma_1, \Sigma_2, \ldots, \Sigma_K$. We now adapt an iterative algorithm, known as Expectation Maximization (EM) algorithm, to extimate the values of μ, Σ and π.

---

[1] Please note that mixture of Gaussian need not be a Gaussian.

We first choose some initial values for these parameters by running K-means clustering algorithm. Then we alternate between two steps known as Expectation(E) Step and the Maximization(M) step to update the values of these parameters until convergence criteria is reached. The EM algorithm for GMM[2] can be summarized as follows:

1. Initialize the parameters μ_k, Σ_k amd π_k by running K-means clustering algorithm and evaluate the log of the likelihood function using Eq.(7)
2. **E Step** Evaluate the responsibilities using Eq.(6) with current parameter values.
3. **M Step** Re-estimate the parameters using the current responsibilities:

$$\mu_k^{new} = \frac{1}{N_k} \sum_{n=1}^{N} \gamma(z_{nk}) \, u_n$$

$$\Sigma_k^{new} = \frac{1}{N_k} \sum_{n=1}^{N} \gamma(z_{nk}) \, (x_n - \mu_k^{new})(x_n - \mu_k^{new})^T$$

$$\pi_k^{new} = \frac{N_k}{N}$$

4. Evaluate the log likelihood:

$$ln \; p(X/\mu, \Sigma, \pi) = \sum_{n=1}^{N} \sum_{k=1}^{K} \pi_k \eta(x_n/\mu_k, \Sigma_k) \tag{7}$$

and check for convergence of log likelihood. If the convergence criterion is not satisfied iterate from step 2.

Using the obtained means of k clusters μ_k, $\forall k = 1, \ldots, K$, first and second order moments are calculated using the Eq. 10. This is used for finding the inclination of the given skewed text line. Figure 1 depicts the mean points obtained for the input skewed document using mixture-of-Gaussians.

2.2   Moments for Skew Estimation

In this section, we present moments based method for the estimation of skew angle. The moments are computed using Eq.(8) and Eq.(9), x and y is the cluster mean points obtained from the GMM, p and q define the order of moments. Angle of each text line present in the document is estimated using Eq.(10). For detailed mathematical derivations, see Ref.[3].

$$m_{pq} = \sum_1^n \sum_1^n x^p \, x^q \tag{8}$$

$$\mu_{pq} = \sum_1^n \sum_1^n (x - \bar{x})^p (y - \bar{y})^q \tag{9}$$

$$\theta = \tfrac{1}{2} tan^{-1} \left[\frac{2\mu_{11}}{(\mu_{20} - \mu_{02})} \right] \tag{10}$$

where θ is the estimated skew angle of the segmented text line.

---

[2] Given a GMM, the objective is to maximize the likelihood function with respect to the parameters comprising μ, Σ and π_k.

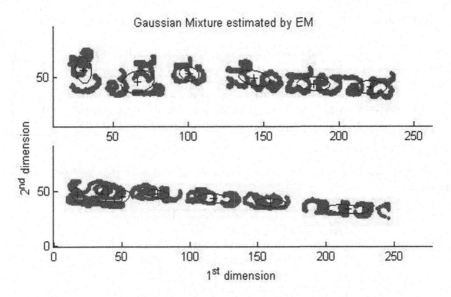

Fig. 1. Illustration of Gaussian Mixture Models for a skewed text line

3 Experimental Results and Comparative Study

This section presents the results of the experiments conducted to study the performance of the proposed method. The method has been implemented in MATLAB 7.0 on a Pentium IV 3.0 GHz with 1GB RAM. For experiment purpose, 20 documents are considered from different sources such as Kannada, Tamil, Telugu, Malayalam, and English. Each document is rotated with four skew angles (3,5,10 and 15). Further to show the superior performance of the proposed method, handwritten English documents and documents with noise are also considered. We have taken two decision parameters such as Mean Skew Angle (M) and Standard Deviation (SD) which are reported in Table 1. In addition, finding optimal number of mixtures to yield best recognition accuracy is highly subjective in nature. Hence, it is empirically fixed to nine mixtures for optimal performance. From Table 1, it is evident that the skew angle obtained by proposed method for English document is better when compared to other south Indian languages.

Table 1. Mean and Standard Deviation obtained by the proposed method

True Angle	Kannada		Telugu		Tamil		Malayalam		English	
	M	SD	M	SD	M	SD	M	SD	M	SD
3	3.12	0.432	2.5	0.3	2.86	0.5	2.86	0.58	3.16	0.18
5	5.05	0.436	5.06	0.46	5.20	0.64	5.80	0.62	5.10	0.28
10	10.5	0.364	9.75	0.61	10.4	0.52	10.5	0.48	10.23	0.15
15	15.02	0.36	14.90	0.37	15.60	0.32	14.62	0.28	15.30	0.17

Table 2. A Comparative study with existing methods for English documents

	Method-1[13]		Method-2[4]		Method-3[6]		Method-4[15]		Proposed Method	
True Angle	M	SD	M	SD	M	SD	M	SD	M	SD
3	3.12	0.34	3.43	0.94	3.86	0.81	3.21	0.51	3.16	0.18
5	5.68	0.456	5.09	1.04	5.71	0.95	5.52	0.68	5.10	0.28
10	10.17	0.42	10.19	0.82	10.73	0.79	9.86	0.54	10.23	0.15
15	15.72	0.51	15.35	0.93	15.53	0.51	15.02	0.32	15.30	0.17

Table 3. Mean and Standard Deviation for clean and noisy handwritten English document

	Handwritten		Noisy Handwritten	
True Angle	M	SD	M	SD
3	3.10	0.20	2.80	0.35
5	4.94	0.32	4.57	0.47
10	10.10	0.10	8.94	0.89
15	15.46	0.37	13.97	0.85

Fig. 2. Clean(top) and noisy(bottom) English handwritten documents

Moreover, amongst four south Indian scripts, our method obtained better results in terms of M and SD for Malayalam documents. A comparative study with other existing methods is carried out to show the performance of our method in terms of accuracy and efficiency. The mean and standard deviation obtained using the proposed method and the other methods are reported in Table 2 for printed English documents. From Table 2 it is clear that the proposed method performs better compared to other existing methods with respect to mean and standard deviation. We extended our experiment for handwritten English documents. For this experiment we considered 10 documents and each are rotated with four mentioned skew angles. Mean and standard deviation obtained for the handwritten English documents are reported in Table 3. It is clear that the proposed method

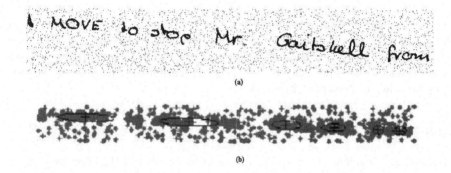

(a)

(b)

Fig. 3. (a): A Noisy handwritten document (b): Its illustration with mixture of six Gaussians

performs better even for handwritten documents. To check the robustness of the proposed algorithm, we tested our method on noisy documents also. For this, we considered five noisy handwritten documents[3] and each were rotated with four true angles. Sample handwritten documents contaminated by noise is as shown in Figure 2. Results obtained from the method are reported in Table 3. Figure 3 shows the illustration of Gaussian mixture for noisy handwritten document.

4   Discussion and Conclusion

Mixture Models are a type of density model which comprise a number of component functions, usually Gaussian. These component functions are combined to provide a multimodal density. GMM is widely used in data mining, pattern recognition, machine learning, and statistical analysis. In many applications, their parameters are determined by maximum likelihood, typically through iterative learning of the EM algorithm. In this paper, an efficient and robust methodology for skew estimation based on GMM is presented. The proposed method is

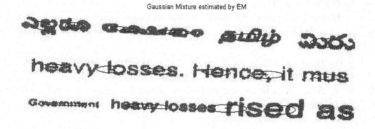

Gaussian Mixture estimated by EM

Fig. 4. Mixture of Gaussians for: (a) a multilingual text line (b) degraded text line and (c) text with varying size of words

---

[3] Here we used Salt-and-Pepper of noise density 0.02

independent of scripts, style and font size. The results for this is illustrated in Fig.4. Extensive experiments have been carried out to study the performance of the proposed method by considering the documents such as printed south Indian scripts, handwritten English and noisy handwritten documents. These experiments revealed the superiority of the proposed method. We plan to extend this work for other Indic scripts in future.

References

1. Basu, S., Chaudhuri, C., Kundu, M., Narsipuri, M., Basu, D.K.: Text line extraction from multi-skewed handwritten documents. Pattern Recognition 40(6), 1825–1839 (2007)
2. Bishop, C.M.: Pattern Recognition and Machine Learning. Springer, Heidelberg (2006)
3. Gonzalez, R.C., Woods, R.E.: Digital Image Processing, 2nd edn. Pearson Education (2002)
4. Yan, H.: Skew correction of document images using interline cross-correlation. Computer Vision, Graphics, and Image Processing 55, 538–543 (1993)
5. Kapoor, R., Deepak, Kamal.: A new algorithm for skew detection and correction. Pattern Recognition Letters 25, 1215 (2004)
6. Lu, Y., Tan, C.L.: A nearest neighbor chain based approach to skew estimation in document images. Pattern Recognition Letters 24, 2315–2323 (2003)
7. Manjunath Aradhya, V.N., Hemantha Kumar, G., Shivakumara, P.: An accurate and efficient skew estimation technique for south indian documents. International Journal of Robotics and Automation (in press)
8. Baird, H.S.: The skew angle of printed documents. In: Proceedings of Conference Society of Photographic Scientists and Engineers, pp. 14–21 (1987)
9. Hou, H.S.: Digital Document Processing. Wisely, New York (1983)
10. Le, D.S., Thoma, G.R., Wechsler, H.: Automatic page orientation and skew angle detection for binary document images. Pattern Recognition 27, 1325–1344 (1994)
11. Srihari, S.N., Govindaraju, V.: Analysis of textual images using the hough transform. Machine Vision and Applications 2, 141–153 (1989)
12. Akiyama, T., Hagita, N.: Automated entry system for printed documents. Pattern Recognition 23(11), 1141–1158 (1990)
13. Pal, U., Chaudhuri, B.B.: An improved document skew angle estimation technique. Pattern Recognition Letters 17, 899–904 (1996)
14. Postl, W.: Detection of linear oblique structures and skew scan in digitized documents. In: Proceedings 8th International Conference on Pattern Recognition, pp. 687–689 (1986)
15. Yang, C., Wang, S., Heng, L.: Skew detection and correction in document images based on straight-line fitting. Pattern Recognition Letters 24, 1871 (2003)
16. Yu, B., Jain, A.K.: A robust and fast skew detection algorithm for generic documents. IEEE Transactions on Pattern Analalysis and Machine Intelligence 29(10), 1599–1629 (1996)

Entropy Based Skew Correction of Document Images

K.R. Arvind, Jayant Kumar, and A.G. Ramakrishnan

MILE Lab, Electrical Engineering,
Indian Institute of Science, Bangalore, India 560012

Abstract. The document images that are fed into an Optical Character Recognition system, might be skewed. This could be due to improper feeding of the document into the scanner or may be due to a faulty scanner. In this paper, we propose a skew detection and correction method for document images. We make use of the inherent randomness in the Horizontal Projection profiles of a text block image, as the skew of the image varies. The proposed algorithm has proved to be very robust and time efficient. The entire process takes less than a second on a 2.4 GHz Pentium IV PC.

1 Introduction

An optical character recognition system takes in the document image generated by a scanner, segments it into words and then to characters, then finally recognizes using some feature based classifier. Although, this procedure seems simple enough, improper feeding of the document into the scanner could produce a skew. This skew could hamper the word/character level segmentation of the document and hence drastically affect the performance the recognition system. Hence we see that, before the processing of the document image, a robust skew detection and correction mechanism is extremely important.

Xiaoyan Zhu and Xiaoxin Yin[1] have proposed a method to correct skew of document images containing both text and non-text. They first divide the image into blocks. Then they classify the blocks into Text/Non-Text using Fourier transform the projection profiles of the blocks as features and Support Vector Machine as the classifier. They determine the skew only for the text blocks by taking the standard deviation of the projection profiles for various angles. They conclude that the angle at which the standard deviation is maximum, is the skew angle of the document image.

Le et al.[2] select a square region dominated by text from the document image and calculate the skew angle by this area. Avanindra and Subhasis Chaudhuri[3] divide the document into blocks and use the median of the cross-correlations of all blocks to determine the skew angle.

Bruno and Rafael[4] have used nearest neighbor clustering approach for skew detection in document images. It works with complex documents and has a range angle detection of 0 to 360 with precision of 0.1. The algorithm starts by boxing

A. Ghosh, R.K. De, and S.K. Pal (Eds.): PReMI 2007, LNCS 4815, pp. 495–502, 2007.

each block of black pixels by using component labeling. Applying the least square method to the middle top point and middle bottom point separately of all blocks of the text-line, it forms two lines located at strategic points. The bottom line is located at the baseline of the text-line and crosses the descending characters. The top line is located above the text-line and crosses the ascending characters. This happens because there are more non-salient characters (e.g. a, c, e, o and u) in the Latin alphabet than ascending and descending ones. If the number of descending strokes is greater than the ascending one then the document is upside-down.

Marisa E. Morita et al.[5] have proposed a morphology-based method to detect and correct handwritten word skew in the treatment of dates written on bank checks. Their aim is to limit the number of parameters and heuristic features necessary for a good skew correction. Their approach is based on the morphological pseudo-convex hull.

In this paper, we propose an entropy based skew correction method. In section 2, we describe our method in detail. In section 3, we discuss the experimental results. Our method can handle document images having multiple skewed blocks. It is time efficient, consuming less than half a second for both skew detection and correction on a block image.

2   System Description

The system is divided into three steps. In the initial step, the document image is cleaned of all noise elements such as spurious dots and lines. Next, it is segmented into its constituent blocks. This is carried out similar to the block segmentation method given by Arvind *et. al.*[6]. Then, the blocks are skew corrected based on their horizontal projection profiles and entropy. The procedure is summarized below:

- Clean the document
- Extract the blocks
- Skew correct the blocks

2.1   Noise Removal

We apply Connected Component Analysis (CCA) and obtain the number of ON pixels and aspect ratios for every component. Then, we find the minimum and maximum values of them. Let them be $minp$, $maxp$ and $mina$, $maxa$ respectively.

$$TP = \frac{np - minp}{maxp - minp} \tag{1}$$

$$TA = \frac{na - mina}{maxa - mina} \tag{2}$$

-where np is number of on pixels in the component and na is aspect ratio of the component

If TP or TA is less than 0.002, which is computed empirically, then we re-move it. We need to remove extraneous dots and especially vertical lines as they could hamper the horizontal projection profiles(HPP) and thereby reduce the effectiveness of the skew correction process.

2.2   Block Segmentation

Arvind *et. al.*[6] have segmented blocks by run-length smoothening the image with the parameter selected such that the intra and inter character gaps, up to a paragraph level, are filled. Then, they apply morphological erosion operator to remove thin joints between blocks. Finally CCA is used to segregate the blocks.

2.3   Skew Detection and Correction

Horizontal Projection Profile. HPP of an image is a vector where each vector element contains the sum of the pixel values in the corresponding row.

$$HPP(i) = \sum_{j=1}^{No.of columns} I(i,j) \tag{3}$$

where $I(i,j)$ is the image.

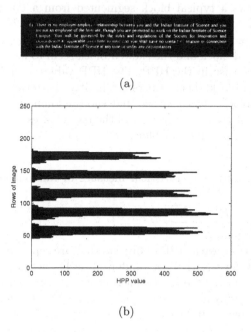

(a)

(b)

Fig. 1. (a) Example of a typical correctly oriented block. (b) HPP of the correctly oriented block.

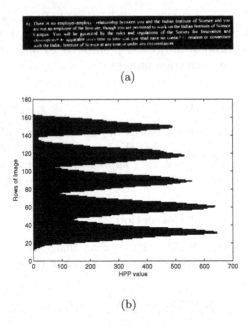

(a)

(b)

Fig. 2. (a) Example of a typically skewed block (b) HPP of the skewed block

Figure 1(a) shows a typical block segmented from a document image and Figure 1(b) its horizontal projection profile(HPP). Figures 2(a) and 2(b) depict the skewed block and its HPP respectively. We observe that the HPPs of both the properly oriented block and the skewed block seem to have a similar pattern of increase and decrease in the HPP. The HPP values actually repeat in case of a skew corrected block. There exists a repetitive pattern as compared to the HPP of a skewed block. Hence we see that the randomness associated with the projection profiles increases as the skew of the block varies. We use this property to detect the skew of a particular block.

Entropy. The entropy is a statistical measure of randomness. It is given by equation 4.

$$E(i) = \sum_i -HPP(i) * log(HPP(i)) \qquad (4)$$

The entropy associated with HPP of objects that are repetitive in nature would be lesser compared to the HPP of random objects such as graphic images or skewed text blocks. i.e. as the randomness increases, the entropy also increases. Assuming that the maximum skew would not be greater than 10 degrees, we rotate the image upto ±10 degrees and obtain the HPPs along with their corresponding entropy values. The resolution of rotation is 1 degree. The angle for which the entropy is minimum, is the rough skew angle of the segmented block. We have taken block image displayed in Figure 4(a) and run the proposed algorithm to correct its skew. Table 1 displays the rotated angles and its corresponding entropy values. It can be observed from the table, that the angle for which entropy value is minimum i.e.1 degree is the coarse skew angle.

Table 1. The angles through which the blocks are rotated and their corresponding entropy values

Angle	Entropy	Angle	Entropy
-10	5.64	10	5.44
-9	5.59	9	5.37
-8	5.53	8	5.30
-7	5.46	7	5.23
-6	5.40	6	5.17
-5	5.33	5	5.11
-4	5.25	4	5.05
-3	5.20	3	4.98
-2	5.11	2	4.87
-1	5.03	1	4.35
0	4.78	0	4.78

After obtaining the coarse skew angle, we do a finer skew angle detection by obtaining the entropy values of the HPP of the block image rotated through ±1 degree with a resolution of 0.1 degrees. Table 2 depicts entropy values obtained during the finer skew angle detection process.

Table 2. The entropy values obtained for fine skew angle detection

Angle	Entropy	Angle	Entropy
0	4.74	1.1	4.402
0.1	4.73	1.2	4.45
0.2	4.66	1.3	4.50
0.3	4.61	1.4	4.56
0.4	4.55	1.5	4.62
0.5	4.49	1.6	4.68
0.6	4.44	1.7	4.74
0.7	4.38	1.8	4.79
0.8	4.34	1.9	4.84
0.9	4.33	2	4.88
1	4.36		

The minimum entropy value is for 0.9 degrees. After obtaining the exact skew angle, the image is rotated by that angle in the opposite direction using bilinear interpolation. Thus the block is skew corrected. Although the skew detection and correction could be done on the entire document image, it is done at block

level since every block with its logos, typing defects etc. could have a different
skew. After skew correction, further processing such as classification and OCR
could be conducted on the block.

3   Experimental Results

3.1   Data Description

Our data consists of 100 document images each of English and Kannada, scanned
at 200 dpi and stored in 1-bit depth monochrome format. These documents contain signatures, logos and other such things along with free-flowing text paragraphs.

3.2   Implementation

The algorithm has been implemented in ANSI-C language and compiled using
GCC compiler. It has been executed on a Pentium IV 2.4 GHz, 512 MB RAM
PC.

3.3   Timing Analysis

Table 3 depicts the timing for various stages in the skew correction process for
100 block images consisting of both English and Kannada images.

Table 3. The Detection and Correction time for 100 block images

Total Detection Time	10.57s
Total Correction Time	33.05s
Total Time	43.62s
Average Time	0.44s

3.4   Time Complexity

During skew detection the entropy calculation is done for a constant number of
times (20 in our case). We find the minimum of all the entropy values calculated.
Since finding the minimum value among all entropy values, involves a constant
number of comparisons, the time complexity of our algorithm mainly depends on
the time complexity of entropy calculation. Entropy calculation for each angle
involves iteratively checking each pixel in the image. Also, we use bilinear interpolation for rotating the image for constant number of times. Hence the overall,
time complexity is O(height*width), which is linear with respect to total number
of pixels in the image.

Figure 3 and 4 depict some of the results we have obtained for English and
Kannada block images.

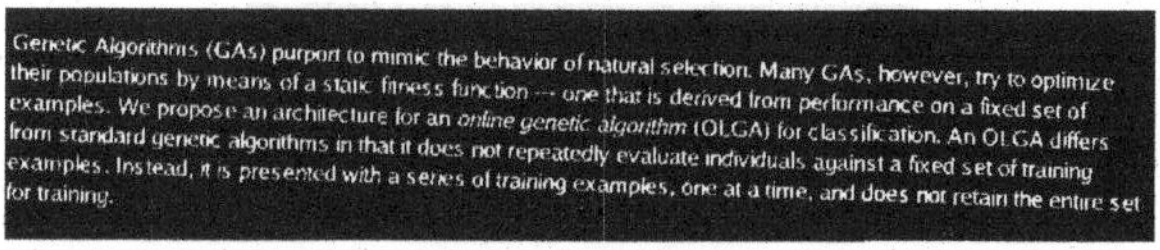

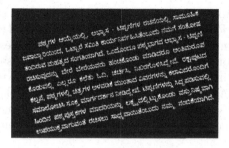

(a)

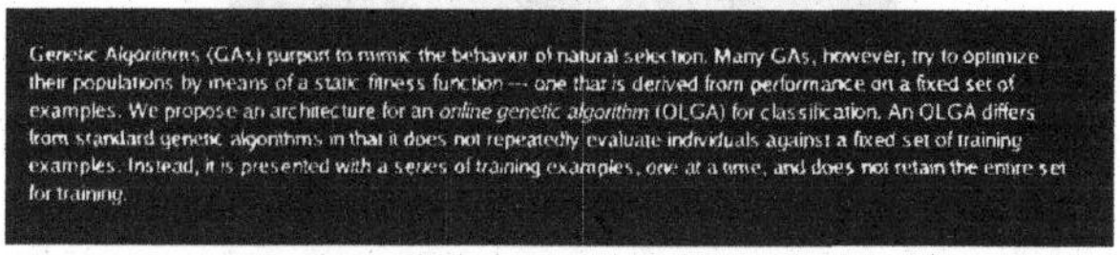

(b)

Fig. 3. (a) Example of a skewed English block image (b)The skew corrected block image

(a)

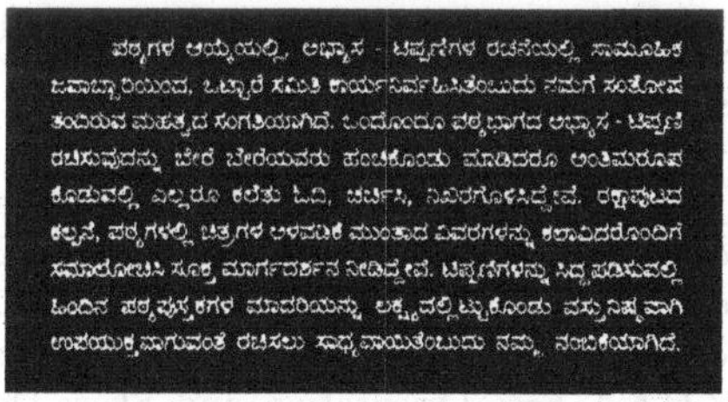

(b)

Fig. 4. (a) Example of a skewed Kannada block image (b) The skew corrected block image

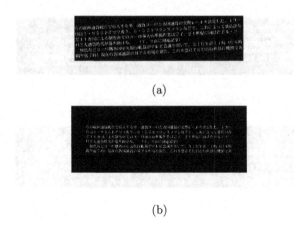

(a)

(b)

Fig. 5. (a) Skewed image of block with Chinese script (b) The skew corrected image

4   Conclusion

Thus we propose a robust and efficient skew detection and correction algorithm based on Horizontal Projection Profiles of a block image and their entropy. In this paper we have shown a range angle detection of ±10 degrees with a precision of 0.1 degrees. Extension to greater angles is possible and is a simple procedure, the drawback being the increase in time consumption. We have tested our algorithm on 100 document images (English and Kannada) i.e. 4421 block images. We have also tested our algorithm on few Chinese documents as shown in Figure 5 and we have observed that the algorithm performs well, independent of script.

References

1. Zhu, X., Yin, X.: A New Textual/Non- textual Classifier for Document Skew Correction. In: Proceedings of the 16th International Conference on Pattern Recognition (ICPR 2002) (2002)
2. Le, D.S., Thoma, G.R., Wechsler, H.: Automatic page orientation and skew angle detection or binary document images. Pattern Recognition 27
3. Avanindra, Chaudhuri, S.: Robust Detection of Skew in Document Images. IEEE Trans on Image Processing 6, 344–349 (1997)
4. Avila, B.T., Lins, R.D.: A Fast Orientation and Skew Detection Algorithm for Monochromatic Document Images. In: Proceedings of the 2005 ACM symposium on Document engineering (2005)
5. Morita, M.E., Bortolozzi, F., Facon, J., Sabourin, R.: Morphological approach of handwritten word skew correction. In: Proceedings of International Symposium on Computer Graphics, Image Processing and Vision (SIBGRAPI 1998)
6. Arvind, K.R., Pati, P.B., Ramakrishnan, A.G.: Automatic text block seperation in document images. In: Proceedings of 4th International Conference on Intelligent Sensing and Information Processing (ICSIP 2006) (2006)

Keyword Extraction from a Single Document Using Centrality Measures

Girish Keshav Palshikar

Tata Research Development and Design Centre (TRDDC),
54B, Hadapsar Industrial Estate, Pune 411013, India
gk.palshikar@tcs.com

Abstract. Keywords characterize the topics discussed in a document. Extracting a small set of keywords from a single document is an important problem in text mining. We propose a hybrid structural and statistical approach to extract keywords. We represent the given document as an undirected graph, whose vertices are words in the document and the edges are labeled with a dissimilarity measure between two words, derived from the frequency of their co-occurrence in the document. We propose that central vertices in this graph are candidates as keywords. We model importance of a word in terms of its centrality in this graph. Using graph-theoretical notions of vertex centrality, we suggest several algorithms to extract keywords from the given document. We demonstrate the effectiveness of the proposed algorithms on real-life documents.

1 Introduction

In information retrieval (IR), given a collection (corpus) of documents, *index terms* are extracted from each document. The set of index terms for each document helps in indexing, searching and retrieving documents relevant to a given query. The automatic index term extraction techniques in IR identify terms that are neither too specific (i.e., occur only within that document) nor too general (i.e., occur in all documents).

There is a related but different problem of automatically extracting keywords from a *single* document, such as an article, a research paper or a news item. A document is characterized by a set of *keywords* (more generally, *key phrases*). Each keyword indicates an important aspect of the subject matter described in the document. Each keyword may describe a major topic discussed in the document. Typically, only a few keywords (10 or 20) are associated with each document, whereas IR associates a large number (hundreds) of index terms with each document in a collection. Moreover, keywords are usually *ordered* in decreasing order of their importance (keywords that are most characteristic of the document occur first). Alternatively, the keywords may also be ordered in increasing order of their generality (most specific terms occur first). Keywords facilitate classification or categorization of a standalone document whereas the index terms facilitate searching the documents within a collection.

A. Ghosh, R.K. De, and S.K. Pal (Eds.): PReMI 2007, LNCS 4815, pp. 503–510, 2007.

The *keyword extraction problem* consists of extracting k keywords from the given document, where $k \geq 1$ is a given integer. We assume that some pre-processing steps are performed on the document before giving it as input to the keyword extraction algorithm. Most approaches to keyword extraction are considered *statistical* in nature. For example, a naive approach is to find the k most frequent words in the document. However, this does not usually work (even after removing stopwords) because keywords are important (meaningful), not necessarily frequent.

In this paper, we propose a graph-theoretical (structural) notion to capture the idea of importance of a word in the given document. We represent the given document as an edge-labeled graph and propose that most keywords would correspond to *central* vertices in this graph. Using appropriate graph-theoretical notions of centrality of a vertex, we then suggest various algorithms to extract keywords from the given document. Since the edge labels are based on the frequency of the words in the document, our approach is not purely structural, but rather a hybrid one, where structural and statistical aspects of the document are represented uniformly in a graph. As a side benefit, the schemes proposed in this paper provide a natural way to order the extracted keywords in terms of their importance (i.e., in terms of the centrality of the corresponding vertices).

Section 2 contains the technical approach and the various keyword extraction algorithms. Section 3 describes results of experiments done to demonstrate the utility of the proposed approach. Section 4 discusses some related work and section 5 contains our conclusions and outlines some further work.

2   Approach

We consider a sentence as an unordered set of words (treating multiple occurrences of a word in a sentence as a single occurrence). Then a document is a collection of sets of words. We simplify the document by applying the following preprocessing steps to it: (i) use of abbreviations (e.g., replace United Nations with UN) (ii) removal of all numbers (a number is rarely a keyword) (iii) removal of stop words (iv) removal of punctuation symbols (except sentence terminators . ? and !) (v) removal of infrequent words (i.e, words that occur less than a specified number of times in the document) (vi) stemming (e.g., using the Porter stemming algorithm).

2.1   Eccentricity-Based Keyword Identification

Definition 1. *A* term graph *is an undirected edge-labeled graph* $G = (V, E, w)$ *where each vertex in V corresponds to a term (i.e, a word) in the document, E is the set of edges and $w : E \rightarrow (0, 1]$ is the edge weight function. There is an undirected edge between terms u and v $(u \neq v)$, with weight $w(u, v)$, iff $0 < w(u, v) \leq 1$, Edge weights indicate dissimilarity (distance) between terms. We assume that w is symmetric i.e., $w(u, v) = w(v, u)$, $\forall u, w$ and $w(u, u) = 0$ $\forall u$.*

One simple scheme to define the weights in the term graph is as follows. Let $c(\{u, v\})$ denote the number of sentences in which terms u and v both occur together. Then

$$w(u, v) = \begin{cases} 0 & \text{if } c(\{u, v\}) = 0 \\ \frac{1}{c(\{u,v\})} & \text{otherwise} \end{cases}$$

If terms u and v do not co-occur in at least one sentence, then the weight $w(u, v) = 0$ and the edge uv is absent in the term graph. Otherwise, the weight $w(u, v) = 1/c(\{u, v\})$. For example, for the collection of sets $\{\{a, b\}, \{a, b\}, \{a, b, c\}, \{a, b\}, \{a, c\}, \{a\}\}$, the $w(a, b) = 0.25$, $w(a, c) = 0.5$ and $w(b, c) = 0.5$. Clearly, lower weight indicates higher strength of the (co-occurrence) relationship between the terms.

In general, the term graph may be disconnected. If the term graph is disconnected, we apply the keyword identification procedure to each component and then return the union of the keywords from each component (alternatively, we may select more keywords from a larger component). To simplify matters, we assume in the following that the term graph is connected.

Consider the following news item posted in TIME magazine's issue for Nov. 21, 2006 (www.timeasia.com), with the headline
Nepal, rebels sign peace accord.

Nepal's government and Maoist rebels have signed a peace accord, ending 10 years of fighting and beginning what is hoped to be an era of peaceful politics in the Himalayan kingdom. In a ceremony, Nepali Prime Minister Girija Prasad Koirala and Maoist leader Prachanda signed the agreement on Tuesday, which brings the rebels into peaceful multiparty democratic politics.

"The politics of violence has ended and a politics of reconciliation has begun," Koirala said after the signing. Last week, the Maoists agreed to intern their combatants and store their arms in camps monitored by the United Nations. Nepal's Maoist rebels have been fighting an armed rebellion for 10 years to replace the monarchy with a republic. More than 13,000 people have been killed in the fighting. According to the agreement, any use of guns by the rebels will be punished. The democratic government and the Maoists have agreed to hold elections in June 2007 for constituent assembly that will decide the fate of the monarchy.

"This is a historic occasion and victory of all Nepali people," Chairman of the Communist Party of Nepal Prachanda said at the signing ceremony, witnessed by political leaders, diplomats, bureaucrats and the media. "A continuity of violence has ended and another continuity of peace has begun," Koirala said. "As a democrat it was my duty to bring non-democrats into the democratic mainstream. That effort is moving ahead towards success. "The peace agreement is an example for the whole world since it is a Nepali effort without outside help," he added. The challenge Nepal now faces is holding constituent assembly elections in a peaceful manner.

Meanwhile, Maoist combatants continued to arrive in seven camps across the country Tuesday, albeit without United Nations monitoring. A tripartite agreement between the government, Maoists and the U.N. has to be signed before the U.N. can be given a mandate to monitor arms and combatants. "I

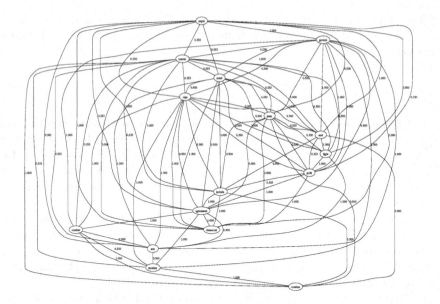

Fig. 1. Term graph for the news item

hope that we will quickly be able to reach tripatriate agreement on the full modalities for the management of arms and armies clarifying essential detail," said Ian Martin, Special Representative of the United Nations Secretary General in Nepal. The Maoists will now join an interim parliament and an interim government, as early as next week, following the agreement.

The pre-processing steps described earlier were applied to the document which reduces the number of words to 97 (from the original 372). The resulting term graph has 97 vertices (one for each word) and 797 edges. The task now is to choose, say 10, keywords from this set of 97 keywords. Our hypothesis is that the keywords are *central* in some sense in the given document. How does one compute the centrality of a word in a document? Since we have represented the document as a graph, we can now appeal to graph-theoretic notions of centrality of vertices. To simplify the graph drawing, Fig. 1 shows another term graph for the same news item, where we have now retained only those words that occur at least 3 times. The resulting term graph has 16 vertices and 84 edges.

Definition 2. *Given a term graph G, the distance $d(u, v)$ between two terms u and v is the sum of the edge weights on a shortest path from u to v in G. Eccentricity $\epsilon(u)$ of a vertex u in G is the the maximum distance from u to any other vertex v in G i.e., $\epsilon(u) = max\{d(u, v)|v \in G\}$.*

Computing the eccentricity of a given vertex is easy. Dijkstra's single source shortest path algorithm [1] efficiently computes the shortest paths from a given vertex u to all other vertices. Then the eccentricity $\epsilon(u)$ of u is the length of the longest path among these paths. Intuitively, one may expect low eccentricity words to be more important. One approach to automatic selection of keywords is now clear. List the terms in increasing order of their eccentricities in G and pick the first k words (having the least eccentricity). Using the proposed algorithm, the eccentricities (and degrees) of the first 16 words (when ordered in terms of increasing eccentricity) in the above news item are as follows:

$\{(nepal, 2.0, 67), (agreement, 2.0, 48), (peac, 2.0, 40), (sign, 2.25, 46),$
$(polit, 2.25, 41), (maoist, 2.333, 54), (rebel, 2.333, 34), (leader, 2.333, 29),$
$(prachanda, 2.333, 29), (ceremoni, 2.333, 29), (end, 2.333, 19), (arm, 2.5, 34),$
$(govern, 2.5, 34), (hope, 2.5, 30), (koirala, 2.5, 23), (fight, 2.5, 21)\}$

While there is no problem in choosing the first 5 keywords, there is some ambiguity in the choice of the last 5 keywords; viz., the 6 words {maoist, rebel, leader, prachanda, ceremoni, end} all have the same eccentricity 2.333. How do we choose 5 words from these 6 words? We prefer keywords which have a high degree in the term graph. Then the next 5 keywords are {maoist, rebel, leader, prachanda, ceremoni} with degrees 54, 34, 29, 29, 29, which are all higher than the degree 19 of the word end. The final set of 10 keywords for the above news item, chosen using this heuristic is: {nepal, agreement, peac, sign, polit, maoist, rebel, leader, prachanda, ceremoni}.

2.2 Using Other Centrality Measures

Apart from eccentricity, betweenness [2] and closeness are two notions of vertex centrality, among others, from social network analysis [7]. Either of these could be used as a measure of centrality to identify keywords from the term graph representation of a given document.

Definition 3. [7] Given an edge-labeled graph $G = (V, E, \lambda)$, the closeness $C(u)$ of a given vertex u is defined as:

$$C(u) = \sum_{v \in V} d(u, v)$$

where $d(u, v)$ is the length of the shortest path from u to v.

Computing closeness for a vertex is similar to computing the eccentricity of that vertex. Vertices with lower value of closeness are more central. So the approach to extract the keywords is same as for eccentricity. We compute the closness value for each vertex, sort the vertices in the ascending order of their closeness values and pick the lowest k vertices as keywords. As earlier, we use the vertex degree to break ties among vertices that have the same closeness value (vertices with larger degree are more preferable). The closeness method yoields the following 10 keywords for the news item: {nepal, peac, maoist, agreement, sign, polit, rebel, arm, govern, leader}. These keywords are similar to the ones produced by the eccentricity method.

2.3 Proximity-Based Keywords Identification

Yet another approach to keyword extraction is possible by applying the link analysis technqiues, such as the cGraph algorithm [3]. Consider a database of research papers, which can be represented as a collection of sets of authors (each paper is represented as a set of authors). Each paper is considered as a *link* among the co-authors of that paper. The database of papers is then represented as a directed, edge-labeled *collaboration graph*, where each vertex corresponds to an author and there is a directed edge from u to v (and from v to u) if they have co-authored at least one paper together. For any vertex u, the sum of the weights on outgoing edges from u must be 1. The weight of every edge is a real number in the interval $[0, 1]$. There are several ways to compute the weight of the edge from u to v; the simplest way is as follows [3]:

$$w(u, v) = \widehat{P}(v|u) = \frac{\sum_{L:(u,v) \subseteq L} \left(\frac{1}{|L|-1} \right)}{\sum_{L:u \in L} 1}$$

Here, the weight $w(u, v)$ is an estimate $\widehat{P}(v|u)$ of the probability that a link that contains u also contains v. The denominator is the count of links that contain u. For example, for the collection of sets
$\{\{a, b\}, \{a, b\}, \{a, b, c\}, \{a, b\}, \{a, c\}, \{a\}\}$,
the weight $w(a, b) = \frac{1+1+0.5+1}{6} = 0.583$. Here, higher weight indicates higher strength of the co-occurrence (co-authorship) relationship between the authors. To make it a suitable dissimilarity measure, we actually use the inverse of $w(u, v)$ as the edge label. Thus, in the example, $w(a, b) = 1/0.583 = 1.715$.

By considering each sentence as a link, we can analogously construct a collaboration graph for a document, with the words as vertices. We can now apply any of the above algorithms to the collaboration graph to obtain a set of keywords. However, there is an additional possibility: that of using the proximity measure [3] between two vertices u and v, defined as:

$$prox(u, v) = \sum_{p \in V(u,v,m)} \prod_{e_i \in p} P(e_i | a_j \forall j < i)$$

where $V(u, v, m)$ is the set of all non-self-intersecting walks p of length $\leq m$ from u to v and e_i is a directed edge in such a walk p. $P(e_i | e_j)$ is the probability that the edge e_i will be added to a path p from u given that the earlier vertices already in p are $a_1, a_2, \ldots$. The function *prox* considers only paths of length at most m steps. All vertices which are not reachable from u in at most m steps are given a very low proximity. Value of m is usually low, say 3 or 4. In the computation of eccentricity, we could use this proximity measure, instead of the distance of the shortest paths. For a vertex u, we consider all vertices $\{v_1, v_2, \ldots\}$ reachable from u in at most m steps and consider the eccentricity of u to be the largest among these values: $\epsilon(u) = max\{prox(u, v_1), prox(u, v_2), \ldots\}$.

3   Experimental Results

We have received very positive feedback on the keywords generated by the algorithms on specific documents. Most users agreed on most of the documents that the keywords generated by the algorithms were good. However, we carried out the following experiment to get a more objective feedback about the quality of the generated keywords. We collected 64 news stories from well-known English news magazines in India, under 5 categories: environment, economy, defence, health and cinema. Each news also had a headline. The everage size of the stories was 1352 words and 8208 characters (without the headline). We generated 10 keywords for each news story (without its headline) and computed how many of these 10 keywords also occurred in the headline. The idea is that the headline would generally contain the main keywords in the news story. For example, the headline `Nepal rebels sign peace accord` of the earlier news story contains 4 keywords `Nepal rebels sign peace`. Note that the a synonym `agreement` of the fifth word `accord` in the headline is a keyword. Nevertheless, we add that the headline is often written to be catchy and exciting, so that it does not always contain the main keywords. Hence we expect only a partial overlap between the generated keywords and the words in the headline. We extracted 10 keywords for each of the 64 news items and compared them with the corresponding headlines. On the average, about 3 to 4 keywords (out of 10) appeared in the headline. We consider this result as rather satisfactory and supporting our keyword extraction algorithms. We also carried out a similar experiment on a set of 300 smaller (one paragraph) financial news items, with similar results.

4   Related Work

[6] presents a closely related approach called KeyGraph for extracting keywords from a document. In Keygraph, a document is represented as a graph, whose vertices are then clustered so that each cluster represents a concept or topic in the document. Then highest ranking words from each cluster are extracted and returned as keywords. In [5], a cognitive process based on priming and activation is used for extracting keywords. The readers's mind is a network of concepts and reading a document activates some concepts, which in turn activates the related concepts and so on. Vertices in a graph represent words and edges denote the associations between words. A mathematical model for activation spreading is specified, based on word frequency and recency of activation. At the end, most highly activated terms are returned as keywords. In [4], a probability distribution of co-occurrences between frequent words and all other words is analysed for bias using χ^2-measure and terms with unusual bias are selected as keywords.

5   Conclusions and Further Work

Keywords are used to characterize or summarize the main topics in a document. Extracting a small set of keywords from a single document is an important problem in text mining. Keyword extraction techniques are beginning to be used in other applications such as web-page clustering and discovering emerging topics by analyzing co-citation graphs. In this paper, we proposed a hybrid structural and statistical approach to extract keywords. We represent the given document as an undirected graph, whose vertices are words in the document and the edges are labeled with a dissimilarity measure between two words, derived from the freuqnecy of their co-occurrence in the document. We then propose that central vertices in this graph are candidates as keywords, where we model the importance of a word in terms of its centrality in this graph. Using appropriate graph-theoretical notions of centrality of a vertex - such as eccentricity, closeness, betweenness and proximity - we suggested several algorithms to extract keywords from the given document. The proposed keyword extraction algorithms appear to be effective when tested on a set of real-life news stories. We found that reducing the number of words (vertices) by retaining only those words that appear at least a user-specified minimum number of times does not decrease the effectiveness of the extracted keywords.

For further work, we are working on refining the proximity-based approach. We also wish to modify our algorithms to take into account other factors such as the position of words, rather than rely purely on their frequencies. We are trying to define an objective measure for the goodness of the extracted keywords, so as to facilitate the comparison of various keywords extraction algorithms. Currently, the number of keywords to be extracted is specified by the user; we are trying to automatically arrive at this number.

Acknowledgements. I thank colleagues in TRDDC for useful discussions and help. Thanks to Dr. Manasee Palshikar for her support and encouragement.

References

1. Cormen, T.H., Leiserson, C.E., Rivest, R.L., Stein, C.: Introduction to Algorithms 2/e. MIT Press, Cambridge (2001)
2. Freeman, L.C.: A set of measures of centrality based on betweenness. Sociometry 40, 35–41 (1977)
3. Kubica, J., Moore, A., Cohn, D., Schneider, J.: Finding underlying structure: A fast graph-based method for link analysis and collaboration queries. In: Proc. 20th Int. Conf. on Machine Learning (ICML 2003) (2003)
4. Matsuo, Y., Ishizuka, M.: Keyword extraction from a single document using word co-occurrence statistical information. Int. Journal on AI Tools 13(1), 157–169 (2004)
5. Matsumura, N., Ohsawa, Y., Ishizuka, M.: Pai: Automatic indexing for extracting assorted keywords from a document. In: Proc. AAAI 2002 (2002)
6. Ohsawa, Y., Benson, N.E., Yachida, M.: Keygraph: automatic indexing by co-occurrence graph based on building construction metaphor. In: Proc. Advanced Digital Library Conference (ADL 1998), pp. 12–18 (1998)
7. Wasserman, S., Faust, K., Iacobucci, D.: Social Network Analysis: Methods and Applications. Cambridge University Press, Cambridge (1995)

A HMM-Based Approach to Recognize Ultra Low Resolution Anti-Aliased Words

Farshideh Einsele, Rolf Ingold, and Jean Hennebert

Université de Fribourg, Boulevard de Pérolles 90, 1700 Fribourg, Switzerland
{farshideh.einsele, rolf.ingold, jean.hennebert}@unifr.ch

Abstract. In this paper, we present a HMM based system that is used to recognize ultra low resolution text such as those frequently embedded in images available on the web. We propose a system that takes specifically the challenges of recognizing text in ultra low resolution images into account. In addition to this, we show in this paper that word models can be advantageously built connecting together sub-HMM-character models and inter-character state. Finally we report on the promising performance of the system using HMM topologies which have been improved to take into account the presupposed minimum length of each character.

1 Introduction

The explosive growth of the World Wide Web has resulted in a colossal data collection with billions of electronic documents. These documents often contain images with textual information providing very high semantic value which could be used for indexing. According to a study done in 2001, of the total number of words visible on a WWW page, 17% are in image form and 76% of these words do not appear elsewhere in the encoded text [2]. Furthermore, the ALT tag, which is recommended for describing an image in HTML language, is frequently not or wrongly used [7]. Web indexation engines would clearly benefit from having access to a so-called Web-image-OCR that could automatically recognize text embedded in images.

One approach is actually to use classical scanner-based OCR systems and to feed them with web images. However, these OCR systems are not trained nor tuned to treat such ultra low resolution images ($<$ 100 dpi) with small point sizes ($<$ 12 points) and with anti-aliasing artefacts. Furthermore, the detection of text area is more difficult in the case of web images as it is frequently surrounded or even superposed to other graphical objects. Acknowledging these difficulties, several related works have been proposing pre-processing methods to transform the image into a representation that is better suited for classical OCR systems [5] [1] [9]. Notwithstanding, most of these studies dealt with bi-level (black and white) images and the proposed methods did not address the specific variabilities of text embedded in web images.

In our approach, instead of focussing on pre-processing methods and relying on classical OCR systems, we propose to build a dedicated characters- and

A. Ghosh, R.K. De, and S.K. Pal (Eds.): PReMI 2007, LNCS 4815, pp. 511–518, 2007.

words-recognizer to handle the specificities of text embedded in web images, i.e. ultra low resolution rendered text images with small point sizes and anti-aliasing artefacts. In this paper, we propose to use a recognition system based on Hidden Markov Models (HMMs). HMMs are well known modelling tools widely used in the fields of handwriting or speech recognition, see for example [10]. HMMs have also been successfully used in classical OCR systems, with the advantage of being able to recognize connected characters such as arabic language [6]. However, they have never been proposed, to the best of our knowledge, for the recognition of text embedded in gray-level images with anti-aliasing. Furthermore, we believe that our approach differs from the classical use of HMMs for OCR tasks since we address the specificities of ultra low resolution rendered text images by applying the following procedures:

- We use a specific feature extraction module based on moments computation instead of the classical projection profiles used in OCR systems. This choice is guided to handle the limited amount of pixels that can be used as input.
- The probability density functions used in the HMMs are trained on ultra low-resolution characters to incorporate the specificities of the features such as the anti-aliasing and downsampling noise, i.e. we don't attempt to remove or reduce the anti-aliasing effect, instead, we model it explicitly.
- The topology chosen for the HMM will model inter-character regions. The associated models will also incorporate the specificities of these regions in terms of anti-aliasing noise due to the close proximity of adjacent characters.

In Section 2 of this paper we give a brief description of the specificities of text images encountered in web images. In section 3, we introduce the word recognizer system we have built and we describe the feature extraction as well as the different HMM topologies that have been used in this work. In section 4 we describe the task used to evaluate the performance of the recognizer and we then report on the results of 8 different configurations of the recognizer. Conclusions and plans for future work are finally presented.

2   Challenges for Web-Image-OCR

Character level. As shown on Fig. 1(a), a character embedded in a web image presents specific characteristics: (1) the character has an *ultra* low resolution, usually smaller than 100 dpi with small point sizes frequently between 6 and 12 points, (2) the character has artefacts due to anti-aliasing filters (3) the same characters can have multiple representations due to the position of the down sampling grid.

In our previous work [4], we have been evaluating a system aiming at the identification of such isolated characters. The objective was to assess the feasibility of recognizing such characters and to evaluate the difficulty of the task. We proposed to use a feature extraction based on first and second order central moments coupled with a statistical model based on multivariate Gaussian density functions. Encouraging results were obtained with overall recognition rates above 99.5%, consistently observed on 168 different fonts. While encouraging, these results do not correspond to a realistic situation as we did the assumption that characters were isolated, which is not the case in practice (see next section).

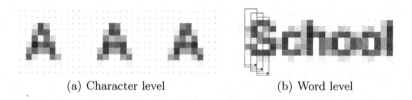

(a) Character level   (b) Word level

Fig. 1. (a) Example of anti-aliased, down sampled character 'A' with different grid alignments. (b) Low resolution version of word 'School' and illustration of the sliding window used for the feature extraction.

Word level. Fig. 1(b) illustrates a more realistic example of characters composing a word in a low resolution image. As can be observed, there are no character interspaces available to segment characters within the word. Furthermore, the anti-aliasing artefact of adjacent characters is superposing. This contextual effect is clearly a new extra source of variability on the character samples.

Therefore, as an extension of our isolated character evaluation presented in the previous section, we wanted to measure the impact of this contextual variability [3]. We used the same classification system on the same task, assuming that character segmentation was known. The only difference was coming from the contextual noise as an extra source of variability. We observed an overall character recognition rate of 98.5% which is less than the former result but still encouraging.

However, the system used to obtain these results is not realistic as it intrinsically implies that the segmentation of characters inside of the word is known. A direct observation of the word "School" in Fig. 1(b) can convince us that well-known pre-segmentation methods used in classical OCR systems can not be applied anymore in our case [8]. Acknowledging this fact, we opted to build a new recognizer system that is able to solve the classification problem simultaneously with the segmentation problem.

3   System Description

In a similar way as what is frequently done in or cursive handwriting or speech recognition modeling, we decided to build our Web-OCR system using a sliding-window feature extraction feeding an HMM based classifier. This combination has the clear advantage that the segmentation and recognition problem can be solved simultaneously. HMMs also bring further advantages by allowing to naturally include linguistic knowledge such as lexical or grammatical models inside of the framework.

Before going to the more complex task of connected word recognition involving more advanced linguistic models, we decided to build and evaluate a HMM system designed for the task of isolated word image recognition. Such a task actually makes sense as we can reasonably assume that words can be isolated with classical image processing techniques, even though they are low resolution and have small font sizes. The details of our system are described hereafter.

3.1 Feature Extraction

HMMs are basically modelling ordered sequences of features that are function of a single independent variable. Inspired by feature extraction used for speech or cursive handwriting recognition, we decided to compute a naturally left-right ordered sequence of features by sliding an analysis window on top of the word. Therefore the independent variable is simply, in our case, the x-axis. As illustrated on Fig. 1(b), we used a 2 pixels length window shifted 1 pixel right. In each analysis window, a feature vector of 8 components including the first and second order central moments is computed. The choice of moments as features was directed by their good discriminative representation as observed in our previous studies (see section 2). The word image is then represented by a sequence of feature vector $X = \{x_1, x_2, ..., x_N\}$ where x_n is a 8 component vector.

3.2 Hidden Markov Models

The HMM will model the likelihood of the observation sequence X given the model parameters associated to a word i. By applying the usual simplifying assumption of HMMs (see for example [10]), the likelihood $P(X|M_i)$ of X given the model M_i can be expressed as the sum, over all possible paths of length N, of the product of emission probabilities and transition probabilities measured along the paths. Alternatively, the Viterbi criterion can also be used, stating that instead of considering all potential paths through the HMM, only the best path is taken into account, i.e. the path that maximizes the product of emission and transition probabilities. In this work, we have based our training and testing approaches using the Viterbi criterion.

For sake of flexibility, we have chosen to associate HMM states to characters. Doing this, any word can be modelled by an HMM where the corresponding states are simply connected together.

Fig. 2 illustrates the single word HMM-recognizer that has been used to perform our evaluation. We have built a vocabulary of 520 words that were different from the words used in the training set. The words were picked from a dictionary in a way to represent all characters according to their natural occurrence. Each test image has been synthetically generated using the *Verdana* font. From a high resolution version, the image was then downsampled to obtain the equivalent of 9 point size with a resolution of 72 dpi, applying anti-aliasing filters and different grid alignments, such as illustrated on Fig. 1(b). The recognizer first computes the feature vectors that are fed to the 520 competing HMMs. The computation of the likelihood for each model is performed using the Viterbi criterion. As, in our test, all words are equally a priori probable, the winning model is simply the one that leads to the maximum likelihood value.

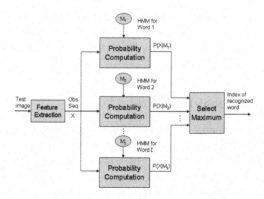

Fig. 2. Single word HMM-Recognizer

4 Topologies

We investigated different topologies for the HMMs.

4.1 Simple Left-Right

This topology was chosen for its simplicity and was, for us, the first configuration to investigate. Fig. 3, illustrates this topology (top). As the performances

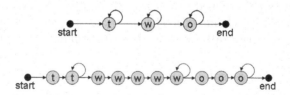

Fig. 3. Simple left-right topology (top) and left-right minimum duration tolopogy (bottom)

obtained with this topology were not so convincing, we inspected some state sequences obtained on mis-recognized words. An example of state sequence for word 'two' calculated with simple left-right topology is reproduced below:

- Genuine word= two
- Recognized word = one
- Best state sequence for word one: o o o o o o o o o o n n e e e e
- Best state sequence for word two: t t t t w w w w w o o o o o o o o
- Ground truth state sequence for word two: t t t t w w w w w w w o o o o o

What is happening can be explained as follows. By nature, the HMM is trying to maximize its likelihood. It will then associate few observations to states that gives low emission probabilities while it will spend more observations in states giving high emission probabilities. The state sequence for the winning word 'one' is clearly spending only two observations in character model 'n'. Character 'n' is of course longer than 2 pixels (actually 3 pixels considering the width of the analysis window) in our images.

4.2   Left-Right Minimal-Duration

In order to avoid phenomena as the one described above, one can introduce so-called minimum duration topologies. This topology is simply obtained by repeating a state a number of time obliging the Viterbi algorithm to spend a minimal amount of observations in the same category of character while the last character state has the possibility to be repeated for an unlimited amount of time. Fig. 3 illustrates such a topology for word 'two' (bottom). In our configuration, the minimum duration values have been obtained from the a priori known font metric information. We have to underline that such minimum duration values are dependent to a given font, leading to a system tuned for a specific font.

4.3   The Inter-character Model

According to our previous study of characters in context (see section 2 and [3]), the left and right borders of characters are influenced by the left and right borders of their adjacent characters. We have observed that this *noisy* zone between two adjacent characters includes up to three pixels for font sizes between 6 to 12 points. We have decided to treat this anti-aliasing noise as an additional character model and perfomed training as for another character. We therefore called this new pseudo character model the *inter-character* ('#' in our figures). Fig. 4(a) illustrates the word 'two' and the corresponding inter-character zones.

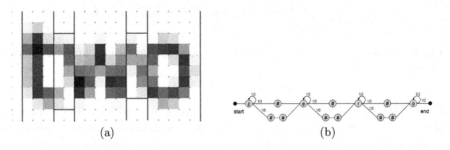

(a)   (b)

Fig. 4. (a) Inter-character zones from word 'two' (b) Inter-Character topology

The previous two HMM-topologies have been modified to take into account the 'inter-character' model. Fig. 4(b) illustrates this modification for the simple left-right HMM-topology.

5   Evaluation Tests

According to the topologies introduced above, different evaluation tests have been performed. We also experimented with two implementations of the Gaussian probability density functions used to estimate the emission probabilities. The first one is using the regular *full* covariance matrix. The second one is using a simplified *diagonal* covariance matrix, making the extra assumption that the components of the feature vectors are de-correlated. While this assumption is potentially too restrictive, it allows a much faster computation of the emission probabilities.

Table1 illustrates the results of theses evaluation tests. The table is divided into two parts. The first part presents results obtained with our previously described two HMM-topologies. The second part presents results obtained from the same HMM-topologies, which have been extended with the inter-character model. Our main observations are the following:

- As expected, we see clearly an improvement coming from the introduction of minimum duration topologies.
- Using full covariance matrices leads to better results than with diagonal covariance matrices.
- The most significant improvement is coming from the introduction of the inter-character model into the HMM topologies.

These encouraging results indicate clearly that HMMs are strongly suitable to simultaneously segment and recognize ultra low resolution words, provided that the right topologies and models are used.

We have to notice that the results are obtained for a test corpus of 520 words. The performance for the HMM using the inter-character model are potentially not anymore significantly different. We plan to use a larger number of test samples and a more difficult task, increasing the test corpus to several thousand words, to better put in evidence system differences.

Table 1. Recognition rates , **without inter-char** (with inter-char)

	diag. covariance	full covariance
Simple left-right	**52%** ($> 99\%$)	**56%** ($> 99\%$)
Minimum duration	**60%** ($> 99\%$)	**79%** ($> 99\%$)

6   Conclusion and Future Work

Most of the approaches to solve the problem of recognizing text embedded in web images are based on pre-processing the images in order to feed them in classical OCR systems. We have presented in this paper a competing approach that is not based on pre-processing of the images but that aims at directly model the specificities of these images: gray-level, ultra low resolution and anti-aliased. An extra piece of difficulty to the task is coming from the fact that adjacent characters cannot be anymore pre-segmented using well-known image processing algorithms. We therefore have proposed to base our system on HMMs that ally powerful statistical models with the ability of performing segmentation in the same time as recognition. While we acknowledge that our experiments are currently performed on synthetic data instead of real-world data, the evaluation tests that we performed with the HMM system brings promising results and clearly show that HMMs are able to simultaneously segment and recognize ultra low resolution words.

In our future works, we plan to improve the HMM system using more powerful emission probability estimators such as multi-Gaussian models. Such models would allow us to relax the assumption that the features are distributed according to a unique Gaussian. In addition to this, we plan to use multi-state character models that will allow us to capture more finely the characteristics of multi-stroke characters. Finally, we plan to move towards more complex recognition tasks involving much larger vocabularies or even open vocabularies, using multi-font context and using data extracted from real-world web images.

References

1. Antonacopoulos, A., Karatzas, D.: Text extraction from web images based on a split-and-merge segmentation method using color perception. In: Proc. of ICPR 2004, Cambridge, UK (August 2004)
2. Antonacopoulos, A., Karatzas, D., Lopetz, J.O.: Accessing textual information embedded in internet images. In: Proc. of Electronic Imaging II, San Jose, California, USA (January 2001)
3. Einsele, F., Hennebert, J., Ingold, R.: Towards identification of very low resolution, anti-aliased characters. In: Proc. of ISSPA 2007, Sharjah, UAE (February 2007)
4. Einsele, F., Ingold, R.: A study of the variability of very low resolution characters and the feasibility of their discrimination using geometrical features. In: Proc. of 4th Enformatika Int. Conf. on Pattern Recognition and Computer Vision, Istanbul, Turkey, pp. 213–217 (June 2005)
5. Lopresti, D., Zhou, J.: Locating and recognizing text in www images. Information Retrieval 2(2/3), 177–206 (2000)
6. Lu, Z., Bazzi, I., Kornai, A., Makhoul, J., Natarajan, P., Schwartz, R.: A robust, language-independent ocr system. In: Proc. 27th IAPR Workshop, vol. 3584, pp. 96–104 (January 1999)
7. Munson, E.V., Tsymbalenko, Y.: Using html metadata to find relevant images on the web. In: Web Document Analysis
8. Nagy, G.: Twenty years of document image analysis in pami. IEEE Transactions on Pattern Analysis and Machine Intelligence 22, 38–62 (2000)
9. Perantonis, S.J., Gatos, B., Maragos, V.: A novel web image processing algorithm for text area identification that helps commercial ocr engines to improve their web recognition accuracy. In: Proc. of the second Int. Workshop on Web Document Analysis, Edinburgh, United Kingdom (August 2003)
10. Rabiner, L., Juang, B.-H.: Fundamentals Of Speech Recognition. Prentice Hall, Englewood Cliffs (1993)

Text Region Extraction from Quality Degraded Document Images

S. Abirami and D. Manjula

Department of Computer Science & Engg, Anna University
Chennai, India.
{abirami@cs.annauniv.edu, manju@annauniv.edu}

Abstract. In this paper we present a well designed method that makes use of edge information to extract textual blocks from gray scale document images. It aims at detecting textual regions on heavy noise infected newspaper images and separate them from graphical regions. The algorithm traces the feature points in different entities and then groups those edge points of textual regions. Finally feature based connected component merging was introduced to gather homogeneous textual regions together within the scope of its bounding rectangles. The proposed method can be used to locate text in-group of newspaper images with multiple page layouts. Initial results are encouraging, then they are experimented with considerable number of newspaper images with different layout structures and promising results were obtained. This finds its major application in digital libraries for OCR where information can be of different quality depending on the age of the scanned paper.

Keywords: Text Extraction, Edge Detection, Block Merging.

1 Introduction

In spite of the wide spread use of computers and other digital facilities, paper document keeps occupying a central place in our everyday life. Segmenting an image into coherent regions is the first step before applying an object recognition method. For example, prior to using an optical character recognizer (OCR), the characters must be extracted from the image. In this paper, the problem of finding text region in a quality degraded newspaper image is discussed. The stated work may be a part of work that facilitates the news article retrieval system.

Newspaper images contain quite a lot of noise, may be one of them was photocopied from the earliest issue of local newspaper dated back or it might have been crushed for some reasons. Isolation of text and graphics is more difficult to do by a segmentation process of a general method. In this paper, we are interested in the preliminary step of document analysis and recognition. Our present goal is to extract textual areas from noisy document images.

We briefly present previous research in the physical segmentation of a document and text detection (in Section 2). Then we describe our work (in Section 3), which mainly consists of detection of zones of interest and discrimination of text areas. Results have been discussed in Section 4. Section 5 shows the performance measures. Section 6 concludes the work.

A. Ghosh, R.K. De, and S.K. Pal (Eds.): PReMI 2007, LNCS 4815, pp. 519–527, 2007.
© Springer-Verlag Berlin Heidelberg 2007

2 Related Works

Techniques for document segmentation and layout analysis are traditionally subdivided into three main categories: bottom-up, top-down and hybrid techniques. Some other up-to-date methods are introduced by recent progresses in this area, so as to expand the scope of above categorization [4]. Bottom-up techniques [5,6,7] progressively merge evidence at increasing scales to form, e.g., words from characters, lines from words, columns from text lines. Top-down techniques [8,9,10] start by detecting the large-scale features of the image (e.g., columns) and proceed by successive splitting until they reach the smallest-scale features (i.e., individual characters, or text lines).

Most methods do not fit into one of these two categories and are therefore called hybrid. Among these we can find methods based on texture analysis and methods based on background analysis [11].

Text / Graphics separation aims at segmenting the document into two layers: a layer assumed to contain text and a layer containing graphical objects. In [1], a consolidated method proposed by Fletcher and Kasturi, with a number of improvements, to make it more suitable for graphics-rich documents is proposed. As a result of these, we want to extract the textual regions successfully from quality degraded document images, which could be useful for text retrieval systems.

3 Text Region Extraction

In this paper, we attempted to extract the text region from quality degraded document images. A high level design of the text region extraction system attempted in this work is shown in figure1. The input to the system could be a color or a gray scale image obtained by scanning the newspapers. Documents containing text, graphics, figures, maps and tables are taken as input. Then the scanned image passes through the following phases.

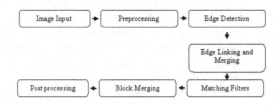

Fig. 1. Text Region Extraction System

3.1 Preprocessing

As a first step, input image undergoes preprocessing. If it is a color image, it has to be converted to a grayscale image. Since newspaper images contain noise, it has to be smoothed out. These two steps of grayscale conversion and smoothing, does forms the basic operation in preprocessing.

3.1.1 Smoothing

This step is used to filter out the noises in the original image before the location and detection of edges. In this system, Gaussian filter has been used for smoothing process. The larger the width of the Gaussian mask, the lower is the detector's sensitivity to noise.

3.2 Edge Detection

Once the preprocessing gets over, the next step is to detect the edges in the image. In this system, Canny Edge Detector Algorithm is applied for edge detection. The various steps involved in this phase is summarized below.

Edge detection is outlined as:

- Storage of smoothed image. S[i,j]   (1)
- Computation of the gradient of smoothed array using 2*2 first difference approximation.

$$P[i,j] = (S[i,j+1] - S[i,j] + S[i+1,j+1] - S[i+1,j])/2 \qquad (2)$$

$$Q[i,j] = (S[i,j] - S[i+1,j] + S[i,j+1] - S[i+1,j+1])/2 \qquad (3)$$

- Calculation of magnitude and orientation for each pixel.

$$M[i,j] = SQRT(P[i,j]^2 + Q[i,j]^2) \qquad (4)$$

$$O[i,j] = arctan(Q[i,j],P[i,j]) \qquad (5)$$

- Apply non-maxima suppression to the gradient magnitude.

$$N[i,j] = nms(M[i,j],L[i,j]) \qquad (6)$$

- Apply suitable threshold to detect and enhance edge.
  - T1 (if needed T2)

To reduce the number of false edge fragments in the non-maxima suppressed gradient magnitude, a threshold has to be applied. All values below the threshold are changed to zero.

3.3 Edge Linking and Merging

Once the edge points are detected, with these edge points, links between these edges has to be identified. This process is called edge linking. Once links are determined, small edge lists need to be merged into longer lines called as edge merging. To perform edge linking, connected components of the edge pixels have to be detected. This is done by applying 8-Adjacnecy Connectivity algorithms.

Once the edge links are determined, calculate the difference between a pair of edge links. If the difference remains to be minimum, then merge it to form a single larger path. Finally, for each path, finds its maximum and minimum point and bound it with a rectangle.

3.4 Matching Filters

Once Edge linking and merging gets over, filter has been applied to remove lines that don't help in finding text components, edges of the bounding box of graphics or text.

First, Pixel Chain code has been computed to match the patterns that could form lines to be removed. Next to that, white pixel gradient average for each component has been computed to remove the things that have very small average (Box Filter).

3.4.1 Pixel Chain Coding

Features of interest range of feature dimensions, noise dimensions, from the set are selected. This is done with knowledge of the characteristics of desired signal or undesired noise. Short isolated lines and spurs that match the pattern are eliminated. In addition to the advantage of feature-based and contextual filtering, pixel chain coding also provides a computational advantage over pixel-based processing because only one directional information is stored than the two coordinate information for each pixel. Figure 2 shows the directional numbers for a pixel P.

5	6	7
4	P	0
3	2	1

Fig. 2. Pixel Chain Coding   **Fig. 3.** Matching Filters

Thus the detected links and their corresponding PCC in the above valid edges in figure 3 are as follows

>Link1:- (0,0)->(7,2) PCC:- 0->0->0->0->0->2->0->0->0->2->0->0->0->0->0
>Link2:- (5,0)->9,0) PCC:- 0->0->0->0->0->0->0->>1->0

With these Chain codes, line patterns are recognized. If the code follows a similar direction value throughout its course, then its pattern does forms a line, which can be deleted. Similarly, if codes with minimal variance match the pattern, they can be deleted making an assumption that the minimal variance could have been caused due to noise. This process is called as line filtering.

3.4.2 Box Filters

As the next step the image is subjected to box filtering. In this step, the white pixel average for each component is calculated. If the average number of white pixels available in a component is less than the threshold average being supplied, then the component is considered to represent a graphic, and it is deleted.

Fig. 4.1. Graphic   **Fig. 4.2.** Text

3.5 Block Merging

At this stage it is considered that all the components so far detected as text. But it is not certain. Instead of analyzing smaller components, they are merged to form bigger blocks by connected component Analysis. Different components of a single block, either graphic or text, closed to each other are connected to form bigger blocks for further analysis.

3.6 Post Processing

In this phase, as a post process, histogram analysis has been made on the image to discriminate the graphical regions, further present on the image. Histogram analysis is similar to box filter, with a variation that, the image being considered is a grayscale image. Average number of white pixel distribution is calculated. Since the range of values is more, we calculated the edge gradient average (considering all the gradients) and normalized number of edge gradient pixel in the component more than the threshold. It is determined that, both the measures remain to be higher than the threshold for text components. These measures are calculated as follows. The first feature, edge gradient average, is defined as:

$$F_a = \frac{\sum_{i=T_h}^{G_m} H(i) \bullet i}{\sum_{i=T_h}^{G_m} H(i)}$$

Where Th is the threshold value, Gm is the maximum value of the gradient, and H(i) is the histogram of the edge gradient. The second feature Fn, normalized number of edge pixels, is defined as:

$$F_n = \frac{\sum_{i=T_h}^{G_m} H(i)}{N_x \bullet N_y}$$

Where Nx, Ny are the width and the height of the block, respectively. This feature represents the density of the edge pixels in a block. In general, the graphic blocks, which contain much blank space, has the lower density of edge pixels in contrast to the text blocks. So, we can use Fn to discriminate graphic block from the text block. Histogram of each component is drawn and analyzed for regularity of extremes. If the regularity is smooth with minimal variance, then it is concluded as text and for irregular extremes, it is considered as graphics and eliminated. Those components that remain form the output. An example is given in the figures 5.1 and 5.2.

Fig. 5.1. Graphic Block

Fig. 5.2. Text Block

4 Experiments and Results

This system has been implemented using Java Advanced Imaging. A set of 1000 images from different newspapers have been scanned at a resolution less than 200

Fig. 6. Preprocessed Image

dots per inch have been tested in this system and promising results were obtained. When an input image is supplied to the system, it undergoes preprocessing, which is shown in figure 6.

Canny edge detection algorithm is applied on the image to enhance the valid edge points. Valid edges are subjected to connectivity algorithm, to find the edge links. After merging, links lying in close proximity with minimal discontinuation are made as a bigger path and bounded by rectangular box. Figure 7 shows the image after edge linking and merging. Connected Component Analysis is performed with the neighboring blocks in horizontal direction and merged together to form a bigger component. Figure 8 shows the image after block merging process.

   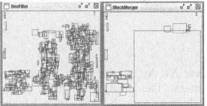

Fig. 7. Edge Linking and Merging   **Fig. 8.** Block Merging

As a final step, the bigger components are subjected to a histogram processing where graphical components are discriminated and removed. Extracted Text region is shown in Figure 9.

Fig. 9. Extracted Text Region

5   Performance Measurements

A set of 1000 images from different newspapers have been scanned at a resolution less than 200 dots per inch have been tested in this system and promising results were obtained. Test Data and Results of some of the images have been projected in the table 1.

Table 1. Test Data and Results

Image Name	Gaussian	LF	PBFT	SBFT	PP (XxY)	HT	TAR	TACR	TAWR	TTANR	TTAR
img 18	Y	N	0.09	0.13	3 x 3	0.5	955	955	0	0	955
img 122	Y	Y	0.1	0.12	3 X 3	0.3	601	601	0	20	621
img 233	Y	Y N		0.13	4 x 4	0.3	3437	3426	11	30	3467
img 419	Y	Y	0.09	0.1	3 x 3	0.5	1663	1602	61	45	1708
img 512	Y	Y	0.14	0.11	2 x 2	0.4	1973	1960	13	40	2010
img 640	Y	Y	0.1	0.11	4 x 4	0.55	678	678	0	30	708
img 67	Y	Y	0.09	0.11	5 x 5	0.3	1666	1666	0	60	1726
img 188	Y	Y	0.09	0.11	4 x 4	0.15	1024	1024	0	45	1069
img 99	Y	N N		0.1	3 x 3	0.2	1109	1109	0	30	1139
img 300	Y	Y	0.13	0.13	4 x 4	0.26	753	463	290	20	773

LF	Line Filter	**TTAR**	Total Text Area To be Retrieved
PBFT	Primary Box Filter Threshold	**Precision = (TACR/TAR)*100**	
SBFT	Secondary Box Filter Threshold	**Recall = (TACR/TTAR)*100**	
PP (X x Y)	Post Process (X x Y)	**TACR**	Total Area Correctly Retrieved
HT	Histogram Threshold	**TAWR**	Total Area Wrongly Retrieved
TAR	Total Area Retrieved	**TTANR**	Total Text C27Area Not Retrieved

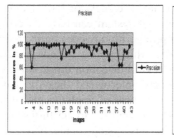

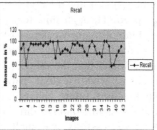

Fig. 10. Precision Graph   **Fig. 11.** Recall Graph

Figure 10 & 11 represents the precision and recall graph, which depicts the text area extracted by the system in percentage.

The precision and recall of those corresponding images tested as per table 1 are given in the following table 2 in measures of percentage. Figure 12 represents various precision and recall measures obtained for various test images with reference to the table 2.

Table 2. Precision Vs. Recall

Image Name	Recall in %	Precision in %
img 18	88	100
img 122	96	100
img 233	97	100
img 419	97	99
img 512	93	96
img 640	98	99
img 67	81	97
img 188	96	97
img 99	94	100
img 300	98	98

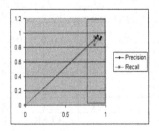

Fig. 12. Precision Vs Recall

6   Conclusion and Future Work

Current OCR techniques and other document segmentation and recognition techniques do not work well for documents with text printed under shaded or textured backgrounds or those with non structural layouts. We proposed a Text region extraction, which works well for the documents even with degraded quality.

The main benefits of our method fall on its efficiency for less memory usage and fast running speed. Through this, text retrieval system can have an efficient access over the text. One of major weakness of our algorithm resides on its assumption that all text is oriented in the same direction, which is by default horizontal. From experiments we found that the ratio of successful detection drops down gradually in exceptional cases when textual paragraphs intertwined heavily with irregular graphical blocks. As a future enhancement, instead of applying global threshold as followed in this system, adaptive thresholds can be used to get further refinement. If these are done successfully, we shall hope this system could play a vital role in newspaper retrieval system and digital library system invariable of the quality of the documents being maintained from past.

References

1. Tombre, K., Tabbone, S., Lamiroy, B.: Text / Graphics Separation Revisited (2002)
2. Chanda, S., Pal, U.: English, Devnagari and Urdu Text Identification (August 2003)
3. Kasthuri, R., O' Gorman, L., Govindaraju, V.: Document Image Analysis – A Primer
4. Chan, C.H., Pau, L.F., Wang, P.S.P.: Handbook of Pattern Recognition & Computer Vision, 2nd edn. (1999)
5. Pavlidis, T., Zhou, J.: Page Segmentation and Classification. Computer Vision Graphics Image Processing 54(6), 484–496 (1992)
6. Wang, D., Srihari, S.N.: Classification of newspaper image blocks using texture analysis. Computer Vision, Graphics, and Image Processing 47, 327–352 (1989)
7. Fletcher, L.A., Kasturi, R.: A robust algorithm for text string separation from mixed text/graphics images. IEEE Transaction on Pattern Analysis and Machine Intelligence 10(6), 910–918 (1988)
8. Le, D.X., Thoma, G.R.: Document classification using connectionist models. In: Proc. of IEEE International Conference on Neural Networks, Orlando, Florida, vol. 5, pp. 3009–3014 (June 1994)
9. Ohya, J., Shio, A., Akamatsu, S.: Recognizing Characters in Scene Images. IEEE Transaction on PAMI 16(2), 214–224 (1994)
10. Baird, W.S., Jones, S.E., Fortune, S.J.: Image segmentation by shape directed covers. In: Proc. Of ICPR, pp. 820–825 (1990)
11. Jain, K., Bhattacharjee, S.: Text Segmentation Using Gabor Filters for Automatic Document Processing. Machine Vision and Applications 5(3), 169–184 (1992)

Offline Handwritten Devanagari Word Recognition: An HMM Based Approach

Swapan Kumar Parui and Bikash Shaw

Computer Vision & Pattern Recognition Unit, Indian Statistical Institute,
203, B.T. Road, Kolkata, India, 700108
swapan@isical.ac.in, bikash_t@isical.ac.in

Abstract. A hidden Markov model (HMM) for recognition of handwritten Devanagari words is proposed. The HMM has the property that its states are not defined a priori, but are determined automatically based on a database of handwritten word images. A handwritten word is assumed to be a string of several stroke primitives. These are in fact the states of the proposed HMM and are found using certain mixture distributions. One HMM is constructed for each word. To classify an unknown word image, its class conditional probability for each HMM is computed. The classification scheme has been tested on a small handwritten Devanagari word database developed recently. The classification accuracy is 87.71% and 82.89% for training and test sets respectively.

Keywords: Hidden Markov Model(HMM), Devanagari Word Recognition, Stroke Primitives.

1   Introduction

Handwriting recognition is one of the challenging problems in Pattern Recognition. The problem has been studied for several decades and many reports on handwriting recognition in the scripts of developed nations are available in the literature. However, only a few works on handwriting recognition in Indian scripts have been reported ([1]-[3]). The present paper deals with recognition of offline handwritten Devanagari words. Works on recognition of handwritten Devanagari characters/numerals exist ([4],[5]). However, no work on handwritten Devanagari word recognition has been reported.

According to literature review there are two approaches for handwritten word recognition: local or analytical approach held at the character level [6] and global approach held at the word level [7]. The first approach deals with the segmentation problem i.e., the words are first segmented into characters or pseudo-characters, then the character model is used for recognition. Since word segmentation is itself a challenging problem, the success of recognition module depends much on segmentation performance. The second approach treats the word itself as a single entity and it goes for recognition without doing segmentation explicitly. However this approach is restricted to applications with small lexicon.

A. Ghosh, R.K. De, and S.K. Pal (Eds.): PReMI 2007, LNCS 4815, pp. 528–535, 2007.

In our present work for word recognition we have applied the second approach because of two reasons: (a) to avoid the overhead of segmentation and (b) due to lack of standard benchmark database for training the classifier. Since a standard benchmark database was not availabe for Indian script so we created a word database for Devanagari to test the performance of our system. In the present report, training and test results of the proposed approach are presented on the basis of this database.

We have used a hidden Markov model (HMM) in the proposed scheme for recognition of handwritten Devanagari words. An HMM is capable of making use of both the statistical and structural information present in handwritten images. This is why HMMs have been used in several handwritten character recognition tasks in recent years [8]. In such HMMs, the states are usually defined as predetermined entities. However, in the present HMM a data-driven or adaptive approach is taken to define the states. The proposed method is robust in the sense that it is independent of several aspects of input such as thickness, size etc.

The next section describes the Devanagari word database followed by Preprocessing and feature extraction in Section 3. Section 4 proposes a HMM classifier. Experimental results are illustrated in Section 5 with conclusions drawn in the last section.

0	आसनसोल	10	हुगली	20	इटावा	30	फरिदावाद	40	लुधियाना
1	औरंगाबाद	11	मैसूर	21	राणाघाट	31	डेहरीओनसोन	41	पोरबंदर
2	कांकीनाड़ा	12	छपरा	22	साहिबगंज	32	गिरिडीह	42	तिसतातोरसा
3	कपूरथला	13	मेरठ	23	अंडमान	33	एटा	43	बिराटि
4	खजूराहो	14	ऊटी	24	भरतपुर	34	उलबेड़िया	44	काकूरगाछी
5	ऋषिकेश	15	झरिया	25	हावड़ा	35	डानकुनी	45	देवघर
6	नैनीताल	16	अहमदाबाद	26	जोधपुर	36	सेवड़ाफुलि	46	चित्रकूट
7	चौरंगी	17	महेषतला	27	पानागढ़	37	थाने	47	कोचीन
8	त्रिवेणी	18	एलौरा	28	विजयवाड़ा	38	वैशाली	48	चंदौसी
9	वाराणसी	19	लक्षमनपुर	29	क्षत्रपतीनगर	39	देहरादून	49	तंजौर

(a)

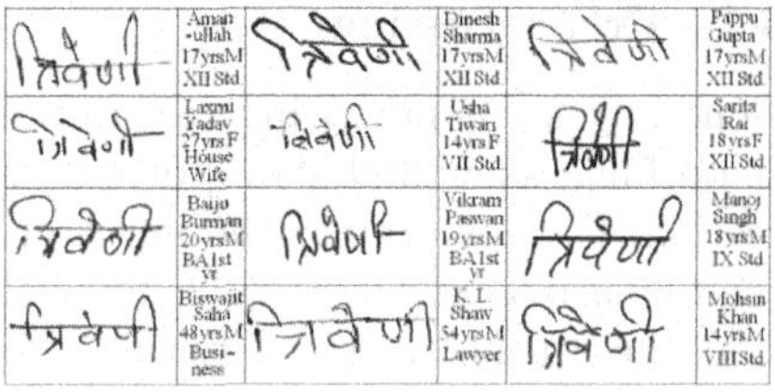

(b)

Fig. 1. (a) Class number and the Devanagari words forming the 50 classes in our database, (b) Several handwritten samples of the same town name "Tribeni", printed form shown in (a) for class number 8, having lots of variation in writing style

2 Handwritten Devanagari Word Database

Handwritten English benchmark word database exist for the research community and CEDAR word database [9] is one of them. But, there does not exist any benchmark word database for any Indic script. Here, we have attempted to create such a database for handwritten Devanagari words. For data collection, we have designed a special kind of form to collect the data. The form contains 50 boxes within which a writer is to write.

The writers were from different classes of society. They were school/college students, business men, housewives, professionals etc. Each writer was asked to fill a form where the word corresponding to each town name is written once.

No restrictions were imposed on the writing style and no handwritten models were provided in order to obtain a heterogeneous database. These handwritten documents were then scanned at 300 dots-per-inch resolution, in 256 levels of gray. For our experiments, we have considered 50 different names of Indian towns, i.e., the number of word classes is 50. Then the whole database of handwritten words is randomly split into two data sets: a training set with 7000 word images, i.e. 140 word images per class and a test set with 3000 words, i.e. 60 words per class.

A few samples from the same town name of this database written by different classes of people are shown in Fig. 1(b), illustrating variation in handwriting style.

3   Pre-processing and Feature Extraction

Variation in handwriting style makes the handwriting recognition problem quite difficult. So to minimize the effect of writing variability related to different writing styles, we take some preprocessing steps. Feature extraction part exploits the global approach for extracting features without explicitly going for word segmentation.

3.1   Preprocessing

Generally for handwriting recognition, the preprocessing stage includes image smoothing, skew and slant correction, image height and pen stroke width correction. For smoothing, the input gray level image is first median filtered and then binarized by Otsu's [10] thresholding method. The binarized image is then smoothed using median filtering. No skew and slant correction is done here. However, our feature extraction method is insensitive to skew/slant within +/-5 degrees. No image height and pen stroke width correction is done since the extracted features are invariant under image height and the extracted strokes are always one-pixel thick irrespective of the stroke width.

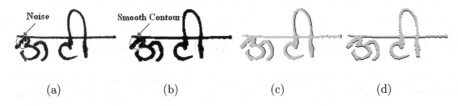

(a)   (b)   (c)   (d)

Fig. 2. (a) An input word image for the word Ooty, (b) Image obtained after thresholding and smoothing, Dark and gray pixels indicate (c) E and A images respectively, (d) S and A images respectively

A sample image from the present database and the same after thresholding and smoothing are shown respectively in Figs. 2(a) and 2(b).

3.2 Extraction of Strokes

Let A be the binarized image. We now describe the process of extraction of vertical and horizontal strokes that are present in A. Let E be a binary image consisting of object pixels in A whose right or east neighbour is in the background. That is, the object pixels of A that are visible from the east Fig. 2(c) form E. Similarly, S is defined as the binary image consisting of object pixels in A whose bottom or south neighbour is in the background Fig. 2(d).

The connected components in E represent strokes that are vertical while the connected components in S represent strokes that are horizontal. Each horizontal or vertical stroke is a digital curve. Shapes of these strokes are analyzed for extraction of features. Very short curves are not considered.

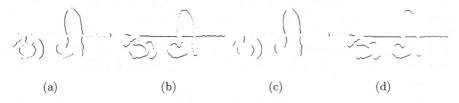

(a)   (b)   (c)   (d)

Fig. 3. (a) E image consisting of vertical strokes obtained from the image in Fig. 2(b), (b) S image consisting of horizontal strokes obtained from the image in Fig. 2(b), (c) Final E image after removal of smaller vertical strokes from the image in (a), (d) Final S image after removal of smaller horizontal strokes from the image in (b)

3.3 Extraction of Features

One of the major factors for the success of any handwritten recognition module is its feature extraction part. The feature should be selected in such a way that it should reduce the intra-class variability and increase the inter-class discriminability in the feature space. From each stroke in E and S, 8 scalar features are extracted. These features indicate the shape, size and position of a digital curve with respect to the word image. A curve C in E is traced from bottom upward. Suppose the bottom most and the top most pixel positions in C are P_0 and P_5. The four points P_1, ... , P_4 on C are found such that the curve distances between P_{i-1} and P_i ($i=1,...,5$) are equal [11]. Let α_i, $i = 1,..., 5$ be the angles that the lines $\overrightarrow{P_{i-1}P_i}$ make with the x-axis. Since the stroke here is vertical, $45^0 \leq \alpha_i \leq 135^0$. α_i's are features that are invariant under scaling and represent only the shape. The position features of C are given by $\overline{X}$,$\overline{Y}$ which are the x and y-coordinates of the centre of gravity of the pixel positions in C. $\overline{X}$ is also useful in arranging the strokes present in an image from left to right. Let L be the length of the stroke C. The 3 features $\overline{X}$, $\overline{Y}$ and L are normalized with respect to the image height. Thus, the feature vector becomes $(\alpha_1, \alpha_2, \alpha_3, \alpha_4, \alpha_5, \overline{X}, \overline{Y}, L)$.

The features extracted from a horizontal stroke C in S are similar. Here C is traced from west to east. The feature vector of a horizontal stroke C is defined as $(\beta_1, \beta_2, \beta_3, \beta_4, \beta_5, \overline{X}, \overline{Y}, L)$ where $-45^0 \leq \beta_i \leq 45^0$, $\overline{X}, \overline{Y}$ and L are defined in the same way as before [11].

4   Proposed HMM Classifier

An HMM with the state space $S = s_1, \ldots , s_N$ and observation sequence $Q = q_1,$ $\ldots , q_T$ is defined as $\gamma = (\pi, A, B)$ where the initial state distribution is given by $\pi = \{\pi_i\}$, $\pi_i = $ Prob $(q_1 = s_i)$, the state transition probability distribution by A $= \{a_{ij}(t)\}$ where $a_{ij}(t) = $ Prob $(q_{t+1} = s_j/q_t = s_i)$ and the observation symbol probability distributions by $B = \{b_i\}$ where $b_i(O_t)$ is the distribution for state i and O_t is the observation at instant t. The HMM here is non-homogeneous.

Here the problem is how to efficiently compute $P(O/\gamma)$, the probability of the observation sequence, given an observation sequence $O = O_1, \ldots , O_T$ and a model $\gamma = (\pi, A, B)$. For a classifier of m classes of patterns, we denote m different HMMs by γ_j, $j = 1, ..., m$. Let an input pattern X of an unknown class have a sequence O. The probability $P(O/\gamma_j)$ is computed for each model γ_j and X is assigned to class c whose model shows the highest probability. That is,

$$c = arg \max_{1 \leq j \leq m} P(O/\gamma_j) \qquad (1)$$

For a given γ, $P(O/\gamma)$ is computed using the well known forward and backward algorithms [12] . Note that the observation sequence $O = O_1, \ldots , O_T$ in our problem is the sequence of feature vectors of the strokes (arranged from left to right) that are present in a handwritten word image. T is the number of strokes in the image. The states here are certain feature primitives (or more specifically, individual 8-dimensional Gaussian distributions in the feature space) that are found below using EM algorithm.

4.1   HMM Parameters

A feature vector $\theta = (\theta_1, \theta_2, \theta_3, \theta_4, \theta_5, \overline{X}, \overline{Y}, L)$ can come either from a vertical or a horizontal stroke. It is assumed that the features follow a multivariate Gaussian mixture distribution. In other words, $\theta = (\theta_1, \theta_2, \theta_3, \theta_4, \theta_5, \overline{X}, \overline{Y}, L)$ has a distribution $f(\theta)$ which is a mixture of K 8-dimensional Gaussian distributions, namely,

$$f(\theta) = \sum_{k=1}^{K} P_k f_k(\theta) \qquad (2)$$

where

$$f_k(\theta) = exp\{-0.5(\theta - \mu_k)^T \Sigma_k^{-1}(\theta - \mu_k)\}/\{(2\pi)^{8/2}|\Sigma_k|^{1/2}\} \qquad (3)$$

and P_k is the prior probability of the k-th component. The unknown parameters of the mixture distribution, namely, $P_k, \mu_k, \Sigma_k, (k = 1, ..., K)$ are estimated using the EM (Expectation Maximization) algorithm ([13],[14]) that maximizes the log likelihood of the training vectors $\{\theta_i, i = 1, ..., n\}$ coming from the distribution given by $f(\theta)$.

The state space of the proposed HMM consists of states which are characterized by the probability density functions $f_k(\theta)$ $(k = 1, ..., K)$. It is assumed that the vertical and horizontal strokes in the word image database are distributed around K different prototype strokes. These are called <u>stroke primitives</u> corresponding to the mean shape vectors $\mu_1, \mu_2, ..., \mu_k$. These K stroke primitives constitute the state space. Thus, the states here are not defined a priori but are determined adaptively on the basis of the training set of word images.

To determine the optimum values of K, we use the Bayesian information criterion (BIC) which is defined as $BIC(K) = -2LL + mlog(n)$, for a Gaussian mixture model with K components, LL is the log likelihood value, m is the number of independent parameters to be estimated, n is the number of observations. For several K values, the $BIC(K)$ values are computed. The first local minimum indicates the optimum K value.

4.2   Estimation of HMM Parameters

In our implementation, $N = K$ and observation symbol probability distribution $b_i(O_t)$ is, in fact, the Gaussian distribution $f_i(\theta) = N(\mu_i, \Sigma_i)$. Thus

$$b_i(O_t) = exp\{-0.5(O_t - \mu_i)^T \Sigma_i^{-1}(O_t - \mu_i)\}/\{(2\pi)^{8/2}|\Sigma_i|^{1/2}\} \qquad (4)$$

The parameters produced by EM algorithm are $P_1, P_2, ..., P_N, \mu_1, \mu_2, ..., \mu_N, \Sigma_1, \Sigma_2, ..., \Sigma_N$. Let, in a word image, the strokes be arranged from left to right on the basis of $\overline{X}$ to generate the observation sequence $O_1, O_2, ... , O_T$. For each O_t, compute

$$h_i(O_t) = p_i b_i(O_t)/\{\sum_{j=1}^{N} p_j b_j(O_t)\} \qquad (5)$$

and O_t is assigned to state k where

$$k = arg \max_{1 \le i \le N} h_i(O_t). \qquad (6)$$

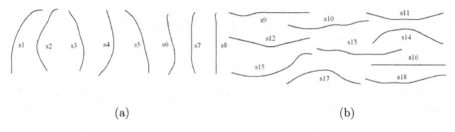

(a)   (b)

Fig. 4. Stroke primitives for (a) vertical and (b) horizontal strokes for Devanagari word Ooty

This assignment to respective states is done for all L observation sequences (L is the number of training images). From these L state sequences, the estimates of the initial probabilities are computed as ($1 \leq i \leq N$) $\pi_i = $ (*number of occurrences of* $\{q_1 \in s_i\}$)/(*total number of occurrences of* $q1$).

The transition probability estimates $a_{i,j}(t)$ are computed as ($1 \leq i \leq N, 1 \leq j \leq N, 1 \leq t \leq T-1$) (*number of occurrences of* $\{q_t \in s_i \& q_{t+1} \in s_j\}$) / (*total number of occurrences of* $\{q_t \in s_i\}$). The above HMM parameter estimates are fine-tuned using re-estimation by Baum-Welch forward-backward algorithm.

5   Experimental Results

The proposed scheme has been tested on the recently developed database of handwritten Devanagari word images. The results of our study are reported below. To the best of our knowledge, there does not exist any other standard database of handwritten Devanagari word images. The training and test datasets here consist of 7000 and 3000 handwritten word images respectively. From these word images, 156906 horizontal and 137250 vertical strokes have been extracted from the training set whereas 67245 horizontal and 58821 vertical strokes have been extracted from the test set. The parameters of an HMM for each of the 50 word classes are determined using the method described in Section 4.

For example, for word class **Ooty**, the K value is found to be 18. The curves corresponding to the 18 mean vectors μ_k are shown in Fig. 4. These represent the 18 HMM states for **Ooty**. For the image shown in Figs. 3(c) & 3(d), the strokes are shown in Fig. 5. The strokes arranged in terms of $\overline{X}$ from left to right are e1, e2, e3, ..., e26. The most likely states of these 26 strokes individually are s15, s4, s13, s17, s7, s15, s17, s6, s4, s16, s17, s15, s3, s15, s2, s14, s15, s4, s1, s14, s16, s3, s18, s6, s16 and s3 respectively. The probability $P(O/\gamma_j)$ is computed

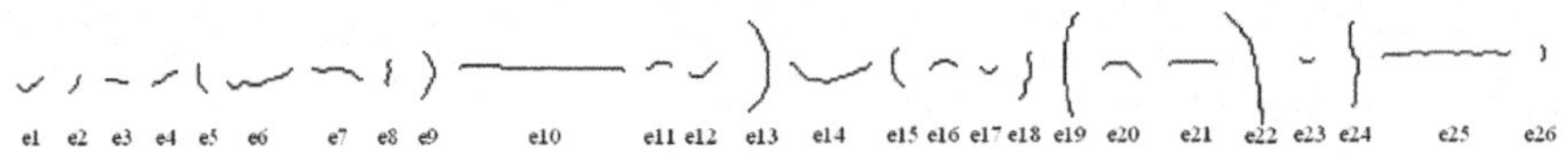

Fig. 5. e1 to e26 represent the strokes arranged from left to right along X-axis

for $j = 1, ..., 50$ and the image is classified as class c where

$$c = arg \max_{1 \leq j \leq 50} P(O/\gamma_j) \tag{7}$$

We have achieved 82.89% correct recognition rate on the test set and 87.71% on the training set.

6   Conclusion

In this paper we have proposed a HMM based approach to recognition of handwritten Devanagari words. The results of our approach are promising for a small Lexicon size. It indicates that it is possible to scale our system to large vocabularies. Our system is based on Global approach, which extracts global features thus reducing the overhead of segmentation. Our future work will be to combine both these local and global approaches resulting in a hybrid approach for more efficiency.

References

1. Rahman, A.F.R., Rahman, R., Fairhurst, M.C.: Recognition of handwritten Bengali characters: a novel multistage approach. Pattern Recognition 35, 997–1006 (2002)
2. Bhattacharya, U., Das, T.K., Datta, A., Parui, S.K., Chaudhuri, B.B.: A hybrid scheme for handprinted numeral recognition based on a self-organizing network and MLP classifiers. Int. J. Patt. Recog. & Art. Intell. 16(7), 845–864 (2002)
3. Bhattacharya, U., Chaudhuri, B.B.: A majority voting scheme for multiresolution recognition of handprinted numerals. In: Proc. of the 7th ICDAR, Edinburgh, Scotland, vol. I, pp. 16–20 (2003)
4. Ramakrishnan, K.R., Srinivasan, S.H., Bhagavathy, S.: The independent components of characters are 'Strokes'. In: Proc. of the 5th ICDAR, pp. 414–417 (1999)
5. Lehal, G.S., Bhatt, N.: A recognition system for Devnagri and English handwritten numerals. Advances in Multimodal Interfaces. In: Tan, T., Shi, Y., Gao, W. (eds.) ICMI 2001. LNCS, vol. 1948, pp. 442–449. Springer, Heidelberg (2000)
6. Kim, G.: Recognition of offline handwritten words and its extension to phrase recognition, PhD Thesis. University of New York at Buffalo, USA (1996)
7. Guillevic, D.: Unconstrained Handwriting Recognition Applied to the Processing of Bank Cheques, Thesis of Doctor's Degree in the Department of Computer Science at Concordia University, Canada (1995)
8. Park, H., Lee, S.: Off-line recognition of large-set handwritten characters with multiple hidden Markov models. Pattern Recognition 29, 231–244 (1996)
9. Hull, J.J.: A database for handwritten text recognition research. IEEE Trans. Patt. Anal. Mach.Intel. 16, 550–554 (1994)
10. Otsu, N.: A threshold selection method from gray-level histograms. IEEE Trans. on Systems, Man, and Cybernetics 9, 62–66 (1979)
11. Bhattacharya, U., Parui, S.K., Shaw, B., Bhattacharya, K.: Neural Combination of ANN and HMM for Handwritten Devnagari Numeral Recognition. 10th-IWFHR, pp. 613–618 (2006)
12. Rabiner, L.R.: A tutorial on hidden Markov models and selected applications in speech recognition. Proceedings of the IEEE 77(2), 257–286 (1989)
13. Fukunaga, K.: Introduction to Statistical Pattern Recognition, 2nd edn. Academic Press, San Diego (1980)
14. Parui, S. K., Bhattacharya, U., Shaw, B., Poddar, D.: A Novel Hidden Markov Models for Recognition of Bangla Characters. 3rd-WCVGIP, pp. 174–179 (2006)

HMM Parameter Estimation with Genetic Algorithm for Handwritten Word Recognition

Tapan K. Bhowmik[1], Swapan K. Parui[2], Manika Kar[3], and Utpal Roy[4]

[1] IBM India Pvt Ltd, BCS Building, Salt Lake, Kolkata - 700091, India
tkbhowmik@gmail.com
[2] Computer Vision and Pattern Recognition Unit, Indian Statistical Institute,
Kolkata-700108, India
swapan@isical.ac.in
[3] Cognizant Technology Solutions, Salt Lake, Kolkata - 700091, India
manika.kar@cognizant.com
[4] Dept. of Computer and System Sciences, Visva-Bharati, Santiniketan, India
roy.utpal@gmail.com

Abstract. This paper presents a recognition system for isolated handwritten Bangla words, with a fixed lexicon, using a Hidden Markov Model (HMM). A stochastic search method, namely, Genetic Algorithm (GA) is used to train the HMM. The HMM is a left-right HMM. For feature extraction, the image boundary is traced both in the anticlockwise and clockwise directions and the significant changes in direction along the boundary are noted. Certain features defined on the basis of these changes are used in the proposed model.

1 Introduction

Off-line handwritten word recognition is the transcription of handwritten data into a symbolic (ASCII) electronic format. It has several applications such as reading addresses on mail pieces [1], reading amounts on bank checks [2], extracting census data on forms, reading address blocks on tax forms etc. There are two major approaches to recognition of a handwritten word image: analytical approach [3] and holistic approach [4]. The idea of analytical approach is to recognize the input word image as a series of segmented sub-images, called primitives. Holistic approach, on the other hand, considers the word image as a single, indivisible entity, and attempts to recognize the word from its over all shape. The present work deals with handwritten word recognition in Bangla with a holistic approach. To the best of our knowledge, no such work is reported on Bangla handwritten word recognition. We consider a set of 117 town names in West Bengal. For feature extraction, the contour of a word image is traced both in clockwise and anticlockwise directions and the points with directional changes are observed. The sequence of such change points represents a basic shape of the word image. Such change points are encoded into certain change codes. These change codes along with their position define the feature vector. A holistic approach based on discrete hidden Markov model (HMM) is used

A. Ghosh, R.K. De, and S.K. Pal (Eds.): PReMI 2007, LNCS 4815, pp. 536–544, 2007.

as the recognition engine here. One HMM is constructed for each word class. The Baum-Welch re-estimation method has traditionally been the first choice for training such an HMM. Yet a problem of over fitting on training samples may arise with this method. To resolve this problem to some extent, genetic algorithm (GA) has been used for optimizing the parameters of HMM.

2   HMM for Word Recognition

An HMM consists of three sets of parameters $\pi = \{\pi_i\}$, $A = \{a_{ij}\}$ and $B = \{b_{jk}\}$, $1 \leq i, j \leq N$, $1 \leq k \leq M$, where π is the initial state probability distribution, A is the state transition probability distribution matrix and B is the observation symbol probability distribution matrix. Here N is the number of states in the model and M is the total number of distinct observation symbols per state. The elements of the matrices $\{a_{ij}\}$ and $\{b_{jk}\}$ always satisfy the following conditions:

$$\sum_{j=1}^{N} a_{ij} = 1, \text{ where } i = 1, 2, \ldots, N \tag{1}$$

$$\sum_{k=1}^{M} b_{jk} = 1, \text{ where } j = 1, 2, \ldots, N \tag{2}$$

The complete notation of HMM is $\lambda = (\pi, A, B)$. Each handwritten word is represented by a sequence of observations $O = O_1, O_2, \ldots, O_T$ where O_t is the observation symbol observed at time t. The isolated word recognition problem can then be regarded as that of computing $\arg\max_i \{P(w_i|O)\}$, where w_i is the ith handwritten word. Now using Bayes' rule:

$$P(w_i|O) = \frac{P(O|w_i)P(w_i)}{P(O)} = \frac{P(O|w_i)P(w_i)}{\sum_j P(O|w_j)P(w_j)}.$$

Thus, for a given set of prior probabilities $P(w_i)$, the most probable word depends only on the likelihood $P(O|w_i)P(w_i)$. However, for a given O the direct estimation of the joint conditional probability $P(O_1, O_2, \ldots, O_T|w_i)$ is not practicable. Instead, for each w_i, we consider $P(O|w_i) = P(O|\lambda_i)$, where λ_i is the HMM corresponding to word w_i. Thus, in the classification stage, given an unknown input sequence, $O = O_1, O_2, \ldots, O_T$ the probability $P(O|\lambda_i)$ is computed for each model λ_i, and O is classified in that class whose model shows the highest likelihood $P(O|\lambda_i)P(\lambda_i)$.

3   Feature Extraction

Feature extraction is one of the key modules for any recognition system. Several features may be extracted from a word image. But the structural feature is the most effective in a problem like the present one. Here the structural feature is extracted as the outer and inner boundaries of a word image traced in both anticlockwise and clockwise directions starting from any boundary point. Suppose such a boundary is represented as $d_1, d_2, \ldots, d_n$, where $d_i \in \{1, 2, \ldots, 8\}$ is a directional code is shown in Fig. 1. The original image is sufficiently smoothed so that the $e_i = d_{i+1} - d_i \pmod{8}$ is +1 or 0 or -1. Now our goal is to determine pixels where there is a change in direction along the boundary and the type of change. Even in a digital straight line, there may be two different directional codes and hence a change in direction. In order to avoid such spurious changes and find the genuine changes in direction, we do the following. Note that in a digital straight line, e_i is always zero except in cases where two consecutive e_i values are either (+1, -1) or (-1, +1). Let $E_j = \sum_{i=1}^{j} e_i$. For any digital straight line, E_j is always zero except in cases where it is either -1 or +1 surrounded by zeros. For a genuine change in direction, the value of E_j will be non-zero for at least two consecutive pixels. Thus, when for the first time both the values of E_j and E_{j+1} are non-zero, we say that there is a change in direction at $(j + 1)^{st}$ pixel. This information of change in direction is encoded as (D, X, Y), where the D represents two consecutive chain codes (Fig. 2) and the coordinates of the $(j + 1)^{st}$ pixel (called a feature point) are given by (X, Y). After a change in direction is encountered, the value of E_j is initialized to zero. This process continues until the starting pixel is reached again. This is repeated for both inner and outer boundaries of the image in both clockwise and anticlockwise directions. All types of possible change codes encountered during traversing the image boundary and their corresponding directional codes are shown in Fig. 2. Note that the change in direction given by the two pairs of directional codes

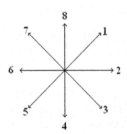

Fig. 1. 8-directional chain codes

(5,4) and (8,1) is the same except that one is obtained from clockwise traversal and the other from the anticlockwise traversal. Thus, these two pairs should result in the same change code. Therefore, all possible changes in direction can be represented by 8 change codes given in Fig. 2. The matrix representation of all possible pairs of directional codes and their corresponding change codes are shown in the Fig. 3. Hence the final feature is (O, X, Y) where $f : D \rightarrow O$, where O is a change code and f is the encoding given by the matrix $(c_{ij})_{8 \times 8}$

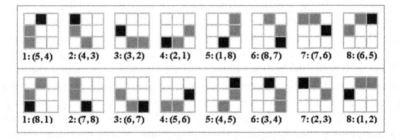

Fig. 2. Change codes corresponding to pairs of directional codes

$$(c_{ij})_{8\times8} = \begin{pmatrix} 0 & 8 & 0 & 0 & 0 & 0 & 0 & 5 \\ 4 & 0 & 7 & 0 & 0 & 0 & 0 & 0 \\ 0 & 3 & 0 & 6 & 0 & 0 & 0 & 0 \\ 0 & 0 & 2 & 0 & 5 & 0 & 0 & 0 \\ 0 & 0 & 0 & 1 & 0 & 4 & 0 & 0 \\ 0 & 0 & 0 & 0 & 8 & 0 & 3 & 0 \\ 0 & 0 & 0 & 0 & 0 & 7 & 0 & 2 \\ 1 & 0 & 0 & 0 & 0 & 0 & 6 & 0 \end{pmatrix}$$

Fig. 3. Matrix representation of change codes

(Fig. 3) and (X, Y) is the positional information associated with D. The contour representations of a word image *"DIGHA"* as well as the observed feature points when traversed in anticlockwise and clockwise directions are shown in Fig. 4.

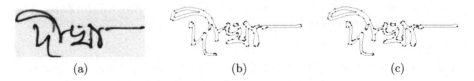

        (a)   (b)   (c)

Fig. 4. (a) Original image (b) Feature points extracted in anticlockwise direction (c) Feature points extracted in clockwise direction

4 HMM State Definition

First, we determine the middle zone of the binary word image by analyzing the horizontal histogram. The histogram is drawn based on the following measurement. Let

$$H_i = \begin{cases} (row_i - \frac{1}{R}\sum row_j), & \text{if } (row_i - \frac{1}{R}\sum row_j) > 0 \\ 0, & \text{otherwise.} \end{cases}$$

where row_i = *No of object pixels in the ith row of the word image*, and R = *No. of object pixel rows of the word image*. The min and max indices of positive H_i give the middle zone boundaries of the word image. $H_i > 0$ indicates that the ith row is significant. The first significant row from the top and the first significant row from the bottom define the middle zone. To define the states of our proposed HMM, the significant vertical strokes [5] are extracted from the word image. A vertical stroke is a connected component of the set of object pixels whose left neighbors are in the background [5]. The vertical strokes whose number of pixels is greater than (height of the middle zone * 0.40) are called significant. On the basis of the significant vertical strokes, which are associated with the middle zone, the whole word image is partitioned into several segments as follows. There is a certain segmenting point between each pair of two consecutive vertical strokes. The segmenting point is determined based on the minimum number of object pixels in the vertical histogram of the middle zone between the two vertical strokes. The vertical strokes and their corresponding segmenting points of a few word images from our database are shown in Fig. 5. The vertical strokes are

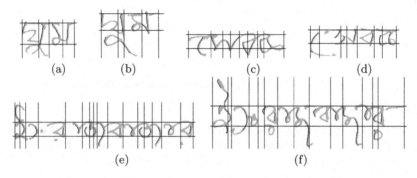

Fig. 5. Vertical strokes and their corresponding segmenting points for Bangla word images:(a), (b) *"GHUM"*; (c), (d) *"SEBAK"* and (e), (f) *"ENGRAJBAZAR"*. The segmenting points are indicated by the vertical straight lines.

arranged from left to right on the basis of $\overline{X}$, where $(\overline{X}, \overline{Y})$ is the centre of gravity of the vertical stroke. Sometimes two consecutive vertical strokes occur very close to each other with one being nearly on the top of the other. In such cases, the larger of the two strokes is retained and the other removed. The strokes are said to be very close if $|\overline{X}_j - \overline{X}_{j-1}| \leq T$, where T is the thickness of the writing pen and $\overline{X}_j$ is the X-coordinate of the center of gravity of the vertical stroke j. In this study, any three consecutive segments are considered as a single state. Starting from left to right the first state is constructed starting from the first segment. For example, the first state is the triplet consisting of segments 1, 2, 3; the second state is the triplet consisting of the segments 2, 3, 4 etc. In Fig 5(a), the number of segments is 4, so the number of states is (4-3+1) = 2. On the other hand, the number of states in Fig 5(c) is (8-3+1) = 6. In case the number of segments is found to be less than 3, the whole word image is considered as a single state. The distribution of the number of segments for one of the most lengthy words *"ENGRAJBAZAR"* from our database is shown in Fig 6 where it is seen that the maximum number of segments is found to be 16.

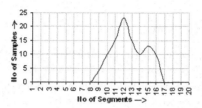

Fig. 6.

So for constructing the HMM for *"ENGRAJBAZAR"*, the number of states is considered $(16-3+1) = 14$. In case the number of segments is more than 16, the last segments are included in the last state (14th state). Similarly, the number of states in the HMM of the word *"GHUM"* is $(6-3+1) = 4$.

5 Genetic Algorithm

Genetic algorithms are an optimization technique [6]. They have wide use in classification problems for determining the optimal class boundaries, the optimal density parameters etc. The GA is also used in HMM based classification problems [7] for estimating the HMM parameters where a certain fitness function is optimized. The fitness function is always chosen in such a way that the most optimum value of the fitness function gives the best possible HMM in which the input observation sequence most closely fits the HMM model. The basic steps involved in an iteration of GA are described in Table 1. Here, L is the population

Table 1. Basic Steps of Genetic Algorithm

Gen	f	Prob.	Selection	Pairing	Crossover	Mutation
$s_1 = 100\ldots$	$f_1 = f(s_1)$	$p_1 = \frac{f_1}{F}$	s_1^*	(t_1, t_2)	t_1^*	s_1^{**}
$s_2 = 001\ldots$	$f_2 = f(s_2)$	$p_2 = \frac{f_2}{F}$	s_2^*	(t_3, t_4)	t_2^*	s_2^{**}
$\vdots$	$\vdots$	$\vdots$	$\vdots$	$\vdots$	$\vdots$	$\vdots$
$s_L = 101\ldots$	$f_L = f(s_L)$	$p_L = \frac{f_L}{F}$	s_L^*	(t_{L-1}, t_L)	t_L^*	s_L^{**}
	$F = \sum\limits_{i=1}^{L} f_i$	$\sum\limits_{i=1}^{L} p_i = 1$				

size, s_i, s_i^*, t_i, t_i^*, s_i^{**} are the chromosomes at different stages and f is the fitness function. There are two aspects in any optimization algorithm, namely, exploration and exploitation. The crossover operation is responsible for exploitation while the mutation operation is responsible for exploration in the search space.

6   HMM Training with GA

Apart from crossover and mutation operations there are two important components in GA : encoding the chromosome and a fitness function.

6.1   Encoding Mechanism

To construct a chromosome, all the parameters of an HMM are arranged in a sequence. In this study, we have used a left-right HMM model with one order jump. A 4-state left-right HMM model with one order jump is shown in Fig 7(a). Note that the initial state distribution here is fixed as $\pi = \{1, 0, 0, \ldots\}$. Thus π is not included in the search space of GA. The observation symbol set here is $\{1, 2, \ldots 8\}$ which is the set of change codes shown in Fig 2. Hence, the number of distinct observation symbols is 8, but the number of states varies from one HMM to another. For example, the number of states of the HMM for "$GHUM$" word image is 4. So the state transition probability distribution is given by $A = \{a_{ij}\}$ in which there are only 9 $a_{ij} > 0$, and the other a_{ij} parameters are always 0 (Fig 7(b)).

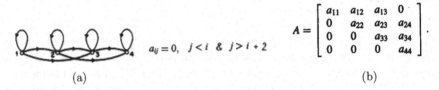

$$a_{ij} = 0, \ j < i \ \& \ j > i + 2 \qquad A = \begin{bmatrix} a_{11} & a_{12} & a_{13} & 0 \\ 0 & a_{22} & a_{23} & a_{24} \\ 0 & 0 & a_{33} & a_{34} \\ 0 & 0 & 0 & a_{44} \end{bmatrix}.$$

(a)   (b)

Fig. 7. (a) LR-HMM model, (b) State transition probability distribution matrix

The observation symbol distribution probability matrix $B = \{b_{jk}\}$ includes 4 x 8 = 32 parameters where b_{jk} is the probability that the kth symbol occurs in the jth state. The total number of parameters of the HMM $\lambda = (\pi, A, B)$ for "$GHUM$" word image is $9 + 32 = 41$. The training of an HMM involves searching for the best values of these 41 parameters each of which ranges from 0 to 1. So, in the GA-HMM training, the chromosome consists of two parts, **A** and **B**, one for matrix $A = \{a_{ij}\}$ and the other for matrix $B = \{b_{jk}\}$. We encode one chromosome ("$GHUM$" word image HMM) as 41 real numbers. The genetic representation of the HMM model for the "$GHUM$" word image is shown in Fig 8.

A						B					
a_{11}	a_{12}	a_{13}	a_{22}	$\cdots$	a_{44}	b_{11}	b_{12}	$\cdots$	b_{18}	b_{21}	$\cdots$	b_{48}

Fig. 8. Genetic representation of HMM for word "$GHUM$"

6.2   Fitness Function

The fitness function is one of the most crucial components in GA. The objective of the re-estimation process is to find the best possible HMM, which the input observation sequence fits closely. For a given model λ, $P(O|\lambda)$ measures how far the input observation sequence $O = O_1, O_2, \ldots$ fits the model. A greater value of $P(O|\lambda)$ gives a closer fit. From this point of view, $P(O|\lambda)$ has been used in the fitness function. Note that L is the number of solutions in the population and $\lambda^{(i)}$ is the ith solution of λ. Then the fitness function is defined as $f_i = \frac{q_i}{\sum\limits_{l=1}^{L} q_l}$, where $q_i = \sum P(O/\lambda^{(i)})$ is the sum of probabilities of all the training observations in a word class for a given $\lambda^{(i)}$. $P(O|\lambda^{(i)})$ has been calculated by the well-known forward algorithm [8].

6.3   Initial Gene Production

For every sample image in a particular word class, a set of feature vectors of the form (O, X, Y) is extracted, where O is a change code and (X, Y) is the position of the corresponding feature point. Let this set be (O_i, X_i, Y_i), $i = 1, 2, \ldots T$. Suppose the number of states is N. Each O_i will belong to one state or two states or three states depending on (X_i, Y_i). For example, if (X_i, Y_i) falls in segment 2, then O_i will belong to both the states 1 and 2. This is repeated for all the training images of a particular word class. For each state, b_{jk} is computed as the relative frequency of the kth change code in the jth state. $B = \{b_{jk}\}$ is the initial estimate of the observation symbol probability distribution, which will be incorporated in each of L chromosomes. Now, no initial estimates are made for $A = \{a_{ij}\}$. Random numbers are generated to compute the initial values of a_{ij}'s such that $\sum\limits_{j} a_{ij} = 1$, for all i. In this experiment, population size L has been taken to be 20.

6.4   Crossover and Mutation

For GA-HMM training, roulette wheel selection scheme has been used to select the chromosome. After selection is over, pairs of chromosomes are picked up randomly from the population to be subjected to crossover. In this study double point crossover mechanism is used. Two points are chosen randomly, one is from part **A** and the other is from part **B**. The portions of two chromosomes between two selected crossover points are exchanged to reproduce two new chromosomes. For mutation operation one chromosome is chosen randomly from the population. From it, two bits, one from part **A** and the other from part **B**, are selected randomly to be flipped. Both the crossover and mutation operations are controlled by two probabilities: crossover probability ρ_c and mutation probability ρ_m. For training an HMM we have used $\rho_c = 0.7$ and $\rho_m = 0.01$. It should be noted that after crossover or mutation, all the HMM parameters of the population are normalized to satisfy the equations 1 and 2.

7   Experimental Results

The experiment has been carried out on a recently developed database of handwritten Bangla words. We have considered 117 town names in West Bengal. Our database contains 14,625 handwritten word samples from 125 writers. Each writer has written only one sample word for each town name. Then it was randomly divided into training and test sets. The training set consists of 11700 word image samples while the test set consists of 2925 word image samples. The recognition accuracy found on test set varies between 71% to 86%.

8   Conclusion

The proposed method is quite general in the sense that it can be applied to other scripts also. Only the state definition is to be changed. Another aspect is that no size normalization is necessary in the proposed approach. Also, it is robust with respect to minor rotations of the input image.

References

1. Dehghan, M., Faez, K., Ahmadi, M., Shridhar, M.: Unconstrained farsi (arabic) handwritten word recognition using fuzzy vector quantization and hidden markov models. Pattern Recognition Letters 22, 209–214 (2001)
2. Guillevic, D., Suen, C.: Hmm word recognition engine. In: Proceedings of the Int. Conf. on Document Analysis and Recognition, pp. 544–547 (1997)
3. Govindaraju, V., Kim, G.: A lexicon driven approach to handwritten word recognition for real-time applications. IEEE Trans. Pattern Analysis Machine Intelligence 19(4), 366–379 (1997)
4. Madhvanath, S., Govindaraju, V.: The role of holistic paradigms in handwritten word recognition. IEEE Trans. Pattern Analysis Machine Intelligence 23(2), 149–164 (2001)
5. Bhowmik, T., Bhattacharya, U., Parui, S.: Recognition of bangla handwritten characters using an mlp classifier based on stroke features. In: Pal, N.R., Kasabov, N., Mudi, R.K., Pal, S., Parui, S.K. (eds.) ICONIP 2004. LNCS, vol. 3316, pp. 814–819. Springer, Heidelberg (2004)
6. Goldberg, D.: Genetic algorithms in search optimization and machine learning. Addison-Wesley, Reading (1989)
7. Won, K., Prugel-Bennett, A., Krogh, A.: Training hmm structure with genetic algorithm for biological sequence analysis. Bioinformatics 20(18), 3613–3619 (2004)
8. Rabiner, L.: A tutorial on hidden markov models and selected applications in speech recognition. Trans. on. IEEE 77(2), 257–286 (1989)

A Hidden Markov Model Based Named Entity Recognition System: Bengali and Hindi as Case Studies

Asif Ekbal and Sivaji Bandyopadhyay

Computer Science and Engineering Department, Jadavpur University, Kolkata, India
asif.ekbal@gmail.com, sivaji_cse_ju@yahoo.com

Abstract. Named Entity Recognition (NER) has an important role in almost all Natural Language Processing (NLP) application areas including information retrieval, machine translation, question-answering system, automatic summarization etc. This paper reports about the development of a statistical Hidden Markov Model (HMM) based NER system. The system is initially developed for Bengali using a tagged Bengali news corpus, developed from the archive of a leading Bengali newspaper available in the web. The system is trained with a training corpus of 150,000 wordforms, initially tagged with a HMM based part of speech (POS) tagger. Evaluation results of the 10-fold cross validation test yield an average Recall, Precision and F-Score values of 90.2%, 79.48% and 84.5%, respectively. This HMM based NER system is then trained and tested on the Hindi data to show its effectiveness towards the language independent abilities. Experimental results of the 10-fold cross validation test has demonstrated the average Recall, Precision and F-Score values of 82.5%, 74.6% and 78.35%, respectively with 27,151 Hindi wordforms.

Keywords: Named Entity (NE), Named Entity Recognition (NER), Hidden Markov Model (HMM), Named Entity Recognition in Bengali.

1   Introduction

Named Entity Recognition is an important tool in almost all NLP application areas. The objective of named entity recognition is to identify and classify every word/term in a document into some predefined categories like person name, location name, organization name, miscellaneous name (date, time, percentage and monetary expressions) and "none-of-the-above". The challenge in detection of named entities is that such expressions are hard to analyze using traditional NLP because they belong to the open class of expressions, i.e., there is an infinite variety and new expressions are constantly being invented.

During the last decade, NER has drawn more and more attention from the NE tasks [1] [2] in Message Understanding Conferences (MUCs) [MUC6; MUC7]. This reflects the importance of NER in information extraction. The problem of correct identification of named entities is specifically addressed and benchmarked

A. Ghosh, R.K. De, and S.K. Pal (Eds.): PReMI 2007, LNCS 4815, pp. 545–552, 2007.

by the developers of Information Extraction System, such as the GATE system [3]. NER also finds application in question-answering [4] systems and machine translation [5].

The current trend in NER is to use the machine learning (ML) approach, which is more attractive in that it is trainable and adoptable and the maintenance of a ML system is much cheaper than that of a rule-based one. Rule-based approaches lack the ability of coping with the problems of robustness and portability. Each new source of text requires significant tweaking of rules to maintain optimal performance and the maintenance costs could be quite steep. The representative machine-learning approaches used in NER are HMM (BBN's IdentiFinder in [6]), Maximum Entropy (New York University's MENE in [7]) and Conditional Random Fields [8].

Among the machine learning approaches, the evaluation performance of the HMM is quite impressive. The main reason may be due to its better ability of capturing the locality of phenomena, which indicates names in text. Moreover, HMM seems more and more used in NER because of the efficiency of the Viterbi algorithm [9] used in decoding the NE-class state sequences. Zhou and Su [10] reported state of the art results on the MUC-6 and MUC-7 data using an HMM-based tagger. However, the performance of a ML system is always poorer than that of a rule-based one. This may be because current ML approaches capture important evidences behind NER problem much less effectively than human experts who handcraft the rules, although machine learning approaches always provide important statistical information that is not available to human experts.

All the works, carried out already in the area of NER, are in non-Indian languages. In Indian languages, particularly in Bengali, the works in the area of named entity recognition can be found in [11] [12]. Other than Bengali, the work on NER can be found in [13] for Hindi. Here, in this paper, an HMM based NER system has been reported that outperforms the systems developed in [11] [12].

The paper is organized as follows. The NER system has been described in Section 2. Experimental results with the 10-fold cross validation tests in terms of three evaluation parameters, Precision, Recall and F-Score, are reported in Section 3 for Bengali and Hindi. Finally, Section 4 concludes the paper.

2   Named Entity Recognition in Bengali and Hindi

Bengali is one of the widely used languages all over the world. It is the seventh popular language in the world, second in India and the national language of Bangladesh. Hindi is the national language of India. Named Entity (NE) identification in Indian languages (ILs) in general is difficult and challenging as there is no concept of capitalization in ILs. A tagged Bengali news corpus, developed from the archive of a widely read Bengali news paper available in the web, has been used in this work for NER in Bengali. At present the corpus contains around 34 million wordforms in ISCII (Indian Standard Code for Information

Interchange) and UTF-8 format. The location, reporter, agency and different date tags in the tagged corpus help to identify some NEs that appear in the fixed places of the newspaper. The training corpus for Hindi was obtained from the SPSAL 20007 Contest [1]. An HMM based named entity tagger has been used in this work to identify named entities in Bengali/Hindi and classify them into person, location, organization and miscellaneous names. Miscellaneous names include date, time, percentage and monetary expressions. To apply HMM in named entity tagging task the NE tags, as shown in Table 1, are defined.

Table 1. Named Entity Tag Set

NE Tag	Meaning	Example
PER	Single-word person name	*sachin*/PER
LOC	Single-word location name	*jadavpur*/LOC
ORG	Single-word organization name	*infosys*/ORG
MISC	Single-word miscellaneous name	*100%*/MISC
B-PER	Beginning, Internal or the	*sachin*/B-PER
I-PER	End of a multi-word	*ramesh*/I-PER
E-PER	person name	*tendulkar*/E-PER
B-LOC	Beginning, Internal or the	*mahatma*/B-LOC
I-LOC	End of a multi-word	*gandhi*/I-LOC
E-LOC	location name	*road*/E-LOC
B-ORG	Beginning, Internal or the	*bhaba*/B-ORG
I-ORG	End of a multi-word	*atomic*/I-ORG *research*/I-ORG
E-ORG	orgnization name	*centre*/E-ORG
B-MISC	Beginning, Internal or the	*10 e*/B-MISC
I-MISC	End of a multi-word	*magh*/I-MISC
E-MISC	miscellaneous name	*1402*/E-MISC
NNE	Not Named Entity	*neta*/NNE, *bidhansabha*/NNE

2.1 Hidden Markov Model Based Named Entity Tagging

The goal of NER is to find a stochastic optimal tag sequence $T = t_1, t_2, t_3, \ldots, t_n$ for a given word sequence $W = w_1, w_2, w_3 \ldots, w_n$. Generally, the most probable tag sequence is assigned to each sentence following the Viterbi algorithm [9]. The tagging problem becomes equivalent to searching for $argmax_T P(T) * P(W|T)$, by the application of Bayes' law ($P(W)$ is constant).

The probability of the NE tag, i.e., $P(T)$ can be calculated by Markov assumption which states that the probability of a tag is dependent only on a small, fixed number of previous NE tags. Here, in this work, a trigram model has been used. So, the probability of a NE tag depends on two previous tags, and then we have,

$$P(T) = P(t_1) \times P(t_2|t_1) \times P(t_3|t_1, t_2) \times P(t_4|t_2, t_3) \times \ldots \times P(t_n|t_{n-2}, t_{n-1})$$

---

[1] http://shiva.iiit.ac.in/SPSAL2007/check_login.php

An additional tag '$\$$' (dummy tag) has been introduced in this work to represent the beginning of a sentence. So, the previous probability equation can be slightly modified as:

$$P(T) = P(t_1|\$) \times P(t_2|\$, t_1) \times P(t_3|t_1, t_2) \times P(t_4|t_2, t_3) \times \ldots \times P(t_n|t_{n-2}, t_{n-1})$$

Due to sparse data problem, the linear interpolation method has been used to smooth the trigram probabilities as follows: $P'(t_n|t_{n-2}, t_{n-1}) = \lambda_1 P(t_n) + \lambda_2 P(t_n|t_{n-1}) + \lambda_3 P(t_n|t_{n-2}, t_{n-1})$ such that the λs sum to 1. The values of λs have been calculated by the following method [14]:

1. set $\lambda_1 = \lambda_2 = \lambda_3 = 0$
2. for each tri-gram (t_1, t_2, t_3) with $freq(t_1, t_2, t_3) > 0$ depending on the maximum of the following three values:
   - case: $\frac{(freq(t_1, t_2, t_3) - 1)}{(freq(t_1, t_2) - 1)}$: increment λ_3 by $freq(t_1, t_2, t_3)$
   - case: $\frac{(freq(t_2, t_3) - 1)}{(freq(t_2) - 1)}$: increment λ_2 by $freq(t_1, t_2, t_3)$
   - case: $\frac{(freq(t_3) - 1)}{(N - 1)}$: increment λ_1 by $freq(t_1, t_2, t_3)$
3. normalize $\lambda_1, \lambda_2, \lambda_3$.

Here, N is the corpus size, i.e., the number of tokens present in the training corpus. If the denominator in one of the expression is 0, then the result of that expression is defined to be 0. The -1 in both the numerator and denominator has been considered for taking unseen data into account.

By making the simplifying assumption that the relation between a word and its tag is independent of context, $P(W|T)$ can be calculated as: $P(W|T) \approx P(w_1|t_1) \times P(w_2|t_2) \times \ldots \times P(w_n|t_n)$.

The emission probabilities in the above equation can be calculated from the training set as, $P(w_i|t_i) = \frac{freq(w_i|t_i)}{freq(t_i)}$.

2.2   Context Dependency

To make the Markov model more powerful, additional context dependent features are introduced to the emission probability in this work. This specifies that the probability of the current word depends on the tag of the previous word and the tag to be assigned to the current word. Now, $P(W|T)$ is calculated by the equation:

$$P(W|T) \approx P(w_1|\$, t_1) \times P(w_2|t_1, t_2) \times \ldots \times P(w_n|t_{n-1}, t_n)$$

So, the emission probability can be calculated as:

$$P(w_i|t_{i-1}, t_i) = \frac{freq(t_{i-1}, t_i, w_i)}{freq(t_{i-1}, t_i)}.$$

Here, also the smoothing technique is applied rather than using the emission probability directly. The emission probability is calculated as:

$P'(w_i|t_{i-1}, t_i) = \theta_1 P(w_i|t_i) + \theta_2 P(w_i|t_{i-1}, t_i)$, where θ_1, θ_2 are two constants such that all θs sum to 1.

The values of θs should be different for different words. But the calculation of θs for every word takes a considerable time and hence θs are calculated for the entire training corpus. In general, the values of θs can be calculated by the same method that is adopted in calculating λs.

2.3   Viterbi Algorithm

The Viterbi algorithm [9] allows us to find the best T in linear time. The idea behind the algorithm is that of all the state sequences, only the most probable of these sequences need to be considered. The trigram model has been used in the present work. The pseudo code of the algorithm is shown bellow.

for $i = 1$ to Number_of_Words_in_Sentence
  for each state $c \in$ Tag_Set
    for each state $b \in$ Tag_Set
      for each state $a \in$ Tag_Set
        for the best state sequence ending in state
        a at time $(i-2)$, b at time $(i-1)$, compute
        the probability of that state sequence going
        to state c at time i.
      end
    end
  end
Determine the most-probable state sequence ending in state c at time i.
end

So if every word can have S possible tags, then the Viterbi algorithm runs in $O(S^3 \times |W|)$ time, or linear time with respect to the length of the sentence.

2.4   Handling the Unknown Words

Handling of unknown words is an important issue in NE tagging. Viterbi algorithm [9] attempts to assign a NE tag to the unknown words. Specifically, suffix features of the words and a lexicon are used to handle the unknown words in Bengali.

For words which have not been seen in the training set, $P(w_i|t_i)$ is estimated based on features of the unknown words, such as whether the word contains a particular suffix. There may be two different kinds of suffixes that could be helpful in predicting the NE classes of the unknown words. The corresponding lists have been prepared. The first category contains the suffixes that could usually appear at the end of different NEs and non-NEs. This list has 435 entries including a null suffix that has been kept for those words that have none of the suffixes in the list. The second category is the set of suffixes that may occur with person names (e.g., -babu, -da, -di etc.) and location names (e.g., -land, -pur, -lia etc.). The person and location lists contain 51 and 46 entries respectively. The probability distribution of a particular suffix with respect to specific tag is generated from all words in the training set that share the same suffix. Two additional features that cover the numbers and symbols are also considered.

To handle the unknown words further, a lexicon [15], which was developed in an unsupervised way from the tagged Bengali news corpus, has been used. Lexicon contains the Bengali root words and their basic part of speech information such as: noun, verb, adjective, adverb, pronoun and indeclinable, excluding NEs. The lexicon has around 100,000 word entries. The heuristic is that *'if an unknown word is found to appear in the lexicon, then most likely it is not a named entity'*.

3   Experimental Results

A portion of the tagged (not NE tagged/POS tagged) news corpus, containing 150,000 wordforms, has been used to train the NER system. The training corpus is initially run through an HMM-based part of speech (POS) tagger [16] to tag the training corpus with the 26 different POS tags [2], defined for the Indian languages. This POS-tagged training set is then manually checked for the correctness. The POS tags representing NEs are replaced by the appropriate NE tags as defined in Table 1, and the rest are replaced by the NNE tags. The training set thus obtained is a corpus tagged with sixteen NE tags and one non-NE tag. In the output, sixteen NE tags are replaced appropriately by the four NE tags, viz., 'Person', 'Location', 'Organization' and 'Miscellaneous'.

The NER system has been evaluated in terms of Recall, Precision and F-Score as defined below:

$$\text{Recall (R)} = \frac{\text{(No. of tagged NEs)}}{\text{(Total no. of NEs present in the test set)}} \times 100\%$$

$$\text{Precision (P)} = \frac{\text{(No. of correctly tagged NEs)}}{\text{(No. of tagged NEs)}} \times 100\%$$

$$\text{F-Score (FS)} = \frac{(2 \times \text{Recall} \times \text{Precision})}{(\text{Recall} + \text{Precision})} \times 100\%$$

The training set is initially distributed into 10 subsets of equal size. In the cross validation test, one subset is withheld for testing while the remaining 9 subsets are used as the training sets. This process is repeated 10 times to yield an average result, which is called the 10-fold cross validation test. The experimental results of the 10-fold cross validation test are reported in Table 2. The NER system has demonstrated an average Recall, Precision and F-Score values of 90.2%, 79.48% and 84.5%, respectively. A close investigation to the experimental results reveals that the precision errors are mostly concerned with the organization names. The lack of robustness of the system to handle the unknown organization names properly might be the possible reason behind the fall in precision of organization names. Unlike person or location names, there is no list of suffixes that could be helpful in predicting the class of the unknown organization names. The other existing Bengali NER systems [11] [12] were also trained and tested with the

---

[2] http://shiva.iiit.ac.in/SPSAL2007/iiit_tagset_guidelines.pdf

same dataset. Evaluation results of these systems demonstrated average F-Score values of 74.5% and 77.9% with the 10-fold cross validation test.

The HMM based NER system is also trained and tested with the Hindi data [3] to show its effectiveness for the language independent nature. The Hindi data was tagged with 26 POS tags. The POS tags representing NEs are replaced appropriately by the NE tags of Table 1, and the rest are replaced by the NNE tags. Evaluation results of the system for the 10-fold cross validation test yield an average Recall, Precision and F-Score values of 82.5%, 74.6% and 78.35%, respectively with a training corpus of 27,151 wordforms. The experimental results are presented in Table 3. The one possible reason behind the poorer performance of the system for Hindi might be the smaller amount of training data in comparison to Bengali. Another reason may be its inability to handle the unknown words as efficiently as Bengali. Unlike Bengali, there are no lists of suffixes or lexicon for Hindi in the system.

Table 2. Results of 10-fold cross validation test for Bengali

Test Set	1	2	3	4	5	6	7	8	9	10	Average
Recall	90.80	90.75	90.63	90.49	90.31	90.25	90.12	89.81	89.72	90.12	90.30
Precision	80.40	80.30	79.15	79.87	79.75	79.52	79.39	78.12	78.27	80.40	79.52
F-Score	85.29	84.98	84.50	84.85	84.70	84.55	84.42	83.56	83.60	84.98	84.50

Table 3. Results of 10-fold cross validation test for Hindi

Test Set	1	2	3	4	5	6	7	8	9	10	Average
Recall	83.10	82.80	82.62	82.91	82.73	82.54	82.31	82.76	82.65	82.58	82.50
Precision	75.34	75.13	74.97	74.81	74.12	73.90	73.73	74.46	74.96	74.58	74.60
F-Score	79.03	78.78	78.61	78.65	78.19	77.98	77.78	77.94	78.16	78.38	78.35

4   Conclusion

In this paper, we have presented a named entity recognizer that uses HMM framework with more contextual information. The system has been evaluated with Bengali and Hindi data. The system uses a HMM based POS tagger for the preparation of training data for Bengali. The evaluation results of 10-fold cross validation test shows that such system has high Recall and good F-Score values for Bengali. The system has also shown good Recall and impressive F-Score values with a relatively smaller Hindi training set. Future works include investigating methods to boost precision of the NER system. Building NER

---

[3] http://shiva.iiit.ac.in/SPSAL2007/check_login.php

systems for Bengali using other statistical techniques like Maximum Entropy Markov Model (MEMM), Conditional Random Fields (CRFs) and analyzing the performance of these systems is another interesting task.

References

1. Chinchor, N.: MUC-6 Named Entity Task Definition (Version 2.1). In: MUC-6, Maryland (1995)
2. Chinchor, N.: MUC-7 Named Entity Task Definition (Version 3.5). In: MUC-7, Fairfax (1998)
3. Cunningham, H.: GATE, a General Architecture for Text Engineering. Computers and the Humanities 36, 223–254 (2002)
4. Moldovan, D., Harabagiu, S., Girju, R., Morarescu, P., Lacatusu, F., Novischi, A., Badulescu, A., Bolohan, O.: LCC Tools for Question Answering. In: Text REtrieval Conference (TREC 2002) (2002)
5. Babych, B., Hartley, A.: Improving Machine Translation Quality with Automatic Named Entity Recognition. In: Proceedings of EAMT/EACL 2003 Workshop on MT and other Language Technology Tools, pp. 1–8 (2003)
6. Bikel, D.M., Schwartz, R.L., Weischedel, R.M.: An Algorithm that Learns What's in a Name. Machine Learning 34(1-3), 211–231 (1999)
7. Borthwick, A.: Maximum Entropy Approach to Named Entity Recognition. PhD thesis, New York University (1999)
8. McCallum, A., Li, W.: Early results for Named Entity Recognition with Conditional Random Fields, Feature Induction and Web-enhanced Lexicons. In: Proceedings of CoNLL (2003)
9. Viterbi, A.J.: Error Bounds for Convolutional Codes and an Asymptotically Optimum Decoding Algorithm. IEEE Transaction on Information Theory 13(2), 260–267 (1967)
10. Zhou, G., Su, J.: Named Entity Recognition using an HMM-based Chunk Tagger. In: Proceedings of ACL, Philadelphia, pp. 473–480 (2002)
11. Ekbal, A., Bandyopadhyay, S.: Pattern Based Bootstrapping Method for Named Entity Recognition. In: Proceedings of ICAPR-2007, Kolkata, India, pp. 349–355 (2007)
12. Ekbal, A., Bandyopadhyay, S.: Lexical Pattern Learning from Corpus Data for Named Entity Recognition. In: Proceedings of 5th International Conference on Natural Language Processing (ICON), Hyderabad, India, pp. 123–128 (2007)
13. Li, W., McCallum, A.: Rapid Development of Hindi Named Entity Recognition using Conditional Random Fields and Feature Induction. ACM Transactions on Asian Language Information Processing (TALIP) 2(3), 290–294 (2003)
14. Brants, T.: TnT a Statistical Parts-of-Speech Tagger. In: Proceedings of the Sixth Conference on Applied Natural Language Processing ANLP-2000, pp. 224–231 (2000)
15. Ekbal, A., Bandyopadhyay, S.: Lexicon Development and POS Tagging using a Tagged Bengali News Corpus. In: Proceedings of the 20th International Florida AI Research Society Conference (FLAIRS-2007), Florida, pp. 261–263 (2007)
16. Ekbal, A., Mondal, S., Bandyopadhyay, S.: POS Tagging using HMM and Rule-based Chunking. In: Proceedings of the IJCAI Workshop on Shallow Parsing for South Asian Languages, Hyderabad, India, pp. 31–34 (2007)

Detecting Misspelled Words in Turkish Text Using Syllable n-gram Frequencies

Rıfat Aşlıyan, Korhan Günel, and Tatyana Yakhno

Dokuz Eylül University, İzmir, Turkey

Abstract. In this study, we have designed and implemented a system which decides whether or not a word is misspelled in Turkish text. Firstly, three databases of syllable monogram, bigram and trigram frequencies are constructed using the syllables that are derived from five different Turkish corpora. Then, the system takes words in Turkish text as an input and computes the probability distribution of words using syllable monogram, bigram and trigram frequencies from the databases. If the probability distribution of a word is zero, it is decided that this word is misspelled. For testing the system, we have constructed two text databases with the same words. One text database has 685 misspelled words. The other has 685 correctly spelled words. The words from these text databases are taken as inputs for the system. The system produces two results for each word: "Correctly spelled word" or "Misspelled word". The system that is designed with monogram and bigram frequencies has 86% success rate for the misspelled words and has 88% success rate for the correctly spelled words. According to the system designed with bigram and trigram frequencies, there is 97% success rate for the misspelled words and there is 98% success rate for the correctly spelled words.

1 Introduction

To detect misspelled words in a text is an old problem. Today, most of word processors include some sort of misspelled word detection. Misspelled word detection is worthy in the area of cryptology, data compression, speech synthesis, speech recognition and optical character recognition [1] [2] [3] [4]. The traditional way of detecting misspelled words is to use a word list, usually also containing some grammatical information, and to look up every word in the word list [6] from dictionary.

The main disadvantage of this approach is that if the dictionary is not large enough, the algorithm will report some of correct words as misspelled, because they are not included in the dictionary. For most natural languages the size of dictionary needed is too large to fit in the working memory of an ordinary computer. In Turkish this is a big problem, because Turkish is an agglutinative language and too many new words can be constructed by adding suffixes.

To overcome this difficulties we have proposed a new approach for detecting misspelled words in Turkish text. We have used Turkish syllable n-gram

A. Ghosh, R.K. De, and S.K. Pal (Eds.): PReMI 2007, LNCS 4815, pp. 553–559, 2007.

frequencies which are generated from several Turkish corpora [5] [7]. From the corpora we have extracted syllable monogram, bigram and trigram frequencies using TASA (Turkish Automatic Spelling Algorithm) [8]. We have used these n-gram frequencies for calculating a word probability distribution. After that the system has decided whether a word is misspelled or not. In this approach we don't need word list. We have only Turkish syllables and their monogram, bigram and trigram frequencies.

The paper is organized as follows. In Section 2, we described the system architecture. We explained how syllable n-gram frequencies and word probability distribution are computed. We discussed the empirical results of the system in Section 3. Future directions and coclusions are described in Section 4.

2   System Architecture

The system consists of three main components. First component is preprocessing which cleans a text. Second component is TASA, and third component is calculating probability distribution of words. As shown in Figure 1, the system takes words in Turkish text as input and gives the result for each word as "Misspelled Word" or "Correctly Spelled Word".

In preprocessing component of the system, punctuation marks are cleaned. All letters in the text are converted to lower case. Blank characters between two successive words are limited with only one blank character.

In second component, TASA [8] takes the Turkish clean text as an input and gives the Turkish syllabified text. The system divides words into syllables putting the dash character between two syllables. For example, the word "kitaplık" in Turkish text is converted into the syllabified word "ki-tap-lık" in Turkish syllabified text.

In third component, the probability distribution is calculated for each syllabified word. The system uses syllable monogram, bigram and trigram frequencies to find this probability distribution. How these n-gram frequencies are computed is explained in detail in the following Section 2.1.

2.1   Calculation of Syllable n-gram Frequencies

We have used the Turkish corpora [5] [7] which includes 304178 Turkish words and the corpora is preprocessed as seen in Figure 1. The system TASA syllabifies all Turkish words in the corpora. We have constructed Turkish syllable corpora from the Turkish word corpora. Turkish syllable corpora contains 900342 Turkish syllables. As shown in Table 1, Table 2 and Table 3, Turkish syllable monogram, bigram and trigram frequencies are calculated. For example, the frequency of the syllable monogram "la" is 21322. In Table 2 and Table 3, "blank" represents only one blank character. We accepted blank character as syllable for the system. Table 2 shows the frequencies of some Turkish syllable bigram.

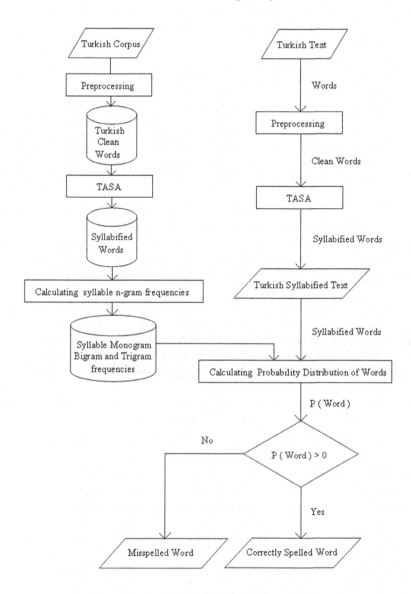

Fig. 1. System architecture

2.2   Calculation of the Probability Distribution of Words

An n-gram is a sub-sequence of n items from a given sequence. n-grams are used in various areas of statistical natural language processing and genetic sequence analysis. The items in question can be letters, syllables, words according to the application.

Table 1. Monogram statistics for Turkish corpus

Monogram	Frequency	%	Monogram	Frequency	%	Monogram	Frequency	%
la	21322	2.37	ra	11611	1.29	ec	5909	0.66
ma	17704	1.97	da	10930	1.21	rih	5186	0.58
li	15124	1.68	ve	10570	1.17	mak	3654	0.41
a	13439	1.49	ri	9618	1.07	di	3022	0.34
ta	13372	1.48	rı	9105	1.01	ne	2788	0.31
i	12827	1.42	me	8776	0.97	si	2741	0.30
de	11699	1.30	e	7312	0.81	mek	2718	0.30

Table 2. Bigram statistics for Turkish corpus

Bigram	Frequency	%	Bigram	Frequency	%	Bigram	Frequency	%
blank i	12880	1.43	blank ka	7907	0.88	rih li	4941	0.55
blank a	11194	1.24	da blank	7429	0.82	blank ko	4681	0.52
blank ve	9793	1.09	ec blank	5857	0.65	lı blank	4572	0.51
blank e	9601	1.07	la rı	5659	0.63	rı blank	4140	0.46
li blank	9410	1.04	le ri	5551	0.62	ma blank	3703	0.41
ve blank	8803	0.98	ta rih	5186	0.58	ma sı	3695	0.41
blank ta	8296	0.92	le blank	5109	0.57	mak blank	3558	0.39

Table 3. Trigram statistics for Turkish corpus

Trigram	Frequency	%	Trigram	Frequency	%
blank ve blank	8566	0.05	e ec blank	3227	0.02
blank ta rih	5238	0.03	blank i le	3037	0.02
ta rih li	4944	0.03	blank bir blank	3000	0.02
rih li blank	4944	0.03	i le blank	2997	0.02
blank i liş	3437	0.02	la rı blank	2979	0.02
i liş kin	3282	0.02	le ri blank	2905	0.02
blank e ec	3228	0.02	blank kon sey	2896	0.02
kon sey blank	2895	0.02	blank i çin	2750	0.01
blank ko mis	2606	0.01	blank ka ra	2578	0.01
ko mis yon	2515	0.01	mis yon blank	2514	0.01

An n-gram of size 1 is a "monogram"; size 2 is a "bigram"; size 3 is a "trigram"; and size 4 or more is simply called an "n-gram" or "$(n-1)$-order Markov model" [9].

An n-gram model predicts x_i based on $x_{i-1}, x_{i-2}, x_{i-3}, \ldots, x_{i-n}$. When used for language modeling independence assumptions are made so that each word depends only on the last n words. This Markov model is used as an approximation of the true underlying language. This assumption is important because it massively simplifies the problem of learning the language model from data.

Suppose that a word W in Turkish syllabified text consists of the syllable sequence $s_1, s_2, s_3, \ldots, s_t$. This word has t syllables. To obtain the n-gram probability distribution [10] of the word W, we have used formula (1).

$$P(W) = P(s_1, s_2, s_3, \ldots, s_t) = \prod_{j-1}^{t} P(s_i \mid s_{i-n+1}, s_{i-n+2}, \ldots, s_{i-2}, s_{i-1}) \quad (1)$$

In n-gram model, the parameter $P(s_i \mid s_{i-n+1}, s_{i-n+2}, \ldots, s_{i-2}, s_{i-1})$ in the formula (1) can be estimated with Maximum Likelihood Estimation (MLE) [8] technique as shown in formula (2).

$$P(s_i \mid s_{i-n+1}, s_{i-n+2}, \ldots, s_{i-2}, s_{i-1}) \approx \frac{C(s_{i-n+1}, s_{i-n+2}, \ldots, s_{i-1}, s_i)}{C(s_{i-n+1}, s_{i-n+2}, \ldots, s_{i-1})} \quad (2)$$

So, we conclude as formula (3).

$$P(W) = P(s_1, s_2, s_3, \ldots, s_t) \approx \prod_{i=1}^{t} \frac{C(s_{i-n+1}, s_{i-n+2}, \ldots, s_{i-1}, s_i)}{C(s_{i-n+1}, s_{i-n+2}, \ldots, s_{i-1})} \quad (3)$$

In formula (2) and (3), $C(s_{i-n+1}, s_{i-n+2}, \ldots, s_{i-1}, s_i)$ is the frequency of the syllable sequence $s_{i-n+1}, s_{i-n+2}, \ldots, s_{i-1}, s_i$. Furthermore, $C(s_{i-n+1}, s_{i-n+2}, \ldots, s_{i-1})$ is the frequency of the syllable sequence $s_{i-n+1}, s_{i-n+2}, \ldots, s_{i-1}$. The frequencies of these syllable sequences can be calculated from the Turkish corpora [5].

For bigram(n=2) and trigram(n=3) models, probability distribution $P(W)$ can be computed as shown in formula (4) and (5) respectively.

$$P(W) = P(s_1, s_2, s_3, \ldots, s_t) = \prod_{i=1}^{t} P(s_i \mid s_{i-1}) \approx \prod_{i=1}^{t} \frac{C(s_{i-1}, s_i)}{C(s_{i-1})} \quad (4)$$

$$P(W) = P(s_1, s_2, s_3, \ldots, s_t) = \prod_{i=1}^{t} P(s_i \mid s_{i-2}, s_{i-1}) \approx \prod_{i=1}^{t} \frac{C(s_{i-2}, s_{i-1}, s_i)}{C(s_{i-2}, s_{i-1})} \quad (5)$$

For example, according to bigram model we can calculate the probability distribution of a word in Turkish text. Assume that we have a text which includes some words as shown in (6). This text is converted to syllabified text as (7). Syllables are delimited with dash character.

$$\text{"}\ldots\text{bu gün okulda şenlik var}\ldots\text{"} \quad (6)$$

$$\text{"}\ldots\text{bu gün o-kul-da şen-lik var}\ldots\text{"} \quad (7)$$

Assume that the word W ="okulda" in (6) is taken for computing its probability distribution and W can be written as the syllable sequence $W = s_1, s_2, s_3 =$ "o","kul","da" as shown in (7). Here, s_1 ="o", s_2 ="kul", s_3 ="da". We accepted blank character as a syllable. We call this syllable as λ. So, assume that syllable monogram frequencies are $C(\text{"}\lambda\text{"})=0.003$, $C(\text{"o"})=0.002$, $C(\text{"kul"})=0.004$ and syllable bigram frequencies are $C(\text{"}\lambda\text{","o"})=0.0001$, $C(\text{"o","kul"})=0.0002$, $C(\text{"kul","da"})=0.0003$. We have calculated $P(\text{ "okulda" })$ using bigram model. We have found that the probability distribution of the word "okulda" is 0.0002475 as shown in (8).

$$P(W) = P(\text{"okulda"}) = P(s_1, s_2, s_3) = P(\text{"o","kul","da"}) \qquad (8)$$

$$= \prod_{i=1}^{3} P(s_i \mid s_{i-1}) \approx \prod_{i=1}^{3} \frac{C(s_{i-1} \mid s_i)}{C(s_{i-1})}$$

$$= \left(\frac{C(\text{"}\lambda\text{", "o"})}{C(\text{"}\lambda\text{"})} \right) \left(\frac{C(\text{"o", "kul"})}{C(\text{"o"})} \right) \left(\frac{C(\text{"kul", "da"})}{C(\text{"kul"})} \right)$$

3   Testing the System

We have designed and implemented two systems to detect misspelled words in Turkish text. One uses monogram and bigram frequencies. The size of monogram database is 41 kilobytes and our monogram consists of 4141 different syllables. The size of bigram and trigram databases are 570 and 2858 kilobytes respectively. While the bigram database includes 46684 syllable pairs, the trigram database consists of 183529 ternary syllables. The other uses bigram and trigram frequencies. We have tested these two systems. To test the systems, we have two Turkish texts. One is correctly spelled text which includes 685 correctly spelled words. The other is misspelled text which has 685 misspelled words. These two texts have same words. Namely, misspelled words are generated with putting errors on the correctly spelled words. These error types are substitution, deletion, insertion, transposition and split word errors. The systems takes correctly spelled and misspelled texts as input and gives the results for each word as "correctly spelled word" or "misspelled word". As it is shown in Figure 1, probability distributions are calculated for each word. If the probability distribution of a word is equal to zero, system decides that the word is misspelled. If it is greater than zero, system decides that the word is correctly spelled.

The system works with Intel-based NT, Windows 2000, XP, Windows 2003 Server systems with 512MB RAM and it has been developed using Borland C++ Builder Professional.

We have first tested the system on the correctly spelled text using monogram and bigram frequencies. The system determines 602 correctly spelled words from the correctly spelled text, so the words are correctly recognized with 88% success rate. Also, 589 misspelled words within the misspelled text are decided successfully by the system. Namely, the system which is tested on the misspelled text correctly recognized the words with 86% success rate.

Finally we have tested the system on the correctly spelled text using bigram and trigram frequencies. The system determines 671 of 685 correctly spelled words from the correctly spelled text. The success rate on correctly recognition of the words is 98%. Furthermore, 664 of 685 misspelled words within the misspelled text are decided successfully by the system. Thus, the system which is tested on the misspelled text correctly recognized the words with 97% success rate. Our system performance is competitive with similar systems. The system checks the approximately 75.000 words per second.

4 Conclusion

In this study, we have presented two systems to detect misspelled words on Turkish text. The system which uses syllable monogram and bigram frequencies is quite successful to find misspelled words. The system's success rate is 86% to detect misspelled words on Turkish text. But, we had better success rate when we develop the system that uses syllable bigram and trigram frequencies. This system reached 97% success rate to detect misspelled words.

For future directions, we plan to develop a system which uses syllable trigram and 4-gram frequencies to find misspelled words. Syllable n-gram frequencies are very important for our system. Therefore, we plan to extract new syllable n-gram frequencies from different Turkish corpora.

References

1. Barari, L., QasemiZadeh, B.: CloniZER spell checker adaptive language independent spell checker. In: AIML 2005 Conference CICC, Cairo, Egypt, pp. 19–21 (2005)
2. Tong, X., Evans, D.A.: A statistical approach to automatic OCR error correction in context. In: Proceedings of the Fourth Workshop on Very Large Corpora, Copenhagen. Denmark, pp. 88–100 (1996)
3. Kang, S.S., Woo, C.W.: Automatic segmentation of words using syllable bigram statistics. In: Proceedings of the Sixth Natural Language Processing Pacific Rim Symposium, Tokyo. Japan, November 27-30, pp. 729–732 (2001)
4. Deorowicz, S., Ciura, M.G.: Correcting spelling errors by modelling their causes. International Journal of Applied Mathematics and Computer Science 15(2), 275–285 (2005)
5. Dalkilic, G., Cebi, Y.: Creating a Turkish corpus and determining word length. DEU Muhendislik Fakultesi Fen ve Muhendislik Dergisi 5(1), 1–7 (2003)
6. Kukich, K.: Techniques for automatically correcting words in text. ACM Computing Surveys 24(4), 377–439 (1992)
7. Kurumu, T.D.: Imla kilavuzu. Ankara (2001)
8. Asliyan, R., Günel K.: Design and implementation for extracting Turkish syllables and analyzing Turkish syllables. In: International Symposium on Innovations in Intelligent Systems and Applications. INISTA (2005)
9. Zhuang, L., Bao, T., Zhu, X., Wang, C., Naoi, S.: A chinese OCR spelling check appoarch based on statistical language models. IEEE International Conference on Systems, Man and Cybernetics, 4727–4732 (2004)
10. Jurafsky, D., Martin, J.H.: Speech and language processing. Prentice-Hall Press, Englewood Cliffs (2000)

Self Adaptable Recognizer for Document Image Collections

Million Meshesha and C.V. Jawahar

Center for Visual Information Technology,
International Institute of Information Technology,
Hyderabad - 500 032, India
jawahar@iiit.ac.in

Abstract. This paper presents an architecture that enables the recognizer to learn incrementally and, thereby adapt to document image collections for performance improvement. We argue that the recognition scheme for a book could be considerably different from that designed for isolated pages. We employ learning procedures to capture the relevant information available online, and feed it back to update the knowledge of the system. Experimental results show the effectiveness of our design for improving the performance on-the-fly.

1 Adaptable OCR System

The success of document image indexing and retrieval in the newly emerging digital libraries considerably depends on the availability of robust OCRs that can take care of the diversity in the document image collections. Performance of the state of the art OCRs are not very encouraging for these collections [1,2]. Recent study by Lin [3] shows that document recognition research is still in great need for better accuracy and reliability, as well as for effective information retrieval and delivery. We need a recognition system that is capable of intelligently adapting to the characteristics of documents of interest, and improving the performance over time. Machine learning offers one of the most cost effective and practical approaches to the design of pattern classifiers for a broad range of pattern recognition applications like character recognition [4]. Learning algorithm could be supervised, unsupervised or reinforcement-based [5]. Supervised techniques have been successfully demonstrated for character recognition application as offline training in OCR systems. However applicability of semi-supervised and reinforcement learning algorithms are not yet explored in their full potential.

Most of the recent research in OCR has been centered around building fully automatic, high performing intelligent classification systems with good generalization capability [6]. Intelligent OCRs with excellent performance on a given page are reported for Latin scripts [7]. However, when it comes to the suitability of converting an old book to text and providing a text-like access, even the present day OCRs are found to be insufficient [1]. The performance of these recognizers decline as the diversity in the collection of documents (with unseen

A. Ghosh, R.K. De, and S.K. Pal (Eds.): PReMI 2007, LNCS 4815, pp. 560–567, 2007.

fonts and poor in quality) increases. We argue that an adaptive recognition system, which can learn from its experience and improve its performance over time, is better suited for such document collections (that vary in printing and quality).

In this paper, we argue that:

1. The technique for recognizing a book (a reasonably large collection of documents in single font and consistent formatting) can be different from OCRs that are designed to be part of scanner drivers or meant for isolated pages.
2. The notion of generalization from a small set of training samples, which is critical to the design of pattern classifiers, need not be the only performance goal. One can positively look for 'overfitting' to a book, as long as the recognition engine can provide better results on that specific collection. In general, the multimodality of the data distribution need not be completely modeled; but can be accepted as a reality.

An approach for designing a semi-automatic adaptive OCR for document images in digital libraries is reported in [8]. The OCR is designed to support an interactive retraining (based on users feedback) for performance improvement. In this paper, we present an intelligent and self adaptable OCR that employs machine learning algorithms for validation, labeling, sampling and incremental learning in a recognition cycle. Post-processor based feedback mechanism is enabled to provide additional knowledge to the system.

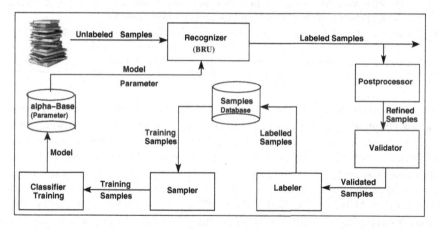

Fig. 1. An architecture of OCR learning framework employed for the recognition of document collection

2   Learning Schemes

There are possibly two ways in which the continuous performance improvements can take place in recognizers for document collections in digital libraries.

1. Learn from the direct or indirect user feedback (as in the case of relevance feedback models, search engines etc.) captured during the search.
2. Improve the performance of recognizers by adapting to a specific collection by absorbing the input from language models (e.g.. Dictionaries).

In this paper, we do not model the user interactions, and follow the second approach, where knowledge from the post-processor is used for developing new labeled examples, and thereby improving the recognition accuracy. In the conventional OCRs also, language models are integrated for improving the performance of the base-classifier. However they are not designed to learn from their experience. If a word which is misclassified once is given to the base classifier, it repeats the mistake again. However, we are interested in a framework, which will make the classifier intelligent and correct the mistake in the future.

Algorithm 1. Recognition of Text in Document Collections

1: Document images are preprocessed and words are extracted for recognition.
2: BRU outputs the recognized text.
3: Post-processor checks the validity of the word based on some language model and outputs the corrected word, if found to be invalid.
4: Validator visually validates the result of post-processing. By matching the rendered text and word image, visual similarity is carefully computed. Errors introduced by the post-processor results in outliers.
5: Those with visually similar appearance are considered as new training samples. Semi-supervised learning is used in labeling the new samples.
6: Bagging is used to create a sample set and classifier is trained (preferably incrementally) to obtain the new set of parameters and stored in α-base.
7: Base-classifier is incrementally learn for knowledge updation.

Overview: Architecture of the proposed recognition system is shown in Figure 1, and also explained in Algorithm 1. Given a word image for recognition, the Basic Recognizer Unit (BRU) converts it into text. A post-processor is then used to rectify/verify the recognized word. In our present implementation, we use a simple dictionary-based post-processor which is designed following a reverse dictionary approach with the help of a trie data structure. Our Basic Recognition Unit is an SVM-based classifier trained offline on few *apriori* available synthetic training examples (that are prepared in single font, style and size) by extracting vector representation of the entire image. In this work, the conventional open-loop system of classifier followed by post-processor is closed by automatically generating training data from the "test" image and retraining (in fact incrementally modifying) the BRU. Before passing mis-recognized samples as new training datasets we run a validator to detect outliers that deviate from the original sample. Samples that are similar to the original ones are labeled using labeler and transferred to the sample database. This system, thus, has a framework for the creation, validation, labeling and use of mis-recognized samples as new training examples for performance improvement over time.

This character recognition system is highly data-driven. It employs feedback for performance improvement. Initial recognition errors are immediately detected and corrected. This technique is also useful when sufficient training samples or *apriori* knowledge is lacking (for example recognizing in a new font/style). The implementation of this approach could be computationally intensive. However, today's OCR environment especially the ones for digitizing books and large document image collections can afford to do so.

Automatic Preparation of Labeled Data: One of the basic problem of the learning process in a dynamic environment is getting high quality new training samples. Once the BRU recognises a word and outputs a textual representation, post-processor verifies the validity of the word with the help of a dictionary (or a language model) and either accepts the word as valid or corrects with an alternative. There are six possible situations.

1. BRU correctly recognizes a word and post-processor accepts it as valid word.
2. BRU correctly recognizes, but post-processor fails to accept it as valid and suggests an alternative.
3. BRU makes a mistake, and post-processor corrects to the right word.
4. BRU makes a mistake, and post-processor corrects/modifies to a wrong word.
5. BRU correctly recognizes, but post-processor modifies to a wrong word.
6. Post-processor fails to suggest an acceptable alternative to a text provided by BRU.

We employ the knowledge from the post-processor in creating a new set of labeled examples. For this, we validate the result of the post-processing, by matching in the image space. *This is more of verification rather than recognition.* Given the text, it is rendered into an image, and matched in the image space. Let $p_1, p_2, \ldots p_m$ and $q_1, q_2, \ldots q_n$ be a set of feature vectors, extracted by scanning (column-wise) the vertical strips of the rendered text and word images, where m and n are width of a given words. They are aligned using dynamic time warping (DTW) for similarity measure. If all the symbols match (one-to-one), then we have achieved the labeling of all the connected components in the document image (situation 1). When only some of the symbols match, they are labeled and the rest of the symbols are treated as unlabeled. In short, at the end of the first phase of validation using a DTW-based algorithm, most of the components from a given document image gets labeled automatically and gets added to the database of training examples. The rest of the samples are treated as unlabeled. Note that, at this stage, our interest is limited to improving the performance of BRU, rather than addressing the problem of merges and splits in the document images.

For each unlabeled data x_i we attach probabilities p_{ij} of belonging to class i, i.e. $p_{ij} = prob(x_i \in w_j)$. This is an initialization. These probabilities are then iteratively improved by an Expectation Maximization (EM) [9] based formulation. The E-Step uses the current parameter estimates to find the best probabilistic labels for class membership using a multivariate normal (Gaussian) distribution. The M-Step then refines the parameters to maximize the total likelihood. The

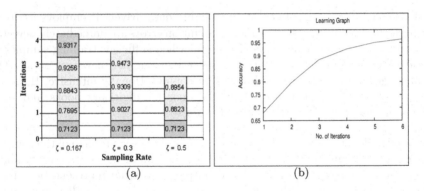

Fig. 2. (a) A change in accuracy and learning rate at different sampling rate (ζ). (b) An improvement in the performance of the recognizer through learning from poor quality documents.

steps are iterated until the change in parameter values falls below some predefined threshold. In a way, it allows to put the unlabeled examples into a cluster with most of the samples already labeled. When most of the samples are labeled, it converges in one step, and assigns label to the unlabeled samples.

Sampling: The ability to actively select the most useful training samples is an important aspect of building an efficient classifier. Boosting and bagging are being increasingly used for this purpose [5,10]. Boosting uses all instances of the datasets at each iteration, but associates a weight for each sample. Bagging, on the other hand, takes the available training samples and generates a new sample set by selecting them randomly and with replacement. In our implementation, training examples are generated throughout the recognition process. We need to use these samples as new training datasets in subsequent iterations; as a result of which we use bagging for sampling. For each session $k = 1, 2, \ldots, n$, new training set of size N is sampled from the database. Bagging works as follows. Consider a training dataset $D = d_1^{c_i}, d_2^{c_i}, \ldots, d_m^{c_i}$ where m is the total samples available in class c_i. Bagging selects a subset of representative samples randomly from the available training sample collections that are accumulated through feedback. In each session k, a re-sampled training set D^k is built for constructing/training a classifier C^k. Classifier C^k has better accumulated knowledge than the previous C^{k-1} increasing its applicability to the given document collection.

Incremental Updation: We use SVM classifier [5] as the basic recognition unit, and employ an incremental learning approach [11] to train it. By identifying the decision boundary with maximal margin, SVM results in better generalization during classification. Margin maximization leads to an optimization problem the solution of which is expressed uniquely in terms of support vectors s_i. An input pattern x is classified into class $y \in \{-1, +1\}$ according to the decision function:

$$y = sgn(\sum_i \alpha_i y_i K(s_i, x)) \tag{1}$$

Table 1. Performance improvement with the use of incremental learning vs. retraining during the learning process.

Iterations	Retraining	Incremental
1	0.652475	0.652475
2	0.867845	0.882446
3	0.898620	0.907383
4	0.901226	0.910813
5	0.929970	0.939294
6	0.941537	0.948294
7	0.943962	0.959026
8	0.952096	0.959026
9	0.952096	0.959026
10	0.952096	0.959026

which takes the form of a linear combination of kernels $K(s_i, x)$ weighted by training labels y_i and coefficients α_i. The coefficients α_i are nonzero only for training data that are support vectors, so that $\sum_i \alpha_i y_i K(s_i, x)$ is sparse and the support vectors capture the relevant information present in the training data.

Incremental SVM works as follows. The representation of the data seen so far for each class is given by the set of support vectors describing the learned decision boundary. These support vectors are combined with the new incoming datasets to provide the training data for the incremental step. The incremental step then updates the solution for addition of a single training sample x_i by incrementing the coefficient i and simultaneously adjusting previously assigned coefficients $\alpha_j (j < i)$ on the present and all previous training samples.

3 Implementation, Results and Discussions

Our implementation in c/c++ is tested on document image collections. The design of our system is based on a multi-core approach [2]. Keeping the futuristic large scale computational applications, it is implemented as layers with plugin interfaces for modules to ease replacing one module with another. SVM based classifier could be modified as a cascaded classifier combination with out major changes in other parts of the code. Each module internally can decide on multiple algorithm implementations of the same functionality that may be interchanged at run-time. This helps in selection and use of an appropriate algorithm or a set of parameters for a document collection or script. The system allows transparent run-time addition and selection of modules thereby enabling the decoupling of the application and the plug-ins.

The learning scenario involves post-processor based feedback mechanism to update the knowledge of the recognizer. Machine learning procedures are integrated for labeling, sampling and validating the new training samples collected online. Scanned images are segmented into words and submitted for recognition in a batch. Recognition results are post-processed to resolve ambiguity among

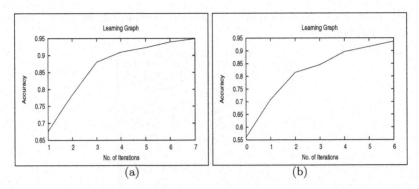

Fig. 3. Recognition accuracy of the OCR through learning in presence of (a) font variations and (b) style variations

the candidate words. Mis-recognized words are validated and fedback as new training datasets for learning incrementally. The labeler is used to split these words into components and assign their membership category (class). The sampler selects subset of these datasets and pass them for learning. Sampling speeds up the learning rate η if proper sampling rate ζ is used. As shown in Figure 2(a) the learning rate and improvements in recognition accuracy are related with the sampling rate used. Initially the accuracy jumps from 71.23% to 76.95%, 90.27% and 88.23% for sampling rate 0.167, 0.3. 0.5 respectivelly. We can observe that the number of iterations decrease as the sampling rate increase, but with better accuracy obtained at sampling rate ζ of 0.3. This shows that sampling at 30% of the original training datasets speeds the learning rate with better improvements in recognition results. This is only an empirical observation. We use this sampling rate for our experimentation.

Once new training samples are selected, incremental SVM classifier is used for updating the knowledge of the recognizer. Table 1 presents performance of incremental learning vis-a-vis retraining. The result shows that the two approaches have comparable performance. However, the use of incremental learning has further advantage in terms of both computational time and space complexity. In a way, it eases the implementation of SVM classifier on large datasets.

We validate the performance of the learning framework on real-life printed document images (of English) that are (i) poor in quality and (ii) varies in fonts and styles as depicted in Figures 2(b) and 3, respectively. The result demonstrate the advantage of our learning strategy for performance improvement. Because of the quality and printing variations in document images, a very low recognition result is obtained at the initial stage, which amounts to less than 70%. Within few iterations of learning, however, the recognition accuracy improved, on the average, to 95%. Most of the recognition errors happen since characters do not always look the way they should because of degradation or printing variations. A character's hole is filled, e.g. 'o', 'u' and 'n' are recognized as 'dot', some part of the character (e.g. dot of 'i') is missing and recognized as '1' or 'l', etc. Different fonts produce a character with different shape that may visually similar with other character. It has been observed that the recognizer is able to adopt to the given document images based on the additional samples of confusing components that are fedback for learning. In this way, the learning framework can enable the OCR to learn from its experience and adapt to varying document image collections in printing and quality. Further work is needed for (i) Extending this framework for many of the complex Indian scripts (ii) Addressing the segmentation errors at various stages. The present design assumes that segmentation (of pages into words as well as words into recognizable symbols) is available. This assumption needs to be relaxed in future.

4   Conclusion

We have presented a novel approach for learning during the recognition of document image collections. The architecture integrates advanced learning procedures that automatically interacts and pass feedback for further learning. This enables the OCR to easily accumulate knowledge for performance improvement. This strategy is promising for the recognition of large digitized document images in applications, like digital libraries. Experiments are ongoing to validate our approach on diverse collections with changes in script, font, etc.

References

1. Feng, S., Manmatha, R.: A hierarchical, HMM-based automatic evaluation of OCR accuracy for a digital library of books. In: Joint Conference on Digital Libraries (JCDL), pp. 109–118 (2006)
2. Sankar, P., et al.: Digitizing a million books: Challenges for document analysis. In: Proc. of the Seventh IAPR Workshop on Document Analysis Systems, pp. 425–436 (2006)
3. Lin, X.: DRR research beyond COTS OCR software: A survey. In: SPIE Conference on Document Recognition and Retrieval XII, San Jose, CA, pp. 16–20 (2005)
4. Xu, Y., Nagy, G.: Prototype extraction and adaptive OCR. IEEE Transactions on Pattern Analysis and Machine Intelligence 21, 1280–1296 (1999)
5. Hastie, T., Tibshirani, R., Friedman, J.: The elements of statistical learning. Springer, Heidelberg (2001)
6. Nagy, G.: Twenty years of document image analysis in PAMI. IEEE Transactions on Pattern Analysis and Machine Intelligence 22, 38–62 (2000)
7. Kahan, S., Pavlidis, T., Baird, H.S.: On the recognition of printed characters of any font and size. IEEE Transactions on Pattern Analysis and Machine Intelligence 9, 274–288 (1987)
8. Rawat, S., et al.: A semi-automatic adaptive OCR for digital libraries. In: Proc. of the Seventh IAPR Workshop on Document Analysis Systems, pp. 13–24 (2006)
9. Ivanov, Y., Blumberg, B., Pentland, A.: Expectation maximization for weakly labeled data. In: Proc. of the Int. Conf. on Machine Learning, pp. 218–225 (2001)
10. Iyengar, V.S., Apte, C., Zhang, T.: Active learning using adaptive resampling. In: Sixth Int. Conference on Knowledge Discovery and Data Mining, pp. 92–98 (2000)
11. Diehl, C., Cauwenberghs, G.: SVM incremental learning, adaptation and optimization. In: Proc. IEEE Int. Joint Conf. Neural Networks, pp. 2685–2690 (2003)

Mixture-of-Laplacian Faces and Its Application to Face Recognition

S. Noushath,[1] Ashok Rao[2], and G. Hemantha Kumar[1]

[1] Dept of Studies in Computer Science,University of Mysore,
Mysore - 570 006, India
nawali_naushad@yahoo.co.in
[2] Dept of Electronics and Communication, SJ College of Engineering,
Mysore - 570 006, India
ashokrao.mys@gmail.com

Abstract. The locality preserving projection (LPP), known as *Laplacianfaces*, was recently proposed as a transformation technique of mapping which optimally preserves the neighborhood structure of the dataset. In this paper, an efficient method for face recognition called *mixture-of-Laplacianfaces* (or LPP mixture model) is proposed, which obtains several sets of Laplacianfaces through Expectation-Maximization (EM) learning of Gaussian Mixture Models (GMM). Experiments carried out by using this on ORL, FERET and COIL-20 indicate superior performance as compared with method based on Laplacianfaces and other contemporary subspace methods.

1 Introduction

The goal of dimensionality reduction algorithm is to map data points $X = x_1, x_2, \ldots, x_m$ in $\Re^N$ to a subspace $\Re^l$, where $l \ll N$. The LPP [9] is a subspace method which also incorporates the neighborhood information while mapping the data points to a subspace. Since it preserves the neighborhood information, its classification performance is much better than other subspace approaches like PCA [6] and FLD [7]. Here, we briefly outline the LPP model [9]. The main objective of LPP is to preserve the local structure of the input vector space by explicitly considering the manifold structure. The first step of this algorithm is to construct the adjacency graph $\mathbf{G}$ of m nodes, such that node i and j are linked if x_i and x_j are *close* w.r.t each other in any of the following two conditions:

1. k-nearest neighbors: Nodes i and j are linked by an edge, if i is among k-nearest neighbors of j or vice-versa.
2. ϵ-neighbors: Nodes i and j are linked by an edge if $\|x_i - x_j\|^2 < \epsilon$, where $\|\cdot\|$ is the usual Euclidean norm.

Next step is to construct the weight matrix $\mathbf{W}$, which is a sparse symmetric $m \times m$ matrix with weights w_{ij} if there is an edge between nodes i and j, and 0 if there is no edge. Two alternative criterion to construct the weight matrix:

A. Ghosh, R.K. De, and S.K. Pal (Eds.): PReMI 2007, LNCS 4815, pp. 568–575, 2007.

1. Heat-Kernel: $w_{ij} = e^{\frac{-\|x_i - x_j\|^2}{t}}$, if i and j are linked.
2. $w_{ij} = 1$, iff nodes i and j are linked by an edge.

The objective function of LPP model is to solve the following generalized eigen-vector problem:

$$XLX^T a = \lambda XDX^T a \tag{1}$$

Where $\mathbf{D}$ is the diagonal matrix with entries as $D_{ii} = \sum_j w_{ji}$ and $L = D - W$ is the laplacian matrix.

The transformation matrix $\mathbf{A}$ is formed by arranging the eigenvectors of Eq.(1) ordered according to their eigenvalues, $\lambda_1 < \lambda_2, \ldots, < \lambda_l$. Thus, the feature vector y_i of input x_i is obtained as follows:

$$x_i \rightarrow y_i = A^T \cdot x_i \ \forall i = 1, 2, \ldots, m \tag{2}$$

Note: The XDX^T matrix is always singular because of high-dimensional nature of the image space. To alleviate this problem, PCA is used as the preprocessing step to reduce the dimensionality of the input vector space.

2   Mixture-of-Laplacianfaces Using GMM

Here, mixture-of-Laplacianfaces obtained through EM learning of the GMM is presented. To implement this, we first explain the concepts of GMM [1][1].

2.1   PCA Mixture Model

The goal of this model is to partition set of all classes into several clusters and to obtain PCA transformation matrix for each cluster. Here, a class and its density function of the N-dimensional data x is represented as $P(x) = \sum_{k=1}^{K} P(x/k, \theta_k)P(k)$. Where $P(x/k, \theta_k)$ and $P(k)$ represent the conditional density and apriori probability of the kth cluster respectively, and θ_k is the unknown model parameters which is to be calculated through EM learning. The multivariate Gaussian distribution function to model $P(x/k, \theta_k)$ is as follows:

$$\eta(x/\mu_k, \Sigma_k) = \frac{1}{(2\pi)^{D/2} |\Sigma|^{1/2}} e^{-\frac{1}{2}(x-\mu_k)^T \Sigma_k^{-1}(x-\mu_k)} \tag{3}$$

Where μ_k and Σ_k are the mean and covariance matrix of kth cluster respectively. A distribution can be written as a linear superposition of Gaussian[2] in the form $P(x) = \sum_{k=1}^{K} \pi_k \eta(x/\mu_k, \Sigma_k)$. Where π_k is called the mixing coefficient that is set to the fractions of data points assigned to k^{th} cluster. Now log of the likelihood function is given by:

$$ln \ P(X/\pi, \mu, \Sigma) = \sum_{n=1}^{N} ln \sum_{k=1}^{K} \pi_k \ \eta(x_n/\mu_k, \Sigma_k) \tag{4}$$

---

[1] It is also referred as probabilistic PCA or PCA mixtures in Refs.[5] and [1].
[2] Note that mixture of Gaussian need not be Gaussian but Fourier of Gaussian is still a Gaussian.

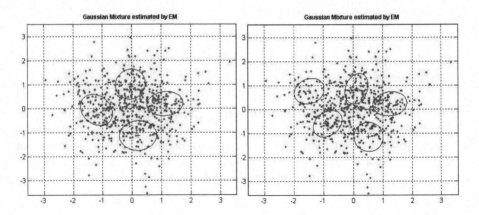

Fig. 1. Illustration of Gaussian Mixture Models

An elegant and powerful method for finding the maximum likelihood solution for GMM is called the EM algorithm [1]. To find a suitable initialization for a GMM, the means μ_k, covariances Σ_k and mixing coefficients π_k are initialized by running K-means clustering algorithm, which are then subsequently estimated using EM. We now alternate between the following two steps which are called Expectation(E) step and the Maximization(M) step. In the E-step, current values of the parameters are used to evaluate the posteriori probabilities given by:

$$\xi(z_{nk}) = \frac{\pi_k \ \eta(x/\mu_k, \Sigma_k)}{\sum_{j=1}^{K} \pi_j \ \eta(u/\mu_j, \Sigma_j)} \tag{5}$$

In the M-step, we use these probabilities to re-estimate the means μ_k^{new}, covariance matrix Σ_k^{new} and mixing coefficient π_k^{new} respectively as follows:

$$\mu_k^{new} = \frac{1}{N_k} \sum_{n=1}^{N} \xi(z_{nk}) \, x_n \tag{6}$$

$$\Sigma_k^{new} = \frac{1}{N_k} \sum_{n=1}^{N} \xi(z_{nk}) \, (x_n - \mu_k^{new})(x_n - \mu_k^{new})^T \tag{7}$$

$$\pi_k^{new} = \frac{N_k}{N} \tag{8}$$

Each update of the parameters resulting from the E-step followed by an M-step is guaranteed to increase the log likelihood function and the algorithm is deemed to converge when the change in the log likelihood function falls below some threshold. Fig.1 illustrates the mixture of four and five Gaussians for a data of 600 random points, where the $+$ mark indicates the cluster means. Now, k^{th} eigen value and eigen vector parameters are obtained using eigenvalue analysis $\Sigma_k W_{kj} = \lambda_{kj} W_{kj} \ \forall k = 1, \dots, K$. The PCA transformation matrix W_k is obtained by selecting l dominant eigen vectors of kth cluster.

2.2   LPP Mixture Model

Since we obtained K number of PCA transformation matrices using PCA mixture model, a feature set for each mixture is obtained in the LPP mixture model. The objective function of the proposed method now becomes as follows:

$$X_k L X_k^T a_k = X_k D X_k^T a_k \ \forall k = 1, 2, \ldots, K \tag{9}$$

Where X_k represents $l \times n$ feature matrix of training samples obtained after the transformation through k^{th} PCA mixture. The $\mathbf{D}$ and $\mathbf{L}$ matrix are obtained as mentioned previously in section 1. The transformation matrices $A_k = (a_1^k, a_2^k, \ldots, a_l^k)$ of LPP mixture model are formed by arranging l eigenvectors of kth LPP mixture in Eq.(9) according to l largest eigenvalues $\lambda_1^k < \lambda_2^k < \ldots, < \lambda_l^k \ \forall k = 1, 2, \ldots, K$. Using the A_ks, features for a training sample x can be obtained as follows:

$$f_i^k = A_k^T x_i \ \forall i = 1, 2, \ldots, m \ and \ \forall k = 1, 2, \ldots, K \tag{10}$$

Since there are K mixtures, K number of features are obtained for a unknown sample $\mathbf{I}$. To combine K classification results of $\mathbf{I}$ from all the mixtures, a distance matrix is constructed and denoted by $D(I) = (d_{ij})_{m \times K}$ where d_{ij} is set to 1 if I is matched to i^{th} training sample after transformation through j^{th} mixture, else it is set to 0. Consequently, the total confidence value that the sample I belongs to the ith class is $TC_I(i) = \sum_{j=1}^{K} d_{ij}$. Finally, identity of the test sample $\mathbf{I}$ is computed as follows:

$$Identity(I) = argmax_i(TC_I(i)) \ 1 \le i \le m \tag{11}$$

3   Experiments

The performance of the proposed method is evaluated using three standard databases namely ORL[3], FERET and COIL-20[4]. The nearest neighbor classifier (Euclidean distance) is used for classification and experiments are carried out on a P4 3GHz PC with 1GB RAM memory and Matlab 7.0 environment.

3.1   Results on the ORL Database

Two types of experiments, i.e. performance under clean and noise conditions, are carried out on this database. Our preliminary experiments suggest that the classification performance of the proposed method is impacted by the number of mixtures learned. Hence, to study the effect of number of mixtures on classification performance, we conduct an experiment by varying both number of training samples and mixtures. This result is depicted in Table 1[5]. It is apparent

---

[3] www.uk.research.att.com/facedatabase.html

[4] www1.cs.columbia.edu/CAVE/research/softlib/coil-20.html

[5] Last row shows the results of conventional LPP method.

Table 1. Recognition Accuracy(%) for varying number of mixtures (40 dimensions)

Number of Mixtures	Number of Training Samples					
	2	3	4	6	7	8
2	86.50	91.25	93.75	98.00	98.75	99.50
3	88.50	91.50	94,25	98.50	99.00	99.50
4	90.00	93.00	94.50	98.00	99.00	99.50
5	90.50	93.00	94.00	98.00	98.75	99.50
LPP method[9]	82.00	82.25	89.75	95.00	96.50	98.25

from table that, mixture of four Gaussians could be a optimal choice for competitive results and reasonable computational burden. Hence in all our subsequent experiments, we use four mixtures. Now, for comparative analysis, an experiment is conducted by training first five samples of each class. Table 2 presents the comparison of different subspace methods on recognition accuracy and running time(s). From the table it is clear that the proposed method significantly

Table 2. Comparison of different subspace methods for 40 dimensions

Parameters	Methods						
	PCA[6]	PCA+DCT[2]	PCA+Wavelet[3]	LDA[7]	LPP[9]	PCA Mixture[5]	Proposed Method
Accuracy(%)	93.25	93.50	92.50	92.25	94.50	95.50	97.00
Time(s)	15.89	13.10	14.44	17.78	16.39	189.44	201.22

outperformed other methods in terms of recognition accuracy. Nevertheless, the drawback of our method is the computational burden involved while EM learning of the GMM. This appears to be the case for mixture models in general.

The issue of noise modeling is crucial to check the robustness of the algorithm under real time pattern recognition and computer vision tasks. In our work, we have modeled five noise environments by using different continuous distributions in their discretized version. Using first image from each class, we generated 10

Table 3. Average recognition accuracy for different noise conditions for 25 dimensions

Noise Conditions	Methods						
	PCA	PCA+DCT	PCA+Wavelet	LDA	LPP	PCA Mixture	Proposed Method
Gaussian	81.50	82.00	68.25	88.00	87.00	93.50	95.25
Salt-and-Pepper	56.50	57.00	46.25	59.50	61.25	63.50	67.50
Exponential	31.50	30.50	78.50	43.00	51.50	71.00	73.50
Weibull	34.50	34.50	68.00	39.50	43.50	55.50	60.25
Beta	86.50	85.25	79.00	87.75	93.50	99.50	100.0

Table 4. Comparing different methods under *SSPP* condition for 20 dimensions

Database Size	PCA	PCA+DCT	PCA+Wavelet	Sampled-FLDA	PCA Mixture	LPP	PC^2A	Proposed
20	100	100	100	100	100	100	100	100
40	87.50	87.50	90.00	90.00	90.00	87.50	87.50	90.00
60	85.00	90.00	86.66	88.33	88.33	84.00	88.33	88.33
80	87.50	90.00	88.75	86.25	88.75	85.40	88.75	85.00
100	85.00	88.00	91.00	89.00	83.33	76.00	87.00	83.33
120	88.33	87.50	87.50	90.00	90.00	86.66	89.16	91.00
140	87.14	89.28	86.42	88.57	89.28	86.42	87.14	89.28
160	87.50	86.25	86.25	83.75	86.25	83.50	88.12	90.00
180	86.11	87.22	88.88	82.22	86.66	78.88	86.66	88.88
200	89.00	88.00	87.50	81.00	81.00	80.50	88.50	89.50

noise images by varying the noise density from $0.1, 0.2, \ldots, 1.0$. Likewise, 50 noise images are created for each class corresponding to 5 different distributions. The 2000 noisy images (40×50) thus created are used as test samples. First five clean samples from each class of ORL are used as training samples. Table 3 presents the average recognition accuracy (average of ten different noise densities) obtained by various algorithm. Some analysis from this experiment are:

1. The *PCA+Wavelet* method is the only robust algorithm under exponential and weibull noise conditions. Since wavelet has the advantage of both space and scale orientation, it is able to represent these noise disturbance in few scales across space and/or few space points across scales.
2. Performances of PCA and LPP mixture models are truly robust and next best performing under weibull and exponential noise because these methods are backed with Gaussian mixture models, which is known to be robust.
3. The LPP algorithm is the next best performing algorithm under all noise conditions. This we believe could be due to the utilization of the neighborhood information in this algorithm.
4. Beta noise model has had very little influence on the performance of algorithms. However, it would be interesting to see their behavior for varying values of control parameters in Beta distribution.
5. Overall the *best* under all conditions is the proposed method.

3.2   Results on the FERET Database

We have selected a partial FERET database [8] to evaluate the performance of the algorithm under *single-sample-per-person* (SSPP) problem. This subset database contains 400 images of 200 persons. Each person has two images (**fa** and **fb**) with different facial expressions. The **fa** are used as gallery for training and **fb** images are used as probes for testing. Some interesting studies have been proposed exclusively to tackle this situation [8,10]. We compare the recognition performance when number of training images is increased gradually from 20 to 200 in steps of 20.

Table 4 suggests that the proposed method can effectively be adapted for SSPP conditions. This is apparent when number of images in the database is increased (from 40 onwards it is as good or better than all other methods). In addition, the proposed model outperformed LPP method in all the cases by a significant margin. However, when images in the database are less(20-100), PCA+DCT and PCA+Wavelet methods outperformed other methods. This is because, for less number of training samples, the possibility of scattering of data is high (as can be seen in Fig.1). Thus GMM models work well under dense data (more number of data points), which is also the type of data where DCT/wavelets could work less better.

3.3 Results on the COIL-20 Object Database

This database contains 1440 gray scale images of 20 different objects, where each object contains 72 different views of varying pose angles. Initially, an ex-

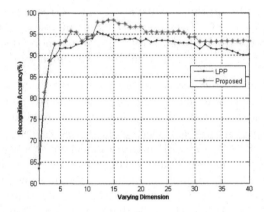

Fig. 2. Recognition accuracy for varying dimensions

periment was conducted by training first 36 samples of each object for varying dimensions. This result is depicted in Fig.2. It is quite evident that the proposed method outperformed the LPP model for all dimensions. Best recognition accuracy obtained by LPP and proposed method are 95.48% and 98.31% respectively. To make full use of the available data, another experiment was conducted by varying the number of training samples and best recognition accuracy for each case is determined. This result is shown in Table 5. It reveals that the proposed method outperformed other two methods in all the cases. This result endorses the applicability of the proposed method for object recognition.

4   Conclusions and Future Scope

The mixture-of-Laplacianfaces has been proposed in this paper. Unlike the conventional Laplacianface method, the proposed method obtains several sets of

Table 5. Best Recognition accuracy(%) on COIL-20 object database

Methods	Number of Training Samples			
	12	18	24	36
PCA[4]	78.89	81.67	88.13	93.69
LPP	80.84	83.43	92.73	95.48
Proposed	81.25	84.51	94.56	98.31

features learned through EM algorithm of GMM. The performance of the algorithm was compared with several contemporary subspace methods under both clean and noise conditions. Besides these, we have also conducted experiments to study the performance of the proposed method under SSPP conditions and also for object images. The proposed algorithm outperformed other subspace methods under these test conditions. However, the main demerit of this algorithm is that it is computationally intensive.

Nevertheless, selecting optimal number of mixtures to yield better performance is highly subjective in nature, which deserves further study. Performance of the proposed algorithm would further improve if wavelet or DCT coefficients are used instead of raw pixel values. Performance for varying percentage of DCT coefficients or different basis of wavelets makes the study a lot more interesting. This gives lot of scope for further study and investigation.

References

1. Bishop, C.M.: Pattern Recognition and Machine Learning. Springer, Heidelberg (2006)
2. Chen, W., Er, M.J., Wu, S.: PCA and LDA in DCT domain. Pattern Recognition Letters 26, 2471–2482 (2005)
3. Chien, Chen.: Discriminant waveletfaces and nearest feature classifiers for face recognition. IEEE Transactions on Pattern Analysis and Machine Intelligence 24(12), 1644–1649 (2002)
4. Murase, H., Nayar, S.K.: Visual learning and recognition of 3D objects from appearance. International Journal of Computer Vision 14(1), 5–24 (1995)
5. Kim, H.C., Kim, D., Bang, S.Y.: Face recognition using the mixture-of-eigenfaces method. Pattern Recognition Letters 23, 1549–1558 (2002)
6. Turk, M., Pentland, A.: Eigenfaces for Recognition. Journal of Cognitive Neuroscience 3(1), 71–86 (1991)
7. Belhumeur, P., Hespanha, J., Kriegman, D.: Eigenfaces vs Fisherfaces: Recognition using Class Specific Linear Projection. IEEE Transactions on Pattern Analalysis and Machine Intelligence 19(7), 711–720 (1997)
8. Wu, J., Zhou, Z.-H.: Face recognition with one training image per person. Pattern Recognition Letters 23, 1711–1719 (2002)
9. He, X., Yan, S., Hu, Y., Niyogi, P.: Face recognition using Laplacianfaces. IEEE Transactions on Pattern Analysis and Machine Intelligence 27(3), 328–340 (2005)
10. Yin, H., Fu, P., Meng, S.: Sampled FLDA for face recognition with single training image per person. Neurocomputing 69(16-18), 2443–2445 (2006)

Age Transformation for Improving Face Recognition Performance

Richa Singh[1], Mayank Vatsa[1], Afzel Noore[1], and Sanjay K. Singh[2]

[1] West Virginia Univeristy, Morgantown WV 26506, USA
richas@csee.wvu.edu, mayankv@csee.wvu.edu, noore@csee.wvu.edu
[2] Purvanchal University, Uttar Pradesh 222001, India
sksiet@yahoo.com

Abstract. This paper presents a novel age transformation algorithm to handle the challenge of facial aging in face recognition. The proposed algorithm registers the gallery and probe face images in polar coordinate domain and minimizes the variations in facial features caused due to aging. The efficacy of the proposed age transformation algorithm is validated using 2D log polar Gabor based face recognition algorithm on a face database that comprises of face images with large age progression. Experimental results show that the proposed algorithm significantly improves the verification and identification performance.

1 Introduction

Human face undergoes significant changes as a person grows older. The facial features vary for every person and are affected by several factors such as exposure to sunlight, inherent genetics, and nutrition. The performance of face recognition systems cannot contend with the dynamics of temporal metamorphosis over a period of time. Law enforcement agencies such as crime and record bureau regularly require matching a probe image with the individuals in the missing person database. In such applications, there may be significant differences between facial features of probe and gallery images due to age variation. For example, if the age of a probe image is 15 years and the gallery image of the same person is of 5 years, existing face recognition algorithms are ineffective and may not yield the desired results.

One approach to handle this challenge is to regularly update the database with recent images or templates. However, this method is not feasible for applications such as border control, homeland security, and missing person identification. To address this issue, researchers have proposed several age simulation and modeling techniques. These techniques model the facial growth over a period of time to minimize the difference between probe and gallery images. Burt and Perrett et al. [1] proposed an age simulation algorithm using shape and texture, and created composite face images for different age groups. They further analyzed and measured the facial cues affected by age variations. Tiddeman [2] proposed wavelet transform based age simulation to prototype the composite face images. Lanitis et al. [3] - [5] proposed statistical models for face simulation. They used

A. Ghosh, R.K. De, and S.K. Pal (Eds.): PReMI 2007, LNCS 4815, pp. 576–583, 2007.

training images to learn the relationship between coded face representation and actual age of subjects. This relationship was then used to estimate the age of an individual and to reconstruct the face at any age. Gandhi [6] proposed Support Vector Regression to predict the age of frontal faces. He further used the aging function with the image based surface detail transfer method to simulate face image at any younger or older age. Wang *et al.* [7] obtained the texture and shape information of a face image using PCA and used this information in reconstructing the shape and texture at any particular age. Recently, Ramanathan and Chellappa [8], [9] proposed a craniofacial growth model that characterizes the shape variations in human faces during formative years. They further developed a Bayesian age difference classifier to verify the identity between two images and to estimate the age difference between them.

In this paper, we propose a novel age transformation algorithm to minimize the age difference between two face images. The proposed algorithm is applicable in practical scenarios such as homeland security and missing person database which operates on face images with large age variations as shown in Fig 1. The proposed age transformation algorithm registers two face images in polar coordinates and minimizes the aging variations. Unlike the conventional method, we transform gallery face images with respect to probe face image and compute the verification and identification performance. Experiments are performed using 2D log polar Gabor transform based face recognition algorithm [10] on a face database of 130 individuals with images of varying age. The proposed algorithm is described in Section 2 and the experimental results are discussed in Section 3.

16 years   29 years   43 years

Fig. 1. Images showing variations in the facial characteristics of an individual at different age [11]

2 Proposed Registration Based Age Transformation Algorithm

In this section, we propose the registration based age transformation algorithm which can be used to recognize face images with significant age difference between them. We first explain the algorithm for face verification (1:1 matching) and then extend it to identification (1:N matching). The algorithm is described as follows:

Step 1: Let F_G and F_P be the detected gallery and probe face images to be matched. $F_G(x, y)$ and $F_P(x, y)$ are transformed into polar form to obtain

$F_G(r,\theta)$ and $F_P(r,\theta)$ respectively. Here, r and θ are defined with respect to the center coordinate (x_c, y_c).

$$r = \sqrt{(x-x_c)^2 + (y-y_c)^2} \qquad 0 \le r \le r_{max} \tag{1}$$

$$\theta = tan^{-1}\left(\frac{y-y_c}{x-x_c}\right) \tag{2}$$

The coordinates of eyes and mouth are used to form a triangle and the center point of this triangle is chosen as the center point, (x_c, y_c), for cartesian to polar conversion. Fig. 2 shows an example of cartesian to polar conversion around the center point. This cartesian to polar conversion eliminates minor variations due to pose and provides robust feature mapping used in the next steps of the algorithm.

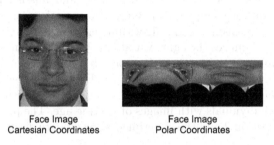

Face Image   Face Image
Cartesian Coordinates   Polar Coordinates

Fig. 2. Example of cartesian to polar coordinate conversion

Step 2: The corner features of $F_G(r,\theta)$ and $F_P(r,\theta)$ are computed using phase congruency based edge and corner detection algorithm [12]. Fig. 3 shows the output of phase congruency based edge and corner detection algorithm on a face image in polar coordinate.

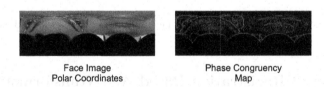

Face Image   Phase Congruency
Polar Coordinates   Map

Fig. 3. Phase congruency map of a polar face image

Step 3: A set of common corner points are selected from the corner features of $F_G(r,\theta)$ and $F_P(r,\theta)$ using the local correlation technique [13]. In the gallery and probe images, a window of size 7×7 is chosen around the corner points. One to many correlation is then performed on both the gallery and probe images and the corner pairs having the maximum correlation output are selected. Also, few corner point pairs could be selected incorrectly. To remove these incorrect pairs, we select only those pairs that have a correlation greater than 0.85. This technique yields a set of common corner point coordinates which are finally used in the registration based age transformation algorithm.

Step 4: The registration based age transformation is performed on a transformation space, S, such that

$$S = \begin{bmatrix} a & b & 0 \\ c & d & 0 \\ e & f & 1 \end{bmatrix} \tag{3}$$

where, a, b, c, d, e, f are the transformation parameters for shear, scale, rotation, and translation. We compute the transformation parameter, S', by using the common feature point coordinates and the transformation space, S,

$$S' = arg \; max_{\{S\}} \left[T\{F_P(r, \theta), S(F_G(r, \theta))\} \right] \tag{4}$$

$T\{\cdot\}$ is the feature point registration algorithm described in [14]. Thus $F_G(r, \theta)$ is registered with respect to $F_P(r, \theta)$ using the transformation parameters, S'.

Step 5: To account for both linear and non-linear variations, we first apply the global transformation followed by the local registration. Global transformation is performed by applying Step 4 on the image to minimize the global variations due to shear, scale, rotation, and translation. Local transformation is then performed on the globally registered images by applying Step 4 in blocks of size 8×8. This local registration compensates for the non-linear variations in facial features. The registered face images are finally transformed back to cartesian coordinates from polar coordinates.

Fig. 4 shows examples of gallery face image, probe face image, and age difference minimized gallery face image. Once the age difference between gallery and probe face image is minimized, face recognition algorithm can be efficiently applied to verify the identity of the probe image. The proposed algorithm can be easily extended for identification. For identifying any given probe image, first the gallery images are transformed with respect to the probe image and then the probe image is matched with all the transformed gallery images.

3   Experimental Results

To validate the performance of the proposed registration based age transformation algorithm, we use face aging database which comprises of 1578 images from 130 individuals or classes. The images are obtained partly from the FG-Net database [11] and partly collected by the authors. We divide the face database

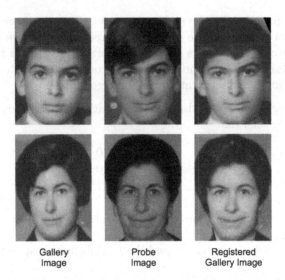

Gallery   Probe   Registered
Image   Image   Gallery Image

Fig. 4. Results of the proposed age transformation algorithm on images from the FG-Net face database [11]

into three age groups: (1) 1-18 years, (2) 19-40 years, and (3) beyond 41 years. The database is divided into these three age groups because we observed that face development depends on the age of a person. For example, development in muscles and bone structure cause significant changes in the face during the age of 1-18 years. From 19-40 years, the growth rate is comparatively lower, whereas after 40 years, wrinkles and skin loosening cause major change in facial features and appearance. It is thus very difficult to accurately model an individual's face for very large age variations such as from age 10 to age 60. Considering these factors, we divide the database into three age groups to evaluate the performance. Details of the images in these three age groups are provided in Table 1. We use one frontal face image per class as the gallery image and the remaining face images as the probe images.

Table 1. Details of the face database used for validation

Age group	Number of class	Total number of images	Total number of images per class	Average age difference in years
1-18 years	59	605	5-16	8
19-40 years	68	735	8-15	10
Beyond 41 years	34	238	4-10	5

We have used 2D log polar Gabor transform based recognition algorithm to match the face images [10]. To validate the proposed age transformation algorithm, we compute the performance of the face recognition algorithm both with and without the transformation algorithm. The experiments are divided into two parts: face verification and face identification.

3.1 Face Verification

To validate the performance of the proposed age transformation algorithm, we first apply the proposed age transformation algorithm on the gallery and probe images. Face recognition algorithm is then applied to match the transformed gallery and probe face images. The performance is computed in terms of verification accuracy at 0.01% False Accept Rate (FAR). Table 2 shows the verification performance of the face recognition algorithm with and without the proposed age transformation algorithm for the three age groups. For the age group of 1-18 years, an improvement of 37.21% is observed with the use of the proposed age transformation algorithm. Since the proposed algorithm minimizes the variations by registering gallery and probe images both locally and globally, the performance of face recognition is greatly enhanced. Similarly, for the age group of 19-40 years and beyond 41 years, the performance of face recognition is improved by 23.55% and 14.49% respectively.

Table 2. Verification results of the proposed age transformation algorithm. Verification performance is computed at 0.01% FAR.

Age group	Verification accuracy (%)		
	Without registration based age transformation	With registration based age transformation	Improvement in verification accuracy
1-18 years	21.14	58.35	37.21
19-40 years	49.87	73.42	23.55
Beyond 41 years	72.60	87.09	14.49

3.2 Face Identification

We next evaluate the performance of the proposed algorithm for face identification. Face identification is more challenging than face verification because of the high false accept rate [15]. Similar to the face verification experiment, we compute the identification accuracies with and without the proposed registration based age transformation algorithm. Fig. 5 shows the Cumulative Match Characteristic (CMC) plots for this experiment. Rank 1 to 10 identification accuracies for 1-18 years age group shown in Fig. 5(a) clearly indicate that with the proposed age transformation algorithm, the performance of face identification is improved by around 7-18%. For 18-40 years, an improvement of 18-28% is observed in the identification performance whereas beyond 41 years an improvement of 7-14% is observed. The identification experiment thus shows that the proposed age transformation algorithm effectively minimizes the age difference between gallery and probe images and improves the identification performance.

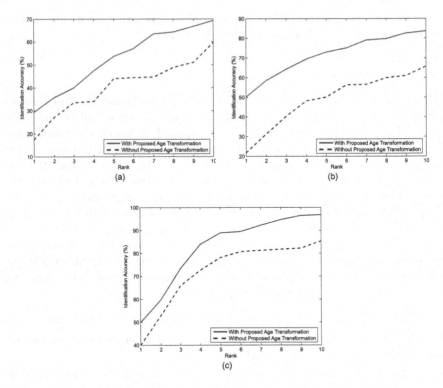

Fig. 5. CMC plot of the proposed age transformation algorithm for (a) 1-18 years age group, (b) 19-40 years age group, (c) Beyond 41 year age group

4   Conclusion

Many law enforcement applications deal with face recognition across age progression. However, existing face recognition algorithms do not perform well in such scenarios. In this paper, we proposed a registration based age transformation algorithm to minimize the age difference between gallery and probe face images. The proposed algorithm registers the gallery and probe face images by transforming them into polar coordinates. The results show significant improvement in the verification and identification performance when the proposed age transformation algorithm is applied. The results also suggest that the proposed algorithm yields better performance for 19-40 years age group and beyond 41 years age group. However, further research is required to improve the recognition performance for 1-18 years age group. Also, additional research is required when dealing with the issue of facial aging along with variations in pose, expression, illumination, and disguise.

Acknowledgment

Authors would like to acknowledge the FG-NET consortium for providing the FG-Net face database used in this research.

References

1. Burt, D.M., Perrett, D.I.: Perception of age in adult caucasian male faces: computer graphic manipulation of shape and colour information. In: Proceedings of Royal Society London, Series B 259, 137–143 (1995)
2. Tiddeman, B., Burt, M., Perrett, D.: Prototyping and transforming facial textures for perception research. IEEE Computer Graphics and Applications 21(5), 42–50 (2001)
3. Lanitis, A., Taylor, C.J., Cootes, T.F.: Toward automatic simulation of aging effects on face images. IEEE Transactions on Pattern Analysis and Machine Intelligence 24(4), 442–455 (2002)
4. Lanitis, A.: On the significance of different facial parts for automatic age estimation. In: Proceedings of International Conference on Digital Signal Processing, vol. 2, pp. 1027–1030 (2002)
5. Lanitis, A., Draganova, C., Christodoulou, C.: Comparing different classifiers for automatic age estimation. IEEE Transactions on Systems, Man, and Cybernactics 34(1), 621–628 (2004)
6. Gandhi, M.: A method for automatic synthesis of aged human facial images, Masters Thesis, Department of Electrical and Computer Engineering, McGill University (2004)
7. Wang, J., Shang, Y., Su, G., Lin, X.: Age simulation for face recognition. In: Proceedings of International Conference on Pattern Recognition, pp. 913–916 (2006)
8. Ramanathan, N., Chellappa, R.: Face verification across age progression. IEEE Transactions on Image Processing 15(11), 3349–3362 (2006)
9. Ramanathan, N., Chellappa, R.: Modeling age progression in young faces. Proceedings of IEEE Computer Vision and Pattern Recognition 1, 387–394 (2006)
10. Singh, R., Vatsa, M., Noore, A.: Face recognition with disguise and single gallery images. Image and Vision Computing (2007)
11. FG-Net Aging Database, http://www.fgnet.rsunit.com/
12. Kovesi, P.D.: Image features from phase congruency, Videre: Journal of Computer Vision Research, MIT Press 1(3) (1999)
13. Singh, R., Vatsa, M., Noore, A.: Improving verification accuracy by synthesis of locally enhanced biometric images and deformable model. Signal Processing 87(11), 2746–2764 (2007)
14. Gonzalez, R.C., Woods, R.E.: Digital image processing, 2nd edn. Prentice Hall, Englewood Cliffs (2002)
15. Li, S.Z., Jain, A.K.: Handbook of face recognition. Springer, Heidelberg (2005)

Recognizing Facial Expression Using Particle Filter Based Feature Points Tracker

Rakesh Tripathi and R. Aravind

Department of Electrical Engineering, Indian Institute of Technology Madras,
Chennai-36, India

Abstract. The paper focuses on an evaluation of particle filter based facial feature tracker. Particle filter is a successful tool in the non-linear and the non-Gaussian estimation problems. We developed a particle filter based facial points tracker with a simple observation model based on sum-of-squared differences (SSD) between the intensities. Multistate face component model is used to estimate the occluded feature points. The important distances are calculated from tracked points. Two kinds of classification schemes are considered, the hidden Markov model (HMM) as sequence based recognizer and support vector machine (SVM) as frame based recognizer. A comparative study is shown in the classification of five basic expressions, i.e., anger, sadness, happiness, surprise and disgust. The tests are conducted on Cohn-Kanade and MMI face expression databases.

1   Introduction

Facial expression is an important cue for humans communication and emotion recognition. Expression recognition [1], [2], [3], [5] provides computers an aid to understanding human behavior. Facial expression is a necessary feedback from the user. It has many applications in behavioral science, medicine, security, man-machine interaction and animation. There are three basic elements of a facial expression recognition system, i.e., feature extraction, data representation and classification. Feature extraction involves the capture of deformation in face during the change of expression. So we need to track the important (key) points with precision. The recognition rate is affected by the goodness of tracked face points. A variety of methods exist for tracking.However, accuracy in the estimation of displacement is always under scrutiny. Here, we present a particle filter [6], [7] based feature tracker for the key points. Motivated from [8] work, the SSD based observation likelihood is formulated. At this stage, we assumed that the key-points are given for expression recognition in first frame. For our implementation, we have manually marked the facial points in the first frame. Our first goal is to track key-points in a face video which must be robust to large deformable motion, low textured region, shadows, small illumination variation (specially eye region with spectacles), occlusion due to some expressions and facial hair. The particle filter based tracker is robust to all above situations except occlusion due to some expressions. The geometric information from face

A. Ghosh, R.K. De, and S.K. Pal (Eds.): PReMI 2007, LNCS 4815, pp. 584–591, 2007.

[3] is used to recover the occluded points. The euclidean distances between face features are computed using these tracked points.

It is shown in [4] that the tracking of feature points are very effective in producing facial action coding system (FACS). Most of the efforts made to know FACS using points tracking [3], [4], [7], and it is believed that the accuracy of FACS detection directly related to facial expression. Though, we are interested to produced rigorous results for expression recognition without using any rule file created from FACS. HMM [1] and SVM [5] are used for evaluation of features obtained from optical flow computation. Motivated from these works, we formed our second research goal as the evaluation of the goodness of tracked points in different classification schemes. The performance of extracted features is evaluated using both the classifiers, i.e., sequence based classifier the HMM, and frame based classifier the SVM. The results are shown on Cohn-Kanade and MMI face expression database [9], [10], and a set of videos (with 8 subjects) captured in our laboratory.

2 Particle Filter Based Facial Points Tracker

The particle filter employs a set of weighted particles $\{\mathbf{x}_k^i; \ i = 1, ..., N_s\}$ at time k with associated weights $\{w_k^i; \ i = 1, ..., N_s\}$ to represent the posterior pdf $p(\mathbf{x}_k|\mathbf{z}_{1:k})$. The weights are normalized such that $\sum_i w_k^i = 1$. If $\{\mathbf{z}_{1:k}^i; \ i = 1, ..., N_s\}$ is a set of all available measurements up to time k, then the posterior density at k can approximated as $p(\mathbf{x}_k|\mathbf{z}_{1:k}) \approx \sum_{i=1}^{N_s} w_k^i \, \delta(\mathbf{x}_k - \mathbf{x}_k^i)$. The sequential importance sampling/resampling particle filter for our application is described below.

1. State Prediction: The state is defined to be the coordinate of particle location. Let the location of ith particle of kth frame corresponding to a feature point is denoted by (x_k^i, y_k^i) and intensity at this point by $I_k(x_k^i, y_k^i)$, so that the state is $\mathbf{x}_k^i = [x_k^i \ y_k^i]$. The transition model is linear with the Gaussian noise of zero mean and known variance σ^2 as process noise $\mathbf{v}_k$. The transition equation is $\mathbf{x}_k^i = \mathbf{x}_{k-1}^i + \mathbf{v}_k$.

2. Formulation of the Observation Model: Our observation model is based upon the sum-of-squared-differences (SSD) described in [8]. We define the observation $\mathbf{z}_k$ at a given $\mathbf{x}_k^i$ to be the image intensities at pixel locations in the window of size $(2N + 1) \times (2N + 1)$. The displacement between i_{th} particle of a feature point in the first frame located at (x_1^i, y_1^i), and in the k_{th} frame at (x_k^i, y_k^i) is $(d_x^i, d_y^i) = (x_1^i - x_k^i, \ y_1^i - y_k^i)$. The SSD is defined as,

$$D(d_x^i, d_y^i) = \sum_{m=-N}^{N} \sum_{n=-N}^{N} \left(\frac{I_1(x_1^i + m, y_1^i + n)}{M_1} - \frac{I_k(x_1^i + d_x^i + m, y_1^i + d_y^i + n)}{M_k^i} \right)^2$$

$$(1)$$

where M_1 is mean of image intensity in the window of marked point by user in the first frame, and M_k^i is mean of intensity in the window of frame k and particle i.

3. Weighting the Samples: The response distribution $R(d_x^i, d_y^i)$ is formulated as, $R(d_x^i, d_y^i) = \exp(-D(d_x^i, d_y^i)/\sigma_s^2)$, where σ_s is a scaling parameter given as $\sigma_s = \tau/M_1$. The normalized weight $w_k^i = R(d_x^i, d_y^i)/ \sum_i R(d_x^i, d_y^i)$.

4. State Update and Estimation: The resampling eliminates particles that have small weights and concentrate on particles with large weights. In our case, SSD is very low at positions of exact match and hence likelihood is very high. Hence, we need to do resampling at each time step [6]. The posterior mean estimator is used for determining the position of feature points. The resampling algorithm produces equally weighted particles, so the posterior mean estimate $\mathbf{S}_k^{Mean}$ at time step k is given by

$$\mathbf{S}_k^{Mean} = \sum_{i=1}^{N_s} w_k^i \mathbf{x}_k^i = \frac{1}{N_s} \sum_{i=1}^{N_s} \mathbf{x}_k^i \tag{2}$$

2.1   Incorporating Face Geometry

The results of tracking using particle filter is shown in Fig. 1. It works well for expressions having enough observations for each points. However, for the tightly closed lip and the closed eye as in anger (see third figure of Fig. 1), particle filter cannot track the exact position of center of upper and lower lip and eyelid. It is because of the SSD based likelihood, partial occlusion of key points make the likelihood very low at true positions. However, if we can know that the mouth (or eye) is tightly closed, then the position of center points of upper and lower lip (or eyelid) is the middle point of two corners of mouth (or eye). To know the state of features (tightly closed or open), we used a variant of multistate facial component model (MFCM) proposed in [3]. A method to detect the state of mouth and eye is described in [3]. The number of lip pixels in tightly closed mouth is very less than open situation. The lip and non-lip pixels are modeled and classified by the Gaussian mixture model. Whereas, the presence

Fig. 1. Tracking using particle filter and MFCM representation

of iris gives important information regarding state of eye. The iris is visible for open eye only. The edginess image [11] followed by binary conversion gives good discriminating information for the presence of iris (Fig. 2(a) and Fig. 2(c)). The iris mask shown in Fig. 2(b) is used to calculate the threshold based on number of white pixels in the region of mask and average iris intensity [3]. The change of states is shown in last two figures of Fig. 1.

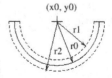

(a) Edginess image for open and closed eyes.

(b) Iris mask, we consider the pixels between r1 and r2.

(c) Corresponding binarized images.

Fig. 2. Iris is searched between left and right eye corners with mask shown in Fig. 2(b)

3 Facial Expression Recognition

3.1 Feature Representation

We calculated the Euclidean distances between tracked feature points. These distances represent the change of facial expression [3], [7]. Upper face consists the motion of eye, brow, cheek and forehead, and gives 9 parameters and lower face consists motion of mouth, and gives 8 parameters listed in Table 1 and shown in Fig. 3.

3.2 Classification of Expressions

Two classifiers are used to evaluate the goodness of extracted features, sequence based and frame based. The sequence based recognizer exploits the temporal relationship between adjacent frames along with the spatial information. The HMM is widely used for making a sequence of decisions from an interconnected observations. Five HMMs are modeled using Baum-Welch algorithm, one for each expression: anger, sadness, happiness, disgust and surprise. The log-likelihood obtained using forward algorithm from each HMM is used to classify an expression corresponding to the highest probability. HMM delays the decisions by 10

Table 1. Important distances. Subscript 0 represents first frame distance.

Inner brow	$\frac{HL2-HL2_0}{HL2_0}$	Lip height	$\frac{(HU+HL)-(HU_0+HL_0)}{HU_0+Hl_0}$
Outer brow	$\frac{HL1-HL1_0}{HL1_0}$	Lip Width	$\frac{W-W_0}{W_0}$
Eye Height	$\frac{(HEL1+HEL2)-(HEL1_0+HEL2_0)}{HEL1_0+HEL2_0}$	Lip corner	$\frac{HLC-HLC_0}{HLC_0}$
Brows distance	$\frac{D-D_0}{D_0}$	Top lip	$\frac{HU-HU_0}{HU_0}$
Cheek motion	$\frac{CL-CL_0}{CL_0}$	Bottom lip	$\frac{HL-HL_0}{HL_0}$
Nose to upper lip width	$\frac{HN-HN_0}{HN_0}$	Chin motion	$\frac{HC-HC_0}{HC_0}$

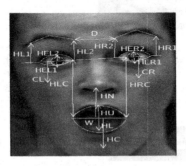

Fig. 3. Important distances between face features

to 15 frame because first it generates all the connected hidden sequence of certain length and then calculates the likelihood for the decisions based on optimal sequence of hidden states.

The frame based recognizer gives decision after each frame independent of previous decisions. Although the frame based recognizer loose the temporal information between adjacent frames, recognition at frame level is specially useful in the situations of mixture of expressions in a single video. We used SVM with Gaussian kernel as frame level classifier. Features calculated from each frame are separately fed to SVM for training and classification. The one-against-all strategy is applied in multiclass classification with SVM.

4   Experimental Results

The database is re-sized to 180 × 280 from cropped facial image using bicubical interpolation. All the frames have frontal view of the face with small in-plane rigid head motion. It is assumed that key-points have limited motion of 20 to 40 pixels. Total 21 facial points are tracked. The SSD window size is 13 × 13. Fifteen important frames per video is chosen, showing the change of expression from the neutral.

For the classification using HMM, each frame is represented by a vector of 17 Euclidean distances, and each video is converted into temporally connected sequence (observable). The statistics regarding the number of training and testing subjects with different kinds of database in shown in Table 2 and Table 3. The subjects used for training is not included to determine the recognition rates (person independent test). The average recognition rates are calculated by comparing human observed correct number of videos classified in rest of the videos. We also captured the expression videos of 8 persons in our laboratory with a motivation of measuring performance with different kind of training sets, keeping the same parameters for feature extraction (different thresholds) Table 4. However, sad expression is felt difficult to produce, so results are shown only with 4 expressions.

It is observed that the features associated with surprise and happiness expressions are much discriminative than other expressions. We can recognize these two expressions from all the videos of Cohn-Kanade database correctly using the lower face features only. Similarly in case of MMI database except one video, all other videos of these two expressions are recognized correctly. There is no ambiguity in the recognition of surprise and happiness expressions using full face features. It is seen that the lower face features for disgust expression is not discriminative. However, the more videos can be recognized by using full face features. Anger and sadness expressions have mutually overlapped features. We felt that the sadness is most difficult expression to recognize. The recognition performance of various training sets with our made database is shown in Table 4. It can be observed that the parameters used for different database are tuned good in comparison to the results from different database.

Table 2. Classification with 3 training and 7 test videos of Cohn-Kanade face expression database using HMM

Face Features	Surprise	Sadness	Anger	Disgust	Happiness	Avg. recognition rates (in %)
Lower	7	4	4	3	7	71.5
Upper+Lower	7	4	5	5	7	80

Table 3. Classification with 3 training and 9 test videos of MMI face expression database using HMM

Face Features	Surprise	Sadness	Anger	Disgust	Happiness	Avg. recognition rates (in %)
Lower	8	4	5	5	8	66.7
Upper+Lower	9	6	6	7	9	82.2

An extra class of neutral face action is considered while classification using SVM because all the videos have first few frames as neutral. The SVM is trained with 120 frames (100 with expression and 20 neutral). The experimental results with different kinds of database is given in Table 5 and Table 6. It should be noted

Table 4. Classification with videos captured in our lab. of 8 persons using HMM

Training set	Face features	Surprise	Anger	Disgust	Happiness	Average recog. rate
Cohn-Kanade	Lower face	8	5	4	7	75
	Full face	8	7	5	7	84.4
MMI	Lower face	7	4	4	7	68.8
	Full face	8	5	5	7	78.1

that these recognition rates are obtained from the confusion matrix using one-against-all strategy. The recognition rate is calculated by observing the correct number of frames within the total number of frames of same expression and given in percentage. The number of samples used for testing are 60 frames for each class with the Cohn-Kanade database and 70 frames with the MMI database. The neutral expression is not considered in determining average recognition rate from SVM because the number of samples for neutral is very high (200 frames in Cohn-Kanade database and 280 frames in MMI database), first few frames of each video, and recognition rate for neutral expression is more than 94%. The major difference in the classification by both classifier in found in the case of disgust and anger. The SVM performs better than HMM for recognizing disgust using only lower face features. Whereas the results from anger is just opposite to that of disgust.

Table 5. Recognition rates (in %) with 120 training and 500 test frames (200 neutral and 60 per expression) of Cohn-Kanade face expression database using SVM

Face Features	Neutral	Surprise	Sadness	Anger	Disgust	Happiness	Avg. recognition rates
Lower	94.6	93.6	76.0	36.0	77.5	84.2	73.5
Upper+Lower	96.6	98.2	84.1	50.1	78.7	79.0	78.0

Table 6. Recognition rates (in %) with 120 training and 630 test frames (280 neutral and 70 per expression) of MMI face expression database using SVM

Face Features	Neutral	Surprise	Sadness	Anger	Disgust	Happiness	Avg. recognition rates
Lower	91.4	89.4	50.2	51.3	71.5	87.4	70
Upper+Lower	92.2	91.2	64.7	58.5	77.4	89.2	76.2

5   Conclusion

In this paper, we evaluated the goodness of features obtained from particle filter based feature points tracker. The particle filter with the Gaussian prior and linear transition model is able to generate particles in interesting regions. The SSD based observation model is enough robust to give the information from low textured regions also. However, the expressions like anger occludes the feature points. So the output from the particle filter is modified using geometrical information from face. The HMM and the SVM can classify the two expressions surprise and happiness clearly. With HMM sad is the most difficult expression to recognize, whereas features associated with anger gives poor performance with SVM. In general full face features gives improvement in the classification. We feel that the classification performance can be improved for the other expressions by using transient features also [3].

References

1. Yacoob, Y., Davis, L.S.: Recognizing Human Facial Expression from Long Image Sequences using Optical Flow. IEEE Trans. PAMI 18(6), 636–642 (1996)
2. Bartlett, M.S., Braathen, B., Littlewort-Ford, G., Harshey, J., Fasel, I., Marks, T., Smith, E., Sejnowski, T., Movellen, R.L.: Automatic analysis of spontaneous facial behaviour: A final project report, Technical Report INC-MPLab, UCSD (2001)
3. Tian, Y., Kanade, T., Cohn, J.: Recognizing action unit for facial expression analysis. IEEE Trans. PAMI 23(2), 97–114 (2001)
4. Cohn, J.F., Zlochower, A.J., Lian, J., Kanade, T.: Automated face analysis by feature point tracking has high concurrent validity with manual FACS coding. Psycophysiology 36, 35–43 (1999)
5. Anderson, K., McOwan, P.W.: A real time automated system for the recognition of human facial expression. IEEE Trans. System, Man, and Cyb.–PartB: Cyb. 36(1), 96–105 (2006)
6. Arulampalam, M.S., Maskell, S., Gordan, N., Clapp, T.: A Tutorial on particle filter for online nonlinear/non-Gaussian Bayesian tracking. IEEE Trans. Signal Processing 50(2), 172–188 (2002)
7. Valstar, M.F., Patras, I., Pantic, M.: Facial action unit detection using probabilistic actively learned support vector machines on tracked facial Point data. In: CVPR 2005
8. Singh, A., Allen, P.: Image flow computation: An estimation-theoretic framework and a unified perspective. CVGIP: Image Understanding 56(2), 152–177 (1992)
9. Kanade, T., Cohn, J.F., Tian, Y.: Comprehensive database for facial expression analysis. In: Fourth International Conference on Automatic Facial and Gesture Recognition, Grenobel, France (2004)
10. Pantic, M., Valstar, M.F., Rademaker, R., Maat, L.: In: Proc. IEEE Int'l Conf. Multmedia and Expo (ICME 2005), Amsterdam, The Netherlands (July 2005)
11. Kumar, P.K., Das, S., Yegnanarayana, B.: One Dimension processing of the images. In: International Conference on Multimedia Processing Systems, Chennai, India (2000)

Human Gait Recognition Using Temporal Slices

Shruti Srivastava and Shamik Sural

School of Information Technology
Indian Institute of Technology, Kharagpur, India
shrutisr@gmail.com, shamik@sit.iitkgp.ernet.in

Abstract. Gait along with body structure has been recognized as a potential biometric feature for identifying human beings. The spatial and temporal shape of motion of an individual is usually the same for all gait cycles and is considered to be unique to that individual. In this paper we introduce a Temporal Slice based approach for gait recognition. Temporal Slices are a set of two-dimensional images extracted along the time dimension of an image volume. They encode a rich set of visual patterns for similarity measure and have been widely used for motion detection. We show that the features obtained from tensor histogram of these temporal slices can be efficiently used as gait features for recognition of human beings.

Keywords: Gait biometrics, Temporal Slices, Tensor Histogram, Multiclass SVM.

1 Introduction

Recognition and verification of human beings is a major security concern in many restricted areas. There are various methods used to address this problem including use of signatures and passwords, biometric feature based methods like facial recognition, iris scan, fingerprints, etc. But none of these methods can be hidden from the suspects or people being monitored. Rather, they require cooperation from the subject in some way or the other. Till now the only perceivable biometric feature that can be captured from a distance is the gait. People walking in a passage can be easily monitored by hidden cameras, that too without hindering the normal activities in that arena.

Gait is the style of walking of a person. It can be termed as the shape of motion. Gait, together with the body structure, can be used as an efficient biometric feature for uniquely identifying a person. In fact, early medical studies show that there are 24 different components to human gait, and that, if all the measurements are considered, the gait of an individual is unique. Statistics show that if there are no external factors involved, the possibility of two individuals having the same pattern formed by their gait and body structure is also less [1,2]. External factors that can affect gait includes prior knowledge of being monitored, voluntary attempt to copy other's style of walking, injury, pregnancy, etc.

The gait of a person is a periodic activity with each gait cycle covering two strides; the left foot forward and the right foot forward. Gait biometrics can be

A. Ghosh, R.K. De, and S.K. Pal (Eds.): PReMI 2007, LNCS 4815, pp. 592–599, 2007.

considered to be derived from the shape and dynamics of the two strides. The gait dynamics is vulnerable to change under varying factors like walking surface, walking speed, etc. So, gait dynamics alone cannot be considered as a stable source of biometric information [3]. Both the shape and the dynamics of a gait cycle for a person forms the gait biometric feature for that person.

Gait volume is a collection of image sequences of an individual in a gait cycle. Temporal slices provide rich visual pattern and have so far been used for motion characterization. It can also have the potential of characterizing human motion as well as distinguishing one subject from another.

2   Related Work

Current gait recognition approaches may be classified into two main classes [1,2,4,5] Model based Approach and Motion based Approach. Both methodologies follow the general framework of feature extraction, feature correspondence and high-level processing. The major difference is with regard to feature correspondence between two consecutive frames. Model based approach aims at modeling human body structure or motion and extracts image features from these models. They generally perform model matching in each frame of a walking sequence to match the model parameters such as trajectories, limb length and angular speed [5]. Motion based or appearance based approach characterizes the whole body movement pattern by a compact representation regardless of the underlying structure. It can be further classified into state-space method and spatio-temporal method [2]. State space method considers gait motion to be composed of a sequence of static body poses and recognizes it by considering temporal variations of observations with respect to that static pose. Spatio-temporal method characterizes the spatio temporal distribution generated by gait motion in its continuum.

Liu and Sarkar [3] categorized gait recognition approaches into the following three types: i) Temporal Alignment Based ii) Static Parameter Based and iii) Silhouette Shape Based. The temporal alignment based approach emphasizes both shape and dynamics. It aims at extracting features from silhouettes, pre-shape representation, silhouette parts and Fourier descriptors. A sequence of these features is aligned with the sequence to be matched by either simple temporal correlation [2], dynamic time warping [6], hidden Markov models [4], phase locked-loops or Fourier analysis. The distance measure is either Euclidean [2], simple dot product-based or Procrustes distance. The Static Parameter Based approach uses parameters to characterize gait dynamics, such as stride length, cadence, stride speed and sometimes static body parameters like ratio of various body parts. The Silhouette Shape Based approach emphasizes on silhouette shape similarity and disregards temporal information. This approach generally transforms the silhouette sequence into a single image representation such as averaged silhouette [5] or an image representation derived from the width vectors in each frame (Freize patterns).

Sagawa et al. [8] applied fourier transforms on the sliced planes of image volume to extract the frequency characteristics of gait for an individual. They used these frequency characteristics for tracking the moving individuals in a video shot. Ngo [7] used temporal histogram to find the distribution of local orientation in the temporal slices obtained from the image volume of a video shot. The feature vector obtained from orientation histogram was used for clustering and retrieval of video shots.

3   Temporal Slice Based Approach

Our approach is based on the analysis and processing of gait patterns in temporal slice images. A temporal slice is a set of two-dimensional (2-D) images extracted along the time dimension of an image volume. The gait volume is the set of temporal images in a gait cycle and the temporal slices are extracted from this gait volume. One dimension of the temporal slice is time and another dimension is the x or y axis of the spatial frames in a gait volume.

Fig. 1. Sequence of Frames in a Gait cycle (here, N=7)

If there are N number of frames in a gait cycle and the image size is height × width (Figure 1), the gait volume will be a 3D representation of dimension height × width × N (Figure 2). Thus, gait volume has one dimension as time while the other dimension has 2D frame sequences in the spatial domain. Figure 2 represents a gait volume of seven frame sequences in time domain. The horizontal slice (Figure 3 (a)) has one dimension as the time and another dimension as x axis (i.e. y value is fixed). In vertical slice (Figure 3 (b)) dimensions are time and y axis (height), value of x (width) is fixed.

We apply Gaussian derivatives of the horizontal slice to obtain the partial derivatives along the x axis ($\mathbf{H_x}$) and the partial derivatives along the time axis ($\mathbf{H_t}$). Using these gradient vectors we calculate the gradient tensor as follows:

$$Q = \begin{pmatrix} \mathbf{H_x} \\ \mathbf{H_t} \end{pmatrix} \begin{pmatrix} \mathbf{H_x} & \mathbf{H_t} \end{pmatrix} \tag{1}$$

$$= \begin{pmatrix} \mathbf{H_x}^2 & \mathbf{H_x H_t} \\ \mathbf{H_x H_t} & \mathbf{H_x}^2 \end{pmatrix}$$

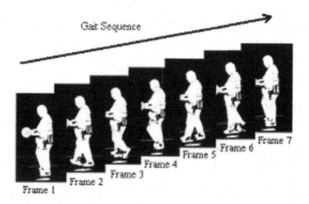

Fig. 2. Gait Volume

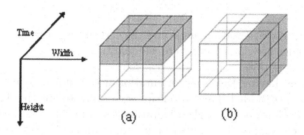

Fig. 3. (a) Horizontal Slice (b) Vertical Slice of a gait volume

Which can be written as:

$$Q = \begin{pmatrix} q_{11} & q_{12} \\ q_{21} & q_{22} \end{pmatrix} \tag{2}$$

Here, H_x and H_t are multiplied pixel by pixel, i.e. $H_x^2(i,j) = H_x(i,j) * H_x(i,j)$ and so on.

After calculating gradient tensor for all the horizontal slices, we get a gradient tensor cube. The gradient tensor cube consists of three cubes q_{11}, q_{12}, q_{22} each of the same dimension as that of the gait volume.

On this gradient tensor cube, we use Sobel averaging over a 3x3 window (on each of q_{11}, q_{12}, q_{22} separately to get S_{11}, S_{12}, S_{22}). The averaging is done in spatial domain, i.e. by fixing the time value. This gives us the structure tensor (τ).

$$\tau = \begin{pmatrix} S_{11} & S_{12} \\ S_{21} & S_{22} \end{pmatrix} \tag{3}$$

Here, S_{11}, S_{12}, S_{22} are the 3D cubes of the same size as that of the gait volume.

The Sobel averaging is done for edge detection; S_{11} gives orientation in x direction, S_{22} in time direction and S_{12} in both the directions.

The angle of rotation of the principle axes of the structure tensor for each pixel is given by:

$$\theta = \frac{1}{2} \arctan \frac{2S_{12}}{S_{11} - S_{22}} \tag{4}$$

The local orientation of a pixel can be written as:

$$\phi = \begin{cases} \theta - \frac{\pi}{2} & \theta > 0 \\ \theta + \frac{\pi}{2} & \text{otherwise} \end{cases}$$

Thus, $\phi = \left[-\frac{\pi}{2}, +\frac{\pi}{2}\right]$

The certainty factor with which a pixel has a particular local orientation is given by:

$$c = \frac{(S_{11} - S_{22})^2 + 4S_{12}{}^2}{(S_{11} + S_{22})^2} \tag{5}$$

After getting the orientation angle and the certainty factor for each pixel, the orientation angles are quantized. Here we quantize the orientation angle into 8 levels between $\left[-\frac{\pi}{2}, +\frac{\pi}{2}\right]$.

All the pixels having the same orientation are grouped together and their certainty factors are added. This is done for each frame to obtain a 1D orientation histogram for each frame. This 1D orientation is summed across the time domain to give a vector (equal to the number of quantization levels of orientation angle), which constitutes a feature vector for the horizontal slices.

Similar procedure is applied on the vertical slices to get another set of features. The feature vector from horizontal slices and the feature vector from vertical slices are clubbed together to give the feature vector of the gait of a given individual.

We use Multiclass Support Vector Machine (SVM)[15] for training and testing our model. SVM uses discriminating method of training and predicts multivariate outputs. Let $S = \{(\boldsymbol{x}_1, y_1), \ldots, (\boldsymbol{x}_m, y_m)\}$ be a set of m training examples. $\boldsymbol{x}_i$ is a vector of the training dataset ($\boldsymbol{x}_i \in \chi$) and y_i is an integer from the set $Y = \{1, \ldots, k\}$, where k is the number of classes. A multiclass classifier is a function H: $\chi \rightarrow Y$ that maps an instance $\boldsymbol{x}$ to an element y of Y. The classifier used by us is of the form:

$$H_M(\boldsymbol{x}) = arg \max_{r=1}^{k} \{M_r . \boldsymbol{x}\} \tag{6}$$

The matrix M is of size k×n, n is the number of features in the vector, M_r is the rth row of M. The predicted label for a probe vector $\boldsymbol{x}$ is the index of the row attaining the highest similarity score with $\boldsymbol{x}$. The empirical error is given by:

$$\epsilon_S(M) = \frac{1}{m} \sum_{i=1}^{m} [\max_r \{M_r . \boldsymbol{x} + 1 - \delta_{y,r}\} - M_y . \boldsymbol{x}] \tag{7}$$

$\delta_{p.q} = 1$ if p=q and 0 otherwise. This bound is zero if the confidence value of the correct label is larger by at least 1 than the confidence assigned to rest of the labels. Otherwise there is a loss which is linearly proportional to the difference between the confidence of the correct label and the maximum among the confidence of other labels.

4 Result and Conclusion

We have used CMU's Mobo dataset [9] for our experiments. It comprises of indoor video sequences of 25 subjects. Different modes of walking are considered i.e. walking on an inclined plane, walking with a ball in one hand, fast walk and slow walk. A person in different modes of walk is considered to be of different class. So, if there are five people in our dataset each recorded for all four modes of walk, there is a total of twenty classes. The binary silhouettes in the dataset are normalized and horizontally aligned. By normalization we mean to proportionally resize each silhouette image so that all silhouettes have the same height. Horizontal alignment is to center the upper half silhouette part with respect to its horizontal centroid.

The CMU dataset contains binary silhouettes of size 640×486 and there are 34 such frames per gait cycle. So, the dimension of gait volume is 640×486×34. For horizontal slices we fix the first dimension, while for the vertical slices we fix the second dimension. Thus, the dimension of a horizontal slice is 486×34 while that for vertical slice is 640×34.

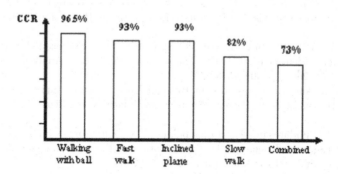

Fig. 4. The Correct Class Recognition (CCR) percentage

We have used 10 fold cross validation technique for testing our model. We tested on all the four types of walk separately. The correct class recognition (CCR) for individuals walking with a ball was 96.5%. For individuals walking at high pace and for individuals walking on an inclined plane, the CCR was 93%. CCR for slow walk was 82%. When our training and testing data contained feature vectors from all the modes of walking, the CCR achieved was 73% (Figure 4).

Algorithm Used	Slow walk	With Ball	Fast walk	Inclined plane
CMU [11]	100	92	76	-
UMD [12]	72	-	32	-
UMD [13]	72	-	12	-
Georgia Tech*	-	50	45	-
MIT [14]	100	50	64	-
Baseline [10]	92	88	72	-
Temporal Slice	**82**	**96.5**	**93**	**93**

* http://www.cc.gatech.edu/cpl/projects/hid/CMUexpt.html
- result unknown

Fig. 5. Comparative Results in CCR of different Algorithms on CMU Dataset (Ref. [10])

A comparative result is shown in Figure 5. It gives the CCR of various algorithms tested on CMU Mobo dataset. We have taken this data from Sarkar's work [10] and included our result. Our algorithm performs better than others for fast walk and walking with a ball. There was no data available for inclined plane walk, so a comparison could not be done. Thus, temporal slice can be potentially used as a feature for gait recognition.

References

1. Wang, L., Tan, T., Ning, H., Hu, W.: Automatic Gait Recognition Based on Statistical Shape Analysis. IEEE Transactions on Image Processing 12(9), 1120–1131 (2003)
2. Wang, L., Tan, T., Ning, H., Hu, W.: Silhouette Analysis Based Gait Recognition for Human Identification. IEEE Transactions on Pattern Analysis and Machine Intelligence 25(12), 1505–1518 (2003)
3. Liu, Z., Sarkar, S.: Improved Gait Recognition by Gait Dynamics Normalization. IEEE Transactions on Pattern Analysis and Machine Intelligence 28(6), 863–876 (2006)
4. Kale, A., Sundaresan, A., Rajagopalan, A.N., Cuntoor, N.P., Roy-Chowdhury, A.K., Kruger, V., Chellappa, R.: Identification of Humans Using Gait. IEEE Transactions on Image Processing 13, 1163–1173 (2004)
5. Han, J., Bhanu, B.: Individual Recognition using Gait Energy Image. IEEE Transactions of Pattern Recognition and Machine Intelligence 28(2), 316–322 (2006)
6. Bolgouris, N.V., Plataniotis, K.N., Hatzinakos, D.: Gait Recognition using Dynamic Time Warping. In: Sixth IEEE Workshop on Multimedia Signal Processing, pp. 969–979 (2004)
7. Ngo, C.W., Pong, T.C., Zhang, H.J.: On Clustering and Retrieval of Video Shots through Temporal Slice Analysis. IEEE Transactions on Multimedia 4(4), 446–458 (2002)
8. Sagawa, R., Makihara, Y., Echigo, T., Yagi, Y.: Matching Gait Image Sequences in the Frequency Domain for Tracking People at a Distance. In: Proceedings of 7th Asian Conference on Computer Vision, vol. 2, pp. 141–150 (2006)

9. Gross, R., Shi, J.: The CMU Motion of Body (MoBo) Database, Tech. Report CMU-RI-TR-01-18, Robotics Institute, Carnegie Mellon University (2001)
10. Sarkar, S., Phillips, P.J., Liu, Z., Robledo-Vega, I., Grother, P., Bowyer, K.W.: The Human ID Gait Challenge Problem: Data Sets, Performance, and Analysis. IEEE Transactions on Pattern Analysis and Machine Intelligence 27(2), 162–177 (2005)
11. Collins, R., Gross, R., Shi, J.: Silhouette-Based Human Identification from Body Shape and Gait. In: Proceedings of International Conference on Automatic Face and Gesture Recognition, pp. 366–371 (2002)
12. Kale, A., Rajagopalan, A., Cuntoor, N., Krueger, V.: Gait-Based Recognition of Humans using Continuous HMMs. In: Proceedings of International Conference on Automatic Face and Gesture Recognition, pp. 321–326 (May 2002)
13. Ben Abdelkader, C., Cutler, R., Davis, L.: Motion-Based Recognition of People in Eigengait Space. In: Proceedings of International Conference on Automatic Face and Gesture Recognition, pp. 267–272 (2002)
14. Lee, L., Grimson, W.: Gait Analysis for Recognition and Classification. In: Proceedings of International Conference on Automatic Face and Gesture Recognition, pp. 155–162 (2002)
15. Crammer, K., Singer, Y.: On the Algorithmic Implementation of Multiclass kernel based Vector Machine. Journal of Machine learning Research, 265–292 (2001)

Accurate Iris Boundary Detection in Iris-Based Biometric Authentication Process

Somnath Dey and Debasis Samanta

School of Information Technology,
Indian Institute of Technology, Kharagpur, India-721302
somnath_dey2003@yahoo.co.in, dsamanta@iitkgp.ac.in

Abstract. This paper presents an efficient technique for accurate detection of iris boundary, which is an important issue for any iris-based biometric identification system. Our proposed technique follows scaling, histogram equalization, edge detection and finally removal of unnecessary edges present in the eye image. Scaling and removing unnecessary edges enables us to reduce the search space for iris boundary. Experimental results show that with our approach it is possible to detect iris boundary as much as 98% of the eye images in CASIA database accurately and it needs only 25% time compared to the existing approaches.

Keywords: Iris recognition; biometric authentication; image segmentation; image processing.

1 Introduction

Now a days, biometric system is widely used for automatic recognition of an individual based on some unique characteristic owned by the individual. Iris recognition is a method of biometric identification based on high-resolution images of the irises of an individual's eyes. Iris is a thin contractile diaphragm, which lies between pupil and the white sclera of a human eye. Different parts of an eye image is shown in Fig. 1.

In iris biometric system, an important task is iris localization. It is observed that the iris localization task is the most computationally intensive task. In iris localization task, we try to locate iris part from an eye image. Iris part

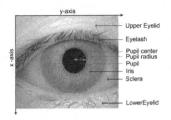

Fig. 1. The typical components in an eye image

A. Ghosh, R.K. De, and S.K. Pal (Eds.): PReMI 2007, LNCS 4815, pp. 600–607, 2007.

localization is necessary to isolate the iris part of the image in between the iris boundary (between sclera and iris) and outside the pupil. This task mainly consists of two sub tasks: detecting pupil boundary (between pupil and iris), and detecting iris boundary (between iris and sclera).

Efficiency of iris recognition in any iris based biometric authentication process greatly relies on how much accurately we detect the iris part from an eye image. The standard procedure to detect an iris boundary is to detect pupil center, the pupil radius, and subsequently use these information to find out iris boundary. Even though a successful detection of pupil helps us to limit the search space significantly, still an issue remains. The pupil and iris center are not necessarily concentric. We may trap in an improper outcome if the issue as stated above is not taken into consideration. Therefore, finding the coordinate of the iris center and radius are must.

In this paper, we detect the pupil boundary and capture pupil information (that is, pupil radius and pupil center) initially. This pupil information is used to detect iris center and radius followed by detecting the iris boundary. We propose an efficient approach to achieve the above mentioned tasks. In addition to these, we also address other issues such as, low contrast between sclera and iris, non uniform illumination, noise due to eye lashes, iris region obscured by eyelids and eye lashes, white spot due to reflection of light etc.

2   Related Work

Daugman's system [1,2] uses integro-differential operator to detect pupil and iris boundary. This operator is sensible to the specular spot reflection of non diffused artificial light. Wildes [3] uses binary edge map and voting each edge points to instantiate particular contour parameter values to detect pupil and iris boundary. Li Ma et al. [4] calculate the summation of intensity value along each row and each column. They choose the particular row and column along which summations are minimum. That row and column are used as approximate x- and y-coordinate of pupil center. Once the approximate pupil center is chosen they apply Canny edge detection [5] and Hough transformation [6] in a rectangular region centered at pupil center to detect pupil and iris boundary circle. In [7,8], Canny edge detection and Hough transform is used to detect pupil and iris boundary. J. M. H. Ali et al. [9] use Laplacian of Gaussian (LOG) operator for edge detection and median filter to remove the garbage pixels contain in edge of iris image. They count the black pixels in each row and column and choose the first pixel last pixel positions of the row and column which contain the maximum number of black pixels to find out pupil center and pupil radius. Mid-point algorithm of circle and ellipse are used to fit pupil boundary. Similarly, boundary fitting technique by using a coarse scale is applied to locate iris boundary. C. Tisse et al. [10] use integro-differential operators [1] with Hough transform strategy. They use gradient decomposed Hough transform to find the pupil center and iris center. In [11], a linear threshold and Freeman's chain code is used to isolate the pupillary region and then central moment is used to find pupil center. In [12], Canny edge detection method [5] and bisection method is used to find pupil center. Theresholding and morphological opening is used to detect pupil region and center mass of pupil region is calculated for pupil center in [13]. They use active contour models (snakes) assuming the constraint that there is no internal energy to detect iris boundary. L. Liam et al. [14] converts the image to a binary image using threshold technique. They find pupil and iris by matching disk shape in the image.

3 Proposed Approach

We consider x- axis of an eye image towards vertical and y- axis toward horizontal as shown in Fig. 1. We use the following conventions in this paper. r_p denotes the radius of a pupil, x_p and y_p denote the x- and y- coordinate of a pupil center, respectively. r_i denotes the radius of an iris, x_i and y_i denote the x- and y- coordinate of an iris center, respectively.

In our approach, we first find pupil boundary information (namely, pupil center and pupil radius) and then use this information for detecting the iris boundary. We detect the pupil radius (r_p) and pupil center (x_p, y_p) using the steps: *(i)* down scaling the eye image. *(ii)* appling power transform and threshold the image. *(iii)* detecting edges. *(iv)* finding all connected element and removing irrelevant connected element. *(v)* finding pupil center and radius using circle fitting. *(vi)* up scaling the pupil information to get actual values. Details of the above mentioned steps are reported in [15].

All the steps in iris boundary detection are discussed in the following sub sections.

3.1 Preprocessing

We propose *histogram equalization* and *median filter* as preprocessing tasks. We use pupil information and down scaled eye image in iris boundary detection step. In an eye image, low contrast between iris and sclera part is present. In this step of iris boundary detection, we use histogram equalization [16] to increase the contrast between iris and sclera part of an eye image. Histogram equalization is the technique by which the dynamic range of the histogram and contrast of an image is increased based on the intensity histogram of the image. The histogram equalization is done on an image using Eq. (1).

$$s_i = \left[\sum_{j=0}^{i} n_i \right] \times \frac{L-1}{N} \tag{1}$$

where s_i represents the new intensity value corresponding to the i-th intensity level. n_i is the number of pixel having intensity value i, $i = 0, 1, 2, \ldots L - 1$. N and $L - 1$ represent the total number of pixel in the image and maximum intensity level, respectively.

We use median filter prior to the edge detection. Median filter removes the noise due to histogram equalization. Median filtering replaces pixels with the median value in the neighborhood. Figure 2 (c) shows the eye image after applying median filter of size 5×5 on the histogram equalized image.

3.2 Edge Detection

We use the Canny edge detection [5] method in our approach. The Canny operator works in a multi-stage process. In its first stage, we smooth the input image using Gaussian smooth operator. Gaussian smoothing reduces the noise in the eye image and also eliminates the chance of producing an edge due to noise. In the next stage, we apply Canny edge detection [5] for both horizontal and vertical edges on smoothed image. In the final stage, we suppress the local maxima. Figure 3 (a) shows the image after the Canny edge detection.

3.3 Removing Unnecessary Edge Information

It is observed that there is a possibility of staying many unnecessary edges present after the above mentioned steps. This is also evident in Fig 3(a). Unnecessary edges appear due to pupillary boundary, eyelids, eye lashes, non uniform illumination and blood vessel in the iris. We focus our interest only on the iris boundary edges. We try to remove above irrelevant edges as much as possible. The removal of irrelevant edges done in the following steps. We represent the pixel value by 1 for edge and 0 for no edge in an edge image.

We find out the edges caused by pupil boundary. To find the pupil boundary we use the knowledge of pupil radius (r_p) and pupil center (x_p, y_p). We remove edge pixels whose distance from pupil center (x_p, y_p) is less than the pupil radius (r_p). Mainly vertical edges form the iris boundary in the edge image. Horizontal edges occurs mainly for the eyelids of the eye images. We remove the horizontal edges of length greater than two pixel. We remove these edges following Eq. (2).

$$O(x,y) = \begin{cases} I(x,y) \oplus (I(x,y-1) \bullet I(x,y) \bullet I(x,y+1)) & \text{if } 0 \le x \le N-1; \\ & \quad 1 \le y \le M-2 \\ 0 & \quad y=1 \text{ or } 0 \text{ ; or} \\ & \quad d_p \le r_p + 20 \end{cases}$$

(2)

Where, $I(x,y)$ and $O(x,y)$ represent the pixel value in edge image and removed edge image at (x,y) position, respectively. M and N are the image width and

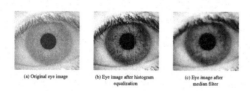

(a) Original eye image   (b) Eye image after histogram equalization   (c) Eye image after median filter

Fig. 2. Eye image in different preprocessing step

image height, respectively. d_p is the distance of a pixel at (x, y) from pupil center (x_p, y_p) i.e. $d_p = \sqrt{(x - x_p)^2 + (y - y_p)^2}$. $\oplus$ and $\bullet$ are two Boolean operators denoting ex-OR and AND, respectively. The value 20 is added with r_p because from our experiments we have observed that no edges are significant as a part of iris boundary which are $r_p + 20$ distance apart from the pupil center (x_p, y_p) in all images. Figure 3(b) shows the image after removing the pupil boundary edge.

We remove the smaller edges from the image produced after horizontal edge elimination. To do this first we find out all connected components and length of the corresponding component in the image. We eliminates those connected components whose length is less than the 50% of the maximum length because edges produce at the iris and sclera region is larger than other edges. Figure 3(c) shows the edge image after removal of smaller length edges.

3.4 Calculating Approximate Iris Radius

Let, $(x_{e1}, y_{e1}), (x_{e2}, y_{e2}), \ldots, (x_{ek}, y_{ek})$ be the edge points present in the edge image and k be the number of edge points. We compute the distance between each edge points and pupil center in Eq. (3).

$$d(i) = \lfloor \sqrt{(x_{e1} - x_p)^2 + (y_{e1} - y_p)^2 + 0.5} \rfloor, i = 1, 2, \ldots, k. \tag{3}$$

The number of distance levels (L) computed using Eq. (4).

$$L = \max_i\{d(i)\} - \min_i\{d(i)\} + 1, i = 1, 2, \ldots, k. \tag{4}$$

We represent each level of distance as d_j, $j = 1, 2, \ldots, L$. Now, we compute the probability of each distance occurrence in Eq. (5).

$$p(d_j) = \frac{h(d_j)}{k}, \ j = 1, 2, \ldots, L. \tag{5}$$

where, $h(d_j)$ is the number of edge points with distance d_j. We determine the approximate radius of iris (r_{approx}) in Eq. (6).

$$r_{approx} = \max_{d_j}\{p(d_j)\}, \ j = 1, 2, \ldots, L. \tag{6}$$

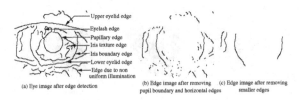

(a) Eye image after edge detection   (b) Edge image after removing pupil boundary and horizontal edges   (c) Edge image after removing smaller edges

Fig. 3. Edge images at different edge removing step

Table 1. Number of search points and search times required to find iris boundaries with and without elimination of irrelevant edges

Image	With eliminating irrelevant edges (our approach)		Without Eliminating irrelevant edges (existing approach)	
	No of search points	Search time (ms)	No of search points	Search time (ms)
Image-1	241	796.559	1412	5501.460
Image-2	213	698.427	1306	4685.471
Image-3	285	965.753	1682	6618.287
Image-4	215	710.220	1318	4989.161
Image-5	235	765.251	1395	5354.782
Average	238	787.242	1422	5429.832

3.5 Finding Iris Center and Radius

Finding the iris boundary is same as finding a circle from an eye image. We have already calculated the approximate radius of iris r_{approx}. Now, to reduce the search space we find the approximate iris center. The actual center of the circle is near to the approximation center. We choose the pupil center (x_p, y_p) as approximate iris center. We consider a small region (7×7) around approximate circle center to find actual iris circle center. To find the actual radius of iris circle, we use all circles for radius from (approximate radius-7) to (approximate radius+7) to fit circle. We consider only those points which are in the circular region with in radius 75% of r_{approx} and 125% of r_{approx} from center (x_p, y_p). We use circle equation $[(x - x_c)^2 + (y - y_c)^2 = r_c^2]$ to count those points which lies on the circle for all values of $(x_c, y_c) \in (7 \times 7)$ around approximate circle center (x_p, y_p) and all values of $r_c \in$ {approximate radius-7,approximate radius-6,..., approximate radius+7}. We choose that (x_c, y_c) and r_c for which count gives the maximum value. We also store corresponding count values as iris center (x_{dn_i}, y_{dn_i}) and iris radius (r_{dn_i}). Table 1 shows the searching time and number of pixels to be searched to fit iris boundary in scaled image for both cases: without eliminating irrelevant edges (existing approaches) and eliminating irrelevant edges (our approach). It may be noted that this iris radius and iris center are corresponding to the down sampled image.

As the processed image is in down scale (with scale factor 0.5 [15]) so we get the iris radius (r_i) and the iris center (x_i, y_i) of the original image by multiplying a factor 2 with iris radius (r_{dn_i}) and iris center (x_{dn_i}, y_{dn_i}) of down scale image.

4   Implementation and Experimental Results

We have implemented our approach of iris detection using C programing language in Fedora Core 5 operating system environment. We use GNU compiler GCC version 4.1.0 for compiling and executing our program.

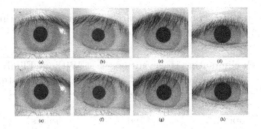

Fig. 4. Successfully detected iris boundary: (a) Eye image with light reflection, (b) Non uniform illumination eye image, (c) Eye image occluded by eyelash (d) Not properly open eye image, (e)-(h) Detected iris boundary corresponding to images in (a)-(d)

Table 2. Comparison of our work with others

Methodology	Accuracy rate
Daugman [2]	54.44%
Wildes [3]	86.49%
Masek [7]	83.92%
Liam and Chekima [14]	64.64%
T. Mäenpää [13]	93%
This Work	97.62%

Our approach has been tested on 108 different set of CASIA iris image database collected by Institute of Automation, Chinese Academy of Science [17]. The database includes 756 iris images from 108 individuals. For each eye, seven images are there which have been captured in two sessions; three samples are collected from first session and four samples are collected in second session. In Fig 4(a)-(d), some degraded sample images from CASIA eye image database are shown. Our approach successfully detect iris boundary though light reflection present at iris boundary (Fig. 4(a)), non uniform illumination in eye image (Fig. 4(b)), iris is occluded by eye lashes (Fig. 4(c)). This approach is also accurately find iris boundary when eye is not properly opened as shown in Fig. 4(d). Figure 4(e)-(h) show the detected iris boundary using our proposed method corresponding to the images in Fig 4(a)-(d).

We compare our results with some best known algorithms [18]. The result of comparisons is shown in Table 2. From Table 2, it is evident that proposed approach has the best performance. The accuracy rate is up to 97.62%. In [8,12], work on iris localization have been reported. These work do not mention on accuracy rate and hence we could not compare those work with our work.

5   Conclusions

We propose an efficient and accurate iris detection method for developing better biometric identification system in wide spread application areas. Proposed approache able to detect iris boundary when eye images are not of good quality, with non-uniform illumination of images, eyes are occluded with eye lashes, only part of irises are visible etc. As a consequence, our approach is more practical than the existing approaches. Significantly our approach is 75% faster than the existing approaches and also detects iris boundary almost 98% accurately as substantiated by experimental results on CASIA eye image database.

References

1. Daugman, J.: How iris recognition works. IEEE Trans. on Circuits and Systems for Video Technology 14(1), 21–30 (2004)
2. Daugman, J.G.: High confidence visual recognition of persons by a test of statistical independence. IEEE Trans.on PAMI 15(11), 1148–1161 (1993)
3. Wildes, R.P.: Iris Recognition: An Emerging Biometric Technology. Proc. of the IEEE 85(9), 1348–1363 (1997)
4. Ma, L., Tan, T., Wang, Y., Zhang, D.: Efficient Iris Recognition by Characterizing Key Local Variations. IEEE Trans. on Image Processing 13(6), 739–750 (2004)
5. Canny, J.: A computational approach to edge detection. IEEE Trans. on PAMI 8(6), 679–698 (1986)
6. Hough, P.: Method and means for recognizing complex patterns. U.S. Patent 3,069,654 (December 1962)
7. Masek, L.: Recognition of human iris patterns for biometric identification (2003), http://www.csse.uwa.edu.au/pk/studentprojects/libor
8. Huang, J., Wang, Y., Tan, T., Cui, J.: A New Iris Segmentation Method for Recognition. In: Proc. of the ICPR, vol. 3, pp. 554–557 (August 2004)
9. Ali, J.M.H., Hassanien, A.E.: An Iris Recognition System to Enhance E-security Environment Based on Wavelet Theory. Advanced Modeling and Optimization journal 5(2), 93–104 (2003)
10. Tisse, C.L., Martin, L., Torres, L., Robert, M.: Person identification technique using human iris recognition. In: Proc. of Vision Interface, Canada, pp. 294–299 (2002)
11. Vasta, M., Singh, R., Noore, A.: Reducing the False Rejection Rate of Iris Recognition Using Textural and Topological Fearures. Int. Jnl. of Signal Processing 2(1), 66–72 (2005)
12. Sung, H., Lim, J., Park, J., Lee, Y.: Iris Recognition Using Collarette Boundary Localization. In: Proc. of ICPR, vol. 4, pp. 857–860 (August 2004)
13. Mäenpää, T.: An Iterative Algorithm for Fast Iris Detection. In: Li, S.Z., Sun, Z., Tan, T., Pankanti, S., Chollet, G., Zhang, D. (eds.) IWBRS 2005. LNCS, vol. 3781, pp. 127–134. Springer, Heidelberg (2005)
14. Liam, L., Chekima, A., Fan, L., Dargham, J.: Iris recognition using self-organizing neural network. In: IEEE, 2002 Student Conf. on Research and Developing Systems, Malasya, pp. 169–172 (2002)
15. Dey, S., Samanta, D.: An Efficient Approach of Pupil Detection in Iris Images. Techinal Report (January 2007)
16. Ritter, G.X., Wilson, J.N.: Handbook of Computer Vision Algorithms in Image Algebra. CRC Press (1996)
17. CASIA iris image database, http://www.sinobiometrics.com
18. Proença, H., Alexandre, L.A.: UBIRIS: A noisy iris image database. In: ICIAP. vol. 1, pp. 970–977 (2005)

Confidence Measure for Temporal Registration of Recurrent Non-uniform Samples

Meghna Singh[1], Mrinal Mandal[1], and Anup Basu[2]

[1] Department of Electrical and Computer Engineering,
[2] Department of Computing Science,
University of Alberta, Edmonton, AB, Canada
{meghna,mandal}@ece.ualberta.ca,
anup@cs.ualberta.ca

Abstract. Temporal registration refers to the methods used to align time varying sample sets with respect to each other. While reconstruction from a single sample set may generate aliasing, registration of multiple sample sets increases the effective sampling rate and therefore helps alleviate the problems created by low acquisition rates. However, since registration is mostly computed as an iterative best estimate, any error in registration translates directly into an increase in reconstruction error. In this paper we present a confidence measure based on local and global temporal registration errors, computed between sample sets, to determine if a given set of samples is suitable for inclusion in the reconstruction of a higher resolution temporal dataset. We also discuss implications of the non-uniform sampling theorem on the proposed confidence measure. Experimental results with real and synthetic data are provided to validate the proposed confidence measure.

Keywords: Recurrent non-uniform sampling, Temporal registration, Confidence measure.

1 Introduction

Temporal registration, which is the computation of a correspondence in time between two sequences, has lately received much attention from researchers in the context of spatio-temporal super-resolution videos [1-3][9]. Temporal registration between sequences becomes imperative when a signal is acquired at a rate much lower than the highest frequency of the signal. Since reconstruction from a single sample set leads to aliasing, one proposed solution to this problem is to acquire multiple low sampled sequences, and compute a temporal relation between them. Fusing multiple low sampled sequences improves the effective sampling rate for reconstruction purposes. Our past work [1-3] dealt with estimating an event dynamics model and using it to register two videos acquired at low sampling rates. The computed registration was then further used to generate a temporally super-resolved video. In [1] we showed that using the global event dynamics for local registration is a better approach than local linear interpolation [9].

A. Ghosh, R.K. De, and S.K. Pal (Eds.): PReMI 2007, LNCS 4815, pp. 608–615, 2007.

A limitation with using temporal registration to generate super-resolved videos [1]-[3][9] is that the registration is based on optimizations, such as linear least squares or projection on to convex sets. Since no time-stamp information is available, these optimizations are only a best estimate of the actual registration. The reconstruction process depends greatly on the computed registration, hence it is better to leave out a sample set for which the registration might be incorrect, rather than including it for reconstruction. In this work, we develop a confidence measure which can be used to determine if a sample set should be included or excluded from the reconstruction process, in order to minimize the reconstruction error. The following assumptions are made in this work: (i) A signal is sampled at a frequency much lower than the Nyquist sampling rate (this happens often in MRI acquisitions of dynamic events and in video acquisitions of fast occurring events), (ii) multiple such sample sets are available, (iii) the temporal relation between these low sampled sets can be computed using methods described in [1], however, there is uncertainty associated with this computed relation. In this work, we develop a confidence measure to answer the following questions: (i) How reliable is the computed temporal relation between the sample sets, (ii) How can we compute a confidence measure to decide if a sample set should be included or excluded from the reconstruction process?

The rest of this paper is organized as follows. In Section 2, we review related work in temporal registration and recurrent non-uniform sampling. In Section 3, we present the proposed confidence measure and discuss the various factors that influence its computation. In Section 4, we present the experimental setup and results of the reconstruction algorithm based on the developed confidence measure. Lastly, we present the conclusions of our work in Section 5.

2   Review of Related Work

Reconstruction from multiple sample sets which are offset from each other by a known time interval has been studied in the past as the domain of recurrent non-uniform sampling and reconstruction [4-8]. The non-uniform sampling theorem from [7] states that *"a function f(t), bandlimited to $-\Omega/2 \leq \omega \leq \Omega/2$, can be uniquely reconstructed from a set of samples which are non-uniformly spaced but satisfy the condition that there be precisely N distinct samples to every interval of length NT, where N is some finite integer."* Reference [7] also provides an interpolation formula for reconstruction of the signal from its non-uniform samples provided the condition in the theorem above is satisfied. Other researchers ([5],[6],[8] and references therein) have developed simplifications to the interpolation, such as global polynomial fitting, trigonometric polynomials, moving least squares, radial basis functions and variational approach using splines. We use the reconstruction algorithm from [10] to evaluate our confidence metric. Feichtinger et al. [10] compute sampling atoms or synthesis functions using approximation operators such that every bandlimited function has a stable summed expansion of the type shown in Eq.1, where e_n are the sampling atoms. Unlike uniform sampling functions the sampling atoms are not necessarily translations of a mother function and are computed using frame theory and adaptive weights to improve the numerical efficiency of the computation.

$$f = \sum_{n \in Z} f(t_n)e_n \tag{1}$$

Confidence measures have been proposed in a variety of fields. In signal processing and pattern recognition, confidence measures have been computed extensively for speech recognition [11][12] . These confidence measures are mostly based on the probability distributions of the likelihood functions of speech utterances, which are derived from Hidden Markov Models. Our proposed work is unique as it introduces the concept of a confidence measure in temporal registration and reconstruction from recurrent non-uniform samples. It also solves a real world problem faced in temporal super-resolution − determining the worthiness of a sample set.

3   Proposed Confidence Metric

Let the recurrent sample sets have a fixed sampling interval T which is much lower than the Nyquist sampling interval, as shown in Fig.1. In Fig.1, we illustrate two sample sets, offset from each other with a time interval t_0 . t_0 is modeled as a discrete uniform distribution as all values within the finite time interval are equally possible. Most reconstruction algorithms assume that the value of t_0 is known *a priori* (i.e. it is a controlled variable) or its exact value can be computed. However, in temporal registration t_0 is the interval that we seek to compute, often with an unknown degree of inaccuracy.

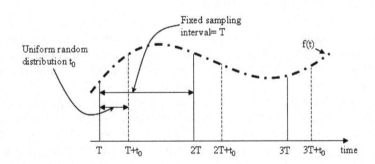

Fig. 1. Illustration of recurrent non-uniform sampling with two sample sets

In previous work [1-3] we computed t_0 by minimizing the registration error in the least square sense. In this work we express the registration error as an aggregate of local registration error and global registration error. Global registration error (E_g) is the error computed from a one-to-one discrete sample correspondence. Local registration error (E_l) on the other hand is computed by first constructing a continuous event model from all the discrete samples of a set, and then registering the event models on a sub-integer time level. Given two sample sets $f(kT)$ and $f(kT + t_0)$, as shown in Fig.1, the global registration error can be computed as follows:

$$E_g = \arg\min_{t_0}(\sum_k \|f(kT) - f(kT + t_0)\|^2) \tag{2}$$

Subsequent to computing the event models $c(t)$ and $c(t + t_0)$ by methods described in [1-3], the local registration error can be expressed as follows:

$$E_l = \arg\min_{t_0}(\int \|c(t) - c(t + t_0)\|^2 dt) \tag{3}$$

The registration errors defined in Eq.2 and Eq.3 give a general idea of the confidence in the temporal registration. However, they do not relate to the confidence measure by a simple proportionality, i.e. a large global registration error does not imply a poor confidence in the registration. In fact, as we will demonstrate in our experiments, a large global registration error indicates a more uniform distribution of the sample sets and therefore a better reconstruction or a higher confidence. For now let us express the confidence measure as a weighted sum of E_g and E_l as follows:

$$c(q, r) = w_g \times (E_g)^q + w_l \times (E_l)^r \tag{4}$$

In Eq.4, w_g and w_l are weights assigned to the contribution of both the registration errors to the overall confidence measure. If no prior information is known about the registration then $w_g = w_l = 0.5$. If prior information is known then the weights can be unequally assigned. The parameters q and r in Eq. 4 are defined as variables in the confidence measure as they can be used to incorporate non-linearity or to invert the proportionality of the registration errors. We present two hypotheses that are supported by experimental results shown in Fig. 2 and by the discussion that follows.

Hypothesis 1: E_g is an indicator of t_0 , and a value of t_0 which places the sample sets as far apart from each other as possible results in a better reconstruction, hence an increase in E_g should have a positive effect on the confidence measure.

Hypothesis 2: E_l is an indicator of the overall error in registration, and a large E_l results in poorer reconstruction, hence an increase in E_l should have a negative effect on the confidence measure.

Figure 2(a) shows the relationship between the reconstruction error and the global registration error. The reconstruction error reported in the figure is the mean of the SSE (sum of squared errors) of all the test cases, hence denoted as the MSE . It can be seen that E_g demonstrates a linear relationship with the reconstruction error. As E_g increases the reconstruction error decreases, i.e. the confidence measure which should be associated with E_g should be in direct increasing proportion. Therefore 'q' in Eq.4 can be approximated to '1'. We also fitted the MSE vs. E_g curve with a quadratic function and it can be seen from Fig. 2(a) that a linear fit is a sufficiently good approximation of the curve. The same analysis is applied to the relationship between the reconstruction error and the local registration error E_l , Fig. 2(b). As E_l increases, the reconstruction error increases. We invert the abscissa to plot MSE vs. E_l^{-1}, to define an inverse relationship between the confidence measure and E_l. Hence, 'r' in Eq.4 can be approximated to '-1'. The proposed confidence measure χ can therefore be expressed as follows:

$$\chi = c(1, -1) = w_g \times (E_g)^1 + w_l \times (E_l)^{-1} \tag{5}$$

It would be interesting at this point to discuss the implications of the non-uniform sampling theorem to temporal registration. For there to be N distinct samples to every interval of length NT, the average non-uniform sampling rate required is the same as the Nyquist rate. The farther the recurrent sample sets are spaced apart in time with respect to each other, the higher is the average sampling rate. This leads to the conclusion that for optimal reconstruction, t_0 should be close to mid-way between kT and $(k+1)T$ samples in Fig.1. Such positioning of t_0 will lead to a large error in the global registration of two sequences, which supports Hypothesis 1.

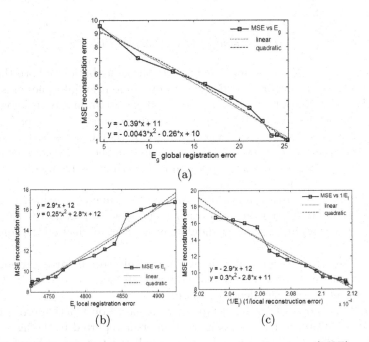

Fig. 2. (a) Relationship between mean square reconstruction error (MSE) and global registration error E_g, (b) Relationship between MSE and local registration error E_l, (c) Relationship between MSE and (E_l^{-1})

4   Experimental Setup and Results

We tested the proposed confidence measure on both synthetic and real data. Samples of synthetic data and real test videos are available for viewing at: `http://www.ece.ualberta.ca/~meghna/Premi_webpage/PReMI_2007.html`. Synthetic data was generated as a high resolution random signal which was band-limited to a user controlled frequency. This high resolution data was then sampled at a very low discrete sampling rate. For example, a 25Hz band-limited signal was sampled at 2Hz. Multiple sample sets at a fixed low sample rate were also generated by initializing the starting point of the sample sets randomly with a uniform distribution. Temporal registration was then computed using methods described in [1]. However, since the samples in each set were limited, the computed registration was not accurate to a sub-frame level. These multiple sample sets were then iteratively fused together, one at a time, based on erroneous time stamp information. Reconstruction algorithm from [10] was used to reconstruct a signal from the fused samples. By adding a sample set at each iteration (see Fig.3 (a-b)), we were able to observe that the MSE falls as the number of sample sets increases. Also, as per our hypothesis, if the sample sets are such that the t_0 is small i.e. the sample sets are closely spaced, then the reconstruction error falls at a rate much slower than if the sample sets were further apart (Fig.3(c)). The relationships developed between E_g, E_l and MSE for the synthetic test cases have already been presented in Fig.2.

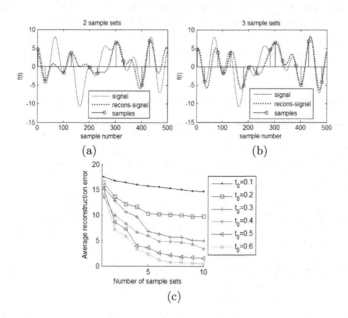

Fig. 3. (a) Reconstruction of the signal $f(t)$ with only two sample sets at very low sampling rate, (b) Reconstruction with three sample sets. (c) Illustration of decrease in reconstruction error with increase in t_0 (reported as a normalized number [0,1]).

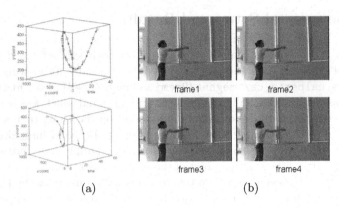

(a)   (b)

Fig. 4. (a) Sample trajectories from real data sequence, (b) Sample frames from real data sequence

For our real test cases, we used video sequences of an individual swinging a ball tied to the end of a string. The video sequences were captured at 30 frames per second and the trajectory of the ball was extracted via background subtraction techniques and motion tracking. This trajectory was then used as a high resolution signal which was further down-sampled at low sampling rates, as shown in Fig.4(a). An event model [1] was used to compute the temporal registration between the low sampled signals. In each experiment, we arbitrarily chose one sample set as the parent against whom two other recurrent sample sets were registered. Based on the local and global registration errors computed using Eq.2 and Eq.3 we determined the confidence measure as per Eq.5. The two confidence measures $\chi(1), \chi(2)$ for each recurrent set were compared to each

Table 1. Experimental results of confidence metric χ and reconstruction error (SSE)

	$E_g(1)$	$E_l(1)$	$E_g(2)$	$E_l(2)$	$\chi(1)$	$\chi(2)$	$SSE(1)$	$SSE(2)$
Seq1	44.0	230.3	35.6	596.7	1	0	1.3	2.6
Seq2	15.7	49.8	4.0	8.9	1	0	4.0	13.0
Seq3	22.5	374.0	16.8	212.4	1	0	205.4	355.3
Seq4	97.2	2959.0	55.1	776.0	1	0	2.4	40.3
Seq5	26.5	22.8	8.7	20.6	1	0	47.3	68.6
Seq6	446.0	41.1	652.0	22.3	0	1	1700	4.2
Seq7	234.0	90.3	652.0	22.3	0	1	81200	42000
Seq8	129.0	4.6	364.0	321.0	0	1	20.0	4.0
Seq9	43.7	7081.5	19.7	2228.7	1	0	8.9	16.1
Seq10	7.8	443.4	28.9	3687.7	0	1	10.9	10.3
Seq11	21.6	7409.3	16.8	5970.7	1	0	8.7	9.3
Seq12	28.9	5823.3	22.3	3121.5	1	0	7.9	10.7
Seq13	16.3	2387.7	17.4	2590.6	0	1	9.7	9.3
Seq14	4.6	169.5	30.3	6190.9	0	1	15.6	9.9

other. If $\chi(1) > \chi(2)$, then $\chi(1)$ was given an absolute value of 1, else it was set to 0. If more than two sample sets are being compared then a normalized value can be assigned to the confidence measure. Table 1 shows the local and global registration errors for a parent set and two recurrent sets, for 14 different signals including both synthetic and real data (real data is from seq6-8). A confidence metric of 1 indicates that the corresponding recurrent set is a better candidate for reconstruction, as proven by the corresponding computed reconstruction error. It can be seen that the proposed confidence metric is a suitable indicator of the reconstruction error.

5   Conclusion

In this work, we presented a confidence measure to determine the suitability of recurrent sample sets for reconstruction purposes based on the computed temporal registration errors. Experiments with real and synthetic data were conducted, and the results support the proposed confidence measure. The confidence measure is able to successfully indicate the sample sets which would lead to a higher error if included in the reconstruction process. As part of future work we plan to use this confidence measure with recurrent sample sets of MRI data.

References

1. Singh, M., Basu, A., Mandal, M.: Event Dynamics based Temporal Registration. IEEE Trans. on Multimedia 9(5), 1004–1015 (2007)
2. Singh, M., Thompson, R., Basu, A., Rieger, J., Mandal, M.: Image based Temporal Registration of MRI Data for Medical Visualization. In: Intl. Conf. on Image Processing ICIP, October 8–11, 2006, pp. 1169–1172
3. Singh, M., Basu, A., Mandal, M.: Temporal Alignment of Time Varying MRI Datasets for High Resolution Medical Visualization. In: Intl. Symposium on Visual Computing ISVC, pp. 222–231 (2006)
4. Kim, S.P., Bose, N.K.: Reconstruction of 2D bandlimited discrete signals from nonuniform samples. IEE Proceedings 137(3) (June 1990)
5. Arigovindan, M., Suhling, M., Hunziker, P., Unser, M.: Variational Image Reconstruction from Arbitrarily Spaced Samples: A Fast Multiresolution Spline Solution. IEEE Trans. on Image Processing 14(4) (April 2005)
6. Micchelli, C.A.: Interpolation of scattered data: distance matrices and conditionally positive definite functions, Constructive Approximation 2 (1986)
7. Freeman, H.: Discrete-Time Systems. John Wiley and Sons Inc, Chichester (1965)
8. Marvasti, F.: Nonuniform Sampling Theory and Practice. Kluwer Academic, New York (2001)
9. Shechtman, E., Caspi, Y., Irani, M.: Space-time super-resolution. IEEE Trans. on Pattern Analysis and Machine Intelligence 27(4), 531–545 (2005)
10. Feichtinger, H.G., Werther, T.: Improved Locality for Irregular Sampling Algorithms. In: Proc. ICASSP (1999)
11. Mortensen, E.N., Barrett, W.A.: A Confidence Measure for Boundary Detection and Object Selection. In: Proc. CVPR (2001)
12. Moreau, N., Jouvet, D.: Use of a confidence measure based on frame level likelihood ratios for the rejection of incorrect data. In: Proc. Eurospeech (1999)

Automatic Detection of Human Fall in Video

Vinay Vishwakarma, Chittaranjan Mandal, and Shamik Sural

School of Information Technology
Indian Institute of Technology, Kharagpur, India
{vvinay,chitta,shamik}@sit.iitkgp.ernet.in

Abstract. In this paper, we present an approach for human fall detection, which has important applications in the field of safety and security. The proposed approach consists of two parts: object detection and the use of a fall model. We use an adaptive background subtraction method to detect a moving object and mark it with its minimum-bounding box. The fall model uses a set of extracted features to analyze, detect and confirm a fall. We implement a two-state finite state machine (FSM) to continuously monitor people and their activities. Experimental results show that our method can detect most of the possible types of single human falls quite accurately.

1 Introduction

Human fall is one of the major health problems for elderly people. Falls are dangerous and often cause serious injuries that may even lead to death. Fall related injuries have been among the five most common causes of death amongst the elderly population. Falls represent 38% of all home accidents and cause 70% of death in the 75+ age group. It is shown in [1] that the number of reported human falls per year was around 60,000 with an associated cost of at least £400 million in the UK.

Early detection of a fall is an important step in avoiding any serious injuries. An automatic fall detection system can help to address this problem by reducing the time between the fall and arrival of required assistance. Here, we present an approach for human fall detection using a single camera video sequence. Our approach consists of two steps: object detection and the use of a fall model. We apply an adaptive background subtraction method to detect a moving object and mark it with its minimum-bounding box. The fall model consists of two parts: fall detection and fall confirmation. It uses a set of extracted features to analyze, detect and confirm a fall. In the fall model, the first two features (aspect ratio, horizontal and vertical gradient values of an object) are responsible for fall detection and the third feature (fall angle) is used for fall confirmation. We also implement a two-state finite state machine to continuously monitor people and their activities.

The organization of the paper is as follows. Section 2 explains the related work on fall detection. Section 3 describes the object detection method. Section 4 elaborates the fall model. In Section 5, we present experimental results followed by conclusion in section 6.

A. Ghosh, R.K. De, and S.K. Pal (Eds.): PReMI 2007, LNCS 4815, pp. 616–623, 2007.

2   Related Work

Primarily, there are three methods of fall detection, classified in the following categories:

1. Acoustics based Fall Detection
2. Wearable Sensor based Fall Detection
3. Video based Fall Detection

In video based fall detection, human activity is captured in a video that is further analyzed using image processing techniques. Since video cameras have been widely used for surveillance as well as home and health care applications, we use this approach for our fall detection method.

Due to the advancements in vision technologies, many individuals and organizations are concentrating on fall detection using video based approaches. In [2], authors have used background modeling and subtraction of video frames in HSV color space. An on-line hypothesis-testing algorithm is employed in conjunction with a finite state machine to infer fall incident detection. However, they only use aspect ratio of a person as an observation feature based on which fall incident is detected. Lue and Hu [3] have presented a fall detection algorithm using dynamic motion pattern analysis. They assume that a fall can only start when the subject is in an upright position and characterizes a big change in either X or Y direction when a fall starts. In [5], Toreyinet al. have used a background estimation method to detect moving regions. Using connected component analysis, they obtain the minimum bounding rectangles (blob) and calculate the aspect ratio. They also use audio channel data based decisions and fuse it with video data based decisions to reach a final decision. In [6], authors subtract the current image from the background image to extract the foreground of interest. To obtain the associated threshold, they consider a subject's personal information such as weight and height. Each extracted aspect ratio is validated with the user's personal information to detect the fall. In [7], McKenna and Nait-Charif [7] have used a particle filtering method to track a person and extract his trajectories using 5-D ellipse parameter space in each sequence. An associated threshold on a person's speed is used to label the inactivity zone and human fall. In [8], authors use 3-D velocity as a feature parameter to detect human fall from a single camera video sequence. At first, 3-D trajectory is extracted by tracking a person's head with the help of a particle filter as it has a large movement during a fall. Next, 3-D velocity is computed from 3-D trajectory of the head.

Most of the existing vision based fall detection systems use either motion information or a background subtraction method for object detection. An abrupt change in the aspect ratio is analyzed in different ways such as Hidden Markov Model (HMM), adaptive threshold and the user's personal information to detect falls in video. A person's velocity is often used to classify a human either as walking or falling in a video. There are also some other existing models, but they work well only in restricted environment.

In our approach, we use an adaptive background subtraction method using a Gaussian Mixture Model (GMM) in YCbCr color space for object detection. We propose a fall model that consists of two steps, first fall detection and then fall confirmation. We extract three features from an object and use the first two features for fall detection and the last for fall confirmation. We implement a simple two-state finite state machine (FSM) to continuously monitor people and their activities.

3   Object Detection

The first and the most important task of human fall detection is to detect humans accurately so we apply an adaptive background subtraction method using a Gaussian mixture model (GMM) and then extract a set of features for fall modeling.

3.1   Background Modeling and Subtraction

A recorded video is used as an input which is stored as a sequence of frames using the Berkeley MPEG Decoder. For every frame, we convert its pixel from the RGB color space to the YCbCr color space. We use mean values of image pixels for further processing.

GMM considers each background pixel as a mixture of Gaussians. The Gaussians are evaluated using a simple heuristic to hypothesize pixels which are most likely to be part of the background process. The probability that an observed pixel has intensity value X_t at time t is modeled by a mixture of K Gaussians as

$$P(X_t) = \sum_{i=1}^{k} w_{i,t} * \eta(X_t, \mu_{i,t}, \Sigma_{i,t}). \tag{1}$$

where

$$\eta(X_t, \mu_{i,t}, \Sigma_{i,t}) = \frac{1}{((2\pi)^{\frac{n}{2}} |\Sigma|^{\frac{n}{2}})} * e^{\frac{1}{2}(X_t - \mu_t)^T \Sigma^{-1}(X_t - \mu_t)}. \tag{2}$$

$$\omega_{i,t} = (1 - \alpha)w_{i,t-1} + \alpha(M_{k,t}). \tag{3}$$

Here m is the mean, α is the learning rate and $M_{k,t}$ is 1 for the model which matches and 0 for the rest.

The background estimation problem is addressed by specifying the Gaussian distributions which have the most supporting evidence and the least variance. Since a moving object has larger variance than a background pixel, in order to represent background process, first the Gaussians are ordered by the value of $\frac{\omega}{\alpha}$ in decreasing order. The background process stays on top with the lowest variance by applying a threshold T, where

$$B = argmin_b(\sum_{k=1}^{b} \omega_k \geq T).\qquad(4)$$

All pixels X_t which do not match any of these components will be marked as foreground.

Pixels are partitioned as being either in the foreground or in the background and marked appropriately. We apply connected component analysis that can identify and analyze each connected set of pixels to mark the rectangular bounding box over an object.

3.2   Feature Extraction

We extract a set of features from each object and its bounding box such as aspect ratio, horizontal (G_x) and vertical (G_y) gradient values and fall angle, which we use further in the fall model.

Aspect Ratio. The aspect ratio of a person is a simple yet effective feature for differentiating normal standing pose from other abnormal pose. In Table 1(a), we compare the aspect ratio of an object in different human pose.

Horizontal and Vertical Gradients of an Object. When a fall starts, a major change occurs in either X or Y direction. For each pixel, we calculate its horizontal (G_x) and vertical (G_y) gradient values. In Table 1(b), we compare horizontal and vertical gradient value of an object's pixel in different human pose.

Fall Angle. Fall angle (θ) is the angle of a vertical line through the centroid of object with respect to the horizontal axis of the bounding box. Centroid (C_x, C_y) is the center of mass co-ordinates of an object. In Table 1(c), we compare fall angles of an object in different pose such as walking and falling.

Table 1. Comparison of features distribution of object in different pose

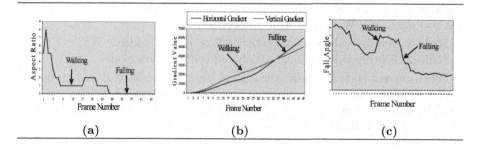

(a)   (b)   (c)

4   Fall Model

Building a fall Model consists of two steps: fall detection and fall confirmation. For the fall detection step, we use aspect ratio and object's horizontal and vertical gradient values. For the fall confirmation, we use fall angle with respect to the horizontal axis of its bounding box. We use rule-based decisions to detect and confirm the fall.

4.1   Fall Detection

1. Aspect ratio of human body changes during fall. When a person falls, the height and width of his bounding box changes drastically.
2. When a person is walking, the horizontal gradient value is less than the vertical gradient value ($G_x < G_y$) and when a person is falling, the horizontal gradient value is greater than the vertical gradient value ($G_x > G_y$).
3. For every feature, we assign a binary value. If the extracted feature satisfies the rules, we assign binary value 1 otherwise 0.
4. We apply OR operation on the feature values. If we get the resultant binary value as 1 then we detect the person as falling, otherwise not.

4.2   Fall Confirmation

When a person is standing, we assume that he is in an upright position and the angle of a vertical line through the centroid with respect to horizontal axis of the bounding box should be approximately 90 degree. When a person is walking, the θ value varies from 45 degree to 90 degree (depending on their style and speed of walking) and when a person is falling, the angle is always less than 45 degrees.

For every frame where a fall has been detected, we apply the fall confirmation step. We calculate the fall angle (θ) and if θ value is less than 45 degree, we confirm that the person is falling. Similarly, we take next few (in our approach, it is seven) frames and analyze their features using the fall model to confirm a fall situation.

4.3   State Transition

To continuously monitor human behavior, which changes from time to time, we implement a simple two-state finite state machine (FSM). As shown in Fig. 1, the two states are 'Walk' and 'Fall' respectively.

Rule 1: Feature values should satisfy fall detection model.

Rule 2: Feature values should satisfy fall confirmation model.

When current state is 'Walk', the system begins to perform Rule 1 testing. If Rule 1 is not satisfied, the state remains unchanged; otherwise the state transits to 'Fall'. When current state is 'Fall', the system begins to perform Rule 2 testing. If Rule 2 is satisfied, the state remains unchanged; otherwise it transits back to 'Walk' state, which is the case when a person has fallen and again started to walk. Alarms will be triggered once a person remains in the state of 'Fall' for a period longer than a pre-set duration.

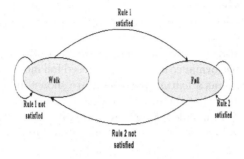

Fig. 1. A finite state machine for human fall detection

5 Experimental Results

We have implemented our proposed approach using C as the programming language in linux and tested it intensively in a laboratory environment. To verify the feasibility of our proposed approach, we have taken 45 video clips (indoor, outdoor and omni-video) as our test target. A handycam (SONY DCR-HC40E MiniDV PAL Handycam Camcorder) was used to capture indoor and outdoor video clips. Video clips contain a number of different possible types of human fall (sideway, forward and backward) and no fall condition. In every video clip, one or more moving object exists in the scene. In this paper, we use a set of criteria to evaluate our system including accuracy, sensitivity and specificity [6].

Table 2. Recognition results

Video Types	Scene Types	Total Frames	Fall Types	TP	FP	FN	TN
Indoor	Single	93	Forward	20	0	0	73
Indoor	Single	216	Backward	56	0	0	150
Indoor	Single	286	Sideway	76	0	0	210
Indoor	Single	100	No Fall	0	0	0	100
Outdoor	Single	87	Forward	47	0	0	40
Outdoor	Single	141	Backward	14	0	0	127
Indoor	Multiple	175	Sideway	80	30	15	50
Outdoor	Multiple	624	Sideway	144	10	120	350
Omni-video	single	376	Forward	100	10	10	256
Omni-video	Multiple	1007	Forward	257	50	440	260

In Table 2, we show results of indoor, outdoor and omni-video clips containing different types of possible human fall. In our experiment, a fixed threshold is set for every feature. For aspect ratio, we set the threshold between 0 and 1. For horizontal (G_x) and vertical (G_y) gradient values of an object, G_x is less than G_y in the case of a walking person and G_x is greater than G_y in the case of a falling person.

For fall angle (θ), θ is between 45 degree to 90 degree in the case of a walking person and θ is less than 45 degree in the case of a falling person. We have selected these thresholds empirically. In our experiments, we first evaluate the system performance (accuracy, sensitivity and specificity) using aspect ratio as a feature. Next, we evaluate system performance using our proposed fall model as shown in table 3. Results of aspect ratio as a feature parameter are shown in parenthesis. The experimental results show that the proposed method can accurately detect most of the possible types of fall in video. Some successful and unsuccessful image frames of human fall detected by our approach are shown in Table 4.

Table 3. System performance of proposed fall model and aspect ratio approach

Video Content	Accuracy(%)	Specificity(%)	Sensitivity(%)
Indoor + Single	100 (95)	100 (97)	100 (90)
Outdoor + Single	100 (93)	100 (90)	100 (95)
Indoor + Multiple	74 (62)	84 (50)	62 (73)
Outdoor + Multiple	79 (64)	97 (85)	54 (37)
Omni-video + Single	94 (89)	96 (92)	90 (81)
Omni-video + Multiple	51 (40)	83 (49)	36 (29)

Table 4. Image frames of fall detected by our proposed approach

Our approach is able to achieve promising results only when there is a single person in the scene. For multiple people in the scene or in a crowd, this approach is not able to detect the fall accurately. For all video, we consider that first few frames contain only background scene.

6 Conclusion

We have presented a method for automatic detection of human fall in video. The proposed approach contains two main components, object detection and the use of a fall model. For object detection, we use an adaptive background subtraction method using a Gaussian Mixture Model in YCbCr color space. For the fall model, we extract a set of features such as aspect ratio, horizontal and vertical gradient values of an object as well as fall angle. In our experiments, we have taken three types of video clips (indoor, outdoor and omni-video) for both single and multiple people in the scene. Our experimental results show that the proposed method can accurately detect a single falling person. In future work, we plan to improve the fall model and apply it for multiple people in the scene.

References

1. Marquis-Faulkes, F., McKenna, S.J., Newell, A.F., Gregor, P.: Gathering the requirements for a fall monitor using drama and video with older people. Technology and Disability 17, 227–236 (2005)
2. Tao, J., Turjo, M., Wong, M., Wang, M., Tan, Y.: Fall incidents detection for intelligent video surveillance. In: Fifth International Conference on Information, Communication and Signal Processing, pp. 1590–1594 (2005)
3. Luo, S., Hu, Q.: A dynamic motion pattern analysis approach to fall detection. In: IEEE International workshop in Biomedical Circuit and Systems (2004)
4. Alwan, M., Rajendran, P.J.: A smart and passive floor-vibration based fall detector for elderly. Information and Communication Technologies 1, 1003–1007 (2006)
5. Ugur Toreyin, B., Dedeoglu, Y., Enis Cetin, A.: HMM based falling person detection using both audio and video. In: Proc. IEEE Conf. Signal Processing and Communication Application (14th) (2006)
6. Miaou, S., Sung, P., Huang, C.: A Customized Human Fall Detection System using omni-camera images and personal information. In: Proc. of the 1st Distributed Diagnosis and Home Healthcare (D2H2) Conference, Arlington, Virginia, USA (2006)
7. McKenna, S.J., Nait-Charif, H.: Summarising contextual activity and detecting unusual inactivity in a supportive home environment. Pattern Analysis Application (14th) 7, 386–401 (2005)
8. Rougier, C., Meunier, J.: Demo: Fall detection using 3D head trajectory extracted from a single camera video sequence. Journal of Telemedicine and Telecare 11(4) (2005)
9. Stauffer, C., Grimson, W.E.L.: Adaptive background mixture model for real time tracking. In: Proc. IEEE Conf. Computer Vision and Pattern Recognition CVPR 1999, pp. 246–252 (1999)

Deformable Object Tracking: A Kernel Density Estimation Approach Via Level Set Function Evolution

Nilanjan Ray and Baidya Nath Saha

Department of Computing Science
University of Alberta, Edmonton, Canada
{nray1,baidya}@cs.ualberta.ca

Abstract. Automated tracking of deformable objects that change shape and size drastically is challenging. For useful results, one needs an efficient deformable object model. In this regard, we propose a novel deformable object model via joint probability density of level set function and image intensity/feature values. Given the delineated object boundary on the first image frame of a video sequence, we learn the aforementioned joint probability density via kernel (Parzen window) method. From the next frame onward, we match this learned probability density with the probability density on the current frame by minimizing Kullback-Leibler divergence. This minimization procedure is cast in a variational framework and a minimizer is obtained by solving a partial differential equation (PDE). A stable and efficient numerical scheme is proposed for solving this resulting PDE. We demonstrate the efficacy of the proposed tracking method on myocardial border tracking from mouse heart cine magnetic resonance imagery (MRI).

Keywords: Kernel density estimation, Parzen window, KL divergence, level set, cine MRI.

1 Introduction

For deformable object tracking, we need an object model that has at the least the following capabilities: *Recognition capability* – the model should recognize the object from one video frame to the next and *discrimination capability* – it should accurately delineate the object boundary from the surrounding. The tracking affair is further complicated when the object being tracked *deforms heavily* from one video frame to the next. Many automated object tracking applications fall into such a problem category. An important application is myocardial border tracking from cine MRI [1].

If the objects to be tracked are not deformable, or, if accurate object boundary delineation is not required, then plenty of fixed template tracking methods can be exploited. Correlation based trackers are famous among them. More powerful methods include mean shift tracker that models the object by its intensity histogram [2].

A. Ghosh, R.K. De, and S.K. Pal (Eds.): PReMI 2007, LNCS 4815, pp. 624–631, 2007.

On the other hand, if the deformation of the object is moderate then we can also employ many powerful deformable object models proposed to date (see [3] for a review of deformable models). Based on the number of training objects required, moderately deformable models can be categorized broadly into two types: (1) models that require more prior knowledge and fewer training objects, (2) models that require many training examples and less prior knowledge about the objects. Examples of the first kind include parametric contour models, such as, affine invariant g-snake [4]. For g-snake even a single example object might be adequate for modeling the object via affine transformation invariant parameters. Note that g-snake can tolerate large deformations, as long as the deformations belong to the family of shapes produced by affine transformations of a base object shape. Other examples in this category are contour models that can be described by geometric primitive shapes (such as circles, ellipses, etc.) parameterized by only a few scalar numbers. An example of the second category of moderately deformable models is principal component analysis (PCA)-based object models, such as active shape and appearance models [5]. PCA -based models use many training objects to learn the object characteristics. It is noteworthy that if the assumption of moderate object deformation holds, moderately deformable models as described above, have adequate recognition and discrimination abilities that can be utilized in object tracking.

Yet a third type of deformable object model is available that only imposes local smoothness/regularity on the object shape and is thus capable of undergoing very large and almost arbitrary deformations. Examples include classical Kass-Witkin-Terzopoulos snake model [6], spline snakes, as well as level set-based object models imposing only local regularization [7]. These large deformation models have seen much of their success in image segmentation because of their excellent ability in object boundary delineation. Typically these models lack the knowledge of object specific characteristics (features) and thus have poor recognition capability. In other words, they are suitable for unsupervised segmentation or clustering, and not as much effective in supervised tasks such as object tracking/recognition.

In this paper we propose a deformable object model that is, to a good extent, capable of undergoing large deformations as well as learning object and background features. As a result we can achieve aforementioned three desirable properties required in tracking, *viz.*, recognition, discrimination and large deformation. Our proposed deformable object model is a joint probability density function (pdf). Given an object boundary contour on the initial video frame, we compute the signed distance function (level set function). Next, the joint pdf of this level set function and image intensities (features) is constructed as the deformable object model. Note that this joint pdf is a function of object boundary. Thus in the subsequent video frames, we search for the right object boundary that results in a joint pdf similar to the joint pdf we learn on the first video frame.

In connection with the proposed deformable object model we mention two closely related previous work as follows. Leventon proposed an object model, which is a joint pdf of level set function (created from object boundary) and image intensity at *every pixel location* in an image for the purpose of segmentation [8]. Note that his model is completely incapable of undergoing severe deformation as, every pixel (inside as well as outside the object) carries its own joint pdf, as opposed to the proposed model that is attached to an entire object. Also note that Leventon's model is not translation invariant - an indispensable property for target tracking. His method requires many training examples to learn this pixel location-based joint pdf and is unsuitable for tracking applications where the object only in the first video frame can be used for learning.

The other related previous work sits at the opposite extreme of Leventon's work in terms of deformation capability. Freedman and Zhang proposed to track the intensity (feature) histogram of an object delineated by a flexible contour [9]. Like the proposed method their method also searches for the best contour location by minimizing the dissimilarity between a model pdf and the current feature pdf. Note that Freedman and Zhang's object model is composed of image intensity/feature only within an object and it has no information of the background at all. In this paper, by experiments, we demonstrate that their model severely lacks object delineation ability.

In the light of the aforementioned discussion on related work, our proposed object model derives desirable properties for tracking from both Leventon's and Freedman and Zhang's models. The proposed object model learns the information about object boundary, object intensities (features), and its background intensity. The proposed model is also highly deformable in nature.

2   Background

This section describes two basic ingredients of our method, *viz.*, level set function and similarity/dissimilarity measures between pdfs.

2.1   Level Set Function

A closed contour on a 2D plane has essentially two representations - parametric and geometric. In the parametric representation the coordinates $(X(s), Y(s))$ of the points on the curve are expressed as functions of a scalar parameter s. This parameter could, for example, denote the length of the curve measured from a particular point on the curve. Geometric representation however avoids this explicit parametric description altogether. It represents the closed curve as an intersection of a plane and a surface.

The set of points on a closed curve can be conceived as the zero level set of the surface, *i.e.*, the curve is the iso-contour with zero height. We refer to this surface as level set function. Given a contour, the construction of a level set function $\phi(x, y)$ that embeds this contour as a zero level set can be obtained by the signed distance transform:

$$\phi(x, y) = \begin{cases} -\sqrt{(x - X(x,y))^2 + (y - Y(x,y))^2}, & \text{if } (x, y) \text{ is inside the object} \\ \sqrt{(x - X(x,y))^2 + (y - Y(x,y))^2}, & \text{otherwise,} \end{cases}$$

where $(X(x,y), Y(x,y))$ is the point on the curve nearest to the (x,y) point (pixel location) in the image domain. We adopt the convention that the signed distance function shown here is negative inside the object and positive outside the object. Thus the curve is the zero level set of this surface $\phi(x,y)$. Refer to [7] for further details on level sets and their applications in image analysis.

2.2 Similarity/Dissimilarity Measures for PDF

Another essential ingredient in the proposed tracking method is a similarity /dissimilarity measure between pdfs. In this paper we use the dissimilarity measure Kullback-Leibler divergence (KL-divergence) [10]:

$$KL(P,Q) = \int Q(z) log(\tfrac{Q(z)}{P(z)}) dz,$$

where $P(z)$ and $Q(z)$ are two pdfs that are being compared. KL-divergence measures the dissimilarity between P and Q. KL-divergence is non-negative and is zero only when P and Q are equal. Larger dissimilarity between P and Q yields larger value of KL-divergence. There is another widely used similarity measure for pdfs called Bhattacharya coefficient [2]:

$$BC(P,Q) = \int \sqrt{Q(z)P(z)} dz,$$

Bhattacharya coefficient is a number between 0 and 1. It achieves a value of 1 when P and Q are equal. The more similar P and Q are, the closer is the value of the Bhattacharya coefficient to unity. Bhattacharya coefficient has been successfully utilized in mean-shift tracking method [2]. By utilizing Jensens inequality [10] we, however, show here that decrease in KL-divergence ends up incrementing Bhattacharya coefficient and we choose to use KL-divergence in the proposed tracking method.

Proposition 1. *Decreases in KL-divergence increases Bhattacharya coefficient.*

Proof: By Jensons inequality we can write the following:

$$log(BC) = log(\int \sqrt{Q(z)P(z)} dz) \geq \int Q(z) log(\sqrt{\tfrac{P(z)}{Q(z)}}) dz = -\tfrac{1}{2} KL.$$

Thus we have: $BC \geq \exp\left(-\tfrac{1}{2} KL\right)$, which shows that decrease of KL-divergence KL (a non-negative number) increases the Bhattacharya coefficient BC.

3   Proposed Method

3.1   Proposed Deformable Object Model

As already mentioned in the Introduction section, the proposed deformable object model is the joint pdf of level set function and image intensity learned on the first frame of a video sequence where tracking begins. Let $Q(l,i)$ denote this joint pdf, where l and i respectively denote level set function value and image intensity value. By means of Parzen window estimate using Gaussian kernel one can express Q as:

$$Q(l,i) = \frac{1}{C} \int \int \exp\left(-\frac{(\psi(x,y)-l)^2}{2\sigma_l^2}\right) exp\left(-\frac{(J(x,y)-i)^2}{2\sigma_i^2}\right) dx dy,$$

where $\psi(x,y)$ and $J(x,y)$ are respectively the level set function and the image intensity on the first image frame. C is a normalization factor that forces $Q(l,i)$ to integrate to unity. σ_i and σ_l are standard deviations of the Gaussian kernels. It is assumed that the object boundary is provided on the first image frame, so the construction of $\psi(x,y)$ is performed by a signed distance transform from the object boundary. On the second frame (or subsequent frames) we construct a similar joint pdf $P(l,i)$ as follows:

$$P(l,i) = \frac{1}{C} \int \int \exp\left(-\frac{(\phi(x,y)-l)^2}{2\sigma_l^2}\right) exp\left(-\frac{(I(x,y)-i)^2}{2\sigma_i^2}\right) dx dy,$$

where $\phi(x,y)$ and $I(x,y)$ are respectively the level set function and the image intensity on the second frame. Note that $\phi(x,y)$ is not known on the second frame and we want to compute $\phi(x,y)$ so that P and Q are as close as possible. Once we find out the desired $\phi(x,y)$, the zero level set of $\phi(x,y)$ would provide us with the object boundary on the second (or the subsequent) image frame.

3.2   Matching Object Models

To compute the desired object boundary on the second or any subsequent frame, we minimize the KL-divergence between P and Q to obtain the desired ϕ^* :

$$\phi^* = arg\min_{\phi} \int \int Q(l,i) log\left(\frac{Q(l,i)}{P(l,i)}\right) dl di = arg\max_{\phi} \int \int Q(l,i) log(P(l,i)) dl di.$$

The last equality follows because Q is not a function of $\phi(x,y)$ and only P is a function of $\phi(x,y)$,i.e.,$P \equiv P(l,i;\phi)$. Thus the energy functional to maximize with respect to ϕ becomes:

$$E(\phi) = \int \int Q(l,i) log(P(l,i;\phi)) dl di.$$

Applying calculus of variations [10] one can show that the gradient ascent partial differential equation (PDE) for the maximizing level set function becomes:

$$\frac{\partial \phi}{\partial t}(x,y) = -\int \int \frac{Q(l,i)}{P(l,i;\phi)} \frac{(\phi(x,y)-l)}{\sigma_l^2} \exp\left(-\frac{(\phi(x,y)-l)^2}{2\sigma_l^2}\right) exp\left(-\frac{(I(x,y)-i)^2}{2\sigma_i^2}\right) dl di,$$

We can compactly express this PDE by using the convolution notation "$*$" :

$$\frac{\partial \phi}{\partial t}(x,y) = -(\tfrac{Q}{P} * g_1)(\phi(x,y), I(x,y)),$$

where g_1 is a function defined as: $g_1(l,i) = \frac{l}{\sigma_l^2} \exp(-\frac{l^2}{2\sigma_l^2}) \exp(-\frac{i^2}{2\sigma_i^2})$ and $\tfrac{Q}{P}$ is simply ratio of Q and P : $\tfrac{Q}{P}(l,i) = \frac{Q(l,i)}{P(l,i;\phi)}$.

3.3   Efficient and Stable Implementation

In order to solve the aforementioned PDE one can use an explicit discretization scheme [11] as follows:

$$\phi^{t+1}(x,y) = \phi^t(x,y) - (\delta t)(\tfrac{Q}{P} * g_1)(\phi^t(x,y), I(x,y)),$$

where the t denotes the iteration number and δt denotes time step size. The explicit scheme is however not suitable for practical purposes as instability develops unless the time step size is very small, in which case the computation takes a long time to converge. Also the solution quality is often not acceptable. Fortunately we can work around this problem using a semi-implicit scheme [11] as follows:

$$\phi^{t+1}(x,y) = \frac{\phi^t(x,y) + (\delta t)(\tfrac{lQ}{P} * g)(\phi^t(x,y), I(x,y))}{1 + (\delta t)(\tfrac{Q}{P} * g)(\phi^t(x,y), I(x,y))},$$

where g is the Gaussian convolution kernel: $g(l,i) = \exp(-\tfrac{l^2}{2\sigma_l^2})\exp(-\tfrac{i^2}{2\sigma_i^2})$ and $\tfrac{lQ}{P}$ is a function defined as: $\tfrac{lQ}{P}(l,i) = \tfrac{lQ(l,i)}{P(l,i;\phi)}$. Note that although one requires two convolutions in each iteration of the semi-implicit numerical scheme as opposed to one convolution in the explicit scheme, the convergence is many times faster in the former as the there is no restriction on the time step size.

4   Results, Comparisons and Discussion

We first illustrate by a straightforward experiment that Kullback-Leibler flow of Freedman and Zhang [9] lacks the object discrimination capability, whereas the proposed method is superior in this respect. The leftmost image of Fig. 1 shows the first frame of a synthetic image sequence where the object is a homogeneous circle on a homogeneous background. The object is shown to be delineated by a boundary for which both Freeman and Zhang's method and the proposed method learn the pdf. Note however that Freedman and Zhang's method learns the intensity histogram within the object boundary, whereas the proposed method learns a joint pdf of level set function and image intensity both inside and outside the object boundary. The middle and the rightmost images of Fig. 1 show respectively the results by Freedman and Zhang and the proposed method on the second frame of the synthetic image sequence. Note that the object in the second frame has grown larger in size and Freedman and Zhang's method could not perceive this change because it can only keep track of what is happening inside the object, not the surrounding. The proposed method is however seen to correctly delineate the object boundary on the second frame even though the object has grown bigger in size.

Next, we apply the proposed method and another popular competing active contour method, *viz.*, gradient vector flow (GVF) snake method [12] on two cine MRI sequences–BSL and PK. The PK sequence shows much rapid and vigorous cardiac motions than the BSL sequence. Fig. 2 shows a few images from the PK

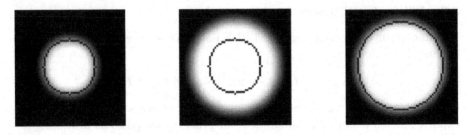

Fig. 1. From left: first image frame and delineated object; second image frame and the result by Freedman and Zhang's method; result by the proposed method

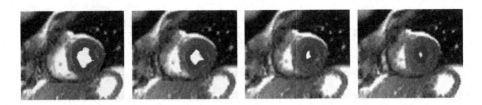

Fig. 2. Tracking myocardial boundary by the proposed tracking method–a few cine MRI frames and delineated boundaries on them

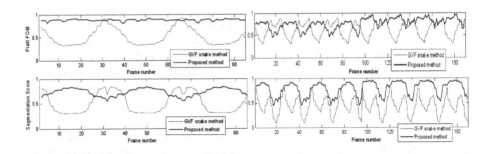

Fig. 3. Comparison of segmentation metrics for GVF snake method and the proposed method on the BSL (left column) and the PK (right column) sequence

sequence where the proposed method is shown to be able track the mouse heart boundary.

Fig. 3 shows comparisons of GVF snake method and the proposed algorithm on two cine MRI sequences. Table 1 summarizes the results by reporting only the mean values of the performance metrics. In these experiments we used two performance metrics: (a) Pratt's figure of merit (FOM) that is widely used for measuring the accuracy of edge detection [13], and (b) segmentation score, which we define as the ratio of intersection over the union of the segmented regions for ground truth segmented image and the segmentation produced by an automated algorithm. For ground truth generations we manually labeled all the

Table 1. Comparisons of mean performance metrics for GVF snake and the proposed method

	Pratt's FOM		Segmentation Score	
	BSL	PK	BSL	PK
GVF Snake Method	0.51	0.62	0.51	0.44
Proposed Method	0.88	0.77	0.74	0.79

frames of the two cine MRI sequences. There are two parameters in the proposed method, *viz.*, the two standard deviations of the Gaussian kernels. We took $\sigma_i = 1$ and $\sigma_l = 0.5$ by cross-validation based on Pratt's FOM. These comparisons demonstrate that the performances of the proposed method on both the rapid and the slow heart rate sequences are superior.

References

1. Janiczek, R., Ray, N., Acton, S.T., Roy, R.J., French, B.A., Epstein, F.H.: Markov chain Monte Carlo method for tracking myocardial borders. Computational Imaging III. In: Bouman, C.A., Miller, E.L. (eds.) Proceedings of the SPIE, vol. 5674, pp. 211–218 (2005)
2. Comaniciu, D., Ramesh, V., Meer, P.: Kernel-Based Object Tracking. IEEE PAMI 25(5), 564–577 (2003)
3. Montagnat, T., Delingette, H., Ayache, N.: A review of deformable surfaces: topology, geometry and deformation. Image and Vision Computing 19(14), 1023–1040 (2001)
4. Lai, C., Chin, R.: Deformable Contours: Modeling and Extraction. IEEE PAMI 17(11), 1084–1090 (1995)
5. Cootes, T.F., Edwards, G.J., Taylor, C.J.: Active appearance models. IEEE PAMI 23(6), 681–685 (2001)
6. Kass, M., Witkin, A., Terzopoulos, D.: Snakes: Active contour models. International J. of Comp. Vis. 1(4), 1405–1573 (1988)
7. Sethian, J.A.: Level Set Methods and Fast Marching Methods: Evolving Interfaces in Computational Geometry, Fluid Mechanics, Computer Vision and Materials Sciences. Cambridge University Press, Cambridge (1999)
8. Leventon, M.: Statistical Models for Medical Image Analysis. Ph.D. Thesis, MIT (2000)
9. Freedman, D., Zhang, T.: Active contours for tracking distributions. IEEE Trans. Image Proc. 13(4), 518–526 (2004)
10. Bishop, C.M.: Pattern Recognition and Machine Learning. Springer, Heidelberg (2006)
11. Ames, W.F.: Numerical methods for partial differential equations. Academic, New York (1992)
12. Xu, C., Prince, J.L.: Snakes, Shapes, and Gradient Vector Flow. IEEE Transactions on Image Processing 7(3), 359–369 (1998)
13. Pratt, W.K.: Digital Image Processing. Wiley InterScience, Chichester (2002)

Spatio-temporal Descriptor Using 3D Curvature Scale Space

A. Dyana and Sukhendu Das

Visualization and Perception Lab, Computer Science and Engineering Deptt.
Indian Institute of Technology Madras, Chennai, India
dyana@cse.iitm.ernet.in, sdas@iitm.ac.in

Abstract. This paper presents a novel technique to jointly represent the shape and motion of video objects for the purpose of content based video retrieval (CBVR). It enables to retrieve similar objects undergoing similar motion patterns, that are not captured only using motion trajectory or shape descriptors. In our approach, both shape and motion information are integrated in a unified spatio-temporal representation. Curvature scale space theory proposed by Mokhtarian is extended (in 3D) to represent shape as well as motion trajectory of video objects. A sequence of 2D contours are taken as input and convolved with a 2D Gaussian. The zero crossings are found out from the curvature of evolved surfaces, which form the 3D CSS surface. The peaks from the 3D CSS surface form the features for joint spatio-temporal representation of video objects. Experiments are carried out on CBVR and results show good performance of the algorithm.

Keywords: Curvature scale space, motion trajectory, content based video retrieval, spatio-temporal descriptor.

1 Introduction

As there is an increase in availability of video data, there is a need to describe video based on its content. MPEG-4 provides access and manipulation of video objects and MPEG-7 describes the features of the multimedia content. Shape and motion are important features in content based video retrieval. Shape descriptors (spatial domain) are classified into two categories: region based and contour based. A contour-based descriptor encapsulates the shape properties of the object's outline (silhouette). Fourier descriptors, Medial axis transform, Shape signature [1], Shape context [2], Geometric representations, Grid representation [3] are some of the contour based techniques to represent object shapes.

Multiscale description of contours is an emerging research area for describing shapes. The Multiscale Fourier descriptors improves the shape retrieval accuracy of the commonly used Fourier descriptors [4]. A multiscale extension to the medial axis transform (MAT) [5] or skeleton is obtained by combining information derived from a scale-space hierarchy of boundary representations with region information provided by the MAT. Two leading approaches emerged: one by Latecki et al. [6] based on the best possible correspondence of visual parts

A. Ghosh, R.K. De, and S.K. Pal (Eds.): PReMI 2007, LNCS 4815, pp. 632–640, 2007.

and a second approach developed by Mokhtarian et al. based on the curvature scale space (CSS) representation [7]. Both descriptors are based on the computation of a similarity measure on the best possible correspondence between maximal convex/concave arcs contained in simplified versions of boundary contours. It is impossible to use the CSS descriptor to distinguish between totally convex shapes. Given the above observations, a new shape description method termed multi-scale convexity concavity (MCC) representation was proposed by T. Adamek [8]. A multiscale, morphological method for the purpose of shape-based object recognition was proposed [9].

Motion trajectory describes the displacements of objects in time, where objects are defined as spatio-temporal regions. For modeling object's trajectory, Dimitrova and Golshani [10] utilized a chain coding scheme to represent different objects' movements. Little and Gu [11] used an interpolation scheme to model object trajectories by using a set of polynomial basis. In addition, Sahouria [12] applied a Harr transform for representing object trajectories in spatial domain through a multiscale analysis. In addition, Dagtas et al. [13] proposed a trajectory-based model and a trail-based model for video retrieval by taking advantages of the Fourier transform and Mellin transform, respectively. Two affine-invariant representations for motion trajectories based on curvature scale space (CSS) and centroid distance function (CDF) was derived.

Spatio-temporal descriptors to represent position and motion of regions has been proposed in [14][15]. The position, motion and color describes the region in a high dimensional space. K nearest neighbor retrieval was used in [14]. In [15], a Graph based description was used to describe relation between regions. In content based video retrieval systems such as Netra-V [16], the shape and motion features are represented individually and the match results are integrated to select a video clip. Chang et al. [17] proposed a VideoQ system for video searching using a set of visual features like color, texture, shape, and motions. In our proposed method, the standard 2D CSS representation is modified to represent both shape and motion of video objects in a unified spatio-temporal representation. The deformed shapes over time are taken as input and a 3D CSS descriptor is generated for each video object.

2   Standard Curvature Scale Space

A CSS image can be considered as a multi-scale organization of the invariant local features of a free-form 2D contour. The CSS image is a multi-scale organization of the inflection points (or curvature zero-crossing points) of the contour as it evolves. Curvature is a local measure of how fast a planar contour is bending. Contour evolution is achieved by first parameterizing by arc length. This involves sampling the contour at equal intervals and recording the 2D coordinates of each sampled point. The result is a set of two coordinate functions (of arclength) which are then convolved with a 1D Gaussian filter of increasing width or standard deviation. Curvatures are computed for all the smoothed contours. As a result, curvature zero-crossing points can be recovered and mapped to the CSS image in which the horizontal axis represents the arclength parameter (u) on the original contour, and the vertical axis represents the standard deviation (scale) of the Gaussian filter. Fig. 1 illustrates an example for 2D CSS representation. Fig. 1b is the CSS image for the fish contour shown in Fig. 1a. The maximas of lower scale are considered to be noise which is ignored in the matching stage. The features recovered from a CSS image for matching are the maxima of its

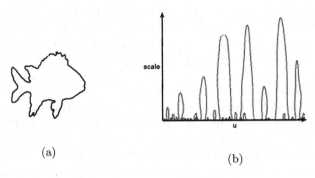

(a)   (b)

Fig. 1. 2D CSS image (b) for the fish shape in (a)

zero-crossing contours [18]. The matching of two CSS images consists of finding the optimal horizontal shift of the maxima in one of the CSS images that would yield the best possible overlap with the maxima of the other CSS image. The matching cost is then defined as the sum of pairwise distances (in CSS) between corresponding pairs of maxima.

3   Proposed 3D CSS Representation

The curvature scale space for 2D contours proposed by Mokhtarian [19] for planar curves, which has been standardized in MPEG-7, is extended in our work. This is a joint spatio-temporal representation of moving objects in a video (shot) and is similar to the 3D surface representation in [7].

Mokhtarian represented 2D contour by a parametric vector,

$$r(u) = ((x(u), y(u))$$ (1)

We represent the input sequence of contours by the following equation which has two parameters: spatial(u) and temporal (v), as

$$r(u, v) = ((x(u, v), y(u, v))$$ (2)

To normalize the arc length (u), the contour is sampled and represented by 200 equally spaced points. Similarly, the temporal parameter (v) is normalized by uniformly sampling the trajectory and representing by 40 points. The formula for computing the curvature is extended from 2D as:

$$\kappa(u,v) = \frac{\dot{x}(u,v)\ddot{y}(u,v) - \dot{y}(u,v)\ddot{x}(u,v)}{(\dot{x}(u,v)^2 + \dot{y}(u,v)^2)^{3/2}} \tag{3}$$

If,

$$\Gamma = \{(x(u,v), y(u,v)) | u \in [0,1], v \in [0,1]\} \tag{4}$$

then the evolved contours when convolved with Gaussian is given by

$$\Gamma_\sigma = \{(X_\sigma(u,v), Y_\sigma(u,v)) | u \in [0,1], v \in [0,1]\} \tag{5}$$

where

$$X_\sigma(u,v) = x(u,v) \otimes g_\sigma(u,v)$$
$$Y_\sigma(u,v) = y(u,v) \otimes g_\sigma(u,v) \tag{6}$$

$$g_\sigma = \frac{1}{2\pi\sigma^2} e^{-u^2+v^2/2\sigma^2} \quad \text{is a 2D Gaussian function.}$$

and $\otimes$ indicates the 2D convolution function.

Using Eqns. 3 and 5, the curvature of the evolved contours is given by,

$$\kappa_\sigma(u,v) = \frac{\dot{X}_\sigma(u,v)\ddot{Y}_\sigma(u,v) - \dot{Y}_\sigma(u,v)\ddot{X}_\sigma(u,v)}{\left(\dot{X}_\sigma(u,v)^2 + \dot{Y}_\sigma(u,v)^2\right)^{3/2}} \tag{7}$$

where

$$\dot{X}_\sigma(u,v) = \frac{\partial}{\partial u}(x(u,v) \otimes g_\sigma(u,v)) + \frac{\partial}{\partial v}(x(u,v) \otimes g_\sigma(u,v))$$
$$= x(u,v) \otimes \dot{g}_\sigma(u,v) \tag{8}$$

and

$$\ddot{X}_\sigma(u,v) = \frac{\partial^2}{\partial u^2}(x(u,v) \otimes g_\sigma(u,v)) + \frac{\partial^2}{\partial v^2}(x(u,v) \otimes g_\sigma(u,v))$$
$$= x(u,v) \otimes \ddot{g}_\sigma(u,v) \tag{9}$$

Similarly,

$$\dot{Y}_\sigma(u,v) = y(u,v) \otimes \dot{g}_\sigma(u,v)$$
$$\ddot{Y}_\sigma(u,v) = y(u,v) \otimes \ddot{g}_\sigma(u,v) \tag{10}$$

The symbol $\otimes$ used in Eqns. 8 - 10 indicates a 2D convolution operation. To implement the same, the numerical addition of two separable derivatives of the gaussian function(g) along x and y is taken, and then convolved with the 2D function (x or y).

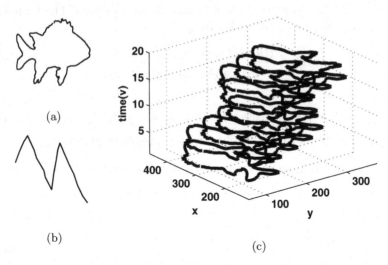

Fig. 2. Input sequence of contours in (c) for the object (a) fish and its trajctory in (b)

The solution of the equation given in 7

$$\kappa_\sigma(u, v) = 0 \tag{11}$$

gives the 2D zero crossing contours.

Fig. 2 shows the input sequence of contours of a fish shown in Fig. 2a which is translated along the trajectory shown in Fig. 2b. The contour of the fish is represented by x and y coordinates and the fish is moved along the trajectory over time. The change in the contour position is depicted along the v axis. The surface thus formed by the stack of 2D contours, is evolved by a 2D Gaussian function at different scales (σ).The surface can be interpreted as a set of normalized 2D contours, where the z coordinate is v. Hence the curvature calculation can be extended from 2D as shown in Eqn. 3. As the contours get evolved it is smoothed in both the dimensions, spatial and temporal. The surface eventually transform into a cylindrical structure. When viewed along the spatial coordinates (x and y), the contour evolves into an elliptical structure.

4   3D CSS Surface

3D CSS surface is a 3D plot of the zero crossings at each level. Axis variables are the spatial parameter (u), the temporal parameter (v) and scale(σ). The surface (see Fig. 3) consists of a series of hills and valleys. Every hill corresponds to the successive pair of occurances of convexity and concavity (or vice-versa) in the sequence of contours (i.e surface). The local peaks are found out by first deleting the global peak at the largest value of sigma, and then recursively traversing down and eliminating its neighbours until a valley is reached. The set of peaks

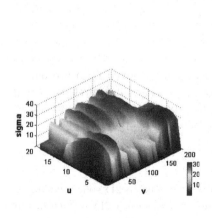

Fig. 3. 3D CSS surface

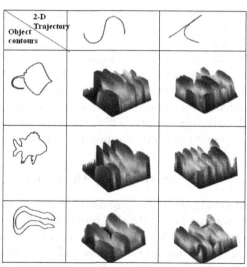

Fig. 4. 3D CSS surfaces for three different objects and two different motion trajectories

form the features to jointly represent shape and motion of the object. The overall algorithm of the proposed method is given in the following section.

4.1   Overall Algorithm

Input for the system is a sequence of 2D contours of video objects. From a set of consecutive frames, contours are taken and stacked over time to form a 3D surface. x and y axis is characterized by spatial parameter (u) and z axis by a temporal parameter (v) (see Eqn. 2). The steps of the overall algorithm are as follows:

1. The 3D surface is smoothed by a 2D Gaussian of minimum σ to obtain an evolved surface (Eqn. 5)
2. Curvature of the evolved surface is computed using Eqn. 3
3. The zero crossings of the curvature in the spatial as well as temporal dimensions are found out and the locations (u, v, σ) of the same are stored.
4. Increment the value of sigma and repeat steps 2 and 3.
5. The set of zero crossings are marked for each sigma which forms a 3D CSS surface as shown in Fig. 3.
6. The peaks from the 3D CSS surfaces are obtained which form the features for the moving object.
7. To match the query with the models from the database the peaks of the query and model are matched by the matching algorithm described in section 4.2

Fig. 4 illustrates the discriminating ability of the 3D CSS representation to distinguish between different combinations of trajectories and objects. A contour of a fish is taken and it is moved or translated along the trajectory to obtain a 3D CSS surface. Three different shape contours along two different trajectories are shown. For the same trajectory (motion) and for different objects, the 3D CSS surfaces have similar structure along v axis (the time axis). For the same object with different motions (trajectories), the 3D CSS surfaces have similar structure along u axis (the parameter for object contours). As observed from the figure, different combinations of trajectories and objects provide different structures of the 3D CSS surface. This shows the uniqueness property of the algorithm to represent the spatio-temporal information of an object shape in a video.

4.2   3D CSS Matching Algorithm

Matching algorithm proposed for matching CSS maxima in 2D space by Mokhtarian has been extended for matching the set of peaks on a 3D surface in our case. Every video object in the database is represented by the locations of peaks (u, v, σ) in its 3D CSS surface. The location of peaks are normalized. The set of peaks which form the features representing the shape trajectory pair for the query is matched with the set of peaks in the database. The pairs are retrieved according to the match-cost. The lower the match-cost, higher is the similarity of the model with the query.

In the standard CSS, the maximas (u, σ) in the CSS image form the features. In our 3D CSS, the location of the peaks represented by triplets (u, v, σ) form the features. The feature list is sorted by the scale coordinate of the peaks. The highest scale maximum of the query is matched with the model whose σ coordinate lies close (within 80%) to that of the highest model peak. For each matched peak, nodes are created with two lists. The first list contains the peaks of the query, the second list contains its corresponding matching peak in the model. The algorithm proceeds in similar to the standard 2D CSS matching algorithm. In the standard 2D CSS matching algorithm, the horizontal distance of the maxima and the height of the peaks contribute to the match-cost. In our algorithm, the euclidean distance (in 2D space) between the peaks and the height of the peaks contribute to the match-cost.

The system has the property of translation, rotation and scaling invariants of shape and trajectories. This is based on the principle that 2D CSS is also invariant to rotation, translation and scaling. It works based on the principle that similar video objects should have similar representations.

5   Experimental Results

Objects used for our experiments are from MPEG7-B dataset [20] and a part of the SQUID database [18] (marine animals). The trajectory set is part of the database downloaded from [21]. For each combination of object and trajectory, a 3D CSS descriptor is generated and stored. For a query, the 3D CSS descriptor is obtained and matched with the models from the database.

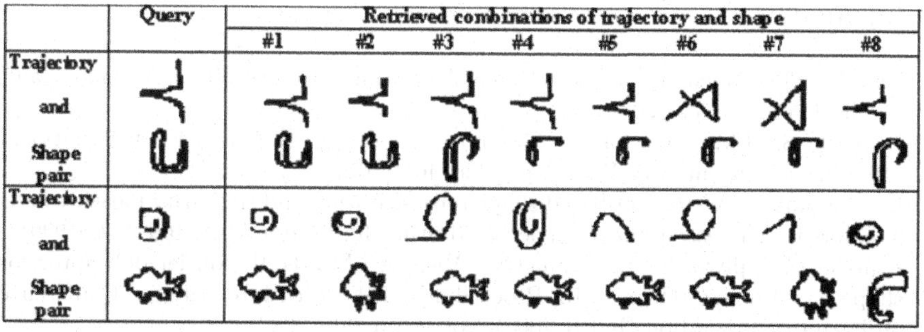

	Query	Retrieved combinations of trajectory and shape							
		#1	#2	#3	#4	#5	#6	#7	#8
Trajectory and Shape pair									
Trajectory and Shape pair									

Fig. 5. Top eight shape and trajectory pairs retrieved for the query in the first row of the table, arranged in descending order of similarity

Table 1. Average Accuracy for retrieval estimated over first 10 outputs

Dataset	Accuracy (in percentage)
MPEG7-B [20]	74.78
SQUID [18]	77.01

One set of experiments were conducted on 100 shapes from MPEG7-B dataset and 24 trajectories from [21]. Other set of experiments involved 24 shapes of "fish" from SQUID database and 36 trajectories from [21]. Fig. 5 shows the results of retrieval for the query "fish" from SQUID, as displayed in the first column. All other entries (columnwise) show the shape and motion pairs retrieved, which are arranged in decreasing order of similarity (ie. ascending order of match-cost) from left to right. From the results, it is observed that, the retrieved pairs are similar to the query. The results show that the proposed method is invariant to translation, rotation and scale for both shape and trajectory. Perspective foreshortening due to sidewise movement of the marine animal, is not considered within the current scope of work. Table. 1 shows the average accuracy of retrieval estimated over 100 queries. Accuracy is calculated as the ratio of the number of correct outputs observed for the first 10 retrieved videos.

6   Conclusion

We have proposed a unique spatio-temporal 3D CSS representation to represent shape and motion of video objects. The deformation of shape is captured in the representation. The algorithm has shown good results for CBVR application. The 3D CSS matching algorithm can be modified to be more computationally efficient.

References

1. Davies, E.R.: Machine Vision: Theory, Algorithms, Practicalities. Morgan Kaufmann, San Francisco (2005)
2. Belongie, S., J.M., Puzicha, J.: Matching shapes. In: Eighth IEEE International Conference on Computer Vision, pp. 456–461 (2001)
3. Lu,, Sajjanhar, A.: Region-based shape representation and similarity measure suitable for content-based image retrieval. Multimedia System 7(2), 165–174 (1999)
4. Kunttu, I., Lepistö, L., Rauhamaa, J., Visa, A.: Multiscale fourier descriptor for shape-based image retrieval. In: Proceedings of the 17th International Conference on Pattern Recognition (ICPR 2004), vol. 2, pp. 765–768.
5. Ogniewicz, L.R.: Skeleton-space: a multiscale shape description combining region and boundary information. In: Proc. Comput. Vision Pattern Recogn.
6. Latecki, L.J., Lakamper, R., Eckhardt, U.: Shape descriptors for non-rigid shapes with a single closed contour. In: CVPR, pp. 424–429
7. Mokhtarian, F., Bober, M.: Curvature Scale Space Representation: Theory, Applications and MPEG-7 Standardization. Kluwer Academic Publishers, The Netherlands (2003)
8. Adamek, T., O'Connor, N.E.: A multiscale representation method for nonrigid shapes with a single closed contour. IEEE Transactions on Ciruits and systems for video technology 14(5), 742–753 (2004)
9. Jalba, A.C., Wilkinson, M.H.F., Roerdink, J.B.T.M.: Shape representation and recognition through morphological curvature scale spaces. IEEE Transactions on Image Processing 15(2), 331–341 (2006)
10. Dimitrova, N., Golshani, F.: Motion recovery for video content classification. ACM Trans. Inf. Syst. 13(14), 408–439 (1995)
11. Little, J.J., Gu, Z.: Video retrieval by spatial and temporal structure of trajectories. In: Proc. SPIE Storage and Retrieval for Media Databases, vol. 13, pp. 408–439 (1995)
12. Sahouria, E.: Video indexing based on object motion. Master's thesis, Dept. Elect. Eng. Comp. Sci, Univ. California, Berkeley (1997)
13. Dagtas, S., et al.: Models for motion-based video indexing and retrieval. IEEE Transactions on Image Processing 9(1), 88–101 (2000)
14. DeMenthon, D., Doermann, D.: Video retrieval using spatio-temporal descriptors. In: ACM Multimedia, pp. 508–517 (2003)
15. Chatzis, S., Doulamis, A., Kosmopoulos, D., Varvarigou, T.: Video representation and retrieval using spatio-temporal descriptors and region relations. In: ICANN (2), pp. 94–103 (2006)
16. Deng, Y., Manjunath, B.S.: Netra-v: Toward an object-based video representation. IEEE Transactions on Ciruits and systems for video technology 8(5), 116–127 (1998)
17. Chang, S.F., et al.: A fully automated content-based video search engine supporting spatiotemporal queries. IEEE Transactions on Ciruits and systems for video technology 8(5), 602–615 (1998)
18. SQUID, http://www.ee.surrey.ac.uk/research/vssp/imagedb
19. Mokhtarian, F., Abbasi, S., Kittler, J.: Robust and efficient shape indexing through curvature scale space. In: British Machine Vision Conference, pp. 53–62 (1996)
20. MPEG7-BDataset, http://csr.bu.edu/asl/data/experiments/cikm2005/database2/
21. TrajectoryDatabase, http://mmplab.eed.yzu.edu.tw/trajectory/trajectory.rar

Shot Boundary Detection Using Frame Transition Parameters and Edge Strength Scatter

P.P. Mohanta[1], S.K. Saha[2], and B. Chanda[1]

[1] ECS Unit, Indian Statistical Institute, Kolkata, India
[2] CSE Department, Jadavpur University, Kolkata, India

Abstract. We have presented a unified model for various types of video shot transitions. Based on that model, we adhere to frame estimation scheme using previous and next frames. The frame parameters accompanied by a scatter measure of edge strength and average intensity constitute the feature vector of a frame. Finally, the frames are classified as no change (within shot frame), abrupt change or gradual change frames using a multilayer perceptron network. The scheme is free from the problems of selecting thresholds and/or window size as used by various schemes. Moreover, the handling of both, abrupt and gradual transitions along with non-transition frames under a single and uniform framework is the unique feature of the work.

Keywords: shot detection, abrupt transitions, cut, gradual transitions.

1 Introduction

Due to the advancement of video technology the volume of digital video data has increased dramatically. But, the tools available for browsing such databases are still primitive in nature. To address the problem, indexing and retrieval has become an active area of research. Video segmentation is the fundamental step for the said application including video indexing, content analysis of video sequence, video accessing, retrieving and browsing, video compression and others. A fast and automatic technique for temporal segmentation of video content is very crucial for accurate content description.

The objective of video shot segmentation is to partition video into meaningful and basic structural units called shots. A shot corresponds to a sequence of frames captured through a continuous record (in time and space) by camera [1]. It describes a meaningful event over a continuous sequence of frames. Once the boundaries of the shots are detected, further analysis of content and interpretation can be performed on such units.

The transition may be of various types and broadly categorized as *abrupt* and *gradual* transition. Abrupt transition is also known as *cut* and it denotes instantaneous transition from one shot to another. On the other hand, a gradual transition is obtained by incorporating photographic effect usually through editing. It can be further classified as fade-in, fade-out and dissolve. Fade-out is

A. Ghosh, R.K. De, and S.K. Pal (Eds.): PReMI 2007, LNCS 4815, pp. 641–648, 2007.
© Springer-Verlag Berlin Heidelberg 2007

a gradual transition of a scene to a constant image (commonly a black frame) and fade-in is reverse transition. Dissolve is gradual super-imposition of two consecutive shot.

It is easy to detect the cuts as it involves two successive frames which are highly uncorrelated and differs significantly. But, during gradual transition, the successive frames may not differ much. Thus, the major challenge becomes to distinguish between the gradual transition and the nominal changes in the scene.

Lots of work has been reported on cut detection. Comparatively, a less amount of work has dealt with gradual transition. The basis of all such algorithms lies in detecting the visual discontinuities in time domain. The schemes extract the visual features and deploys a similarity between the frames. Most of the cut detection schemes identifies the transition if the difference between two consecutive frames exceeds a certain threshold. Similarity (difference) of the frames are measured in terms of features computed from the frame. A wide variety of features have been reported in various works. The simplest one is pixel wise difference [2,3]. But as it is very sensitive to motion of objects, grayscale/colour histogram based features are also tried in [4] though the histograms lack spatial information. As an alternative, features based on motion vector analysis [5], edge tracking [6], edge changes [7], entropy measures [8] are also used. In order to detect gradual transitions, twin comparison method [2] deals with two threshold values to detect cuts and gradual transition. Yeo and Lin [3] proposed plateau detection technique where difference between current frame and kth frame that follows is considered. But, in this case, proper selection of k is a non-trivial task. A method called chromatic scaling has been discussed in [9]. In another approach, transitions are detected by counting the entering/exiting edge pixels [10]. Machine learning and multi resolution concept [11] are also reported for dissolve identification. Algorithm evolved by combining the concept of object tracking and feature based approaches was also tried for dissolve detection [12]. In [13], the variance of pixel intensity of a sequence is modeled as a parabolic curve and based on that model a detection scheme is presented.

It appears that although a lot of schemes have been tried but they have their own merits and demerits and almost none of them has tried to detect all kinds of shot boundaries in a comprehensive way. In this work, we present a parametric model of the shot transitions of various types which, in turn, will be used to detect and classify the shot. The paper is organized as follows. Section 2 presents the formulation of the problem and the details of the proposed scheme. Experimental results are presented in section 3 and concluding remarks are put in section 4.

2   Proposed Methodology

In this section, a general framework is presented to describe the transition of various types and it will act as the basis for the proposed shot detection scheme.

In case of abrupt transition, last frame of a shot and first one of the following shot are uncorrelated. A cut is generated by the natural process of capturing video data through the camera. On the contrary, gradual transitions (*fade-in, fade-out and dissolve or cross-fading*) are generated through editing. Dissolves are generated by super-imposing the boundary frames of two successive shots over a duration. In such cases of gradual transition, intensity of one boundary frame gradually decreases and that of other one increases during the phase of transition. For *fade-out* (or *fade-in*) the intensity of boundary frames are gradually reduced (or increased) and last (or first) frame of such transitions is commonly a black frame. Thus, unlike abrupt transitions, gradual transitions spans over a range of frames. It is also obvious that presence of motion and activities usually are very insignificant in the frames of such edited transitions.

The successive frames within a shot and also those within the span of a gradual transition show little differences. Thus, the ability to distinguish the two situation controls the false and misclassification rate. It may be noted that, differences between the successive frames within a shot is mostly caused by camera and/or object motion keeping the background otherwise unaltered. But, for gradual transition, it mainly comes from the editing process.

2.1   Problem Formulation

With the background idea of natural and edited transitions, we formulate the scenario as follows.

Let, $f_1, f_2, \ldots, f_n$ denotes a sequence of frames in a video. Suppose, f_{l_1} and f_{l_2} are the last representative frame of a shot and the first representative frame of the following shot, where, $0 \leq l_1 < l_2 \leq n$. The frames in transitions are denoted by f_i where i varies from l_1 to l_2. Such frames may be represented as

$$f_i = A_i f_{l_1} + B_i f_{l_2} \qquad (1)$$

where, $0 \leq A_i, B_i \leq 1$ and $A_i + B_i = 1$. Basically, A_i's and B_i's modulates the intensities of the frames being super-imposed.

In case of a *cut*, $l_2 = l_1 + 1$. From equation 1, it is obvious that there will be two transition frames with $A_i = 1, B_i = 0$ for one and reverse for the other.

For *gradual transition*, A_i gradually decreases from 1 to 0 and B_i successively increases from 0 to 1. It may be noted that, in case of *fade-in*, f_{l_1} is the black frame and for *fade-out*, f_{l_2} is the black frame.

Thus, the model shown in equation 1 can represent all sorts of transitions. But, during the process of boundary detection, the representative frames (f_{l_1} and f_{l_2}) are not available. As a matter of fact, it is the task of detection process to find out these frames along with the boundaries. Even then, model described in equation 1 provides the underlying structure for our scheme. Based on the model of equation 1, f_{i-1}, f_i and f_{i+1} can be represented as follows.

$$f_{i-1} = A_{i-1} f_{l_1} + B_{i-1} f_{l_2} \qquad (2)$$

$$f_i = A_i f_{l_1} + B_i f_{l_2} \qquad (3)$$

$$f_{i+1} = A_{i+1} f_{l_1} + B_{i+1} f_{l_2} \tag{4}$$

By manipulating the equations 2, 3 and 4, f_i can be represented in terms of its previous and following frame as shown in equation 5.

$$f_i = a_i f_{i-1} + b_i f_{i+1} \tag{5}$$

where $a_i = \frac{A_i B_{i+1} - B_i A_{i+1}}{A_{i-1} B_{i+1} - A_{i+1} B_{i-1}}$ and $b_i = \frac{B_i A_{i-1} - A_i B_{i-1}}{A_{i-1} B_{i+1} - A_{i+1} B_{i-1}}$, and $A_i + B_i = 1$ implies that $a_i + b_i = 1$. This model is also valid for the frames within a shot (*i.e.* no change frames) and in that case, ideally, it will be $a_i = b_i = 0.5$. Thus, all types of frames can be estimated from its previous and following frame. Ideally, the characteristic pattern of (a_i, b_i) is similar to that of (A_i, B_i) and can be used for shot detection and classification purpose. This has motivated us to go for frame parameter estimation for shot boundary detection and classification. As we adhere to the model of equation 5 instead of that in equation 1, there is no need to consider a sliding window of suitable size for studying the characteristics of (a_i, b_i). Thus, our methodology will remain free from the burden of selecting the window size as it deals with only previous and next frame.

2.2   Computation of Frame Transition Parameters

A frame (or image) in a video sequence consists of two major types of components: background and foreground objects. Over the frames background is either static (no change) or may undergo little motion due to camera pan and tilt. On the other hand, foreground objects exhibit activities including significant motion. To incorporate both types of characteristics in the frame transition parameter estimation we use both global and local (edge scatter) features.

Estimation based on global feature: Let f_{i_e} denotes the estimate for i-th frame using the equation 5 with appropriate parameters (a_i, b_i). As mentioned earlier, a frame consists of background and foreground or active objects. Consistency of the background can better be represented in terms of global features. Since we have to deal with a huge amount of data, it is advisable to use some features which need as little computation as possible. In the proposed method we have used gray level histogram of the frames.

For the time being, let us consider that the frames are continuous domain containing continuous value of intensity. Thus gray level histogram may be treated as probability density function (p.d.f.) and let the p.d.f. of frame f_i is denoted by $p_i(v_i)$, where $v_i = f_i(x, y)$; so that the Jacobian of the linear transformation may be applied to estimate the p.d.f. of the candidate frame from the p.d.f's of the subsequent and the previous frame. Equation 5 suggests that the intensity of i-th frame is obtained by linear transformation of the $(i - 1)$-th and $(i + 1)$-th frames. Thus, the p.d.f. of frame f_i may be obtained from $p_{i-1}(v_{i-1})$ and $p_{i+1}(v_{i+1})$. To make the formulation mathematically tractable and computationally efficient we assume that v_{i-1} and v_{i+1} are independently distributed. Thus, the joint distribution of v_{i-1} and v_{i+1} is defined as $p(v_{i-1}, v_{i+1}) = p_{i-1}(v_{i-1}) \times p_{i+1}(v_{i+1})$. To derive the distribution of v_{i_e} let

$v_{i_e} = a_i v_{i-1} + b_i v_{i+1}$ and $u = v_{i+1}$ which implies $v_{i-1} = \frac{v_{i_e} - b_i u}{a_i}$. Now, the Jacobian of the transformation is $|J| = \left| \frac{\delta(v_{i-1}, v_{i+1})}{\delta(v_{i_e}, u)} \right| = \frac{1}{a_i}$

Thus, the joint distribution of v_{i_e} and u is $p(v_{i_e}, u) = p_{i-1}(\frac{v_{i_e} - b_i u}{a_i}) \times p_{i+1}(u) \times \frac{1}{a_i}$
The distribution of v_{i_e} is

$$ p_{i_e}(v_{i_e}) = \frac{1}{a_i} \int p_{i-1}(\frac{v_{i_e} - b_i u}{a_i}) \; p_{i+1}(u) \; du \tag{6} $$

Hence, by minimizing the error between the actual p.d.f. $p_i(v_i)$ and the estimated p.d.f. $p_{i_e}(v_{i_e})$ we obtain the appropriate values of parameters (a_i, b_i) which characterize the frame transition. In this work the said error is measured as the Bhattacharya distance.

Actually, the estimation process is carried out based on the intensity histogram of the frames. Thus, p_{i_e}, p_i, p_{i-1} and p_{i+1} represent the intensity histogram of respective frames. We try to find out the values of a_i and b_i to obtain the best estimate for p_i. Along with those, E_i, the error of estimation is also taken as a feature of estimate. The computation steps are as follows.

- p_{i-1}, p_i and p_{i+1}, the normalized 256 bin intensity histograms are computed.
- p_{i-1} and p_{i+1} are shifted to make $\mu_i = \mu_{i-1} = \mu_{i+1}$, where μ_i is the average intensity value of the i-th frame.
- Exhaustive search for a_i, b_i is employed to attain optimum E_i.
- $E_i = dist(p_{i_e}, p_i)$, the error between the frames f_{i_e} and f_i.
- a_i, b_i and E_i, the estimated error for the best (a_i, b_i), are taken as transition characteristic features for the i-th frame.

It may be noted that in case of cut f_i and f_{i-1} or f_i and f_{i+1} are highly uncorrelated suggested by either a_i or b_i equal to zero. On the other hand, f_{i-1}, f_i and f_{i+1} are strongly correlated in case of *no change* and *gradual transition*. This is revealed by non-zero values of both a_i and b_i in estimating these frames. This may lead to confusion in identifying *within shot* frames and *dissolve* frames. However, it may be shown that the *within shot* frames can be estimated from either previous or next frame only, which is not possible in case of *dissolve* frames. This suggests a strong distinction between them. So, in order to reduce the conflict in detection and classification, we further consider the estimation process using only the previous frame and only the next frame. The equations for such estimation are as follows.

$$ f_{i_e} = a_{i_p} f_{i-1} \tag{7} $$

$$ f_{i_e} = b_{i_n} f_{i+1} \tag{8} $$

As in earlier case, the parameters a_{i_p}, b_{i_n} and error of estimation e_p, e_n are used as features. Average intensity (I_{avg}) of the frame is also taken as a parameter. In identification of *fade-in* and *fade-out*, I_{avg} contributes significantly. For *fade-in*, I_{avg} will show a gradual increase and it is reverse for *fade-out*.

Estimation based on local features: Object motion as well as the gradual shift (if any) of the background may be arrested by detecting shift of edge points over the frames. This information may be extracted using neighborhood operators and are treated as local features. Edge strength (gradient magnitude at edge points) between two successive frames change due to change of a_i and b_i of equation 5 or motion or both. Thus scatter matrix of edge strength of two successive frames provides a representation of intensity transition as well as motion. It may be noted that if there is no motion, the scatter matrix for two successive *within shot* frames is, ideally, a diagonal matrix and it is a banded diagonal in case of gradual transition only. In case of abrupt transition the scatter matrix deviates significantly from the diagonal form. Non-diagonal elements would be loaded even more if motion is included. However, edge strength of *within shot* frames or *gradual change* frames, where motion is small and regular, is usually aligned or accumulated along the diagonal of the scatter matrix if the shifted position of the edge points can be found. This calls for solving the well known correspondence problem. If motion shifts a point at most by K in any direction and object width is more than $2K$, then we simply solve the correspondence problem by taking two similar edge points (in terms of relative magnitude and direction) of two frames within $K \times K$ window as the original and the shifted edge point.

Implementation of Scatter matrix and computation of local feature S_m as follows. Let, g_{i-1}, g_i and g_{i+1} denote the gradient images corresponding to $(i-1)$-th, ith and $(i+1)$-th frame. The gradient images are subjected to a 5×5 max filter to solve correspondence problem upto some extent. Then the scatter matrices S_1 and S_2 of dimension 256×256 are formed corresponding to (g_{i-1}, g_i) and (g_{i+1}, g_i) respectively. In order to obtain S_m, two vectors S_{v_1} and S_{v_2} corresponding to S_1 and S_2 are formed, where the elements in S_{v_i} are the normalized sum of the values along the diagonal and its parallels in S_i. Finally, the Bhattacharya distance between S_{v_1} and S_{v_2} is taken as S_m.

2.3   Detection and Classification

Thus, 9 features $< a_i, b_i, E_i, a_{i_p}, e_{i_p}, b_{i_n}, e_{i_n}, I_{avg}, S_m >$ corresponding to each frame is obtained. As the model of equation 5 is being used in our frame estimation process, a classification scheme is required to detect and classify the shot boundaries. Here, we have relied on neural network based approach and a Multilayer Perceptron (MLP) network has been used. MLP network is a multi-class classifier consisting of several layers of neurons of which first one is the input layer and the last one is the output layer, remaining layers are called hidden layers. In the architecture employed here, there are complete connections between the nodes in successive layers but there is no connection between the nodes within a layer. Corresponding to each frame an input vector is provided to the network. Corresponding to i-th frame, the input feature vector is of 27-dimensions which is formed by putting together the features of $(i-1)$-th, i-th and $(i+1)$-th frames. Thus, the feature vector for a frame also relies on previous and following frames. During training phase, along with the input vector, a label denoting the class

Table 1. Confusion Matrix and Accuracy for Training Data

Actual	Recognized Class			Classification
Class	nc	gc	ac	Accuracy
nc	7051	112	0	98.44%
gc	141	537	0	79.2%
ac	0	0	57	100.0%

Table 2. Confusion Matrix and Accuracy for Test Data

Actual	Recognized Class			Classification
Class	nc	gc	ac	Accuracy
nc	7028	133	2	98.12%
gc	160	517	0	76.37%
ac	1	0	56	98.25%

of the frame (*i.e. no change, abrupt change* or *gradual change*) is also provided. Then the connection weights are set such that the error between the network output and the target output (*i.e.* the classification error) becomes minimum.

3   Experimental Results

The frames are manually groundtruthed and the feature vectors are labeled accordingly. The frames are classified into three categories such as *no change (nc)*, *gradual change (gc)* and *abrupt change (ac)*. The frames within a shot belong to *nc*. The successive frames where abrupt change (cut) occurs are marked as *ac* and the frames under gradual transition *(dissolve, fade-in, fade-out)* are labeled as class *gc*. The dataset for this experiment consists of 15,765 frames collected from various video files like BOR03, BOR19, UGS04, UGS09 etc. present in *TRECVID 2001* test database downloaded from http://www.open-video.org. It contains different types transitions like abrupt change(cut), fade-in, fade-out and dissolve.

In our experiment, MLP network has only one hidden layer. Number of hidden nodes is chosen experimentally and set to 15. It relies on back-propagation learning. The learning rate is 0.6 and the number of iteration used for training is 15,000. The training dataset is generated by randomly choosing 50% frames of each category present in the dataset and the rest are used as the test dataset. The experiment is repeated several times by selecting different training and test dataset. The average result in the form of confusion matrix is shown in tables 1 and 2. The overall training and testing accuracy achieved are 96.79% and 96.25% respectively. Thus, it can be argued that the proposed methodology is capable enough to classify the frames reliably. As it was indicated, confusion occurs between gradual transitions (more specifically, the dissolve) and frames within the shot. It has occurred mostly because of the presence of camera motion, zooming effect, very slow paced gradual transitions etc.

4 Conclusion

We have presented and adhered to a unified model of shot transitions. Considering the model as underlying framework, a scheme is proposed to identify a frame based on its previous and following frames. The transition parameters along with the scatter matrix of edge strength and average intensity describes a frame. For classification, we have employed a Neural Network with back-propagation. It classifies the frame into one of the three categories: *no change, gradual change* or *abrupt change*. Thus, a unified model based scheme is presented which is free from the critical issues like various threshold or window size selection.

References

1. Cabedo, X.U., Bhattacharjee, S.K.: Shot detection tools in digital video. In: Proc. Non-linear Model Based Image Analysis, pp. 121–126. Springer, Heidelberg (1998)
2. Zhang, H.J.: Automatic partitioning of full-motion video. ACM/Springer Multimedia Systems 1(1), 10–28 (1993)
3. Yeo, B., Liu, B.: Rapid scene analysis on compressed video. IEEE Trans. on Circuits and Systems for Video Technology 5(6), 533–544 (1995)
4. Patel, N.V., Sethi, I.K.: Video shot detection and characterization for video databases. Pattern Recognition 30(4), 583–592 (1997)
5. Huang, C.L., Liao, B.Y.: A robust scene-change detection method for video segmentation. IEEE Trans. on Circuits and Systems for Video Technology 11(12), 1281–1288 (2001)
6. Zabih, R., Miller, J., Mai, K.: A feature based algorithm for detecting and classifying scene breaks. In: Proc. ACM Multimedia 1995, pp. 189–200 (1995)
7. Hanjalic, A.: Shot-boundary detection: Unraveled and resolved? IEEE Trans. on Circuits and Systems for Video Technology 12(2), 99–104 (2002)
8. Cernekova, Z., Pitas, I., Nikou, C.: Information theory-based shot cut/fade detection and video summarization. IEEE Trans. on Circuits and Systems for Video Technology 16(1), 82–91 (2006)
9. Hampapur, A., Jain, R., Weymouth, T.: Production model based digital video segmentation. Multimedia Tools and Applications 1, 1–38 (2002)
10. Zabih, R., Miller, J., Mai, K.: A feature based algorithm for detecting and classifying production effects. Multimedia Systems 7(2), 119–128 (1999)
11. Lienhart, R.: Reliable dissolve detection. In: Proc. SPIE conf. on SRMD(4315), pp. 219 – 230 (1999)
12. Porter, S., Mirmehdi, M., Thomas, B.: Detection and classification of shot transitions. In: Proc. 12th British Machine Vision Conference, pp. 73–82. BMVA press (2001)
13. Yoo, H.W., Ryoo, H.J., Jang, D.S.: Gradual shot boundary detection using localized edge blocks. Multimedia Tools and Applications 28, 283–300 (2006)

Improved Tracking of Multiple Vehicles Using Invariant Feature-Based Matching

Jae-Young Choi, Jin-Woo Choi, and Young-Kyu Yang

College of Software, Kyungwon University,
Seongnam, Gyeonggi, 461-701, Republic of Korea
{jychoi,jwchoi,ykyang}@kyungwon.ac.kr

Abstract. In case of monitoring road traffic, the image based monitoring system is more useful than any other system such as GPS or loop detector because it can give the whole picture of the two-dimensional traffic situation. The idea of this paper is that the quad-tree scheme segments MBR following from the background subtraction process. Then the segmented and detected vehicle regions, ROIs, are tracked by SIFT algorithm. Our method succeeded detecting and tracking multiple moving vehicles accurately in sequence frame. The proposed method is very useful for the video based applications such as automatic traffic monitoring system.

1 Introduction

Image based monitoring and surveillance system becomes popular due to its excellent performance against the installation and maintenance cost. Moreover, the output of image based system such as a number of vehicles, a class of vehicles, distribution of vehicles, and a speed of car can be used for automatic routing and control traffic as well as traffic statistics [5]. Especially, the emergency situation can be solved by image based surveillance system quickly since the user who monitors the display device can intervene in the system at any time during traffic observation.

However, detecting and tracking objects in images taken by mounted camera on the street lamp or pedestrian bridge across a road have some errors due to ambient illumination, changing the shape or size of moving car.

We have developed an image based system that extracts moving vehicles using quad-tree segmentation, and tracks multiple vehicles from the sequences of images using the Scale-invariant Feature Transform to improve the tracking performance which is robust to changing the intensity, shape, and size of vehicle.

This paper starts by introducing an overview of vehicle tracking in vision based aspect and scale invariant feature transform method in Section 2. Section 3 describes our tracking technique. Experimental results are reported in Section 4 and summarizes conclusion.

A. Ghosh, R.K. De, and S.K. Pal (Eds.): PReMI 2007, LNCS 4815, pp. 649–656, 2007.

2   Background

2.1   Vehicle Tracking

To tracking the vehicles, it is important process to extract vehicle in advance. The common method for object detection and extraction in sequence frames is that compares contiguous images and subtracts the image from the previous image in order to eliminate background and get moving region within two images. It is robust and easy to take change object. The results of subtraction, however, are influenced by environmental change and various speed of vehicle in spite of these advantages. That is, change the illumination and shadows allows the background to update frequently. Thus, often update of background causes accumulation of update error. Too slow or too fast speed also affects the extraction of vehicle region. Another method is template matching technique which uses a template to compare the features such as intensity, shape, and so on. Due to the fact that vehicles have different shape and size, it is difficult to choose a proper template to find car object in image.

Once the moving vehicles are detected in the current image, the tracking process tracks vehicles during a tracking interval over further input frames. In the literatures, there are many methods in the tracking of moving object. 3D model based vehicle tracking system has previously been investigated by several researchers, but the most serous weakness of this approach is the reliance on detailed geometric object model like template matching method. Region based tracking is popular technique if background subtraction method was used for detecting vehicle. This process, however, makes the task of segmenting individual car difficult in case of under congested traffic conditions, vehicles partially occlude each other instead of being spatially isolated. Feature based tracking method tracks subfeatures such as distinguishable points on the object. The advantage of this approach is that even in the presence of partial occlusion or deformation of shape, it could detect some of the remains of visible features on the moving object. In this paper, the feature based tracking algorithm is used to complementary to region based object extraction.

2.2   Invariant Feature-Based Matching

It is necessary to compare images and match the same object to track the object from previous image to next image. Early work in image matching has two types; direct and feature based. Feature-based methods try to extract salient features such as edges and corners and use a small amount of local information, for example, correlation of a small image patch, to establish matches. Direct methods attempt to use all of the pixel values using template in order to iteratively align images. At the intersection of these approaches there are invariant features which are robust to image scale, rotation, and partially invariant to changing viewpoints, and change in illumination [1],[3].

The interest point detector must select image locations that contain a high degree of information content. Interest point detectors range from classic feature detectors such as Harris corners or derivative of Gaussian maxima to more elaborate methods such as maximally stable regions and stable local phase structures [4],[6]. Several other scale invariant interest point detectors have been proposed.

Previous approaches using corner detectors have a serious defect which is that they examine an image at only a single scale. This means the detectors respond to different image points as the change in scale become large [9]. SIFT is an efficient method to identify stable key locations in scale space. Therefore, the different scales of an image will have no effect on the set of key locations selected.

The scale invariant feature transform which combines a scale invariant region detector and a descriptor based on the gradient distribution in the detected regions [7]. The descriptor is represented by a 3D histogram of gradient locations and orientations which makes the descriptor robust to small geometric distortions and errors in the region detection. The first stage identifies key locations in scale space by looking for locations that are maxima or minima of a difference of Gaussian function. Each point is used to generate a feature vector that describes the local image region sampled relative to its scale space coordinate frame. The resulting feature vectors are called SIFT keypoints which are used in a nearest neighbor approach to indexing to identify candidate object models. Keypoints are first identified through a Hough transform hash table, and then through a least squares fit to a final estimate of model parameters.

3   Tracking of Multiple Vehicles

In recent years, the image-based traffic monitoring system is a remarkable alternative for magnetic loop detectors because it is easy to install and has more abilities such as vehicle class, vehicle path, queue length, etc as well as vehicle speed and count. Image-based traffic monitoring system, however, has some problem with respect to the error of the vehicle detection and tracking due to the variety of input environment. Especially, the shape of vehicle is deformed because of the aspect ratio regarding to view point as the moving object comes to the camera.

For vehicle matching and tracking in above situation, SIFT is useful because it provides robust matching across a substantial range of affine distortion, addition of noise, and partially change in illumination.

This section describes a proposed approach to estimate traffic parameters as shown in Fig. 1.

3.1   Multiple Vehicle Detection

When the first frame is input together with reference (background) image, the proposed algorithm subtracts the intensity value of each pixel in the image $I_k(x, y)$ from the corresponding value in the reference scene $I_{ref}(x, y)$, and applies region segmentation technique, in this paper the quad-tree segmentation, to the subtracted image $I_D(x, y)$.

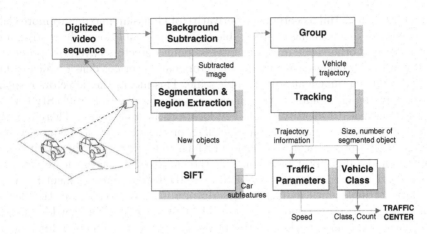

Fig. 1. Blockdiagram for tracking multiple vehicles and estimation of traffic parameters

Image segmentation is essential in the implementation of feature-based techniques because effective segmentation will isolate the important homogeneous regions of the image.

Using a quad-tree decomposition, features can be extracted from spatial blocks. A quad is a tree data structure in which each internal node has up to four children. Quad-tree is most often used to partition a two dimensional space by recursively subdividing it into four quadrants or geometric regions. The regions may be square or rectangular, or may have arbitrary shapes [8]. The suggested algorithm uses bottom-up construction which consists of binary decisions to merge, where construction begins with the smallest possible block size in the quad-tree. If all relevant subblocks have been combined into a larger block, then a decision is made whether to combine the larger regions into a yet larger region.

After quad-segmenting the algorithm classifies adjacent blocks which group homogenous properties according to the type of detected data, and makes it region of interest (ROI). If there are many segment regions more than certain threshold value, it means that the difference of background between compared images is large. In such case, changing the reference frame using background update is required.

The advantage of using quad-tree segmentation is that the output of segmentation can be minimum boundary rectangle (MBR), and reduce the cost of SIFT

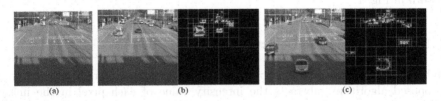

Fig. 2. Quad-tree segmentation. (a) Reference image, (b) and (c) Input image and object region detection using quad-tree segmentation.

because it does not need to check features on the whole image in order to make SIFT keypoints. Furthermore, specific threshold value is not needed for extracting object from a subtracted image since quad-tree eliminates the isolate and small leaf as a noise of background. Fig. 2 shows the input image and result of object region detection using quad-tree segmentation on subtracted image from reference image.

3.2 Matching and Tracking

The segmented vehicle objects, ROIs, are detected continuously via moving object extraction and tracking using SIFT algorithm. SIFT can extract distinctive features from image to be used to matching different views, color, and shapes.

The SIFT descriptors are constructed from two scale spaces; the Gaussian scale space of the input image $I(x, y)$ as in (1) and difference of Gaussian as in (2), where g_σ is an isotropic Gaussian kernel of variance $\sigma^2 I$. Scale space is function $F(x, y, \sigma) \in R$ of a spatial coordinate $x, y \in R^2$ and a scale coordinate $\sigma \in R_+$, Since a scale space $F(\cdot, \sigma)$ typically represents the same information at various scales $\sigma \in R$, its domain is sampled in a particular way in order to reduce the redundancy.

$$G(x, y, \sigma) \cong (g_\sigma * I)(x, y) \tag{1}$$

$$\begin{aligned} D(x, y, \sigma) &= G(x, y, k\sigma) - G(x, y, \sigma) \\ &\cong (k - 1)\sigma^2 \nabla^2 G \end{aligned} \tag{2}$$

Using scale-space method the image is progressively Gaussian blurred (smoothed) in level σ_n, and produces a new series of spaces with the difference of Gaussians (DOG). It provides a close approximation to the scale-normalized Laplacian of Gaussian as shown in above (2) and below Fig. 3.

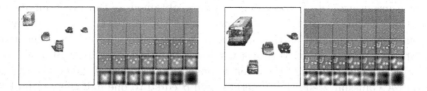

Fig. 3. Vehicle object and its scale spaces

Input image will produce several thousand overlapping features such as Fig. 4(a) to identify potential interest points (keypoints) that are invariant to the scale and orientation. From the extrema in scale space the keypoints are chosen and assigned orientation as shown in Fig. 4(b). In order to detect the local maxima and minima of $D(x, y, \sigma)$, each sample point is compared to its eight

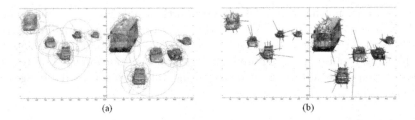

Fig. 4. Feature descriptor and its orientation. (a) Local image descriptor, (b) Orientations of keypoints.

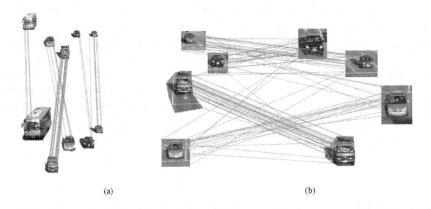

Fig. 5. Example of matching features and tracking vehicle. (a) feature tracking in case of lane cross, (b) feature matching in case of different scale.

neighbors in the current image and nine neighbors in the above and below scale. It is selected only if it is larger than all of these neighbors or smaller than all of them. The cost of this check is reasonably low due to the fact that most sample points will be eliminated following the first few checks.

The orientation θ of a keypoint (x, σ) is obtained as the predominant orientation of the gradient in a window around the keypoint. The predominant orientation is obtained as the maximum of the histogram of the gradient orientations $\angle \nabla G(x_1, x_2, \sigma)$ within a window. The SIFT descriptor of a keypoint (x, σ) is a local statistic of the orientations of the gradient of the Gaussian scale space. A Gaussian weighting function with σ equal to one half the width of the descriptor window is used to assign a weight to the magnitude $|\nabla G|$ of each sample point. The purpose of this Gaussian window is to avoid sudden changes in the descriptor with small changes in the position of the window, and to give less emphasis to gradients that are far from the center of the descriptor. To reduce the effects of illumination change, the feature vector is also normalized to unit length. Fig. 5 illustrates the best matching keypoints which are compared between descriptors with minimum Euclidean distance for the invariant descriptor vector [2].

Once vehicle features group the same region in quad tree, the grouper uses a common motion constraint to collect features into a vehicle: corner features that are seen as moving rigidly together probably belong to the same object. In other words, features from the same vehicle will follow similar trajectory and two such features will be offset by the same spatial translation in every frame. Two features from different vehicles, on the other hand, will have distinctly different trajectories and their spatial offset will change from frame to frame. If the object split across two similar amount of feature groups during a tracking, the algorithm ascribe the divided objects to the different object. Therefore the system decides to the implicit occlusion, and generates two tracking trajectories for multiple vehicles even if those were occluded each other at initial point.

4   Experimental Results and Conclusion

The proposed algorithm was tested off-line using sequential image from video stream which are often characterized by multiple moving vehicle, vehicle changed lane, variable illumination condition, and so on.

The ranges of widths and lengths are set according to the prior knowledge of the road. Especially, the speed error is large when the location of the car is far from the camera if incorrect information is used for camera calibration.

It is not difficult to detect the features (SIFT keypoints) since vehicles have a number of corner features. Furthermore, average amount of processing time can be reduced by using only interest (detected) region, whereas the traditional SIFT used entire image for searching keypoints as show Fig. 6. As one would expect from SIFT tracking, the suggested method is robust to track and occlusion even if the vehicles are overgrouped or oversegmented.

We are developing a feature based tracking instead of tracking entire region using SIFT for estimating traffic parameters on image based system. Experiments give satisfying results to validate the proposed algorithm, especially for

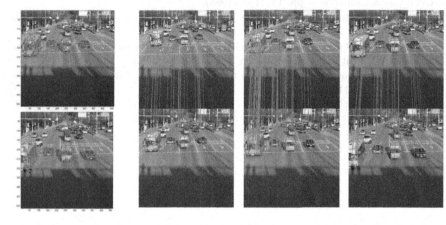

Fig. 6. Example of detecting keypoints and tracking vehicles between frames

invariant vehicle size, illumination changes, lane changes, and partial occlusion of vehicles. This paper suggested multiple vehicles detection and tracking method using scale invariant feature transform to improve the performance of tracking for extracting traffic parameter such as vehicle count, speed, class, and so on from video stream. The experimental result presents the proposed method is effective and robust on tracking multiple vehicles, especially in cases that a vehicle changes a lane, vehicle is occluded by another object, and deformation of vehicle is occurred by moving car.

Acknowledgements

This research was supported by the MIC(Ministry of Information and Communication), Korea, under the ITRC(Information Technology Research Center) support program supervised by the IITA(Institute of Information Technology Assessment). (IITA-2006-C1090-0603-0040)

References

1. Mikolajczyk, K., Schmid, C.: A performance evaluation of local descriptors. IEEE Trans. Pattern Analysis and Machine Intelligence 27(10), 1615–1630 (2005)
2. Gepperth, A., Edelbrunner, J., Bucher, T.: Real-time detection and classification of cars in video sequences. In: Proc. Intelligent Vehicles Symposium, pp. 625–631 (2005)
3. Bay, H., Tuytelaars, T., Gool, L.V.: SURF: Speeded up robust features. In: European Conf. Computer Vision, pp. 404–417 (2006)
4. Carneiro, G., Jepson, A.: Multi-scale phase-based features. In: Int'l Conf. Computer Vision and Pattern Recognition, pp. 736–743 (2003)
5. Ha, D.-M., Lee, J.-M., Kim, Y.-D.: Neural-edge-based vehicle detection and traffic parameter extraction. Image and Vision Computing 22, 899–907 (2004)
6. Harris, C., Stephens, M.: A combined corner and edge detector. In: Proc. of the Alvey Vision Conference, pp. 147–151 (1988)
7. Lowe, D.G.: Distinctive image features from scale-invariant keypoints. Int'l J. Computer Vision 60(2), 91–110 (2004)
8. Smith, J.R., Chang, S.-F.: Quad-tree segmentation for texture-based image query. In: Proc. ACM Int'l Conf. Multimedia, pp. 279–286 (1994)
9. Lindeberg, T.: Feature detection with automatic scale selection. Int'l J. Computer Vision 30(2), 77–116 (1998)

An Adaptive Bayesian Technique for Tracking Multiple Objects

Pankaj Kumar, Michael J. Brooks, and Anton van den Hengel

University of Adelaide
School of Computer Science
South Australia 5000
pankaj.kumar@adelaide.edu.au, michael.brooks@adelaide.edu.au,
anton.vandenhengel@adelaide.edu.au

Abstract. Robust tracking of objects in video is a key challenge in computer vision with applications in automated surveillance, video indexing, human-computer-interaction, gesture recognition, traffic monitoring, etc. Many algorithms have been developed for tracking an object in controlled environments. However, they are susceptible to failure when the challenge is to track multiple objects that undergo appearance change to due to factors such as variation in illumination and object pose. In this paper we present a tracker based on Bayesian estimation, which is relatively robust to object appearance change, and can track multiple targets simultaneously in real time. The object model for computing the likelihood function is incrementally updated and uses background-foreground segmentation information to ameliorate the problem of drift associated with object model update schemes. We demonstrate the efficacy of the proposed method by tracking objects in image sequences from the CAVIAR dataset.

1 Introduction

Reliably tracking an object through an extended image sequence remains a fundamental and challenging problem in computer vision. While considerable progress has been made, robust tracking in unconstrained environments remains an unsolved problem [1]. Some of the reasons for the difficulty are:

- noise in image data and camera artifacts
- unpredictable and nonlinear object motion
- articulated nature of some objects, e.g., humans
- partial or full occlusion of objects by other objects or background elements
- illumination changes affecting both background and object
- real time processing requirements
- change in detail or appearance of the object as it moves within the Field of View (FoV) of the camera

In this paper we present a novel solution to the problem of multiple-object tracking in the presence of fluctuating object appearance. This variation in appearance might be due to one of many factors, including change in shape, orientation, pose, depth, or illumination. Figure 1 shows some examples of changing object appearance under variation in illumination, orientation, depth and shape.

A. Ghosh, R.K. De, and S.K. Pal (Eds.): PReMI 2007, LNCS 4815, pp. 657–665, 2007.
© Springer-Verlag Berlin Heidelberg 2007

Fig. 1. The image windows show two examples of change in object appearance exhibited in video from a single camera

Fig. 2. These frames show errors in multi object tracking as a result of the drift problem

Our approach to multiple object tracking in the presence of appearance change is to employ a real time adaptive tracking algorithm that incrementally updates the object model in a novel and effective manner. The update method incorporates information from a background-foreground segmentation process, and this serves to ameliorate the drift problem. Drift is a problem commonly associated with object tracking and is a significant nuisance for methods which perform object model updates. It is manifest by the accumulation of small errors in the target model that result in mis-tracking whereby the object is lost or an alternative region is incorrectly tracked. Figure 2 shows an example of mis-tracking of multiple objects due to drift. In the past this drift problem has typically been ameliorated by carefully choosing the initial parameters, using tracking parameters which are invariant to change, and updating the target model or template in a sophisticated way. As the environmental constraints are relaxed it becomes difficult to choose features which are invariant to change in the object and environment. Sophisticated update techniques cannot alone completely ameliorate the drift problem. Rather, the key here is to use additional information of the right kind.

2   Related Work

We briefly review developments in object tracking most relevant to this work. In a landmark paper, Comaniciu and Meer developed a kernel based object representation and applied mean shift to tracking objects [2]. Mean shift trackers employ a single hypothesis and tend to be less suitable for multiple object tracking.

An object update model is difficult to incorporate and tracking may fail under significant illumination or colour changes. Ross *et. al.* [3] proposed an adaptive probabilistic real time tracker that updates the model using an incremental update of a so-called eigenbasis. They demonstrated the efficacy of their method on single object tracking. Nummiaro *et. al.* [4] developed an adaptive particle filter tracker, updating the object model by taking a weighted average of the current and a new histogram of the object. There is no explicit discussion on how to solve the drift problem.

R. Collins and Y. Liu [5] and B. Han and L. Davis [6] presented methods of online selection of the most discriminative feature for tracking objects. Here again there is no explicit solution for adapting to changes in the object model. In [7], Han *et. al.* presented a kernel based Bayesian filtering framework which adopts an analytic approach to better approximate and propagate a density function. This formulation helps in better tracking objects with high dimensional state vectors. In [8] an on-line density based appearance model was presented in which the density of each pixel was composed of a mixture of Gaussians and the parameters of the mixture were determined by the mean shift algorithm. This method works well for updating the model of the target but the algorithm tracks a single object only. Perez *et. al.* [9] proposed a multiple cue tracker for tracking objects in front of a web cam. They introduced a generic importance sampling mechanism for data fusion and applied it to fuse various subsets of colour, motion, and stereo sound for tele-conferencing and surveillance using fixed cameras. Appearance update is not factored into the approach.

The novelty of the work presented in this paper is that it uses background-foreground segmentation information in updating the object model incrementally thus preventing the background regions from becoming a part of the model. This mechanism reduces the tendency for corruption of the object model and thus acts to offset the drift problem.

3   An Adaptive Bayesian Tracker

We now present a novel multiple object tracker after first detailing some preparatory material.

3.1   Bayesian State Estimation

The aim of the Bayesian estimation process is to compute the posterior probability density function (pdf) $p(X_t|Z^T)$ of the state vector X_t given a set of measurements $Z^T = (z_1, \cdots, z_t)$ from the sensor, which in visual tracking is a camera. In Bayesian tracking, we adopt a process model

$$X_{t+1} = F_{t+1}(X_t, v_t), \tag{1}$$

where $F_{t+1} : R^{n_x} \times \mathcal{R}^{n_v} \to \mathcal{R}^{n_x}$ is a nonlinear function of the state, and v_t is independent identically distributed (i.i.d.) process noise. Further we adopt a measurement model

$$Z_t = h_t(X_t, n_t), \tag{2}$$

where $h_t : \mathcal{R}^{n_x} \times \mathcal{R}^{n_n} \rightarrow \mathcal{R}^{n_z}$ is a possibly nonlinear function and n_t is i.i.d. measurement noise. In Bayesian estimation the problem is then to recursively calculate the degree of belief in the state X_t given all the measurements $Z_{1:t}$; that is, we are required to construct the pdf $p(X_t|Z_{1:t})$. This is done in two stages of prediction and update. The prediction stage involves the use of the process model (1) in the Chapman-Kolmogorov equation

$$p(X_{t+1}|Z_{1:t}) = \int p(X_{t+1}|X_t)p(X_t|Z_{1:t})dX_t. \tag{3}$$

This is followed by an update step when the measurement at $t + 1$ becomes available. The update is done using Bayes' theorem [10]

$$p(X_{t+1}|Z_{1:t+1}) = \frac{p(Z_{t+1}|X_{t+1})p(X_{t+1}|Z_{1:t})}{p(Z_{t+1}|Z_{1:t})}. \tag{4}$$

The likelihood function $p(Z_{t+1}|X_{t+1})$ is defined by the measurement model, and the normalizing constant is obtained using the total probability theorem [10].

3.2   Particle Filter

The particle filter is a special case of Bayesian estimation process (see [11] for a tutorial on particle filters incorporating real-time, nonlinear, non-Gaussian Bayesian tracking). Particle filters were first used in [12] to track objects in video. The key idea of a particle filter is to approximate the probability distribution of the state X_t of the object with a set of samples/particles and their weights $\{X_t^i, w_t^i\}_{i=1}^{N_s}$. Each sample/particle can be understood to be a hypothetical state of the object and the weight/belief for this hypothetical state is computed using the likelihood function. The particles at each iteration are computed using the system model (1).

Motion Model. In real life scenes, and especially with humans walking, it is very difficult to know the motion model a priori; also human movements and interactions can result in very unpredictable motions. Therefore we use a random walk motion model, where the next particle vector is obtained by adding random noise to the current particle. Given the state vector $X_t = [x_c, y_c, b, h]^T$, where x_c, y_c are the co-ordinates of the centroid of the object and b, h are the breadth and height of the object, the update is given by:

$$X_{t+1} = X_t + v_t. \tag{5}$$

The state vector defines a window in the image frame, which is the measurement obtained from the camera.

Likelihood Function. The weight of a particle is computed using a likelihood function, which is equivalent to the measurement model of the Bayesian estimation process

$$\mathcal{L}(Z_t|X_t) = \mathcal{L}_{colour}(Z_{colour,t}|X_t). \tag{6}$$

This function is colour based. We compute the likelihood measure using a non-parametric representation of the colour histogram of the object $P = p^{(u)}{}_{u=1...m}$ and particle/candidates $Q = q^{(u)}{}_{u=1...m}$, where m is the number of bins in the histogram. Colour histograms change with variation in illumination, object pose, etc. However, they are (1) relatively robust to partial occlusion, (2) rotation invariant, (3) scale invariant, and (4) efficient to compute. The disadvantages of colour histograms are ameliorated by intelligently updating the object model and incorporating foreground segmentation information.

It has been argued in previous works [4], [2] that not all pixels in the object or candidate region are equally important in describing the object or candidate. Pixels on the boundary of a region are typically more prone to errors than the pixels in the interior of the region. The general trend in the solution has been to use a kernel function like the Epanechnikov kernel [2] to weight the pixels' contribution to the histogram. The same kernel function is applied irrespective of the position of the region. Our contention is that this blind application of the kernel function can accentuate the drift problem when the object model is updated. Small errors can accumulate to the point where the target model no longer reflects the appearance of the object being tracked. Our strategy in building the object and candidate histogram is to weight the pixel contribution by the background-foreground segmentation information. In our implementation we have used the fast queue based background-foreground segmentation method [13]. The foreground segmentation result is cleaned up using morphological operations. The Manhattan distance transform [14], [15] is then applied to get the weights of the pixels for their contribution to the object/candidate histogram. In a binary image the distance transform replaces intensity of each foreground pixel with the distance of that pixel to its nearest background pixel. Figure 3 shows the weights of the pixels scaled to $[0-255]$ (for their contribution in building histogram model of the object)computed using the Manhattan distance transform. Scores of the bins of the histogram are computed using the following equation

$$p^{(u)} = \sum_{\mathbf{x_i} \in Foreground\ Region} w(\mathbf{x_i})\ \delta(g(\mathbf{x_i}) - u), \qquad (7)$$

where δ is the Kronecker delta function, $g(\mathbf{x_i})$ assigns a bin in the histogram to the colour at location $\mathbf{x_i}$, and $w(\mathbf{x_i})$ is the weight of the pixel at location $\mathbf{x_i}$ obtained on application of the distance transform to the foreground segmented image. The weights for background pixels are always zero, which makes it nearly impossible for the tracker to shift to background regions of the scene. When two or more objects merge, it is effectively detected using a merge-split algorithm [16]; the update of the object model is temporarily halted when the objects have merged.

3.3   Model Update

To handle the appearance change of the object due to variation in illumination, pose, distance from the camera, etc., the object model is updated using the auto-regressive learning process

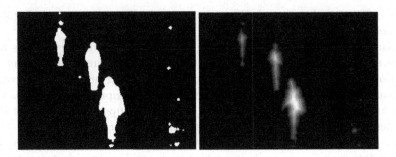

Fig. 3. The weight image generated by application of the distance transform to the foreground segmentation image, showing three people

$$P_{t+1} = (1 - \alpha)P_t + \alpha P_t^{est}. \qquad (8)$$

Here P_t^{est} is the histogram of the region defined by the mode of the samples used in tracking the object, and α is the learning rate. The higher the value of α the faster the object model will be updated to the new region. The model update is applied when the likelihood of the current estimate of the state of the object X_t^{est}, with respect to the current measurement Z_t, given by

$$\mathcal{L}_{colour}(Z_t|X_t^{est}) = exp(-d(P_t, P_t^{est})/\sigma_z) \qquad (9)$$

is greater than an empirical threshold. The quantity $d(P,Q) = \sqrt{1 - \rho(P,Q)}$ is the Bhattacharyya distance based on the Bhattacharyya coefficient, $\rho(P,Q) = \sum_{i=1}^{m} \sqrt{p^{(i)} q^{(i)}}$

4   Results

The tracker has been tested on a number of real video sequences. Figure 4 shows the tracking result in an image sequence from the CAVIAR data set. The objects were tracked using 100 particles per object. In some cases the tracker was successful in tracking objects with as few as 40 particles. Each object was modelled in *RGB* colour space. The tracker was able to track the objects in spite of changes in pose, illumination and scale. There is significant illumination

Fig. 4. These images show the tracking of three objects simultaneously form a video in the CAVIAR data set

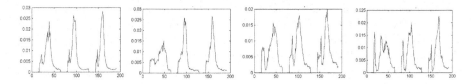

Fig. 5. These images shows the object model of the person being tracked with solid bounding box. Y-axis are the bin scores and X-axis are different bins of the *RGB* channels. Notice the significant change in object model for different instances of tracking.

Fig. 6. These images show the tracking result of three objects on an another sequence from CAVIAR data set

change for example the lights coming from the shop windows. The illumination change is well captured in the object model shown in Figure 5. This figure shows the *RGB* reference histogram model of the person being tracked with solid bounding box for the time instance for which tracking results are shown in Figure 4. The change in reference model reflects the change in the appearance of the object. Because of the large depth of the corridor there is significant change in the object scale as well. Figure 6 shows tracking result on another video from the CAVIAR data set.

Each object was manually initialised and the algorithm is robust to small errors in initialisation.

Figure 7 shows successful tracking of two persons as they cross each other. One target completely occludes other at one point in this sequence. This video is from

Fig. 7. These images show the tracking of two objects when they cross each other. One occluding the other almost completely.

the data set of our vision lab. A drawback of the tracker presented here is that, when shadows are detected as foreground then there are errors in localisation of the object being tracked. This can be improved by using a shadow detection algorithm along with foreground detection algorithm.

5   Conclusion

An enhanced particle filter system was developed for robust, multiple-object tracking. Key to the approach is the use of new object model and update method for the model that incorporates foreground information obtained via a background subtraction process. This provides improved handling of object models undergoing change, rendering the system less susceptible to the drift problem. As a consequence, the tracker gives improved performance in the presence of changes in object appearance due to partial occlusion, variation in illumination, pose, and scale. Experimental results suggest the method holds promise.

References

1. Yilmaz, A., Javed, O., Shah, M.: Object tracking: A survey. ACM Comput. Surv. 38(4), 13 (2006)
2. Dorin, C., Visvanathan, R., Peter, M.: Kernel-based object tracking. IEEE Transactions on Pattern Analysis and Machine Intelligence 25(5), 564–577 (2003)
3. Ross, D., Lim, J., Yang, M.-H.: Adaptive probabilistic visual tracking with incremental subspace update. In: Proceedings of the Eighth European Conference on Computer Vision (ECCV 2004)
4. Nummiaro, K., Koller-Meier, E., Gool, L.J.V.: Object tracking with an adaptive color-based particle filter. In: Proceedings of the 24th DAGM Symposium on Pattern Recognition, pp. 353–360. Springer, London, UK (2002)
5. Collins, R.T., Liu, Y., Leordeanu, M.: Online selection of discriminative tracking features. IEEE Transactions on Pattern Analysis and Machine Intelligence 27(10), 1631–1643 (2005)
6. Han, B., Davis, L.: Object tracking by adaptive feature extraction. In: ICIP 2004. International Conference on Image Processing, vol. 3, pp. 1501–1504 (2004)
7. Han, B., Zhu, Y., Comaniciu, D., Davis, L.: Kernel-based bayesian filtering for object tracking. In: CVPR 2005. Proceedings of the 2005 IEEE Computer Society Conference on Computer Vision and Pattern Recognition, vol. 1, pp. 227–234, Washington, USA (2005)
8. Han, B., Davis, L.: On-line density-based appearance modeling for object tracking. In: Proceedings of the Tenth IEEE International Conference on Computer Vision, pp. 1492–1499. IEEE Computer Society, Washington, DC, USA (2005)
9. Perez, P., Vermaak, J., Blake, A.: Data fusion for visual tracking with particles. Proceedings of the IEEE 92(3), 495–513 (2004)
10. Papoulis, A.: Probablilty Random Variables and Stochastic Processes. In: S.W. (ed.), 3rd edn., McGraw-Hill Internationals, New York (1991)
11. Arulampalam, S., Maskell, S., Gordon, N., Clapp, T.: A tutorial on particle filters for on-line non-linear/non-gaussian bayesian tracking. IEEE Transactions on Signal Processing 50(2), 174–188 (2002)

12. Isard, M., Blake, A.: Condensation – conditional density propagation for visual tracking. International Journal of Computer Vision 29(1), 5–28 (1998)
13. Kumar, P., Ranganath, S., Huang, W.: Queue based fast background modelling and fast hysteresis thresholding for better foreground segmentation. In: Proceedings of the Fourth International Conference on Information, Communications and Signal Processing (2003)
14. Rosenfeld, A., Pfaltz, J.: Distance functions in digital pictures. Pattern Recognition 1, 33–61 (1968)
15. Jain, A.K.: Fundamentals of Digital Image Processing. In: Kailath, T. (ed.) Prentice Hall International, Englewood Cliffs (1989)
16. Kumar, P., Ranganath, S., Sengupta, K., Huang, W.: Cooperative multitarget tracking with efficient split and merge handling. IEEE Transactions on Circuts and Systems for Video Technology 16(12), 1477–1490 (2006)

Computationally Efficient MCTF for MC-EZBC Scalable Video Coding Framework

A.K. Karunakar and M.M. Manohara Pai

Department of Information and Communication Technology,
Manipal Institute of Technology
Manipal 576 104 India
{karunakar.ak,mmm.pai}@manipal.edu

Abstract. The discrete wavelet transforms (DWTs) applied temporally under motion compensation (i.e. Motion Compensation Temporal Filtering (MCTF)) has recently become a very powerful tool in scalable video compression, especially when implemented through lifting. The major bottleneck for speed of the encoder is the computational complexity of the bidirectional motion estimation in MCTF. This paper proposes a novel predictive technique to reduce the computational complexity of MCTF. In the proposed technique the temporal filtering is done without motion compensation. The resultant high frequency frames are used to predict the blocks under motion. Motion estimation is carried out only for the predicted blocks under motion. This significantly reduces the number of blocks that undergoes motion estimation and hence the computationally complexity of MCTF is reduced by 44% to 92% over variety of standard test sequences without compromising the quality of the decoded video. The proposed algorithm is implemented in MC-EZBC, a 3D-subband scalable video coding system.

Keywords: Motion Estimation, Motion Compensated Temporal Filtering, Temporal Filtering, MC-EZBC.

1 Introduction

The Scalable Video Coding (SVC) is one of the most important features of modern video communication system. For a truly scalable coding, the encoder needs to operate independently from the decoder, while in predictive schemes the encoder has to keep track and use certain information from the decoder's side (typically, target bit-rate), in order to operate properly.

The 3D sub-band video coding has appeared recently as a promising alternative to hybrid DPCM video coding techniques; it provides high energy compaction, scalable bit-stream for network and user adaptation and resilience to transmission errors. While early attempts to apply separable 3D wavelet transform directly to the video data didn't produce high coding gains, it was soon realized that, in order to fully exploit inter-frame redundancy, the temporal part of the transform must compensate for motion between frames. In one of the first attempts to incorporate motion into 3D wavelet video coding, Taubman and

A. Ghosh, R.K. De, and S.K. Pal (Eds.): PReMI 2007, LNCS 4815, pp. 666–673, 2007.

Zakhor [1] pre-distorted the input video sequence by translating frames relative to one another before the wavelet transform so as to compensate for camera pan. Wang et. al. [2] used mosaicing to warp each video frame into a common coordinate system and applied a shape-adaptive 3D wavelet transform on the warped video. Both of these schemes adopt a global motion model that is inadequate for enhancing the temporal correlation among video frames in many sequences with local motion. To overcome this limitation, Ohm [3] proposed local block-based motion, similar to that used in standard video coders, while paying special attention to covered/uncovered and "connected/unconnected" regions. Failing to achieve perfect reconstruction with motion alignment at 1/2-pixel resolution, Ohm's scheme showed no significant performance improvement. Only recently it has been generalized to sub-pixel accuracies. This paper proposes a technique that will apply motion estimation only to those blocks which has undergone some motion and hence increase the speed of the scalable encoder significantly.

The rest of the paper is organized as follows. In section II, MCTF is explained. The section III discusses the proposed technique. Section IV and V is simulation results and conclusion respectively.

2 Motion Compensated Temporal Filtering

The idea of using motion-compensated temporal DWT (MCTF) was introduced by Ohm [3] and developed by Choi and Woods [4]. A motion compensated lifting framework for TDWT is proposed by several researchers [5],[6],[7],[8],[9],[10]. Temporal decomposition using any desired motion model and any desired wavelet kernel with finite support is possible using lifting framework for MC TDWT. The results reported in [11,12]indicate superior performance with the bi-orthogonal 5/3 wavelet kernel, compared to conventional Haar wavelet transform. As discussed in LIMAT framework [10], MC TDWT is accomplished through a sequence of temporal lifting steps and the motion compensation is performed inside each lifting steps. Let $M_{k1 \to k2}(f_{k1})$ denote a motion-compensated mapping of frame k1 onto the coordinate system of frame k2. Using this notation we can implement motion compensated lifting steps for 5/3 analysis as below.

$$h_k[x] = f_{2k+1}[x] - \frac{1}{2}(f_{2k}[M_{2k \to 2k+1}(x)] + f_{2k+2}[M_{2k+2 \to 2k+1}(x)]) \qquad (1)$$

$$l_k[x] = f_{2k}[x] + \frac{1}{4}(h_{k-1}[M_{2k-1 \to 2k}(x)] + h_k[M_{2k+1 \to 2k}(x)]) \qquad (2)$$

The motion compensation of lifting steps effectively causes the temporal subband analysis filters to be applied along the motion trajectories induced by motion compensation operators, M. These temporal lifting steps are shown in Fig. 1. Equation (1) is commonly known as **prediction** step it produces the high pass temporal frames h_k, as the residual left after bi-directional motion compensation of the odd indexed frames based on the even indexed frames. In region where the motion model captures the actual motion, the energy in the high pass frames will be close to zero. Motion model failure, however causes multiple edges and increased energy in the high pass temporal frames. Equation (2) is commonly known as the update step. Its interpretation is not as immediate as that of the prediction step, but it servers to ensure that frame l_k corresponds to a low pass filtering of the input frame sequence along the motion trajectories using the transforms five-tap low-pass analysis filter. Regardless of the motion

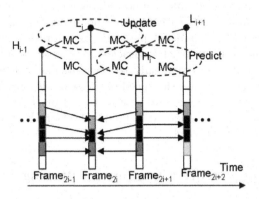

Fig. 1. One level lifting based MCTF using bi-orthogonal 5/3 filter

model used for the M operators, the temporal transform can be trivially inverted by reversing the order of the lifting steps and replacing addition with subtraction as follows.

$$f_{2k}[x] = l_k[x] - \frac{1}{4}(h_{k-1}[M_{2k-1\to 2k}(x)] + h_k[M_{2k+1\to 2k}(x)]) \qquad (3)$$

$$f_{2k+1}[x] = h_k[x] + \frac{1}{2}(f_{2k}[M_{2k\to 2k+1}(x)] + f_{2k+2}[M_{2k+2\to 2k+1}(x)]) \qquad (4)$$

3   Computationally Efficient MCTF

The temporal decomposition of a video sequence using any wavelet filter without motion compensation causes blurriness in the region wherever there is a motion, as shown in Fig. 2(a). In order to remove the blurriness in the low frequency frames temporal filtering is done along the motion trajectory (i.e. MCTF) as shown in Fig. 2(b) and the energy in the high frequency frame is also reduced, thus supports for compression efficiency. The MCTF framework consumes dominant portion of the encoding time due to its bi-directional motion estimation (four times motion estimation is to be performed in order to decompose a pair of frames using 5/3 filters)[10], in which the motion estimation is performed for all the blocks in the frame irrespective of motion presence.

In slow motion videos (e.g. Akiyo, News, etc.,) and the video with fixed background (e.g. Claire) most of the blocks in a frame remains stationary. These observations on several test video sequences motivates us for proposing a novel technique to identify zero motion blocks (ZMB), without actually carrying out the motion estimation and hence contributes to reduce the computationally complexity of MCTF. The temporal filtering is applied along the corresponding blocks of the frames without motion compensation.

(a)   (b)

Fig. 2. (a)Blurred image with out MCTF Clear image after MCTF

In a region where there is no motion and pixel values also remain same, the temporal filtering results in zero energy in high frequency frame. If there is no motion, the pixel value does not vary much due to inherent nature of video, thus there will be very less energy in high frequency frame.

In the proposed technique the "predict" step of lifting is applied to the input frames without any motion compensation and the high frequency frame is obtained (i.e. residual energy). During this step sum of the pixel values for each block in high frequency frame is computed. The zero motion blocks are detected using sum of the pixel values of high frequency frame. If the sum is less than 512 (in case of block size 16 X 16), that block is considered as zero motion block otherwise as a block with motion. The threshold value for motion detection is taken as 512, since there are total 256 pixels in each block and the technique empirically (Table-1) decides average value of each pixel as two when there is no motion.

Our assumption is empirically proved by applying the assumption on all classes of test videos and the results is shown in Fig. 3, Fig. 4, Fig. 5 and Fig. 6. In all the test videos actual number of zero motion blocks and the estimated approximate number of zero motion blocks using our assumptions are coinciding with each other on almost all the frames. Hence our assumption to detect zero motion blocks is true and gives expected results in complexity reduction of MCTF. The technique computes the sum of pixel values of a residual block. In general for zero motion blocks this sum will be less than n^2 (where n X n is the block size) and for remaining block MCTF is done as usual.

Algorithm

Step 1. Apply "prediction" operation of lifting wavelet transform on the input frames and obtain residual (or high frequency) frame using following equation.

$$h_k[x] = f_{2k+1}[x] - \frac{1}{2}(f_{2k}(x)] + f_{2k+2}(x)]) \tag{5}$$

During this process calculate the sum of the pixel values of the individual blocks.
Step 2. For each block
   If (SUM > n X n)// when n X n is block size
   {
   Consider that as a block with motion do motion, estimation for that block and obtain   motion vector. Again carry out temporal filtering along the motion vector using Equ. (1) and update the corresponding block in the residual frame obtained in step 1.
   }
Endif
Step 3. For each block apply following "update" step of lifting to obtain low frequency frame.
If (a block undergoes motion)
          Equ. (2)
Else

$$l_k[x] = f_{2k}[x] + \frac{1}{4}(h_{k-1}(x)] + h_k(x)]) \tag{6}$$

Endif

4   Simulation Results

During simulation of the proposed technique we have considered full search motion estimation with 1/8 pixel accuracy, 5/3 wavelet transform for temporal filtering, Debauchees 9/7 wavelet filter for spatial wavelet transform, window size 15 X 15 and block size 16 X 16. Standard test sequences like Akiyo, News, Foreman, etc., of QCIF resolution at 30 frames per second showing all varieties of motions are considered.

The Fig. (3) shows the actual blocks with zero motion(standard count) found after motion estimation and calculated number of blocks with zero motion (proposed technique count) without computing motion estimation. For various values of SAE the number of computed blocks with zero motion is shown for various types of videos. Hence we have chosen a threshold of 512 for the SAE, which compromises with complexity and quality of the decoded video.

Standard test sequences	Standard count/ PSNR(dB)	Proposed Technique count/PSNR(dB)				
		SUM (SAE)				
		0	256	512	1024	2048
Akiyo	8610	4414	6982	7728	8211	8558
	55.92	55.92	55.92	55.92	55.92	55.92
News	8137	2124	6171	6928	7587	8092
	40.00	40.00	40.00	40.00	39.98	39.99
Claire	8207	0	6778	7610	8138	8513
	51.39	51.39	51.40	51.40	51.40	51.40
Container	8416	0	5049	6763	7969	8554
	50.75	50.75	50.75	50.75	50.75	50.75
Foreman	2857	0	359	1269	2915	5055
	32.30	32.30	32.33	32.14	32.15	31.88
Table Tennies	2203	0	521	1361	2791	4829
	30.08	30.08	30.08	30.08	28.69	32.20
Car Phone	3929	0	1344	2894	4844	6959
	40.30	40.30	40.31	40.27	40.29	39.74
Coast Gaurd	1673	0	11	154	1269	4069
	33.79	33.79	33.79	33.79	33.79	32.97

Fig. 3. The actual number of blocks with zero motion and identified blocks as zero motion blocks using proposed technique with qualtiy of the decoded video in terms of PSNR (dB)

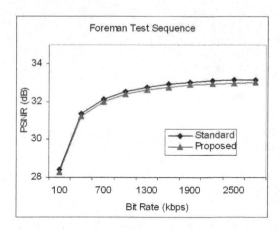

Fig. 4. Comparison of objective quality PSNR (dB) for Foreman sequence at various bit rates

The quality of the decoded video at various bit rate for standard and proposed techniques are shown in Fig. 3, Fig. 4, Fig. 5 and Fig. 6. The objective and subjective quality of the proposed technique is same as the standard techniques.

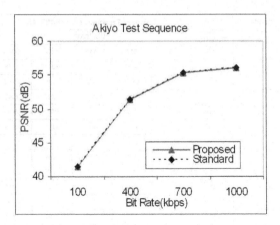

Fig. 5. Comparison of objective quality PSNR (dB) for Akiyo sequence at various bit rates

Fig. 6. Decoded 30th Foreman video frame at 2800kbps (a) Proposed Technique (b) Standard Technique

5   Conclusion

This paper proposed a novel idea to reduce the computational complexity of MCTF by effectively applying MCTF to the blocks having some motion. For the remaining blocks not having any motion, simply temporal filtering is applied. Hence unnecessary motion estimation for most of the blocks is avoided and complexity of the entire MCTF framework is reduced significantly. The results obtained from the MC-EZBC framework show that subjective and objective quality of the decoded video remains almost the same as that of the standard.

References

1. Taubman, D., Zakhor, A.: Multirate 3-D subband coding of video. IEEE Trans. Image Process. 3, 572–588 (1994)
2. Wang, A., Xiong, Z., Chou, P., Mehrotra, S.: Three-dimensional wavelet coding of video with global motion compensation. In: Proc. Data Compression Conference, pp. 404–413 (March 1999)
3. Ohm, J.: Three-dimensional subband coding with motion compensation. IEEE Trans. Image Process. 3, 559–571 (1994)
4. Choi, S., Woods, J.: Motion compensated 3d subband coding of video. IEEE Trans. Image Proc. 8, 155–167 (1999)
5. Pesquet-Popescu, B., Bottreau, V.: Three dimensional lifting schemes for motion compensated video compression. In: IEEE Int. Conf. Accoust. Speech and Signal Proc., pp. 1793–1796 (2001)
6. Bottreau, V., Benetiere, M., Felts, B., Pesquet-Popescu, B.: A fully SCalable 3d subband video codec. In: IEEE Int. Conf. Image Proc., pp. 1017–1020 (2001)
7. Luo, L., Li, J., Li, S., Zhuang, Z., Zhang, Y.-Q.: Motion compensated lifting wavelet and its application in video coding. In: IEEE, Int. Conf. on Multimedia and Expo, pp. 481–484 (2001)
8. Secker, A., Taubman, D.: Motion-compensated highly scalable video compression using an adaptive 3d wavelet transform based on lifting. In: IEEE Int. conf. Image Proc., pp. 1029–1032 (2001)
9. Secker, A., Taubman, D.: Highly scalable video compression using a lifting-based 3d wavelet transform with deformable mesh motion compensation. In: IEEE Int. conf. Image Proc., pp. 749–752 (2002)
10. Secker, A., Taubman, D.: Lifting based invertible motion adaptive transform (LIMAT) framework for highly scalable video compression. IEEE Trans. Image Proc. 12, 1530–1542 (2003)
11. Secker, A., Taubman, D.: Motion-compensated highly scalable video compression using an adaptive 3d wavelet transform based on lifting. In: IEEE Int. conf. Image Proc., pp. 1029–1032 (2001)
12. Secker, A., Taubman, D.: Highly scalable video compression using a lifting-based 3d wavelet transform with deformable mesh motion compensation. In: IEEE Int. conf. Image Proc., pp. 749–752 (2002)
13. Woods, et al.: Bi-Directional MC-EZBC with lifting implementation. IEEE Transaction of Circuits, Systems and Video Technology 14(10) (October 2004)
14. Choi, S., Woods, J.W.: Motion-compensated 3-D subband coding of video. IEEE Trans. Image Processing 8, 155–167 (1999)
15. Antonini, M., Barlaud, M., Mathieu, P., Daubechies, I.: Image coding using wavelet transform. IEEE Trans. Image Processing 1, 205–220 (1992)
16. Woods, et al.: Embedded image coding using zeroblocks of subband/wavelet coefficients and context modeling. Presented at the MPEG-4 Workshop and Exhibition at ISCAS 2000, Geneva, Switzerland (May 2000)

Author Index

Lecture Notes in Computer Science

Sublibrary 6: Image Processing, Computer Vision, Pattern Recognition, and Graphics

Vol. 4292: G. Bebis, R. Boyle, B. Parvin, D. Koracin, P. Remagnino, A. Nefian, G. Meenakshisundaram, V. Pascucci, J. Zara, J. Molineros, H. Theisel, T. Malzbender (Eds.), Advances in Visual Computing, Part II. XXXII, 906 pages. 2006.

Vol. 4291: G. Bebis, R. Boyle, B. Parvin, D. Koracin, P. Remagnino, A. Nefian, G. Meenakshisundaram, V. Pascucci, J. Zara, J. Molineros, H. Theisel, T. Malzbender (Eds.), Advances in Visual Computing, Part I. XXXI, 916 pages. 2006.

Vol. 4245: A. Kuba, L.G. Nyúl, K. Palágyi (Eds.), Discrete Geometry for Computer Imagery. XIII, 688 pages. 2006.

Vol. 4241: R.R. Beichel, M. Sonka (Eds.), Computer Vision Approaches to Medical Image Analysis. XI, 262 pages. 2006.

Vol. 4225: J.F. Martínez-Trinidad, J.A. Carrasco Ochoa, J. Kittler (Eds.), Progress in Pattern Recognition, Image Analysis and Applications. XIX, 995 pages. 2006.

Vol. 4191: R. Larsen, M. Nielsen, J. Sporring (Eds.), Medical Image Computing and Computer-Assisted Intervention – MICCAI 2006, Part II. XXXVIII, 981 pages. 2006.

Vol. 4190: R. Larsen, M. Nielsen, J. Sporring (Eds.), Medical Image Computing and Computer-Assisted Intervention – MICCAI 2006, Part I. XXXVVIII, 949 pages. 2006.

Vol. 4179: J. Blanc-Talon, W. Philips, D. Popescu, P. Scheunders (Eds.), Advanced Concepts for Intelligent Vision Systems. XXIV, 1224 pages. 2006.

Vol. 4174: K. Franke, K.-R. Müller, B. Nickolay, R. Schäfer (Eds.), Pattern Recognition. XX, 773 pages. 2006.

Vol. 4170: J. Ponce, M. Hebert, C. Schmid, A. Zisserman (Eds.), Toward Category-Level Object Recognition. XI, 618 pages. 2006.

Vol. 4153: N. Zheng, X. Jiang, X. Lan (Eds.), Advances in Machine Vision, Image Processing, and Pattern Analysis. XIII, 506 pages. 2006.

Vol. 4142: A. Campilho, M. Kamel (Eds.), Image Analysis and Recognition, Part II. XXVII, 923 pages. 2006.

Vol. 4141: A. Campilho, M. Kamel (Eds.), Image Analysis and Recognition, Part I. XXVIII, 939 pages. 2006.

Vol. 4122: R. Stiefelhagen, J.S. Garofolo (Eds.), Multimodal Technologies for Perception of Humans. XII, 360 pages. 2007.

Vol. 4109: D.-Y. Yeung, J.T. Kwok, A. Fred, F. Roli, D. de Ridder (Eds.), Structural, Syntactic, and Statistical Pattern Recognition. XXI, 939 pages. 2006.

Vol. 4091: G.-Z. Yang, T. Jiang, D. Shen, L. Gu, J. Yang (Eds.), Medical Imaging and Augmented Reality. XIII, 399 pages. 2006.

Vol. 4073: A. Butz, B. Fisher, A. Krüger, P. Olivier (Eds.), Smart Graphics. XI, 263 pages. 2006.

Vol. 4069: F.J. Perales, R.B. Fisher (Eds.), Articulated Motion and Deformable Objects. XV, 526 pages. 2006.

Vol. 4057: J.P.W. Pluim, B. Likar, F.A. Gerritsen (Eds.), Biomedical Image Registration. XII, 324 pages. 2006.

Vol. 4046: S.M. Astley, M. Brady, C. Rose, R. Zwiggelaar (Eds.), Digital Mammography. XVI, 654 pages. 2006.

Vol. 4040: R. Reulke, U. Eckardt, B. Flach, U. Knauer, K. Polthier (Eds.), Combinatorial Image Analysis. XII, 482 pages. 2006.

Vol. 4035: T. Nishita, Q. Peng, H.-P. Seidel (Eds.), Advances in Computer Graphics. XX, 771 pages. 2006.

Vol. 3979: T.S. Huang, N. Sebe, M. Lew, V. Pavlović, M. Kölsch, A. Galata, B. Kisačanin (Eds.), Computer Vision in Human-Computer Interaction. XII, 121 pages. 2006.

Vol. 3954: A. Leonardis, H. Bischof, A. Pinz (Eds.), Computer Vision – ECCV 2006, Part IV. XVII, 613 pages. 2006.

Vol. 3953: A. Leonardis, H. Bischof, A. Pinz (Eds.), Computer Vision – ECCV 2006, Part III. XVII, 649 pages. 2006.

Vol. 3952: A. Leonardis, H. Bischof, A. Pinz (Eds.), Computer Vision – ECCV 2006, Part II. XVII, 661 pages. 2006.

Vol. 3951: A. Leonardis, H. Bischof, A. Pinz (Eds.), Computer Vision – ECCV 2006, Part I. XXXV, 639 pages. 2006.

Vol. 3948: H.I. Christensen, H.-H. Nagel (Eds.), Cognitive Vision Systems. VIII, 367 pages. 2006.

Vol. 3926: W. Liu, J. Lladós (Eds.), Graphics Recognition. XII, 428 pages. 2006.

Vol. 3872: H. Bunke, A.L. Spitz (Eds.), Document Analysis Systems VII. XIII, 630 pages. 2006.

Vol. 3852: P.J. Narayanan, S.K. Nayar, H.-Y. Shum (Eds.), Computer Vision – ACCV 2006, Part II. XXXI, 977 pages. 2006.

Vol. 3851: P.J. Narayanan, S.K. Nayar, H.-Y. Shum (Eds.), Computer Vision – ACCV 2006, Part I. XXXI, 973 pages. 2006.

Vol. 3832: D. Zhang, A.K. Jain (Eds.), Advances in Biometrics. XX, 796 pages. 2005.

Vol. 3736: S. Bres, R. Laurini (Eds.), Visual Information and Information Systems. XI, 291 pages. 2006.

Vol. 3667: W.J. MacLean (Ed.), Spatial Coherence for Visual Motion Analysis. IX, 141 pages. 2006.

Vol. 3417: B. Jähne, R. Mester, E. Barth, H. Scharr (Eds.), Complex Motion. X, 235 pages. 2007.

Vol. 2396: T.M. Caelli, A. Amin, R.P.W. Duin, M.S. Kamel, D. de Ridder (Eds.), Structural, Syntactic, and Statistical Pattern Recognition. XVI, 863 pages. 2002.

Vol. 1679: C. Taylor, A. Colchester (Eds.), Medical Image Computing and Computer-Assisted Intervention – MICCAI'99. XXI, 1240 pages. 1999.